Trajectory of 75 years of Indian Agriculture after Independence

P. K. Ghosh • Anup Das • Raka Saxena •
Kaushik Banerjee • Gouranga Kar • D. Vijay
Editors

Trajectory of 75 years of Indian Agriculture after Independence

Springer

Editors
P. K. Ghosh
National Institute of Biotic Stress Management
Raipur, Chhattisgarh, India

Anup Das
ICAR Research Complex for Eastern Region
Patna, Bihar, India

Raka Saxena
ICAR- National Institute of Agricultural Economics and Policy Research
New Delhi, India

Kaushik Banerjee
National Research Centre on Grapes
Pune, Maharashtra, India

Gouranga Kar
Central Research Institute on Jute & Allied Fibre
Barrackpore, West Bengal, India

D. Vijay
Indian Agricultural Research Institute
New Delhi, India

ISBN 978-981-19-7996-5 ISBN 978-981-19-7997-2 (eBook)
https://doi.org/10.1007/978-981-19-7997-2

This Springer imprint is published by the registered company Springer Nature Singapore Pte Ltd.
The registered company address is: 152 Beach Road, #21-01/04 Gateway East, Singapore 189721, Singapore

Foreword

Agriculture being the primary source of income for the largest section of India's population, sustaining high productivity, ensuring resource use efficiency and securing profitability assume critical importance in catering to the country's growing food demand. Since independence, agricultural output has increased dramatically placing India as one of the top five producers in respect of 80% of global agri-produce. The year 2021–2022 closed with a record food grain output of 314.51 million tons and record horticultural output of 335 million tons. The year also saw agri-exports surpassing USD 50 billion. Notwithstanding this, our food system continues to fall short of meeting the nutritional security besides suffering from a huge deficit of certain commodities like oilseeds and millets.

The nation's agrarian structure has however been acquiring the needed dynamism in recent times. As markets are evolving, farmers are gradually shifting from production focus to demand-led market focus in practicing agriculture. The tentative changes in this direction need to be consolidated, which entail a fundamental rethink of the role of agricultural knowledge, science and technology.

The economic, social and environmental dimensions are closely inter-linked in the domain of agriculture. This calls for reconciling the contradictions that would emerge between economic growth and ecological sustainability, and therefore, necessitating broad-basing the growth triggers across all the sub-sectors including dairy, livestock and fisheries besides agronomic and horticultural crops. Logically, this orientation warrants a new policy framework and appropriate investments in the agriculture sector. The intensive nature of agriculture adopted since the mid-1960s in the irrigated regions of the country has resulted in certain challenges to sustaining the productivity. These include loss of soil organic carbon, increase in problematic soils and decline in groundwater table. Further, the climatic change-linked weather aberrations are complicating the production system across both rainfed and irrigated systems. Yield gaps vis-a-vis the technical potential exist across various agricultural crops and states, minimizing which will require sharing of technology with all farmers that promotes sustainability of higher yield based on total factor productivity, climate resilience and feasibility for efficient market integrations. Given the current need for generating profitability, farmers will need to be repurposed to practise agricultural value systems based on healthy value chains.

The moral and foremost responsibility of agriculture is to ensure food security of the people. So far, this has meant carbohydrate security. It is important that the production system is remandated to produce a cafeteria of agri-produce that will meet the nation's nutrient requirements for a balanced diet and healthy living. Broad-based production systems should also focus on outputting safe and healthy commodities by adopting Good Agriculture Practices. It cannot be "anyhow and at any cost" system anymore. This new approach also bears the potential that can generate spin-off positive advantages like minimizing production and income risks.

I am happy to learn that Dr. P. K. Ghosh and a group of scientists have thoroughly examined the development of Indian agriculture since independence and shared with readers the agricultural landscape over the last 75 years. The book *Trajectory of 75 Years of Indian Agriculture After Independence* addresses many issues in a pragmatic way and documents various facets of Indian agriculture from different perspectives. The choice of themes and topics. well researched content and logical structuring of the book makes it a useful reading for all those interested in agriculture including researchers, academics, policy makers and programme executives.

That the book is being brought out during the "Azadi Ka Amrit Kaal" of India, and when the country looks forward to a new trajectory of growth after its 75 years of meaningful transformation, is more than a fortuitous coincidence.

I congratulate the authors and editors for contributing richly to the knowledge domain of the agriculture sector through this book.

Department of Agriculture & Farmers Welfare Ashok Dalwai
National Rainfed Area Authority (NRAA)
New Delhi, India

Preface

The majority of India's rural population relies on agriculture as their main source of income. This sector provides employment to a sizable population in the country. Realizing the importance of agriculture, the Government of India made several attempts to evolve appropriate agricultural policies for producers' and consumers' welfare. Adoption of improved HYV seeds, coupled with efficient irrigation, other inputs and price support led to a notable increase in agricultural production and productivity making it possible to feed the growing human population. The approach to agricultural reforms was largely production centric, but several concerns related to farmers' distress and low incomes emerged in the recent decade. India is the land of diverse agro-ecologies that covers many typologies and is dominated by various crops and animal-based systems. These diverse typologies allow the country to produce a number of major and minor agri-based commodities. The crop sector contributes maximum to the value of output realized from agricultural and allied activities. Livestock and fisheries sectors are growing appreciably and are regarded as high growth sectors. To provide a boost, the Government has created a separate ministry to focus on these sectors. The outputs have increased manifolds since independence, and we are now the largest exporter of bovine meat and crustaceans.

Currently, the country produces more than 1.2 billion tonnes of agricultural produce. However, the commodities vary in terms of their importance for food, feed and fodder, clothing along with being used as raw and intermediate products for industries. The country has now emerged as a significant global player particularly in crustaceans, basmati rice, cotton, bovine meat and many other commodities. Agricultural exports crossed USD 50 billion in 2021–2022. India's export basket is now diversified with non-traditional items, and differential products are also gaining importance. However, the country has been deficit in production of oilseeds, and the major chunk of edible oil requirement in the country is met through imports. Our imports of palm oil have reached the level of approximately 10 million tonnes. A number of steps have been taken by the Government of India to boost the productivity and production of domestic oilseeds in the country.

The agrarian structure in India has also undergone significant change. Corresponding to the growth trajectory, the cropping pattern has also gone through significant changes over time. The Government of India is places much emphasis on the promotion of coarse cereals, which are now emerging as important for assured health benefits and termed as nutri-cereals. Much emphasis has also been placed on pulses and oilseeds for attaining self-sufficiency. There is a clear shift from food grains towards fruits and vegetables. Due to the shift in demand patterns, farmers have also responded to market signals and are gradually shifting the production mix to meet the growing demand for these commodities. Increasing demand of high value commodities like fruits and vegetables in coming years could be tapped by shift in policy focus as well as investment towards allied sectors for improving productivity, quality and efficiency.

The nutrition transition is reported to be an important consequence of the economic and structural transformation. The economic growth, changes in tastes and preferences and urbanization have resulted in changing consumption pattern away from traditional food commodities to processed and high value commodities. There is a clear consumption shift from food grains towards high value products like fruits and vegetables, livestock and fisheries. This is manifest in increasing area coverage and production in respect of these sectors and reflects in the increasing value share of these items over time. The increasing demand for high value commodities like fruits and vegetables, livestock products and fisheries can be tapped better, by a shift in policy to focus on what would constitute the growth drivers in the coming years.

Further, Indian agriculture is challenged owing to climate variability, occurrence of natural disasters, poor logistics, price uncertainties, etc. In many cases, the crop productivity in India is much lower than other producing nations. We need to ensure that the maximum potential of any crop variety is harvested at the farmers' field. Significant yield gaps exist across various crops in different states, as well as within states. Bridging these yield gaps will not only increase the production but also help to improve the efficiency of land and labour, reduce production cost and add to food security. Improving the crop yields is also important from the perspective of releasing land for other productive uses, by diversifying into high value activities and commodities, and to allow farmers to scale up integrated farming practices.

Raipur, Chhattisgarh, India P. K. Ghosh
Patna, Bihar, India Anup Das
New Delhi, India Raka Saxena
Pune, Maharashtra, India Kaushik Banerjee
Barrackpore, West Bengal, India Gouranga Kar
New Delhi, India D. Vijay

Contents

About the Editors

P. K. Ghosh works as Founder Director and Vice Chancellor of ICAR-National Institute of Biotic Stress Management (NIBSM), Raipur, earlier demonstrated leadership as National Coordinator of National Agricultural Higher Education Project-World Bank-aided project, ICAR, New Delhi, Director of Indian Grassland and Fodder Research Institute, Jhansi, and Head of Crop Production Division, Indian Institute of Pulses Research. His pioneering works are on climate change, carbon sequestration, resource use efficiency, conservation agriculture, crop diversification, soil water conservation and integrated farming system. He has been identified as one of the two percent of agricultural scientists at the global level and is a member of the Royal Society of Research, London. He is a recipient of 19 national awards, including the M.S. Randhawa Memorial Award for best Administration, and is also a Fellow of the National Academy of Sciences, India, and National Academy of Agricultural Sciences (NAAS), and three professional societies. He served as Chairman by organizing the 23rd International Grassland Congress (IGC) held for the first time in India, an executive member of the IGC continuing committee representing South-East Asia, and the Sectional President of the Agricultural section of Indian Science Congress. He is the editor-in-chief of the National Academy of Science, India. Considering his academic, administrative and research excellence, he was selected as a co-opted member of the DFI Committee, constituted by the PMO and contributed to bring 14 volumes on DFI. He also formulated four policy papers for the country. He published 20 books and 167 research papers in high impact factor journals with a total citation of 7947, an h-index of 40 and an i-10 index of 90 and a total publication of 304.

Anup Das presently serves as Director, ICAR Research Complex for Eastern Region, Patna, Bihar, India, and also served as Principal Scientist (Agronomy) at ICAR Research Complex for NEH Region, Tripura Centre, India. His extensive research works led to the development of package of practices for no-till production of pulses and oilseeds in rice and maize fallow areas; organic production packages of 33 important crops on cropping system approach; standardization of Modified System of Rice Intensification and Raised and Sunken Bed land configuration, watershed-based Integrated Farming System models, produce quality, input use efficiencies, soil properties and C-pools under different land uses and management practices. He has guided/co-guided 17 master's and two Ph.D. students and taught six courses to PG students. He has published 200 research articles, 74 books, compendium and technical bulletins, and many other documents with a total citation of 4000 and an h-index of 36 to his credit. He is actively involved in technology transfer for livelihood improvement of farmers through large-scale capacity building programmes and demonstrations benefitting about 50,000 farmers in northeast India. He has received several national awards like the ICAR-Lal Bahadur Shastri outstanding young scientist award, Swami Sahajanand Saraswati outstanding extension scientist award, Fakhruddin Ali Ahmed award for improving tribal farming system, ICAR-Inter-disciplinary team research award in natural resource management and First Dr. H.K. Jain—CAU, Imphal Award for research excellence in northeast India among many others. Dr. Das is a recipient of the prestigious Fellowships of National Academy of Agricultural Sciences (2019), Indian Society of Agronomy (2017), Indian Association of Hill Farming (IAHF-2017) and Society for Biotic and Environmental Research (2020).

Raka Saxena is currently working as a Principal Scientist at ICAR-National Institute for Agricultural Economics and Policy Research (NIAP), located in New Delhi, India. Her areas of interest in research include agricultural trade, animal science and agricultural economics. She has more than 15 years of research experience and has published more than 50 papers in prestigious publications on a national and worldwide scale. In addition, she has written more than 20 book chapters and eight policy papers, briefs and working papers. She has contributed via innovative, multi-disciplinary and multi-institutional initiatives that are founded on evidence-based policy insights. She maintained her research agenda with a focus on the urgent need for prioritizing and evaluating agricultural research in order to directly benefit the nation's farmers. She focused on two important aspects of the country's agriculture, notably those that dealt with market intelligence for handling price volatility and enhancing farmers' incomes. Her most recent study focuses on methods for maximizing the potential of the agricultural trade to raise farm incomes.

Kaushik Banerjee is a Principal Scientist from ICAR-National Research Centre for Grapes, Pune. He is heading the National Reference Laboratory on pesticide residues and mycotoxins in India. His area of research focuses on the development of efficient analysis methods for the sensitive and confirmatory estimation of pesticide residues and mycotoxins in agricultural and food matrices and risk assessment studies for the fixation of crop-specific maximum residue limits. Dr. Banerjee's extensive contributions to science and the community have earned him numerous national and international laurels. He received the prestigious Harvey W. Wiley Award of AOAC INTERNATIONAL in 2017 and the Recognition Award of the National Academy of Agricultural Sciences (NAAS) in 2019. Earlier, he was named as a Fellow by the Royal Society of Chemistry (FRSC), National Fellow-Indian Council of Agricultural Research and Fellow-NAAS.

Gouranga Kar is presently the Director of ICAR-Central Research Institute for Jute and Allied Fibres (ICAR-CRIJAF), Barrackpore. Earlier he served as Principal Scientist, ICAR-Indian Institute of Water Management, Bhubaneswar. He has done pioneering contribution in quantification of water footprint in India on farm scale and designed and implemented many national/international projects related to water and watershed management, crop growth modelling, climate change research, mitigation and adaptation, land use and cropping system using RS and GIS. He developed GIS-based decision rules to prepare sustainable alternative land-use plan on watershed basis. To increase crop productivity and cropping intensity of rainfed rice agro-ecosystem, he has developed crop diversification technology along with in-situ soil conservation, technologies for modification of microclimate, low cost soil amelioration of acidic land and irrigation scheduling of crops of post-rice season. He has designed innovative flood water harvesting system, on-dyke agro-forestry, pond-based farming and standardized cultivation of salt-resistant crops for coastal agro-ecosystem, designed site-specific groundwater recharge structures and validated in different agro-climatic regions of the country. He also significantly contributed in standardizing dug well-based drip-fertigation system to obtain higher land and crop productivity of vegetable crops. He has published more than 200 research papers, review paper and book chapters of immense academic as well as practical importance. He is also a fellow of different scientific societies like National Academy of Agricultural Sciences (FNAAS), Indian Society of Soil Science (FISSS), Indian Association of Soil and Water Conservationists (FIASWC), Association of Agro-meteorologists (FAAM), West Bengal Academy of Science and Technology (FWAST) and Indian Society of Coastal Agricultural Research (ISCAR). Also he is a recipient of many prestigious awards, namely ICAR Vasant Rao Naik Award 2004; IARI Sukumar Bosu Memorial Award; ASWC Dr. K. G. Tejwani Award 2006; ICAR Hari Om Ashram Trust Award 2007-08; USDA Norman Borlaug International Agricultural Science and Technology Fellowship Award 2008; NAAS Recognition Award 2012; ICAR-Swami Sahajananda Saraswati Award 2019; and Rajbhasa Gaurav Puraskar 2019.

D. Vijay is a Principal Scientist at ICAR-Indian Agricultural Research Institute, New Delhi, and has more than 17 years of research, teaching and extension experience in seed science and technology. He specialized in seed physiology and has good expertise in seed production of various crops, including grass species. He worked as a zonal seed production officer and nodal officer (seed) in different institutes for more than 9 years. He published more than 30 research articles in various national and international journals, two books, ten book chapters and several bulletins, compendium chapters and training manuals. He is a gold medallist in UG and PhD and received several young/best scientist awards from different academic societies and is an associate fellow of the Range Management Society of India. He is an editorial board member of four journals and delivered more than 30 invited lectures in several national and international training programmes. He guided three MSc students and presently guiding four PhD students at ICAR-IARI. He got nominated as a Regional Representative, India, of the International Society for Seed Science, UK.

Part I

Transformation of Indian Agriculture

Trends and Trajectory of Indian Agriculture 1

Raka Saxena, Sonia Chauhan, Dilip K. Ghosh, Rohit Kumar,
Dilip Kumar, and Anup Das

Abstract

The Government of India made several attempts to evolve appropriate agricultural policies. The approach to agricultural reforms was largely production-centric, but concerns related to farmers' distress and low incomes emerged in the recent decade. Realizing those, the Government constituted a committee for Doubling of Farmers' Income (DFI) by 2022–2023. This chapter studies the detailed trajectory of Indian agriculture in terms of changes in macroeconomic indicators, particularly agricultural growth, employment, agricultural trade, changes in agrarian structure, cropping pattern, production profile, and changes in consumption pattern during various growth phases. The crop sector contributed the maximum to agricultural growth. Livestock and fisheries sectors have grown at an appreciable growth. The number of marginal and small holdings and the area under such holdings have increased while the number of semi-medium, medium, and large holdings and the area under such holdings has reduced. There is a clear consumption shift from food grains toward high-value products like fruits and vegetables, livestock, and fisheries. This manifests in increasing area coverage and production in respect of these sectors and reflects in the

---

R. Saxena (✉) · S. Chauhan · D. Kumar
ICAR-National Institute of Agricultural Economics and Policy Research, New Delhi, India
e-mail: raka.saxena@icar.gov.in

D. K. Ghosh
ICAR-National Institute of Biotic Stress Management, Raipur, India

R. Kumar
Department of Dairy Business Management, Sanjay Gandhi Institute of Dairy Technology, Bihar Animal Sciences University, Patna, India

A. Das
ICAR Research Complex for Eastern Region, Patna, Bihar, India

P. K. Ghosh et al. (eds.), *Trajectory of 75 years of Indian Agriculture after Independence*, https://doi.org/10.1007/978-981-19-7997-2_1

increasing value share of these items over time. The exports of agricultural commodities have jumped significantly. The country needs to handle the challenges related to climate uncertainties and other productivity barriers.

Keywords

Agricultural growth · Employment · Agrarian structure · Cropping pattern · Consumption pattern · Exports

1.1 Chronology of Policy Reforms in Agriculture

Agriculture has been a significant component of the Indian economy and supports the mainstay of the majority of the rural population in India. Though the share of agriculture and allied in total gross value added (GVA) has declined from 60% in 1950–1951 to 20% in 2019–2020, the agricultural output has been consistently increasing. India is one of the largest producers of rice, wheat, pulses, sugarcane, cotton, groundnuts, fruits, vegetables, milk, and fisheries. This sector provides employment to a sizable population in India. Realizing the importance of agriculture, the Government of India made several attempts to evolve appropriate agricultural policies for producers' and consumers' welfare. To start with, the Royal Commission was set up in 1928 to examine the conditions of the agricultural and rural economy and recommend measures for agricultural improvement in India. The first Food Grains Policy Committee was set up in 1943 to tackle the ill effects of the Second World War on the availability, supply, distribution, and prices of food products. Severe food shortages were faced due to partition and lower productivity immediately after attaining independence. With the aim to provide food to all, the government appointed Foodgrain Policy Committee in 1947 to study the food distribution aspects. The main feature of this policy was the gradual withdrawal of control and restrictions on the food grains movement. Subsequently, to examine food problems and their solutions, various other committees like Maitra Committee (1950), Mehta Committee (1957), and Venkatappaiah Committee (1966) were constituted. The recommendations of these committees were vital in the formulation of agricultural policies for the distribution of food grains, minimum support prices, subsidies, public storage, procurement, and trade protection measures (Government of India 2018).

The first comprehensive agricultural policy in the country was brought out by the National Commission on Agriculture (1976) with a major emphasis on agricultural production along with natural resource management including land, water, and soil conservation. The outcomes were realized in the suggestion/initiation of various development schemes to improve the condition of natural resources and the income of farmers. Another committee chaired by Shri Bhanu Pratap Singh in 1990 made recommendations covering all major sectors of the agricultural economy and provided the base for the first draft of agricultural policy resolution, which resulted in the first ever comprehensive National Agricultural Policy introduced in 2000 (Government of India 2018). Further, the National Commission on Farmers (NCF) was set up in 2004 and studied the major concerns pertaining to Indian farming and the farmers, largely emphasizing agricultural production, productivity, food availability, and sustainability and also addressed other major agricultural dimensions related to land reforms, irrigation, infrastructure, along with agricultural research and development.

The approach to agricultural reforms was largely production-centric, but several concerns related to farmers' distress and low incomes emerged in the recent decade. Realizing those, the government constituted a committee for Doubling of Farmers' Income (DFI) by 2022–2023; the agenda focused on holistic solutions and suggested the major reforms required in the agriculture sector. DFI Committee also considered and suggested the market reforms in a big way and also focused on the allied sectors like animal husbandry, poultry, fisheries, and secondary agriculture.

Considering this chronology, this chapter studies the detailed trajectory of Indian agriculture in terms of changes in macroeconomic indicators, particularly agricultural growth, employment, agricultural trade, changes in agrarian structure, cropping pattern, production profile, and changes in consumption pattern during various growth phases. For analyzing the changes, the entire period during the last 75 years has been classified into four broad phases: pre-green revolution phase (1950–1951 to 1966–1967), green revolution phase (1967–1968 to 1987–1988), diversification and economic reforms phase (1988–1989 to 2009–2010), and market and income reforms (2010–2011 to 2019–2020).[1] We followed a systematic approach and identified the structural breaks using the Bai-Perron methodology.[2] The breaks were identified in the GVA of agriculture at 2011–2012 prices. The analysis of selected parameters (for continuous series) has been done according to the identified structural breaks.

1.2 What Comprises Indian Agriculture?

Agriculture and its allied sectors play an important role in the Indian economy. Agriculture and allied sector consists of four broad subsectors, namely crop sector, livestock, forestry, and fisheries. The crop sector contributes the maximum output value realized from agricultural and allied activities. Livestock and fisheries sectors are growing appreciably and are categorized as high-growth sectors. India is the land of diverse agroecologies and covers various typologies, namely arid, coastal, hilly, mountainous, irrigated, and rainfed. These diverse typologies allow the country to

[1] Chand and Shinoj (2012) delineated six phases in agricultural development, namely pre-green revolution period (1950–1951 to 1966–1967), early green revolution period (1968–1969 to 1975–1976), period of wider technology dissemination (1975–1976 to 1988–1989), period of diversification (1988–1989 to 1995–1996), post-reform period (1995–1996 to 2004–2005), and period of recovery (2004–2005 to 2010–2011).

[2] The models for a structural break can integrate the structural changes in selected variables through the model parameters. Bai and Perron (1998) provide the standard framework for the structural breaks model in which some, but not all, of the model parameters, are allowed to break at m possible break points (Bai and Perron 1998).

produce a number of major and minor agri-based commodities. Box 1.1 provides the total production of all major agri-commodities in India. Currently, the country produces approximately 1.2 billion tonnes of agricultural produce. However, the commodities vary in terms of their importance for food, feed and fodder, and clothing, along with being used as raw and intermediate products for industry. Being the largest sector in agriculture, the crop sector holds a large number of contributors. Field crops include cereals, pulses, oilseeds, sugars, and fibers. Among the cereals, paddy and wheat are the major staple crops and comprise the maximum share. Recently, the Government of India accorded greater emphasis on the cultivation of coarse cereals/millets owing to their nutritional benefits. Gram, pigeon pea, black gram, and green gram are the most important pulses. The country attained an appreciable increase in the production of pulses, and the total pulse production increased from 16–17 million tonnes (2014–15) to around 24–25 million tonnes (2019–20).

Box 1.1 Volumes of Different Commodities Produced in India, TE 2019–2020 (000′ Tonnes)

Cereals								
Paddy 116035		Wheat 103775	Sorghum 4350	Pearl millet 9412	Barley 1712	Maize 28411	Finger millet 1660	Small millets 381
Pulses								
Gram 10798.6	Pigeon pea 3832.3		Black gram 2877.9	Green gram 2328	Lentil 1318			
								Oilseeds
Soybean 11809	Linseed 131	Sesamum 701	Groundnut 8644	Rapeseed and mustard 8936	Castor seed 1535	Niger seed 52	Safflower 41	Sunflower 217
Sugar and jaggery								
Sugarcane 385,274								
Fibers								
Cotton 32,304		Jute 9511	Mesta 399					
Indigo, dyes and tanning material, drugs and narcotics								
Tea 915		Coffee 311	Cocoa 23					
Condiments and spices								
Cardamom 24		Dry chilies 1692	Black pepper 58	Ginger 1817	Turmeric 999	Areca nut 1110	Garlic 2986	Coriander 668
Fennel 134		Cumin 767	Ajwain 26	Fenugreek 193	Tamarind 193	Nutmeg 15	Cloves 1	
Fruits								
Banana 31288	Cashew Nut 754	Mango 21155	Grapes 3047	Papaya 5940	Apple 2486	Mosambi 3524	Lemon 3439	Orange 5827
Walnut 295	Litchi 711	Pineapple 1716	Sapota 1047	Almonds 12	Jack fruit 1778	Watermelon 2399	Muskmelon 1198	Pear 303
Guava 4223	Gooseberry 1095	Passion fruit 73		Peach 113	Pomegranate 2791	Strawberry 6	Other fruits 2295	Plum 87
Vegetables								
Potato 50,021	Sweet potato 1266	Tapioca 5329	Onion 24,057	Cabbage 9145	Cauliflower 8897	Okra 6209	Tomato 19,980	Green peas 5611

(continued)

Box 1.1 (continued)

Radish 3129	Beans 2301	Bitter gourd 1203	Bottle gourd 2933	Capsicum 452	Carrot 1790	Cucumber 100	Pumpkin 1936	
Other Crops								
Rubber 685,667		Mushroom 293						
Livestock								
Milk group 176,467		Meat 7733	Eggs 104 (billion nos.)	Wool 4242 (million kg)				
Fisheries								
Inland fish 8806		Marine fish 3675						

Source: Directorate of Economic and Statistics, Government of India.

Major oilseeds include soybean, rapeseed and mustard, groundnut, and coconut. The country has been in deficit in the production of oilseeds, and the major chunk of edible oil requirement in the country is met through imports. Our imports of palm oil have reached the level of approximately ten million tonnes. A number of steps have been taken by the Government of India to boost the productivity and production of domestic oilseeds in the country. The country is also the leading producer of sugarcane. The recent sugar policy aims at enhancing sugar exports and also aims to diversify its usage for strengthening domestic producers and sugar mills. A number of fruits and vegetables are grown under diverse agroclimatic environments. Horticultural output is increasing at an encouraging pace leading to greater participation of the country at the international level. Livestock and fisheries are the high-growth sectors. To provide a boost, the Government has created a separate ministry to focus on these sectors. The outputs have increased manifolds since independence, and we are now the largest exporter of bovine meat and crustaceans. Subsequent chapters delve more into the technological and other growth aspects of each of these subsectors.

1.3 Phases in Indian Agriculture

Several efforts have been made since independence to evolve appropriate agricultural policies impacting the agricultural systems in India. Given the food deficiency that prevailed and continued since independence, the growth strategy was largely production-centric. The agricultural sector received the much-needed attention that helped the nation achieve high levels of production for achieving food security. To have a more systematic portrait, we delineated the growth phases in agriculture and identified the structural breaks (Fig. 1.1).

To realize the goal of self-sufficiency in the agricultural sector, the Green Revolution was initiated in 1966–1967, which led to appreciable productivity

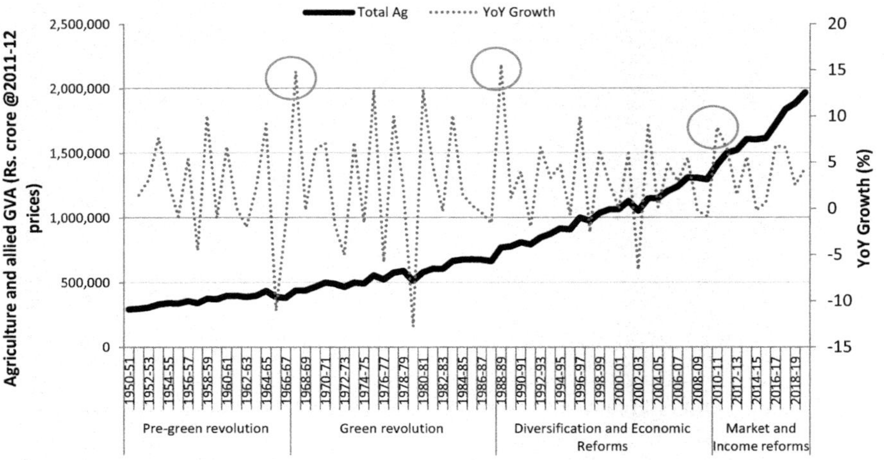

Fig. 1.1 Various phases in the growth of Indian agriculture. Source: Central Statistics Office, MoSPI

growth in major grains, mainly rice and wheat. The adoption of improved HYV seeds coupled with efficient irrigation, other inputs, and price support led to an increase in productivity, making it possible to feed the growing human population. This led to a considerable rise in agricultural production, making India self-sufficient in food grains. During the decade of the 1980s, efforts were made to spur agricultural growth in low productivity and stagnant states and regions, and special packages were launched to disseminate improved agricultural technology in underdeveloped states. The strategy focused on improved crop varieties and technologies suitable for dry-land regions to improve agricultural productivity. The productivity enhancements were realized through an increase in public investment in agriculture, research & development, and infrastructure. Agricultural research, extension, input supply, credit, marketing, price support, and the spread of technology were the prime concern of policymakers (Rao 1996).

Expansion of area was the main source of growth during the 1950s and 1960s; productivity became the main source of growth in agricultural production afterward. This phase also witnessed agrarian reforms, a number of institutional reforms, and the strengthening of agricultural credit institutions. Land reforms were important in increasing agricultural production during this phase. For price support to boost the adoption of improved technology, Agricultural Price Commission was set up in 1964, and an agricultural price policy was formulated to ensure minimum support prices to the farmers for selected agricultural commodities.

The second phase in Indian agriculture resulted in the attainment of self-sufficiency in food grains. Meanwhile, the Operation Flood Program launched by the Government of India with international cooperation in different phases resulted in enhanced milk output. A 'National Milk Grid' was formed that connected milk producers throughout the country along with the consumers. This led the country to become the highest milk producer in the world. Indian agriculture witnessed diversification toward nonfood grains output like milk, fishery, poultry, vegetables, fruits, etc. which accelerated growth in agricultural GDP during the 1980s (Chand 2003). The Blue Revolution concentrates mainly on enhancing the production and productivity of aquaculture and fisheries both from the inland and marine sources to meet the food and nutritional security of the nation. Economic reforms initiated in 1991 focused on deregulation, reduced government participation in economic activities, and liberalization. Opening up the economy for agricultural trade was a significant development and impacted agriculture. The country signed a number of multilateral trade agreements to boost our agricultural exports. The comprehensive e New Agricultural Policy was launched in 2000, aiming at 4% per annum growth in agriculture while meeting sustainability and equity goals. The recent phase focuses on various reforms for enhancing farmers' income in the country.

1.4 Agricultural Growth

This section presents the trends and growth in Gross Value Added (GVA) across various subsectors at 2011–2012 prices. The contribution of the agricultural and allied sectors to the overall GVA has changed intensely over the years. The contribution of the agricultural sector has declined from approximately 56% in pre-green revolution phase to about 16% in the recent phase, which was mainly dominated by several market and income reforms. The share of the service sector has enhanced drastically during this growth trajectory. Though the share of agriculture has declined, the growth momentum has picked up. The recent phase has witnessed an appreciable growth of over 3.5% per annum (Table 1.1).

During the recent phase, livestock and fisheries registered a growth of 1.54, 7.48, and 8.56% per annum, respectively (Fig. 1.2). The fisheries sector is growing at an appreciable and sustainable rate and is ahead among all sub-sectors. The livestock sector is likely to emerge as an engine of growth in the agricultural sector and can be relied upon for risk mitigation and minimizing losses to the farmers. The studies show that livestock is the best insurance against agrarian distress as the sector is the source of sustained income and generates income more frequently than the crop sector (Fig. 1.3).

1.5 Cropping Pattern

Corresponding to the growth trajectory, the cropping pattern has gone significant changes over time. Table 1.2 provides insight into the area share of different crop categories to gross cropped area (GCA) over time. The area under cereals has increased over time; paddy and wheat have been the principal cereals and accounted for more than one-third share of GCA. The Government of India places a lot of emphasis on the promotion of coarse cereals, which are now emerging as important for assured health benefits and are termed as nutri-cereals. A lot of emphasis is also

Table 1.1 Trends in GVA across sectors

Sector	Pre-green Revolution 1950–1951 to 1966–1967	Green Revolution 1967–1968 to 1987–1988	Diversification and Economic Reforms 1988–1989 to 2009–2010	Market and Income Reforms 2010–2011 to 2019–2020
Average GVA@2011–2012 prices (Rs billion)				
Agriculture and allied	3631	5585	10,328	16,691
Agriculture	2956	4589	8683	14,367
Industry	960	2401	8314	21,314
Services	2287	5452	21,868	64,999
Share (%)				
Agriculture and allied	55.89	43.12	25.83	16.20
Agriculture	45.51	35.43	21.72	13.95
Industry	14.78	18.53	20.80	20.69
Services	35.21	42.09	54.70	63.10
GVA growth (% p.a.)				
Agriculture and allied	2.02	2.30	2.70	3.55
Agriculture	2.00	2.52	2.77	3.24
Industry	6.33	4.88	6.25	6.21
Services	4.87	4.70	7.45	7.52

being accorded to pulses and oilseeds for attaining self-sufficiency in both subsectors. There is a clear shift from food grains toward fruits and vegetables. Because of the shift in demand pattern, the farmers have also responded to market signals and gradually shifted the production mix to meet the growing demand for these commodities, as indicated by the increasing value share of these items over time. Increasing demand for high-value commodities like fruits and vegetables in coming years could be tapped by a shift in policy focus as well as an investment toward allied sectors for improving productivity, quality, and efficiency (Fig. 1.4).

1.6 Dependence on Agriculture and Agrarian Structure

The dependence on agriculture can be assured from the fact that how many workers in the form of cultivators and agricultural laborers are employed in agriculture. Such data on agricultural workers are available in the population census on decadal bases. Besides, the National Sample Survey Office (NSSO) conducts surveys on employment and unemployment on a quinquennial basis. Table 1.3 presents the data on agricultural workers in the economy since 1951. Except for 1971, a continuous increase has been noticed in total workers engaged in agriculture. However, the number of cultivators declined between 2001 and 2011. The number of agricultural

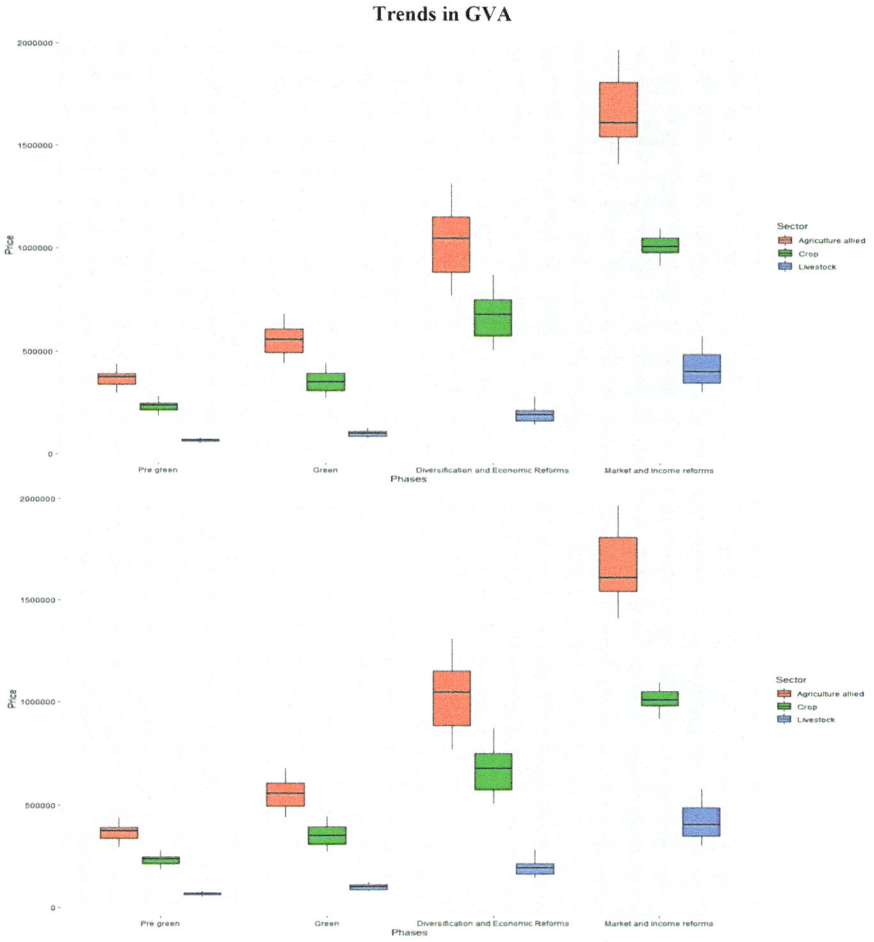

Fig. 1.2 Growth rates in GDP across subsectors at 2004–2005 prices. Source: Central Statistics Office, Ministry of Statistics and Programme Implementation, Govt. of India

laborers increased from 107 million to 144 million. The data indicate that the interest of the farming community in agriculture is declining, and consequently, the self-employed workers in agriculture are leaving the industry. This reported shift is good, provided the workers, who left the sector, are productively and gainfully employed in alternate sectors/industries.

The agrarian structure in India has also undergone significant change in India. The average land holding has witnessed a continuous decline since independence, and it declined from 2.69 hectares in 1960–1961 to 1.08 in 2015–2016. The total number of operational holdings in the country increased from 71.011 million in 1970–1971 to 146.45 million in 2015–2016. Small and marginal farm holders now

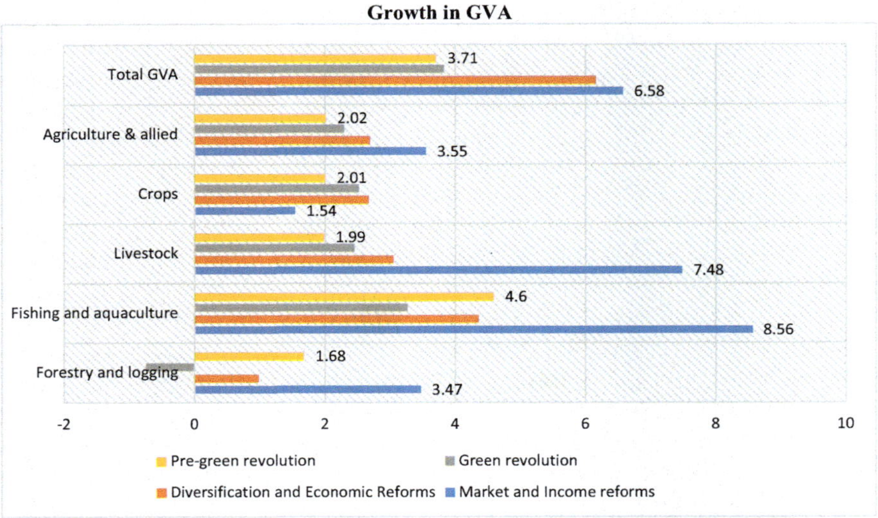

Fig. 1.3 Growth rates in GDP across subsectors. Source: Central Statistics Office, MoSPI, Govt. of India

Table 1.2 Area shares of crop categories to gross cropped area (GCA)

Crops	Pre-green revolution period	Green revolution period	Diversification and economic reforms	Market and income reforms
Paddy and wheat	30.44	35.12	36.84	37.48
Pulses	15.19	13.39	12.04	13.38
Oilseeds	8.60	9.72	18.51	17.55
Sugar	1.49	1.69	2.17	2.49

Source: Agricultural Statistics at a Glance, Various issues

cultivate 47% of total operated land and constitute 86% of the total number of landholdings. Fragmented and scattered holdings (as is the case of most of the marginal and small farms in India) do not allow better utilization of farm resources and technology adoption by the farmers; as a result, it reduces productivity. Moreover, this also hinders the diversification process, which is considered a key in enhancing the income of farmers. However, studies have also reported that small farms are more efficient in terms of technology adoption and better resource use productivity. A study by Chand et al. (2011) indicates that the smaller the holding size, the higher the use of inputs, crop intensity, and coverage under HYVs, reflecting technology. The study concluded that agriculture productivity in marginal and smallholdings was found to be much higher than the average productivity for all size categories; however, per capita output is low in smallholdings despite higher productivity due to lower per capita availability of land (Table 1.4).

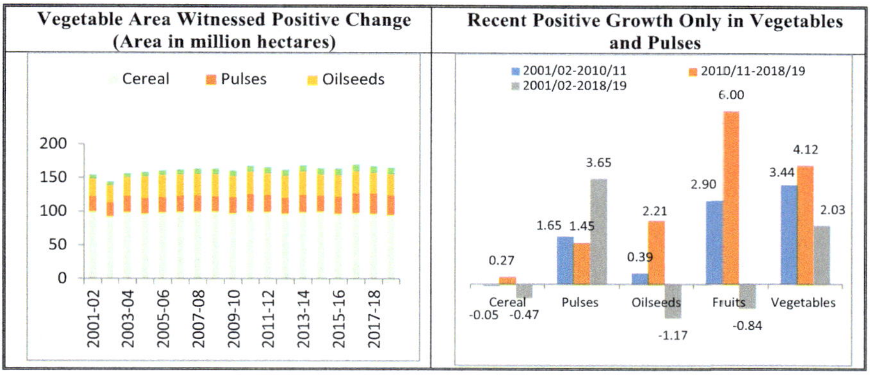

Fig. 1.4 Recent changes in cropping pattern. Source: Agricultural Statistics at a Glance, Various issues

1.7 Domestic Marketing and Trade

The bulk of the agricultural produce was marketed through the traditional spot markets. The Agriculture Produce Marketing Committee (APMC) was established to regulate agricultural marketing and promote efficient trade. The Government of India established the Agricultural Prices Commission (APC) in 1964 to create a stable and remunerative price environment. Agricultural price policy in India started with the basic intention of incentivizing Indian farmers to adopt the improved package of technology effectively. The model Agricultural Produce Marketing (Development and Regulation) Act, 2003 was formulated and shared with all the states for implementation. Only a few states could completely modify the existing Acts and implement the amendments. Newer marketing platforms were also created, such as eNAM.

Contrary to the overall trade, India has always been a net exporter in the case of agriculture despite the initial phases of attaining self-sufficiency in most of the commodities. The exports of agricultural commodities picked up after 1970–1971; however, a kick start was attained only after 1994–1995 with the launch of global trade reforms and trade integration with the establishment of the World Trade Organization. Over the last 25 years since India's liberalization, foreign trade has expanded multifold and seen significant structural shifts in product mix as well as geographic spread. Liberalization in trade policies related to easing several trade restrictions, reduction in tariff levels across different products, and other trade reforms have assisted the growth of foreign trade, especially in the first two decades post-liberalization. India's exports increased from Rs. 0.32 lakh crores in 1990–1991 to Rs. 23.08 lakh crores in 2018–2019, and imports also enhanced from Rs. 0.43 lakh crores in 1990–1991 to Rs. 35.95 lakh crore in 2018–2019 (Table 1.5). The composition of exports has gone substantial changes post-liberalization. India's

Table 1.3 Workers in the agricultural sector

Year	Total workers	Agricultural workers			Share in total workers (%)		
		Cultivators	Agricultural laborers	Total	Share of cultivators in total workers	Share of agriculture labor to total worker	Share of agriculture workers in total worker
1951	139.5	69.9	27.3	97.2	50.11	19.57	69.68
1961	188.7	99.6	31.5	131.1	52.78	16.69	69.48
1971	180.4	78.2	47.5	125.7	43.35	26.33	69.68
1981	244.6	92.5	55.5	148	37.82	22.69	60.51
1991	314.1	110.7	74.6	185.3	35.24	23.75	58.99
2001	402.2	127.3	106.8	234.1	31.65	26.55	58.20
2011	481.9	118.8	144.3	263.1	24.65	29.94	54.60

Source: Census of India, Various issues

Table 1.4 Changing the agrarian structure: number and size of holdings

	Below 1.0	1.0–2.0	2.0–4.0	4.0–10.0	Above 10.0	All
Number of holdings (thousands)						
1960–1961	19,900	10,900	9200	6600	2300	48,900
1970–1971	35,682	13,432	10,681	7932	2766	70,493
1976–1977	50,122	16,072	12,455	8068	2166	88,883
1980–1981	56,147	17,922	13,252	7916	1918	97,155
1985–1986	63,389	20,092	13,923	7580	1654	106,637
1990–1991	71,179	21,643	14,261	7092	1404	115,580
1995–1996	76,222	22,814	14,067	6868	1230	120,822
2001–2001	83,694	23,930	14,127	6375	1096	129,222
2005–2006	64,316	18,776	11,218	5336	1003	100,650
2006–2007	92,826	24,779	13,896	5875	973	138,348
2010–2011	92,688	24,746	13,869	5854	953	138,110
2015–2016	100,251	25,809	13,993	5561	838	146,454
Average holding size (hectares)						
1960–1961	0.44	1.47	2.85	6.08	17.57	2.69
1970–1971	0.41	1.44	2.81	6.08	18.10	2.30
1976–1977	0.39	1.41	2.77	6.04	17.57	2.00
1980–1981	0.39	1.44	2.78	6.02	17.41	1.84
1985–1986	0.39	1.43	2.77	5.96	17.21	1.69
1990–1991	0.39	1.43	2.76	5.90	17.33	1.57
1995–1996	0.40	1.42	2.73	5.84	17.21	1.41
2001–2001	0.40	1.42	2.72	5.81	17.12	1.33
2005–2006	0.38	1.38	2.68	5.74	17.08	1.23
2006–2007	0.42	1.42	2.72	5.78	15.57	1.30
2010–2011	0.39	1.42	2.71	5.76	17.38	1.15
2015–2016	0.38	1.40	2.69	5.72	17.07	1.08

Source: Agricultural Census, Various Years

export basket is now diversified, with nontraditional items and differential products also gaining importance. Agricultural exports and imports have also increased considerably during the last 25 years. Since the year 2005–2006, there can be seen marked surge both in the export and import of agricultural commodities. The absolute agricultural trade has expanded, and also the share of agricultural exports has increased in recent years (Fig. 1.5).

Table 1.5 Patterns and contribution of agricultural trade from India

Years	Agriculture exports (Rs. crore)	Total exports (Rs. crore)	Agriculture exports to total exports (%)	Agriculture imports (Rs. crore)	Total imports (Rs. crore)	Agriculture imports to total imports (%)	Net trade (Rs crore)	Net agricultural trade (Rs crore)
1965–1966	335	806	42	536	1394	38.43	–589	–201
1970–1971	565	1535	37	604	1642	36.80	–107	–39
1975–1976	1686	4042	42	2142	5265	40.68	–1223	–457
1980–1981	2376	6683	36	2300	12,549	18.32	–5866	76
1985–1986	3271	10,895	30	3889	19,657	19.78	–8763	–618
1990–1991	6013	32,527	18.49	1206	43,171	2.79	–10,644	4807
1995–1996	20,398	106,353	19.18	5890	122,678	4.8	–16,325	14,508
2000–2001	28,657	201,356	14.23	12,086	228,307	5.29	–26,950	16,571
2005–2006	45,711	456,418	10.78	15,978	660,409	3.26	–203,991	29,733
2010–2011	113,047	1,136,964	10.28	51,074	1,683,467	3.41	–546,503	61,973
2015–2016	215,396	1,716,378	12.55	140,289	2,490,298	5.63	–773,920	75,107
2017–2018	251,563	1,956,514	12.86	152,095	3,001,033	5.07	–1,044,514	99,469
2018–2019	274,571	2,307,726	11.90	137,019	3,594,674	3.81	–12,869	137,551

Source: Agricultural Statistics at a Glance, (various issues)

Recent Marketing Reforms

In order to revive the Indian economy, the Centre government has announced the Atma Nirbhar Bharat Abhiyan. It includes the announcement of Rs 1 lakh crore fund to finance agriculture infrastructure projects at the farm gate and produce aggregation points. Also, Rs 500 crore has been allocated to extend Operation Greens, which comprises Tomatoes, Onion, and Potatoes (TOP) to ALL fruits and vegetables (TOTAL). Apart from it, the government has also shown the intention to usher in agricultural marketing reforms. These reforms relate to amending the Essential Commodities Act (ECA), 1955, bringing Central legislation to allow farmers to sell their products to anyone outside the APMC mandi yard and creating a legal framework for contract farming.

1.8 Shift in Consumption Pattern

India is a country with around 1.38 billion population (almost 17% of the world) with changing food habits. The agricultural systems and policies in the initial phases largely catered to meeting the basic objective of reducing hunger and poverty. Breakthroughs in production technology for enhancing the food grain yield, named the Green Revolution, led to self-sufficiency in food production. Rising income further facilitated consumption diversification leading to greater demand for the consumption of high-value products. The nutrition transition is reported to be an important consequence of economic and structural transformation. Economic growth, changes in tastes and preferences, and urbanization has resulted in changing consumption pattern away from traditional food commodities to processed and high-value commodities (Murty 2000; Meenakshi 1996; Rao 2000) (Fig. 1.6).

There is a clear consumption shift from food grains toward high-value products like fruits and vegetables, livestock, and fisheries. This is manifest in increasing area coverage and production in respect of these sectors and reflected in the increasing value share of these items over time. The increasing demand for high-value commodities like fruits and vegetables, livestock products, and fisheries can be tapped better by a shift in policy to focus on what would constitute the growth drivers in the coming years. This potential can be better harvested by reorienting the policy to enhance investments in these allied sectors for improving productivity, quality, and efficiency. Over the past two decades, the share of food in total expenditure (as explained by the monthly per capita expenditure, MPCE) has fallen in rural India, signaling a clear shift in expenditure behavior. An increasing trend toward nonfood expenditure is clearly visible, and one would expect the trend to continue in the near future. The physical consumption quantities display a consistent decline in cereals in both rural and urban India, and the trend holds true for pulses as well. Around 15% of the household expenditure budget has shifted toward nonfood

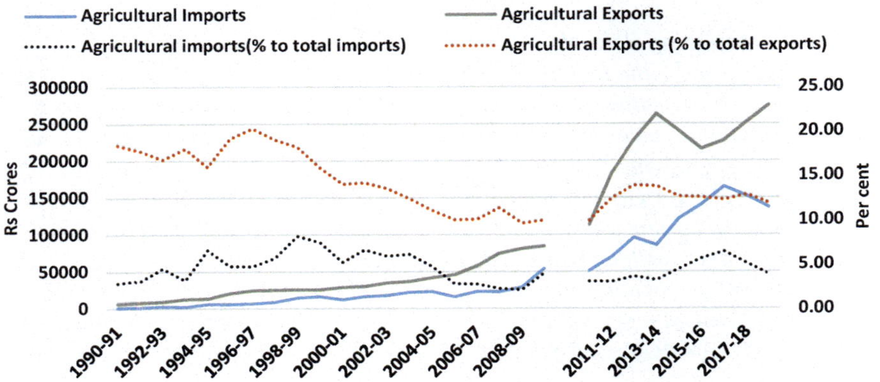

Fig. 1.5 Contribution of agricultural trade to total. (Source: Central Statistics Office, MoSPI)

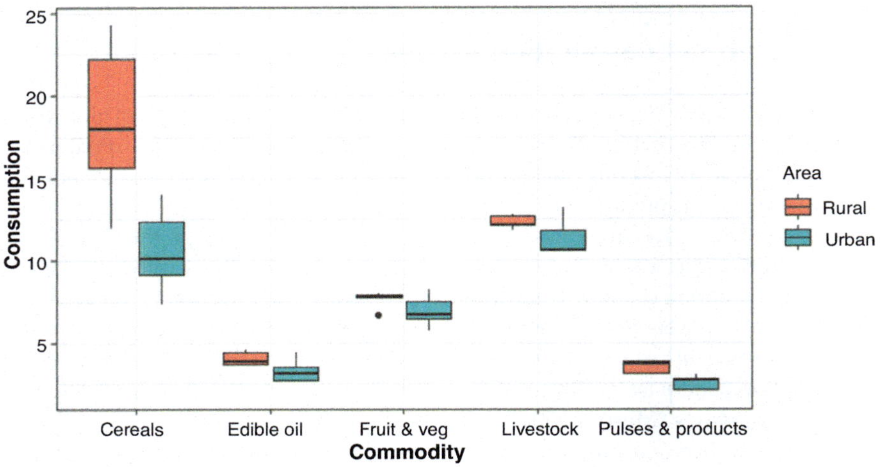

Fig. 1.6 Consumption pattern of food items in India. (Source: NSSO Rounds on Household Consumer Expenditure Survey, Govt of India)

expenditure. While expenditures on fuel, light, and other items have also registered consistent and marginal improvements, doubling expenditure on durable goods can be appreciated as it reflects household welfare. More than 9% increase in expenditure on other goods and services also indicates an increasing preference toward nonfood items than food items (Table 1.6).

Table 1.6 Trends and patterns in consumption

Item group	Share in total consumption expenditure (%)									
	Rural					Urban				
	1993–1994	1999–2000	2004–2005	2009–2010	2011–2012	1993–1994	1999–2000	2004–2005	2009–2010	2011–2012
Consumption pattern of major items (per person per month)										
Cereals (kg)	13.4	12.72	12.12	11.35	11.22	10.6	10.42	9.94	9.37	9.28
Pulses (kg)	0.76	0.84	0.71	0.65	0.78	0.86	1.00	0.82	0.79	0.90
Milk (liter)	3.94	3.79	3.87	4.12	4.33	4.89	5.10	5.11	5.36	5.42
Egg (number)	0.64	1.09	1.01	1.73	1.94	1.48	2.06	1.72	2.67	3.18
Fish (kg)	0.18	0.21	0.20	0.27	0.27	0.20	0.22	0.21	0.24	0.25
Mutton (kg)	0.06	0.07	0.05	0.05	0.05	0.11	0.10	0.07	0.09	0.08
Chicken (kg)	0.02	0.04	0.05	0.12	0.18	0.03	0.60	0.85	0.18	0.24
Consumption expenditure on major categories (MPCE value shares)										
Cereals	24.2	22.2	18.0	15.6	12.0	14.0	12.4	10.1	9.1	7.3
Pulses and products	3.8	3.8	3.1	3.7	3.1	3.0	2.8	2.1	2.7	2.1
Milk and products	9.5	8.8	8.5	8.6	9.1	9.8	8.7	7.9	7.8	7.8
Edible oil	4.4	3.7	4.6	3.7	3.8	4.4	3.1	3.5	2.6	2.7
Egg, fish, and meat	3.3	3.3	3.3	3.5	3.6	3.4	3.1	2.7	2.7	2.8
Vegetables	6.0	6.2	6.1	6.2	4.8	5.5	5.1	4.5	4.3	3.4
Fruits and nuts	1.7	1.7	1.9	1.6	1.9	2.7	2.4	2.2	2.1	2.3
Sugar	3.1	2.4	2.4	2.4	1.8	2.4	1.6	1.5	1.5	1.2
Salt and spices	2.7	3.0	2.5	2.4	2.4	2.0	2.2	1.7	1.5	1.7
Beverages, etc.	4.2	4.2	4.5	5.6	5.8	7.2	6.4	6.2	6.3	7.1
Food (total)	**63.2**	**59.4**	**55.0**	**53.6**	**48.6**	**54.7**	**48.1**	**42.5**	**40.7**	**38.5**
Nonfood (total)	**36.8**	**40.6**	**45.0**	**46.4**	**51.4**	**45.3**	**51.9**	**57.5**	**59.3**	**61.5**
Total expenditure	**100.0**	**100.0**	**100.0**	**100.0**	**100.0**	**100.0**	**100.0**	**100.0**	**100.0**	**100.0**

Source: MOSPI, Various Reports

Box 1.2 Projected Demand of Major Food Commodities (Million Tonnes)

Commodities	2030*	2050**
Cereals	284	359
Pulses	26.6	46
Edible oils	21.3	39
Vegetables	192	342
Fruits	103	305
Milk	170.4	401
Sugar	39.2	58
Meat	9.2	14
Egg	5.8	10
Fish	11.1	22

Source: *Kumar et al. (2016) for projected demand in 2030, **NCAP Vision 2050 for projected demand in 2050.

Owing to the increasing population pressure over the years, demand for food is naturally expected to increase in the coming years. Various studies have projected the demand for food grains under alternative assumptions of income growth, distribution of income, and future dynamics of rural and urban populations. Box 1.2 presents the projected national demand for major food commodities for the years 2030 and 2050, as estimated in different studies. Substantial increases in the consumption of high-value food commodities like fruits, vegetables, milk, meat, fish, and eggs have been projected. It is encouraging to note that the outputs have risen more than the projections because of technological improvements and better logistics. The rising food demand needs to be accompanied by increasing demand for its safety and quality owing to the rising health consciousness of the masses. The major challenge lies in developing technologies, practices, varieties, and breeds that are high-yielding as well as safe for human health.

1.9 Conclusions

The growth trajectory of the agricultural sector has moved through different phases and evidenced varied growth patterns along with changes in other related sectors. The crop sector contributed the maximum to agricultural growth, and overall growth in agriculture was largely determined by growth in the crop sector. Livestock and fisheries sectors have grown at an appreciable growth. Fisheries growth has outpaced growth in other sectors. Since independence, food grains production enhanced by six times, fish by 19 times, milk 12 times, and eggs 67 times since 1950–1951, thus making a visible impact on national food and nutritional security. However, the interest of the farming community in agriculture is declining, and consequently, the agricultural workers who are self-employed in agriculture are leaving the industry. This reported shift is good, provided the workers, who left the sector, are productively and gainfully employed in alternate sectors/industries. Trends in components of nonfood items remain more or less equal in rural and urban domains. Rather, while food and nonfood expenditure are converging in the rural sphere, urban India shows a clear divergence, with a sharp fall in food expenses and a corresponding increase in nonfood expenses.

The country is the largest producer of pulses in the world and the second-largest producer of paddy and wheat. India is also an important producer of commercial crops like cotton, sugarcane, and tobacco. There is a need to sustain the enhanced outputs and maximize land productivity. In many cases, the crop productivity in India is much lower than in other producing nations. We need to ensure that the maximum potential of any crop variety is harvested at the farmers' field. Significant yield gaps exist across various crops in different states, as well as within states. Bridging these yield gaps will not only increase production but also help to improve the efficiency of land and labor, reduce production costs, and add to food security. Improving the crop yields is also important from the perspective of releasing land for other productive uses, by diversifying into high value activities and commodities, and to allow farmers to scale up integrated farming practices.

The Government is placing a strong emphasis on neglected sectors. India suffers from a huge deficit of edible oils and has emerged as the major importer of food oil in the world. As against the domestic requirement of about 23 million tonnes of vegetable oils, the country's domestic output from both primary and secondary sources is about 8.5 million tonnes. The oilseeds output is 32 million tonnes, far short of the requirement and potential. The yields can be optimized, and there is scope to increase the net sown area by improving both cultivable wasteland and fallow. The gross sown area can be increased through intensive cropping.

The number of marginal and small holdings in the country and the area under such holdings have increased, while the number of semi-medium, medium, and large holdings and the area under such holdings has reduced. The area under nonfood crops as a proportion of the total cropped area is increasing, but still, there is a dominance of food crops. This shift in the allocation of the area from food crops to nonfood crops reflect a change from subsistence cropping to commercial cropping. Furthermore, the area under fruits, vegetables, and oilseeds is gradually increasing. This is because the consumption pattern is shifting from cereals to noncereals. Policy reforms are continuously facilitating the growth of agriculture and will help sustain its growth trajectory.

References

Bai J, Perron P (1998) Estimating and testing linear models with multiple structural changes. Econometrica 66:47–78

Chand R (2003) Government intervention in Foodgrain Markets in the Changing Context, policy paper 19. National Centre for Agricultural Economics and Policy Research, New Delhi

Chand R, Shinoj P (2012) Temporal and spatial variations in agricultural growth and its determinants. Econ Polit Wkly 47(26–27):55–64

Chand R, Prasanna PAL, Singh A (2011) Farm size and productivity: understanding the strengths of smallholders, and improving their livelihoods. Econ Polit Wkly 46(26 & 27):5–11

Government of India (2018) Report of the committee on doubling farmers' income, volume I march of agriculture since Independence and growth trends. Department of Agriculture, cooperation and farmers' welfare, Ministry of Agriculture & farmers' welfare, New Delhi

Meenakshi JV (1996) How important are changes in taste? A state-level analysis of food demand. Econ Polit Wkly 31:3265–3269

Murty KN (2000) Changes in taste and demand pattern for cereals: implication for food security in semi-arid tropical India. Agric Econ Res Rev 13(1):25–51

Rao CH (2000) Declining demand for foodgrains in rural India: causes and implications. Econ Polit Wkly 35:201–206

Rao VM (1996) Agricultural development with a human face. Econ Polit Wkly 31(26):A-50–A-62

Food Supply and Security

2

Anugu Amarender Reddy, Suresh Babu, Parmod Kumar,
and Soora Naresh Kumar

Abstract

India's total food grain production in 1950–1951 was low at 50.8 million tonnes, with a population of 361 million. Thus, the food grain production in 1950–1951 was 140.7 kg per person per annum or 0.39 kg per day. Thanks to Indian farmers and agricultural scientists who worked hard to increase the food grain production through new crop varieties and production technologies, along with the supportive policies of the governments that paved the way for the Green Revolution in Indian Agriculture. Achievements of the green revolution further led to achievements in other agricultural and allied sectors like the white revolution with substantial gains from milk production, followed by the yellow revolution with a significant increase in edible oilseed production, and the pink revolution with an increase in meat and poultry production to a significant extent.

This chapter mainly discusses where does India stand today in terms of its agriculture when compared to its independence in 1947? As the data for 1947 for most of the indicators is not available, 1951 is considered the base year and compared the various indicators for the year 2021.

A. A. Reddy (✉)
Section of Design & Analysis, ICAR-Central Research Institute for Dryland Agriculture, Hyderabad, India
e-mail: amarender.reddy@icar.gov.in

S. Babu
International Food Policy Research Institute (IFPRI), Washington, DC, USA

P. Kumar
Giri Institute of Development Studies (GIDS), Lucknow, India

S. N. Kumar
ICAR-Indian Agricultural Research Institute, New Delhi, India

P. K. Ghosh et al. (eds.), *Trajectory of 75 years of Indian Agriculture after Independence*, https://doi.org/10.1007/978-981-19-7997-2_2

2.1 Introduction

India's total food grain production in 1950–1951 was low at 50.8 million tonnes, with a population of 361 million. Thus, the food grain production in 1950–1951 was 140.7 kg per person per annum or 0.39 kg per day. Thanks to Indian farmers and agricultural scientists who worked hard to increase food grain production through new crop varieties and production technologies, along with the supportive policies of the governments that paved the way for the Green Revolution in Indian Agriculture. Achievements of the green revolution further led to achievements in other agricultural and allied sectors, like the white revolution with substantial gains from milk production, followed by the yellow revolution with a significant increase in edible oilseed production, and the pink revolution with an increase in meat and poultry production to a significant extent.

This chapter mainly discusses where does India stand today in terms of its agriculture when compared to its independence in 1947? As the data for 1947 for most of the indicators is not available, 1951 is considered the base year, and compared the various indicators for the year 2021.

2.2 Period-Wise Food Security Strategies and Approaches

Broadly, the overall period from independence to 2021 is divided into: Phase-I, Pregreen revolution period (1947–1965); Phase-II, Green Revolution Period (1965–1980); Phase-III, Post-Green Revolution Period (1980–1991); Phase IV, Economics reforms period (1991–2015); and Phase V, One nation, One market (2015 onwards). The problems faced, strategies, and approaches followed in different periods are presented in Table 2.1.

2.2.1 Pre-Green Revolution Period (1947–1965)

When India became an independent nation on August 15, 1947, India's population was just 330 million, mostly living in villages with poverty and low life expectancy. Agriculture had very low productivity, mostly led by self-sustaining small farmers. The marketed surplus was very limited, and mostly marketed in weekly markets along with other daily necessities. The share of coarse cereals and pulses was much higher in the food basket. The trading was mostly done by informal and unregulated petty village traders, who purchased from the farmers and sold to aggregators or in weekly markets with limited postharvest processing like cleaning, grading, and packaging. Slowly different state governments started regulating the agricultural markets through the formation of Agricultural Produce Market Committees (APMCs) at the block level with twin objectives of regulating the malpractices of traders and commission agents to ensure reasonable prices to farmers and also to maintain reasonable prices for consumers. This period is characterized by frequent famine-like situations. Traders frequently hoarded and speculated on food grains to

Table 2.1 Problems, strategies, and approaches followed since independence

Phase	Status and approach
Phase-I: Pre-Green Revolution Period (1947–1965)	Problem: Deficit in food production
	Strategy: "Grow more food" campaign and improved food security
	Approach: Marketing system designed to the handle deficit, regulate interstate trade, manage food security, APMC Acts and ECA Act, 1955, PL-480
Phase-II: Green Revolution Period (1965–1980)	Problem: Self-sufficiency in food grains, ushering in "green revolution" (wheat and rice) and "operation flood" (milk sector)
	Motive: Ensure food security
	Approach: Usage of technology and high yielding varieties (HYVs) to boost production and distribution through procurement. Number of important institutions set up (FCI, APC, MSP, CWC, and SAUs)
Phase-III: Post-Green Revolution Period(1980–1991)	Status: Diversification toward high-value crops.
	Strategy: Enhance the value of the output
	Approach: Focus on commercial agriculture, setting up National Horticultural Board (NHB) and agricultural and processed products export development authority (APEDA)
Phase IV: Economics reforms period (1991–2015)	Problem: Approaching surplus
	Strategy: Improving the functioning of markets and greater international market access for export and imports
	Approach: Signing of Agreement of Agriculture of WTO; rapid growth of poultry and milk production; Initiation of consultations on Market reforms: Report of Committee on Strengthening and Developing of Agricultural Marketing under the chairmanship of Shri. Shankerlal Garu submitted on 29.06.2001, Model APMC Act 2003 to increase private sector participation in marketing and processing; Model APMC Rules, 2007. Some states adopted the Model Act.
Phase V: One nation, one market (2015 onwards)	Problem: Food secure, but the problem of plenty emerges especially in cereals
	Strategy: Enhance farmers' income with freedom to market access with one nation, one market strategy
	Approach: Toward a National unified Electronic National Agricultural Market (e-NAM), the Model Agricultural Produce and Livestock Marketing (Promotion & Facilitation) Act, 2017 (APLM) allows for the operation of alternative markets and unified national markets; GST rollout, streamline interstate trade

Source: Ministry of Agriculture and Farmers Welfare (2018) Report of the "Committee on Doubling on Farmer's Income," Ministry of Agriculture and Farmers Welfare, Government of India https://agricoop.nic.in/en/doubling-farmers

take advantage of high price fluctuations. To control speculative and large-scale hoarding of food grains, the government of India passed the Essential Commodities Act (ECA Act), 1955, to impose stock limits maintained by the middlemen and traders and release excess stocks into the market for sale in the times of famines and food shortages. Further, during this period, India relied on supplies of food grains from the United States under Public Law 480 (PL-480) against rupee payments, as India did not have much foreign exchange to buy large quantities of food in international markets. During the peak years, under PL-480, India imported 10 mt of food grains, this huge import dependence on basic necessities like food was a humiliation to the government; hence the government started taking steps to become food self-sufficient.

2.2.2 Green Revolution Period (1965–1980)

During the late sixties, the government gave special emphasis on food self-sufficiency to reduce dependency on PL-480 imports. India imported 18,000 tonnes of high-yielding varieties (HYV) of wheat from Mexico in 1966 and ushered in the Green Revolution. Improved technologies (HYV seeds, water, and fertilizers) and innovation in institutional development (like the cooperative movement ushered in the white revolution) have been the driving force of the green revolution. Although seeds originally came from outside the country (as with the short duration high yield varieties of paddy and wheat). But the farmer-entrepreneur takes the risk of adopting these seeds and technologies, puts in his/her best efforts, and the nation reaps a rich harvest to feed its citizens. Measures were taken during the green revolution (1960s to 1980s): like FCI, PDS, and their impact.

2.2.3 Post-Green Revolution Era (1980–1991)

The last phase could be the period of the post-green revolution, where a lot of focus was being provided on improving productivity and quality with the objective of enhancing farmers' income. Green revolution of paddy and wheat spread to other states like Andhra Pradesh, Karnataka, Tamil Nadu and also western states like Gujarat. To some extent, there was a shift to commercial crops and increased use of fertilizers for increasing productivity. This period saw the success of the yellow revolution meant for self-sufficiency in oilseeds (the Technology Mission on Oilseeds).

2.2.4 Economic Reform Period (1991–2015)

The growing private sector participation especially in the seed sector and with Bt cotton seeds from large private-sector companies such as Mahyco and Monsanto and the spread of basmati rice varieties like Pusa basmati and several hybrids of maize became popular among farmers during this phase. The exports of many agricultural commodities started picking up, especially basmati rice, oil cake (meal), meat, and fish.

2.2.5 One Nation, One Market (2015 Onwards)

The Government of India focuses on one nation, one market with sweeping reforms in agricultural markets, strengthening market infrastructure, and encouraging private markets and public–private partnerships. With all these efforts, the agricultural sector showed astonishing resilience during the COVID pandemic. Agriculture was the only sector that grew at a comfortable rate of 3.5% during 2020—the COVID year, while all other sectors suffered negative growth. Even during the strict lockdown, there was no shortage or no price hike for essential food items like rice, wheat, fruits, and vegetables. The supply chains evolved themselves to overcome logistical bottlenecks. It is now the responsibility of the policymakers and government to build and support these resilient supply chains.

2.3 Evolution of Technological and Institutional Provisions (1951–2021)

Technology has been the main driver of change over the years, whether it is the green revolution led by wheat and rice, or maize productivity led by hybrids or cotton production led by Bt technology. Fig. 2.1 shows the changes in the yield index with 1951 as the base year. From the last 70 years, the highest yield increase was recorded in cotton (yield index increased from 100 to 585), indicating 5.8 times increase in yield, mostly after 2001 with the introduction of Bt cotton varieties; the next highest

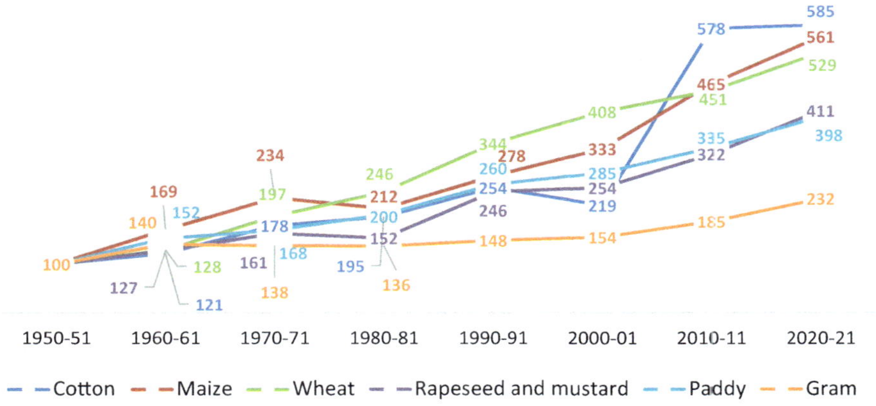

Fig. 2.1 Index of yields of major crops from 1951 to 2021. Source: Agriculture at a glance, various issues

increase was in maize (from 100 to 561), indicating 5.61 times increase over 70 years, followed by wheat 5.29 times, rapeseed and mustard (4.11 times), paddy 3.98 times and 2.32 times in gram. These are some of the success stories from the last 70 years. Although yield played a major role in the overall increase in production and availability of food items, the increase in the area also played a major role. For example, there is a significant increase in cotton areas in response to increased demand and prices.

2.4 Changes in Food Availability (1951–2021)

An increase in food production is the first indicator of food available for consumption by the population. While the population has grown from 361 million in 1947 to almost 1.403 billion in 2021, that is by almost 3.9 times; our cereal production has increased by almost 6.73 times (from about 42.4 mt in 1951 to 285.3 mt in 2021) (Table 2.2). Pulses production has gone up by three times (from 8.4 mt in 1951 to 25.5 mt in 2021), oilseeds production by 6.9 times (from 5.2 mt to 36.1 mt), cotton by more than 12 times (from 3.0 million bales to 36.1 million bales), milk by 12.4 times (from 17 mt to 210 mt), meat by more than 7.3 times (from 1.2 mt to 8.8 mt), poultry by 43 times (from 0.1 mt to 4.3 mt), eggs by 64 times (from 2 billion to 128 billion), and fish by 18.1 times. It indicates India performed in terms of the availability of various food items even after discounting for population growth (Fig. 2.2).

It indicates the highest growth in production was in eggs, followed by poultry, fish, milk, and cotton with a more than tenfold increase, while the increase is limited in the case of pulses, cereals, oilseeds, and meat. However, in the case of pulses, production growth is less than the population growth, resulting in a decrease in per capita availability of pulses in 2021 compared to 1951, even after accounting for import growth.

2.5 Changes in Per Capita Availability (1951–2021)

Except for pulses and coarse cereals, the net availability of all other food items increased over the last 70 years. Per capita availability of rice increased from 58 kg to 73 kg (25.9%), wheat increased from 24 kg to 65 kg (171% increase), and total food grains (cereals plus pulses) increased from 144 kg to 187 kg (an increase of 30%). Edible oil availability increased from 3 kg to 19 kg (533% increase), milk per capita availability increased from 130 g/day to 427 g/day (228% increase), and meat availability increased from 3.3 kg to 6.5 kg (97% increase). The spectacular performance was seen in poultry and eggs from 0.2 kg to 3.1 kg (1450% increase) and from 5 to 91 (1720% increase), respectively (Table 2.3).

Table 2.2 Annual production (mt) from 1951 to 2021

Year	Cereal	Pulse	Oilseed	Cotton*	Milk	Meat	Poultry	Egg (billion nos)	Fish	Population (million)
1951	42.4	8.4	5.2	3.0	17.0	1.2	0.1	2	0.8	361
1961	63.4	10.4	7.0	5.6	17.5	1.7	0.1	3	0.9	439
1971	79.2	9.3	9.6	4.8	19.6	2.1	0.1	7	1.6	548
1981	105.2	9.0	9.4	7.0	29.3	2.7	0.1	10	2.2	683
1991	142.7	12.1	18.6	9.8	45.8	3.7	0.5	21	3.4	846
2001	165.3	12.0	18.4	9.5	66.1	4.3	1.0	37	5.1	1029
2011	185.8	17.3	32.5	33.0	98.2	5.2	2.2	62	7.2	1211
2021	285.3	25.5	36.1	36.0	210.0	8.8	4.3	128	14.5	1403
Increase (times)	6.73	3.0	6.9	12.0	12.4	7.3	43.0	64.0	18.1	3.9

Source: Agricultural statistics at a glance, various issues. Note: *Million bales of 170 kg each

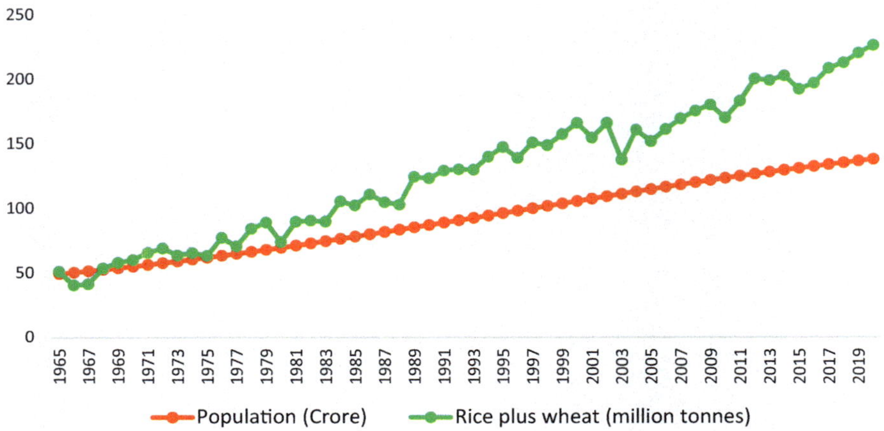

Fig. 2.2 Trends in India's population (crore) and food grain production (million tonnes). (Source: Agricultural statistics at a glance, various issues)

2.6 Evolution of Institutional Framework

India has transformed from a famine-like situation in the 1940s and 1950s to a food-surplus economy over the years. Now it is a leading agricultural exporting country. During the shortage period, the government of India came out with new regulations and institutional framework to ration food grains so that there would not be any spike in prices of essential food items. First among the regulations is the establishment of regulated markets at the block level under Agricultural Produce Market Committee (APMC) Act to eliminate speculation and malpractices by the traders and commission agents and to provide a platform for farmers to sell their produce at reasonable prices. The Essential Commodities Act of 1955 was to impose stock limits on traders to curb hoarding and speculation. With the onset of the green revolution, to procure grain in green revolution states and distribute in food deficit states, a plethora of institutions evolved, of which the Food Corporation of India (FCI) in 1964, the Warehousing Corporation of India (WCI) 1962, Agricultural Price Commission (APC) in 1965 were established. However, after this initial phase, there was stagnation in reforms in the agricultural sector. Then after a long gap, Model APMC Act 2003 was distributed to states to reform their agricultural markets through (i) the introduction of private markets, (ii) contract farming, (iii) electronic national markets, (iv) direct marketing, etc. This model act is changed as per the new information and consensus (Box 2.1 for details of institutional and legislative developments related to agriculture and food).

Table 2.3 Per capita net availability of food grains (per annum) in India (kilograms per year)

Year	Rice	Wheat	Other cereal	Cereals	Pulses	Food grains	Edible oil	Cotton (meters)	Milk (g/day)	Meat	Poultry	Egg (number)	Fish
1951	58	24	40	122	22	144	3	8	130	3.3	0.2	5	2.1
1961	73	29	44	146	25	171	3	15	126	3.9	0.2	7	1.9
1971	70	38	44	152	19	171	4	16	110	3.8	0.2	13	3.0
1981	72	47	33	152	14	166	4	17	128	3.9	0.2	15	3.2
1991	81	60	29	171	15	186	6	24	176	4.4	0.6	25	4.1
2001	70	50	21	141	11	152	8	31	217	4.1	0.9	36	5.0
2011	66	60	24	150	16	171	14	44	263	4.3	1.8	51	5.9
2021	73	65	31	170	18	187	19	47	427	6.5	3.1	91	10.3
% increase	25.9	171	−23	39	−18	30	533	488	228	97	1450	1720	390

Source: Agricultural statistics at a glance, various issues. Note: *Million bales of 170 kg each

Box 2.1 Historical Legal and Regulatory Framework Related to Food Markets (Laws and Orders)
Big role for the government

- Berar Cotton and Grain Market Act of 1887.
- Agricultural Produce Marketing (Grading and Marketing), Act, 1937.
- AGMARK, 1937.
- Public Distribution System(PDS), 1945.
- The Essential Commodities Act, 1955.
- State Trading Corporation of India, 1956.
- National Cooperative Development Act, 1962.
- The Warehousing Corporations Act, 1962.
- Food Corporation of India Act, 1964.
- Agricultural Price Commission, 1965.
- Prevention of Black Marketing and Maintenance of Supply of Essential Commodities Act, 1980.
- The Bureau of Indian Standards Act 1986.

Liberalization of agricultural markets

- Model APMC Act (Agricultural Produce Marketing Committee) Act 2003.
- Forward Contract (Regulation) Amendment Act 2006.
- The (Warehousing Development and Regulation) Act, 2007.
- National Food Security Act, 2013.
- Model Agriculture Land Leasing Act, 2016.
- Model Agricultural Produce and Livestock Marketing (promotion and facilitation) Act, 2017.
- Electronic national agriculture markets (eNAM).
- Model Contract Farming Act, 2018.

2.7 Demand and Supply Projections 2033

With the current trend of growth in technology and incentive structure, by 2033, there will be a huge surplus in food grains, especially cereals, i.e., wheat and rice. There will be a surplus also in milk and its products and vegetables. But there is likely a deficit in oilseeds, coarse cereals, pulses, and fruits (Fig. 2.3).

2.8 Evolution of Public Distribution System (PDS)

It is not the availability of food grains, but its distribution to the most vulnerable population, low-income households, and the poorest of the poor and disabled people that cause persistent hunger and malnutrition, and thereby their proper distribution is

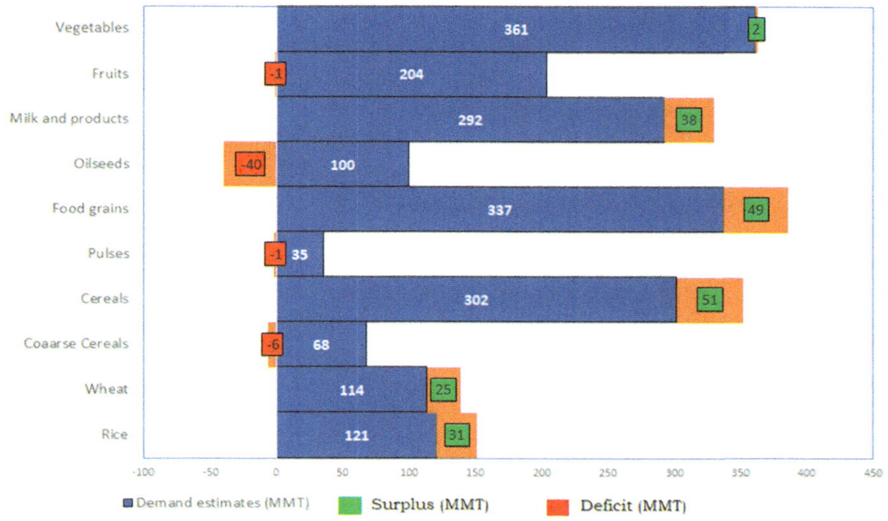

Fig. 2.3 Aggregate demand and supply projections, 2032–33. (Source: NITI Aayog 2018)

Table 2.4 Net availability, procurement, and public distribution of food grains (in million tonnes)

Year	Net production of food grains	Net imports	Net availability of food grains	Procurement	Public distribution
1951	48.1	4.8	52.4	3.8	8.0
1961	72.0	3.5	75.7	0.5	4.0
1971	94.9	2.0	94.3	8.9	7.8
1981	113.4	0.7	114.3	13.0	13.0
1991	154.3	(−)0.1	158.6	19.6	20.8
2001	172.2	(−)2.9	156.9	42.6	13.2
2011	213.9	(−)2.9	203.1	64.5	47.9
2021	276.4	−20.6	253.4	56.8	53.8 (in addition 29.9 mt under PMGKAY) = 83.69

Source: https://www.indiabudget.gov.in/budget_archive/es2000-01/app1.19.pdf
Source: Loksabha unstarred question No. 1614 answered on 8th December 2021

crucial. Keeping the recurrent famines during and after the second world war, the Public Distribution System (PDS) scheme was first started on 14 January 1945 and was launched in its current form in June 1947. The introduction of rationing in India dates back to the 1940s Bengal famine. This rationing system was revived in the wake of acute food shortage during the early 1960s, before the Green Revolution. It involves two types, RPDS and TPDS. In 1992, PDS became RPDS (Revamped PDS), focusing the poor families, especially in far-flung, hilly, remote, and inaccessible areas. In 1997 RPDS became TPDS (Targeted PDS) which established Fair Price Shops for the distribution of food grains at subsidized rates (Table 2.4).

Another major push for food security and PDS was in 2013 with the enactment of The National Food Security Act 2013 (also the 'Right to Food Act'). It is an Indian Act of Parliament that aims to provide subsidized food grains to approximately two-thirds of the country's 1.4 billion people. It was signed into law.

The National Food Security Act, 2013 (NFSA 2013) converts into legal entitlements for existing food security programs of the Government of India. It includes the Midday Meal Scheme, Integrated Child Development Services scheme, and the Public Distribution System. Further, the NFSA 2013 recognizes maternity entitlements. The Midday Meal Scheme and the Integrated Child Development Services Scheme are universal in nature, whereas the PDS will reach about two-thirds of the population (75% in rural areas and 50% in urban areas). As per the latest data, 79.72 crore persons were covered under the NFSA. A total of 5.44 lakh fair-price shops exist across the country for the distribution of food grains to the NFSA beneficiaries.

Under the provisions of the bill, beneficiaries of the Public Distribution System (PDS) are entitled to 5 kg per person per month of cereals. Rice is sold at ₹3 per kg; wheat at ₹2 per kg, and coarse grains (millet) at ₹1 per kg. Pregnant women, lactating mothers, and certain categories of children are eligible for daily free cereals.

To meet the COVID situation, Pradhan Mantri Garib Kalyan Anna Yojana (PMGKAY) was launched to ensure that the targeted beneficiaries (including migrant workers) do not face any issues of food security on account of the pandemic. PMGKAY provided for additional allocation of food grains @ 5 kg per person per month free of cost for all the beneficiaries covered under the Targeted Public Distribution System (Antyodaya Anna Yojana & Priority Households). This was over and above the monthly benefit provided under the National Food Security Act 2013 (Table 2.4).

2.9 Malnutrition

To address anemia and micronutrient deficiency in the country, the Government of India approved a Centrally Sponsored Pilot Scheme on "Fortification of Rice and its Distribution under Targeted Public Distribution System (TPDS)" for a period of 3 years beginning in 2019–2020. The Pilot Scheme is intended to focus on 15 Districts. Under this scheme, state governments have started the distribution of fortified rice. Mid-Day Meal (MDM) and ICDS are covered under the provisions of NFSA.

The government has notified Food Security Allowance Rules, 2O15, under this, pregnant women and lactating mothers and children in the age group of 6 months to 14 years are entitled to meals as per prescribed nutritional norms under Integrated Child Development Services (IGDS) and Mid-Day Meal MDM) schemes. Higher nutritional norms are prescribed for malnourished children up to 6 years of age.

2.10 Children

As per the World Bank (2013), approximately 60 million children in India were underweight, about 45% were stunted (too short for their age), 20% were wasted (too thin for their height, indicating acute malnutrition), 75% were anemic, and 57% were vitamin A deficient. The numbers keep increasing over the years. The majority of them belong to the most vulnerable sections of society living in rural areas. To cater to this section, Anganwadi Centres were created.

It was also observed that the AWCs, especially the ones in tribal hamlets, lacked proper Teaching and Learning Materials, had no LPG cylinder connection, no drinking water facility, no mats or furniture, etc., and were in deplorable conditions.

Now is the time to change the narrative and reimagine the whole framework of working of AWCs in light of digital interventions. Technology has already paved its way in AWCs. The AWWs have been provided with smartphones, and the supervisors with tablets for recording the data accurately and digitally. In 2015, the NITI Aayog recommended providing better sanitation and drinking water facilities, improved power supply, and basic medicines for the AWCs.

Women empowerment is a prerequisite for child development. Many studies on intra-household food consumption pointed out that economic contributions, physically strenuous work, cultural beliefs and social status, concerns about fair allocation, decision-making power, household-level food sufficiency, wealth, and seasonality will determine the food and nutrition status of a child. A number of studies pointed out that women and men have different priorities when controlling income, with women more likely to invest in children's education, nutrition, and health (Masamha et al. 2018; Agarwal 2021).

The increasing presence of undernutrition, obesity, and micronutrient deficiencies simultaneously in the same households results in a triple burden of malnutrition (Béné et al. 2015). In this context, it would be wise to reconsider the inclusion of healthy food commodities such as pulses, nutri-cereals, biofortified rice, etc., in the PDS basket.

2.11 Increasing Exports

There was a significant improvement in the food surplus situation and exports after the 1990s in India. Globally more countries are opening up their domestic economies, providing greater opportunities for India's exports. Exports of agricultural commodities steeply increased since 1991 and started rapidly increasing after 2006–2007. Overall exports of India also increased during this period with the industrial liberalization and free export and import policies to align in line with WTO commitments. Although agricultural exports increased steeply, their relative share decreased from the peak of 20% of total national exports to 12% in 2019–2020. At the same time, India's agricultural imports remain very low, below 10% of total national imports. With the growing surplus of many agricultural commodities, there is a need for increasing agricultural exports by identifying new export commodities

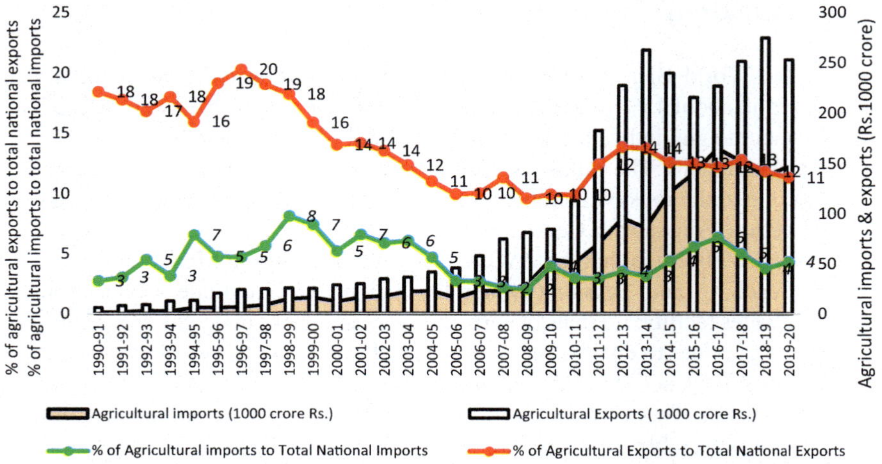

Fig. 2.4 Agricultural exports and imports from 1990–1991 to 2019–2020. Source: Agricultural statistics at a glance, various issues

and countries and strengthening existing commodity groups to the existing destinations. This can be done through long-term export contracts or by joining different trade groups like SAAFTA (Fig. 2.4).

2.12 Inefficiencies in Procurement and PDS

Every year, the government of India spends huge amounts of money, up to 2–4 lakh crore, toward food subsidies and public distribution systems. Although procurement operations at Minimum Support Price (MSP) of paddy and wheat and distribution through ration shops for the needy served its purpose of remunerative prices to farmers and food security for the majority of Indians, inefficiencies have built up in the system over the years.

The buffer stock of food grains in 2020 is more than three times the existing buffer stocking norm (Fig. 2.5). Except for a few years, stocks have been much higher than buffer norms in the recent past. There is about 50 mt excess stock worth about 1.5 lakh crore blocked with no use to either farmers or consumers.

On the supply side, owing to the rise in the MSP, the government has become the sole buyer of paddy and wheat in states like Punjab, Haryana, and Telangana. On the demand side, the expanding food grain commitments under the National Food Security Act pressure the FCI to intensify its procurement drive. Consequently, the government procured a major share of paddy and wheat in the market.

The grain management cost includes acquisition cost, which covers pooled cost, procurement incidentals, and distribution cost. Combining the MSP with the cost of logistics, storage, handling, and distribution, the economic costs for rice and wheat are estimated at ₹37,026 and ₹27,026 per tonne, respectively (Fig. 2.6). The cost of

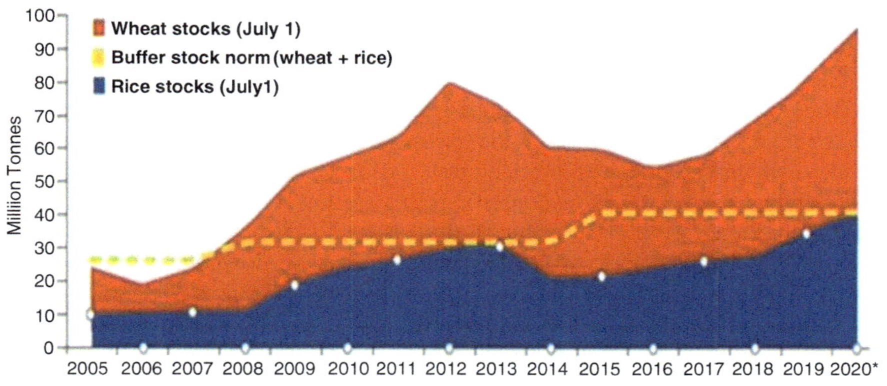

Fig. 2.5 Open-ended procurement operations. (Source: Reddy (2021a, 2021b))

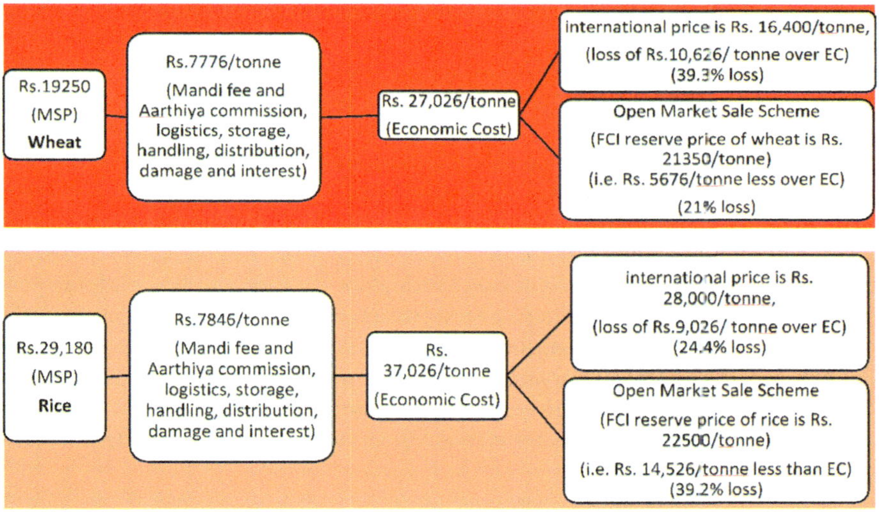

Fig. 2.6 Economics of buffer stock operations for the year 2021. Source: Reddy (2021a)

maintaining strategic stocks or buffer carrying costs is rising, which is attributed to increased procurement incidentals (such as storage, handling, transportation, and other charges), and acquisition and distribution costs. Prices of both rice and wheat in the Open Market Sale Scheme (OMSS) of FCI came down because of excess supply. The price of rice in OMSS is ₹22,500 per tonne in 2020 against ₹27,850 per tonne earlier 3 years back. In the same way, the OMSS price of wheat has also come down to ₹21,350 per tonne against ₹22,450 per tonne earlier. These OMSS prices of both rice and wheat are significantly lower than the economic cost. The reduced sale price of grains for OMSS is due to the excess buffer stock and the shortage of storage. The poor demand for grains in the open market in recent years is also another reason. Therefore, the increasing gap between costs and revenues has strained the finances of the FCI as they lose 39.2% for rice and 21% for wheat in OPSS. The same thing happens for international prices also (Fig. 2.6).

The practice of procuring more than stipulated buffer quantities leads to certain imbalances. It requires offloading the excess stock in the international or domestic market, which could lead to a price crash. Also, it could lead to a loss when the international market prices or domestic prices are lower than the actual economic cost of the stock.

Table 2.5 indicates the vast difference in the share of the procurement of grains against the total production in the case of rice, wheat, and pulses. In case of pulses, procurement is generally below 10% of the production, while in the case of paddy and wheat, it is 43% and 36%, respectively. The neglect of pulses, oilseeds, and nutri-cereals with a focus on only paddy and wheat disrupts a balanced diet among the poor consumers who depend on PDS for their consumption requirements. This needs to be rectified with more procurement of pulses, oilseeds, and other crops to enhance balance in food systems at the national and household levels.

2.13 FAO-Based Food and Nutrition Security Indicators

According to a currently accepted definition (FAO 2000), "Food Security" is achieved when it is ensured that "all people, at all times, have physical, social and economic access to sufficient, safe and nutritious food which meets their dietary needs and food preferences for an active and healthy life." Food is here defined as any substance that people eat and drink to maintain life and growth. As a result, safe and clean water is an essential part of food commodities. The nutrition focus adds the aspects of caring practices and health services and healthy environments to this definition and concept. This aims at what is more precisely called "Nutrition Security," which can be defined as adequate nutritional status in terms of protein, energy, vitamins, and minerals for all household members at all times (Quisumbing et al. 1995 p. 12).

Now, after incorporating nutrition aspects, "Food and nutrition security is achieved, if adequate food (quantity, quality, safety, sociocultural acceptability) is available and accessible for and satisfactorily used and utilized by all individuals at all times to live a healthy and active life." This definition combines food and nutrition security (FNS) and emphasizes several aspects, i.e., "Availability," "Accessibility," and "Use and Utilization" of food. The inclusion of the use and utilization aspect underscores the fact that "Nutrition Security" is more than "Food Security."

2.14 Aspects of Food and Nutrition Security

The conceptual framework of food security Fig. 2.7 illustrates the relationship among the various elements of food security. Two factors influence the framework: a physical and a temporal factor. The physical determinant is the food flow:

Table 2.5 Procurement of rice, wheat, and pulses

Year	Rice		Wheat		Pulses	
	Production (LT)	Procurement (% of production)	Production (LT)	Procurement (% of production)	Production (LT)	Procurement (% of production)
2015–2016	1044.1	32.8	922.9	24.9	16.3	0.0
2016–2017	1097.0	34.7	985.1	31.3	23.1	0.0
2017–2018	1127.6	33.9	998.7	35.8	25.4	6.4
2018–2019	1164.8	38.1	1036.0	32.9	22.1	18.9
2019–2020	1179.4	43.3	1071.9	36.4	23.2	6.5

Source: http://164.100.24.220/loksabhaquestions/annex/178/AU3447.pdf
Source: Lok Sabha Unstarred Question No.3447 Answered On 23rd March,2022

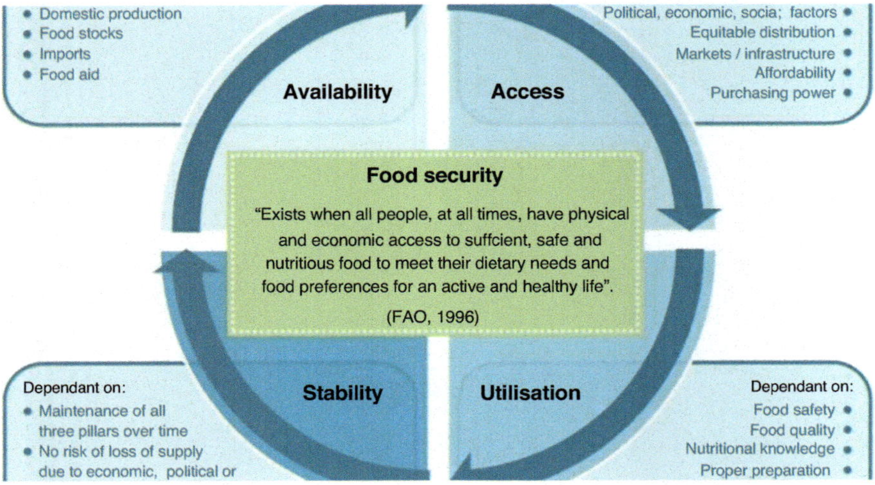

Fig. 2.7 Conceptual framework for food security. Source: FAO (1996)

Availability, Accessibility, Use, and Utilization. The temporal determinant of FNS refers to stability, which affects all three physical elements. In this section, different pillars of food security for India were analyzed from the year 2000–2019.

2.14.1 Food Availability

Availability refers to the physical existence of food, be it from its own production or from the market. On a national level, food availability is a combination of domestic food production, commercial food imports, food aid, and domestic food stocks, as well as the underlying determinants of each of these factors. The use of the term availability is often confusing since it can refer to food supplies available at both the household level and at a more aggregate (regional or national) level. However, the term is applied most commonly in reference to food supplies at the regional or national level. Indicators of availability are explained in Fig. 2.8.

Average dietary supply adequacy, the share of dietary energy supply derived from cereals, roots, tubers, and average protein supply are the indicators of availability. Based on FAO data, the national average per capita food energy supply is calculated. Rice and wheat are the major sources of dietary energy in India. In the early 2000s, the supply of dietary energy was less stagnant (Fig. 2.8) and gradually increased after 2007. Major sources of protein for human consumption are pulses, meat, fish, and milk products. The average protein supply is measured in g/cap/day. Protein supply was somewhat stagnant or even declining in the early 2000s, but from 2005, it shows an increasing trend year-on-year in the same way the share of dietary energy supply derived from cereals, roots, and tubers shows an increasing trend.

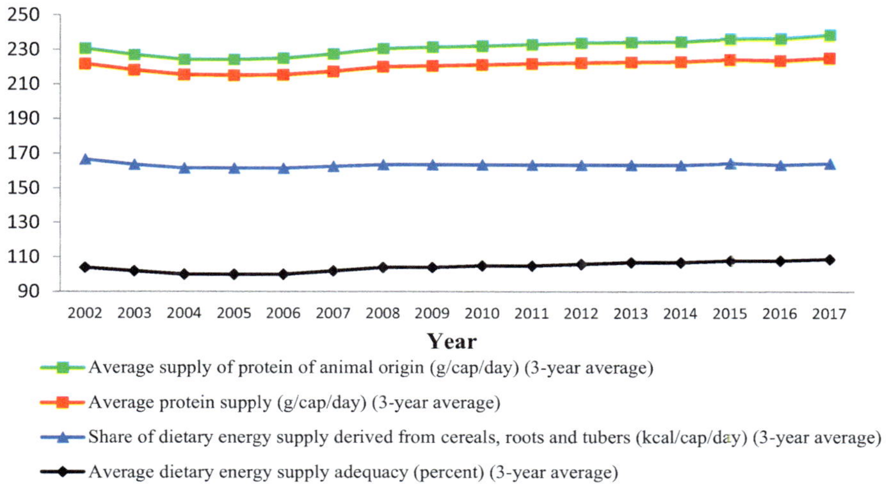

Fig. 2.8 Indicators of food availability. (Source: FAOSTAT 2022)

2.14.2 Food Access

Access is ensured when all households and all individuals within those households have sufficient resources to obtain appropriate foods for a nutritious diet. It is dependent on the level of household resources—capital, labor, and knowledge—and prices.

Figure 2.9 depicts the rail line density and prevalence of undernourishment. The rail line density was lower in 2002, but it slowly improved after 2011. The prevalence of undernourishment is an outcome indicator of food access. It shows the probability that a randomly selected individual from the population consumes a number of sufficient calories to cover her/his energy requirement for an active and healthy life. Undernourishment was high in 2005 (22.1%) but decreased steeply.

GDP per capita was based on PPP$. PPP-GDP is GDP that is converted into international dollars using PPP rates. It was measured in constant 2011 international dollars. Fig. 2.10 depicts the Gross domestic product per capita, PPP, dissemination from 2002 to 2018; the GDP per capita was lowest in 2002 at about 2711.5 dollars and gradually showed an increasing trend, indicating increased purchasing power and food access.

2.14.3 Food Utilization

The use of food refers to the socioeconomic aspect of household food security. If sufficient and nutritious food is both available and accessible, the household has to make decisions concerning what food is to be purchased, prepared, and consumed (demanded) and how the food is allocated within the household. In households

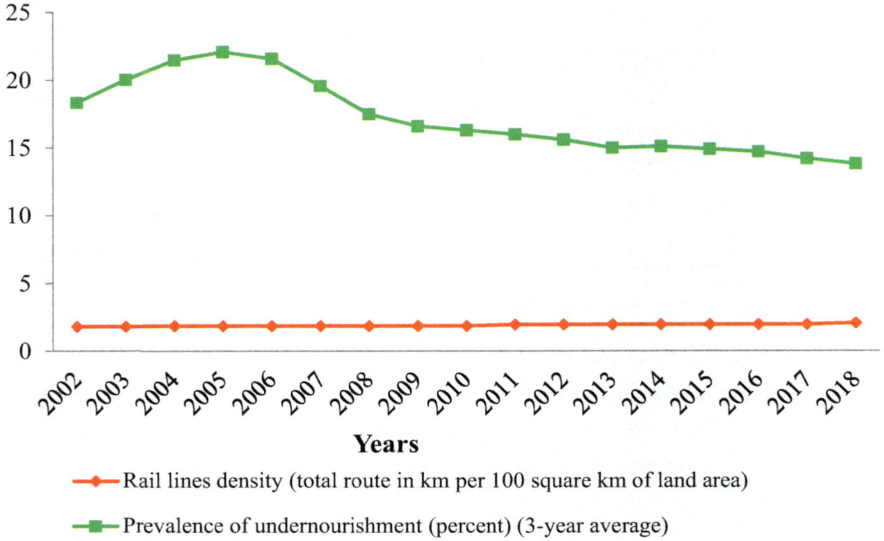

Fig. 2.9 Indicators of food access (Source: FAOSTAT 2022)

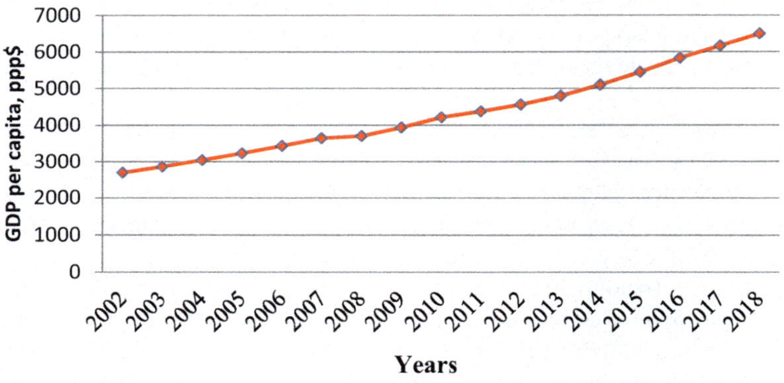

Fig. 2.10 Gross domestic product per capita, PPP$ (FAOSTAT 2022)

where distribution is unequal, even if the measured aggregate access is sufficient, some individuals may suffer from food deficiency. The same is true if the composition of the consumed food is unbalanced.

Focusing on the individual level of food security also requires taking the biological **utilization** of food into consideration. This refers to the ability of the human body to take food and convert it into energy which is either used to undertake daily activities or is stored. Utilization requires not only an adequate diet but also a healthy physical environment, including safe drinking water and adequate sanitary

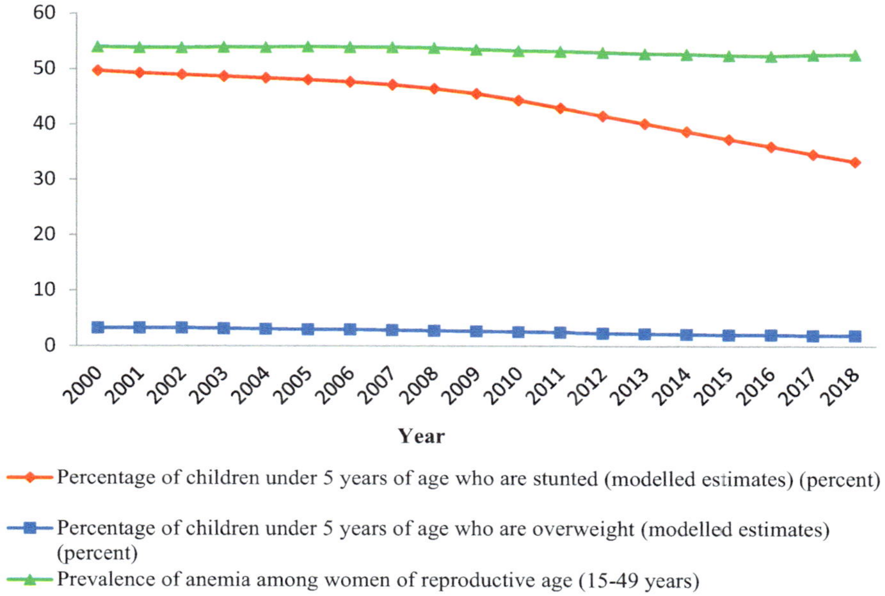

Fig. 2.11 Indicators of utilization (Source: FAOSTAT 2022)

facilities (so as to avoid disease) and an understanding of proper health care, food preparation, and storage processes.

The percentage of children under 5 years of age who are stunted and overweight are outcome indicators in utilization. Children under 5 years of age who are stunted and overweight were higher in 2000 at about 49.8 and 3.3%, respectively; year-on-year, there is a steep decrease in the percentage of children under 5 years of age who are stunted and overweight (Fig. 2.11).

2.14.4 Food Stability

Food stability or sustainability refers to the temporal dimension of nutrition security, i.e., the time frame over which food security is being considered. In much of the food security literature, a distinction is made between chronic food insecurity—the inability to meet food needs on an ongoing basis—and transitory food insecurity, when the inability to meet food needs is of a temporary nature (Maxwell and Frankenberger 1992). Transitory food insecurity is sometimes divided into two subcategories: (i) cyclical, where there is a regular pattern to food insecurity, e.g., the "lean season" or "hungry season" that occurs in the period just before harvest, and (ii) temporary, which is the result of a short-term, exogenous shock such as droughts or floods.

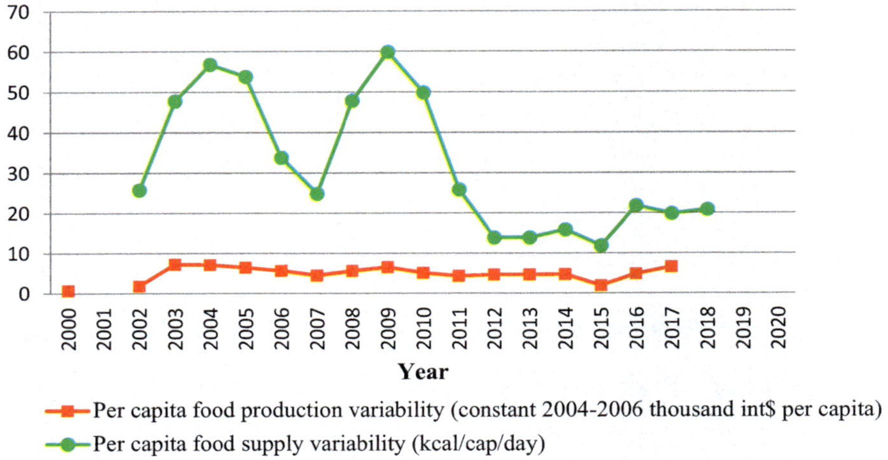

Fig. 2.12 Indicators of stability

Per capita food production variability and Per capita food supply variability (kcal/cap/day) are depicted in Fig. 2.12 that, per capita food production variability corresponds to the variability of the net food production value in constant 2004–2006 international dollars divided by the population number as from UN 2010 estimates. Variability is based on the trend of the net food production index number per capita over the period of 2002–2018 and corresponds to the standard deviation of the deviation from the trend over a period of 5 years. Per capita food production variability was much lower in 2002, which gradually increased in 2003. After 2003 there was a steep decrease in Per capita food production variability, whereas it was much higher in 2017, and it is the same in the case of per capita food supply variability.

2.15 The Problem Analysis

The above analysis showed some undernourishment, especially among women and children, from both secondary sources, which was supported by primary data collection from a few village studies. A detailed analysis of the problem was done by drawing a fishbone diagram of problem analysis, diagnosis, and dissection (Fig. 2.13). The major causes of undernourishment are poverty, irregular incomes, low crop yields, unhealthy food habits like an excessive proportion of rice compared to fruits and vegetables and infectious diseases.

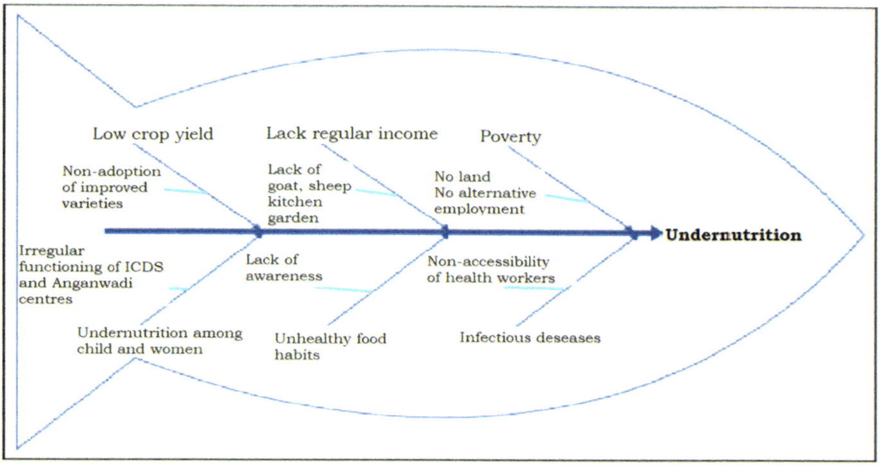

Fig. 2.13 Root cause analysis of undernourishment through fishbone diagram (Source: Focus group discussion in a village in Telangana by the authors)

2.16 Conclusions

This chapter has attempted to analyze data from three dimensions of food availability, access to nutritious food and utilization of food supplies in India. The results and analysis indicate that there is very good progress in food availability with increased production of all food items, and there is also a visible improvement in food access with the implementation of the Food Security Act. There is a need to focus on agriculture diversification, enhanced productivity, incentives for promotion of cultivation, marketing, and demand generation of nutri-rich foods like coarse cereals and pulses. There is a need to focus on strengthening safety net programs like PDS, improving Child and women nutrition programs like ICDS and mid-day meal programs to reduce undernourishment and anemia, and reducing wasting prevalence among children and women. There is a need for improvement in water, sanitation, and hygiene practices and changing consumer behavior toward healthy food.

References

Agarwal B (2021) Livelihoods in COVID times: gendered perils and new pathways in India. World Dev 139:105312

Béné C, Barange M, Subasinghe R, Pinstrup-Andersen P, Merino G, Hemre GI, Williams M (2015) Feeding 9 billion by 2050–Putting fish back on the menu. Food Sec 7:261–274

FAO (2000), The state of food insecurity in the world 2000. Rome

FAO (1996) Declaration on world food security. World food summit. FAO, Rome

Food and Agriculture Organization of the United Nations (2022) FAOSTAT Statistical Database. FAO, Rome

Masamha B, Thebe V, Uzokwe VN (2018) Mapping cassava food value chains in Tanzania's smallholder farming sector: the implications of intra-household gender dynamics. J Rural Stud 58:82–92

Maxwell S, Frankenberger T (eds) (1992) Household food security: concepts, indicators, and measurements: a technical review. UNICEF/IFAD, New York/Rome

NITI Aayog (2018) Demand and supply projections towards 2030: the working group report chaired by Dr. Parmod Kumar. NITI Aayog. Government of India

Quisumbing AR et al (1995) Women: the key to food security. IFPRI Food Policy Report, Washington

Reddy AA (2021a) Assuring farmers income in the context of new farm Laws: issues and the way forward. In: 2021 conference, august 17–31, 2021, virtual (no. 315243). International Association of Agricultural Economists, Toronto, Ontario

Reddy AA (2021b) Boost price stabilisation fund for pulses, oilseeds, the tribune. July 20 2021. https://www.tribuneindia.com/news/features/boost-price-stabilisation-fund-for-pulses-oilseeds-285149?faodatalab=2021-07-19-1

World bank (2013) Helping India Combat Persistently High Rates of Malnutrition May 13, 2013 https://www.worldbank.org/en/news/feature/2013/05/13/helping-india-combat-persistently-high-rates-of-malnutrition

Changing Structure of Rural Employment 3

S. K. Srivastava, Jaspal Singh, and Raka Saxena

Abstract

This chapter empirically examined the long-run gender and occupation-wise changes in rural employment in India using the nationally representative surveys of the National Sample Survey Office (NSSO). Also, studied the implications of changes in labor supply in agriculture on the farm economy. The estimated workforce in rural areas increased from 191 million in 1972–1973 to 361 million in 2019–2020, however, the growth in the workforce remained less than the population growth indicated by the declining worker-population ratio (WPR). Despite a declining share, rural areas still engage 70% of the country's workforce. The results reveal a significantly higher rate of participation in the male workforce as compared to female counterparts in rural areas. The share of females in the rural workforce declined from 36% in 1972–1973 to 24% in 2017–2018. It is worth noting that since 2018–2019, the rural workforce is witnessing a reversal in the declining trend in female participation. Interestingly, the Period Labor Force Survey (PLFS) conducted during 2019–2020 revealed a significant increase in the rural workforce and growth in the agricultural workforce. Withdrawal of agricultural labor affects the farm economy either by creating a physical scarcity of labor or through the rise in farm wages. Because of the inelastic demand for labor, an increase in wages could not bring a proportionate decrease in labor use and result in an increase in labor cost in crop cultivation. Thus, the extent of the decline in labor use is found to be insufficient to negate wage-push cost inflation. This warrants concerted efforts to accelerate the pace of farm mechanization and its economic access to farmers to partially substitute labor.

S. K. Srivastava (✉) · R. Saxena
ICAR-National Institute of Agricultural Economics and Policy Research, New Delhi, India

J. Singh
NITI Aayog, New Delhi, India

P. K. Ghosh et al. (eds.), *Trajectory of 75 years of Indian Agriculture after Independence*, https://doi.org/10.1007/978-981-19-7997-2_3

Keywords

Rural employment · Workforce · Cultivators · Agricultural laborers · Labor demand · Elasticities

3.1 Introduction

Despite rising urbanization, rural areas contribute significantly to the output and employment of the Indian economy. According to the 2011 population Census, 68.8% of the country's population resided in rural areas. Further, 70.9% of the workforce and 46.9% of the output were contributed by rural areas in 2011–2012 (Chand et al. 2017a). Population projections also indicate that India will continue to be predominantly rural till the year 2050, after which the urban population is estimated to overtake the rural population (United Nations 2012). Therefore, developments in rural areas play a critical role in the overall economic development of the nation. Chand et al. (2017b) have observed that an urban worker earns about three times or more than the income of a rural worker. Given the high rural–urban disparity in income and thus unplanned rural-to-urban migration, there is a strong need to generate productive employment in rural areas.

Employment diversification from agriculture to more productive nonfarm sectors has been considered a major source of improving worker productivity in rural areas (Kumar et al. 2011; Himanshu et al. 2011; Chand et al. 2017a). Many scholars have observed that such transitions in rural employment are taking place but at a slower pace than in rural output (Aggarwal and Kumar 2012; Maurya and Vaishampayan 2012; Papola 2012). It is essential to understand the nature of such transitions at a disaggregate level so that suitable strategies can be devised for accelerating the pace of employment diversification toward nonfarm sectors.

Although employment diversification (away from agriculture) is desirable from an economic development point of view, it may lead to physical and economic scarcity of labor for the agriculture sector. The physical scarcity of labor due to their outward movement may hamper the timely completion of farm operations, particularly during the peak season. Further, it may increase labor wages and push cost inflation. As labor constitutes a predominant share of production costs (Raghavan 2008; Srivastava et al. 2017), an increased wage bill can also adversely affect farm profitability. It is therefore required to assess the implications of structural changes in rural employment on the farm economy and devise coping strategies.

With this background, the chapter has empirically examined long-run gender and occupation-wise changes in rural employment in India using the nationally representative surveys of the National Sample Survey Office (NSSO). Further, the implications of changes in labor supply in agriculture on the farm economy have been studied using the Cost of Cultivation (COC) Surveys of the Directorate of Economics and Statistics, Ministry of Agriculture and Farmers Welfare. The findings provide useful insights for developing technological and policy measures for efficient management of labor resources in agriculture.

Table 3.1 Worker–population ratio (WPR) and rural share in the total workforce

			Rural share in total workforce (%)		
Year	Rural workforce (million)	WPR	Agriculture	Nonagriculture	Total
1972–1973	191	46	97	47	84
1983	226	45	96	48	81
1993–1994	293	44	96	47	78
2004–2005	343	44	96	47	75
2011–2012	336	40	96	49	71
2017–2018	316	35	96	49	69
2018–2019	327	36	96	51	70
2019–2020	361	39	96	49	70

Source: Authors' estimates based on NSS estimates, Census population, and National Account Statistics

3.2 Changing Rural Employment

The evidence from successive employment surveys of NSSO reveal that majority of the workforce (based on usual activity status) is employed in rural areas in India. The estimated workforce in rural areas increased from 191 million in 1972–1973 to 361 million in 2019–2020 (Table 3.1). But, the growth in the workforce remained less than the population growth which is indicated by the declining worker–population ratio (WPR). The rural workforce constituted 84% share in the total workforce of India in 1972–1973. Temporally, the rural share in total employment has declined because of relatively higher growth in the urban workforce. Relatively higher growth in the workforce in urban areas could be due to rural-to-urban migration and better employment opportunities in these areas. Despite a declining share, rural areas still engage 70% of the country's workforce. Chand et al. (2017a) observed that the share of rural areas in national output declined at a sharper rate than the decline in rural share in employment during 1970–1971 to 2011–2012. This implies that a major portion of the overall economic growth in the country came from the capital-intensive sectors in urban areas without generating significant employment.

Occupation-wise disaggregation shows that almost entire agricultural activities are taken up in rural areas only. Moreover, about half of the nonagricultural workforce is employed in rural areas. In 2011–2012, 49% of the total nonagricultural workforce engaged in rural areas produced only 35% of the total nonagricultural output in the country (Chand et al. 2017a). This implies that economic activities in rural areas are more labor-intensive, with less worker productivity as compared to urban areas.

3.3 Long-Run Changes in the Structure of Rural Employment

3.3.1 Gender-Wise Changes in the Rural Workforce

The estimates from NSSO surveys on employment reveal a significantly higher rate of participation in the male workforce as compared to female counterparts in rural areas (Table 3.2). This led to the domination of males in the total rural workforce, constituting a 64% share of the total rural workforce in 1972–1973. Over time, the participation of females in the workforce further declined and reached 18% by the year 2017–2018. Consequently, the share of females in the rural workforce declined from 36% in 1972–1973 to 24% in 2017–2018. These evidence reveal a de-feminizing trend in the rural workforce during the period under consideration. The withdrawal of female workers also contributed to the decline in the absolute number of the rural workforce between 2004–2005 and 2017–2018. It is worth noting that since 2018–2019, the rural workforce is witnessing a reversal in the declining trend in female participation.

3.3.2 Occupation-Wise Changes in the Rural Workforce

Rural areas are witnessing a structural change in employment over the years. During 1972–1973, 86% of the rural workforce (based on usual status) was engaged in the agriculture sector (Fig. 3.1). With the economic development, rural areas started diversifying toward nonagricultural activities, and by the year 1993–1994, the share of agriculture in the rural workforce declined to 78% (Table 3.3). The pace of employment diversification further accelerated with significantly higher growth in nonagricultural activities (3.99%) as compared to agriculture activities (0.82%) between 1993–1994 and 2004–2005. Interestingly, the agricultural workforce declined by about 34 million at an annual growth rate of 2.06% between 2004–2005 and 2011–2012. On the other hand, 28 million workers joined

Table 3.2 Gender-wise changes in workforce structure in rural areas

	WPR (usual status)		Share in total workforce		
Year	Male	Female	Male	Female	Total (million)
1972–1973	55	32	64	36	191
1983–1984	55	34	63	37	226
1993–1994	55	33	64	36	293
2004–2005	55	33	64	36	343
2011–2012	54	25	70	30	336
2017–2018	52	18	76	24	316
2018–2019	52	19	74	26	327
2019–2020	54	24	70	30	361

Source: Authors' estimates based on NSS estimates, Census population, and National Account Statistics

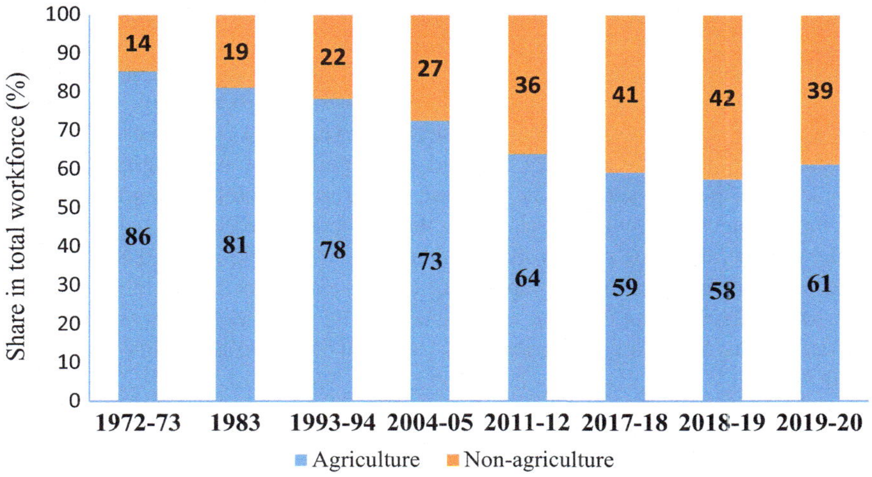

Fig. 3.1 Distribution of rural workforce between agriculture and nonagriculture sectors

Table 3.3 Changes in the agricultural and nonagricultural workforce in rural India

	Agriculture			Nonagriculture			Total		
Year	Male	Female	Total	Male	Female	Total	Male	Female	Total
Number (million)									
1993–1994	139	90	229	49	14	63	188	105	293
2004–2005	146	103	249	73	21	93	219	124	342
2011–2012	139	76	215	95	25	121	234	102	336
2017–2018	131	56	187	108	21	128	239	77	315
2018–2019	129	60	188	114	24	137	243	84	326
2019–2020	141	81	222	113	26	139	254	107	360
Growth rate (%)									
1993–1994 to 2004–2005	0.45	1.35	0.82	4.19	3.66	3.99	1.55	1.70	1.59
2004–2005 to 2011–2012	–0.64	–4.26	–2.06	3.81	3.00	3.70	0.98	–2.80	–0.28
2011–2012 to 2017–2018	–0.96	–4.94	–2.29	2.06	–3.49	1.01	0.32	–4.57	–1.04
2017–2018 to 2019–2020	3.47	20.33	8.80	2.55	12.58	4.08	3.05	18.31	6.91

Source: NSSO surveys on employment and unemployment (various years)

nonagricultural jobs during the same period. This shows a clear preference of workforce for nonagricultural jobs, and the movement of workers outside agriculture is influenced by a complex set of factors such as the pattern of economic growth, inter-sectoral differences in wage rate and worker productivity, government programs, education and sociocultural factors prevailing in rural areas (Chand and Srivastava 2014). The majority of workers withdrawing from agriculture were female. It is to be noted that the increase in the nonagricultural workforce could not be commensurate with the decline in the agricultural workforce, which resulted in the net decline in the rural workforce between 2004–2005 and 2011–2012. Such evidence also reveals a limited capacity of nonagricultural activities in rural areas to provide productive employment to the incoming workforce. Efforts are, therefore, needed to augment the nonagricultural jobs and improve the skills of the rural workforce.

The withdrawal of the agricultural workforce further accelerated, and additional 28 million workers left agriculture between 2011–2012 and 2017–2018. This was accompanied by a slowdown in nonagricultural employment, and the overall rural workforce declined by 21 million. The share of agriculture in the rural workforce reached 59% by the year 2017–2018. The Period Labor Force Survey (PLFS) conducted during 2019–2020 revealed a significant increase in the rural workforce, and growth in the agricultural workforce between 2017–2018 and 2019–2020 was more than doubled as compared to the nonagricultural workforce. The increase in the agricultural and nonagricultural workforce was majorly accounted for by female workers.

The agricultural workforce comprises of both cultivators and agricultural labors. The changes in cultivators and agricultural labors in the country are presented in Table 3.4. The estimates using NSS and census population show that cultivators constituted about 60% of the total agricultural workforce in 1993–1994. Between 1993–1994 and 2004–2005, the number of cultivators increased by 22 million at an annual growth rate of 1.47%. On the other hand, agricultural labor registered a marginal decline during this period. The subsequent period between 2004–2005 and 2011–2012 witnessed a decline in the number of cultivators at an annual growth rate of 1.83%. The decline in the absolute number of cultivators occurred first time in the history of agricultural development since independence. The decline in agricultural labor also accelerated in this period. The decline in the agricultural workforce, along with the simultaneous increase in nonagricultural workforce, implies employment diversification toward more productive nonagricultural sectors, which is desirable from the economic development point of view. Notwithstanding, the withdrawal of agricultural workers was primarily accounted for by female.

The evidence from subsequent NSSO surveys, however, revealed a deceleration in the withdrawal of cultivators between 2011–2012 and 2017–2018. In fact, male cultivators increased at an annual growth rate of 1.76%, and withdrawal was restricted only to female cultivators during this period. Agricultural laborers continued to withdraw from agriculture at a higher rate (−6.56%) during this period as compared to the previous period. The deceleration in the decline of cultivators could be due to the recent government's emphasis on improving farmers' income

Table 3.4 Changes in agricultural workforce in rural India

	Cultivator			Agricultural labor			Agricultural workers		
Year	Male	Female	Total	Male	Female	Total	Male	Female	Total
Number (million)									
1993–1994	85	53	138	54	37	91	139	90	229
2004–2005	93	67	160	53	37	89	146	103	249
2011–2012	91	49	140	48	27	75	139	76	215
2017–2018	101	37	138	30	20	50	131	56	187
2018–2019	100	41	140	29	19	49	129	60	188
2019–2020	108	58	165	33	24	57	141	81	222
Growth rate (%)									
1993–1994 to 2004–2005	0.89	2.32	1.47	−0.27	−0.18	−0.23	0.45	1.35	0.82
2004–2005 to 2011–2012	−0.24	−4.32	−1.83	−1.41	−4.34	−2.56	−0.64	−4.26	−2.06
2011–2012 to 2017–2018	1.76	−4.75	−0.30	−7.54	−5.07	−6.56	−0.96	−4.94	−2.29
2017–2018 to 2019–2020	3.06	25.50	9.47	5.05	10.10	6.94	3.47	20.33	8.80

Source: NSSO surveys on employment and unemployment (various years)

and/or the limited capacity of nonagricultural sectors to provide productive employment to the incoming workers. Interestingly, the recent period between 2017–2018 and 2019–2020 has witnessed a significant increase in the number of both cultivators and agricultural laborers.

3.4 Implications of Changes in Agricultural Labor on the Farm Economy

The decline in the supply of agricultural labor is expected to create physical and economic scarcity of labor. Physical scarcity adversely affects the timely completion of farm operations, particularly during the peak season. On the other hand, economic scarcity implies an upward movement in farm wages because of a contraction in labor supply which in turn adversely affects farm profitability and farmers' income. The magnitude of this effect would depend on the relative changes in wages and labor use and labor share in the cost of cultivation. Presently, labor constitutes 46% share of the average cost of cultivation (CostA_1 + FL) of principal crops (Table 3.5).

Table 3.5 Changes in average labor use and cost of cultivation of major crops during 1993–1994 to 2016–2017

Year	Average labor use (hrs/ha)			Average real labor cost (Rs/ha)	Average real labor wages (Rs/hr)	Average real cost A_1 + FL (Rs/ha)	Share of labor cost in cost A_1 + FL[a]
	Male	Female	Total				
Absolute numbers							
1993–1994	455	246	701	4367	6.2	10,585	41.3
2004–2005	419	223	642	4971	7.7	12,938	38.4
2011–2012	412	220	632	7205	11.4	15,651	46.0
2016–2017	366	189	555	7218	13.0	15,705	46.0
Growth rate (%)							
1994–2005	–0.89	–1.13	–0.97	1.30	2.10	2.03	–
2005–2012	–0.26	–0.15	–0.22	5.45	5.82	2.76	–
2012–2017	–2.34	–2.99	–2.56	0.04	2.67	0.07	–

Source: Srivastava et al. (2017)
[a]At current prices

According to the COC surveys, average labor use[1] in the cultivation of major crops declined by 8% between 1993–1994 and 2004–2005. But, despite the reduction in labor use, labor costs at real prices increased by 14% on account of 33% rise in real wages. Incremental labor cost contributed 26% of the total increase in Cost A_1 + FL during this period. Nevertheless, the share of labor in CostA_1 + FL reduced from 41.3% in 1993–1994 to 38.4% in 2004–2005 because of a relatively higher increase in the cost of other factors of production. The subsequent period till 2011–2012 witnessed a significant rise in real labor wages which resulted in 45% increase in labor cost (despite a decline in labor use). This inflated real cost A_1 + FL by 82%, and the share of labor in a cost increased to 46% by the year 2011–2012. Interestingly, the decline in labor use accelerated during the latest period 2011–2012 to 2016–2017, which negated the effect of rising wages on labor costs.

This evidence indicates that despite the reduction in labor use in crop cultivation, labor costs could not be saved during the past 24 years. This phenomenon is explained by the inelastic nature of the demand for labor in crop cultivation.

[1]An aggregate time series of labor use (family and hired) was constructed using COC summary data on 10 principal crops in 19 states. The selected crops include paddy, wheat, maize, jowar, gram, arhar, rapeseed and mustard, groundnut, sugarcane, and cotton, which covered 64.58% of the Gross Cropped Area (GCA) in the country in 2015–2016. The cultivated area under each crop in the respective state was used as a weight to construct aggregate time series.

Table 3.6 Estimated elasticities of labor demand in selected crops in India

Crops	Price elasticity of labor demand
Paddy	−0.20
Wheat	−0.27
Sorghum	−0.25
Maize	−0.22
Pigeon pea	−0.22
Gram	−0.16
Groundnut	−0.16
Rapeseed and mustard	−0.23
Cotton	−0.20
Sugarcane	−0.20
Overall	−0.21

Source: Srivastava et al. (2017)

The estimated price elasticities of labor demand were negative and less than one in all the selected crops, with an average value of −0.21 (Table 3.6). This implies that in the situation of wage rise, labor use in crop cultivation reduces less than proportionately, resulting in rising labor costs. As the magnitude of reduction in labor use is insufficient to negate the wage-push cost inflation, it is necessary to promote farm mechanization and improve its economic access to farmers through institutional innovations (e.g., custom hiring centers). Srivastava et al. (2017) have observed that the present level of farm mechanization is inadequate to offset the wage-push cost inflation in Indian agriculture.

3.5 Conclusions

Despite declining contributions to total employment, rural areas still engage majority of the country's workforce in agricultural and nonagricultural activities. The economic activities in rural areas are more labor-intensive, with less worker productivity as compared to urban areas. The rural workforce is dominated by male workers, and the evidence reveal a de-feminizing trend in rural employment between 1972–1973 and 2017–2018. However, the participation of females in the workforce has been increasing in recent years. Further, rural areas are witnessing a structural change in employment away from agriculture toward nonagricultural activities. Consequently, the share of agriculture in the total rural workforce has been declining over the years. From 2004 to 2005, both cultivators and agricultural labors are withdrawing from agriculture. The recent NSSO survey, however, reveals a deceleration in the withdrawal of cultivators and an acceleration in the withdrawal of agricultural labors. The slowdown in the withdrawal of cultivators could be due to the effect of ongoing agricultural reforms raising their expectations about remunerative returns from farming. This could also imply the limited capacity of nonfarm sectors to absorb the incoming workforce and necessitates the strengthening of rural nonfarm sectors so as to generate gainful employment opportunities.

Withdrawal of agricultural labor affects the farm economy either by creating a physical scarcity of labor or through the rise in farm wages. Because of the inelastic demand for labor, an increase in wages could not bring a proportionate decrease in labor use and result in an increase in labor cost in crop cultivation. Thus, the extent of the decline in labor use is found to be insufficient to negate the wage-push cost inflation. This warrants concerted efforts to accelerate the pace of farm mechanization and its economic access to farmers to partially substitute labor.

References

Aggarwal A, Kumar N (2012) Structural change, industrialization and poverty reduction: the case of India. Development Papers 1206, Economic and Social Commission for Asia and the Pacific (ESCAP). United Nations, South and South-West Asia Office, New Delhi

Chand R, Srivastava SK (2014) Changes in rural labour market and its implications for Indian agriculture. Econ Polit Wkly 49(10):47–54

Chand R, Srivastava SK, Singh J (2017a) Changing structure of rural economy of India: implications for employment and growth. Discussion Paper, NITI Aayog, New Delhi

Chand R, Srivastava SK, Singh J (2017b) Changes in rural economy in India, 1971 to 2012: lessons for job-led growth. Econ Polit Wkly 52(52):64–71

Himanshu PA, Lanjouw R, Mukhopadhyay, Murgai R (2011) Non-farm diversification and rural poverty decline: a perspective from Indian sample survey and village study data. Working paper 44, Asia Research Centre

Kumar A Kumar S, Singh DK, Shivjee (2011) Rural employment diversification in India: trends, determinants and implications on poverty. Agric Econ Res Rev 24: 361–372

Maurya NK, Vaishampayan JV (2012) Growth and structural changes in India's industrial sector. Inter J Econ 6(2):321–331

Papola TS (2012) Structural changes in the Indian economy: emerging patterns and implications, ISID Working Paper No 12, Institute for Studies in Industrial Development, New Delhi

Raghavan M (2008) Changing pattern of input use and cost of cultivation. Econ Polit Wkly 43(26–27):123–129

Srivastava SK, Chand R, Singh J (2017) Changing crop production cost in India: input prices, substitution and technology effects. Agric Econ Res Rev 30:171–182

United Nations (2012) World urbanization prospects: the 2011 revision. ST/ESA/SER.A/322, Department of Economic and Social Affairs, Population Division, New York

Agricultural Technologies

4

Anup Das, Saurav Saha, Jayanta Layek, Subhash Babu, R. Saxena, and G. I. Ramkrushna

Abstract

Provisioning food and nutritional security for India's 1.4 billion population is a challenge and requires robust agricultural research and extension mechanism to address the complex issues. Further, India has only 2.4% and 4% of global land and water availability, respectively with which it should support 17.5% of the world population. From the position of 'begging bowl' for its 330 million population in the 1950s, India not only became food secure for its ~1.4 billion current population but became a net exporter of food grains. While there was a 5.6 times increase in food grain production since independence, there was also a substantial increase in the production of pulses (3 times), oilseed (5 times), cotton (11.4 times), sugarcane (6.6 times), milk (9.7 times), fish (15.2 times), egg (47.5 times), etc. Except for the British period, there was a steady increase in the development of agricultural technology and subsequent production of agricultural and allied enterprises over the centuries. Scientific integration of crops and

A. Das (✉)
ICAR Research Complex for Eastern Region, Patna, Bihar, India
e-mail: anup.das2@icar.gov.in

S. Saha
ICAR Research Complex for NEH Region, Sikkim Centre, Gangtok, Sikkim, India

J. Layek
ICAR Research Complex for NEH Region, Umiam, Meghalaya, India

S. Babu
ICAR-Indian Agricultural Research Institute, New Delhi, India

R. Saxena
ICAR-National Institute of Agricultural Economics and Policy Research, New Delhi, India

G. I. Ramkrushna
ICAR-Central Institute for Cotton Research, Nagpur, Maharashtra, India

P. K. Ghosh et al. (eds.), *Trajectory of 75 years of Indian Agriculture after Independence*, https://doi.org/10.1007/978-981-19-7997-2_4

livestock, along with the concept of manuring and irrigation, has existed since Indus Valley Civilization. After the Independence, there was enormous growth in the agricultural sector and food production due to different pioneer schemes of the Government of India like Grow More Food Campaign and Integrated Production Program. The introduction of high-yielding varieties (HYVs), fertilizer, provision of irrigation, and improved agronomic practices resulted in the Green revolution in the 1960s and enhanced food grain production significantly over the years. The creation of the Department of Agricultural Research and Education (DARE), Indian Council of Agricultural Research (ICAR), and Krishi Vigyan Kendra (KVK) are the important milestones in Indian agriculture to get operation flood, yellow revolution, blue revolution, etc. Development of HYVs and hybrids of different agricultural and horticultural crops, a scientific package of practices for cultivation of crops, livestock, fisheries, and allied enterprises, availability of farm implements, post harvesting and processing technologies, supply on quality inputs like manures, fertilizer, pesticides, etc. and availability of institutional credit and procurement of produce further strengthen the Indian agriculture. Remote sensing technology is widely used for crop yield estimation, area expansion under specific crops in order to forecast the district/ regional level productivity, and also used to provide crop insurance to farmers for crop loss due to climatic or biotic stress. The use of no-till technology along with drip or sprinkler irrigation is the need of the hour for conserving our scarce natural resources and enhancing productivity and income. While biotechnology and genetic engineering have resulted in the development of varieties with pest resistance and increased crop yields, farm mechanization has led to efficient tilling, harvesting, water use, and a reduction in manual labor requirements. The development of location-specific different integrated farming systems (IFS) or integrated organic farming system (IOFS) models also helped enhance resource use efficiency, system productivity, and income of farmers. The use of information and communication technology and weather-based agro-advisories also help farmers' preparedness for forthcoming sporadic extreme weather events such as flash floods, drought, or pest outbreaks. The Indian population is slated to grow to about 1.7 billion by 2050. Food production has to be increased in the context of worsening land and water scarcity, soil degradation, multinutrient deficiency, and climate-change-related extreme weather events. The challenge is to find ways and means to produce enough quality to ensure food and nutritional security without deteriorating the environment. The challenge of producing more from reduced area and water under agriculture and minimizing post-harvest losses will have a major impact on agricultural development. Digital smart agriculture with the use of artificial intelligence, blockchain, remote sensing and GIS technology, use of drones and robots, etc. may take Indian agriculture to the next paradigm.

Keywords

Food security · Technological advances · Integrated farming system · Sustainable agriculture · Smart agriculture

4.1 Introduction

Feeding an ever-growing population and maintaining food and nutritional requirements warrants enhancement in agricultural productivity using modern technologies. Slow agricultural growth concerns policymakers as some two-thirds of India's population depends on rural employment for a living. The technology gap needs to be filled on a fast-track basis to match productivity ratios with the rest of the world. Thanks to technological advancements, our food production, which was merely 50 million tonnes at the time of independence, will have reached more than 315 million tonnes by 202-22. Similarly, in the case of livestock, India is the leading producer of milk in the world and the largest producer of pulses. This is also happening at a time when we have to achieve the Sustainable Development Goals. Food production has to be increased in the context of worsening land and water scarcity and climate-change-related weather shocks. India's current water use efficiency is barely 36–40%, which may lead to a "water scarce" situation by 2030 (NITI Aayog 2018). Further, there is a high deficiency of nitrogen (N), phosphorus (P), potassium (K) and increasing deficiencies of zinc (Zn), boron (B), iron (Fe), manganese (Mn), and sulfur (S), with very low nutrient use efficiency (36–40% for N and 15% for P). There is a constant decline of soil organic matter, with soil organic carbon (SOC) reaching to 0.25% in Indo Gangetic plains (IGPs). The groundwater table depletion in non-traditional rice production areas and arsenic pollution in the eastern region of India are further challenges. Soil and nutrient loss through runoff and anthropogenic activities also requires massive attention to sustain agriculture for posterity. Hence, a significant paradigm shift of adopting integrated soil, water, and biodiversity-centric approach is required to increase and sustain agriculture-food system productivity and growth, through conserving natural resources and mitigating climate change, and moving from Green to Evergreen Revolution to Evergreen Economy (Swaminathan 2007; Singh 2011). Off late emphasis is also given on natural farming, organic farming, regenerative agriculture and other alternative ecofriendly practices (to cover ~10 m ha) to reduce pressure on land, revitalizing environment and promote resilient agriculture with focus on benefiting small and marginal farmers. Thus, while promoting modern farming practices and technologies, the learnings from traditional practices must also be integrated for a sustainable development.

4.2 Evolution of Agricultural Technologies in India

Agriculture has always been accorded an important profession in Indian society. There is endless evidence that indicates the flourishing state of our agriculture in the pre-British period. Many of the indigenous practices which were perfected centuries ago, such as the rotation of crops, the practice of drill husbandry, etc., were relatively unknown in seventeenth-century Europe and are often cited as the major advances achieved during the eighteenth century 'agricultural revolution' in Europe. The all-around excellence of Indian agriculture was perhaps first documented in detail by Alexander Walker in a report written around 1820 (CFPS 2011).

Agriculture in British India

- Agriculture was almost stagnant during this period
- Emphasis was on cash and plantation crops than food crops.
- Production of food crops decreased, mass impoverishment and destitute farmers, trade was major focus.
- Favorable policies for the rulers
- Population increased but production decreased
- 1943 famine in Bengal due to less production, wrong policy and diversion of food for soldiers

Indian agriculture, which grew at the rate of about 1.0% per annum during the 50 years before independence, has grown at the rate of about 2.6% per annum in the post-independence era. Expansion of area was the main source of growth in the period of 1950s and 1960s. After that, the contribution of increased land area under agricultural production declined over time, and an increase in productivity became the main source of growth in agricultural production. Success in eradicating dependence on imported food grains was an important factor of progress in agriculture. Land reforms, the inauguration of the Agricultural Price Commission to ensure remunerative prices to producers, a new agricultural strategy, investment in agricultural research and extension services, provision of credit facilities, and improving rural infrastructure are some of the steps initiated by Government of India which contributed to agricultural development and growth in India (Tripathi and Prasad 2009). Indigenous farming practices with a focus on intensive tilling, recycling on-farm resources, use of FYM/Composts/green manure, use of tank silts, indigenous traditional knowledge base (ITKs), traditional long-duration varieties including primitive practices like shifting cultivation (reduced from more than two million ha at the end of 2000 to less than 0.75 million ha by 2020), cattle shed rotations in fields, existed in India long before independence.

Agriculture in Post-independence

- Till mid 1960s India relied on imports and food aids to meet domestic requirements
- Drought in 1965 and 1966 led to reforms in agriculture policies
- Food security through Green revolution
- High yielding disease resistance wheat varieties along with improved agronomic practices
- HYVs needed high chemical fertilizers, pesticides and irrigation

The fragility of the Indian food production and distribution system was exposed after the disastrous Bengal famine in 1943. In order to increase the food grain production levels, the cultivated area under food crops was expanded through special programs, like Grow More Food Campaign. Further, on the basis of the recommendations of the Food Grain Policy Committee in 1947, the reclamation of 25 million acres (ten million hectares) of land was undertaken to expand the cultivated area under food crops. But this singular focus on food grain production led to an acute shortage in cotton and other fiber production in the late 1940s, and this problem was addressed through an Integrated Production Programme initiated in 1950. At this stage, the vital role of technology was realized, particularly the land-augmenting technology that enhances the yield per unit of land. The introduction of technical changes through improved seed, fertilizer, irrigation, mechanization, and plant protection has brought dramatic changes in agricultural production since the 1950s. These technical changes in agriculture have proceeded in a step-by-step manner; initially, the focus was on the development of land, irrigation, and other inputs; subsequently, the emphasis was on high-yielding varieties (HYVs) and improved "package of practices," and finally, postharvest and marketing aspects were explored.

4.2.1 Pre-green Revolution Period

To address the problems of food shortage and the resultant rise in food prices, the agriculture sector was given the highest priority in India's first 5-year plan (FYP). The focus was placed on the expansion of irrigation, land reclamation, and the domestic production of inorganic fertilizers. A number of major irrigation projects were completed during the first FYP (1951–1956), including the Tungabhadra Project (1956), Gandhi Sagar (1960), Ghataprabha Left Bank (1956), and Hirakhud Dam (1956), and many others. In addition to surface irrigation, groundwater irrigation through tube wells gained popularity in the 1950s. Further, the mechanization of irrigation operations, as reflected by the steep increase in the number of diesel engines and electric pumps, picked up momentum in the late 1950s. As a result, the extent of irrigated area in the total cultivated area went up from about 17% in the 1950s to about 20% in the 1960s. The number of electric pumps for irrigation purposes increased steeply, from about 56,000 at the end of the first FYP period to more than 500,000 by the end of the third FYP (1961–1966) period, as a result of the expansion in rural electrification launched during the first plan period.

4.2.2 Fertilizer Use

Though the role of fertilizers as an important source of nutrients and a substitute for organic manures was recognized, their use was not widespread and was restricted to commercial crops in southern regions of the country until 1950. In order to create awareness among farmers about the importance and methods of fertilizer application, Agricultural Technical Assistance Programme (1951) and the Fertilizer Demonstration Programme (1954–1956) were initiated. Soil testing laboratories were also established to collect and analyze soil samples and to prepare a nutrient status map of the villages. As a result, the application of fertilizers increased, with an average annual growth rate of about 20% in the 1950s and 25% in the 1960s. The consumption of phosphatic and potassic fertilizers was almost negligible until the 1950s but started rising in the subsequent period; the growth in phosphatic and potassic fertilizer application outpaced that of nitrogenous fertilizers during the 1950s and 1960s, although the use was judicious in absolute quantities.

4.2.3 Package of Practices

Based on the recommendations of the Ford Foundation Team, the Intensive Agricultural Development Programme (IADP) was initiated, which featured a "package of practices" for each crop based on recent research findings. Though the program was successful in increasing yields to some extent during the early 1960s, the success was restricted to resource-endowed areas that accounted for only 5% of the total cultivated area in the country. Because of successive droughts, food grain imports reached more than 9.8 million tonnes (ten million tonnes) in 1966–1967. The reason for the program's limited success was recognized to be an inherent problem associated with the traditional improved varieties of wheat and rice. After the application of high doses of fertilizers, those varieties grew taller, with slim stems, and had a tendency to fall over at the time of maturity, notwithstanding the grain weight.

4.2.4 Green Revolution

India had imported the dwarf genes varieties of wheat (Lerma Rojo 64 and Sonara 64) from Mexico, and Indian scientists developed high-yielding, pest-resistant, and input-responsive dwarf wheat varieties (Sonalika, Sharbati Sonara) by crossing these lines with local high-yielding varieties. In rice, the first semidwarf *indica* rice variety, Taichung Native 1, was developed by crossing a semidwarf *indica* variety, Dee-gee-woo-gen, with a drought-resistant variety, 'sai-Yuan-Chung, at the Taichung District Agricultural Improvement Station in Taiwan in 1956. India, being the world's largest source of rice, had actively participated and cooperated in developing new rice varieties with the rice breeders at the International Rice Research Institute by making several crosses of Dee-gee-woo-gen and local varieties. The high-yielding semidwarf rice breeding lines were introduced, and numerous improved varieties were developed with the active participation of the Central Rice Research Institute, the All-India Coordinated Rice Improvement Program, and various state agricultural universities. Finally, high-yielding rice varieties were released for large-scale commercial cultivation all over India in 1966. In addition, high-yielding varieties of coarse cereals, including sorghum, *bajra*, and maize, were also developed through hybridization and were released for large-scale commercial cultivation during the late 1960s.

Thus, began the golden period in Indian agriculture called the "Green Revolution." The area under HYVs of wheat reached more than 60% of total wheat cultivated by the mid-1970s, while it took 15 more years (until 1990–1991) to reach the same level (64%) for rice. Consequently, the growth of food grain production was the highest during the 1980s compared to any other period.

Considering the frequency of drought, a characteristic of Indian agriculture, assured irrigation has become a prerequisite for intensifying agricultural production in India, particularly following the introduction of HYVs. Apart from major irrigation projects, emphasis was also placed on increasing minor irrigation projects through groundwater exploitation. Consequently, the total area under irrigation grew from about 55.8 million acres (22.6 million hectares) in 1950–1951 to about 123 million acres (50 million hectares) in 1979–1980. Considering the limitations to expanding irrigation through canals as well as groundwater, agricultural scientists and planners started promoting rain-fed agriculture through a promising "watershed" technology during the 1980s. The HYVs, due to their responsiveness to fertilizers, particularly to nitrogenous fertilizers, have a tendency to grow succulently and become susceptible to pests and diseases. With increasing irrigation and fertilizer application, the growth of weeds also increased. As a measure of plant protection, the first Pest and Disease Surveillance Service was organized in 1969 in selected districts of the Intensive Agriculture Development program (IADP). The application of pesticides became an important component of crop production, particularly in commercial crops like cotton, tobacco, and sugarcane. Mechanization of irrigation operations expanded rapidly and resulted in a multiplier effect on crop production through crop intensification.

4.2.5 Yellow (Oilseed) Revolution

In view of the dismal performance of oilseeds during the 1960s and 1970s, the Government of India created the Technology Mission on Oilseeds (TMO) in May 1986. The objective of the TMO was to achieve self-sufficiency in edible oils; to achieve this, the TMO implemented the introduction of HYVs of oilseeds, together with the adoption of improved production technology, a better supply of inputs, extension services, and postharvest technologies. Consequently, a major increase in oilseed production was achieved; the production of oilseeds increased from 10.7 million tons in 1985–1986 to 24.6 million tons (25 million tons) in the late 1990s. Following the success in oilseed production, other crops, including pulses, oil palms, and maize, were brought under the TMO in 1990, 1992–1993, and 1995–1996, respectively.

4.2.6 White Revolution

The government initiated a massive dairy development program, popularly known as Operation Flood, in 1970. Verghese Kurien was the principal architect of the program, which was implemented in three phases from 1971 to 1996, with support from the European Economic Community, the World Food Programme, and a soft loan from the World Bank.

The aim of the program was to expand milk production through processing and marketing facilities. Toward this objective, milk producers' cooperatives were organized to collect, process, and sell milk to achieve a secured market and remunerative prices. Further, to enhance milk production, inputs such as better feed and fodder, breed improvement through artificial insemination, and disease control measures were also arranged through cooperatives. This coordinated effort led to an increase in milk production and ushered in an era of the "White Revolution." During this period, milk production increased at an annual average growth of about 5%.

4.2.7 Blue Revolution

The Indian Council of Agricultural Research (ICAR) implemented an All India Coordinated Research Project on Brackishwater Fish Farming (1973–1984). The objective of the program was to develop and test various farming technologies under different agroclimatic conditions in the country. The main center of the project was located in West Bengal; other centers were located in Orissa, Andhra Pradesh, Tamil Nadu, Kerala, and Goa for demonstrating the technologies to small-scale farmers. In addition, shrimp hatchery technology was also introduced into the country, and two commercial hatcheries were established in the late 1980s with an initiative from the Marine Products Export Development Authority. As a result, the production of fisheries, particularly inland fisheries, increased significantly, at an average annual growth rate of 7.1% and 6.2% during the 1980s and 1990s, respectively. By 2002, India had become the third-largest producer of fish.

4.2.8 Pulses revolution India's

Pulses production underwent a significant transformation after the observance of the International Year of Pulses in 2016. As one of the world's leading producers and consumers of pulses, the Indian government implemented a range of initiatives and programs in mission mode with an aim on promoting production. These included the National Food Security Mission, the Pradhan Mantri Fasal Bima Yojana, and the Pradhan Mantri Krishi Sinchai Yojana, which provided farmers with high yielding varieties through seed minikits, resource conservation technologies, efficient water application tools, and insurance coverage (Tiwari and Shivhare 2018; Gaur 2021). As a result of these efforts, the production of pulses in India rose from 16.35 million tonnes in 2015–16 to a record high of 23.13 million tonnes in 2016–17, a 43% increase. This trend continued in the following years, with pulses production remaining consistently above 23 million tonnes (25.58 million tonnes in the year 2020–21). The surge in production had numerous positive impacts, such as enhancing food security, improving nutrition, and raising farmers' incomes. It also helped

India reduce its dependence on imports of pulses, which had previously put a significant strain on the country's foreign exchange reserves. To sustain the production, focus on farmers' preference to competitive crops over pulses should be studied, emphasis on research on developing varieties with biotic and abiotic stress tolerant with early maturity, location specific package of practices, improved mechanization, etc. shall take into future planning.

4.2.9 Agroprocessing

In order to encourage food processing, the Government of India has taken a number of steps, first creating the Ministry for Food Processing Industries in 1988. The ministry has taken many initiatives, such as integrated food law, the creation of cold storage facilities, and the establishment of food parks (centers for the distribution of all processed grains and horticultural products across the country). As a result, the food processing industry expanded significantly in the 1990s. Recognizing the potential and comparative advantage that India has in the agroprocessing industry, a number of agro-export zones were established across the country. Thus, it is the changes in postharvest technology that have dominated Indian agriculture since the 1990s.

4.2.10 Remote Sensing Technology

In the early 1970s, India initiated the first practical use of remote sensing in the agriculture sector for early detection of coconut wilt disease in Kerala through air-born Infrared Cameras. In the early 1980s, the extensive use of space technology in policy planning was started through satellite remote sensing with the development of Indian Remote Sensing (IRS) satellites under the Joint Experiment Programme (JEP). At present, remote sensing technology is widely used for crop yield estimation and area expansion under specific crops to forecast the district/regional level productivity under Forecasting Agricultural Output using Space, Agro-meteorology and Land-based Observations (FASAL) project at a national level with considerable accuracy. At present, the satellite sensor-based Normalized Difference Vegetation Index (NDVI) is widely used for real-time drought monitoring under National Agricultural Drought Management and Monitoring System (NADAMS) across India. Remote sensing became the indispensable tool for mapping wasteland, wetland ecosystems, soil resources (land use and land cover, land degradation studies, monitoring of horticultural inventory under the Mission for Integrated Development of Horticulture (MIDH) program, the extent of problem soils (soil salinity, alkalinity, water logging, etc.), Disaster Monitoring (locust invasion), forest vegetation cover assessment, and Climate Change studies across India. The use of microwave remote sensing (L or C bands) enabled the monitoring of soil moisture from space using modern satellites (RADARSAT-2, MODIS, etc.).

4.2.11 Genetically Engineered Crop Varieties (Gene Revolution)

Near stagnation in yields of all field crops in the 1990s created the need to look for alternative technology that could further boost crop yields. In this context, genetic engineering appeared to be a promising choice. In India, the efforts initiated in the late 1990s to release the first genetically engineered crop, "Bt cotton," were not successful until 2002 because of strong protests from environmentalists. The first genetically modified (GM) crop was approved in the year 2002 as Bt cotton and its adoption has been one of the most successful stories in the country's agricultural history. Bt cotton was genetically modified to resist bollworm infestation, which had significantly reduced the use of pesticides and increased cotton yield. Three Bt hybrids were approved by Genetic Engineering Appraisal Committee (GEAC), Ministry of Environment, Forest and Climate Change, Govt. of India, namely MECH 12, MECH 162, and MECH 184, for cultivation in India initially carrying gene Cry 1AC. A quantum jump in the yields of cotton was realized after the introduction of Bt cotton in India. On the other hand, several research units under ICAR and state agricultural universities have been working to develop genetically modified crops suitable for Indian conditions. However, the dilemmas concerning the commercial cultivation of genetically engineered crop varieties still persist, owing to the serious concerns of biosafety and biodiversity.

4.3 Technological Advancements

Indian agriculture is still a technology deficit. Our productivity of food grains, fruits, and vegetables is still far below global averages. The current is about one-third of China's and about half of Vietnam's and Indonesia's. The productivity of pulses and oilseeds in India has the potential to increase by 2.3–2.5 times through attention to seeds, soil health, pest management, crop lifesaving irrigation, and post-harvest technology (Gautam and Kumar 2014). India's population is expected to reach 1.7 billion by 2050, making food security the most important social issue, and food production will have to be increased considerably to meet the needs of a growing population. As per FAO, India is still home to a quarter of all undernourished people globally. There is an urgent need to embrace new technologies like biotechnology, nanotechnology, high-tech protected cultivation, and modern irrigation methods to accelerate agriculture production. There is also a growing realization that previous strategies of generating and promoting technologies have contributed to serious and widespread environmental and natural resource degradation problems. This implies that in the future, the technologies that are developed and promoted must result not only in increased productivity levels but also ensure that the quality of the natural resource base is preserved and enhanced (Kumari 2014). Agriculture also includes livestock, apiculture, pisciculture, apiary, goatery, forestry, etc. has to undergo a significant transformation in order to meet the above-related challenges. Smallholder and marginal farmers, accounting for nearly 86.25% of Indian farmers, 47.38% of the cultivated land, and over 50% of the total agricultural production, are vital not only for India's agrarian economy (tenth Agriculture Census 2015–2016) but also for achieving alleviation of hunger and poverty. Thus, the new agriculture paradigm must ensure that the small and marginal farmers be at the center stage of any technological interventions (Table 4.1).

Innovative technologies are a must to push the yield frontiers further, utilize inputs more efficiently and diversify to higher-value sustainable cropping systems. These are all knowledge-intensive technologies that require both a strong research and extension system and skilled farmers but also a reinvigorated interface where the emphasis is on the mutual exchange of information bringing advantages to all. At the same time potential of less-favored areas must be better exploited to meet the targets of growth and poverty alleviation.

Efficient utilization of resources is the central driving force behind the use of all agricultural technologies. Several resource conservation technologies are being used, including no-till and minimum tillage, green manuring, crop rotations/diversifications, etc. The no-till (NT) in wheat is reported to have reduced the cost of production by Rs. 2000–2500 per ha and saved 15–20% of irrigation water (Tripathi and Prasad 2009). Similarly, by using drip and sprinkler type of irrigation methods, more area can be brought under irrigation by saving 50–80% of water. The use of farm yard manure (FYM), compost, and biofertilizers helps reduce over-dependence on the chemical fertilizers that lead to intensive cultivation. Boosting agricultural growth is essential for inclusive growth because this sector sustains the livelihood of 65% of the population. However, agriculture contributes only 14% to Gross Domestic Product (GDP). Several revolutions in agriculture have taken place to boost the sector. These include the Green Revolution, Evergreen Revolution, Blue Revolution, White Revolution, Yellow Revolution, Bio-technology Revolution, and ICT Revolution. While nearly all relevant technologies and know-how are available to us, the extension of these developed systems to the field is required. Agriculture Extension combined with adequate infrastructure is the key to agricultural growth. The involvement of the private sector would help the absorption of technologies in the fields faster.

Technology in agriculture affects many areas of agriculture, such as sowing/planting practices, fertilizers, pesticides, seed technology, etc. Biotechnology and genetic engineering have resulted in pest resistance and increased crop yields. Mechanization has led to efficient tilling, harvesting, water use, and a reduction in manual labor. Irrigation methods and transportation systems have improved, processing machinery has reduced wastage, etc., and the effect is visible in all areas. New-age technologies should focus on machine learning, robotics, precision agriculture, aeroponics, hydroponics, artificial intelligence, blockchain technology, and more. Some technological advancements that have innovated agriculture:

- **Improved productivity from the mechanization of agriculture**: Indian farming is characterized by small landholdings, and the need is to partner with others to take advantage of modern machines. Combine harvesters are finding greater use to minimize manual labor and make the processes quicker.

Table 4.1 Chronology of some important technological development in Indian agriculture

9000 BC	Agriculture is born Wheat, barley, and jujube were domesticated in the Indian subcontinent
3000–2000 BC	Indus Valley Civilization Irrigation concept developed Cotton, rice, fruits, vegetables Poultry, buffalo Cattle drawn plough
1000 BCE	Sugar processing in India
1000–500 BC	Cultivation of a wide range of cereals, vegetables, and fruits Cow dung provided the manure Irrigation was practiced
100–200 AD	Water storage systems for irrigation Spice trade (cinnamon, black pepper, etc.)
British period (1857–1947)	Agriculture is more or less stagnant Emphasis on cash and plantation crop rather than food crop Policy more for the rulers than the ruled Imperial Council of Agricultural Research (ICAR) established
1940	Grow More Food Campaign
1950	Integrated Production Program
1952	Simple Fertilizer Trial on Cultivators field
1965–1970	High yielding varieties, fertilizer application, need-based irrigation and improved agronomic practices leads to green revolution
1973	Creation of Department of Agricultural Research and Education (DARE) in the Ministry of Agriculture
1979	Launching of Lab-to-Land Program and the National Agricultural Research Project (NARP)
1989–2010	Experiments on cropping systems research
1992–2006	Microirrigation concept initiated (1992) and recognized this as a thrust area (2006)
2002	Transgenic crop (Bt cotton)
2003	System of Rice Intensification (SRI)
2004–2020	Development of IFS and IOFS models for enhancing system productivity and income
2005	Conservation agriculture
2011	National Innovations in Climate Resilient Agriculture (NICRA)
2015	Soil health card to promote Integrated Nutrient Management (INM)
2015	Neem Coated Urea made mandatory (New Urea policy)
2017	MOVCDNER: Mission organic value chain development for North Eastern region
2004–2022	Scientific organic farming and promotion
2021	21 climate-resilient and biofortified special varieties of crops
2021	Digital agriculture mission (artificial intelligence, block chain, remote sensing and GIS technology, use of drones and robots, etc.)

- **Climate/weather prediction through artificial intelligence**: A major advance in agriculture is the use of artificial intelligence (AI). Modern equipment and tools based on AI enable data gathering and assist in precision farming and informed decision-making. Drones, remote sensors, and satellites gather 24/7 data on weather patterns in and around the fields, providing farmers with vital information on temperature, rainfall, soil, humidity, etc. However, AI finds slow acceptance in a country like India where marginal farming, fragmented landholdings, and other socioeconomic reasons act as impediments. But there is no doubt that technologies based on AI can bring precision to large-scale farming and lead to an exponential rise in productivity.
- **Resilient crops developed via the use of biotechnology**: It refers to a wide resource of methodologies that include traditional breeding methods, genetic engineering, and the development of microorganisms for agriculture. Genetic engineering uses the understanding of DNA to identify and work with genes to increase crop resistance to pests, and the development of high-yielding varieties also makes improvements to livestock. The introduction of Bt technology has reduced crop insecurity by 15–20% among Indian cotton growers (Gautam and Kumar 2014). But, the use of genetically modified crops was restricted to cotton only due to concerns echoed by various environmentalist groups. Currently, the Government of India has allowed the trials of other GM crops, which will also give a momentum for adopting other GM crops. The country has also developed golden rice, which is rich in β-carotene. This is a great solution for India as nearly 5000 children go blind every year because of β-carotene deficiency.
- **Improving farm yields and supply chain management using Big Data**: The collection and compilation of data and its further processing to make it useful for decision-making/problem-solving are expanding the way big data functions. Big data is slated to play a major role in smart farming, and the benefits percolate across the entire supply chain and the markets. Agriculture is becoming larger, and it depends on a large number of variables.
- **Information and communication technology (ICT) in agriculture:** In developing countries like India, ICT in agriculture often provides farmers with vital information on sowing, crop protection, and measures for improving soil fertility to improve agricultural productivity. Weather-based agroadvisories issued under Gramin Kishi Mausam Seva (GKMS, real-time SMS-based messaging services) with sufficient lead time often open up the scope for advance farmers' preparedness toward forthcoming sporadic extreme weather events, viz. flash floods, drought, or pest outbreaks in order to minimize significant crop loss. This results in the greater collection and use of complex data, which has to be meaningfully interpreted and managed. Data can be from external sources such as social media, supplier networks, markets, or sensor/machine data from the fields. Transformation of agriculture using big data is taking place that affects crop yield, supply chain management, yield prediction, etc.

4.3.1 High-Yielding Varieties and Technologies

The green revolution of the 1960s would not have occurred without the HYV of Wheat and rice. These HYVs and the area under irrigation and fertilizers saw India becoming a bread basket from once being labeled as a begging bowl. If Indian agriculture is to remain competitive with global agriculture, it has to increase yield per unit area through the development and production of seeds that have more yields, are resistant to multiple diseases/stresses, and can withstand the environmental extremities. Already wide such varieties are developed under National Innovations in Climate Resilient Agriculture (NICRA). National Rice Research Institute (NRRI), Cuttack, Indian Institute of Wheat and Barley Research (IIWBR), Karnal, Indian Agricultural Research Institute (IARI), New Delhi, Indian Institute of Maize Research (IIMR), Ludhiana, and others ICAR institutes have come out with biofortified varieties and multiple stress-tolerant crop varieties.

4.3.2 Modern Technologies

Modern machinery such as laser land levelers, self-propelled sprayers, precision seeders and planters, transplanters for rice and vegetable seedlings, multicrop threshers, and harvesters for cereals and sugarcane today allow technically highly efficient farming and resource conservation. The single operation of laser leveling can reduce water application charges by 25–30% with greater water use efficiency. The precision farming approach recognizes site-specific differences within fields and adjusts management actions accordingly adopting the concept of "doing the right thing in the right place at the right time." Moving forward, this will be the new normal, addressing: (1) increased land and labor productivity by means of gender-neutral technologies, (2) intensification, diversification, and off-farm employment, (3) institutional arrangement to equitable rights, and (4) balanced agroecological settings compatible with minimum risk (Gatzweiler and Von Braun 2016). Coupled with precision farming, the resource conservation technologies like zero tillage and residue management can reduce the cost of cultivation by 25–30% over conventional farming practices. System of Rice Intensification (SRI), which improves the productivity of rice by 30–35% in conventional varieties and over 50% in hybrids, is a potential source of technological revolution for small and marginal farms. It is a skill-intensive technology that cuts the need for inputs such as seeds and fertilizers while raising yields per ha. Aerobic rice production and direct seeding in rice are two breakthrough technologies for saving water and enhancing water productivity. It is predicted that in the next 10 years, the nanotechnology-led application will play a critical role in agriculture. These applications are related to the release and efficient dosage of water and fertilizers, drugs (for livestock), and herbicide delivery. The nanoparticles for new pesticides, insecticides, and insect repellents may also come to play an important role. Land shaping, furrow irrigated raised bed planting (FIRB), crop diversification, sustainable intensification, irrigation at critical stages like CRI stage in wheat, mid-season drainage, use of leaf color chart, green seekers, high-tech poly houses, soil health cards, biofertilizer, Free Air Temperature Enhancement (FATE), Carbon-dioxide and temperature Gradient Chamber (CTGC), Growth chambers, Open Top Chambers (OTCs), eddy-covariance towers, micro-irrigations, the use of information and communication technology (ICT) in agriculture like models, robotics, SMS services, weather forecasting, etc. are some of the modern technologies for efficient resource uses. Fiber extraction technologies, fodder seed production and multiplication technologies, biochar technology, sugarcane production technologies, and integrated weed management using new-generation herbicides, are some of the other promising technologies. Planting soybean on broad bed furrow (BBF) has helped conserve water and raise productivity. Micro irrigation via sprinklers and drips brought a dramatic change in several pockets of the country, especially in undulating topography and sand dunes areas where no other method of irrigation can work. Hi-tech horticulture, like poly house cultivation of vegetables, flowers, medicinal plants, and fruits, constitutes one of the most technologies and skill-intensive agricultural practices. The emphasis should be on informing farmers of the opportunities new technologies offer, improving access to credit, and creating an enabling policy environment for their adoption without major direct financial commitments. Similar great development has also occurred in horticulture, animal husbandry, and fishery sector in India with a quantum jump in productivity and production.

4.3.3 Genetically Modified Organisms

The conventional technologies will not be able to meet the food and nutrition requirements. GM crops are produced by the transfer of genes between organisms for specific traits using laboratory techniques. Modern biological advancement, especially in the fields of biotechnology and molecular biology, offers many advantages when applied in conjunction with traditional plant breeding techniques. Plants obtained through this method are called GMOs (Genetically Modified Organisms) or genetically engineered or transgenic plants. The development of genetically modified crops expressing a variety of novel traits, such as insect resistance, disease resistance, herbicide tolerance, improved nutritional quality, etc. has led to large-scale cultivation of GM crops (Lucht 2015). GM crops commercialized worldwide in the past two decades include tomato, corn, soybean, cotton, canola, rice, potato, squash, melon, and papaya. Genetically engineered cotton (known as Bt cotton) for insect resistance was released for commercial cultivation in India in 2002. During the next few decades, GM crops will play a significant role in ensuring food security and addressing concerns of biotic and abiotic stress. Concerns about environmental risks of GM crops include the impact of introgression of the transgenes into the natural landscape, the impact of gene flow, the effect on nontarget organisms, the evolution of pest resistance, and the loss of biodiversity.

4.3.4 Nanotechnology

Nanotechnology can help in sustaining soil fertility and balanced crop nutrition; effective weed control; enhancing seed emergence using carbon nanotubes; delivery of agriculture chemicals, field-sensing systems to monitor the environmental stresses and crop conditions, and improvement of plant traits against environmental stresses and diseases. It offers a very good opportunity to develop innovative products and applications for agriculture, water treatment, food production, processing, preservation, and packaging, and its use may bring potential benefits to farmers, food industry, and consumers alike. Nanotechnology has a huge potential to revolutionize food packaging. Nanoparticles, such as titanium dioxide, zinc oxide, and magnesium oxide, as well as a combination of them, once functionalized, can be efficient in killing microorganisms and are cheaper and safer to use than metal-based nanoparticles.

4.3.5 Protected Cultivation/Automated Structures

Protected or greenhouse cultivation is the most promising area where the production of horticultural crops has improved qualitatively and quantitatively the world over in the last few decades. New-generation poly houses/greenhouses are automated with temperature, humidity, light intensity, irrigation, and fertigation control system with need-based sensors. In India, the area under protected cultivation is presently around 25,000 ha, while the greenhouse vegetable cultivation area is about 2000 ha. Faced with constraints of land holdings, rapid urbanization, declining crop production, declining biodiversity and ever-increasing population, demand for food, especially vegetables, has increased manifold, and protected cultivation has offered a new dimension to produce more in a limited area.

4.3.6 Farm Mechanization

India has a very high share of labor (55%) with a lesser contribution to farm mechanization (40%). While the USA (2.5%) and Western Europe (3.9%) have a very low share of labor in comparison to a 95% share of mechanization. Power is the major crunch in mechanization as only 1.36 kw/ha of power is used in India in comparison to 8.75 kw/ha in Japan. One of the major bottlenecks in farm mechanization in India is 138 million land holdings which are very large in comparison to only 2–3% of the population having landholdings in the USA. Efficient farm mechanization can help in 15–20% savings in seeds, 15–20% savings in fertilizers, 5–20% increase in cropping intensity, 20–30% savings in time, 20–30% reduction in manual labor, and 10–15% overall increase in farm productivity (Guru et al. 2018). Happy seeder, no-till seed drills, paddy threshers, seed drills, and solar sprayers, among many others, are very efficient technology that revolutionized Indian agriculture.

4.3.7 Laser Land Leveling

The use of modern technology in land leveling, like laser land leveling with suitable instruments, has helped in reducing the time and cost of irrigation. Its simple operation to prepare the land before sowing can reap massive returns, such as increasing yields, saving water, and reducing greenhouse gas emissions. Laser leveling in rice fields reduced irrigation time by 47–69 h/ha/season and improved yield by approximately 7% compared with traditionally leveled fields. In wheat, irrigation time was reduced by 10–12 h/ha/season, and yield increased by 7–9% in laser-leveled fields (Aryal et al. 2015).

4.3.8 Modern Irrigation Methods

In India, around 78% of water is consumed in the agriculture sector, while the remaining part is diverted between drinking, industry, and other uses. Dry land agriculture covers about 60% of cultivated area and hence should be the main focus for efficient water conservation and management. The water use efficiency under the conventional flood method of irrigation is very low because of substantial conveyance and distribution losses. There is a need to adopt modern methods of irrigation, like drip and sprinkler irrigation. Drip methods can save up to 70%, while increasing crop yields by 20–90%. India now leads the world, with nearly two million hectares (about five million acres) under microirrigation methods. But, still, there is tremendous potential to go away from using underground water to adopting such methods by harnessing the vast potential of rainwater. To produce 'More Crop Per Drop,' i.e., enhancing water productivity, a five-pronged strategy (based on ecology) would be helpful: (1) precision water management practices (micro-irrigation, laser leveling, automation), (2) reduced water wastage by discouraging flooding, (3) cropping systems optimization and diversification, (4) induction of solar pumps, and (5) on-farm rain-water harvesting (Singh et al. 2021).

4.3.9 Post-Harvest Technologies

. Post-harvest infrastructure plays an important role in Indian agriculture. In India, less than 10% of farm products are processed. Approximately 2% of fruits and vegetables, 8% marine, 35% milk, and 6% poultry are processed. A considerable proportion of our produce goes wasted in the absence of suitable post-harvest infrastructure. Suitable postharvest infrastructure in terms of cold storage, processing units, road networks in inaccessible areas, and establishment of local regulated markets at the Panchayat (village) levels can boost the agriculture sector by promoting value addition and food processing.

4.3.10 Technologies for Drudgery Reduction

Drudgery is the main bottleneck that keeps youths away from agriculture. Thus, appropriate mechanization and smart farming practices will allow for retaining more youth and talent in agriculture. The ergonomics and landscapes are also given adequate emphasis while developing machinery for a particular area in addition to socioeconomic conditions and other factors.

4.3.11 Biofortified Crops

Biofortification" or "biological fortification" refers to nutritionally enhanced food crops with increased bioavailability to the human population that is developed and grown using modern biotechnology techniques, conventional plant breeding, and agronomic practices. The biofortified varieties are considered to be 1.5–3.0 times more nutritious than the traditional varieties. *It's* a sustainable and cost-effective solution to alleviate malnutrition. India is home to the world's highest number of under-nourished people (194.6 million), where 38.4% of them are children. It is estimated that every $1 invested in a proven nutrition program offers benefits worth $16 (Yadava et al. 2017). The Prime Minister of India dedicated 17 biofortified varieties of eight crops to the nation on the occasion of the 75th Anniversary of the Agriculture and Food Organization (FAO), the United Nations on October 16, 2020. The rice variety CR DHAN 315 is enriched with zinc; the wheat variety HD 3298 is rich in protein and iron, while DBW 303 and DDW 48 are rich in protein and iron. The maize hybrids 1, 2, and 3 are enriched with lysine and tryptophan, and the finger millet varieties CFMV 1 and 2 are rich in calcium, iron, and zinc. The CCLMV1 variety of small millet is rich in iron and zinc. The Pusa Mustard 32 is enriched with low Araucic Acid, while Girnar 4 and 5 varieties of peanuts are rich in increased Oleic Acid and Yam's Shri Neelima and DA 340 varieties are enriched with Anthocyanin. Such varieties will contribute to improving the nutrition of the masses.

4.3.12 Sustainable Technologies

To address various issues related to the fatigues of green revolution technologies, more integrated and regenerative technologies like conservation agriculture (CA), integrated farming systems, organic farming, and natural farming are being emphasized in the country. The emphasis is on rehabilitating degraded lands, achieving land degradation neutrality, enhancing farm productivity, reducing the cost of cultivation, making farming climate resilient, ecosystem restoration, efficient on-farm resource recycling, balancing nutrition and producing healthy foods, biorational pest and disease management and blending ITKs with modern practices. Apart from institutional innovations and the right policies, specific actions are needed for: (1) strengthening scientific manpower through more investment in R&D and by forging viable partnerships with other players in the public and private sectors, (2) revisiting rules and regulations for speeding the transfer of technology, including paid extension, and (3) fostering partnership with the private sector for rapid commercialization of available technologies (Singh et al. 2021). The agricultural research in India gradually moved from crop-based to farming system-based agriculture integrating horticulture, livestock, poultry, and fisheries, and secondary agricultural activities as essential components to achieve sustainable food and nutrition security and to maximize farm income. For the economic and nutrition security of a largely vegetarian population, dairy animals are of prime importance; and to improve their productivity and climate resilience, a two-pronged approach will be required—strengthening the availability of superior germplasm and enhancing fertility through biotechnological augmentation of reproduction (NDRI, Vision 2030).

4.3.13 Technology Transfer Tools

Technology transfer needs effective interactive groups like Self Help Groups (SHGs) and Farmers Clubs, which should become tools for disseminating information about various government-sponsored schemes, and these entities will help in liaising with various line Departments for various activities. Internet and mobile phones are potential tools to Impart knowledge on new developments and improved methods of cultivation/technologies in the field of agriculture. These tools are useful in the dissemination of weather data and agroclimatic conditions' latest information on prices of agricultural produce to farmers. Krishi Vighyan Kendras (KVKs) have been established in each district of the country, and now these are the backbone of technology dissemination in the country. Fast technology transfer will certainly reduce the knowledge deficit among the farmers and will help accelerate the stagnant growth of agriculture, realizing the higher potential of our land and the hard work of our farmers.

There are a number of innovative technologies ready to be scaled up and scaled out, and farmer-led innovations also need to be assessed, validated, refined, and out-scaled to harness their full benefits and adopted in a participatory mode involving all stakeholders (Paroda 2019). Hybrid technology (maize, bajra, sorghum, and rice), biotechnology—GM crops (soybean, mustard, maize, and brinjal), conservation agriculture (3.5–20 m ha), protected cultivation (expand area from 50,000 ha to 2.00 m ha), microirrigation (discourage flood irrigation)—at least 10 m ha, bioenergy/biofuel (use of sugarcane and maize - initially 20%), biofortified crops (QP maize, Fe and Zn rich rice, Fe rich bajra, Zn rich wheat) and ICT for knowledge sharing, ex. e-Chaupal have been identified as potential areas for immediate up-scaling in India (Paroda 2019).

4.3.14 Scientific Organic Farming and Export of Organic Produce

Organic farming is a system that avoids or largely excludes the use of synthetic inputs (such as fertilizers, pesticides, hormones, feed additives, etc.) and, to the maximum extent feasible, rely upon crop rotations, crop residues, animal manures, off-farm organic waste, mineral grade rock additives and biological system of nutrient mobilization and plant protection. An inherited tradition of organic farming in several states of the country is an added advantage. This holds the promise for organic producers to tap the market which is steadily growing (15–25%) in the domestic market related to the export market. In order to develop a technological package of organic farming, including plant protection for different crops and cropping systems, a Network Project on Organic Farming (NPOF) was initiated during 2004–2005 by the Indian Council of Agricultural Research (ICAR), New Delhi with Indian Institute for Farming Systems Research (IIFSR), Modipuram, Uttar Pradesh as a nodal institute with now having 20 centers across India. More than 45 packages of practice for the organic production of crops in cropping systems across India have been developed so far. An integrated organic farming system (IOFS) model was also developed (Umiam, Meghalaya) to meet the diverse requirement of the farm household while preserving the resource base and maintaining the ecology. The model has diversified farming components like field crops (cereals, pulses, and oilseeds), horticultural crops (vegetables, fruits), livestock, and duckery, along with perennial fodder crops, composting units, and central water harvesting pond for composite fish culture and as a source for irrigation during the lean season. As of March 31, 2021 total area under the organic certification process (registered under National Programme for Organic Production) is 4,339,185 ha (2020–2021). This includes 2,657,889 ha cultivable area and another 1,681,296 ha for wild harvest collection. Among all the states, Madhya Pradesh has covered the largest area under organic certification, followed by Rajasthan, Maharashtra, Chhattisgarh, Himachal Pradesh, Jammu and Kashmir, and Karnataka. The total volume of export during 2020–2021 was 888,179.68 MT. The organic food export realization was around INR 707,849.52 lakhs (1040.95 million USD). Organic products are exported to the USA, European Union, Canada, Great Britain, Korean Republic, Israel, Switzerland, Ecuador, Vietnam, Australia, etc. In terms of export value realization, processed foods, including soya meal (57%), lead among the products, followed by oilseeds (9%), cereals and millets (7%), plantation crop products such as tea and coffee (6%), spices and condiments (5%), medicinal plants (5%), dry fruits (3%), sugar (3%), and others.

4.4 Conclusions

The India's population is estimated to be about 1.7 billion by 2050. To meet the challenge of producing enough quality food to assure food and nutritional security without harming the environment, India must find ways to increase food production while dealing with worsening land and water scarcity, soil degradation, and climate-change-related extreme weather events. The increasing role of technology in addressing these issues is the only way forward to a food-secure future. The recent example of growth in pulses sector making India self sufficient is a glaring example of technological advancement in the country. Technology can help save foreign exchange, increase productivity, and lead to an improvement in the overall standard of life, especially for peasants. Innovative approaches such as digital smart agriculture, artificial intelligence, blockchain, remote sensing and GIS technology, drones, and robots could take Indian agriculture to the next level. The pace of adopting modern farming practices is slow, and path-breaking efforts need to be made to educate farmers about the benefits to be had with technology.

References

Aryal JP, Mehrotra M, Jat ML, Sidhu HS (2015) Impacts of laser land leveling in rice–wheat systems of the north–western indo-gangetic plains of India. Food Secur 7(3):725–738

CFPS, Centre for Food and Policy Studies (2011) Science and society: an Indian perspective science and technology under the British rule. Retrieved from http://cpsindia.org/index.php/art/114-science-sustainability-and-indian-nationalresurgence/d-science-and-technology-under-the-british-rule/158-d1-indian-agriculture-before-modernisation. Accessed 16 April 2022

Gatzweiler FW, Von Braun J (2016) Technological and institutional innovations for marginalized smallholders in agricultural development. Springer. ISBN 978-3-319-25718-1; XV, p 435

Gaur P (2021) Can India sustains high growth of pulses production? J Food Legumes 34(1):1–3

Gautam HR, Kumar R (2014) Agricultural development - the road ahead. Kurukshetra 62(8):3–6

Guru PK, Borkar N, Debnath M, Chatterjee D, Sivashankari M, Saha S, Panda BB (2018) Rice mechanization in India: key to enhance productivity and profitability. In: Pathak et al (eds) Rice research for enhancing productivity, profitability and climate resilience. ICAR-National Rice Research Institute, Cuttack, pp 337–351

Kumari S (2014) News ways of improving agriculture. Kurukshetra 62(8):7–10

Lucht JM (2015) Public acceptance of plant biotechnology and GM crops. Viruses 7(8):4254–4281

NDRI, Vision (2030) National Dairy Research Institute, Karnal, Harynana, India. https://ndri.res.in/design/document/vision/pdf

NITI Aayog (2018) NITI Aayog report on composite water management index. Government of India, New Delhi

Paroda RS (2019) Report on policies and action plans for a secure and sustainable agriculture. In: Committee report submitted to the principal scientific adviser to the government of India, New Delhi, p 198

Singh RB (2011) Towards an evergreen revolution – the road map. National Academy of Agricultural Sciences, New Delhi, p 61

Singh RB, Paroda RS, Malavika D (2021) National dialogue indian agriculture towards 2030 pathways for enhancing farmers' income, nutritional security and sustainable food systems thematic session: science, technology and innovation discussion paper: science, technology and innovation, NITI Ayog & FAO, Ministry of Agriculture & Farmers Welfare, Govt of India. Retrieved from https://www.fao.org/fileadmin/user_upload/FAO-countries/India/docs/Full_Paper-8.pdf. Accessed 16 Apr 2022

Swaminathan MS (2007) Agriculture cannot wait – new horizons in Indian agriculture. In: Swaminathan MS (ed). Academic Foundation and National Academy of Agricultural Sciences, New Delhi, p 550

Tiwari AK, Shivhare AK (2018) Pulses in India: retrospect and prospects. Directorate of pulses development, ministry of agriculture and farmers welfare, government of India, Bhopal, Madhya Pradesh, India

Tripathi A, Prasad AR (2009) Estimation of agricultural supply response by cointegration approach. Ind Econ J 57(1):106–131

Yadava DK, Choudhury PR, Hossain F, Kumar D (2017) Biofortified varieties: sustainable way to alleviate malnutrition. Indian Council of Agricultural Research, New Delhi

Indian Seed Sector

5

D. Vijay, H. P. Vijayakumar, C. T. Manjunath Prasad, and P. R. Choudhury

Abstract

The Indian seed sector is one of the most vibrant industries in the country. It is unique with a strong public and private industry. The evolution of this sector is an interesting story and illustrates the role of various contributors, including the policymakers, researchers, departmental personnel, and the private sector. A categorical institutionalization and growth of this sector were initiated before the independence with the suggestions of the Royal Commission on Agriculture (1925). The availability of quality seed played an important role in the green revolution and further contributed to the increase in food grain production and productivity. The time-to-time recommendations of various commissions, committees, and review teams helped in shaping a strong seed sector in India. The foundation of an organized seed industry was laid with the creation of a public sector organization, the National Seeds Corporation, in 1963, which mentored the private sector initially and helped in developing a systematic, independent seed certification system in the country. The National Seed Project, with World Bank financial assistance and other seed projects, helped in the development of infrastructure in both public and private seed organizations. The landmark seed legislations, viz., Seeds Act (1966), Seed Rules (1968), The Seeds (Control) Order (1983), Protection of Plant Varieties and Farmers Rights Act (2001), and several policies like New Policy on Seed Development (1988), National Seed Policy (2002), Exim policy (2002–2007) helped in the overall growth of the industry and paved the way for a competitive private sector. The

D. Vijay (✉) · H. P. Vijayakumar · C. T. Manjunath Prasad
ICAR-Indian Agricultural Research Institute, New Delhi, India
e-mail: dunna.vijay@icar.gov.in

P. R. Choudhury
Indian Council of Agricultural Research, New Delhi, India

P. K. Ghosh et al. (eds.), *Trajectory of 75 years of Indian Agriculture after Independence*, https://doi.org/10.1007/978-981-19-7997-2_5

seed production and supply system showed continuous growth, and the surplus production facilitated the increase in seed replacement rates of various crop groups over a period. The seed exports are also increasing and have a potential for further growth. The increase in the varietal replacement rate of important crops supports the inclusion of climate-resilient varieties in the seed chain. To become the Asian seed hub, the Indian seed sector shall emphasize the incorporation of the latest technological innovations and upgrade to the digital mode to cater the future needs.

Keywords

Indian seed industry · Seed policies · Seed availability · Seed imports and exports · Seed replacement rates · Seed requirement · Varietal replacement rate

5.1 Introduction

Seed is one of the primary inputs for successful and sustainable agriculture. It is a vehicle through which various technologies can reach farmers to bring out a phenomenal change in the agricultural situation of a country or region. Seed quality is of utmost importance for realizing the yield potentials of high-yielding varieties, transforming Indian agriculture and its dependence on food imports. Our ancestors realized the fundamental importance of using quality seed, and the quote "सुबीजम सुस्क्षत्रे जायथे संपधथे" present in 'Manu smriti' which means good seed in good land yields abundance, is an undeniable proof of it. The supply of quality seeds to the farming community helps in achieving higher crop productivity and enhances farm income. Quality seeds possess specific standards for genetic and physical purity, germination, moisture content, and seed health. The legal requirement of meeting these standards through certification and quality control regulation has revolutionized the Indian seed sector. It is well-established that quality seed alone contributes about 15–20% to crop productivity by ensuring good germination, early vigor, and plant stand. The Indian seed industry is unique globally, with strong public and private sectors. The review and recommendations of various committees during the pre-independence era have paved the way for developing a robust public seed sector, which further helped in the birth of the private seed sector through various schemes. India is home to more than 500 private seed companies, including local and multinational enterprises. The success of the green revolution in mitigating food imports started with the import of desired seeds and their multiplication and distribution. The bold steps taken during the last 75 years of Indian independence made the seed industry a viable system that helped the country to achieve self-sufficiency in food grains and aided in venturing into food exports. The vastness of the seed production and supply system and its quality regulation over the periods has revolutionized this sector and made it one of the fastest-growing agricultural input industry in the country. This chapter will unfold the saga of the Indian seed industry from the pre-independence era and its transformation during the last 75 years of independence, along with its contribution to the development of the agriculture sector in India.

5.2 Status of the Seed Sector in the Pre-independence Era (up to 1947)

The development of improved varieties in India was initiated with the commencement of agriculture research at Agricultural Colleges and Research Stations established in the early years of the twentieth century. In the pre-independence era, the state department of agriculture was bestowed with the distribution of quality seeds of the improved varieties. The state department used to adopt two different methods for distributing seeds. In the first method, seeds multiplied at one location were delivered over a larger area. In the second method, small quantities of seeds were supplied to many farmers for further multiplication and dissemination among themselves. However, the second method was ineffective in the spread of improved varieties, and the state departments concentrated on the first method by setting up several seed multiplication centers. The seed produced at these centers was further multiplied with the help of landlords and then distributed to the region's farming community. The Royal Commission on Agriculture, constituted in 1925, examined the progress of seed distribution and the spread of improved varieties and suggested constituting a separate organization within the agriculture department to deal with seed distribution and seed testing. It also advised to encourage new seed merchants, since only flower and vegetable seed merchants were present in India. The commission predicted that it would take several years to develop exclusive seed merchants and suggested that until then, the government shall be involved in seed multiplication and distribution. Later, several other commissions like Sir John Russel Commission (1937), the Famine Enquiry Commission (1944), and Food Grains Policy Committee (1943) reviewed the seed status in India and emphasized seed multiplication and distribution by the agriculture department.

The second world war resulted in several hardships to the country's food availability. After the separation of Burma (presently Myanmar) in April 1937, India imported 1.5–2 million tonnes of rice per annum from it. Due to the second world war, this source of rice import got suddenly cut off, and no alternate source could be tapped. At the same time, wheat supplies from the USA, Canada, and Australia could not be obtained due to the war. This situation led to the emphasis on increasing food production, which ultimately resulted in Grow More Food campaign in 1942. One of the main action lines of this campaign is the intensive cultivation of cultivated lands, making available better seeds and manures, and following better farming practices. The second world war also brought several difficulties for the seedsmen of vegetable crops in India because most of the vegetable seed requirement was met through imports. The private vegetable seed companies were the first to develop seed production facilities for temperate vegetables in Kashmir and Quetta by 1945. The seedsmen of vegetable seeds organized into an association, "The All India Seed

Growers, Merchants and Nurserymen's Association," in 1946 for the rapid development of the vegetable seed industry in India.

5.3 Seed Sector Development in the Pre-green Revolution Period (1947–1966)

The partition of the country in 1947 led to a loss of 7–8 lakh tons of food grains to India. Due to partition, India received 82% of population and 71% of the total irrigated area of the undivided country. This led to only 75% of total cereal production, 65% of wheat, and 68% of rice production of the undivided nation (Fig. 5.1). The partition reduced percent of the total irrigated area to the total cropped area from 24 to 19 in India (Fig. 5.2). These losses led to more emphasis on agriculture to meet the food requirement of the growing population in the consequent 5-year plans. Along with agriculture, the seed sector has also grown during the post-independence period. The Grow More Food Enquire Committee (GMFEC) was constituted in 1952 to review the Grow More Food (GMF) campaign implemented in April 1949 to enhance the production of food crops. One of the important schemes of the GMF campaign is the supply and distribution of seeds and fertilizers. The GMFEC pointed out that the GMF campaign could not achieve its target, and one of the reasons was a lack of seed purity and unsatisfactory seed multiplication. It recommended establishing nucleus seed production farms and strengthening seed multiplication. ICAR constituted an expert standing committee in 1952 to formulate a sound seed improvement program. Due to the recommendations of these committees, a system for distributing improved seeds of food grains came into existence. Despite this, the progress made during the 1950s was poor, and the seed programs were mainly confined to seed distribution.

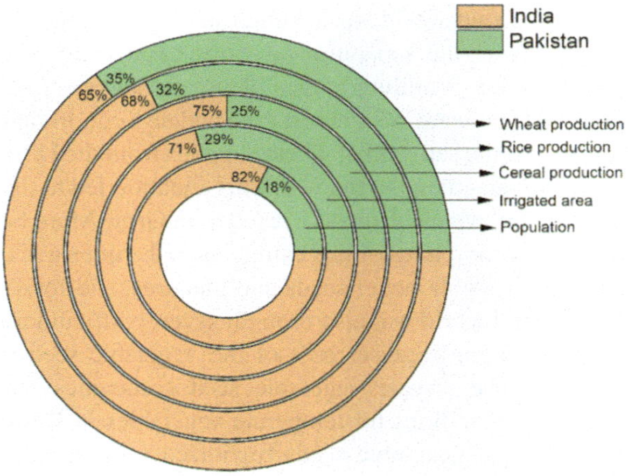

Fig. 5.1 Percent population, irrigated area, and crop production of India and Pakistan due to the partition of the country in 1947. (Data source: GMFEC Report 1952)

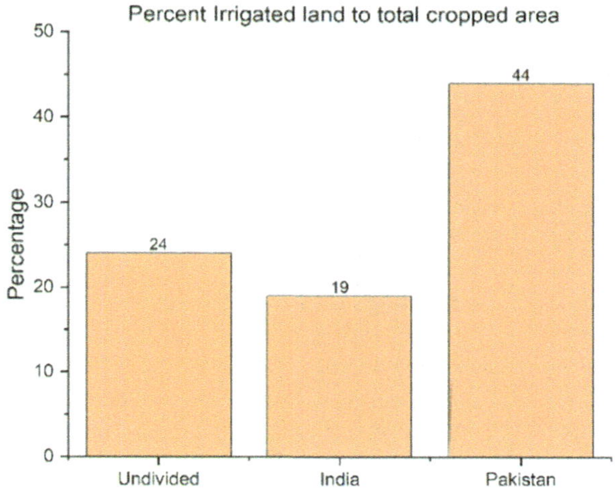

Fig. 5.2 Percent irrigated land to the total cropped area of India and Pakistan due to partition of the country in 1947. (Data source: GMFEC Report 1952)

During the Second 5-Year Plan (1956–1961), special importance to seed multiplication from the nucleus seed stage to the foundation seed stage given by the Indian government by setting up seed farms of 10 ha each. Even though only 59% of seed farms (2551 Nos.) could commence functioning of the targeted 4328 farms during this period, it paved the way for systematic seed production on a large scale. In 1961, the Program Evaluation Organization (PEO), at the instance of the planning commission of India, conducted a study to assess the operation of the seed multiplication and distribution program from 1952–1953 to 1959–1961. The study found that the production and multiplication of nucleus and breeder seed in the first two 5-year plans were not given proper attention even though the Royal Commission on Agriculture (1925) emphasized the importance of these two stages for further maintenance of quality during seed multiplication. The PEO study also observed that only 5% of seed farms had taken all necessary precautions to maintain seed purity, and 90% of farms were running at a loss. The PEO (1961) made a few suggestions, viz. (1) formulation of annual seed plans by state governments, (2) extending the seed production and multiplication to food crops other than paddy and wheat, (3) strengthening of seed farms, and (4) training to human resources of seed farms on quality seed production. During the same period, the first Indo-American Agricultural production team (1959) headed by Dr. Sherman E. Johnson from the Ford foundation examined India's food production problem and suggested several critical seed quality and distribution aspects. The team recommended (1) educating farmers to use improved seed through the extension workers at the village, block, and district level, (2) Involvement of State Agriculture Departments in seed certification and involvement of cooperatives and private seed growers in seed supply, and (3) to set up seed testing laboratories, development of uniform seed certification standards and enactment of seed laws. These recommendations from PEO and Agricultural Production Team laid the foundation for the seed quality control system in the country.

The All India Coordinated Maize Improvement Programme (AICMIP), initiated in 1957 by the Indian Council of Agricultural Research (ICAR) in collaboration with the Rockefeller Foundation, was one of the most significant turning points in developing suitable varieties/hybrids in Indian Agriculture. It marked the beginning of an intensive, multidisciplinary, integrated approach to crop improvement. Four years of its inception, four maize hybrids were released in 1961. Encouraged by the progress of this approach, ICAR simultaneously launched similar projects in 1960 for sorghum and bajra, which released their first hybrids in 1964 and 1965, respectively. The Third 5-Year Plan (1961–1966) has seen some serious efforts to overcome the shortcomings of the seed programs. The year 1961, celebrated as 'World Seed Year' by Food and Agricultural Organization (FAO), is also a remarkable year for the Indian seed sector. This year, the first seed testing laboratory (STL) was established at Indian Agricultural Research Institute (IARI) with the help of the Rockefeller Foundation. This STL became a member laboratory of the International Seed Testing Association (ISTA), Zurich, Switzerland, for uniformity of testing procedures. The newly developed hybrids under All India Coordinated Crop Improvement Programmes necessitated the creation of a separate organization for seed production, resulting in the birth of a Central seed corporation, viz. National Seeds Corporation (NSC), in 1963. The NSC was incorporated under the Ministry of Agriculture to ensure the rapid multiplication and distribution of hybrid seeds to farmers. It paved the way for developing an organized seed industry in India. Its responsibilities in producing high-quality seeds, establishing quality control inspection through seed certification, scientific seed processing, storage, and marketing marked the beginning of systematic seed production on scientific principles. NSC was also involved in developing field inspection methods and seed standards for seed certification and labeling; multiplication of prereleased varieties of all-India importance; assisting in setting up seed processing plants; and training individuals involved in seed production.

5.4 Development of the Seed Sector During the Green Revolution Era (1966–1985)

The world witnessed a tremendous increase in food crop productivity in the last six decades due to the green revolution (1966–1985), particularly in the developing world. The global population increased by 2.5 times, whereas cereal production rose by almost 3.5 times in the last 60 years (Fig. 5.3). The green revolution rapidly increased agricultural output due to increased yield levels. Between 1960 and 2000, the yield levels of developing countries rose by 208% for wheat, 109% for rice, and 157% for maize (FAO 2004). Developing countries in Southeast Asia and India were the first to reap the benefits of the green revolution in cereal food crops (Pingali Prabhu 2012). Agricultural Trade Development and Assistance Act (commonly known as Public Law-480 or "Food for Peace"), enacted by the United States of America (USA) in 1954, enabled food-deficit friendly countries to purchase US agricultural commodities with their local currency instead of USD to save their

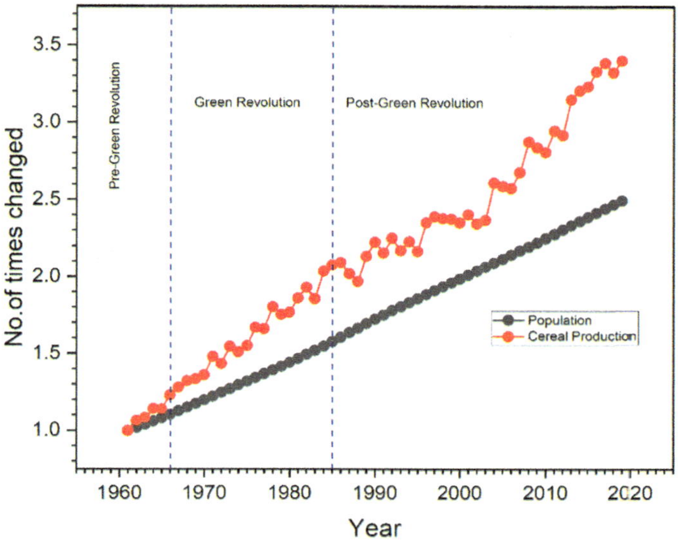

Fig. 5.3 Number of times change in the global population and cereal production with 1961 as base year. (Data source: https://ourworldindata.org)

precious foreign exchange reserves. India signed an agreement in 1956 to import food grains from the USA under PL-480 to meet the food grain requirement of the growing population. The imports under PL-480 masked the necessity and deferred the initiation of programs to enhance the country's agricultural productivity during the Second and Third Five-Year Plans (Hatti 1977).

In India, the population increased by 10% between 1961 and 1966, which coincided with a decline in cereal food grain production. The back-to-back drought in 1965 and 1966 and the USA's delayed supply of food grains due to political differences led to the 'ship to mouth' situation. The green revolution was ushered by importing 18,000 tonnes of wheat seeds in 1966 from CIMMYT, Mexico, under the High Yielding Varieties Programme (HYVP). The HYVP launched in 1966 is a significant milestone in developing the country's seed industry. The HYVP is a comprehensive program consisting of research, irrigation, inputs, extension services, training, and supervision in five crops, viz. paddy, wheat, maize, sorghum, and pearl millet. The output of these five crops taken together increased by 16% during 1964–1965 to 1969–1970. The collaborated efforts of the NSC, state agriculture departments and private entrepreneurs in seed production, processing, certification, testing, storage, and distribution ensured the supply of the required quantity of seed for implementing the HYVP. In the last 60 years, the population of India increased by three times, but due to the green revolution, the cereal food grain production increased by 3.75 times (Fig. 5.4) and changed the food grain situation from deficit to surplus. Crop production and yield trajectory changed due to the green revolution in India and the world (Figs. 5.3 and 5.4). Several drivers led to the success of the

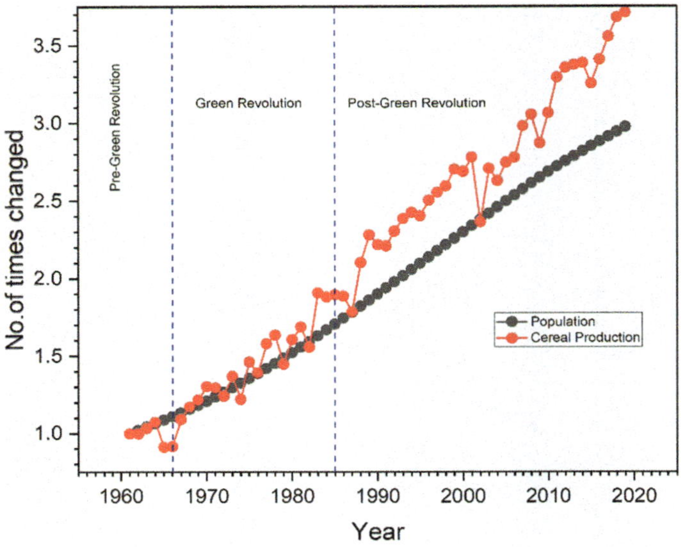

Fig. 5.4 Number of times change in the Indian population and cereal production with 1961 as base year. (Data source: https://ourworldindata.org)

green revolution, and one among them is the production, multiplication, and supply of quality seeds of high-yielding varieties.

The progress of seed development work was very rapid, with many new initiatives on the seed front taken during the green revolution period. To ensure seed quality, the Parliament passed The Seeds Act in 1966, and Seed Rules were framed in 1968. These two played a significant role in providing the legal sanctity to the seed quality assurance and its maintenance. The Seeds Act is essentially a truthful labeling law that empowers the government to notify the seed varieties for specific areas. This act also provides setting up certification agencies for seed certification. However, seed certification is made voluntary, and labeling is compulsory for the seed sold in the market. In the second half of the 1960s, seed testing laboratories were created in almost all states, and several processing plants were also established. To review the rapid changes in the Indian seed sector and to develop a robust seed production system in India, the Government of India constituted a Seed Review Team (SRT) on October 10, 1967 under the leadership of Mr. I.J. Naidu, Joint Secretary to the Govt. of India, Department of Agriculture and members from Research Institutes, state agriculture department, private seed industry and NSC. The SRT submitted its report on 14th June 1968. It made several recommendations on various aspects, including seed production, processing, storage, certification, the seeds act, quality control, the role of multiple agencies, and the seed industry.

Some of the important recommendations made by SRT (1968) were (1) registration of crop varieties for marketing seeds; (2) elimination of varieties with doubtful value; (3) adequate arrangements for prerelease seed multiplication of promising varieties; (4) involvement of Agricultural Universities in foundation seed production in addition to NSC; (5) assigning a significant role to the cooperative and private sectors in seed production, processing, and marketing; (6) once the cooperative and private organizations are developed, governmental agencies should be relieved gradually from these activities and focus on program planning and extension activities; (7) NSC should support State Governments to set up the certification agencies and transfer its certification work to them. During the Fourth Five-Year Plan (1969–1974), GOI planned to transform Indian agriculture from a traditional way of life into an industry-based system by efficiently using available resources and providing economic incentives for higher investment. For this purpose, areas with assured rainfall were selected and suggested to follow an improved package of practices, including seeds of improved varieties, and fixed the particular food production targets for such areas. One such organization is Tarai Development Corporation (TDC), established in 1969 with the World Bank's assistance under National Seed Program. It was another important landmark in developing the seed industry in India. TDC was a role model for seed corporations set up in other states. It aimed at the integrated agricultural development of the Tarai area through the production and supply of quality seeds. Later it was renamed UP Seeds and Tarai Development corporation w.e.f. July 1, 1978.

The National Commission on Agriculture (NCA) reviewed Indian Seed Industry and submitted its final report in 1976 with recommendations for expanding the seed industry on commercial lines. Its suggestions include (1) the development of a national registry of varieties by ICAR and the central seed committee; (2) the creation of a network of seed processing and storage facilities to the extent required for the seed industry; (3) the manufacture of processing equipment; (4) storage of nucleus, and breeder seed in controlled conditions; (5) undertake grow-out test as an integral part of seed testing. It also emphasized selecting congenial areas for seed production and advised introducing seed production technology courses in agricultural universities. It also encouraged small seed growers to form compact areas for seed production. Based on the Seed Review Team (1968) and the National Commission of Agriculture (1971 and 1976) recommendations and the encouraging experience with the Tarai Seeds Project, a working group was established to prepare the proposals for National Seeds Programme. Under this program, a project proposal was submitted to World Bank with components including reorganization of NSC to coordinate and develop the seed industry in the country and interstate seed marketing; expansion of seed storage capacity, establishment of state seed corporations, improvement of breeder seed production facilities at agricultural universities and ICAR institutes; development of research component of seed technology at agricultural universities; provision of facilities and equipment for private seed processing units; expansion of seed quality control through support to seed certification agencies and seed testing laboratories; and providing training and technical assistance. The National Seeds Project (NSP) was launched initially in two phases, NSP-I (1976–1984) and NSP-II (1978–1985), to expand the seed industry with a loan from the World Bank. NSP Phase-I was launched in four states, viz. Punjab, Haryana, Maharashtra, and Andhra Pradesh, and Phase-II in five states, viz. Karnataka,

Rajasthan, Uttar Pradesh, Bihar, and Odisha. The World Bank Supervisory Mission examined the progress of NSP Phase-I and II in 1981 and observed that only 1.72 lakh metric tonnes of seed were produced against the target of 3 lakh metric tonnes (Agrawal 1995). It also found that most of the sanctioned amount was not utilized by the Seed Corporations. However, the Projects have created an excellent infrastructure to meet future seed demands. Even though both projects could not generate the economic returns envisaged in the proposal, they helped to identify the critical issues of public sector seed industry governance and formulate appropriate policies for the development of the seed industry.

Further, the government of India passed the Seeds (Control) Order (1983) under the Essential Commodities Act of 1955 to regulate the seed business, including the import and export of seed, and to stop illegal seed hoarding. This order made compulsory licensing for seed sale, import, and export. Seed quality assurance was entrusted to seed inspectors. It empowered them to draw the samples of seeds meant for sale, import or export, check their quality, and seize any seed if a contravention of this order was committed.

5.5 Seed Sector During the Post-green Revolution Era (1985 Onwards)

The Indian seed sector has seen a categorical change after the green revolution, followed by various policy decisions by the government to encourage the seed industry. During the Seventh Five-Year Plan (1985–1990), seed corporations were established in another four states, viz. Assam, West Bengal, Madhya Pradesh, and Gujarat, under the sponsorship of NSP Phase-III. Further, seed production infrastructure in the private sector also considerably improved under this Phase-III as about 95% of the funds were loaned to private companies. The introduction of the New Seeds Policy in 1988 revolutionized the Indian seed industry by liberalizing the seed trade and granting access to the best quality seed or planting material available anywhere in the world to Indian farmers. Under this policy, the import of seeds under an Open General License (OGL) for sowing purposes both by the public and private seed industry was allowed. A timebound (30 days) permission or rejection by the regulating body was emphasized on the import requests. Separate guidelines were issued under this policy for strengthening the plant quarantine. This policy paved the way for several incentives like reducing import duty on seeds and machines used for seed processing and quality control; preshipment credit; cash compensation on the export of seed; income tax rebate; and backward area subsidy for the growth of the seed industry in the country. These steps by the government have resulted in the import of hybrid seeds of vegetables, flowers, and some food crops by private companies and boosted the private seed sector in the country. With the growth of the seed sector, the research and development activities of the private sector increased, and several national and multinational companies started seed production of their hybrids in several crops. With the growth in the seed sector, along with the commitment by India as a member of the World Trade Organization (WTO) to the

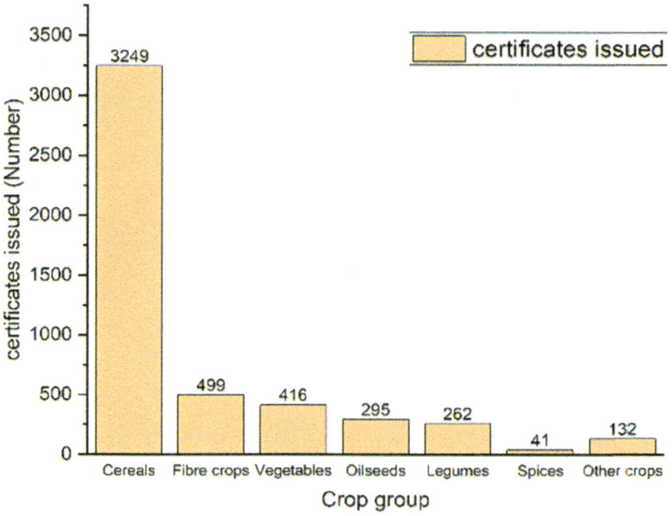

Fig. 5.5 Crop group wise number of plant registration certificates issued by PPV&FRA till March 2022. (Data source: https://plantauthority.gov.in)

Trade-Related Aspects of Intellectual Property Rights (TRIPS) agreement, the Government of India enacted the Protection of Plant Varieties and Farmers' Rights (PPV&FR) Act, 2001. To safeguard the farmers' rights and plant breeders' rights, India adopted a sui-generis system instead of joining the UPOV (International Union for the Protection of New Varieties of Plants) convention to protect new varieties. The PPV&FR Act strengthens the seed industry by providing necessary protection to their intellectual property so that farmers get access to seeds developed by MNCs in other countries. This act also encourages investment in research and development both by the public and private sectors. To date, the Distinctness, Uniformity, and Stability (DUS) testing guidelines for new varieties of 171 crop species were notified by the PPV&FR Authority. The authority has issued 4894 plant variety registration certificates till March 2022. Maximum certificates were issued in cereal crops (3249) and least in spice crops (41) (Fig. 5.5). The certificates were issued under five categories, viz. extant varieties—notified and of common knowledge, farmers' varieties, new varieties and essentially derived varieties. The maximum number of certificates issued belongs to the farmers' varieties category (39.4%), indicating the need for farmers' rights in our country and the role farmers play in maintaining the seed of indigenous/traditional varieties (Fig. 5.6).

The National Seeds Policy 2002 was formulated to enhance the seed replacement rate, which means enhanced availability of quality seed to the farming community. It also intends to create a congenial environment for the faster growth of local seed industries, encourage germplasm import and boost exports and strengthen the monitoring system for the availability of high-quality seed material to the farmers. This policy also emphasized enhancing India's share of global seed export from 1%

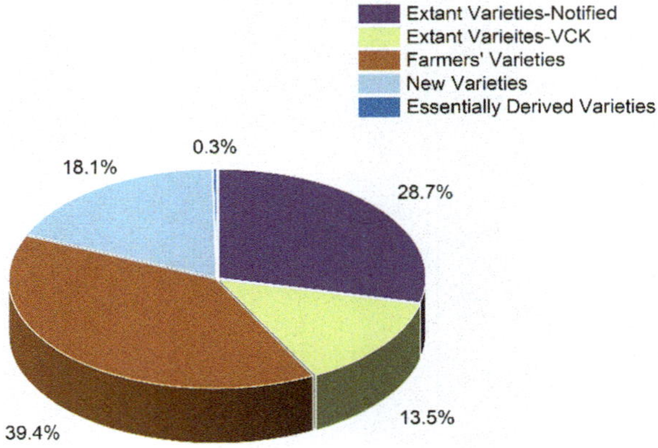

Fig. 5.6 Percentage of different types of plant registration certificates issued by PPV&FRA till March 2022. (Data source: https://plantauthority.gov.in)

to 10% and developing a national seed grid to provide information on the availability of seeds initially by the public sector and later extended to the private sector. Further, to streamline the import and export of seed, the export/import (EXIM) policy 2002–2007 was formulated. This policy is being followed by including the amendments made from time to time. The EXIM Policy reiterates that Plant Quarantine Order 2003 regulates all seeds and planting material imports. The Import licenses will be granted by the Directorate General of Foreign Trade (DGFT) under the Ministry of Commerce and Industry only based on the recommendations of the Department of Agriculture and Farmers Welfare of Ministry of Agriculture and Farmers Welfare. A small quantity of seeds sought to be imported would be given to ICAR, or farms accredited by ICAR, for trial and evaluation for one crop season. The restrictions on the export of all seeds except breeder seed, foundation seed, seeds of wild varieties, and seeds of a few specific crops, viz. onion, berseem, cashew, rubber, nux vomica, pepper cuttings, sandalwood, saffron, neem, forestry species, and wild ornamental plants, were removed from 2004. The EXIM Committee meets every month subject to the tendency of proposals for the import/export of seeds and planting material and furnishes its recommendation to DGFT for necessary permissions.

The government proposed to revise the Seeds Act of 1966 to get it harmonized with other acts and rules and combine it with the Seed (Control) Order 1983 and other policies. The New Seed Bill was proposed in 2004 and later modified in 2019 and is under parliament's consideration. In the new bill, apart from agriculture and horticulture, forestry, cultivation of plantation, medicinal and aromatic plants are brought under its purview. Registration of varieties sold in the country has been made compulsory, and the registration body will maintain a 'National Register of Seeds'. A provision is made under this new seed bill to evaluate the registered varieties and compensate the farmers for the non-performance of varieties under said conditions. The registration of seed dealers, processing units, and seed producers is also compulsory. Seed certification by accredited government organizations apart from certification agencies is permitted. Seed health is given importance in the New Seed Bill in coherence with the international seed trade. Apart from these regulations and policies, from time to time, several schemes, viz., technology mission on oilseeds; subsidy on seed production and distribution; subsidy on seed transport; and seed bank scheme, were initiated by the government to encourage and strengthen the seed sector in the country during post green revolution period.

5.6 Role of Seed Sector in Agricultural Growth

Seed is a low-cost input and is the cheapest means of enhancing agricultural production. The quality of seed and its timely availability in sufficient quantities are the major requirements for the success of any agriculture-based technology. The seed sector has played a tremendous role in increasing agricultural output in terms of production and productivity. With the advent of the science of genetics, plant breeding, and biotechnology, new vistas are opened in the development of varieties and hybrids with resistance to pests and diseases, enhanced food quality, increased yield, and other desirable characteristics by manipulating the genetic composition of the plants. A robust seed system is the foremost step toward the country's food security and acts as a driver of growth in agriculture. In the early nineteenth century, modern India recognized the significance of seeds and emphasized several seed improvement programs. Accordingly, several suggestions were made by the Royal Commission on Agriculture (1925), Grow More Food Enquiry Committee (1952), Indo-American Agricultural Production Team (1959), Seed Review Team (1968), and National Commission on Agriculture (1976). With the government's emphasis on formulating several seed policies and infrastructure development, a robust seed sector has emerged. The Indian seed sector is unique, with strong public and private sector seed companies. The public seed sector mainly concentrates on high-volume and low-value food crops. Important public sector organizations include 50 ICAR Research Institutes of crop science and horticulture; 35 crops related All India Coordinated and Network Projects, 46 State Agricultural and Horticultural Universities (SAUs), 3 Central Agricultural Universities, 722 Krishi Vigyan Kendras (KVKs), 1 National Seeds Corporation (NSC); 15 State Seeds Corporations (SSC) and 24 State Seed Certification Agencies (SCA). The ICAR Institutes and Universities are mainly concerned with breeder seed production, and SAUs and KVKs are involved in the foundation seed production. The seed corporations and department farms are involved in the foundation and certified seed production. The private seed sector also plays an equally important role in seed production but is usually restricted to high-value crops such as vegetables, hybrids of corn, cotton, pearl millet, sunflower, and other similar crops. The private seed sector comprises 540 seed companies, including multinational companies. Among these, 80 companies have their own independent research and development programs to develop new varieties. After the green revolution, the demand for hybrid seeds in the country increased due to the increasing pressure for higher food grain production to meet the requirement of the growing population in the country. The strong seed sector could meet the rising demand and help achieve food security.

5.6.1 Impact of Quality Seed on Food Grain Yield and Production

The green revolution was instrumental in increasing food grain production, changing the 'ship to mouth' condition to surplus food grains with overflowing granaries. Of the several factors that made this possible, significant ones are (1) the photo insensitive, fertilizer responsive, dwarf, high yielding varieties, (2) the availability of adequate inputs like quality seed and fertilizers, and (3) increased area under assured irrigation. In the last 70 years, the area under food grains has increased only by 30 million hectares (~30%), whereas production has increased by 245 million tonnes (~481%). This increase in production is due to a yield increase, which grew from 522 kg/ha to 2325 kg/ha (~445%) in the previous 70 years (Fig. 5.7). Seed is the most essential and primary input of the several inputs required for crop production. The quality seed alone contributes 15–20% of the total production (Chauhan et al. 2015). Thus, the availability of quality seed (certified seed) played an essential role in increasing crop yield. The correlation analysis between the amount of certified seed distributed and crop yield during the last 34 years showed a strong positive correlation with a significant positive Pearson correlation (r) value of 0.918 and an adjusted r^2 value of 0.839 between these two parameters (Fig. 5.8). Similarly, we observed a strong positive correlation ($r = 0.928$, adjusted $r^2 = 0.858$) between certified seed distribution and food grain production per se (Fig. 5.9). These findings irrevocably explain the importance of quality seeds to enhance crop yield and production and justify the government's priority on various seed programs.

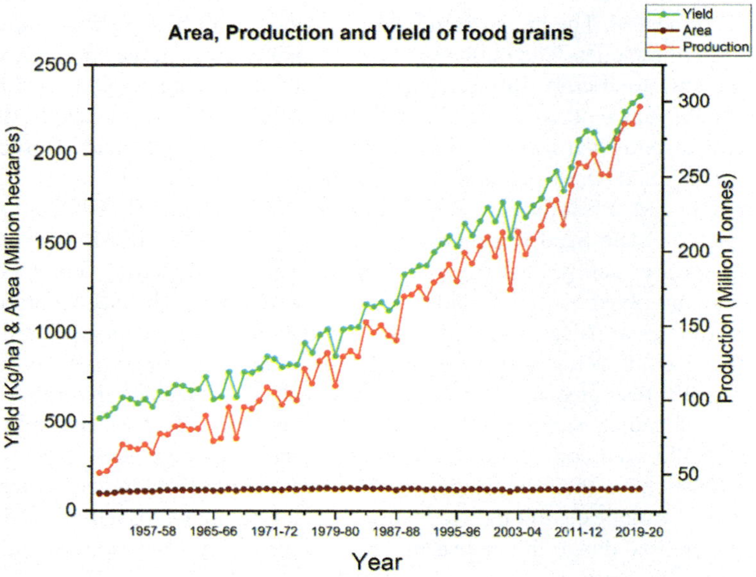

Fig. 5.7 Status of area, production, and yield of food grains in last 70 years (Source: Anonymous 2021)

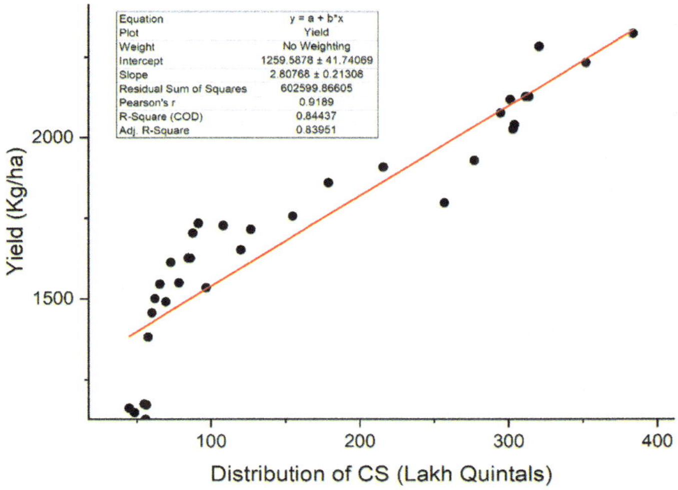

Fig. 5.8 Relation between yield and distribution of certified seed (CS) during the last 34 years (Vijay et al., unpublished data)

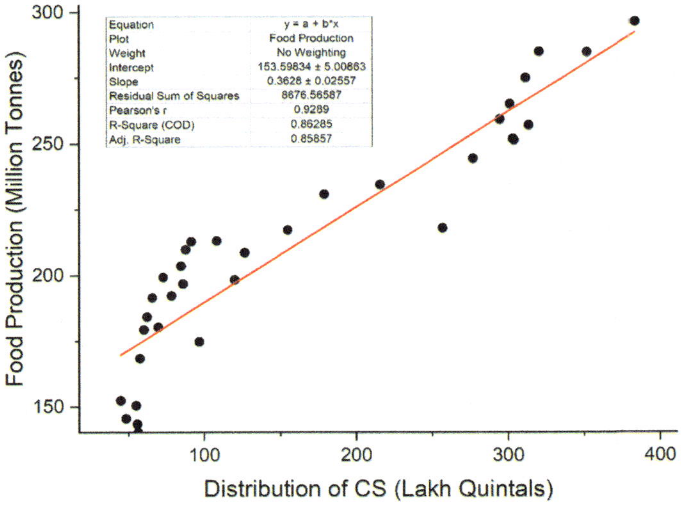

Fig. 5.9 Relation between food grain production and distribution of certified seed (CS) during the last 34 years (Vijay et al., unpublished data)

5.6.2 Status of Seed Requirement and Availability

The success of the seed program depends on proper planning of the seed production of different classes of seeds. In India, the three-generation seed multiplication system

Table 5.1 Status of certified/quality seed requirement and availability

Year	Requirement (million tonnes)	Availability (million tonnes)	Percent surplus (+) or deficit (−)
2004–2005	1.11	1.32	+19.34
2005–2006	1.07	1.41	+31.22
2006–2007	1.29	1.48	+15.08
2007–2008	1.81	1.94	+7.51
2008–2009	2.07	2.50	+20.77
2009–2010	2.49	2.80	+12.28
2010–2011	2.91	3.21	+10.52
2011–2012	3.30	3.54	+7.02
2012–2013	3.15	3.29	+4.25
2013–2014	3.35	3.47	+3.59
2014–2015	3.44	3.52	+2.39
2015–2016	3.37	3.44	+1.91
2016–2017	3.53	3.80	+7.60
2017–2018	3.71	4.19	+12.93
2018–2019	3.54	3.99	+12.82
2019–2020	3.87	4.31	+11.28

Source: Anonymous (2021), Agarwal et al. (2019), and Chauhan et al. (2016)

consisting of the breeder, foundation, and certified seed is followed. The breeder seed is produced in the variety originating research center, whereas the foundation and certified seed production at seed corporations, SAUs, and the private sector. Based on the information available for the area under a particular crop, the seed requirement can be calculated using the standard seed rate for that crop. Based on the last 16 years' data available from the Ministry of Agriculture and Farmers' Welfare, there is always a surplus situation for the available seed ranging from 2% to 30% in different years (Table 5.1). This self-sufficiency in the primary input has helped achieve the country's food security. The certified seed is produced from foundation seed, which in turn is grown from breeder seed. Thus, the current certified seed production cycle begins 3 years earlier with breeder seed production. The generation-wise seed requirement will be calculated based on the crop area, seed multiplication ratio, and ideal seed replacement rate (SRR), and a seed rolling plan will be developed for a successful production program. The Department of Agriculture and Farmers' Welfare at the center and its counterpart in the state set the seed rolling plan for every financial year consisting of Kharif and Rabi seasons. Apart from the certified seed, the breeder and foundation seed also steadily increased during the last 30 years. In the previous 15 years, there was a 50% increase in breeder seed and a more than 100% increase in foundation seed production (Fig. 5.10). This increase could be made possible due to the implementation of the ICAR Seed Project in 2005 at the National Agricultural Research and Education System, consisting of ICAR institutes and SAUs, and with the help of various seed schemes at the central and state levels.

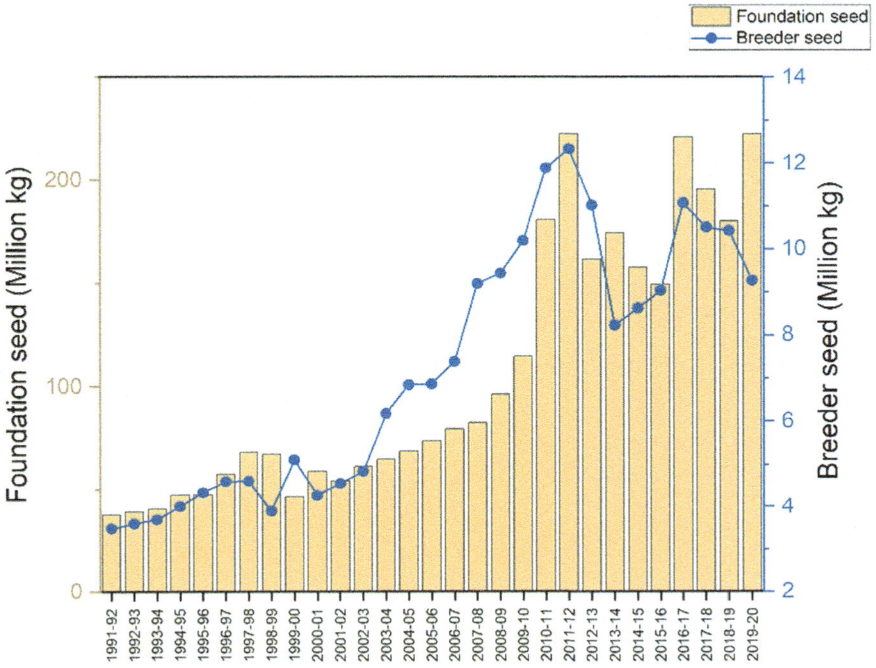

Fig. 5.10 Breeder and foundation seed production during the last 30 years (Source: Anonymous 2013, 2021)

5.6.3 Contribution of the Private Sector

The private sector was in the nascent stage during the green revolution period. However, due to consistent government policy, financial support, and technical help from National Seeds Corporation, the private industry started growing during the post-green revolution era. The National Seed Program gave substantial support to the private sector under NSP Phase III. Further, with the implementation of the New Policy on Seed Development (1988), there was a surge in the private sector with the liberalization of imports and foreign investment/technology support. The growth of the private sector is visible in its contribution to quality seed, which increased from 39% to 62% of the total quantity of quality seed supplied in the last decade (Table 5.2).

The detailed analysis of the contribution of public and private sectors in major crop groups shows that the increase in private sector contribution is not uniform across all crops, and there is a lot of variation in the percent contribution. In cereal crops, including maize, the percent contribution of the private sector increased from 53% in 2012–2013 to 61% in 2019–2020 (Fig. 5.11). Among the cereals, the presence of hybrids in most crops like maize, pearl millet, and sorghum, whose

Table 5.2 Public and private seed sectors share in the supply/availability of quality seed

Year	Quality seed supplied/available (million tonnes)		Percent share of quality seed supplied/available	
	Public sector	Private sector	Public sector	Private sector
2009–2010	1.71	1.09	61.06	38.94
2010–2011	1.65	1.56	51.48	48.52
2011–2012	1.81	1.73	51.08	48.92
2012–2013	1.61	1.67	49.05	50.95
2013–2014	1.68	1.80	48.30	51.70
2014–2015	1.51	2.01	42.95	57.05
2015–2016	1.47	1.96	42.87	57.13
2016–2017	1.84	1.96	48.44	51.56
2017–2018	1.79	2.40	42.76	57.24
2018–2019	1.71	2.28	42.77	57.23
2019–2020	1.64	2.67	37.98	62.02

Data source: Anonymous (2015, 2018, 2021) and Agarwal et al. (2019)

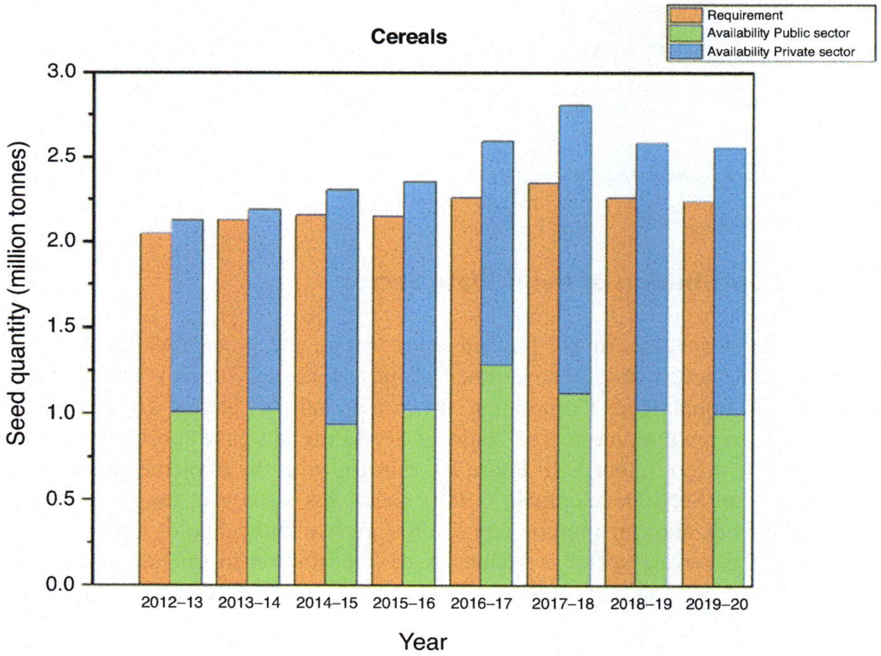

Fig. 5.11 cereal seed requirement and availability with contribution from public and private sectors (Data source: Anonymous 2016, 2018, 2021)

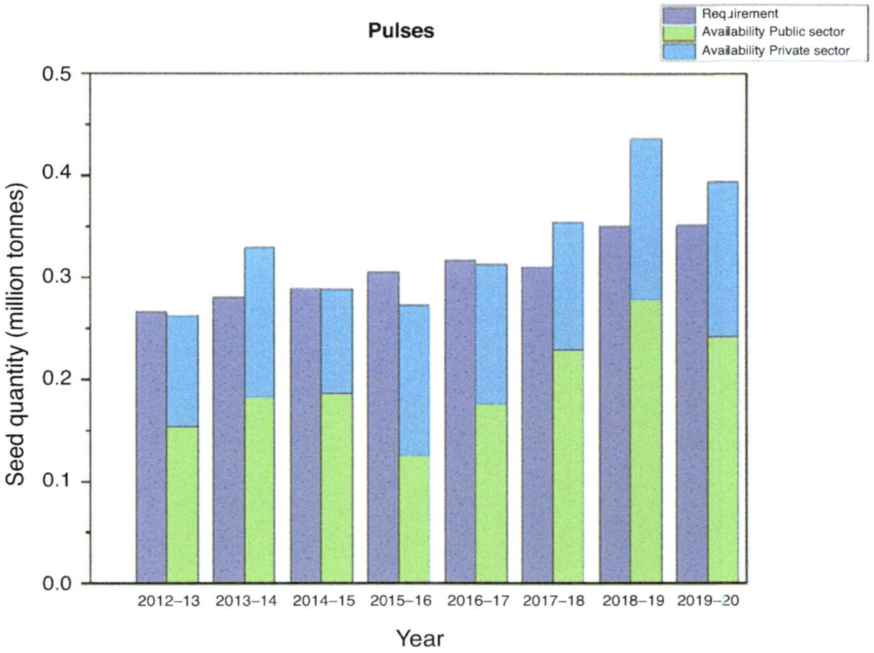

Fig. 5.12 Legume seed requirement and availability with contribution from public and private sectors (Data source: Anonymous 2016, 2018, 2021)

seed replacement rate (SRR) is high, has attracted the private sector to contribute more. The private sector's contribution to pulse crop seed supply is almost static at around 40% (Fig. 5.12). In pulses, the increase in the supply of quality seeds is mainly due to the increased contribution by the public sector (~60%). The situation in oilseed crops is similar to cereals, where the private sector's contribution is steadily growing. The presence of commercially important crops like groundnut, soybean, and castor attracted the private sector's contribution to quality seed supply in recent years, leading to the increase of its share from 43% in 2012–2013 to 57% in 2019–2020 (Fig. 5.13). The private sector is the leader in the quality seed supply of fiber crops, viz., cotton, and jute, with more than 80% contribution. In this crop group, the public sector's contribution decreased from 18% in 2012–2013 to 11% in 2019–2020 (Fig. 5.14). Thus overall, the share of the private sector in the quality seed supply is steadily increasing and is a positive sign of the growth of the seed industry in the country.

Even though the private sector's contribution is increasing in quality/certified seed supply, the breeder seed indents are almost static for cereals and decreasing for oilseeds. In the case of legumes, a sudden surge in seed indent was observed from 2017 to 2018 (Fig. 5.15). The overall private sector's breeder seed indent is less than 20% (Agarwal et al. 2019). These low BS indents, coupled with an increase in

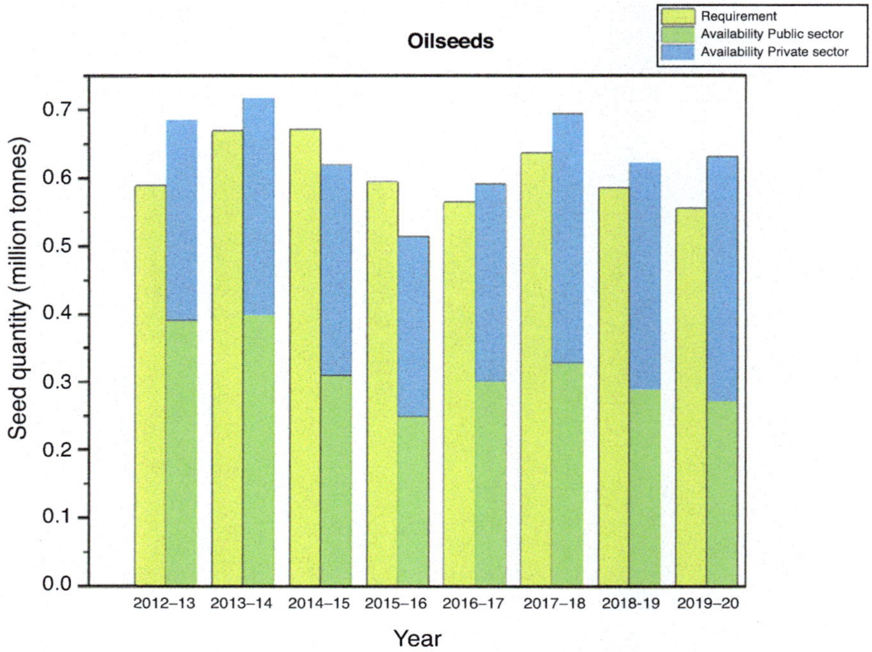

Fig. 5.13 Oilseed requirement and availability with contribution from public and private sectors (Data source: Anonymous 2016, 2018, 2021)

truthfully labeled seed supply, indicate that the private industry's use of public sector varieties is limited. However, in recent years there has been increased collaboration between the public and private sectors. Several seed companies are taking up the seed production of recently developed varieties from the public sector on a non-exclusive basis.

5.6.4 Seed Replacement Rates

The proportion of cropped area covered with quality seed purchased or procured during the crop period to the total cultivated area is called the seed replacement rate (SRR). Seed replacement is necessary for quality maintenance and yield assurance. The degeneration of varieties occurs either due to natural events like mutation, cross-pollination, pests, and diseases or by human activities that cause mechanical mixtures, developmental variations, and minor genetic variations. Therefore, it warrants the replacement of seeds in a stipulated period. The National Seeds Policy 2002 emphasizes enhancing SRR to achieve food production targets as SRR has a well-known strong positive correlation with crop production and productivity. Based on the type of pollination, the ideal SRR of different crops was finalized. The self-

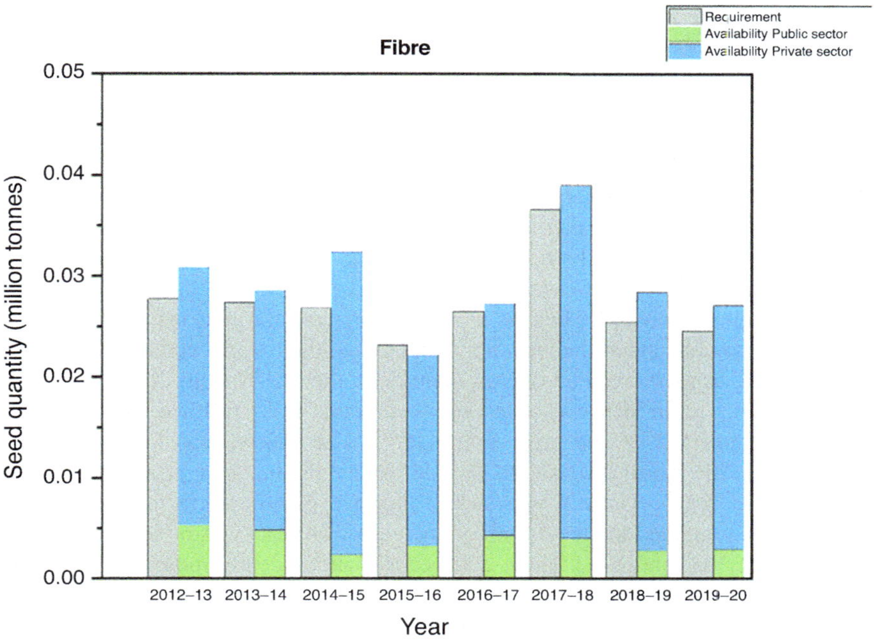

Fig. 5.14 Fiber seed requirement and availability with contribution from public and private sectors (Data source: Anonymous 2016, 2018, 2021)

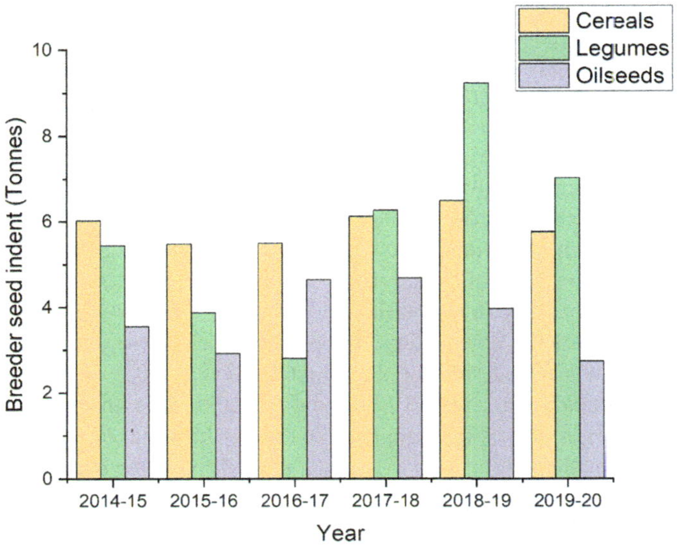

Fig. 5.15 Breeder seed indent by private sector in different crop groups (Agarwal et al. 2019)

pollinated crops require 33% SRR, which means after every 3 years, the seed should be changed; similarly, for cross-pollinated crops, 50% SRR means that fresh seed should be used after every 2 years. In the case of hybrids (F_1) of both self- and cross-pollinated crops, SRR is 100% as segregation occurs in the F_2 generation. The SRR over the last 20 years (2001–2020) varied from state to state and crop to crop. The SRR of important cereals showed an increasing trend in India except in pearl millet (Fig. 5.16). There was a sharp rise in the SRR of maize (62%), followed by wheat (29%). Even though a 22% increase in SRR was observed in rice from 2001 to 2020, the actual growth was only up to 2011 (21%); later, it fluctuated without any significant increase. In the case of sorghum and pearl millet, the SRR fluctuated with a marginal increase. There is a need to analyze the reasons for these SRR fluctuations to understand the farming practices and the role of seed in a better way.

The pulse crops showed an overall increase in their SRR from 2001 (Fig. 5.17). The maximum rise in SRR was observed in red gram (47%), followed by black gram (34%), chickpea (22%), and green gram (19%). The oilseed crops also showed an increase in SRR from 2001 to 2020, with many fluctuations in sunflowers and mustard (Fig. 5.18). There was a steady increase in groundnut up to 2010, but afterward, it remained static. A similar increase was observed in soybean, except for a surge in 2011 and 2012. There is a need to make in-depth studies on SRR based on area, cropping pattern, purchase practices, and its influence on production and productivity to understand the fluctuations and develop guidelines for its steady increase.

5.6.5 Varietal Replacement Rates

The replacement of old varieties with new ones is an essential requirement for enhancing production and productivity. It provides new and upgraded genetic material to the farming community as the new varieties are bred mainly for climate resilience, pest and disease tolerance, improved quality, and increased yield. Quality seeds play an essential role in varietal replacement, and the green revolution was initiated with the supply of quality seeds to the farming community. The Government of India is pushing to replace the old varieties with the newly developed better varieties through Minikit trials, seed rolling plans, and many demonstrations. The varietal replacement rate of major food crops from 2017–2018 to 2019–2020 showed that less than 5-year-old varieties occupy 23% of the total indent and less than 10-year-old varieties around 54%. Crop-wise details show that less than 5-year-old varieties of wheat and soybean occupied more than 40% of their total indent (Table 5.3). Similarly, less than 10-year-old varieties have the highest indent in wheat crop (74%), followed by mungbean (64.4%), chickpea (61.4%), and soybean (55.6%). Based on the number of varieties in the seed chain, the maximum percent of less than 5-year-old varieties are present in soybean (34.3%), followed by wheat (28.5%) and rice (23.2%). At the same time, less than 10-year-old varieties are also maximum in soybean (57.1%), followed by groundnut (56.5%), wheat (54.4%), and rapeseed and mustard (53.5%). Thus, it indicates that a higher percentage of the

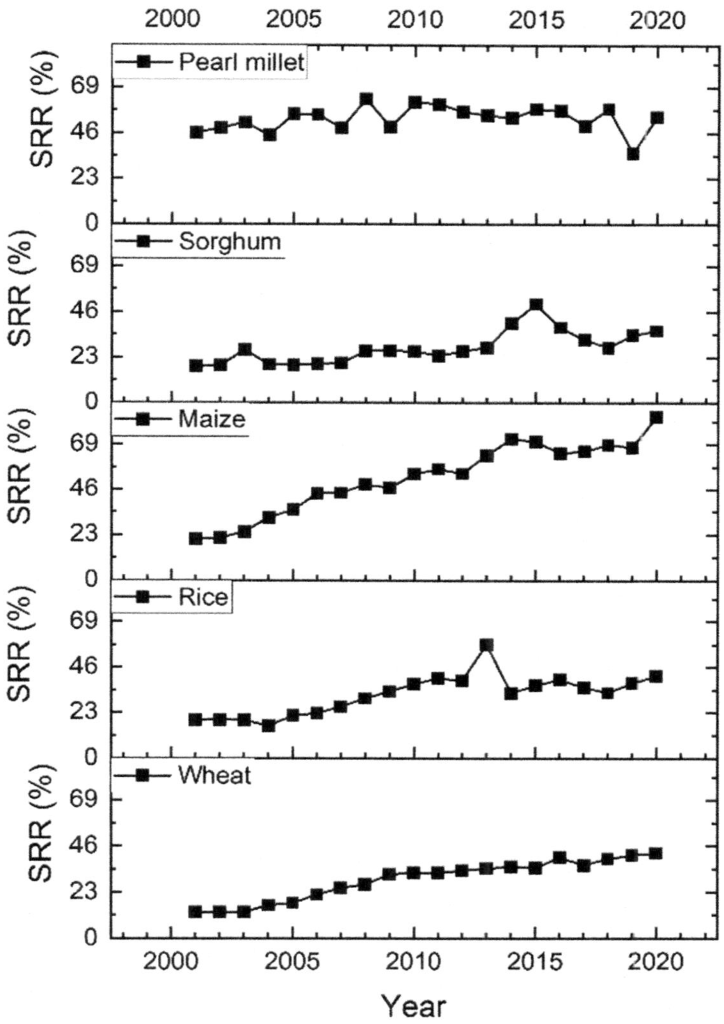

Fig. 5.16 Changes in the seed replacement rate of important cereal crops, wheat, rice, maize, sorghum, and pearl millet from 2001 to 2020 in India. (Data source: Ministry of Agriculture and Farmers' Welfare, GOI)

number of varieties in the seed chain does not mean higher indents. The requirement for climate-resilient varieties is one of the primary reasons for the increased VRR. In India, from 2014 to 2019, of the total varieties released, 83% were climate-resilient. In some crop groups like pulses, oilseeds, and forage crops, all the varieties released have one or the other climate resilience character (Table 5.4).

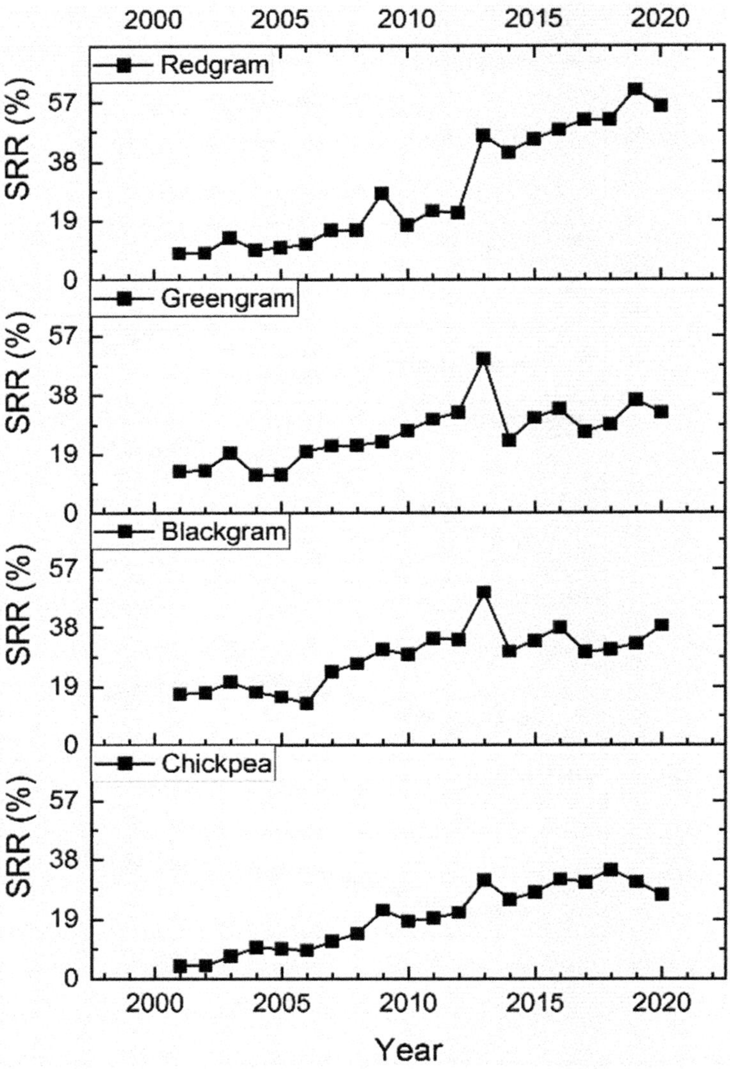

Fig. 5.17 Changes in the seed replacement rate of important pulse crops, chickpea, black gram, green gram, and red gram from 2001 to 2020 in India. (Data source: Ministry of Agriculture and Farmers' Welfare, GOI)

5.6.6 Seed Imports and Exports

The global seeds market size during 2020 was almost US$ 65.4 billion, and it may reach a value of nearly US$ 98.2 billion by 2026. The seeds industry is expected to grow at a CAGR of 7.1% (www.expertmarketresearch.com). India occupies fifth

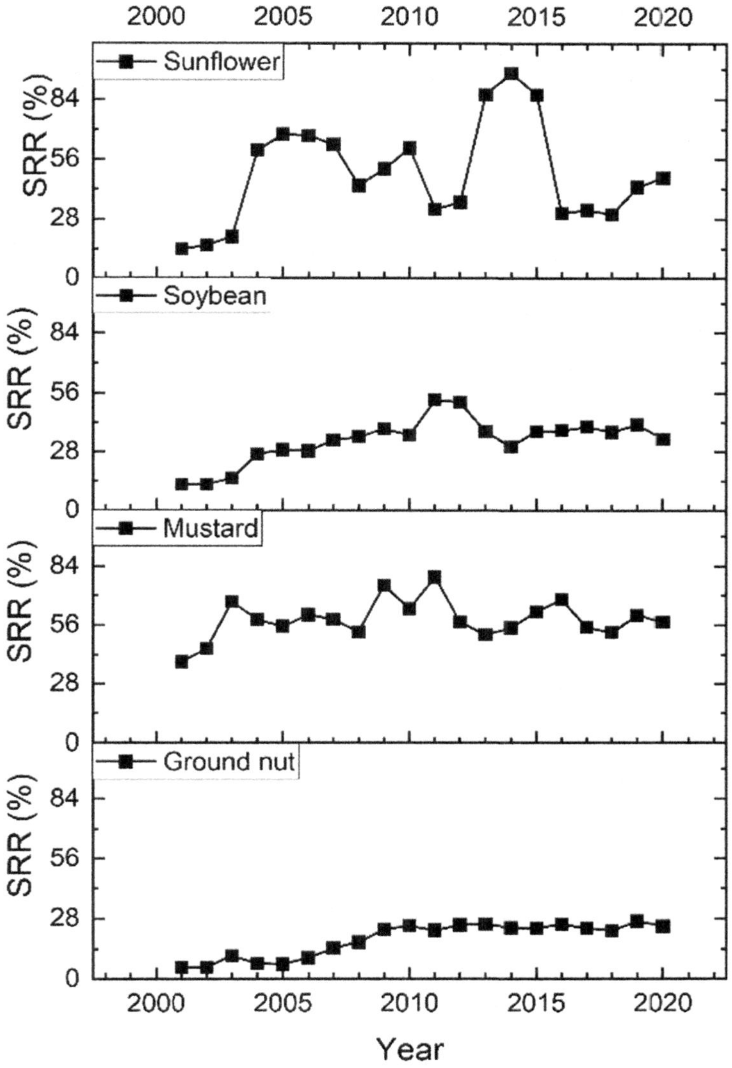

Fig. 5.18 Changes in the seed replacement rate of important oilseed crops, ground nut, mustard, soybean, and sunflower from 2001 to 2020 in India. (Data source: Ministry of Agriculture and Farmers' Welfare, GOI)

position in the global seed market with approx. 5.8% share (Vijay 2021). The Indian seed market reached a value of USD 4.9 billion in 2020 (https://www.imarcgroup.com/seed-industry-in-india). The total quantity of seed exported was in the surplus condition in the last 7 years except in 2015, when the quantity imported was more

Table 5.3 Varietal replacement rate of major field crops in the last 3 years (2017–2018 to 2019–2020)

Crop	No. of varieties in seed chain	Total indent (T)	Varieties <5 years old				Varieties <10 years old			
			No's	Indent (T)	% Share in total indent	% Share in seed chain	No's	Indent (T)	% Share in total indent	% Share in seed chain
Rice	293	472.00	68	70.53	14.8	23.2	129	226.41	47.7	44.0
Wheat	158	2187.33	45	997.70	45.3	28.5	86	1619.03	74.0	54.4
Pigeon pea	55	31.27	8	1.51	4.9	14.5	18	12.09	39.0	32.7
Chickpea	73	1004.53	15	284.59	28.4	20.5	34	602.80	60.1	46.6
Mungbean	53	84.64	5	14.37	16.9	9.4	20	54.67	64.4	37.7
Soybean	35	1818.97	12	748.64	41.5	34.3	20	1008.46	55.6	57.1
Groundnut	46	1057.12	8	123.59	11.5	17.4	26	409.51	38.5	56.5
Rape seed and mustard	43	6.36	9	0.78	12.4	20.9	23	3.27	51.6	53.5

Source: modified from Singh et al. (2020)

Table 5.4 Crop group-wise number of climate-resilient varieties released in the last 5 years (2014–2015 to 2019–2020)

Crops	No. of varieties released	Climate-resilient varieties	Per cent of total varieties
Cereals	645	446	69.15
Oilseeds	183	183	100.00
Pulses	187	187	100.00
Fiber crops	92	89	96.74
Forage crops	81	81	100.00
Sugar crops	44	32	72.73
Others	2	2	100.00
Grand Total	1234	1020	82.66

Source: presentation by Joint Secretary of seeds accessed from https://agricoop.nic.in/sites/default/files/Quality%20Seeds%20JS%20Seeds.pptx

than that exported. However, the trade balance was always negative up to 2017, and it tilted to positive only in 2019 with a phenomenal increase in export quantities (Fig. 5.19). The three major groups of seeds (field crops, vegetables, and flower crops) that India exports and imports showed different trends over the years. The difference in export and import quantities of field crop seeds is only marginal except in 2019 when there was a significant leap in export quantity. Similarly, vegetable crop exports were higher than imports except in 2015 and 2016. The export quantity of flower seeds is substantially higher than the import quantities (Fig. 5.20).

Nevertheless, the trade balance shows a different scenario with a trade deficit in all three categories until the latest exports (Fig. 5.21). There was a shift in trade balance from 2018 onwards when the total trade value of vegetable seeds was surplus, and in 2019 trade surplus was achieved both in food crops and vegetables. Still, we have a substantial negative trade balance in the case of flower seeds, even though our export quantities are higher than our import quantities. Thus, choosing high-value crops, particularly flower seeds, is needed to change the scenario. This positive sign of seed trade will help in increasing our global seed market share. India is bestowed with diverse agro-climatic zones and skilled human resources. Thus, there are many opportunities to support large-scale export-oriented seed production. In addition to vegetable seeds, seeds of hybrid corn, paddy, pearl millet, and cotton have promising export prospects to Asian and African countries.

5.7 Future Thrust Areas

The Indian seed sector has the potential to make the country the "Asian seed hub," provided emphasis was given on certain aspects as detailed below.

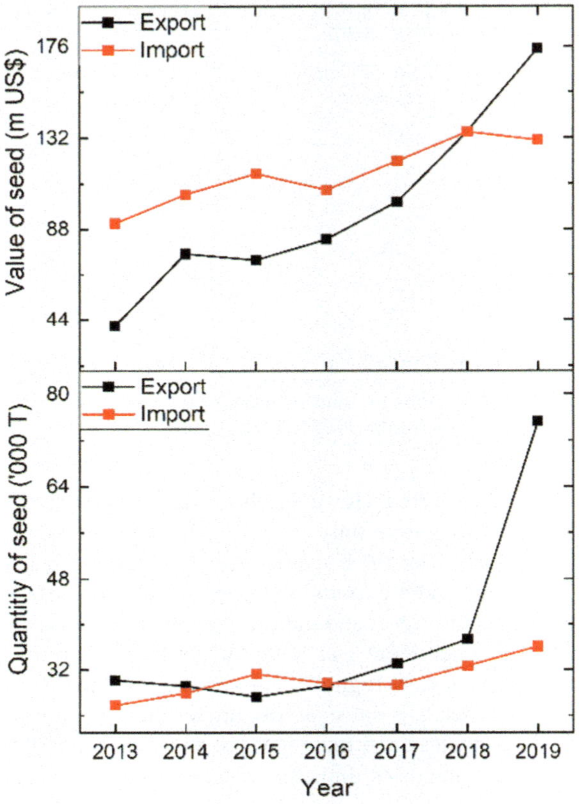

Fig. 5.19 Total quantity ('000 t) and value (m USD) of exported and imported seed by India in the last 7 years (2013–2019) (Data source: ISF 2022)

5.7.1 Seed Production of Biofortified Varieties

Biofortification of plant varieties is the best method for ensuring the country's nutritional security. Recently, 17 biofortified varieties of eight crops were dedicated to the nation by the Prime Minister; with this, the total number of biofortified varieties available in the country has increased to 34. Most of these varieties are under the seed chain; however, there is a lot of export potential for the seed of these varieties because of their nutritional superiority. There is a need to concentrate on the seed production and export of biofortified varieties.

5.7.2 Seed Production of GM/GE Crops

Genetically modified (GM) and/or genome-edited (GE) crops are inevitable in the future. Standard policy on GM seeds is required with prioritization of crops and traits to stop the illegal spread of untested technology. Specified production zones can be

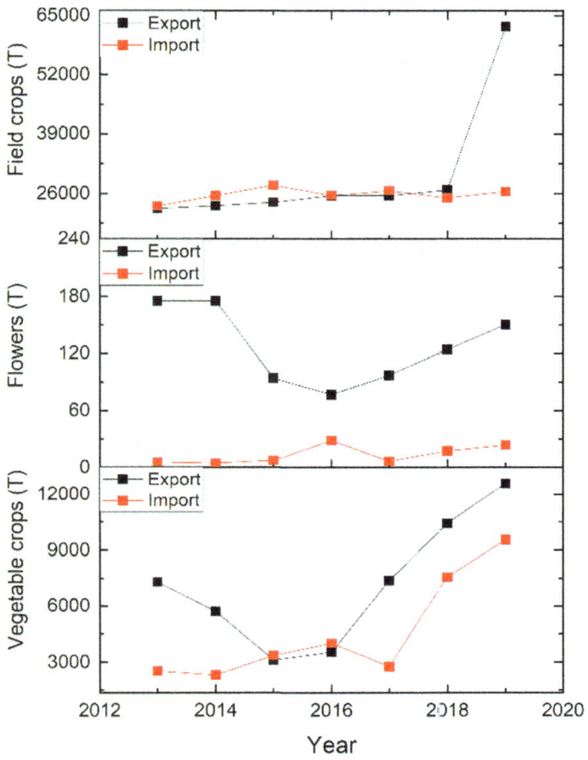

Fig. 5.20 Export and import quantity (t) of seed of different crop groups in the last 7 years (2013–2019) (Date source: ISF 2022)

developed for GM crops for better containment with a regulated seed supply and procurement system. GM seed labeling and testing shall be pushed on a large-scale basis with quality regulation. GM crops have occupied the central part of the crop belt in several countries due to their many advantages. However, *Bt*-cotton is the only GM crop allowed for open-field cultivation in India. Recently the government has exempted genome-edited crops free from exogenous introduced DNA from the biosafety assessment rules of genetically engineered organisms. This will usher in the developing of new varieties with quality traits, pest and disease resistance, and increased yield. Seed production of these new varieties is of utmost importance for their success.

5.7.3 Seed Traceability

Quality assurance is one of the critical parameters in the success of any supply chain, and seed is not an exemption. A mechanism for seed traceability at the national level is to be developed and implemented. It will help to maintain and assure quality and

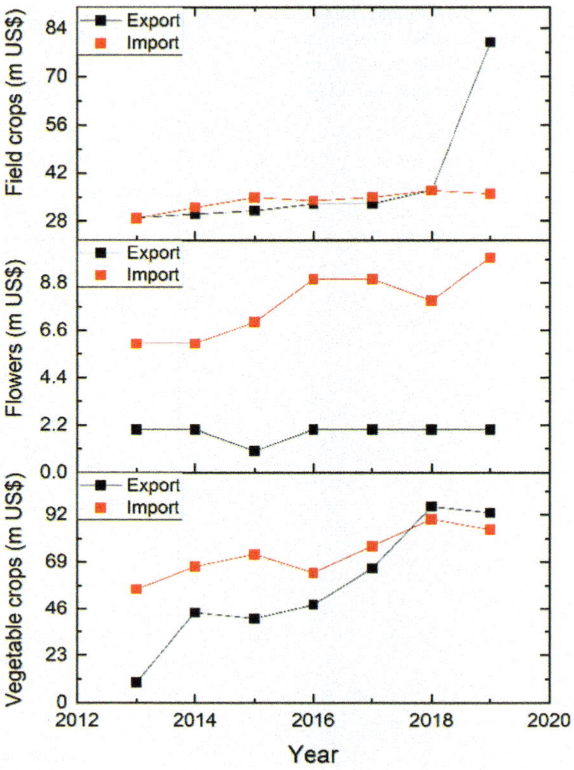

Fig. 5.21 Export and import value (m USD) of seed of different crop groups in the last 7 years (2013–2019) (Data source: ISF 2022)

assist in identifying the bottlenecks in the supply chain. Traceability empowers farmers and agricultural companies in tracing the history of origin, processing, and storage of that particular seed lot. Digitization of information in the seed industry is still in the initial stages, and there is an urgent need to adopt IoT technology.

5.7.4 Networking of Seed

The seed production of different classes occurs in various organizations of the country. It includes more than 800 institutes/organizations in the public sector and more than 500 private companies. There is a need to connect all these production centers to use their seed efficiently. Interlinking all the production organizations to have updated information about seed availability is a big task. Developing a national seed grid for live updating of the seed availability is the need of the hour. Initially, it can be started with public sector organizations and later can be extended to the private sector.

5.7.5 Climate-Resilient Seed Production Research

Climate change will affect seed production on a massive scale. The research on the effect of climate change on seed development and quality is scanty. There is a need to start network projects with specific objectives in this line to cope with the changes. The research output will be helpful to the seed industry in identifying new places for seed production and adopting mitigation strategies. Thus, there is a need for substantial public and private collaboration on this aspect.

5.7.6 OECD Seed Schemes/EU Equivalence

The OECD seed schemes assure seed quality by providing a global framework for seed certification in international trade. India is a participating member of OECD Seed Schemes and started exporting seeds under this scheme. National Designated authority nominated nine designating authorities across all states. A total of 246 varieties in 31 crop species are enlisted in the OECD list of varieties eligible for OECD certification. The scope of OECD seed schemes shall be extended to a greater number of varieties as most of the Indian varieties have the potential to get exported to South Asian and African countries due to similar climatic conditions and food habits. Similar to OECD seed schemes, the EU equivalence provides access to European markets. The non-EU member countries must offer the same quality as EU seed, and the packing shall be as per OECD seed schemes. By gaining EU equivalence, India can export seeds to EU countries easily.

5.7.7 Harmonization of Seed Legislations

Harmonizing the seed rules and regulations is the need of the hour. With the adoption of OECD seed schemes, and the wish to join EU equivalence, India shall synchronize its rules and regulations with international regulations. Harmonizing the procedures and standards in line with the global systems will help ease of movement of seeds to other countries and make India the seed hub of South Asia. The New Seed Bill needs further deliberations as specific contentions exist between PPV&FR Act, 2001, and the proposed New Seed Bill, 2019.

5.7.8 Strengthening of Seed Chain

The seed chain acts as a foundation for the growth and availability of quality seeds in the country. There is a considerable deficit in the efficiency of the seed chain multiplication, resulting in the production of excessive quantities of precious breeder and foundation seed. By ensuring 100% conversion, the energy and resources utilized for breeder and foundation seed production can be diverted to other seed developmental activities.

5.8 Conclusions

The Indian seed industry has evolved under the recommendations of various committees and review teams. The seed sector played a crucial role in realizing the green revolution and increasing the food grain production and yield during post-green revolution period. The systematic development of the industry, with the help of several government policies and funding through National Seed Projects, resulted in the tremendous growth of the seed sector. With the changing scenario of crop improvement and the development of new technological innovations, the seed industry needs to adopt those changes. The age-old act and rules need to be upgraded and shall be harmonized with the national and international regulations for the industry's future growth. The Indian seed industry has a lot of export potential. With surplus production of quality seed, avenues for export like OECD seed schemes, EU equivalence, and ISTA accreditation shall be strengthened and implemented across the country. Identifying potential crops and varieties for export and developing a national seed grid aid in the further development of the industry. Thus, the most vibrant Indian seed sector shall become future-ready by metamorphosing itself with the combined efforts of renewed government policies, innovative research inputs, and dedicated public and private sectors in the most coherent manner.

References

Agarwal DK, Uday Bhaskar K, Sripathy KV, Vijayakumar HP, Kamble UR, Pal G, Ramesh KV, Jeevan Kumar SP (2019) Indian seed statistics. ICAR-Indian Institute of Seed Science, Mau

Agrawal RL (1995) Seed technology. Oxford & IBH Publishing Co. Pvt. Ltd.

Anonymous (2013) Agricultural statistics at a glance. Directorate of Economics and Statistics, DAC &FW, Ministry of Agriculture, Government of India

Anonymous (2015) Agricultural statistics at a glance. Directorate of Economics and Statistics, DAC &FW, Ministry of Agriculture, Government of India

Anonymous (2016) Agricultural statistics at a glance. Directorate of Economics and Statistics, DAC &FW, Ministry of Agriculture, Government of India

Anonymous (2018) Agricultural statistics at a glance. Directorate of Economics and Statistics, DAC &FW, Ministry of Agriculture, Government of India

Anonymous (2021) Agricultural statistics at a glance. Directorate of Economics and Statistics, DAC &FW, Ministry of Agriculture, Government of India

Chauhan JS, Singh AL, Rajendra Prasad S, Pal S (2015) Quality seed: a mega factor in enhancing crop productivity. In: Singh AL (ed) Recent advances in crop physiology, vol II. Daya Publishing House, New Delhi, pp 357–426

Chauhan JS, Prasad SR, Pal S, Choudhary PR, Bhaskar KU (2016) Seed production of field crops in India: quality assurance, status, impact and way forward. Indian J Agric Sci 86(5):563–579

EXIM Policy (2002-2007) The Export and Import Policy for the period 2002-2007. Ministry of Commerce and Industry, Government of India

Famine Enquiry Commission (1944) The Famine Inquiry Commission Final Report, 1945. Manager of Publications, Delhi

FAO (2004) The state of food and agriculture 2003–2004. Food and Agriculture Organization of the United Nations, Rome, p 208

Food Grains Policy Committee (1943) Report of the Foodgrains Policy Committee 1943. Manager of Publications, Delhi
GMFEC Report (1952) Report of the grow more food enquiry committee. Ministry of Agriculture, Government of India, p 128
Grow More Food Enquiry Committee (1952) Report of the Grow More Food Enquiry Committee, 1952. Ministry of Food and Agriculture, Government of India
Hatti N (1977) Impact of assistance under P.L. 480 on Indian economy 1956–1970. Economy and History 20:23–40
https://ourworldindata.org. Accessed 04 Mar 2022
https://plantauthority.gov.in. Accessed 01 Apr 2022
https://www.expertmarketresearch.com. Accessed 30 Nov 2021
https://www.imarcgroup.com/seed-industry-in-india Accessed 30 Nov 2021
https://www.mordorintelligence.com/industry-reports/indian-seed-sector-analysis. Accessed 30 Nov 2021
Indo-American Agricultural Production Team (1959) Report on India's Food Crisis and Steps to Meet it, 1959. Ministry of Food and Agriculture and Ministry of Community Development and Cooperation, Government of India
ISF (2022) International seed federation. www.worldseed.org/isf/seed_statistics.html. Accessed 04 Mar 2022
National Commission of Agriculture (1971, 1976) Report of the National Commission on Agriculture, 1976. Ministry of Agriculture and Irrigation, Government of India
National Seed Policy (2002) National Seed Policy, 2002. Ministry of Agriculture, Government of India
New Policy on Seed Development (1988) New Policy on Seed Development, 1988, No. 11-71/88. SD.1. Ministry of Agriculture, Government of India
PEO (1961) Study of the multiplication and distribution programme for improved seed, 1961 (Study No. 40). Program Evaluation Organization, Planning Commission, Government of India
Pingali Prabhu L (2012) Green revolution: impacts, limits, and the path ahead. Proc Natl Acad Sci 109:12302–12308
Protection of Plant Varieties and Farmers Rights Act (2001) The Protection of Plant Varieties and Farmers Rights Act, 2001 (No. 53 of 2001), dated 30th October 2001. Ministry of Law, Justice and Company Affairs, Government of India
Royal Commission on Agriculture (1925) Royal Commission on Agriculture in India Report, 1928. Government of India
Seed Review Team (1968) Seed Review Team Report, 1968. Ministry of Food, Agriculture, Community Development and Cooperation, Government of India
Singh RP, Chintagunta AD, Agarwal DK, Kureel RS, Kumar SPJ (2020) Varietal replacement rate: prospects and challenges for global food security. Glob Food Sec 25:100324
Sir John Russel Commission (1937) Report on the work of the Imperial Council of Agricultural Research in applying science to crop production in India, 1937. Government of India
SRT (1968) Seed review team report. Government of India, p 234
The Seeds (Control) Order (1983) The Seeds (Control) Order, 1983 of Essential Commodities Act, 1955 (No. 10 of 1955). Ministry of Agriculture, Government of India
The Seeds Act (1966) The seeds Act, 1966 (No. 54 of 1966), dated 29th December 1966. Ministry of Food, Agriculture, Community Development and Cooperation, Government of India
The Seeds Rules (1968) The Seeds Rules, 1968 of The Seeds Act, 1966 (No. 54 of 1966). Ministry of Food, Agriculture, Community Development and Cooperation, Government of India
Vijay D (2021) Innovations in seed sector and its regulations for Atmanirbhar Bharat. In: Ghosh PK, Kumar P, Chakraborty D, Debashis M, Sivalingam PN (eds) Innovations in agriculture for a self-reliant India. New India Publishing House, New Delhi, pp 57–68

Part II

Food, Nutritional and Income Security: Role of Priority Crops and Allied Sectors

Rice

6

Somnath Roy, N. P. Mandal, M. S. Anantha, B. C. Verma, Amrita Banerjee, C. Gireesh, P. Senguttuvel, R. Abdul Fiyaz, Suneetha Kota, Jyothi Badri, L. V. Subba Rao, A. S. Hariprasad, A. V. S. R. Swamy, Mangal Deep Tuti, R. Mahender Kumar, Dipankar Maiti, and R. M. Sundaram

Abstract

A detailed account of rice research in India starting from the inception of the Indian Council of Agricultural Research (ICAR) in 1929, its achievements, and its role in securing the rice productivity of the country is provided. A trend analysis of rice yield gains during the green revolution (GR), and post-GR era is provided. Development and adoption of improved cultivars with higher yield potential and crop management technologies have contributed to continued productivity gain and steered India towards an evergreen revolution. The future strategies for rice improvement, considering the major challenges imposed by global climate change, are discussed in light of genomics-assisted breeding and cutting-edge crop management practices.

Keywords

Rice · India · Green revolution · Productivity · Varietal development

S. Roy · N. P. Mandal (✉) · B. C. Verma · A. Banerjee · D. Maiti
ICAR-National Rice Research Institute, Central Rainfed Upland Rice Research Station, Hazaribagh, Jharkhand, India
e-mail: NP.Mandal@icar.gov.in

M. S. Anantha · C. Gireesh · P. Senguttuvel · R. A. Fiyaz · S. Kota · J. Badri · L. V. S. Rao · A. S. Hariprasad · A. V. S. R. Swamy · M. D. Tuti · R. M. Kumar · R. M. Sundaram
ICAR-Indian Institute of Rice Research, Hyderabad, Telangana, India

P. K. Ghosh et al. (eds.), *Trajectory of 75 years of Indian Agriculture after Independence*, https://doi.org/10.1007/978-981-19-7997-2_6

6.1 Introduction

Rice is the staple crop of Asia and is the key to the food security of around half of the world's population. Rice is also the major source of livelihood for about 140 million rice farming households and millions of rural poor associated with rice farming. For poor consumers, rice accounts for nearly a fifth of the total household expenditure on average. Hence, rice is a strategic commodity as a country's overall economic growth and stability depend on an adequate and cheap supply of this crop. India has the largest area (approximately 43.78 million hectares, M ha) under rice and is second only to China in rice production (118.43 million tonnes, Mt). Over the last 70 years, rice production has increased manifold from 20.58 Mt (in 1950–1951) to 118.43 Mt (in 2019–2020) while the total area under rice has increased marginally from 30.81 Mha to 43.78 Mha during the period (Fig. 6.1) (www.agricoop.nic.in, Agricultural Statistics at a Glance 2020). This remarkable improvement in rice production played a pivotal role in the food security of the country and was achieved through the coordinated efforts from rice scientists of Indian and international institutes under the guidance of the Indian Council of Agricultural Research (ICAR) and the International Rice Research Institute (IRRI), Philippines. The wide adoption of the high-yielding semi-dwarf variety IR8 carrying the semi-dwarf gene, *sd1*, derived from a Taiwanese rice variety Dee-geo-woo-gen, marked the Green Revolution in the country in 1966–1967. Before that, ICAR established the All India Coordinated Rice Improvement Project (AICRP) in Hyderabad in 1965 to fast-forward the varietal development and screening for wider adoption in different regions of the country. The collaborative research activities with IRRI in the subsequent years led to the exchange of genetic material between IRRI and ICAR

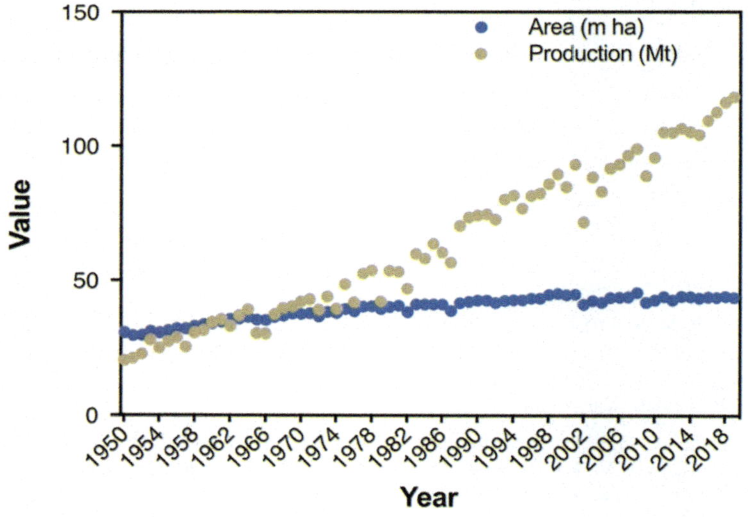

Fig. 6.1 Growth in rice production and area in 70 years (1950–1951 to 2019–2020) in India

(AICRIP), and the researchers in India started using high-yielding disease and insect-resistant varieties such as IR20, IR22, IR26, IR36 and IR72 in their breeding programs. Later the mega variety IR64 possessing excellent grain quality and high yield was developed and released in the Philippines in 1985. IR64 was adopted rapidly by the farmers in most rice-growing countries and was cultivated in over 10 Mha worldwide. The semidwarf high-yielding variety "Jaya" was released by AICRIP in 1968, ushering in the green revolution. This miracle rice variety transformed the country into a state of self-sufficiency by the mid-eighties and reduced rice imports while beginning an era of exporting rice, earning high foreign exchange for the government by the early 1990s.

The Indian Council of Agricultural Research, formerly known as the Imperial Council of Agricultural Research, was established on July 16, 1929 as a registered society under the Societies Registration Act, 1860, in pursuance of the report of the Royal Commission on Agriculture. Since its inception, ICAR has acted as an apex body for coordinating, guiding, and managing research and education in agriculture in the country. The ICAR has played a pioneering role in ushering Green Revolution and subsequent agricultural development in India. Rice being the most important food crop, the Council devoted considerable attention to its improvement from a very early period. An account of significant research and development in rice research and development under the guidance of ICAR is presented in the following sections. The growth rates in rice yield have been analyzed based on data on the productivity of crops (www.agircoop.nic.in) starting from 1950. The period of 70 years (1950–1951 to 2019–2020) has been partitioned into four phases: Pre-Green Revolution (GR) era (1950–1966) is referred to as phase I, the GR era (1967–1983) as phase II, the post-GR era of 1984–2000 as phase III, and the post-GR era of 2001–2020 as phase IV following (Yadav et al. 2019).

6.2 Pre-independence and Pre-green Revolution Era (1940–1966)

In the "pre-Green Revolution" (pre-GR) era (1940–1966), rice improvement focused mainly on the selection of traditional cultivars and the introduction of rice varieties from China, Japan, Russia, and other countries to select new and local cultivars. Throughout the country, indigenous rice germplasm was collected, evaluated, and improved through pure-line and mass selection, and to some extent, through hybridization and mutation. Several coordinated trials were operative in various regions of India. During the pre-GR era, the average yield was 888.71 $\pm$ 114.13 kg/ha, with an annual improvement of 15.7 kg/ha (Fig. 6.2) (Yadav et al. 2019). The rate of increase in rice production was 0.89 Mt/yr and it increased steadily during 1950–1966 (Fig. 6.2).

During this period, a large number of schemes for rice research were funded by ICAR in different rice-growing areas of the country. As a result, wide varieties suitable for varying environments and resistant to pests and diseases were developed. Under the Kashmir scheme, which started in 1940–1941, exotic rice varieties

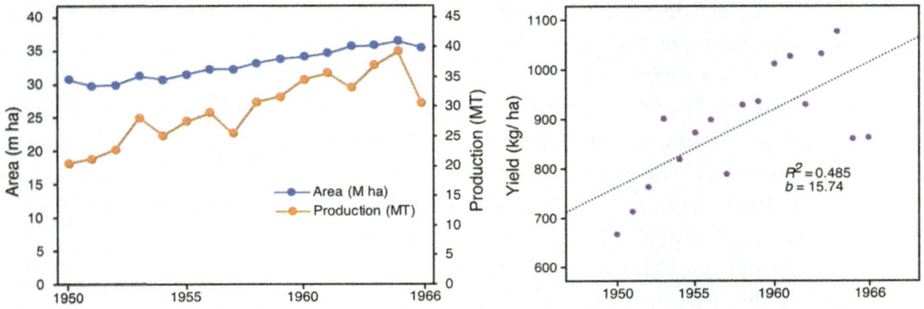

Fig. 6.2 Rice area, production, and productivity (kg/ha) trend in the pre-GR era (1950–1966). The coefficient b is the rate of increase in yield (kg/ha/yr) in each phase, and R^2 is the coefficient of determination

from China, Russia, America, Japan, and Australia were evaluated along with indigenous varieties for yield and earliness to identify promising high-yielding exotic varieties. Vernalization experiments were conducted at Calcutta university, Bengal, to induce flowering in foreign varieties. Due to higher yield potential (5–6 t ha-1), Chinese rice varieties became popular among the growers in Kashmir Valley (ICAR Annual Report 1951–1952). The Chinese varieties such as China 1039, China 988, China 972, and China 1007 helped in solving the food problem in the Kashmir valley during 1945–1950. Looking into the success of the Chinese rice varieties in Kashmir, subsequent trials were conducted starting from 1950–1951 at Orissa, Assam, Uttar Pradesh, and Punjab to test the suitability of Chinese varieties in other ecologies. However, these varieties did not perform satisfactorily in other parts of the country. In 1950, the Russian Dry Land Paddy was introduced in India. During 1952–1953, in the Jammu and Kashmir state, exotic varieties from America, Australia, Italy, and Japan produced around 3.5 t ha^{-1} yields and showed promise for cultivation in high-altitude areas where no cereals are grown except low-yielding small millets and buckwheat. The Kashmir scheme for plant introduction and acclimatization was terminated in 1952, and a new scheme for developing high-yielding non-shattering varieties was started in 1954. In 1962–1963, India began importing rice varieties from temperate countries through IRRI, the Philippines and the Japanese Agricultural Research Institute, Hiratsuka, for testing in temperate regions in Himachal Pradesh, Jammu and Kashmir, and Punjab.

The Bombay scheme was operative during 1945–1952 with an objective of developing improved cultivars with better yield and quality from the rice germplasm of Gujarat. Many promising cultivars such as Sukhwel, Sutarsal, Jirasal, Pankhali, White Kada, Red Kada, early Kolam, and T22-A (a selection from the Basni variety of Uttar Pradesh) were identified and subjected to pure-line selection in this scheme. Under the Travancore scheme, mass selection of the most popular local cultivars, Kallada Samba and Chettivirupu, resulted in the selection of Ptb10 and Cochin 3 with higher yield and early maturity. Research on late and very late varieties was done at Hyderabad, focusing primarily on agronomic and manurial management.

In 1946, the Government of India set up Central Rice Research Institute (now National Rice Research Institute, NRRI) at Cuttack, Orissa, with Dr. K. Ramaiah as the first Director to enable India to produce more and better rice. Both short- and long-term research projects were undertaken to achieve this objective. The *indica-japonica* hybridization scheme was sanctioned during 1950–1951. The *japonica* varieties, despite having higher yields in their own habitat, responsiveness to heavy fertilizer application, short and non-lodging plant structure, and adaptability to tropical conditions, were not suitable for introduction in the *indica* zone. On the other hand, indica varieties are fast germinating with high tillering and are resistant to drought and diseases. Therefore, a comprehensive hybridization scheme was started under the sponsorship of the Indian Council of Agricultural Research and Food and Agriculture Organization (FAO) for the benefit of India and the southeast Asian countries. Under this scheme, much exotic rice germplasm was assembled from different parts of south and southeast Asia. The promising *indica* types selected by the states were sent to CRRI for crossing with selected *japonica* varieties. Around 710 *japonica-indica* cross combinations were made, and seeds of the hybrids were distributed to the participating states and Asian countries for testing. Numerous promising selections were made, showing around a 25% yield advantage. Studies on the spikelet fertility in the *indica-japonica* hybrids showed that *japonicas* gave greater fertility with *indicas* from northern and north-western India than with south Indian cultivars. The promising high-yielding *indica-japonica* hybrids (IJ hybrids) were selected by the states and tested through pilot projects. During 1962–1963, IJ hybrid Np. 27, giving a 30–50% higher yield than Jhona 349, was released in Punjab. IJ hybrid No. 2914 was released as ADT27 in Andhra Pradesh and showed superiority over TKM6 and ART20.

The scheme for the botanical survey and collection of landraces from the Jaypore tract of Orissa, the secondary center of origin of rice, was operative during 1952–1959. This mission, popularly known as Jaypore Botanical Survey (JBS), was the first of its kind ever organized in the world to collect rice germplasm (Chang 1989). Under this scheme, a significant collection of 1930 rice cultivars (1766 cultivated and 164 wild types) have been made. The collection included glutinous, semi-glutinous, scented varieties with long sterile glumes, double grain, and clustered grain types. Grains measuring >12 mm in length have also been collected from the Jaypore tract.

In 1959–1960, a total of 455 improved rice varieties were released, which was the outcome of various schemes operating in different states of India (ICAR 1959). Under Madras schemes, blast-resistant early maturing variety development work started in Coimbatore, Aduthurai, and Buchired-dipalam research stations in 1953. Blast-resistant long-duration cultivars Co25 and Co26 were released for the Madras region in 1951–1952. Quality rice improvement work was undertaken in the Punjab scheme. Several other rice cultivars such as T1 (basmati group), Kele, ADB-4, ADL-4, Bhutmuri, Chakala, S-20 (Jhona Kasawala group), NP125, and T-43 (Palman group) were tested and register higher yield. Rice breeding for special conditions such as flood, deep water, drought, salinity, hilly conditions, and resistance to blast and stem borer was strengthened during 1955–1960. Work for flood and salinity resistance work was started in Andhra, Madras, Kerala, Uttar Pradesh, Orissa, and West Bengal. The rice landraces collected from Assam, Orissa, and West Bengal were characterized for submergence tolerance in 1956. The flood-resistant lines developed at NRRI, Cuttack showed a considerable level of tolerance, and among them, FR43B could withstand more than 21 days of submergence to a depth of 4 feet. In the Assam collection of *bao* and *Sali* rice, Negheri Bao was found to be the best submergence-tolerant landrace at Chinsurah Rice Research Station, West Bengal. Breeding for salt and flood resistance work at West Bengal and Andhra identified several salinity/flood tolerant cultivars such as Bheral, Nona bhelki, Bokra, Rupral Lalgenti, Sadamota, Pathacuchi, Talmugur, Kumargore, Tilakkanchan, Patnai 23 in West Bengal and Kolappala, Pokkali, and Karivennel in Kerala, and Red Kesari, Buddamolakolu, and Kulu in Andhra Pradesh. It was noted that variety 1131 could withstand submergence under 5 ft. for more than 2 months (ICAR Annual Report 1959–1960). In the Coimbatore research station, *Oryza minuta* was identified as a deeper and more vigorous root system. A scheme for research in the mountainous region was started in Kalimpong, West Bengal, in 1953–1954. The selection was made on hill paddy germplasm and local cultivars such as Jasuda, Adday, Krishnabhog, Kalomaishi, and Ramtulasi, along with the Chinese variety CNAB4 were identified as high yielding. Under the West Bengal scheme, the improvement of *boro* rice was targeted. *Boro* landraces were collected and characterized for yield attributes. Local variety Chinsurah Boro 33/138, along with DI3, DI4, and Bhutmuri were found promising. Later in 1955, Chinsurah Boro I was identified, and it was at par in yield (0.7 t ha^{-1}) with Co13, CH45, and Borosali.

The Uniform Blast Nursery Scheme of the Food and Agriculture Organization (FAO) was launched in the year 1962–1963. Under this scheme, the seeds of blast-resistant varieties were shared among the countries, and testing of a global set was started at CRRI, Cuttack, and Lonavala, Maharashtra. During this year, ICAR planned an All-India Co-ordinated Trial with varieties having wide adaptability so that the number of improved varieties under distribution in different states can be reduced. The All India Co-ordinated Rice Improvement Project started on April 1, 1965.

The success of rice breeding in India lies in the coordinated efforts of hundreds of rice breeders working under research institutes of ICAR, State- and Central- Agricultural universities, and Private seed enterprises through the All India Coordinated Rice Improvement Project (AICRIP). It was initiated to conduct multilocation trials to identify suitable widely adaptive genotypes of high yield potential along with appropriate crop management practices. It is the largest national network for rice breeding in the world. Under this project, a nucleus coordinating cell was established at Ranjendranagar, Hyderabad, in 1965 to improve rice production, productivity, and profitability in India by organizing multilocation testing and interdisciplinary research on a national basis. Until the 1970s, the support was provided by the Rockefeller Foundation, International Rice Research Institute, and US Agency for International Development (USAID) through financial assistance and technical support to strengthening the research activities.

During 1965–1966, IR-8, the 'Miracle Rice' bred at IRRI, Philippines, was approved for release and cultivation in single and multicrop areas in Andhra Pradesh, Assam, Bihar, Kerala, Madras, Madhya Pradesh, Orissa, Uttar Pradesh, and West Bengal. IR-8 was found to be less susceptible to bacterial leaf blight (BLB) than Taichung Native 1 and registered an average of 27% more yield than TN1, which was included in the High-yielding Varieties Program of the Government of India. The widespread adoption of IR-8 throughout the country marked the dawn of the Green Revolution in the following decades.

6.3 Green Revolution Era (1967–1983)

In the past 70 years, although the human population has more than doubled, the production of staple crops has tripled during the period, with only around a 30% increase in cultivated lands (Wik et al. 2008). The developing world was able to overcome its chronic food deficit mainly due to collaborative efforts of different nations and international bodies in prioritizing investment in crop research, infrastructure and market development, and appropriate policy support during 1966–1985, known as the Green Revolution era (Pingali 2012). The Green Revolution strategy was based on the foundation that, given appropriate institutional mechanisms, technology spillovers across geo-political boundaries could be captured. The central part of the GR strategy was to develop the necessary institutional capacity, particularly in plant breeding. In the case of rice, the IRRI in the Philippines led to the early development of improved modern production technologies. Based on the success of IRRI and the International Maize and Wheat Improvement Center (CIMMYT) in Mexico, the Consultative Group on International Agricultural Research (CGIAR) came into force specifically to channel technology spillovers in those countries that underinvest in the agriculture sector. The knowledge, inventions, and products (primarily the modern varieties and breeding lines) developed at IRRI, and other CGIAR centers were made publicly available in rice. Different countries responded positively by adopting, disseminating, and delivering the technologies which paved the path to the first Green Revolution. The rice germplasm improvement played a central role in enhancing agricultural production and reducing food prices and poverty. It was estimated that the productivity gain from crop germplasm improvement was 0.8% for rice, while it was 1.0% for wheat across all regions (Evenson and Gollin 2003a). Note that the national agriculture research systems (NARS), in association with international agricultural research centers (IARCs), had played a crucial role in developing, testing, and selecting suitable high-yielding varieties for the farmers. The agricultural research system of India, led by ICAR, played a leading role in the science and techniques of rice breeding and in realizing the effect of the Green Revolution (Evenson and Gollin 2003b).

Before 1965, the growth rate of food grain output was around 2.4% per annum, and after 1965 the food grain production grew at the rate of 3.5% per year. Since the introduction of the high-yielding variety (HYV) technologies, the necessity of

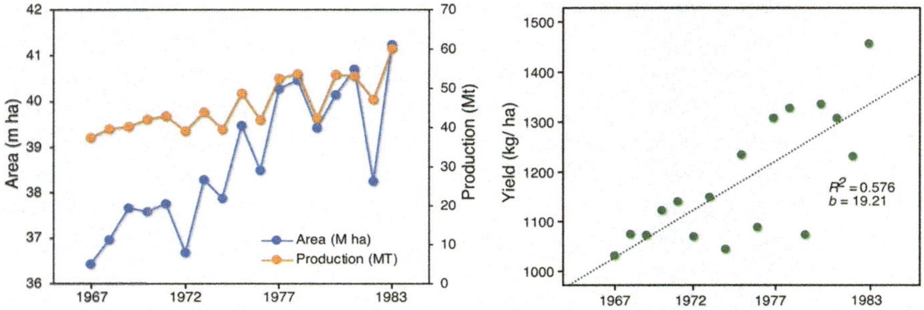

Fig. 6.3 Rice area, production, and productivity (kg/ha) trend in the Green Revolution era (1967–1983). The coefficient b is the rate of increase in yield (kg/ha/yr) in each phase, and R^2 is the coefficient of determination

importing food grains has declined considerably. The major boom in food grain production was due to wheat, followed by rice. Food production increased from 50 million tons in 1950–1979 million tons in 1964. In 1968, it was 95 million tons and reached a quantity of 120 million tons by 1976. During the GR era, the high-yielding dwarf and semidwarf rice varieties made a significant impact, and rice productivity increased by 19.2 kg/ha/yr (Fig. 6.3). The average rice productivity during this period was 1163.69 ± 109.66 kg/ha. The rice production showed an increase of 1.04 Mt/yr.

The early success of breeding improved rice varieties reflected the advanced state of research in the late 1950s in developed countries and India. In Asia, IRRI has played a crucial role in rice germplasm improvement and the development of important high-yielding varieties. The relationship between IRRI and Asian NARS was very fruitful in the early years. After the 1970s, IRRI's role was increasingly that of a germplasm supplier, producing breeding lines for NARS breeders. This role was facilitated by an international network for germplasm exchange that provides NARS breeders with ready access to breeding materials (Evenson and Gollin 2003b). IRRI's initial efforts were concentrated in the irrigated rice environments and then extended to favorable rainfed environments. However, the success of the collaboration between IRRI and NARS did not extend to any significant degree to upland rice environments nor to deep-water environments. The breeders incorporated dwarfing genes that allowed the development of shorter lodging-resistant varieties. The new varieties devoted much of their energy to producing grains and relatively little to straw or vegetative parts. The semidwarf varieties were responsive to better fertilizer than traditional varieties. The farmers, particularly in irrigated and rainfall-sufficient areas, rapidly adopted the new varieties, and the new varieties yielded substantially more than traditional varieties (Dalrymple 1986). The early success of these high-yielding semidwarf varieties was widely referred to as the Green Revolution, which coincided with the initial wave of modern variety releases during 1966–1975. Over the following years, the Green Revolution achieved broader and deeper impacts on

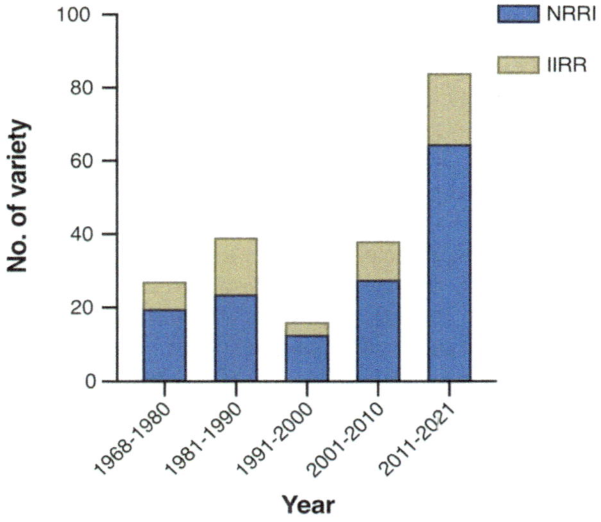

Fig. 6.4 Rice varieties released from two national institutes on rice, NRRI, Cuttack, and IIRR, Hyderabad through CVRC during 1968–2021

the economy and social structure of the country (Bowonder 1979; Evenson and Gollin 2003a).

Following the introduction of the semidwarf variety T(N)1, breeding work was started to develop high-yielding improved plant-type varieties, side by side, work on crop management, primarily fertilizer management, has also been intensified to harness the high-yield potential of new varieties (ICAR 1967). Trials were conducted with several IRRI-bred lines such as IR8-288-3, IR5-114-3-1, IR4-93-2, IR5-198-1-1, IR-177-3-3-2, and IR5-264-1-3-2. The culture IR8-288-3 was subsequently released as IR8, which showed superiority over T(N)1 in yield and seed dormancy. However, with the popularization of semidwarf varieties, the outbreak of bacterial leaf blight (BLB) and bacterial leaf streak (BLS) increased, and these became identified as major diseases of rice. Trials were started at CRRI, Cuttack, to screen for BLB-resistant lines in the late 1960s. In 1968, the rice variety Padma, bred at CRRI, was released along with the variety Jaya by AICRIP. Both these varieties played an important role in enhancing rice yield during the GR era. In this period, both T(N)1 and IR8 were extensively used as high-yielding parents in several breeding programs, and around 15 varieties, including Bala, Ratna, Vijaya, Saket 4, and Kalyani II, were released from CRRI during the GR era. Altogether, 28 varieties were released by two prominent national rice research institutes: NRRI (earlier CRRI) and IIRR (earlier Directorate of Rice Research, DRR) (Fig. 6.4).

To enhance genetic gain, the All India Coordinated Rice Improvement Project (AICRIP), since its inception in 1965, relied on the rigorous testing of breeding lines based on set criteria such as high mean yield, yield stability, pest and disease resistance, agronomic superiority, and grain quality parameters. Initially, AICRIP work was carried out in seven zones. There were 22 centers, and the zonal headquarters were Khudwani, Jorhat, Faizabad, Patna, Hyderabad, Cuttack, and Coimbatore. Twelve regional stations viz., Palampur, Pantnagar, Kapurthala, Chinsurah, Sambalpur, Raipur, Maruteru, Karjat, Nawagam, Mandya, Aduthurai, and Pattambi were established in the major rice-growing states of the Country. Upper Shillong, Kalimpong, and Imphal were identified as testing centers. During 1980–1985, eight more subcenters were added, raising the total to 53. In August 1975, AICRIP was elevated to the status of Directorate of Rice Research (DRR), wherein lead research targeting mainly the irrigated ecosystem was included in the mandate. The Directorate of Rice Research (DRR) was upgraded to national institute status and renamed as "Indian Institute of Rice Research (IIRR)" during the golden jubilee year, from December 15, 2014. In order to resolve the research problems of rice cultivation in India, coordination of AICRIP is now carried out jointly by three ICAR research institutes, namely ICAR-IIRR, ICAR-NRRI, and ICAR-Indian Agricultural Research Institute, with the support of state agricultural universities, state departments, and private seed industry.

Later in the 1980s, yield and quality improvement received major attention in AICRIP. This led to the development of varieties possessing significant biotic stress tolerance with desirable quality. In the 1990s, developing suitable hybrid rice technology and multilocation testing were the major goals. During the 2000s, emphasis was on aerobic, soil stress, and hill trials. In recent times, the objective was on the near-isogenic lines (NILs) introgressed with genes for biotic (blast, bacterial blight, and gall midge) and abiotic (drought and submergence) stresses. Nutritional security is also stressed by enriching the grain with micronutrients such as zinc and iron along with protein. While the emphasis on boro rice improvement continues, efforts to identify material suitable for newly emerging challenges like limited water situations are also being addressed through aerobic trials. Additionally, trials have been initiated on New Plant Type (NPT), Nutrient Use Efficiency (NUE)—Nitrogen and Phosphorous, and *japonica* trial. Summarizing the achievements of AICRIP is a herculean task. Since 1966, a total of 29,023 elite breeding lines developed by different rice research centers were tested in multilocation trials across the country under the umbrella of AICRIP, which culminated in the release of 1436 varieties, including 127 hybrids. Of these, 230 pure line varieties and 83 hybrids were released by the central variety release committee (CVRC). Whereas 1079 pure line varieties and 44 hybrids were released by state variety release committees (SVRC) from different states. With the increase in the export quality of rice, special attention was given to enhancing the quality of rice in the country, leading to the release of around 30 export-quality basmati, and short grain rice varieties. It is worth mentioning that the AICRIP system of multilocation testing of improved genotypes in India has become a role model for establishing similar institutional frameworks in other countries of Africa and Asia.

From 1975–1976, breeding rice for rainfed conditions was prioritized. The short duration cultures developed from *indica-japonica* hybridization at CRRI, Cuttack were tested and appeared to be highly suitable for direct seeded upland conditions. Other cross combinations involving drought-tolerant donors such as Bala, Dular, Lalnakanda, etc., which included CR143-2-2, CR141-1-191, CR141-2-192, CR125-

12-6, and CR124 series along with the early maturing varieties developed at AICRIP such as IET1444 (Rasi), RP79-14, RP79-5 were selected as high drought tolerant and showed potential for zones having rainfall of 900–1100 mm. Other high-yielding varieties such as UPRI-71-12, Karjat-14-7, Ratnagiri, Co38 (Bhagavathy), Co39 (Amarareathy), No.103, MR272, and others were also found suitable for rainfed conditions. Subsequently, Rasi (IET1444) and Akashi (IET2914) were released in 1979 for rainfed upland conditions as a substitute for Bala. Rasi is suitable for problem soils, particularly for low soil phosphate in rainfed upland conditions. In the late 1970s, efforts were made to develop insect and disease-resistant varieties. Breeding for rice tungro disease (RTD) resistance was started in CRRI, and resistant genotypes such as Kataribhog, Pankhari 203, IR28, IR9-4, Pusa 2-21, and Ambemohar 102 were selected. Rice cultures developed from TKM6 were resistant to RTD (ICAR 1975). Interdisciplinary research was started to develop a broad spectrum of resistance to blast, BLB, RTD, stem borer, gall midge, and leaf and plant hoppers, in addition to good plant habit and acceptable grain type. Mutation breeding was also started during 1975–1976 for developing varieties with semidwarf characters and high yield potential using Mashuri, GEB24, and Kaminisaru cultivars. It was also noted that genetic improvement of grain protein content was started at CRRI in the mid-1970s and lines with 10% protein content such as CRHP-1, CRHP-8, and CR198-13. CRHP-8 has developed from the cross Gaisen mochi Pirurutong x Padma, recorded 10–11% protein content. Considerable progress was made in understanding the physiological basis of response to low sunlight during monsoon season and for submergence tolerance (up to 15 days of complete submergence). For adaptation to low light, genotypes having high photosynthetic rates, viz. Ptb10, MTU15, Co29, TKM6, AC4491, Vijaya, IR8, Saket4, and Jamuna, were identified. Studies on submergence tolerance at CRRI helped in identifying varieties (Tilakkacherry, CR1030, CR540, and FR13A) with high survival and regeneration capacity under complete submergence for 15 days (ICAR 1983). The mega rice variety Swarna (MTU 7029), developed at Agriculture Research Center, Maruteru, was released in 1982. This variety spread countrywide. The coordinated efforts of CRRI, Cuttack, AICRIP, Hyderabad and a network of 50 research centers at agricultural universities and other institutions throughout the country led to a significant development of some remarkable HYVs in the Green Revolution era targeting diverse rice ecologies of India.

6.4 Post-GR Era (1984–2020)

In the post-GR era, greater emphasis was given to the development of cultivars with adaptation to specific niche environments, with the result that the rate of productivity increased further to 31.45 kg/ha/yr, which is 68% higher than the gain achieved during the GR era (Yadav et al. 2019). The average productivity during this period has been 2030.5 ± 342.9 kg/ha. Total rice production increased from 58.3 Mt in 1984 to 118.43 Mt in 2019 with a rate of 1.49 Mt/yr. In the first half of the post-GR period (1984–2000), the yield increased at a rate of 31.45 kg/ha/yr, while in the

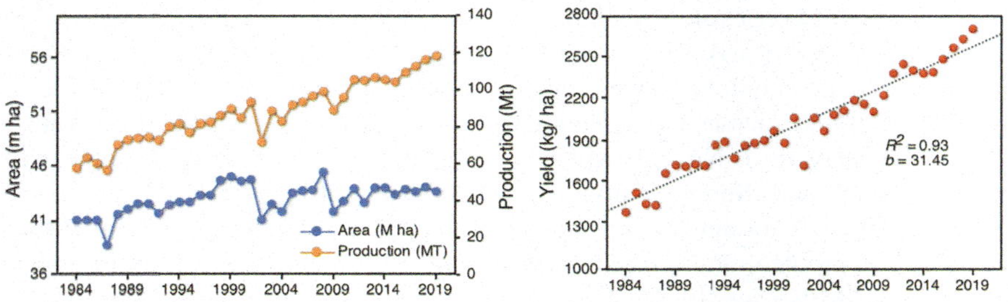

Fig. 6.5 Rice area, production, and productivity (kg/ha) trend in the post-GR era (1984–2019). The coefficient b is the rate of increase in yield (kg/ha/yr) in each phase, and R^2 is the coefficient of determination

second phase (2001–2019) it increased further to 40.93 kg/ha/yr (Fig. 6.5). From early 2000, rice improvement program emphasized the incorporation of multiple genes for resistance to diseases or insect pests, and tolerance to abiotic stresses like submergence and drought using conventional and molecular techniques, leading to the development of climate-resilient varieties. Hybrid development and amalgamation of high yield, short duration, and superior grain quality were given top priority (Yadav et al. 2019).

The varietal improvement program continued to identify suitable genotypes for various agroecological situations, emphasizing rainfed environments. Equal priority was given to isolating varieties with a high degree of resistance to insect pests and diseases for increased and stable yields. Drought affected agricultural production to a great extent in consecutive years from 1984 to 1988. Such environmental phenomenon necessitated developing rice varieties for specific niche environments under rainfed ecologies. Drought-tolerant varieties for rainfed uplands got priority, and varieties were developed using tolerant donors such as Rasi, Dular, Cauvery, and Annapurna. In 1988, the early maturing (90–100 days) variety Annada (CR222-MW10) was released for rainfed uplands. Heera (CR544-12), an extra early (70 days) variety, was released in 1989 for eastern India. Subsequently, wide very-early and early maturing drought-tolerant varieties, such as Kalinga III, Sneha, Vandana, Khandragiri, Ghanteswari, and Vanaprabha were released in the 1990s. Breeding work for varietal development suitable to rainfed shallow lowland, semi-deep, and waterlogged ecologies also produced many important varieties. For rainfed lowland conditions, Pankaj was used in several crossing programs at CRRI. Early maturing varieties were evaluated for post-flood conditions. The development of high-yielding Basmati rice varieties with superior grain quality was strengthened at IARI during the late 1980s. Pusa Basmati 1 (IET10364) was identified in 1989 and gave a considerably higher yield than Basmati 370 and Ranbir Basmati. In 1989, the crossing program with *tropical japonica* (*bulus/javanica*) varieties from Indonesia was started to transfer high tillering ability, lodging resistance, and long-dense panicle characters into HYVs. It was observed that most of the varieties developed from IR8 and Jaya have weaker culms and are thus prone to lodging.

The varieties bred at IRRI played a vital role during the Green Revolution period. Although IR8 had a very high grain yield, it has several defects, most importantly, poor grain quality, lack of disease and insect resistance, and late maturity. The varieties subsequently developed and released since the 1980s improved greatly on these traits (Khush 1999). During the early 1980s, one of the most popular IRRI varieties grown was IR36 which was early maturing (111 days), high yielding, and resistant to insects and diseases (Khush and Virk 2005). IR36 became very popular in India during the 1980s. Later, IR64 was released by IRRI in the Philippines in 1985. This variety represented a breakthrough in combining the excellent palatability of cooked rice with the other traits found in earlier IRRI HYVs. Because of its wide adaptation, early maturity, and improved quality, it became highly popular and was also highly desired by the rice industry. IR64 has also been used extensively as a parent in breeding programs and to develop populations for genetic analysis (Mackill and Khush 2018).

Hybrid rice research in India was first started at CRRI, Cuttack, in the mid-1950s (Sampath and Mohanty 1954; Richharia et al. 1962). Based on the identification of the wild abortive (WA) cytoplasm and the miraculous success and adoption of hybrid rice technology in China, ICAR- Indian Agricultural Research Institute (ICAR-IARI) launched a national program in 1987 with 12 network centers for the development of suitable hybrids for Indian ecologies. This program was further strengthened with technical support from the IRRI, Philippines, and financial support from FAO, UNDP (United Nations Development Programme), Mahyco, World Bank, and Asian Development Bank (ADB) consortium. With concerted efforts, APRH-1, the first rice hybrid, was released by APRRI, Maruteru, in 1994. Immediately after the release of APRH-1 from Andhra Pradesh, the University of Agricultural Sciences, Mandya, released KRH 1 and KRH 2 hybrids. The KRH 2 became highly popular throughout India and widely grown in Southern and Eastern parts of the country. Excellent progress has been made in hybrid rice research and development in India for the past three decades. However, first-generation hybrids focused only on heterotic yield but failed in quality traits. Keeping with past experience, second and third-generation hybrids focused on grain quality in comparison with popular Indian cultivars like Samba Mahsuri and Ponni, resistance to pests and diseases in addition to per se yield heterosis, to have marketability and credibility among farmers and consumers to a larger extent (Janaiah and Xie 2010). Altogether, 121 hybrids with high yield potential and better grain quality were released for commercial cultivation in different rice-growing states across the country (Senguttuvel et al. 2019). All the released hybrids are based on three lines (A, B, and R) breeding systems with WA cytoplasm, except SAVA124 and SAVA134, which are based on two line breeding approach (Senguttuvel et al. 2019). Of the 121 hybrids, 38 were developed by the public sector, and the remaining 83 by the private sector (AICRIP 2020). Some handful of hybrids from the public sector that finds a place in the market are DRRH2, DRRH3, KRH2, KRH4, Pusa RH 10, Rajalaxmi, Ajay, Sahyadri 4, CORH4, CRHR 32, PSD 3, Indira Sona and JRH 8; important hybrids from the private sector are PHB 71, PA 6129, PA 6201, Arize6444 Gold, JKRH 401, 27P31, 27P37, Suruchi, GK 5003, DRH 775, HRI

157, PAC 835, PAC, 837, US 312, Indam 200-017, NK 5251 and 27P11. United Andhra Pradesh, popularly called the "Seed capital of India," where more than 80% of the private seed sector plays a pivotal role in hybrid seed research and production. During the year 2020, hybrid rice was planted in an area of 3 m ha (6.8% of total rice area), and more than 80% of the total hybrid rice area is in Eastern and Central Indian states of Uttar Pradesh, Jharkhand, Bihar, Chhattisgarh, Madhya Pradesh, and Odisha with lesser area in states like Madhya Pradesh, Assam, Gujarat, Punjab, and Haryana. As rice is a key source of livelihood in eastern India, a considerable increase in yield through hybrid rice technology will have a major impact on household food and nutritional security, and income generation, besides an economic impact in the region.

The Biotechnological work in terms of tissue culture and production of doubled haploids (DH) through rice anther culture was started in 1988 at CRRI. Initially, heterotic cross combinations were selected from the crosses of IR64, IR50, PR106, P269, and other elite lines. The earlier DH cultures, AC561 and AC570 showed around a 20% yield advantage over the checks. Tissue culture techniques are being adopted to establish rice suspension cultures from different plant parts in vitro and assess the regeneration capacity of calli obtained from different rice genotypes. Transformation experiments were initiated using a biolistic particle-delivery system. Protocols for producing embryogenic calli from different explants and preculturing or isolated immature embryos were standardized for successful transformations. Molecular tools being used for identifying pathotypes of BLB in 1995–1996. Pathogen diversity was analyzed using virulence gene probes in combination with the classical method of identifying pathotypes. The population structure of blast pathogen *Magneporthe oryzae* in the uplands of Eastern India was characterized using Random amplified polymorphic markers (RAPD) in early 2000 (Chadha and Gopalakrishna 2005). The effectiveness of blast resistance genes such as *Pi-1(t)*, *Pi-2(t)*, *Pi-3(t)*, *Pi-4a(t)*, and *Pi4b(t)* was assessed in different rice growing parts of the country during this period (ICAR 2000). Work on pyramiding BLB resistance genes *xa5*, *xa13*, and *Xa21* in the background of Swarna and IR64 was also started in 2000 (ICAR 2000). Introgression of *xa5*, *xa13*, and *Xa21* through marker-assisted selection (MAS) in BPT5204 and Triguna was reported subsequently (Sundaram et al. 2008). DNA fingerprinting of rice germplasm lines was mostly started during 2002–2003 at National Research Center (NRC) for DNA fingerprinting, NBPGR, New Delhi, using various molecular profiling techniques (ICAR 2003). In the following years, with the advent of agricultural biotechnology, molecular tools were applied to a greater extent in developing improved rice varieties with abiotic and biotic stress tolerance.

6.5 Rice Development Programs: Government Schemes, Salient Achievements

The country's food production after independence and before the green revolution (1947 to 1960) was so bad that there was a risk of famine. The lack of proper technology, awareness, etc., brought India to the verge of massive famine (Eliazer

Nelson et al. 2019). People cannot get adequate food; therefore, starvation and malnutrition are very common. The new agricultural strategy was initiated to achieve the goal of self-sufficiency in agriculture or food production in the 1960s, and the basic principle of this strategy was to apply science and technology to enhance productivity (Tripathi and Prasad 2010). Therefore, the Green Revolution was initiated in the 1960s to increase food production with the objective of alleviating poverty and malnourishment in the country (Eliazer Nelson et al. 2019). To achieve these targets, attention was given to producing and developing high-yielding varieties and hybrids and promoting mono-cropping. This is based on the use of high-yielding varieties and hybrids which are responsive to high doses of fertilizers, particularly nitrogen, and the improved package of practices. The Govt. also launched several programs in sequence to achieve these targets.

The Central Sector Rice Seed Minikit Program has played a major role in improving rice production and productivity in the country by increasing the area under high-yielding varieties, which also includes demonstrating improved crop production technologies to the farmers. The High-Yielding Varieties (HYVs) Program was started during 1966–1967, and the Directorate of Rice Development (DRD) under the Ministry of Agriculture, Govt. of India, monitored its implementation from 1970 onward. The successful implementation of the program significantly increased the area under HYVs from 4.34 million ha (1969–1970) to 33.10 million ha (1999–2000). During 1969–1970 only 16 HYVs were released/notified for cultivation, and thereafter, with the concerted efforts of breeding, the country has released more than 1200 varieties and 127 hybrids of rice so far. The popularization of HYVs in farmers' fields through the rice Minikit program was initiated in 1971–1972 and continued up to the Ninth Plan period. During the Ninth Plan, more emphasis was given to the popularization of location-specific HYVs released/notified during the previous 3 years for favorable areas and 5 years for problematic areas, and a total of 33 lakh seed mini kits of 419 location-specific HYVs were distributed.

A full-fledged Centrally Sponsored "Special Rice Production Programme-SRPP" was started from 1985–1986 with the objective of substantially increasing the productivity of low-productive areas by supplying critical inputs like quality seeds, fertilizers, agrochemicals, implements, equipment, improved package of practices, etc. The program was implemented in 439 blocks of seven Eastern India States with the funding pattern of 50:50 sharing basis between the Govt. of India and the concerned State Government. Quality seeds having high yield potential were provided under the seed Minikit program. After that, the "Special Food grains Production Program (SFPP)" was also launched to achieve the minimum food production of 166 million tonnes during 1988–1989 and 175 million tonnes in the year 1989–1990. This program was implemented in 106 districts of 13 rice-growing states of the country and was 100% funded by the Government of India.

An "Integrated Programme for Rice Development (IPRD)" was also implemented from 1990 to 1991 by unifying the two schemes, SRPP and SFPP-Rice, and four additional states were covered under this scheme. This scheme was implemented in all the districts of the states covered under the program with a funding pattern of 75:

25 between GOI and the concerned State Government. Besides field demonstrations, training programs for farmers were also included under the scheme for the effective transfer of crop production technology. Keeping in view of the specific constraints to rice production, the State Governments were given a choice to include the most suitable components of rice production from an approved list.

The National Development Council (NDC), in its 53rd meeting held on 29th May 2007, adopted a resolution to launch a Food Security Mission on rice, wheat, and pulses to increase the annual production of rice by ten million tonnes, wheat by eight million tonnes and pulses by two million tonnes by the end of the Eleventh Plan (2011–2012). Accordingly, a Centrally Sponsored Scheme, "National Food Security Mission" (NFSM), was launched in October 2007. The main objectives of this scheme are to increase the production of rice, wheat, and pulses through area expansion and productivity enhancement in a sustainable manner in the identified districts of the country; restore soil fertility and productivity at the individual farm level; creation of employment opportunities; and enhancing farm level economy to restore confidence amongst the farmers. NFSM-Rice is being implemented in 25 states covering 194 potential districts, selected based on rice area coverage (>50,000 ha) with a productivity level less than the state average. The scheme is being implemented by organizing cluster demonstrations of varieties that are less than 10 years old following the latest crop production technologies and mechanization to reduce the cost of cultivation. The NFSM met with overwhelming success and achieved the targeted additional production of rice, wheat, and pulses. Because of its success, the scheme continued during the 12th Plan with new targets of additional production of ten million tonnes of rice, eight million tonnes of wheat, four million tonnes of pulses, and three million tonnes of coarse cereals, which were also accomplished. The average rice area of the country during the XII plan period decreased marginally by 0.11 million ha as compared to the XI plan period. At the same time, production has increased by 9.14 million tons due to a productivity hike of 215 kg/ha. Rice production can be further increased by expansion of the area, increasing productivity, and reducing the yield gap. Based on the performance of the 12th Plan and earlier tenure, the Central Government decided to continue the program beyond the 12th plan, i.e., from 2017–2018 to 2019–2020, which is co-terminus with the Fourteenth Finance Commission (FFC) period, with a new target to achieve 13 million tonnes of additional food grain production comprising of five million tonnes of rice, three million tonnes of wheat, three million tonnes of pulses, and two million tonnes of coarse cereals.

Following the recommendation of the Task Force constituted by the Government of India to maximize agricultural production on a sustainable basis, the program 'Bringing Green Revolution to Eastern India' (BGREI) was launched in the year 2010–2011, which intended to address the constraints limiting the productivity of "rice-based cropping systems" in Eastern India comprising seven (7) States namely, Assam, Bihar, Chhattisgarh, Jharkhand, Odisha, Eastern Uttar Pradesh, and West Bengal. The program aims to harness the water potential for enhancing rice production in Eastern India, which was hitherto underutilized. During 2010–2011, the major focus of the states was on the promotion of improved crop production technologies of rice, water harvesting measures, and their utilization for overall agriculture development. The demonstrations included the introduction of new HYVs (emphasis was on stress-tolerant varieties, STRVs) and hybrids, farm machines and implements, nutrients, pesticides, etc. During 2013–2014, based on the experience of implementation, the intervention of Marketing Support, including post-harvest technology, was also included. From 2015 to 2016, the program was modified, including a few interventions like seed distribution of new rice varieties, seed production incentives for newer varieties/hybrids of rice, micronutrients, soil ameliorants, and plant protection chemicals, machines like laser land levelers, etc. A three-tier monitoring structure involving the Central Steering Committee (CSC) under the chairmanship of the Secretary (DAC & FW); State Steering Committee (SSC) for each state under the chairmanship of Agriculture Production Commissioner/Principal Secretary (Agriculture) and District Steering Committee (DSC) headed by District Magistrate/Chief Development Officer is created to monitor the program. ICAR-National Rice Research Institute (NRRI), Cuttack is the nodal agency for monitoring the programme, and scientists of ICAR institutes and SAUs are also allotted one district each for extending technical backstopping and monitoring of the program. All the BGREI states have made a lot of progress in improving rice production and productivity by introducing improved production technologies and infrastructure development in the adopted districts. The impact of various interventions of block demonstrations to drive growth in rice is reflected in changes in yield rates; some of the states even reached the national average in terms of productivity. With the implementation of BGREI and NFSM programs, rice production has increased in seven Eastern India States from 45.65 million tonnes during 2009–2010 to 54.93 million tonnes during 2015–2016.

All these programs focus on increasing productivity to achieve the target food grain production. These lead to more emphasis on monoculture with over-exploitation of natural resources. With the use of HYVs, hybrids, chemical fertilizers, agrochemicals, groundwater, etc., productivity has increased, and since independence, the country has made tremendous progress in food grain production as well as in the overall agricultural sector. India has progressed from 'hand to mouth' conditions to self-sufficiency. We are not only self-sufficient but acquired sufficient resilience to tide over the adverse conditions, and now the country is exporting food grains to other countries. Since the green revolution, the consumption of chemical fertilizer has been growing very fast, and Government is also encouraging the use of fertilizer through heavy subsidies to get high productivity. The total consumption of fertilizer was 70,000 tonnes during 1950–1951 and went to 19,145,000 tonnes during 1999–2000 (270 times). The fertilizer consumption per hectare of the gross cropped area has also increased steadily, from 0.50 kg in 1950–1951 to 74 kg in 1995–1996 (128 times) (Tripathi and Prasad 2010). The use of high-yielding varieties, hybrids, and over-application of fertilizers and other agrochemicals leads to the deterioration of natural resources. Among the different natural resources, water is the most important. In rice, the primary method of cultivation is transplanting, and the conventional puddled transplanting system requires a large quantity of irrigation water. These practices not only lead to the use of a high amount of water but also break to use of a high amount of water requirement but also break capillary pores, destroy soil aggregates and form hard pans. Since the water resources (both surface and underground) are shrinking day by day (Farooq et al. 2009) and the profit margins are decreasing in transplanting mainly because of high labor costs and water requirements (Pandey and Velasco 1999), there is a need to move on from the transplanted method of rice cultivation. SRI was introduced to reduce the water requirement of the rice cultivation process. The practices that culminated in SRI began in the 1960s with the principle of applying a minimum quantity of water and transplanting individual plants in the square. These techniques reduce the water requirement to some extent as it involves intermittent irrigation that keeps the field moist but not inundated. However, water savings from SRI cultivation cannot be to a large extent as it still involves puddling, which is the major water-consuming process of rice cultivation, and it has been reported that water up to 2500–5000 L is required to produce 1 kg of rough rice (Bouman 2009). Due to water scarcity and high labor wages, there was a need for alternative crop establishment methods with less water requirement and higher water productivity. Direct-seeded rice (DSR) is one of the viable options to reduce the water requirement. The development of improved short-duration high yielding varieties and improved package of nutrient and weed management practices encouraged the farmers to shift from traditional transplanting to DSR culture in many rice growing areas including Punjab. Direct seeding offers different advantages like saving irrigation water, labor and energy, time, reduction of greenhouse gases, non-disturbance of soil, maintaining the soil structure, better crop growth, etc.

The progressive development of the different packages of practices and methods of cultivation, overexploitation of natural resources, mismanagement and excessive use of chemical fertilizers, pesticides, and lack of crop rotation caused several natural resources problems. It reduces soil fertility and crop productivity and leads to land deterioration, imbalance of nutrients, deficiency of micronutrients, and depletion of groundwater. The quantum of chemical fertilizers and agrochemicals used during and after the Green Revolution was very high. The overuse of these agrochemicals to achieve high production led to physical, chemical, and biological degradation of the soil (Singh 2000). The excessive use of groundwater for irrigation depleted the water table in many parts of the country. This way, the farmers shifted to unsustainable practices to achieve more yields, and due to the highly demanding nature of the agriculture production system, farmers are unable to withstand the increasing expenses for farming and fall into debt.

Now there is a need to conserve our natural resources for future generations and to sustain the yield. Different technology has been evolved for the sustainable use of resources. With increasing water shortage and labor scarcity, Dry-DSR with minimum or zero tillage (ZT) was adopted to reduce the cost of cultivation. Further, conservation agriculture has been adopted to utilize the different resources, in which zero tillage (ZT) or reduced tillage (RT) was applied, followed by row seeding. This technology is based on the principle of fewer disturbances to the soil and proper utilization of farm resources. Crop residues are being used to cover the soil surface as mulch. When this technology utilizes crop residue as mulch with improved crop and resource management methods, it is termed conservation agriculture (CA) or integrated crop and resource management (ICRM) (Asian Development Bank 2009).

6.6 Future Strategies and Conclusions

India's projected demand for rice is about 197.4 million tons by 2050. Annual genetic yield gains in rainfed rice in India are about 0.68–1.9%, primarily due to conventional breeding and some marker-assisted breeding products released during the last quarter of the previous decade. Recently, an analysis of genetic gain for yield in rice breeding and rice production in India indicated that since the introduction of plant-type-based HYVs (1995–2013), rice breeding research had not increased grain yield substantially by the genetic gain (Muralidharan et al. 2019). The yield stability has increased through the selection of traits, especially to withstand abiotic and biotic stress conditions. The increase in the national rice production of India is also the outcome of continued improvement in crop production skills at the farm level, effective use of irrigation, and several other factors. A long-term (from 1988 to 2019) analysis of AICRIP yield data of hybrids and inbred cultures in irrigated ecosystems across India revealed that the genetic gain or loss in yield of hybrid genotypes estimated over the test duration was non-significant (Muralidharan et al. 2022). Only in <20% of test locations, hybrid F_1 genotypes produced 10% more grain yields than inbred varieties. Therefore, much work is needed to make hybrid rice technology profitable to farmers.

The grain yield in rice is a complex trait determined by the quantitative component traits. The significant developments in rice genomics have provided powerful tools for understanding the genetic and molecular basis of complex traits. The genomic selection strategies, where individuals could be precisely selected based on Genomic Estimated Breeding Values (GEBVs) that are determined by genome-wide high-density markers, are expected to play a vital role in developing new-generation rice varieties with enhanced genetic gains. Genomic selection is now widely adopted to meet the future demand for rice. Through functional characterization/haplotype analysis of genes governing traits of economic importance, a haplotype-based breeding strategy can be implemented for combining superior haplotypes from elite backgrounds using targeted markers to develop improved varieties.

Rice productivity, in general, is increasingly challenged by the ensuing global climate change. The adverse effects of climate change are more evident in rainfed ecologies. To accelerate the yield gain, the integration of physiological traits along with agronomic traits needs to be considered in rice breeding. For example, to maximize the improved performance under drought stress, several independent compatible traits need to be introgressed into elite genetic backgrounds. Recent development in genomics makes it possible to tag any traits with reliable markers for use in the selection of a large population. The International Wheat Improvement Network demonstrated that physiological breeding has led to a significant increase in genetic gains in Australia and several other developing countries. Similar initiatives must be undertaken in the rice breeding programs to improve climate resilience under stress-prone ecologies. Considering the challenge of the future food demand and mitigating the adverse effects of climate change, the priority should be to address the yield gap in available high-yielding varieties and vertically raise the genetic yield by integrating genomic technologies in conventional breeding strategies.

References

AICRIP (2020) Progress Report 2020, Vol I, All India Coordinated Rice Improvement Project. Rajendranagar, Hyderabad

Asian Development Bank (2009) Integrated crop and resource management in the rice-wheat system of South Asia. Asian Development Bank

Bouman BAM (2009) How much water does rice use? Rice Today:28–29

Bowonder B (1979) Impact analysis of the green revolution in India. Technol Forecast Soc Chang 15:297–313. https://doi.org/10.1016/0040-1625(79)90023-4

Chadha S, Gopalakrishna T (2005) Genetic diversity of Indian isolates of rice blast pathogen (Magnaporthe grisea) using molecular markers. Curr Sci 88:1466–1469

Chang TT (1989) The management of rice genetic resources. Genome 31:825–831. https://doi.org/10.1139/g89-145

Dalrymple DG (1986) Development and spread of high-yielding rice varieties in developing countries. Bur Sci Tech, Agency for Int Dev, Washington, DC

Eliazer Nelson ARL, Ravichandran K, Antony U (2019) The impact of the Green Revolution on indigenous crops of India. J Ethnic Foods 6:8. https://doi.org/10.1186/s42779-019-0011-9

Evenson RE, Gollin D (2003a) Assessing the impact of the green revolution, 1960 to 2000. Science 300:758–762. https://doi.org/10.1126/science.1078710

Evenson RE, Gollin D (eds) (2003b) Crop variety improvement and its effect on productivity: the impact of international agricultural research. CABI Pub, Wallingford, OX, 522 p. ISBN: 9780851995496

Farooq M, Wahid A, Lee DJ et al (2009) Advances in drought resistance of rice. Crit Rev Plant Sci 28:199–217. https://doi.org/10.1080/07352680902952173

ICAR (1959) Annual Report 1959–60. Indian Council of Agricultural Research, New Delhi

ICAR (1967) Annual report 1967–68. Indian Council of Agricultural Research, New Delhi

ICAR (1975) Annual report 1975–76. Indian Council of Agricultural Research, New Delhi

ICAR (1983) Annual report 1983–84. Indian Council of Agricultural Research, New Delhi

ICAR (2000) Annual report 2000–01. Indian Council of Agricultural Research, New Delhi

ICAR (2003) Annual report 2003–04. Indian Council of Agricultural Research, New Delhi

Janaiah A, Xie F (2010) Hybrid rice adoption in India. Farm level impacts and challenges. IRRI Technical Bulletin No 14, pp 1–20

Khush GS (1999) Green revolution: preparing for the 21st century. Genome 42:646–655

Khush GS, Virk PS (2005) IR varieties and their impact. Los Baños

Mackill DJ, Khush GS (2018) IR64: a high-quality and high-yielding mega variety. Rice 11:18. https://doi.org/10.1186/s12284-018-0208-3

Muralidharan K, Prasad GSV, Rao CS, Siddiq EA (2019) Genetic gain for yield in rice breeding and rice production in India to meet with the demand from increased human population. Curr Sci 116:544–560. https://doi.org/10.18520/cs/v116/i4/544-560

Muralidharan K, Prasad GSV, Rao CS et al (2022) Grain yield performance of hybrid rice in relation to inbred cultivars in long-term multi-environment tests in India. Crop Sci 63:1133–1148. https://doi.org/10.1002/csc2.20747

Pandey S, Velasco LE (1999) Economics of alternative rice establishment methods in Asia: a strategic analysis. In: Social sciences division discussion paper. International Rice Research Institute, Los Baños

Pingali PL (2012) Green revolution: impacts, limits, and the path ahead. Proc Natl Acad Sci U S A 109:12302–12308. https://doi.org/10.1073/pnas.0912953109

Richharia RH, Misro B, Rao RK (1962) Sterility in the rice hybrids and its significance. Euphytica 11:137–142. https://doi.org/10.1007/BF00033785

Sampath S, Mohanty HK (1954) Cytology of semi-sterile rice hybrids. Curr Sci 23:182–183

Senguttuvel P, Revathi P, Kemparaju KB, et al (2019) Rice hybrids released in India. Compendium No.103/2019. Rajendranagar, Hyderabad

Singh RB (2000) Environmental consequences of agricultural development: a case study from the Green Revolution state of Haryana, India. Agric Ecosyst Environ 82:97–103. https://doi.org/10.1016/S0167-8809(00)00219-X

Sundaram RM, Vishnupriya MR, Biradar SK et al (2008) Marker assisted introgression of bacterial blight resistance in Samba Mahsuri, an elite indica rice variety. Euphytica 160:411–422. https://doi.org/10.1007/s10681-007-9564-6

Tripathi A, Prasad AR (2010) Agricultural development in India since independence: a study on progress, performance, and determinants. J Emerg Knowled Emerg Mark 1:1–31. https://doi.org/10.7885/1946-651x.1007

Wik M, Pingali P, Broca S (2008) Global agricultural performance: past trends and future prospects. In: Background paper for the world development report, pp 1–39

Yadav OP, Singh DV, Dhillon BS, Mohapatra T (2019) India's evergreen revolution in cereals. Curr Sci 116:1805–1808. https://doi.org/10.18520/cs/v116/i11/1805-1808

Wheat

7

S. K. Singh, Satish Kumar, P. L. Kashyap, R. Sendhil, and O. P. Gupta

Abstract

Wheat, a preeminent crop as a food of mankind, is originated in the Central Asian region, which has 225 million ha with global production of 750 million tons. India holds the second position with 109.52 million tons of production and contributes 13% share to the global wheat basket. Wheat acreage has five zones in India where bread, durum, and dicoccum wheat are grown during the rabi season under different production conditions. Wheat improvement was initiated in 1905, and since then, it has reached present-day advancements after its journey through the green revolution. There have been many research advancements in the last 75 years, especially in the post-green revolution period, in the form of the development of more than 500 cultivars for commercial cultivation, different resource conservation and plant protection technologies, crop diversification strategies, product and nutritional quality traits, etc. Nowadays, stagnating yield potential, unavailability of sufficient quantity of quality seeds, low seed replacement, biotic and abiotic stresses in climate change conditions, restrictions to germplasm exchange in new IPR regime, reduced total factor productivity, imbalanced use of fertilizers and yield gaps at farm level have been identified as major challenges to wheat production in the country. The current pace of research efforts needs to be maintained to meet our future demands of 140 million tons by 2050, for which future research efforts would be focused on evolving new and innovative production technologies which can fit into the framework of changing wheat production scenarios. Some key strategic issues for enhanced

S. K. Singh (✉)
Genetics Division, ICAR - Indian Agricultural Research Institute, New Delhi, India
e-mail: Sanjay.singh4@icar.gov.in

S. Kumar · P. L. Kashyap · R. Sendhil · O. P. Gupta
ICAR-Indian Institute of Wheat & Barley Research, Karnal, Haryana, India

P. K. Ghosh et al. (eds.), *Trajectory of 75 years of Indian Agriculture after Independence*, https://doi.org/10.1007/978-981-19-7997-2_7

production levels include breaking yield barriers through genetic enhancement, molecular approach for precision breeding, tailoring wheat genotypes with cropping system perspective, improved varieties for abiotic stresses, focused disease resistance breeding, access to quality seeds of wheat, conservation agriculture technologies, integrated input management, crop diversification, quality improvement, strengthening inter-institutional linkages and support in policy issues. Bridging the yield gap between experimental and farmers' fields can solve the problem to a considerable extent. With support from farmers, policymakers, and extension units, it is expected that present technologies can be further refined and popularized so that wheat production can be enhanced to fulfill future demand for ensured national food security.

Keywords

Wheat · Bread wheat · Durum wheat · Dicoccum wheat · Improvement · Production technologies · Challenges · Strategies

7.1 Introduction

Among the world's crops, wheat (*Triticum species*) is preeminent both in regard to its antiquity and its importance as a food of mankind. In prehistoric times, it was cultivated throughout Europe and was one of the most valuable cereals of ancient Persia, Greece, and Egypt. According to paleo-botanists and archaeologists, the modern domesticated form of wheat originated in South-eastern Anatolia, around the region of Diyarbakir Province in present-day Turkey, at about 8500 BC. In India, evidence from Mohen-Jo-Daro excavations indicated wheat cultivation more than 5000 years ago. Globally wheat is cultivated in approx. 225-million-ha area with production of more than 750.0 million tons of grain. India, one of the greatest success stories of historical Green Revolution in India, made it the second largest producer of wheat in the world after China, with approximately 13% share of global wheat production, and recognized it as the wheat surplus nation. On the other hand, India is also the second largest wheat consumer after China and thus, wheat and its various products play an increasingly important role in managing India's food security.

7.2 Trends in Wheat Area, Production, and Productivity

Wheat is the second most important crop after rice in India and occupies an approximately 31.76-million-ha area. Wheat production in India touched a new height of 109.52 million tons in 2020–2021 (DAC & FW 2021). The major wheat-growing states in India are Uttar Pradesh, Punjab, Haryana, Madhya Pradesh, Rajasthan, and Bihar, which contribute approx. 91.5% of India's wheat production. The trend of area, production, and productivity indicated a significant quantum jump in wheat in post-independence period (Table 7.1, Fig. 7.1), where an increment of 2.23 times in area under wheat (+21.91 million ha: 222.50%), 15.49 times in wheat production (+102.15 million tons: 1548.73%), and 4.11 times in wheat productivity (+2754 kg/ha: 411.23%) was recorded. This gargantuan jump in production, post-independence, is attributed to the increased crop productivity followed by area. Precisely, the impact was more evident since the inception of the All India Coordinated Research Project leading to semidwarf wheat-based Green revolution. Despite recent risks of climate change, decade-wise analysis indicated increasing trends in area, production, and productivity in recent decades.

Globally, wheat is the most widely cultivated and traded cereal. In India, owing to surplus production, wheat imports have reduced considerably from 37.6 lakh tons (triennium ending 1970) to 0.4 lakh tons (triennium ending 2000). In contrary, the exports witnessed a surge from 0.01 to 5.56 lakh tons for the same period. Wheat is one of the major food staples for a majority of Indians; the consumption also registered an increase from 21 to 97.08 million tons between the triennium ending 1970 and 2020 (Sendhil et al. 2021).

7.3 Wheat Species and the Zones

The three species of wheat, namely *Triticum aestivum* (Bread wheat), *Triticum durum* (Macaroni wheat), and *Triticum dicoccum* (Emmer or Khapli), grown on a commercial basis in India are of spring type but cultivated during the winter season. Of these species, *T. aestivum* is the most important, accounting for about 90–95% of the total wheat area of the country, and is grown in almost all the wheat-growing states. *T. durum* occupies approximately 5% of the total wheat area and is confined mostly to central and southern parts of India. The cultivation of *T. dicoccum* is confined largely to the southern region, mainly Karnataka and southern Maharashtra. Another wheat species, *T. sphaerococcum* (Indian dwarf wheat), has now almost vanished and cultivated in some pockets in Gujarat. Wheat in India is cultivated in almost every state except Kerala, thus representing diverse crop-growing conditions and situations. Wheat cultivation in India extends from 9°N (Palni hills) to above 35° N (Srinagar valley of J & K), thus the wheat crop is exposed to a wide range of agroclimatic changes such as humidity, temperature, photoperiod during crop season, soil types, altitudes, latitudes and cropping systems. Based on the agroclimatic conditions and varying agroecological production conditions, the country is broadly divided into five wheat growing zones, namely, Northern Hills Zone (NHZ), North Western Plain Zone (NWPZ), North Eastern Plain Zone (NEPZ), Central Zone (CZ), and Peninsular Zone (PZ). NWPZ, NEPZ, and Central Zones are the main contributors to wheat production. The growing period of wheat is variable from one agroclimatic zone to another, which affects the vegetative and grain filling duration leading to differences in attainable yield. The maximum wheat growing duration is in Northern Hill Zone and the minimum in Peninsular Zone.

Table 7.1 Growth in national area, production, and productivity in wheat

Period	Quantum change			Percent change (%)		
	Area (million ha)	Production (million tons)	Productivity (kg/ha)	Area	Production	Productivity
Pre-Independence	−1.42	−1.48	−47	−12.60	−18.35	−6.58
Post-Independence	21.91	102.15	2754	222.50	1548.73	411.23
• 1947–1948 to 1959–1960	3.53	3.72	102	35.87	56.46	15.26
• 1960–1961 to 1969–1970	3.70	9.09	357	28.62	82.64	41.95
• 1970–1971 to 1979–1980	3.93	8.00	129	21.55	33.57	9.87
• 1980–1981 to 1989–1990	1.22	13.54	491	5.48	37.29	30.12
• 1990–1991 to 1999–2000	3.32	21.23	497	13.74	38.50	21.79
• 2000–2001 to 2009–2010	2.73	11.12	131	10.61	15.96	4.84
• 2010–2011 to 2020–2021	2.69	21.88	436	9.26	25.18	14.57

Pre-independence: 1885 1886 to 1946–1947 and Post-independence: 1947–1948 to 2020–2021

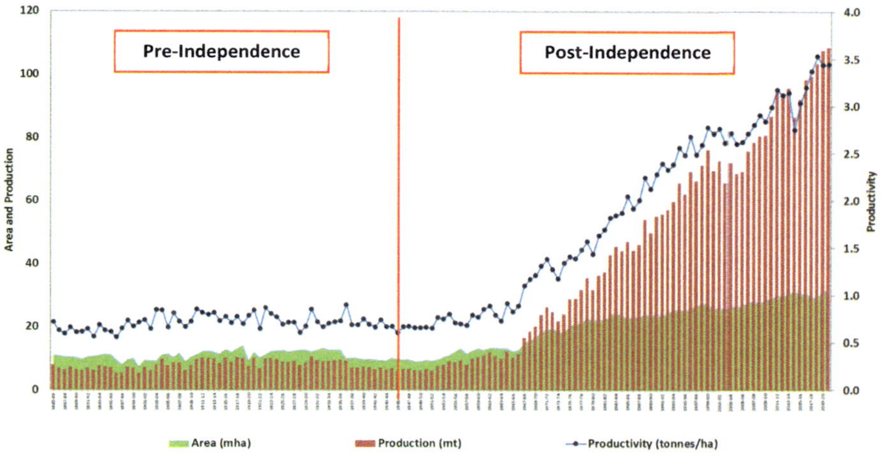

Fig. 7.1 Trends in wheat area, production, and productivity (pre- and post-independence)

7.4 Wheat Growing Season and Cultural Condition

Wheat is cultivated during the winter season from mid-October to April (except in higher hills of north India, where harvesting of wheat is done in the month of May) Sowings of wheat are initiated when the average day-night temperatures are equal to 23 °C. The months of December and January remain to be the coldest, followed by comparatively warmer and higher temperatures in the months of March-April, coinciding with later growth stages of the crop till maturity. Wheat is mainly grown under three production conditions: timely sown, irrigated; late sown, irrigated; and timely sown, restricted irrigation. Nearly, 89% of the wheat area in the country is irrigated and most of it lies in north India. The central, peninsular, and hilly areas have comparatively lower coverage of area under irrigation and grow mostly rainfed. In recent years, a new situation of timely sown, restricted/limited irrigation has emerged in some of the areas of the central and peninsular parts where water for irrigation is not available in sufficient quantity, and thus, the wheat crop is grown with one to two irrigations only.

7.5 Wheat Research in India: Chronological Perspective

7.5.1 Pre-independence Era

Wheat grown in India prior to the twentieth century mostly consisted of mixtures of various botanical forms usually referred to as 'sorts' among which Sharbati, Dara, Safed Pissi, Chandausi, Karachi, Choice White, Hard Red Calcutta, Lal Kanak, Lal Pissi Jaipur Local, Kharchia Local, Mondhya 417, Muzaffar Nagar White, Buxar

White, etc. were some prominent local bread wheat sorts utilized nationally and internationally. A systematic bread-wheat improvement work was initiated by Sir Albert Howard and his wife Lady G.L.C. Howard in 1905 at Imperial (now Indian) Agricultural Research Institute (IARI) at Pusa in Bihar. At the same time, Government Agricultural Colleges and Research Institutes were established at Lyallpur (now in Pakistan), Kanpur, Sabour, Coimbatore, and Pune to undertake research on crops, especially wheat, and later wheat improvement was also undertaken at Powarkheda (Madhya Pradesh), Nagpur, Akola, and Niphad (Maharashtra). In the era of pre-semidwarf bread wheat varieties (1905–1962), the selection from the mixtures resulted in some important pure lines like NP4, NP6, NP12, NP22, Type9, Type11, Pb type8A, C13, C46, AO13, AO49, AO68, AO69, AO85, AO88, AO90, etc. Some of these varieties earned international recognition due to their excellent grain appearance and are still considered a valuable genetic resources. Many Indian local collections, viz. Indian, Indian A, Indian B, Indian 4E, Indian F, Indian G, Indian H, Indian 8, Indian 9, Indian 17, Nabob, Sindhi, Etawah, Indian Pearl, Indian Dwarf, Ranjit, Khapli, Kaali, Muzzafar Nagar Variety, Hindi D, Hindi 7, Hindi 39, Hindi 62, Hindi 90, Hindi 144, Hard Red Calcutta, etc. were utilized in Australia, Egypt, Canada, South Africa, USSR, France, Brazil, and other countries (Jain 1994) for wheat improvement. Hybridization among these varieties and the exotic wheat was utilized in wheat breeding from the early years of the twentieth century which resulted in varieties like NP52, NP80-5, NP120, NP165, NP710, NP770, NP783, NP824, PbC518, PbC591, Niphad4, AO113, etc. Exotic sources such as Ridley, Padova I, and Padova II gained acceptance, and Federation 41, Kononso, Thatcher, Kenya C 10854, Kenya 48F, Democrat, Spalding's, Ridit, Reliance, prolific, Gabo, Timstein, Bobin, Gaza, Regent, etc., contributed significantly in the Indian wheat breeding program as donor lines for incorporating rust resistance leading to the development of disease-resistant cultivars like K53, K54, RS31-1, and Kenphad25. The varieties developed until the early 1960s were tall in stature and prone to lodging and therefore, nonresponsive to high fertilizer usage, e.g., K 65, K 68, Hyb 65, etc. Similar to those of the bread wheat improvement program, the durum improvement was initiated by H.M. Chibber in 1918–1919 in the erstwhile Bombay Presidency by making single plant selections out of the cultivated landraces like Bansi, Kathia, Gangajali, Haura, Jalalia, Jamli, Khandwa, Malvi, etc. These selections yielded superior lines like Bansi103, Bansi 162, Bansi 168, Bansi 224, and Baxi 228-18. The durum wheat evolved in India is a good source of genes for drought and heat tolerance, and they are mostly used as chapati or Dalia due to their unique quality characteristics. Subsequent efforts using exotic germplasm led to the development of improved durums like A 206, A 624, Amrut, N 59, NP 404, A-9-30-1, N 5749, Hybrid 23, Ekdania 69, Narasingarh 111, Bijaga yellow, and Bijaga red.

7.5.2 Post-independence Era

The real breakthrough in productivity occurred from the introduction of the semi-dwarf Mexican wheat in the early 1960s when India took advantage of Norin (having *Rht* dwarfing gene) based germplasm including four semidwarf wheat varieties, viz, Lerma Rojo 64-A, Sonora 63, Sonora 64, and Mayo 64 possessing a unique set of high productivity traits such as non-lodging habit, high fertilizer responsiveness, and resistance to rust and foliar diseases. After extensive evaluation of these varieties during 1963–1964, Lerma Rojo 64-A and Sonora, 64 carrying the genes for dwarfism and resistance to rusts, were released by the Central Varietal Release Committee in 1965, which laid the foundation for increased wheat production. At the same time, the establishment of the All India Coordinated Wheat Improvement Project (AICWIP) in 1965 was an important milestone for systematic developments in wheat research resulting in a major breakthrough in wheat production and productivity. Later on, the advanced lines from CIMMYT provided the base material for the development and commercial release of important amber-colored varieties, namely S227 (Kalyansona), S307, S308 (Sonalika), S331, Chhoti Lerma, and Safed Lerma. Among these, Kalyansona and Sonalika were very popular, occupying nearly 10-million-ha area because of their high yield, rust resistance, amber grains, and adaptability to different agroclimatic conditions of the country and became the harbingers of the 'Wheat Revolution' in India, which was later termed as the "Green Revolution." The red color of Sonora 64 was improved through mutagenesis at IARI, New Delhi, and released as "Sharbati Sonora." Afterward, hybridizations between Mexican and Indian varieties led to the release of WG357, WG377, WL711, HD2009, WH147, and HD1981 for NWPZ; HD1982, UP262, HP1102, HUW12, K7410, and HP1209 for NEPZ; HD2189 and NI5439 for PZ and many others which not only sustained the impact of green revolution but also took the wheat revolution to newer heights. India was the first country to release three multiline wheat varieties, namely MLKS11, KSML3, and KML7406 (Bithoor), for commercial cultivation in 1978. Varieties such as Lok1, HUW234, HD2285, HD2329, HD2189, and some derivatives of *Veery* of CIMMYT, viz. HUW206 and HS207, were very popular among the farmers during the 1980s and provided the necessary boost to wheat production and productivity. The development of wheat varieties CPAN3004, WH542, PBW343, and PBW373 having 1B/1R translocation through the utilization of winter wheat gene pool provided a quantum jump in wheat productivity by way of resistance to diseases, enhanced morphophysiological traits, and wider adaptability. Later HD2687, UP2338, RAJ3765, K9107, NW1014, HP1744, GW273, GW322, GW366, and MACS2496 covered significant wheat acreage in different parts of the country. During the early 2000s, PBW343 was the dominant wheat variety and after its susceptibility to yellow rust, DBW17 caught the farmers' attention as a suitable replacement for PBW343 in NWPZ. The release of mega varieties such as HD2967 in 2011 and HD3086 in 2014 has brought out a revolution in the highly productive environments of northern India. The robust and tall plant types with thick stems have led to changes in the plant type of the varieties, with the newer genotypes being more efficient and adaptive. The

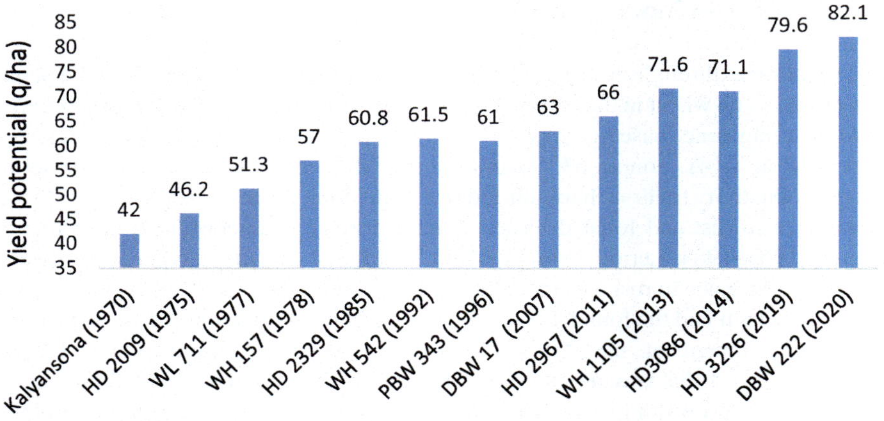

Fig. 7.2 Improving yield potential (q/ha) in high fertility conditions of NWPZ

improvement in yield potential of wheat cultivars under high fertility conditions of the north-western wheat zone—the wheat bowl, is shown in Fig. 7.2. This decade also saw the release of two zinc bio-fortified wheat varieties (HPBW01 and WB2) for cultivation at the farmers' fields in 2017 hence enhancing the nutritional status of farm families. Also, for the first time, the wheat variety PBW723, developed through marker-assisted selection (MAS), was released in 2017. Recent releases DBW187, DBW222, and DBW303 have revolutionized wheat cultivation in the breadbasket of India with a yield potential of more than 8.0 tons/ha.

7.5.3 Wheat Improvement for Biotic Constraints

The most important fungal disease that badly hampered wheat production in India are rust diseases, namely stem or black rust (*P. graminis* f. sp. *tritici*), leaf or brown rust (*Puccinia triticina* Eriks), and stripe or yellow rust (*Puccinia striiformis* f. sp. *tritici*). Historical rust epidemic accounts revealed the first stem rust epidemics in 1786, 1829, and 1946–1947 in the central parts of India (Sleeman 1839; Nagarajan and Joshi 1975; Joshi et al. 1986). The research advances in the intensification of the agriculture system in post green revolution resulted in the rise of pest and pathogen incidences as indicated by brown and yellow rust epidemics in the north-western part of India during 1971–1973 and 1993 (Nayar et al. 1997). At present, yellow (stripe) rust is a potential threat to 10 million ha area of northern India, whereas stem (black) rust is to ~7 million ha of Central and Peninsular areas of India. Leaf or brown rust is of major concern in all the wheat-growing zones of the country. Although virulent pathotypes of yellow rust were identified on Yr9 in 1996 and Yr27 in 2002 (Prashar et al. 2007), cultivars possessing 1BL.1RS wheat-rye (*Secale cereale*) translocation (Singla and Krattinger 2016) like PBW373, PBW343, PBW175, UP2338, UP2425, UP2418, UP2382, and CPAN3004 showed resistance towards multiple fungal diseases including yellow rust (*Yr9*), leaf rust (*Lr26*), stem rust (*Sr31*) and powdery mildew (*Pm 8*). The emergence of 78S84 virulence for *Yr9* in stripe rust led to the elimination of mega wheat variety PBW343. Later virulent pathotypes 46S119, 110S119, 110S84, and 238S119 were detected, and currently, 238S119 has become the most prevalent. It is important to mention that Indian wheat germplasm remained resistant to rusts due to the presence of multiple genes complexes for stem (*Sr31* + *Sr24*, *Sr2*, Sr5, and Sr8), leaf (*Lr26* + Lr13, Lr23, Lr34, *Agropyron* segment carrying Lr24/Sr24), and stripe (Yr9 + Yr2, Yr18, and some unknown adult plant resistance genes) rusts. At present, about 10 million tons of wheat are protected every year from fungal diseases by deploying strategic and integrated disease management modules (Bhardwaj et al. 2019). Efforts have been made to decipher the genome structure, molecular basis of variation, and pathogenicity of rust fungus in wheat, and the genome of three races (K, 31, and 46S119) of *P. striiformis* f. sp. *tritici* (wheat yellow rust fungus) has been unzipped (Kiran et al. 2017). Recently, in India, a major rust resistance gene *Lr80* was identified from the local wheat landrace, Hango-2, and mapped on the 2DS chromosome (Kumar et al. 2021).

Another important foliar disease, Spot blotch (*Bipolaris sorokiniana* (Sacc.) Shoemaker Syn. *Helminthosporium sativus*), is reported to affect nearly 9 million ha of the rice–wheat belt in Northern plains with an average yield loss of about 15–20% (Chand et al. 2003; Duveiller and Sharma 2009). Recent studies have revealed that the resistance to spot blotch is imparted by the presence of at least two or more genes or QTLs and so far, eight quantitative trait loci (QTLs) linked to spot blotch resistance have been identified (Kumar et al. 2010). Besides, powdery mildew (*Erysiphe graminis* DC. f. sp. *tritici* Em. Marchal syn. *Blumeria graminis* (DC.) E. O. Speer), flag smut (*Urocystis agropyri* (G. Preuss) Schrot.), loose smut (*Ustilago tritici* (Pers.) Rostr. Syn. *U. segetum tritici*), head scab (*Fusarium graminearum* Schwabe), etc., are diseases of minor importance due to their localized effects. Powdery mildew hindered wheat cultivation in the north-western plain zone, northern and southern hill zone and was documented to cause 10–15% yield losses under congenial conditions (Kashyap et al. 2021). It has been observed that head scab disease is on the rise under the influence of global warming and the fast adoption of reduced tillage practices in North-western parts of India (Saharan et al. 2021). Flag smut and Loose Smut are also gaining concern in the present climate change scenario.

Karnal bunt (*Tilletia indica* Mitra syn. *Neovossia indica* (M. Mitra) Mundk.) is another disease that has importance in trade perspectives. It was first identified by *Manoranjan Mitra* in the year 1931 and remained a little-known wheat disease localized in North-West India. With the introduction of semidwarf wheat varieties, KB incidences became unusually frequent in North-West India (Joshi et al. 1983). Moreover, intense irrigation and large-scale fertilizer application also added to the aggravation of the intensity and number of KB incidences in the entire Northern India. The KB-infected grains are of low quality as they harbor an unacceptable smell, color, and taste and at as low as 1% infection, the grains/flour become unpalatable and categorized as a disease of quarantine significance (Bishnoi et al. 2020).

A particular concern is the threatening emergence of wheat blast disease (*Magnaporthe grisea* pathotype tritici) in our neighboring country Bangladesh. To prevent its entry in Indian territory, especially in the bordering districts of West Bengal, the most proactive role has been made by ICAR-Indian Institute of Wheat and Barley Research (IWBR) under the aegis of the Indian Council of Agricultural Research (ICAR) to thwart any scope of wheat biosecurity in the county. Under this initiative, a wheat crop holiday i.e. banning wheat cultivation in Murshidabad and Nadia districts of West Bengal, India, with support and incentive for alternate crops (non-poaceaous crops like gram, *urad*, oilseed crops such as rapeseed, mustard, and potatoes) and "no wheat zone" within 5 km along the border area have been recommended. In addition to this, research and development activities related to the characterization of Indian wheat lines for WB resistance in collaboration with CIMMYT, Mexico, have been initiated since 2016 and successfully identified and recommended five resistant/tolerant, high-yielding wheat varieties such as HD3249, DBW187, HD2967 for irrigated timely sown conditions and HD3293, DBW252, and HD3171 for restricted irrigation timely sown conditions. Apart from this, the available potential donors (BH1146, Milan, SHA7, *Aegilops tauschi* derivatives, varieties carrying *Lr34*, and genotypes possessing 2NS translocation) are being exploited for introgression of WB resistance in Indian cultivars as part of the WB anticipatory breeding research initiative.

7.6 Advances in Technological Development

7.6.1 Cultivar Development

The agroclimatic conditions, local preferences, and wheat-based food habits, prevalence of diseases and pests, wheat-based cropping systems, availability of irrigation, and related input factors have a direct bearing on the types of wheat varieties to be developed for commercial cultivation in the country. The farmers of the country have been provided with a choice of varieties during the last 60 years since the inception of the All India Coordinated Wheat and Barley Improvement Program in 1965. The Indian wheat improvement program has significantly contributed to the release of 501 wheat varieties, including *Triticum aestivum*, *T. durum*, *T. dicoccum*, and triticale, through Central Variety Release Committee (CVRC) or State Variety Release Committee (SVRC) for different agroclimatic zones along with relevant

Table 7.2 Released wheat varieties in India (1965–2021)

Crop species	CVRC	SVRC	Total
Bread wheat (*T. aestivum*)	267	155	422
Durum wheat (*T. durum*)	45	23	68
Dicoccum wheat (*T. dicoccum*)	6	1	7
Triticale	4	–	4
Total	322	179	501

production technology (Table 7.2). This included 422 bread wheat, 68 durum, 7 dicoccum, and 4 triticale varieties.

Some varieties released in the recent past have contributed significantly to enhancing wheat production and among them, DBW17, PBW550, HD2967, HD3086, WH1105, K0307, GW 322, GW 366, HI 1544, and MACS 6222 are notable in different areas. The recently released cultivars are presented in a table which is becoming popular among the farmers (Table 7.3). Besides this, more than 300 wheat genetic stocks have been registered with the National Bureau of Plant Genetic Resources (NBPGR), New Delhi, for various biotic and abiotic stress tolerance, yield and quality component traits for their further utilization in wheat improvement programs.

7.6.2 Resource Conservation Technologies

Among the wheat-based cropping systems in India, the rice–wheat cropping system was adopted on a significant area of more than 12 million ha. The intensive tillage under this system needed to optimize energy usage by improving tillage practices and developing efficient machinery. The focused research efforts have resulted in the development of eco-friendly resource conservation technologies, viz., zero tillage, bed planting, rotary tillage, and rotary disc drill, to reduce the cost of cultivation, increase productivity, and improve the soil health and environment. The zero tillage technology, covering about 2.0 million ha area for seeding wheat without any field preparation, besides saving energy, reduces the cost of cultivation, advances the time of wheat sowing by 4–5 days, requires less water for the first irrigation, and results in less infestation of the notorious weed *Phalaris minor* as compared to the conventional tillage. The furrow irrigated raised bed-planting system (FIRBS) is a promising resource conservation tillage technology that saves about 25% seed and fertilizer along with reduced water usage by 25–40% depending upon the soil type and agro-climatic conditions. Rotary tillage technology is another option that completely pulverizes the top 10 cm soil with simultaneous placement of seed and fertilizer and gives the highest productivity with the least specific energy requirement. In some part of the country, like Punjab and Haryana, the burning of crop residues often lead to environmental pollution, loss of soil organic carbon, depletion of plant nutrients such as nitrogen, and loss of soil microflora. To avoid residue burning, a new machine "Rotary Disc Drill" has also been developed at the IIWBR which is capable of seeding under loose residue conditions.

7.6.3 Crop Diversification

Diversification of wheat-based cropping system, especially rice–wheat systems, by introducing short-duration legume crops for grains or green manuring, helps in restoring soil health by enhancing the organic matter content and improving the soil physicochemical properties. These systems also help in controlling weeds

Table 7.3 Latest wheat varieties for diverse agroclimatic and production conditions

Production condition		Improved cultivars
North Western Plains Zone (NWPZ)		
Early sown, irrigated	:	DBW 303, WH 1270
Timely sown, irrigated	:	DBW 222, DBW 187, HD 3226, WB 02, HPBW 01, PBW 723, DBW 88, HD 3086, WH 1105, HD 2967, DPW 621-50, DBW 17, WHD 943 (d), PDW 314 (d), PDW 291 (d)
Late sown, irrigated	:	PBW 752, DBW 173, DBW 90, WH 1124, DBW 71, HD 3059, PBW 590
Very late sown, irrigated	:	PBW 757, HI 1621, HD 3271
Timely sown, rainfed	:	PBW660, PBW644, WH1080
Timely sown, restricted irrigation	:	NIAW 3170, HD 3237, HI 1620, WH 1142, HD 3043
Sodic soils/others	:	KRL-19, KRL 210, KRL 213
North Eastern Plains Zone (NEPZ)		
Timely sown, irrigated	:	HD 3086, DBW 187, K 1006, HD 2967, NW 5054, DBW 39, RAJ 4120, K 0307
Late sown, irrigated	:	DBW107, HD3118, HD2985, HI1563
Very late sown, irrigated	:	HI1621, HD3271
Timely sown, rainfed	:	K1317, HD3171
Timely sown, restricted irrigation	:	DBW252, HI1612
Sodic soils/others	:	KRL-19, KRL 210, KRL 213
Central Zone (CZ)		
Timely sown, irrigated	:	HI 1544, GW 366, GW 322, HI 8759 (d), HD 4728 (d), HI 8737 (d), HI 8713 (d), MPO 1215 (d)
Late sown, irrigated	:	HI 1634, CG 1029, RAJ 4238, MP 3336, MP 1203
Timely sown, rainfed	:	HI 1500, MP 3173, MP 3288
Timely sown, restricted irrigation	:	DBW 110, DDW 47 (d), UAS 466 (d), HI 8627 (d)
Sodic soils/others	:	KRL-19, KRL 210, KRL 213
Peninsular Zone (PZ)		
Timely sown, irrigated	:	DBW 168, MACS 6478, UAS 304, MACS 6222, DDW 48 (d), MACS 3949 (d), WHD 948 (d), UAS 428 (d), UAS 415 (d), HW 1098 (dic), MACS 2971 (dic), DDK 1029 (dic), DDK 1025 (dic)
Late sown, irrigated	:	HI1633, HD3090, AKAW4627, HD2932
Timely sown, rainfed	:	NIAW 1415, PBW 596, UAS 375, UAS 347, MACS 4028 (d), HI 8777 (d), UAS 446 (d) NIDW 1149 (d), GW 1346 (d), HI 8802 (d), HI 8805 (d), MACS 4058 (d)
Timely sown, restricted irrigation	:	NIAW 3170, HI 1605, DBW 93
Northern Hills Zone (NHZ)		
Timely sown, irrigated	:	HS 562, HPW 349, VL 907, HS 507
Late sown, irrigated/ rainfed	:	HS 490, VL 892
Timely sown, rainfed	:	HS 542, HPW 251, VL 829
For high altitude areas	:	VL 832

without herbicide application as the wheat is sown late when the environment is not conducive for the germination of Phalaris minor. Among various diversified cropping sequences, rice–vegetable peas–wheat rotation gave the highest net return per unit area, and it appears to be the best option for maximization of net returns and sustainability of the rice–wheat system.

7.6.4 Plant Protection Technologies

Indian wheat production has been free from disease epidemics during the last four decades mainly because of the systematic deployment of rust resistance genes in high-yielding varieties. The survey and surveillance of rust virulence and identification of resistance donors against important diseases and pests were very crucial activities in this direction. Pest risk analysis has been carried out for Karnal bunt to safeguard the national export potential. Integrated Pest Management (IPM) modules have been developed and validated for cost-effective and eco-friendly control of pests and diseases of wheat. In post green revolution period, a large number of research initiatives have been taken to diagnose and monitor pathogen inocula inside plants and soil with special reference to Karnal bunt, flag smut, leaf rust, and head scab diseases in wheat (Kashyap et al. 2020; Gurjar et al. 2017; Manjunatha et al. 2018; Gupta et al. 2020). The long-term experiments under AICRP on Wheat and Barley have recommended the application of fungicides such as Propiconazole 25% EC (Tilt), Tebuconazole 25% EC (Folicur), Triadimefon 25% EC (Bayleton), and Difenaconazole for control of initial inoculum load or high disease pressure of fungal pathogens and their effective management. Efficient protection technologies have also been developed to control aphids (spray of imidacloprid @ 20 g a.i./ha) and termites (endosulfan, chlorpyriphos, and carbosulfan). Besides, several advanced breeding lines have been evaluated and screened under AICRP on wheat and barley against multiple pathogens to support the breeding program at various hot spot locations under artificially inoculated conditions in different agroecological zones, and potential resistance donors were identified for further utilization (Table 7.4).

7.6.5 Quality of Indian Wheat

A number of physical and biochemical parameters are associated with the quality of wheat. Targeting the surplus wheat and export potentials, the quality parameters have been standardized, and special emphasis was given to identify specific varieties of *chapati*, bread, biscuit, and pasta products. The quality requirements differ for various products like *chapati*, bread, biscuit, and pasta. Hard wheat with strong gluten (>60 mL sedimentation value), >12.0% protein, 5 + 10 high molecular weight glutenin subunit with 9 or 10 Glu-1 scores is required for making good bread, whereas weak and soft wheat with <10.0% protein, weak gluten of <30 mL sedimentation value, and ~50% alkaline water retention capacity (AWRC) is required for biscuits. Good pasta products can be prepared from hard durum wheat

Table 7.4 Current spectrum of wheat diseases in different agroecological zones and identified potential resistance donors

Diseases	Distribution	Potential disease-resistance donors
Stripe rust (yellow rust)	NHZ and NWPZ	DBW 187, DBW 237, DBW 302, DBW 303, HI 1628, HPW 467, HS 660, HS 661, PBW 752, PBW 757, PBW 763, PBW 771, PBW 796, PBW 797, PBW 800, PBW 801, PBW 820, PBW 821, PBW 822, PBW 823, PBW 825, UP 3043, VL 3020, VL 3021, WH 1270, DDW 47 (d), DDW 48 (d), GW 1339 (d), GW 1348 (d), HI 8800 (d), HI 8802 (d), HI 8805 (d), HI 8807 (d), HI 8808 (d), HI 8811 (d), HI 8812 (d), MPO 1336 (d), NIDW 1149 (d), NIDW 1158 (d), WHD 963 (d)
Leaf rust (brown rust)	All the six agroecological zones but more prevalent in NEPZ, CZ and PZ, whereas in NWPZ, it appears late	AKAW 4924, CG 1029, DBW 187, DBW 237, DBW 302, DBW 303, GW 491, GW 492, GW 509, HI 1624, HI 1628, HI 1633, HI 1634, HPW 451, HPW 459, HPW 467, HS 660, HS 661, MACS 5051, NIAW 3171, PBW 752, PBW 757, PBW 763, PBW 771, PBW 796, PBW 797, PBW 800, PBW 801, PBW 820, PBW 821, PBW 822, PBW 823, PBW 825, UP 3016, UP 3043, VL 3020, VL 3021, WH 1270, DDW 47 (d), DDW 48 (d), GW 1339 (d), GW 1346 (d), GW 1348 (d), HI 8800 (d), HI 8802 (d), HI8805 (d), HI 8807 (d), HI 8808 (d), HI 8811 (d), HI 8812 (d), MACS 4059 (d), MPO 1336 (d) NIDW 1149 (d), NIDW 1158 (d), WHD 963 (d), DDK 1054 (dic)
Stem rust (black rust)	PZ and CZ	AKW4924, CG 1029, DBW 302, GW 491, GW 492, GW 509, HI 1624, HI1628, HI 1633, HI 1634, HPW 451, HPW 459, MACS 5051, NIAW 3171, PBW 820, UP 3016, GW 1346 (d), GW 1348 (d), HI 8802, HI 8807 (d), HI 8811 (d), MACS4059 (d), NIDW 1149 (d), NIDW 1158 (d), DDK 1054 (dic)
Karnal bunt	Major problem in NHZ, NWPZ, minor in NEPZ	AKAW4924, ALDAN, Altar 84, CG 1029, CPAN 3045, DBW 110, DBW 187, DBW 237, DBW 302, DBW 303, GW 491, GW 492, GW 509, HD 29, HI 1624, HI 1628, HI 1633, HI 1634, HP 1531, HPW 451, HPW 459, HS 660, HS 661, KBRL 22, KBRL57, MACS 5051, NIAW

(continued)

Table 7.4 (continued)

Diseases	Distribution	Potential disease-resistance donors
		3171, PBW 34, PBW225, PBW 752, PBW 757, PBW 763, PBW 771, PBW 796, PBW 797, PBW 800, PBW 801, PBW 820, PBW 825, UP 3016, UP 3043, VL 3021, W48, W285, W382, W485, DDW 47 (d), DDW 48 (d), GW 1339 (d), GW 1346 (d), GW 1348 (d), HI 8800 (d), HI 8802 (d), HI8805 (d), HI 8807 (d), HI 8811 (d), MACS4059 (d), MPO 1336 (d) NIDW 1149 (d), NIDW 1158 (d), DDK 1054 (dic)
Spot blotch	NEPZ, moderate in CZ, PZ and NWPZ	HI1628, NIAW 3171, PBW 763, PBW 800, UP 3016, VL 3020, WH 1270, HI 8805 (d), DDK 1054 (dic)
Powdery mildew	NHZ and NWPZ but occurrence is erratic	DBW 187, DBW 237, GW 491, GW 492, HI 1628, HPW 451, HPW 459, HS 660, HS 661, MACS 5051, NIAW 3171, PBW 757, UP 3016, GW 1339 (d), GW 1346 (d), HI 8800 (d), MACS 4059 (d), MPO 1336 (d)
Head scab	NWPZ, NEPZ	AKW4924, DBW 187, DBW 237, GW 491, GW 492, HI 1624, HI 1628, HS 660, HPW 451, HPW 459, HS 661, MACS 5051, NIAW 3171, PBW 757, PBW 763, PBW 797, PBW 800, PBW 801, UP 3016, GW 1339 (d), GW 1346 (d), HI 8800 (d), MACS 4059 (d), MPO 1336 (d), DDK 1054 (dic)
Loose Smut	NWPZ, NHZ, and NEPZ	PBW 752, UP 3043
Flag smut	NWPZ	AKAW 4924, CG 1029, DBW 187, DBW 237, DBW 302, DBW 303, GW 491, GW 492, GW 509, HI 1624, HI 1628, HI 1633, HI 1634, HPW 451, HPW459, HPW 467, HS 660, HS 661, MACS 5051, NIAW 3171, PBW 752, PBW 757, PBW 763, PBW 771, PBW 796, PBW 797, PBW 800, PBW 801, PBW 820, PBW 821, PBW 822, PBW 823, PBW 825, UP 3016, UP 3043, VL 3020, VL 3021, WH 1270, DDW 47 (d), DDW 48 (d), GW 1339 (d), GW 1346 (d), GW 1348 (d), HI 8800 (d), HI 8802 (d), HI8805 (d), HI 8807 (d), HI 8808 (d), HI 8811 (d), HI 8812 (d), MACS4059 (d), MPO 1336 (d) NIDW 1149 (d),

(continued)

Table 7.4 (continued)

Diseases	Distribution	Potential disease-resistance donors
		NIDW 1158 (d), WHD 963 (d), DDK 1054 (dic)
Wheat blast	Not yet reported in India	BH 1146, DBW 187, DBW 252, HD 2967, HD 3171, HD 3249, HD 3293, Milan, SHA7 and *Aegilops tauschi* derivatives

d durum, *dic* dicoccum, *NHZ* Northern Hills Zone, *NWPZ* North Western Plains Zone, *NEPZ* North Eastern Plains Zone, *CZ* Central Zone, *PZ* Peninsular zone

Table 7.5 Product-specific varieties

Product	Varieties
Chapati (score > 8.0/10.0)	DBW 71, DBW 303, HD 2888, HD 2967, HD 3086, HD 3237, HI 1500, HI 1634, K 0307, MP 3288, NIAW 1415, PBW 757, PBW 771, WH 1124
Bread (loaf volume >600 mL)	DBW 71, DBW 93, DBW 173, DBW 187, DBW 222, HD 2733, HD 3059, HD 3226, HD 3298, NIAW 1415, WH 1080, WH 1105, WH 1124, WH 1254
Biscuit (spread factor >10.0)	DBW 168, HS 490, NIAW 3170
Durum for pasta (Yellow pigment >7.0/9.0)	UAS 446 (d), DDW48 (d), DDW 47 (d)

with >12.0% protein and strong gluten strength of >40 mL sedimentation value. The yellow berry incidence below 10%, >7 ppm β-carotene content, and γ-gliadin 45 is also required for good quality pasta products. A large number of genotypes were evaluated for product quality and promising varieties were identified as mentioned in Table 7.5. Quality analysis was carried out on large scale with samples collected from different markets and warehouses across the country for the purpose. A wheat quality atlas of the country was also developed, and potential regions have been identified for product-specific varieties.

7.7 Challenges Ahead

7.7.1 Stagnating Yield Potential

After realizing the benefits of the green revolution, steep growth in wheat productivity in frontline states of north-western India was achieved from 1975 to 1995 through the churning of the gene pool and deployment of rust resistance genes in better agronomic backgrounds that resulted in the release of some of the landmark varieties. The increasing trend of wheat productivity in all the wheat areas has reached a kind of saturation. Recently released varieties in NWPZ have shown yield superiority, but the cultivars in other zones are not coming with desired yield superiority which is one of the major concerns for the enhancement of wheat productivity. India is expected to produce 140 million tons of wheat by 2050 (ICAR 2018). Progress made in irrigated and high-fertility wheat regions is significantly higher than that in marginal areas experiencing water stress conditions coupled with heat stress during different crop growth stages. Since significant scope exists for improvement in these new areas, one of the major challenges is to develop high-yielding varieties having tolerance to abiotic stresses, especially heat, drought, salinity, and waterlogging.

7.7.2 Unavailability of Quality Seeds and Low Seed Replacement

Physical and genetic purities of seed are of utmost importance for realizing the actual yield potential of the variety. The rate of seed replacement of newer varieties to the older ones is also an important factor to put the yield levels in high momentum. The major constraints in most of the area are the unavailability of pure-quality seeds and a substantially lower seed replacement rate at the farm level. An enhanced rate of seed replacement with advanced varieties will certainly result in a significant increase in total wheat production in the country. Thus, there is an urgent need to strengthen the seed production system and efficient distribution channel to bring more area under new wheat varieties in a lesser time period.

7.7.3 Global Climate Change

An impact of changing climate on crop production is expected for various latitude limits for all the crop seasons, and the wheat crop is the most affected during the winter season. Wheat is sensitive to high temperatures (both early and late heat), but the magnitude of damage depends on the existing ambient temperature, stage of crop development, and variety. The rise in temperature during December, the period of tillering, and subsequently higher temperatures above 30 °C during the February and March at the stages of anthesis, grain formation, and filling has affected the productivity of varieties having high yield potential. Effects of increased CO_2 on wheat yields will normally be positive but the benefits vary with the prevailing temperature regime and availability of other inputs (water and nutrients). It is predicted that with the doubling of CO_2, the ambient temperature in India would increase by 3 °C and will affect both the area and productivity of wheat. The encounter of negative effects of varying temperature regimes with the benefits of increased CO_2 activity is, therefore, a critical issue for any assessment of wheat production under changing climate. It has been observed that an increase in temperature (about 2 °C) reduced potential wheat grain yields at most places. A net reduction in wheat production is anticipated due to a reduction in the growth period as a result of increased temperature. Besides, increased water requirements may be anticipated in all regions, which will highlight the importance of irrigation management in mitigating climate change.

7.7.4 Restrictions to Germplasm Exchange in New IPR Regime

The success of the green revolution in the early sixties and the later varietal improvement programs were dependent on the exchange of germplasm lines from exotic sources especially International Maize and Wheat Improvement Center (CIMMYT) that were used directly as a variety or as a donor parent in wheat improvement programs. However, due to the emergence of Intellectual Property Rights (IPR) as one of the major global issues, germplasm exchange between the countries is likely to be restricted, and this may influence the pace of genetic improvement in wheat. In the post-GATT scenario, the issue of Intellectual Property Rights gained significant importance and has become a major hurdle in germplasm exchange. Therefore, the Indian wheat program has to focus on prebreeding activities by utilizing unexploited elite sources like landraces, synthetics, and other available wheat genotypes for broadening the genetic diversity for higher yield potential.

7.7.5 Reduced Total Factor Productivity and Imbalanced Use of Fertilizers

The intensive tillage and over-exploitation of the natural resources i.e. soil and water, resulted in the situation where the benefits per unit input used in wheat cultivation are continuously declining, as evident from experimental as well as on farm trials. The earlier fertilizer recommendation of 120:60:40 kg NPK/ha has been enhanced to 150:60:40 or 150:60:60 kg NPK/ha in some soils. Over-mining of essential plant nutrients and burning of crop residues have increased widespread incidences of nutrient deficiencies. The deficiency of Zn in the rice–wheat system and S, Fe, Mn, and Bo from various pockets in the intensive cropping areas are severely reported. Farmers, in general, are not applying potash, and in some areas, soil status has dropped to such an extent that further nutrient mining may change soil nutrient status substantially. Potash and, to some extent, zinc are going to play a major role in sustaining and enhancing production and productivity. Further, an increase in fertilizer requirement in the future to get the same productivity level is also predicted in case of continuing unbalanced fertilization. The depleting soil organic carbon due to intensive tillage is also a very crucial factor at present that has reduced the water and nutrient-holding capacities of the soils. As a consequence, the frequency of irrigation has increased in those areas where water is not a limiting factor. This is further leading to deep percolation of water that often leaches down plant nutrients such as nitrogen and ultimately pollutes groundwater.

7.7.6 Yield Gaps at the Farm Level

The yield gaps were observed between the frontline demonstrations (FLDs) and the farmers' practice at their own fields. The zone-wise analysis indicated the maximum

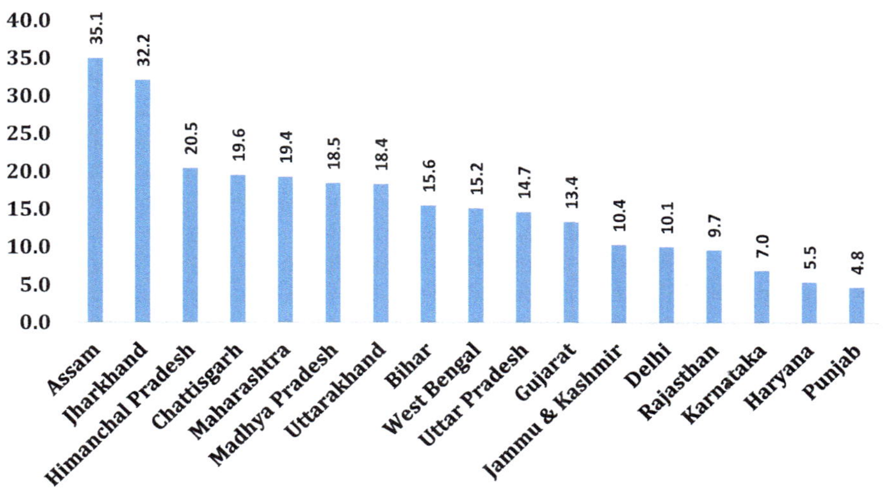

Fig. 7.3 Wheat yield gaps (%) in different states. (Source: AICRPW & B Progress report (Social Sciences), 2019–2020)

yield gap in CZ (13.26 q/ha) followed by NEPZ (9.46 q/ha), PZ (7.6 q/ha), NWPZ (7.35 q/ha), and NHZ (6.18 q/ha). The state-wise yield gap scenario (Fig. 7.3) indicated the highest yield gap in Assam (35.1%) and Jharkhand (32.2%), whereas a low gap was observed in the states of Haryana (5.5%) and Punjab (4.8%). This gap may be due to many related factors to seed, water, and technological adoptions, which can be bridged for enhanced overall production.

7.8 Future Strategies for Enhanced Production

The major thrust for increased wheat production will be in the agroclimatic regions of Indo-Gangetic plains, especially the northeastern plains zone, the plateau region of central India, and the hill regions. In potential regions, the production is expected to increase at a higher rate, whereas a slight increase in yield levels may be achieved in the remaining regions. These expectations are possible with technologies available at present; however, future research efforts would be focused on evolving new and innovative production technologies which can fit into the framework of changing wheat production scenario. Some key strategic issues are discussed for enhanced production levels.

7.8.1 Breaking Yield Barriers Through Genetic Enhancement

The existing yield plateau in the main wheat growing areas is a major concern, and enhanced efforts are being made to break the yield barriers. Genetic diversity will continue to be the key factor, and new approaches need to be adopted to enhance the yield potential of wheat genotypes. In this regard, the future strategy will be aimed to introgress desirable gene complexes from unexploited germplasm and wild progenitors for creating new variability. There is ample scope in prebreeding activities, which has been evident in the better performance of synthetic wheat genotypes and their hybridization with bread and durum wheat. The winter × spring wheat hybridization has been an attractive tool to introgress genes for disease resistance, and resistance/tolerance to various abiotic stresses like drought and cold, etc. Among the other approaches, the exploitation of hybrid vigor through the development of hybrids is very promising for yield gains in marginal climatic conditions.

7.8.2 Molecular Approach for Precision Breeding

One of the important goals of molecular biology is to expedite the breeding program with precision and accuracy to develop cultivars with specific traits within a short period of time. Significant achievements were made in the development of diverse types of molecular markers, the construction of molecular genetics, and physical maps with a reasonably high density of markers. The new molecular tools are potential armaments to revisit the origin and evolution of the wheat genome for the development of more precise QTL/gene analysis for the identification of markers associated with major economic traits for integration of marker-assisted selection (MAS) as a complementary strategy to conventional breeding methodologies of wheat improvement. The transformation approach, including newer techniques of CRISPER/Cas, can also be explored for improving popular varieties for missing traits like disease resistance, abiotic stress tolerance, and quality traits.

7.8.3 Tailoring Wheat Genotypes in Cropping System Perspective

Among the wheat-based cropping systems, the most extensively followed is rice–wheat cropping system. The continuity of this cropping sequence over the years showed a tendency for a decrease in wheat yield though the total productivity per year has increased. Other cropping systems like maize-wheat, potato-wheat, sugarcane-wheat, green pea-wheat, and cotton-wheat are being used in different parts. New resource conservation technologies have been developed for high output from wheat cultivation in a profitable and sustained manner. This needs to be accomplished with specific genotypes for cultivation under different RCTs. In this regard, the experiments showed a significant interaction between varieties and RCTS, suggesting differential behaviors of wheat genotypes. The hybridization and handling of segregating populations under different tillage options is the future thrust for developing tillage-specific genotypes for different agro-climatic zones.

7.8.4 Improved Varieties for Abiotic Stresses

The major abiotic stresses affecting wheat crops are heat, drought, salinity-alkalinity, waterlogging, etc. The wheat crop in the northern plains is exposed to higher ambient temperatures at the time of grain filling, whereas in central and peninsular parts, the crop is exposed to both early and late heat stress which significantly reduces productivity. Hence, breeding for heat tolerance is one of the major issues under prevailing stress environments. Most of the areas in the central and peninsular parts also experience water stress condition and suffers due to moderate to extreme drought conditions. The salinity and alkalinity condition is another threat that severely affects the productivity of wheat. Although resistant genotypes like KRL 19, KRL 210, and KRL 21319 have been released for these suppressive soils, more efforts are needed to reclaim and provide highly tolerant genotypes for these conditions. It has also been observed that a large area sown to wheat (about 3–6 million ha) is subjected to irregular waterlogging at early stages of growth, particularly where soils are sodic. Genotypes tolerant to waterlogging have been identified, which need to be incorporated in high-yielding background.

7.8.5 Disease-Resistance Breeding

Among biotic stresses, rusts pose a major threat to wheat production in India. A limited number of known resistance genes are being used in our breeding programs so far. Also, isolated efforts have been made in characterizing the phenomenon of slow rusting (a trait of durability) in Indian wheat germplasm. At this juncture, none of the genes or their compatible combinations in use is capable of providing complete protection from stripe and leaf rusts. Therefore, it becomes imperative to make use of other unexploited genes, especially those derived from alien sources, to keep the rust epidemics under check in the Indian subcontinent. Although the resistant sources have been identified in international collaboration, the need is to develop high-yielding varieties with resistance to these races. Leaf blight and flag smut, Karnal bunt are other diseases of importance that needs to be taken care of for wheat production and exports.

7.8.6 Access to Quality Seeds of Wheat

The varietal diversity under commercial cultivation plays a barrier to any probable pathogenic hazard. From 1967 to 1977, there were only two varieties, namely Sonalika and Kalyansona, which occupied the largest area in the country. In 1984–1985, there were 62 wheat varieties under the breeder seed program, of which indents of only 11 varieties were higher than 50 quintals, and a similar trend was continued till 2000 AD, although there was a mosaic of many promising varieties like HUW 234, HD 2285, HD 2329, WH 542, UP 2338, Raj 3765, PBW 343, and PBW 373. There was a paradigm shift in seed demand with the development of new varieties like GW 322, DBW 17, PBW 550, HD 2967, HD 3086, among which PBW 343, HD 2967, and HD 3086 have larger areas and are considered mega varieties. Nowadays, although the seed indent of new varieties, including DBW 187 and DBW 222, is more than 1500 q, a large area under few varieties is also a concern in the event of the spread of disease. For quicker replacement of old, susceptible, and otherwise uneconomical varieties, the research institutes of the Indian Council of Agricultural Research (ICAR), State Agricultural Universities (SAUs), and State Departments of Agriculture should have tuned to develop a strategy to distribute and popularize a large number of newly released cultivars. To provide quality seed to the farmers, an extensive breeder seed production program has been taken up under coordinated projects every year, and it has been further strengthened through the ICAR initiative of mega seed project involving SAUs and ICAR institutes to enhance seed research and production capabilities.

7.8.7 Conservation Agriculture

The new resource conservation technologies have shifted the thrust to conservation agriculture. These technologies are capable of placing seeds of wheat directly in the presence of residues left after the harvest of rice/wheat by the combines, which at present is being burnt. This can tremendously contribute to conserving soil health by preventing the loss of nutrients. Conservation agriculture will help to retain the residue on the surface of the soil which will act as biological tillage, conserve the soil moisture, release nutrients to the plant as and when required, reduce soil erosion, keep a check on weeds and provide enabling environment that will ultimately help in the sustenance of soil and crop productivity.

7.8.8 Integrated Water, Nutrient, and Weed Management

Water management is the key to the development of sustainable agriculture for both irrigated as well as rainfed areas for which issues like water conservation, watershed management, sprinkler, and drip irrigation and FIRB (Furrow Irrigated Ridge-till and Bed-planting) system of wheat cultivation needs to be addressed with respect to increased nutrients and moisture use efficiency, avoiding lodging, reducing seed rates, minimizing weed infestation, better residue management and thus reducing cost of wheat cultivation without affecting the productivity. Attention is also needed for the balanced use of chemical fertilizers, the use of biofertilizers, and other sources for improving soil structure and texture. Sustained yield and soil productivity can be accomplished with balanced nutrient addition using animal manures/ locally available organic manures with commercial fertilizers. Some of the measures suggested to improve the micronutrient status of soils under intensive cultivation include the use of a leaf color chart for saving nitrogen, in situ green manuring without additional water, diversification with pulses, residue incorporation/retention, balanced application of fertilizer including micronutrients and development of efficient micronutrient cultivars. Weeds play an important role in realizing the yield potential of any crop, and unfortunately, they are more resistant to abiotic stresses and their nutrient absorption capacity is also more than the wheat crop. As most of the area follows the wheat-based cropping system, there is a need to focus on integrated weed management and succession of weeds in a cropping sequence. Besides physical, cultural, and chemical means, biological weed control in wheat crops using plant pathogens, especially in the form of mycoherbicides, needs to be focused on in the future that will reduce the hazards of groundwater contamination and promote the food safety and protection of endangered species.

7.8.9 Diversification/Intercropping/Companion Cropping

Since the area under wheat is not going to expand further, there is a need to evolve suitable genotypes and production technologies for various synergistic and parallel intercropping/companion-cropping systems. Under irrigated conditions, opportunities for intercropping of wheat exist with autumn-planted sugarcane and potato. This could be achieved by establishing inter-institutional linkages and an effective extension network. Continuous use of rice–wheat system is depleting the soil health and lowering the water table. Hence there is also a need to opt for the possible replacement of rice by some other remunerative crops i.e., maize, baby corn, soybean, etc., through FIRB technology so that burden on the water is reduced. Alternatively, there is a possibility to intensify the system through the introduction of leguminous crops like moong bean, which can also improve soil nutrient status. This will reduce the cost of cultivation and enhance crop productivity and water use efficiency. The present efforts to diversify the rice–wheat system need to be enhanced through support from policymakers and extension workers.

7.8.10 Quality Improvement

In changing socioeconomic scenarios and wheat consumption patterns, high demand is expected for value-added wheat products of bread, biscuit, and pasta. With tremendous human resources and emerging food processing technologies, India has a large scope to develop instant food industries; thus, the Indian wheat program has to strengthen to meet the quality requirement of the domestic and international markets. Increasing global demand, value addition potential, better price in the market, and resistance to Karnal bunt make durum an export commodity. Although high-yielding varieties with better quality are available, efforts are in progress to create variability for beta-carotene, protein content, semolina recovery, and hectoliter weight, besides enhancing resistance against stripe, leaf, and stem rust. A systematic breeding program is also required to improve the nutritional and industrial quality of bread wheat. Food processing industries can take maximum benefit from superior varieties of wheat if procurement and processing of grains at the market level are attended to with special care. Small-scale industries for wheat-based value-added products needs to be encouraged in the rural sector to improve the livelihood of farmers and especially of the rural woman.

7.8.11 Strengthening Linkages and Policy Issues

Strengthening research capabilities of centers for efficient transfer of technology through frontline demonstrations is crucial to bridge the yield gap between the experimental field and farmer's field that can increase an additional wheat production to the tune of approximately 30.0 million tons by the adoption of currently available technologies. Cooperation from state agricultural extension units and Agricultural Universities would be a key factor in this process. Krishi Vigyan Kendras (KVKs) can also play a significant role in the rapid transfer of technologies and needs to be strengthened and their activities should be strongly linked to national wheat research centers. The availability of agricultural inputs in time for the timely sowing of wheat can lead to a remarkable increase in wheat production. Policy support is needed to popularize and spread eco-friendly production technologies such as zero tillage, crop diversification, production and marketing of high-quality product-specific wheat, installation of modern silos to prevent post-harvest losses of grains, and development of wheat-based rural industries and cooperatives for producing and marketing value-added products. Support for human resource development through training at national and international institutes to acquire expertise in advanced science and technologies is also crucial.

7.9 Conclusions

The present yield plateau in most of the wheat zones except NWPZ is the major concern for achieving the targeted wheat production of 140 million tons by 2050. However, bridging the yield gap between experimental and farmers' fields can solve the problem to a considerable extent. For future needs, the genetic enhancements of yield potential through the integration of conventional and molecular approaches with special reference to the changing climatic conditions, integrated management of resources, and incorporation of resistance genes to various biotic and abiotic stresses are very significant factors. The quality of wheat will continue to be one of the major foci of research. With support from farmers, policymakers, and extension units, it is expected that present technologies can be further refined and popularized so that wheat production can be enhanced to fulfill future demand.

References

Bhardwaj SC, Singh GP, Gangwar OP, Prasad P, Kumar S (2019) Status of wheat rust research and progress in rust management-Indian context. Agronomy 9(12):892

Bishnoi SK, He X, Phuke RM, Kashyap PL, Alakonya A, Chhokar V, Singh RP, Singh PK (2020) Karnal Bunt: a re-emerging old foe of wheat. Front Plant Sci 11:569057. https://doi.org/10.3389/fpls.2020.569057

Chand R, Pandey SP, Singh HV, Kumar S, Joshi AK (2003) Variability and its probable cause in natural populations of spot blotch pathogen (*Bipolaris sorokiniana*) of wheat (*T. aestivum* L.) in India. J Plant Dis Prot 110(1):27–35

DAC & FW (2021) Advance estimates of APY of food grain crops 2020-21. www.agricoop.nic.in

Duveiller E, Sharma RC (2009) Genetic improvement and crop management strategies to minimize yield losses in warm non-traditional wheat growing areas due to spot blotch pathogen Cochliobolus sativus. J Phytopathol 157:521–534

Gupta S, Saharan MS, Gurjar MS, Singh J, Bashyal BM, Aggarwal R (2020) Molecular detection of *Fusarium graminearum* causing head blight of wheat by loop mediated isothermal amplification (LAMP) assay. Ind Phytopathol 73(4):667–672

Gurjar MS, Aggarwal R, Jogawat A, Sharma S, Kulshreshtha D, Gupta A, Gogoi R, Thirumalaisamy PP, Saini A (2017) Development of real time PCR assay for detection and quantification of teliospores of *Tilletia indica* in soil. Indian J Exp Biol 55:549–554

ICAR (2018). https://www.thehindubusinessline.com/economy/agri-business/wheat-demand-to-touch-140-mt-by-2050-icar/article9677426.ece. Accessed 11 Jan 2018

Jain KBL (1994) Wheat cultivars in India: names, pedigrees, origins and adaptations. Research Bulletin No. 2. Directorate of Wheat Research, Karnal. 72 p

Joshi LM, Singh DV, Srivastava KD, Wilcoxson RD (1983) Karnal bunt - a minor disease that is now a threat to wheat. Bot Rev 49:309–330

Joshi LM, Singh DV, Srivastava KD (1986) Wheat and wheat diseases in India. In: Joshi LM, Singh DV, Srivastava KD (eds) The threat of Ug99 in South Asia. Problems and progress of wheat pathology in South Asia. Malhotra Publishing House, New Delhi, pp 41–68

Kashyap PL, Kumar S, Kumar RS, Sharma A, Jasrotia P, Singh DP, Singh GP (2020) Molecular diagnostic assay for rapid detection of flag smut fungus (*Urocystis agropyri*) in wheat plants and field soil. Front Plant Sci 11:1039. https://doi.org/10.3389/fpls.2020.01039

Kashyap PL, Kumar S, Sharma A, Kumar RS, Jasrotia P, Singh GP (2021) Diversity and genetic architecture of resistance to powdery mildew in wheat. In: Bisnoi SK, Khan H, Singh SK, Singh GP (eds) Breeding frontiers in wheat. Agrobios, Jodhpur, pp 207–239

Kiran K, Rawal H, Dubey H, Jaswal R, Bhardwaj S, Prasad P, Pal D, Devanna BN, Sharma T (2017) Dissection of genomic features and variations of three pathotypes of *Puccinia striiformis* through whole genome sequencing. Sci Rep 7:1–16

Kumar U, Joshi AK, Kumar S, Chand R, Röder MS (2010) Quantitative trait loci for resistance to spot blotch caused by *Bipolaris sorokiniana* in wheat (*T. aestivum* L.) lines 'Ning 8201' and 'Chirya 3'. Mol Breed 26(3):477–491

Kumar S, Bhardwaj SC, Gangwar OP, Sharma A, Naeela Q, Kumaran VK, Khan H, Prasad P, Miah H, Singh GP, Sharma K, Verma H, Forrest KL, Trethowan RM, Bariana HS, Bansal UK (2021) Lr80: a new and widely effective source of leaf rust resistance of wheat for enhancing diversity of resistance among modern cultivars. Theor Appl Genet 134:849–858

Manjunatha C, Sharma S, Kulshreshtha D, Gupta S, Singh K, Bhardwaj SC, Aggarwal R (2018) Rapid detection of *Puccinia triticina* causing leaf rust of wheat by PCR and loop mediated isothermal amplification. PLoS One 13(4):e0196409

Nagarajan S, Joshi LM (1975) A historical account of wheat rust epidemics in India and their significance. Cereal Rust Bull 3(2):29–33

Nayar SK, Prashar M, Bhardwaj SC (1997) Manual of current techniques in wheat rusts. Research Bulletin No. 2. DWR RS, Shimla, p 32

Prashar M, Bhardwaj SC, Jain SK, Datta D (2007) Pathotypic evolution in *Puccinia striifromis* in India during 1995-2004. Aust J Agric Res 58:602–604

Saharan MS, Kumar HMA, Gurjar MS, Aggarwal R (2021) Fusarium head blight of wheat in India-variability in pathogens associated and sources of resistance: an overview. Ind Phytopathol 74: 345–353. https://doi.org/10.1007/s42360-021-00358-8

Sendhil R, Kumari B, Khandoker S, Jalali S, Acharya KK, Gopalareddy K, Singh GP, Joshi AK (2021) Wheat in Asia – trends, challenges and research priorities. In: Kashyap PL, Gupta V, Gupta OP, Sendhil R, Gopalareddy K, Jasrotia P, Singh GP (eds) New horizons in wheat and barley research: global trends, breeding and quality enhancement. Springer, New Delhi. https://www.springer.com/gp/book/9789811644481

Singla J, Krattinger SG (2016) Biotic stress resistance genes in wheat. In: Wrigley CW, Faubion J, Corke H, Seetharaman K (eds) Encyclopedia of food grains. Elsevier, Oxford, pp 388–392

Sleeman WH (1839) Extracts from Major Sleeman's diary. Trans Agric Hortic Soc India 6:77–79

Maize and Millets

8

Sujay Rakshit, Prabhakar, and Pardeep Kumar

Abstract

Maize is an important cereal crop in India, which has the highest production and productivity in the world after rice and wheat. In India, cultivation is done on an area of 197.20 million ha with a global production of 1148 million tons and average productivity of 5.82 tons/ha productivity. The maize production in the country has increased from 2.0 to 30.24 million metric tons from 1949 to 2020 with an average annual growth rate of 3.90% and presently it occupies a 9.7-million ha area with a mean yield of 3.12 tons/ha. Similarly, various millet including sorghum, pearl millet, finger millet, and small millets (barnyard millet, proso millet, kodo millet, and foxtail millet) are grown throughout the year in the whole country. Millets can be grown in arid and semi-arid types of climate due to their less water requirement. Like maize, it can be used for feed, food and different industrial applications and due to its nutritional value it is called nutri-cereal. The historical view of production, area, and productivity of different millets and maize during the green revolution and post-green revolution indicated that there are tremendous changes. Production and productivity increased during the time. The present book chapter will cover the historical changes in terms of production, productivity and area for millets and maize with current technologies and future perspectives.

S. Rakshit (✉) · P. Kumar
ICAR-Indian Institute of Maize Research, Ludhiana, India
e-mail: s.rakshit@icar.gov.in

Prabhakar
AICRP on Small Millets, Bengaluru, India

P. K. Ghosh et al. (eds.), *Trajectory of 75 years of Indian Agriculture after Independence*, https://doi.org/10.1007/978-981-19-7997-2_8

Keywords

Maize · Barnyard millet · Proso millet · Kodo millet · Foxtail millet

8.1 Introduction

More crop diversity is present in India due to variations in soil, rainfall, and climate. Different coarse cereals like maize, sorghum, pearl millet, finger millet, and small millets (barnyard millet, proso millet, kodo millet, and foxtail millet) are grown throughout the country in diverse ecology. Further, sorghum, pearl millet, finger millet, maize, and small millet are also called nutri-cereals due to their better nutritional value. Among the coarse cereals, maize is the most important cereal crop in the world with huge yield potential among all the cereal crops; therefore, it is called as queen of cereals. Maize, with its wider adaptability, is cultivated in all parts of the country during all three seasons. It is mainly a kharif crop but in some states, viz., Bihar, West Bengal, and Peninsular India, it is grown as a winter/rabi crop and in some states such as Punjab, Haryana, and western Uttar Pradesh as a summer crop also (Rakshit and Chikkappa 2018). It is used as both food and feed. Further, every part of maize has some economic value like grain, leaves, stalk, tassel, and cob which can be used to produce a large number of food and non-food products. In addition, maize is an industrial crop and is used as raw material in several industries like starch, oil, alcoholic beverages, food sweeteners, pharmaceuticals, cosmetics, textile, paper, film, tyre, food processing, and bio-fuel to develop thousands of products. Apart from these specialty corn becoming more popular in the last few years, it includes sweet corn, baby corn for ensuring livelihood and green fodder security in peri-urban areas, and popcorn as a nutritional alternative snack (Kumar et al. 2012).

Millets refer to cereal grasses grown primarily as grain and feed crops. Millets require less water than any other grain crop and provide assured harvests in arid, semi-arid, and mountainous regions of tropics and sub-tropics where monsoon failure and droughts are frequent, soil fertility is poor, and land terrain is challenging. Millet grain forms the main staple for farm households in several old-world countries and among the poorest people. Millet straw is a valuable livestock feed, besides having other uses as a building material and fuel, in those farming systems. There are major millets—sorghum and pearl millet—and several small grain millets—finger millet (ragi), foxtail millet (kangni), kodo millet (kodo), proso millet (cheena), barnyard millet (sawan), brown top millet, and little millet (kutki). These are suitable for fragile and vulnerable environments and sustainable and green agriculture. Promoting these crops can lead to more efficient natural resource management and a holistic approach to sustaining precious agro-biodiversity.

Maize and millet grains are nutritious with good quality protein, rich in minerals, dietary fibre, phytochemicals, and vitamins. A cursory look at the proximate composition of various food grains (Table 8.1) would reveal the distinct nutritional superiority of millets over major food grains such as wheat and rice. Millet, by its

Table 8.1 Nutritional composition of millet grains vis-a-vis fine cereals

Crop	Protein (g)	Carbohydrates (g)	Fat (g)	Fibre (g)	Minerals (g)	Calcium (mg)	Phosphorus (mg)
Wheat	11.8	71.2	1.5	12.9	1.5	41	306
Rice	6.8	78.2	0.5	5.2	0.6	10	160
Maize	11.1	66.2	3.6	2.7	1.5	12.4	348
Sorghum	10.4	72.2	1.9	12.0	1.6	25	222
Bajra	11.6	67.5	5.0	16.0	2.3	42	296
Finger millet	7.3	72.0	1.3	18.8	2.7	344	283
Proso millet	12.5	70.4	1.1	14.2	1.9	14	206
Foxtail millet	12.3	60.9	4.3	14.0	3.3	31	290
Kodo millet	8.3	65.9	1.4	15.0	2.6	27	188
Little millet	8.7	75.7	5.3	12.0	1.7	17	220
Barnyard millet	11.6	74.3	5.8	13.5	4.7	14	121

Source: Nutritive value of Indian foods, NIN 2007; *MILLET in your meals* unique grain properties, exhibits considerable opportunities for diversification of its food use through processing and value addition. The regular use of millet can lead to significant health benefits across all sections of society. The area, production and productivity of coarse cereals also underwent a significant change since independence in India (Table 8.2).

8.2 Maize

Maize is the third most important food crop after rice and wheat, with the highest production and productivity globally (Kumar et al. 2012). It is more versatile due to its diverse applications such as food, feed, fodder and in the recent past as a source of bio-fuel (Rakshit and Chikkappa 2018). Further, maize has wider adaptability and is grown in more than 170 countries worldwide, including tropical, sub-tropical, and temperate regions. Globally, maize is being cultivated in an area of 197.20 million ha with 1148 million thousand tons of production and 5.82 tons/ha productivity (FAOSTAT 2019). The production of maize increased from 265 to 1148 million thousand tons from 1970 to 2019, with an average annual growth rate of 3.41% (FAOSTAT 2019). Out of world maize production, India produces 28.8 million tons in an area of 9.5 million ha with 3.0 tons/ha productivity. Among the total production of maize, 60% is used for feed, 14% for industrial application, 13% for direct food, 6% for processed food, and around 7% for export and other purposes (Rakshit and Chikkappa 2018). India at sixth rank in terms of production after the United States (346 metric tons), China (260 metric tons), Brazil (102 metric tons), Argentina (51 metric tons), and Ukraine (35.9 metric tons). Both area and production of maize have increased in India from 1950 to 2019 with 3.3–9.5 million ha area and 1.7 metric tons to 28.8 metric tons, respectively. This impressive growth of maize in India is the result of higher productivity and cultivation area expansion in different states like Uttar Pradesh, Bihar, Rajasthan, and Madhya Pradesh. However, the maize area increased in the peninsular region after 1980s which represents around 40% and 52% in terms of area and production, respectively. In India, maize is a *kharif* season crop but its area of cultivation is increasing in *rabi* and *spring* seasons also (Yadav et al. 2015). Maize has recorded the highest Compound Annual Growth Rate (CAGR) of 0.68% in yield and 0.73% in production among all the cereals.

Area and Production status of maize The total maize production of India was 1.73 million tons in 1950–1951 which reached double to 3.46 million tons by 1958–1959 on account of a 35% increase in area and 48% in yield (Yadav et al. 2015). Further, the annual maize area increased at 109 ha/year while the yield increased by 24.7 kg/ha/year and this figure reached in 1960–1969 at 168 ha/year and 7.4 kg/ha/year in terms of area and productivity, respectively (Fig. 8.1). But in the 1970s, the productivity gain was negative at −6.9 kg/ha/year while yield enhanced significantly again 29 kg/ha/year in the 1980s with a stable area in both decades. After that, the yield reached maximum growth of 37 kg/ha/year in the 1990s and 46 kg/ha/year in the next decade. Presently yield enhancement is over 60 kg/ha/year.

Table 8.2 Area (million ha), production (million tons) and productivity (kg/ha) of coarse cereals in India

Crop/year	Category	1955–1956	1965–1966	1975–1976	1985–1986	1995–1996	2005–2006	2015–2016	2019–2020
Maize	Area	3.70	4.80	6.03	5.80	5.98	7.59	8.69	9.57
	Production	2.60	4.82	7.26	6.64	9.53	14.7	21.80	28.77
	Productivity	7.0	10.1	12.0	11.5	16.0	19.4	25.10	30.06
Sorghum	Area	17.36	17.68	16.09	16.10	11.33	8.68	6.08	4.48
	Production	6.73	7.58	9.50	10.20	9.33	7.63	4.24	4.38
	Productivity	387	429	591	633	823	880	697	998
Pearl millet	Area	11.34	11.97	11.57	10.65	9.32	9.58	7.13	6.77
	Production	3.43	3.75	5.74	3.66	5.38	7.68	8.07	8.90
	Productivity	302	314	496	344	577	802	1132	1315
Finger millet	Area	2.30	2.70	2.63	2.41	1.77	1.53	1.02	0.97
	Production	1.85	1.33	2.80	2.52	2.50	2.35	1.82	1.68
	Productivity	800	492	1064	1049	1410	1534	1784	1731
Small millets	Area	5.34	4.56	4.67	3.16	1.66	1.06	0.65	0.46
	Production	2.07	1.56	1.92	1.22	0.78	0.47	0.39	0.34
	Productivity	388	341	412	386	469	443	600	740
All millets[a]	Area	36.34	36.91	34.92	32.32	24.08	20.85	23.57	22.25
	Production	14.08	14.22	19.96	17.60	17.99	18.13	36.32	44.19
	Productivity	469	394	641	603	820	915	1541	1986

[a] AgriculturalCensus, Directorate of Economics and Statistics, Department of Agriculture & Cooperation, Government of India

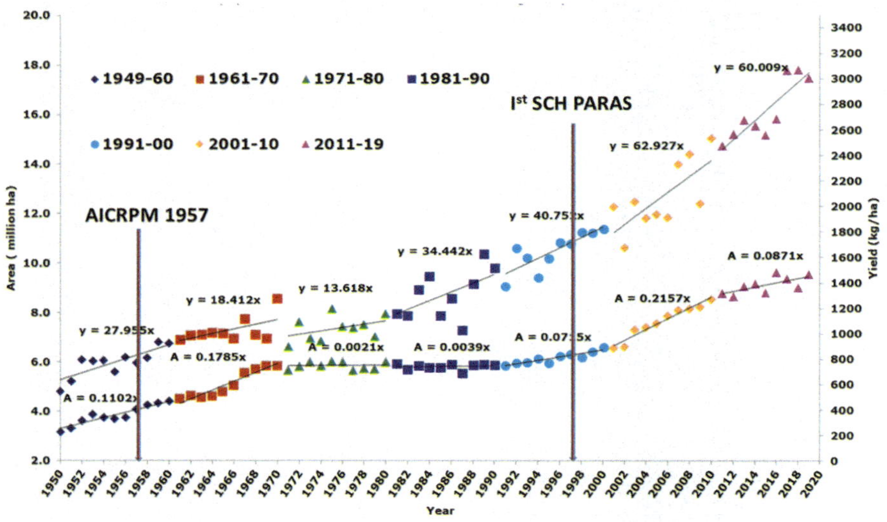

Fig. 8.1 The decadal growth rate of maize area and productivity in India during 1950–2020

8.2.1 Historical Perspective of Maize Improvement in India

8.2.1.1 Green Revolution Period

Maize has the highest productivity among cereals. Considering this fact and with the support of the Rockefeller Foundation, ICAR established the All India Coordinated Maize Improvement Project (AICMIP) in 1957. AICMIP was the first project of its kind among all the crops. The major objective of AICMIP was the development and release of widely adapted high-yielding cultivars, which are resistant to major diseases and insect pests, through extensive multidisciplinary and multi-location testing. Further, low temperature during the winter season in the north Indian centre was the major hindrance to the advancement of breeding material. To address this issue, Winter Nursery Center was established in Hyderabad in 1962 for the advancement of the breeding material in off-season specifically in the winter (rabi) season. Based on the AICMIP network, a total of five agro-climatic zones were established in the entire country.

8.2.1.2 Improvement of Maize

After the establishment of AICMIP in 1957, major emphasis was given to the introduction of exotic germplasm to strengthen the maize improvement programme. In 1961, four double-cross hybrids namely Ganga 1, Ganga 101, Ranjit, and Deccan were released for cultivation using the exotic inbred lines. After that, a set of double top cross hybrids were released, viz., Hi-starch and Ganga Safed 2, Ganga 5, etc. to make seed production easier. The increased demand for maize seeds and some limitations of hybrid seed production in the Indian maize programme realized to took some strong decisions focusing on base germplasm with two major objectives

(1) to release synthetic/composite varieties as cultivars for commercial cultivation where farmers could use seeds from his harvest up to 3–4 years with a minimum compromise with productivity, and (2) to improve the population base enabling extraction of more productive inbreds towards hybrid development. Using the exotic and native germplasm populations through the systematic crossing, six composite cultivars, viz. Vijay, Kisan, Amber, Vikram, Sona, and Jawahar were released for cultivation in 1966. Among these Vijay was more popular not only in India but also in adjoint countries like Pakistan and Nepal with different names (Dhillon and Malhi 2006). In the 1960s two mutants namely, *Opaque-2* (O_2) with high lysine and tryptophan and *flory-2* (fl_2) with high methionine were introduced which opened the research area on quality protein maize (QPM). Using the *Opaque-2* (O_2) mutant, three QPM composites with soft endosperm were released, viz. Shakti, Rattan, and Protina but not became popular due to more susceptibility of diseases and storage pests. Later on, Shakti 1 was released with hard endosperm.

8.2.1.3 Post-green Revolution Period

The AICMIP project was upgraded into the Directorate of Maize Research in 1994 and then full-fledged institute the Indian Institute of Maize Research in 2015. Presently institute carries all the basic, strategic, and applied research for the overall improvement of maize in India and big AICRP on maize network with 30 regular, 2 ICAR, and 30 voluntary centres over the country (Annual maize progress report 2020).

Hybrid Development in Maize

Towards the development of productive hybrids work focus was made on heterotic pools to extract the promising inbred lines from different pools to develop high-yielding single cross hybrids. In this series introgression of the temperate germplasm into tropical material was also carried out. As a result, a large number of inbred lines were derived and screened for yield and adaptability. Using stable and high-yielding inbred lines pair, the first single cross hybrid "Paras" by Punjab Agricultural University. Further, it was observed that single cross hybrids were more productive than DC or DTC hybrids. During the twentieth century, the major focus was given to developing the single cross hybrid for normal grain, QPM, and other specialty corns. Maize breeders developed several normal maize single cross hybrids and dozens of QPM using hard endosperm QPM inbred lines during that period. Around 50,000 hybrids were developed through the public/private sector till 2019 for different states. Among the 500 cultivars released till 2019, a very less number of OPVs, and currently, 65% of the maize area is covered under hybrids. In 2008, the first marker-assisted selection (MAS) derived QPM hybrid, QPM Vivek-9 was released. The series of MAS-derived QPM hybrids, viz., Pusa HM-8 Improved (AQH-8), Pusa HM-9 Improved (AQH-9), and Pusa HM-4 Improved (AQH-4) were released by IARI, New Delhi. Other than these specialty corn like sweet corn, popcorn, and baby corn is one of the important aspects whose demand is increasing every year. Further, there is a green fodder shortage in India and this can be addressed through the cultivation of baby corn and sweet corn as by-products and no additional land required for it (Kumar et al. 2012). The food habit of humans is changing due to urbanization and improved economic status, in this context specialty corn has become more popular in peri-urban areas of the country.

Productivity Challenge

The productivity of maize growing in the rabi/spring season (4.1 tons/ha) is nearly double of kharif season (2.3 tons/ha). However, the area under kharif maize represents 82.3% and among this 75% area is under rainfed conditions. While rabi and spring maize is predominantly grown in irrigated conditions. The kharif maize is adversely affected by the changes that occurred in the climate and extreme weather conditions like uneven rainfall, drought, flooding, and high temperature. Other than these some biotic stresses also reduced the kharif maize productivity, viz. leaf blights, post-flowering stalk rot (PFSR), banded leaf and sheath blight (BLSB), ear rots (ER), downy mildew (DM), borers, and weeds.

Hybrid Production Technology

The maize production has been increased significantly through an increase in area and yield gains. Further, yield gains in any crop can be achieved through both genetic gains and better agricultural practices. Hence, location-specific agronomic practices have been developed with an emphasis on sowing date, plant population, and fertilizer application to enhance maize productivity. The optimum plant population is 74–80,000/ha, 66,000/ha, and 80–90,000/ha for kharif irrigated, rainfed and rabi maize respectively. Nutrient management such as N-P-K @ 150:75:75 in irrigated, 120:40:40 for rainfed, and 250:105:105 for rabi season along with 10 tons/ha FYM reported getting more yield. On the other hand, to apply nitrogen in two split doses by the traditional method, application 4–5 times give the best result. Further, resource conservation technologies (RCT) reduced the timeline window between kharif harvest and sowing of rabi crop, enhancement of organic carbon, nutrient use efficiency, and conserving moisture. After harvesting kharif crop, farmers waste their time preparing land but zero tillage technology provides the opportunity for maize sowing without land preparation by a maize bed planter and ferti-cum-seed planter.

Plant Protection

There are several chemical control measures available to protect maize crops from diseases and insect pests but now time to include eco-friendly control measures. Eventually, releasing of pest resistance cultivars is one of the major priority areas since the beginning and continued up to date. In maize, biopesticides are relatively effective to control pests but it is not a regular practice like chemical control.

Mechanization in Maize

Due to the continuous shortage of agricultural labour, mechanization help with timely farm operations, saving natural resources, low cost of cultivation, good quality product, and improving the living standard of farmers. In Indian conditions land preparation activities done by machinery and some extent sowing also but harvesting and post-harvest handling are mainly manually based.

Post-harvest Handling

Around 20% of losses are reported at the storage level and 5% during harvesting, threshing, cleaning, and harvesting, etc. Further, the post-harvest maize quality is majority dependent on weather and storage conditions. The proper grain storage and drying facility for maize crops are not available in India. Hence, maize grains are damaged during normal storage condition. However, some secondary plant metabolites have been identified to control storage pests such as leaves of *Ixora coccinea*, *Ageratum conyzoides*, and *Erythrina indica*.

Maize Value-Addition

Dry milling, wet milling, and alkali processing are the three major processes of maize grain as raw materials. After processing the end products use to produce more than thousands of products for daily life by different enterprises. The products prepared after dry milling are grids, coarse meal, germ, a fine meal, flour, and hominy feed. The different ready-to-eat snacks such as cornflakes, porridges, and wallpaper paste are prepared using the grids. There are a series of maize-based ready-to-cook and ready-to-eat products have been developed by the Rajendra Prasad Agricultural University, Dholi and the University of Agricultural Sciences, Manyda. Thus, maize is having huge potential to use in processing and value addition by involving women and youth. Many products like pasta, vermicelli, cornflakes, Jelly, ice cream cones, sewaian, flour, dalia, suji, multigrain flour, maize grid, namkeen products, etc. can be developed.

Potential Scaling Up of Proven Technologies

The main challenge to enhancing maize productivity is that the developed technologies are not reaching the stakeholders. Further, youth is the biggest strength in the Indian economy but the biggest challenge is to retain them in agriculture. However, they can be potentially utilized for the effective dissemination of knowledge and out-scaling innovations to fulfil the productivity gap. For this, a detailed action plan needs to be developed involving all stakeholders, viz., planners, researchers, farmers, processors, and traders in Public-Private-Producer-Partnership (PPPP) mode. The challenge of increased productivity may be attained by enhancing breeding efficiency, seed production of single-cross hybrids, production and protection technologies, development of maize value chain, and policy interventions. The strategies to enhance the breeding efficiency through strengthening the pre-breeding activity, genetic enhancement for stress tolerance, use of frontier technologies for enhancing the genetic gain like doubled haploids, MAS, and development of genetically modified maize with their adoption.

Future Strategy

There is tremendous maize growth gained in terms of production after the green revolution through the adoption of improved technologies. Further, maize demand is expected to rise by 45 million metric tons by 2030 and the major users of this demand would be the poultry and piggery sector, processed industry, dairy sector, and biofuel production. Currently, high-yielding single cross hybrids are available to achieve the production target in 2025 to fulfil the demand. In addition, focussed research programmes are needed on the diversification of germplasm, development of climate-resilient hybrids, application of advanced tools, extension, and popularization of improved technologies and production and protection practices to touch the target production.

8.3 Sorghum

Sorghum [*Sorghum bicolor* (L.) Moench] is the fifth most important cereal crop after wheat, rice, maize, and barley across the world. It is mostly cultivated in the arid and semi-arid tropics for its better adaptation to drought, heat, salinity, and flooding. It is the main staple food for the poorest and most food-insecure people of the world. Sorghum is reported to be cultivated across 105 countries representing 41.1 million ha with an average production of 58.6 million tons.

Sorghum is the fourth most important cereal crop in India. This crop was one of the major cereal staples during 1950s and occupied an area of 17.36 million ha but has come down to 4.48 million ha in 2020. The decline has serious concerns about the cropping systems and the food security of these dryland regions of the country. However, the productivity has increased from 387 kg/ha in 1955–1956 to 1018 kg/ha in 2019–2020, with a threefold increase. Though kharif sorghum yield growth rates were relatively higher, it could not offset the declining growth rates in production, as the growth rates in kharif sorghum area were negative and high. Just the opposite is true in the case of rabi sorghum where the area decline was not sufficient to undermine the yield growth, thus resulting in positive production growth rates. The overall increase in productivity of kharif is far more than rabi sorghum. However, the loss in both area and production is greater in kharif sorghum than in rabi. The coverage with high-yielding varieties (HYVs) of sorghum is nearly 80% in kharif and the potential under moderate input is also high (4–6 tons/ha) (Figs. 8.2 and 8.3).

Sorghum genetic resources At the global level, sorghum germplasm collections consist of approximately 168,500 accessions; the largest collection (21% of global total) is held at the International Crops Research Institute for the Semi-Arid Tropics (ICRISAT), Patancheru, India. The total accessions consist of *18%* landraces, old cultivars, *21%* advanced cultivars breeding lines, *and 60%* mixed categories of unknown material, while very few are wild relatives (Upadhyaya et al. 2009). ICAR-Indian Institute of Millets Research (IIMR) is one of the National Active Germplasm Sites (NAGS) to act as a national repository for sorghum germplasm in India. The objectives of the Millets Gene bank are collection, augmentation, conservation, characterization, evaluation, documentation, distribution, and utilization of millets genetic resources (Elangovan 2020).

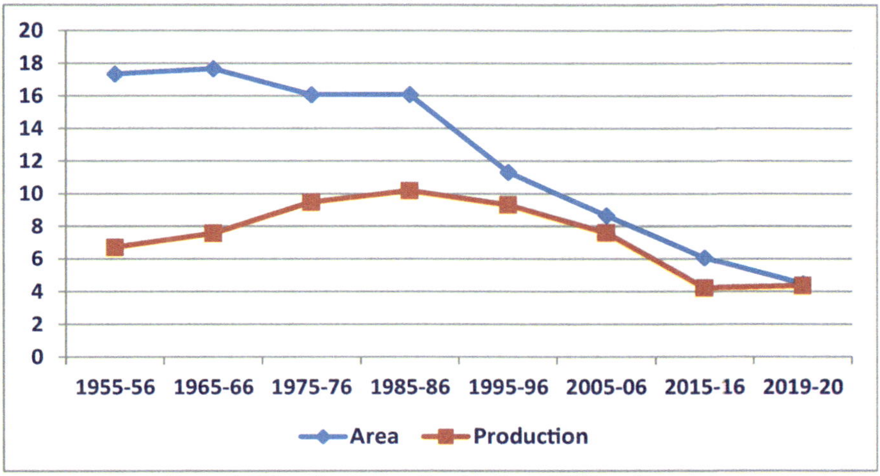

(Area: Million hectares; Production: Million Tonnes)

Fig. 8.2 Area and production of sorghum in India from 1955–1956 to 2019–2020 (Area: million ha; Production: million tons)

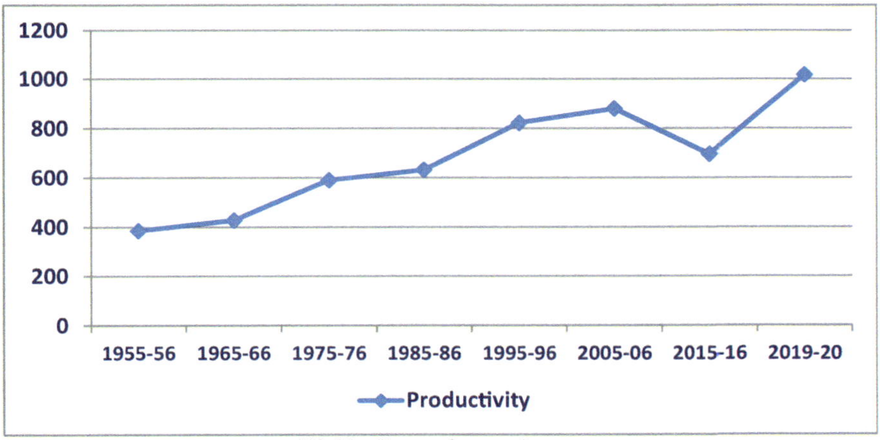

(Yield : Kg per hectare)

Fig. 8.3 Yield of sorghum in India from 1955–1956 to 2019–2020 (Yield: kg/ha)

8.3.1 Historical Perspective of Sorghum Improvement in India

8.3.1.1 Green Revolution Period

Focused sorghum research in India started with the establishment of the Project on Intensification of Regional Research on Cotton, Oilseeds and Millets (PIRRCOM) in 1958. Under the PIRRCOM sorghum research was led by the Indian Agricultural Research Institute (IARI), New Delhi. In 1966 the sorghum research was shifted from New Delhi to Hyderabad as a part of the IARI Regional Research Station. Realizing the success of hybrid sorghum in the United States of America, in 1962, the Indian Council of Agricultural Research (ICAR) launched the Accelerated Hybrid Sorghum Project. In December 1969, All India Coordinated Sorghum Improvement Project (AICSIP) was launched from the existing IARI RRS in Hyderabad.

8.3.1.2 Post-Green Revolution Period

Subsequently in 1987 a full-fledged "National Research Centre for Sorghum (NRCS)" was established which has evolved into the Indian Institute of Millets in 2015. Currently, AICSIP functions with a total of 20 centres spread across nine states.

Grain Sorghum Improvement

Sorghum improvement till the 1960s focused on selections from local landraces, which were tall with low harvest index, photosensitive, late maturing after cease of monsoon, and with localized adaptation. With the launching of the Accelerated Hybrid Sorghum Project through the Rockfellar Foundation a wide range of germplasm was made available in India. This led to significant improvement principally through the manipulation of plant height and maturity. Through these coordinated efforts, the Indian National programme over the years released 41 hybrids and 42 open-pollinated varieties under the kharif sorghum (Rainy season), Rabi sorghum (Post-rainy season), sweet sorghum, and forage sorghum.

Hybrid Sorghum Development in Kharif

Sorghum is one of the crops in which heterosis could be exploited. The Adoption of the first commercial hybrid (CSH 1) in India over much of the rainy season sorghum area, while local varieties confined to fairly narrow specific environmental niches is a testimony to the wide adaptability of hybrids over the varieties (Rao 1982). Indian public sector agricultural research agencies have been breeding improved sorghum genotypes since the early part of the twentieth century. Hybrid sorghum research in India started in the early 1960s. The major achievement of the Accelerated Hybrid Sorghum Project was the availability of male sterile (MS) lines from the USA and diverse germplasm from Africa and southeast Asia. Using these MS lines, three hybrids, viz., CSH 1, CSH 2, and CSH 3 were released during the first decade of hybrid sorghum research. Out of these CSH 1, released in 1964, gained much popularity among farmers. In subsequent decades limitations of hybrid seed production and low stover yield of released hybrids were addressed with the release of CSH 4, CSH 5, and CSH 6. Out of these, CSH 6 with its early maturity gained much popularity. Among the kharif hybrids CSH 1, CSH 5, CSH 6, CSH 9, CSH 14, and CSH 16 need special mention as CSH 5 and CSH 6 had a yield potential of 3.4 t/ha which was raised to 40 t/ha in CSH 9 and further raised to 41.0 t/ha in CSH 16, CSH 23, CSH 25 and CSH 30 with distinct superiority in grain and fodder quality. The release of more than 40 hybrids at the national level and several at state level is a standing testimony to the success of the Indian sorghum improvement programme

(Prabhakar et al. 2015). So far, outstanding and significant progress has been achieved in the case of kharif hybrids, but a great deal is yet to be done to exploit heterosis in breeding hybrids for rabi cultivation.

Varietal Improvement in Kharif Sorghum

The first outcome of focused sorghum breeding in India was the release of an early maturing variety, CSV 1 in 1968. The next decade witnessed the release of five varieties, viz. CSV 2 to CSV 6. Out of these CSV 2 and CSV 3 were of early type, while CSV 4 and CSV 5 were dwarf, and CSV 6 was a relatively tall variety. Zera-zera landrace from Ethiopia was brought into use to incorporate resistance against biotic stresses. Further tan plant pigment got incorporated in all *kharif* nurseries to confer resistance against leaf diseases. During the 1980s, five varieties, like SPV 462, CSV 9 to CSV 13 (K and R) were released. Further using SPV 462 and CSV 13 in crossing programme a high-yielding dual-purpose variety, CSV 15 was released in 1996. Two new varieties were developed from the derivatives of crosses involving this variety, viz. early maturing CSV 17 for light soils and drought-prone areas and improved dual-purpose variety CSV 20. This was followed by the release of two more improved dual-purpose varieties, viz. CSV 23 and CSV 27. Besides, central releases, there are several varieties released for specific agroecologies of different states (Prabhakar et al. 2015).

Varietal Improvement in Rabi Sorghum

The most popular *rabi* variety, M35-1 was released in 1969 from Mohol, Maharashtra. It has remained popular among farmers over the past five decades for its stable performance under rainfed situations with above-average yield and bold lustrous grains. Most of the present-day improved varieties are the result of pure-line selection practiced among the local/popular varieties and their crosses. The popular varieties have lustrous, bold, and globular grains. At the national level, the first *rabi* variety CSV 7R was released in 1974. Subsequent releases are CSV 8R, Swati, CSV 14R, Sel 3, Phule Yashoda (CSV 216R), CSV 18R, CSV 22R, CSV 26R, and CSV 29R. Besides these, the varieties have also been released by states at the regional level. Soil depths play an important role in *rabi* sorghum growing ecologies. In the recent past efforts have been made to develop varieties adapted to specific soil situations (shallow, medium, and deep) (Prabhakar et al. 2015).

Hybrid Development in Rabi Sorghum

Some of the early *rabi* hybrids released are CSH 7R and CSH 8R in the year 1977 and CSH 12R in 1986. Lack of acceptable grain quality with adaption to different *rabi* agro-ecological situations among release hybrids was the principal reason behind the low popularity of *rabi* hybrids among farmers. The second phase of *rabi* sorghum breeding with emphasis on hybrid cultivars, initiated in the late 1980s, resulted in central release of CSH 13R, CSH 15R, CSH 19R in 1990s and CSH 39R in 2019. None of the hybrids gained popularity due to inferior grain and fodder quality as compared to M 35-1. Lack of diversity, grain lustre, and sensitivity to low temperatures and terminal moisture stress of the parental line are the principal bottleneck in *rabi* hybrid breeding (Rao 1982).

Forage Sorghum Improvement
Forage sorghum, with an area of 3.0 million ha area in India, is principally cultivated in Punjab, Haryana, Delhi, western and central Uttar Pradesh, and adjoining areas of Madhya Pradesh. In these states, it is grown during *kharif* and summer seasons, either as single-cut (mostly in *kharif*, as rainfed) or as a multi-cut (summer and *kharif*) forage crop. The major objectives in forage sorghum breeding are to develop varieties both for single-cut and multi-cut with high tonnage, better quality, good seed yield, and resistance to insect pests and diseases. Concerted breeding efforts for the improvement of forage sorghum were initiated in 1970 under AICSIP. Multi-cut hybrids developed at Pantnagar, CSH 20MF and CSH 24MF have improved fodder yield and quality. Some of the recent single-cut forage sorghum varieties are CSV 30F and CSV 32F. The latest forage cultivars possess improvement in terms of resistance to leaf spot diseases, stem borer, and seed yield. Besides these several private-sector hybrids are also popular for multi-cut forage.

Sweet Sorghum Improvement
Sweet sorghum, similar to grain sorghum except for its juice-rich sweet stalk, is considered to be a potential bio-ethanol feedstock, and is expected to meet food, feed, fodder, fuel, and fibre demands. Some sweet sorghum lines contain 15–23% soluble fermentable sugar (by comparison, sugarcane has 14–16%). Early efforts towards sweet sorghum improvement were made at Nimbkar Agricultural Research Institute (NARI), Phaltan, Maharashtra. Sweet sorghum research under AICSIP started in 1989. Concerted research efforts during the last two decades have resulted in excellent sweet sorghum varieties and hybrids for use in ethanol production and for use as green/dry fodder. The productivity ranged between 40 and 50 tons/ha. Stalk yields obtained during *rabi* are 30–35% less with reduced sugar content than *kharif* and summer-grown crops.

8.3.2 Future Strategy

Sorghum improvement efforts have succeeded in increasing the productivity of *kharif* sorghum but could not impact much in *rabi* sorghum. Sorghum as a health food is another area that needs concerted efforts to diversify sorghum uses. Sorghum production can only be enhanced with appropriate policy support by the government. The various factors leading to the decline of sorghum crop in Indian agriculture is a matter of concern. Sorghum-based agricultural systems need to withstand biotic and abiotic stresses because of their cultivation mostly in unfavourable soil and climatic conditions (Prabhakar et al. 2015). Further, they also need to adjust to changing economic (prices and income) and policy-induced stresses as has been the case in India where subsidized wheat and rice are supplied through the public distribution system. With this background in view, there is a need to embark on future approaches for improvement. Promotion of genetic diversity, cropping system stability and economic advantage or parity will become the major criteria. The genetic approaches should promote genetic diversity and cropping system performance and stability.

Climate change is expected to influence the sorghum area and its importance globally, in addition to the biotic and abiotic challenges. Considering all these points, crop improvement research in sorghum needs to be oriented towards genetic and cytoplasmic diversification for high yield and large grain, shoot fly and grain mould resistance, drought and salinity tolerance, post rainy season adaptation, sweet stalk traits, and grain micronutrient density. Grain and stover quality needs special attention to enhance the market value. Concerted efforts to diversify sorghum as health food are another area that needs attention. Sorghum production can only be enhanced with appropriate policy support by the government. In India, with the inclusion of sorghum and other millets in the national food security mission, there is a ray of hope that sorghum will regain a new place in the food basket of the country.

8.4 Pearl Millet

Pearl millet [*Pennisetum glaucum* (L.) R. Br.] is the most widely grown drought-tolerant warm-season coarse grain cereal cultivated on 30 million ha worldwide. It is grown mainly in the arid and semi-arid tropical regions of Africa (18 million ha) and Asia (10 million ha) with India accounting for the largest area (>9 million ha). It is also consumed as feed and fodder for livestock and is the sixth most important cereal crop in the world next to maize, rice, wheat, barley, and sorghum. In India, pearl millet is the fourth most widely cultivated food crop after rice, wheat, and maize. It occupied an area of 6.93 million ha with an average production of 8.61 million tons and productivity of 1243 kg/ha during 2018–2019.

Pearl millet is cultivated in regions with characteristically low and erratic rainfall, high mean temperature, high potential evaporation, and infertile, shallow soils with poor water holding capacity, i.e. under the most adverse agro-climatic conditions where other major crops like maize and sorghum fail to produce economic yields. Despite this, pearl millet has a remarkable ability to respond to favourable environments because of its short developmental stages and capacity for high growth rate, thus making it an excellent crop for the short growing season and under improved crop management.

Pearl millet is critically important for food and nutritional security. It possesses several advantages such as early maturing, drought tolerance, minimal inputs, and is mostly free from biotic and abiotic stresses. Its grains have high protein content, a balanced amino acid profile, and high levels of iron, zinc, and insoluble dietary fibre. Pearl millet is gluten-free and retains its alkaline properties even after being cooked, which is ideal for people suffering from gluten allergy and acidity.

8.4.1 Coordinated System of Testing

8.4.1.1 Green Revolution Period

Research on pearl millet improvement in India is carried through the All India Coordinated Research Project on Pearl Millet (AICRP-PM) which is a continuing central plan project established in 1965 by Indian Council of Agricultural Research (ICAR) under the name of All India Coordinated Millet Improvement Project (AICMIP) with its headquarters at the Indian Agricultural Research Institute, New Delhi. The headquarters of the project were shifted to Pune in 1977.

8.4.1.2 Post-green Revolution Period

Later on, pearl millet was separated from the rest of the millet crops and the All India Coordinated Pearl Millet Improvement Project (AICPMIP) was established in 1985 with its headquarters at Pune as an independent Coordinated project. The ICAR shifted, in July 1995, the headquarters of AICPMIP to Jodhpur in the state of Rajasthan, the state which occupies nearly half of the pearl millet area of the country. The AICRP on Pearl millet headquarters is located at ARS, Mandor, Jodhpur which now comes under Agriculture University, Jodhpur. The AICRP on Pearl millet has a network of 13 AICRP centres in Rajasthan, Gujarat, Maharashtra, Karnataka, Andhra Pradesh, Madhya Pradesh, Punjab, Haryana, and Tamil Nadu (AICRP-PM 2021).

The AICPMIP has played a pioneering role in developing a diverse range of improved breeding lines, parental lines, and commercial hybrids. These hybrids are currently cultivated on approx. 50% of the total pearl millet area of 9–10 million ha. Hybrids maturing in 80–85 days, when cultivated as an irrigated rains/summer season crop in parts of Rajasthan, Gujarat, and Uttar Pradesh have been reported to give as high as 4000–5000 kg/ha of grain yield. With adaptive and nutritional features of pearl millet combined with high yield potential make it an important cereal crop that can effectively address the emerging challenges of global warming, water shortages, land degradation, and food-related health issues. AICPMIP has also developed production-protection technologies specific to agro-eco regions of different states. The HYVs cover about 50% of the total pearl millet area, which is the highest among coarse cereal crops. The area under HYVs is highest in Gujarat where the almost whole area (>90%) has come under hybrid coverage. Although Rajasthan has the highest area under pearl millet, the adoption of HYVs in this state has been very low (25–30%).

8.4.2 Current Status of Pearl Millet Improvement in India

Through ICAR - All India Coordinated Research Project on Pearl millet, a total of 175 hybrids (public 101 and private sectors 74) and 62 varieties were identified and released for cultivation in different agro-ecological zones of the country. These hybrids are cultivated on 60% of the total pearl millet area, leading to increase in crop productivity from 305 kg/ha during 1951–1955 to 1243 kg/ha during 2018–2019.

Pearl millet is the first crop where MAS (marker-aided selection) strategies and tools have been applied to develop "Improved HHB 67". Since its inception in 1974, the Coordinated Project has developed several production and protection technologies specific to different agro-ecological regions which proved useful in enhancing the productivity of improved cultivars to commercial farming scales and thus increased the profitability of pearl millet cultivation. With the release of new high-yielding hybrids and varieties, production has enhanced through increased seed replacement rate and higher productivity (AICRP-PM 2021).

8.4.3 Post-green Revolution Era: Impact of the Cultivation of These Crops and Consequences on Natural Resources and Environment

Enormous progress has been made in India to improve productivity by developing high-yielding cultivars and their improved agronomic management during the last six decades (Jukanti et al. 2016). The accomplishments of pearl millet breeding are often referred to as one of the greatest success stories in Indian agriculture (Yadav et al. 2019, 2021). However, the biological potential of pearl millet has not been fully realized as indicated by the current 1.2 tons/ha national productivity in comparison to the productivity level of 4–5 tons/ha in the summer season in northwestern India.

Indian pearl millet cultivation area has been divided into three mega environments (designated as A1, A, and B zones) considering the geographical location, rainfall pattern, local adaptation, and other environmental conditions (Gupta et al. 2013). The A zone consists of parts of northern India receiving >400 mm of annual rainfall. The A1 zone consists of parts of northwestern India receiving <400 mm of annual rainfall, whereas the B zone accounts for the area in peninsular India receiving more than >400-mm annual rainfall. At present, ~75% of the pearl millet is grown in A and A1 zones and 25% in B zone. Different pearl millet breeding programmes in India have developed their product profiles, depending on the need of their target mega environment.

Early breeding efforts in genetic improvement of pearl millet, which started as early as the 1930s, attempted to capitalize on such existing genetic variation within traditional landraces by subjecting them to simple mass selection (Athwal 1961). There were some innovative attempts in the 1950s to exploit heterosis through developing "chance hybrids" that included growing two parental populations of similar maturity in the mixture and allowing them to cross-pollinate to produce seed that contained ~40% hybrid seed. The chance hybrids out yielded local varieties by 10–15% but could not become popular because of a lack of efficient seed production programmes and their limited genetic superiority. The greater urgency for population improvement programmes started with the acquisition of a diverse range of germplasm from across the world in the 1970s at the International Crops Research Institute for the Semi-Arid Tropics (Witcombe 1999). Eventually, a large number of populations and trait-based composites of the broad genetic base were established, and a diverse range of elite breeding materials was developed (Rai et al. 2012).

Research programmes in India have started breeding both A and R lines and developing hybrids based on these CMS systems. The understanding of the genetics of A4 and A5 CMS (Gupta et al. 2018) helped in the well-organized and efficient utilization of these CMS sources. A range of germplasm material from India and Africa with diverse phenotypic characteristics, such as tillering, panicle size, earliness, grain size, grain colour, and so on, was strategically exploited to diversify the genetic base of both seed and restorer parents (Patil et al. 2020). In the last four decades, hybrid breeding has received a very high priority in India using genetically diverse parental lines targeting various production ecologies that have helped to intensify the genetic gains (Yadav et al. 2012).

It is quite interesting to compare the improvement in pearl millet productivity achieved during the last 28 years (1985–2013) to that achieved during 1960–1985 period. During the first 25 years (1960–1985) of hybrid development, the crop productivity increased @ 6.3 kg/ha/year. This increase went up to over 20 kg/ha/year in the next 25 years (1985–2013). The greater rate of improvement in pearl millet productivity during the last 25 years is due to several reasons (Yadav et al. 2021). Firstly, a far greater number of availabilities of pearl millet cultivars provided a wide range of choice for their cultivation in various agro-ecological regions. A total of 43 cultivars were released during 1960–1985 in comparison to 112 hybrids/varieties released to date. As a result, there have been no major disease epidemics during the last 25 years against quite a few prior to 1985. Secondly, the involvement of the private sector in seed production, distribution, and marketing has helped provide quality seed hybrids to farmers. Thirdly, greater adoption of production technology along with high-yielding hybrids proved synergistic in further augmenting productivity gains.

8.4.4 Future Strategy

Pearl millet is becoming an indispensable food crop that provides calories, nutrition, and livelihood security to the poor and marginal people living in the fragile ecosystem of arid and semiarid regions of South Africa and Sub-Saharan Africa Pearl millet is a crop of choice because of its critical role in enhancing the resilience to climate change. The past breeding priorities and strategies have been able to deliver significant productivity growth realized in pearl millet. Greater use of hybrid technology, employing modern tools, wider inter-institutional, and inter-sectoral partnerships, and improved crop management practices would play a greater role to accomplish much higher genetic gains for yield and nutritional traits for growing populations in SA and SSA. The success would depend upon a deeper understanding of new germplasm, genome, and trait-specific genes for novel traits through an amalgamation of conventional and modern tools and rapid generation techniques in national and international pearl millet breeding programmes.

Pearl millet has shown impressive genetic gains in India for the past seven decades (Yadav et al. 2019, 2021). The crop now is poised to take the next quantum leap in genetic gains. Sustaining pearl millet production and productivity in India have been shown to further enhance the bioavailability of these micronutrients. These technologies need to be tested for their commercial feasibility. Thus, an integrated approach of crop improvement backed with improved crop management, grain processing, and food product development and appropriate policy support will enable pearl millet to play its rightful role in enhancing food and nutritional security.

8.5 Small Millets

Small millets are a group of seven crops comprising finger millet, kodo millet, little millet, foxtail millet, barnyard millet, brown top millet, and proso millet. These are crops of antiquity having a long history of cultivation of more than 5000 years and are known for their suitability to dry lands, hill, and tribal agriculture contributing to food, fodder, and nutritional security at the farm and regional levels. They require a small quantity of water, mature early and are well suited for cultivation under scarcity conditions. They are less prone to pests and diseases. Unlike the major crops, viz. rice, wheat, and maize, the resilience exhibited by the small millets is helpful in their adjustment to different ecological situations and makes them ideal crops for climate change and contingency plantings. They have longer storage life and hence can be termed as 'famine reserves'.

Small millets are known for their unique nutritional properties, particularly for high fibre content, quality protein and mineral composition and contribute significantly to the nutritional security of some of the most disadvantaged groups. They contain 7–12% protein, 75–85% carbohydrates, 1–4% fat, 2–3% minerals, and 15–20% dietary fibre, besides being rich sources of phytochemicals and micronutrients and hence they are aptly termed as 'nutri-cereals'. They are known for their nutraceutical qualities and health benefits besides their nutritional advantages.

Globally, small millets are grown in an area of about 15 million ha and India is the leading producer. Annual planting area under small millets in India is around 2.0 million ha, of this nearly 1.2 million ha is under-finger millet comprising 40–50% of the global area under the crop. During the last three decades the area under finger millet has declined but with a significant improvement in productivity (1650 kg/ha), the annual production is maintained around 2.0 million tons. In contrast, the area under other millets has steeply declined and stagnating productivity resulted in declined production.

The major Finger millet growing states are Karnataka, Tamil Nadu, Andhra Pradesh, Odisha, Jharkhand, Maharashtra, Chhattisgarh and Uttarakhand. Karnataka is the largest grower of ragi accounting for nearly 60% of area and 70% of production in the country. The cultivation of kodo millet, little millet and foxtail millet is more in Madhya Pradesh, Chhattisgarh, Odisha, Tamil Nadu, Jharkhand, Karnataka and Andhra Pradesh. The cultivation of barnyard millet and proso millet is largely seen in Uttarakhand, northeast regions, western U.P., and north Bihar. The utilization of these grains is mainly as food and straw is a valued fodder for bovines. The grain is processed and consumed in traditional way and almost the entire production is utilized at the village/farm level. Despite the superior nutritive value of grains, their use is largely confined to rural areas and very little produce finds its way to urban markets.

8.5.1 Crop Improvement in Early Years

Small millets improvement efforts have been in progress since the beginning of the twentieth century (Seetharam 1998). But, the launching of coordinated crop improvement programmes during the late 1950s and 1960s has contributed significantly by way of developing new superior varieties and concomitant production and protection technologies in all small millets. The release of improved varieties and production packages for general cultivation has helped in threefold increase in grain production in the country. Finger millet among small millets has received a little more attention than the rest.

In the early 1950s and 1960s; crop improvement was confined to fewer states such as Tamil Nadu, Andhra Pradesh, Karnataka, and Uttar Pradesh. The emphasis was on varietal improvement through the selection of better types from local cultivars. In Tamil Nadu, Millet Research Station was established in 1923 at Coimbatore under the erstwhile Madras Presidency. Finger millet work in Karnataka dates back to 1900, initiated at Bangalore especially on finger millet and in Uttar Pradesh at Kanpur and Gorakhpur in 1944. The first finger millet variety released in the country was H 22 as early as 1918 in Karnataka. The other finger millet varieties released were Co 6 (1935); R 0870, ES13, K1, ES11 (1939); Hagari1 (1941), Co1, Co2, Co3, Co4 (1942), VZM 1, VZM 2 (1958) and T36 B (1949).

8.5.2 Green Revolution Period

Finger millet improvement got a fillip in Karnataka during 1950–1960s and several new varieties such as Aruna, Udaya, K1, Purna, ROH 2, and Cauvery were released. Similarly, many varieties were released in other small millets and also in many states (Seetharam 1998). This included little millet variety Co 1 (1954); foxtail millet varieties Co1, Co2, Co 3 (1943), H1, H2 (1948), T 4 (1949); kodo millet varieties PLR 1 (1942), T 2 (1949), Co 1 (1953), proso millet variety Co 1 (1954) and barnyard millet varieties T 46, T25 (1949).

8.5.3 Post-green Revolution Era: Impact of the Cultivation of These Crops and Consequences on Natural Resources and Environment

8.5.3.1 Crop Improvement Efforts During Coordinated Project Era

Millets in general started receiving attention with the launching of the All India Coordinated Millets Improvement Project (AICMIP) in 1969. Small millets also started receiving some attention at a selected few centres. Small millets improvement received a major boost during 1978–1979 with the establishment of five crops specific lead research centres in the country under IDRC assistance. They were Almora in Uttarakhand (barnyard millet), Dholi in Bihar (proso millet), Dindori in Madhya Pradesh (Kodo millet), Semiliguda in Orissa (Little millet) and Nandyal in Andhra Pradesh (foxtail millet). The IDRC project continued till 1985 and the "All India Coordinated Small Millets Improvement Project" (AICSMIP) was established in the year 1986. The centres that were functioning under the IDRC project became part of AICSMIP. Small millets research is focused for developing varieties and other agro-production and protection technologies suitable to different regions. There are 14 centres functioning under AICSMIP spread all over the country to address to the research needs of small millets. The research in the project is focused to state/regional needs from the point of developing appropriate varietal and agro production technology for maximizing production/productivity. The work is multi-disciplinary and applied in nature.

The crop improvement is aimed at developing high-yielding varieties with resistance to blast disease quality fodder, early and medium maturity and white seed in finger millet, resistance to head smut in kodo millet and resistance to shoot fly in foxtail, kodo, proso and little millets. So far, a total of 294 varieties in 6 small millets with finger millet having 136 varieties, have been released in the country from 1986 to 2021. Out of these 86 varieties were released before 1986 (pre-coordinated era) and 208 during 1986–2021 (post-coordinated project era).

Genetic Resources Management

In the past, small millets scientists had limited access to germplasm and worked with a handful of local collections that lacked diversity. This situation was to some extent rectified in 1960s, when first attempts were made by ICAR to pool the collections under PL 480 project. The conservation activities further gained momentum with NBPGR, New Delhi, playing a key role in augmenting the small millets collection. Recognizing the importance and conservation and greater access to germplasm, the All India Coordinated small millets Improvement Project (AICMIP) established a germplasm unit in Bangalore in 1979. This unit, since then, has been making efforts to collect, and pool the available germplasm from various sources and make it available to breeders. This unit is also recognized as National Active Germplasm Site (NAGS) by ICAR/NBPGR and has the mandate to assist in the collection, conservation, evaluation, and documentation of small millets germplasm in the country (Seetharam 2006). Presently the Unit at Bangalore is maintaining one of the largest collections of more than 10,000 accessions of all small millet crops.

Utilization of Germplasm

During the last 25 years, the majority of the accessions have been screened for agronomic, physiological, pathological, and even important grain quality parameters. There is a good database available for most accessions conserved (Seetharam et al. 2006). In order to improve the efficiency for the utilization of germplasm, core subsets have been formed and made available to breeders working at different centres. Identification of several sources of stable resistance to blast disease of finger millet and their deployment in breeding research has been highly rewarding in the evolution of high-yielding blast resistant cultivars in finger millet in the country.

Seed Production

The success story of seed production of improved varieties in Finger millet in the 1990s helped farmers to get more production and productivity of the crop and benefit them. However, in recent years, the seed production of other small millet crops has improved in order to make small millet cultivation more competitive and economically viable.

Crop Production and Protection Technologies

A package of practices such as time of sowing/planting, choice of varieties, time, and method of application of fertilizers have been developed for different regions of the country. Management practices for aberrant weather conditions for mitigating early, mid, and late season drought have been worked out. Remunerative cropping systems involving different pulse crops in millet for different regions have evolved. Technology transfer attempted through frontline demonstrations on the farmer's field and large-scale station demonstrations have helped in narrowing down the yield gap that exists between farmers' fields, demonstration plots and research station trials (Prabhakar 2017; Seetharam and Krishne Gowda 2008; Seetharam 2015a, b). Plant protection measures to control economically important diseases and pests have evolved (Seetharam 2015a, b). Several blast-resistant lines were identified from the germplasm available at NAGS and crop protection for the management of blast diseases has been recommended.

8.5.4 Sub-Mission on Nutri Cereals

A Sub-Mission on Nutri Cereals under NFSM started with an outlay of Rs. 300.00 crores for 2018–2019. Sorghum (Jowar), Bajra, Finger Millets (Ragi/Mandua) and Minor Millets, Foxtail millets (Kangani/kakun), Proso millet (Cheena), Kodo Millet (Kodo), Barnyard millet (Sanwa/Jhangora), Little millet (Kutki), Pseudo Millets (Buck wheat-kuttu, Ameranthus-chaulai) are termed as Nutri-Cereals for production, consumption and trade point of view. The major objectives of the Sub-Mission on Nutri Cereals are development of a strategy for addressing issues concerning production, demand, and research with market-oriented approach. The National Year of Millets was celebrated in 2018 and programmes were implemented in 202 districts of 14 states covering all Nutri cereals. A wide range of activities related to nutri cereals production, awareness by way of Millet campaign, Seminar/symposium/workshop at district/state/national level and Value Chain Integration and Millet Entrepreneurship were undertaken.

8.5.5 Future Strategy

After years of neglect, small millets are finding a place in agricultural research, agendas of large private companies, and from there into our supermarkets. They are increasingly being recommended by doctors and nutritionists as being important for the health and wellness of people and helpful in preventing many diseases related to modern lifestyle including obesity, and diabetes. A lot of elite food chains have begun selling millet and millet-based products on their shelves as health food. The productivity could be increased by more than 50% by adopting improved production practices as indicated by the results of frontline demonstrations.

8.6 Conclusions

By 2050 India has to achieve 3.5 more maize production, 2.2 increase in productivity and 1.4 times in terms of area compared to the current scenario. During the last seven decades (1949–2020), India has increased its maize production from 2.0 to 30.24 million metric tons. Eventually, the challenge is very typical but can be achieved through strong policy support and by cultivating higher yielder single cross hybrids with yield potential of 6–7 tons/ha during kharif season and 9–10 tons/ha during the rabi season. But the major challenge is to adopt the available technologies like single cross hybrids with 100% area and ensure the availability of quality seeds. In addition to that, there is a need to focus on speciality corn like popcorn, sweet corn, and baby corn for livelihood security. Similarly for millets, like a need to focus on rabi sorghum production, promotion of hybrid technology in pearl millet, and popularizing the small millets and their benefits with more production practices. In addition to that, there is need to focused research on germplasm collection and diversification, development of hybrids or varieties suitable for adverse climate conditions, accelerated development of varieties/hybrid cultivars through the application of advanced tools like marker-assisted selection techniques, promotion, and popularization of improved released hybrids/varieties. Further, sustainable production and productivity may be achieved through the adoption of improved production and protection practices.

In recent years milling technology has been improved to enhance grain quality. Millet mills are available for cottage-level and large-scale processing. Millets can be further processed towards various foods such as flakes, quick food cereals, ready-to-eat snacks, supplementary foods, extrusion cooking, malt-based products, weaning foods, and more importantly health foods. The R & D efforts made in the area of grain processing and value addition through the development of novel diversified foods especially in sorghum, finger millet, and other millets are opening new avenues for expanding consumer base, and enhancing absorption for food use. Small millet crops are viewed as important for the health and wellness of people and can help in preventing many kinds of diseases related to modern lifestyle including obesity, diabetes. Of late, plenty of elite food chains have begun selling millet and millet based products on their shelves as health food. Grains can be further processed towards various foods such as flakes, quick food cereals, ready-to-eat snacks, supplementary foods, extrusion cooking, malt-based products, weaning foods, and more importantly health foods. The importance of regular food use of nutrient-dense millet for achieving holistic food and nutritional security is getting widely recognized.

Keeping the above in view the research activities are to be restructured and should have a very high degree of location specificity. The millet R & D should not be viewed from crop and productivity angles only; but holistically on a wider perspective of other advantages accruing through millet-based cropping systems such as ensuring greater ecological balance with the environment, efficient soil and moisture conservation and utilization, adjustments for climate change, minimum pest/disease load, assured harvests and more so the food, nutrition, and health security to people at large.

References

AICRP (2021) Pearl millet. https://aicrpsm.res.in, https://icar.org.in/

Athwal DS (1961) Recent developments in the breeding and improvement of bajra (pearl millet) in the Punjab. Madras Agric J 48:18–19

Dhillon BS, Malhi NS (2006) Maize breeding in India: retrospective analysis and prospects. Ind J Plant Genet Resour 19:327–345

Elangovan M (2020) Millets genetic resources management. In: Elangovan M, Ganapathy KN, Amasiddha B, Singode A, Tonapi VA (eds) Online refresher course on recent advances in millets research. Compendium of lecture notes. Society for Millets Research, ICAR–Indian Institute of Millets Research (IIMR), Hyderabad, pp 1–4

FAOSTAT (2019). http://www.fao.org/faostat/en/#data/QC

Gupta SK, Rathore A, Yadav OP, Rai KN, Khairwal IS, Rajpurohit BS, Das RR (2013) Identifying mega-environments and essential test locations for pearl millet cultivar selection in India. Crop Sci 53:2444–2453

Gupta SK, Yadav DV, Govindaraj M, Boratkar M, Kulkarni VN, Rai KN (2018) Inheritance of fertility restoration of A5 cytoplasmic- nuclear male sterility system in pearl millet [*Pennisetum glaucum* (L.) R. Br.]. Ind J Genet Plant Breed 78:228–232

Jukanti A, Gowda CLL, Rai KN, Manga VK, Bhatt RK (2016) Crops that feed the world 11. Pearl Millet (Pennisetum glaucum L.): an important source of food security, nutrition and health in the arid and semi-arid tropics. Food Secur 8:307–329

Kumar RS, Kumar B, Kaul J, Karjagi CG, Jat SL, Parihar CM, Kumar A (2012) Maize research in India-historical prospective and future challenges. Maize J 1(1):1–6

Patil KS, Gupta SK, Marathi B, Danam S, Thatikunta R, Rathore A, Das RR, Dangi KS, Yadava OP (2020) African and Asian origin pearl millet populations: genetic diversity pattern and its association with yield heterosis. Crop Sci 60:3035–3048

Prabhakar (2017) Small millets: climate resilient crops for food and nutritional security. Mysore J Agric Sci 51(1):52–62

Prabhakar, Aruna C, Venkatesh B, Umakanth AV (2015) Genetic improvement in Sorghum. In: Tonapi VA, Patil JV (eds) Millets-ensuing climatic resilience and nutritional security. ICAR-IIMR, Hyderabad, pp 109–196

Rai KN, Yadav OP, Gupta SK, Mahala RS, Gupta SK (2012) Emerging research priorities in pearl millet. J SAT Agric Res 10:1–4

Rakshit S, Chikkappa GK (2018) Perspective of maize scenario in India: way forward. Maize J 7(2): 49–55

Rao NGP (1982) Transforming traditional sorghum in India. In: Sorghum in the Eighties: Proceedings of the International Symposium on Sorghum, 2–7 Nov 1981, ICRISAT, India, vol 1. International Crops Research Institute for the Semi-Arid Tropics, Andhra Pradesh, pp 39–59

Seetharam A (1998) Small millets - achievements during 1947-97. Indian J Agric Sci 68 (Suppl):47–54

Seetharam A (2006) Millets. In: Hand book of agriculture. ICAR, New Delhi, pp 892–912

Seetharam A (2015a) Small millets research: achievements during last 50 years, In "Millets: Promotion for food, feed, fodder, nutritional and environment security", edited by VA Tonapi, B. Dayakara Rao and JV Patil, ICAR-IIMR, Hyderabad, p 68-90.

Seetharam A (2015b) Genetic improvement in small millets, In Millets-Ensuing climatic resilience and nutritional security: edited by VA Tonapi, B. Dayakara Rao and JV Patil, ICAR-IIMR, Hyderabad, p 233-260.

Seetharam A, Krishne Gowda KT (2008) Production and utilization of small millets in India. In: Food uses of small millets and avenues for further processing and value addition. Project coordination cell, All India Coordinated Small Millets Improvement Project, ICAR, Bangalore

Seetharam A, Patel DP, Halaswamy BH (2006) Small millets. In: Dhillon BS, Saxena S, Agrawal A, Tyagi RK (eds) Plant genetic resources-food grain crops. Narosa Publishing House, Delhi, pp 204–223

Upadhyaya HD, Pundir RPS, Dwivedi SL, Gowda CLL, Reddy VG, Singh S (2009) Developing a mini core collection of sorghum for diversified utilization of germplasm. Crop Sci 49:1769–1780

Witcombe JR (1999) Population improvement. In: Khairwal IS, Rai KN, Andrews DJ, Harinarayana G (eds) Pearl millet breeding. Oxford and IBH, New Delhi, pp 213–256

Yadav OP, Rajpurohit BS, Kherwa GR, Kumar A (2012) Prospects of enhancing pearl millet (Pennisetum glaucum) productivity under drought environments of north-western India through hybrids. Ind J Genet Plant Breed 72:25–30

Yadav OP, Hossain F, Karjagi CG, Kumar B, Zaidi PH, Jat SL, Chawla JS, Kaul J, Hooda KS, Kumar P, Yadava P (2015) Genetic improvement of maize in India: retrospect and prospects. Agric Res 4(4):325–338

Yadav OP, Singh DV, Dhillon BS, Mohapatra T (2019) India's evergreen revolution in cereals. Curr Sci 116:1805–1808

Yadav OP, Gupta SK, Govindaraj M, Sharma R, Varshney RK, Srivastava RK, Rathore A, Mahala RS (2021) Genetic gains in Pearl millet in India, insights into historic breeding strategies and future perspectives. Front Plant Sci 12:645038. https://doi.org/10.3389/fpls.2021.645038

Pulses

9

K. K. Hazra and Partha Sarathi Basu

Abstract

Pulses are nutrient-dense crops and the cheapest source of protein. Among the countries in the world, India produces and consumes the most pulses. These are the third most important crops after cereals and oilseeds. More than 80% of pulses are grown in rainfed and semi-arid regions across diverse climatic conditions. In the era of the Green Revolution, cereals were prioritized which brought self-sufficiency in food production. However, being exhaustive crops and high nutrient responsive, the demand for chemical fertilizers and ground water increased tremendously resulting in land degradation, a decline in the water table, and depletion in the soil fertility level. For the past several decades, pulses were largely ignored and these precious crops were pushed more towards marginal lands. As a result, the productivity of pulses stagnated and the demand-supply gap kept on widening. Many pulse-based technologies have been developed in the past, including region-specific varieties, advanced crop production and protection measures, crop diversification, micro-nutrient supplementation, promoting efficient *Rhizobium* strains, phosphate solubilizing bacteria (PSB) and hybrid technologies together have boosted pulse production in India. Along with scientific advancement supported by policy support such as the launching of various Government schemes such as the creation of pulse villages, accelerated pulse programmes under NFSM, increasing MSP and creation of seed hubs as per seed demand in different locations have brought the country self-sufficiency in pulse production. During this period both vertical and horizontal expansion of pulses including crop diversification in non-traditional areas took place which enabled further scope to expand pulse cultivation in the country. With the development of improved short-duration varieties and the use of genomic-assisted selection,

K. K. Hazra · P. S. Basu (✉)
ICAR-Indian Institute of Pulses Research, Kanpur, Uttar Pradesh, India

P. K. Ghosh et al. (eds.), *Trajectory of 75 years of Indian Agriculture after Independence*, https://doi.org/10.1007/978-981-19-7997-2_9

breeding programs have been accelerated. Whole-genome sequencing of major pulses such as pigeon pea and chickpea provided vast information on specific gene-location imparting resistance towards biotic and abiotic stress. The present article analysed the critical gap in our understanding and explored further possibilities for enhancing pulse production in the country.

Keywords

Pulses · Nutritional security · Sustainability · Trend analysis · Research and development · Climate resilience

9.1 Introduction

India is home to 20 agro-climatic zones and holds the second largest agricultural land in the world with 157.4 million ha of land under cultivation. Thus, agriculture plays an important role in 58% of rural households. Although India is self-sufficient in food production, its food production between 1947 and 1960 was so poor that there were risks of famine. Hence, the Green Revolution was the key to increasing food production in the country, reducing extreme poverty and malnutrition and feeding millions of people. Despite these measures, India has a quarter of the world's hungry population, with 224.3 million malnourished people lacking enough food to meet their daily nutritional needs (Fig. 9.1); 31% of children under 5 years of age are stunted and widespread prevalence (53%) of anaemia among women of reproductive

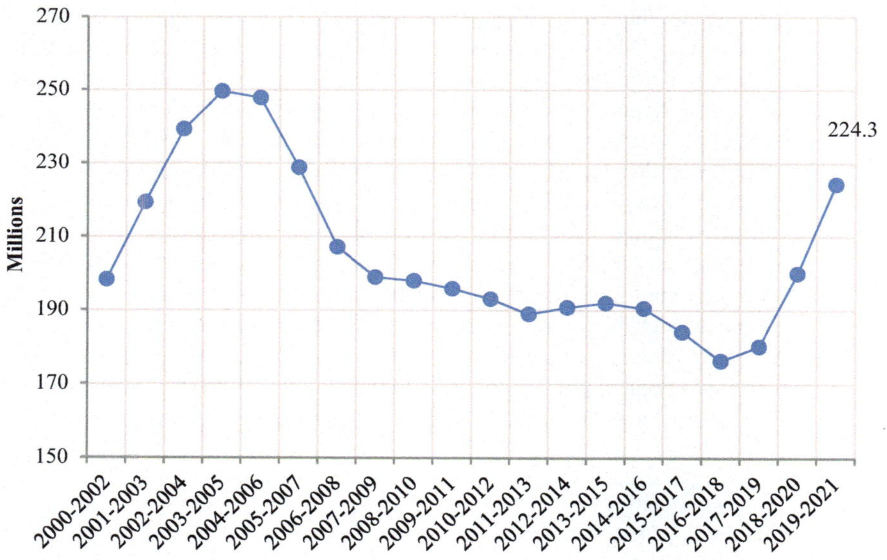

Fig. 9.1 Number of people undernourished (millions) (3-year average) in India (FAOSTAT 2020)

age (15–49 years) (FAOSTAT 2020). Pulses are the cheapest source of dietary proteins for a large section of people. Worldwide pulses are cultivated covering around 80 million ha with an annual production of about 92–96 million T and average productivity of about 900–950 kg ha^{-1}. India is the largest pulse producer accounting for 27–28% of global pulse production sharing about 90% of the global pigeon pea, 65% of chickpea, and 37% of lentil area of the total global pulse production. Chickpea (*Cicer arietinum* L.), pigeon pea (*Cajanus cajan* L.), green gram (*Vigna radiata* L), black gram (*Vigna mungo* L.), and lentils (*Lens culinaris* L.) are the most important among pulses. The net per capita availability of pulses was about 61 g day^{-1} $person^{-1}$, which declined to about 32 g day^{-1} $person^{-1}$, though medical experts recommend 70 g day^{-1} $person^{-1}$. Consistent growing demand for pulses followed by inadequate production and supply has led to a major gap in pulse production in India where a majority of the people are dependent on pulses as the cheapest source of protein. Pulses continue to be an integral part of sustainable crop-production systems because of their biological nitrogen fixation (BNF) ability, low water requirement, and ability to withstand unusual weather conditions.

Pulses provide the right mix of vegetarian proteins of high biological value, with a decline in pulse consumption leading to an increase in malnutrition and a decline in protein intake (Shalendra et al. 2013). About 24% of the world's malnourished people are reported in India (Sharma et al. 2016). If the level of production of pulses in India is not increased, the demand–supply gap is expected to widen. The production of pulses in India has been low and yields have been unpredictable. Low returns per hectare have resulted in the least preference for farmers to grow pulses on irrigated and fertile lands, leading to unstable and low yields (Joshi and Saxena 2002; Tulsi 2015). Pulses are predominantly rainfed crops and inadequate rainfall often causes drastic declines in the productivity of pulses, as there are no alternative mechanisms to cope with the situation. Temperature extremes and drought are the two major abiotic stresses for winter pulses such as chickpeas, lentils, field peas, and rajmash. Cold temperatures and frost often cause substantial yield losses in pigeon peas. Pulses are mostly rainfed, but as the irrigated area increases, pulses are shifted to rainfed areas and their area is replaced by cereals or some cash crop (Singh and Singh 1995). In India, the irrigated area under pulses was only 12%, while under wheat and paddy; it was more than 60% of the total area (Reddy and Reddy 2010). Among biotic stresses, wilt and root rot complexes are the most prevalent diseases in pulses. In addition, pod borer infestation leads to a reduction of grain yield in pulses to a considerable extent. There is a huge gap between supply and demand for pulses. Therefore, the government has taken major steps to scale up pulse production in the country. These steps include increasing support prices, creating seed hubs across the country, and developing pulse villages with technological support. Production of pulses has stagnated for the past several decades largely due to the lesser importance given by policymakers and agri-planners. It is only in the last decade that pulse production in India has seen breakthroughs. This is due to strong seed chain links, expansion of pulse cultivation in new niches, crop diversification, and stronger crop protection and production technologies.

There has been a significant increase in the production of pulses during the last decade. In this chapter, we describe how India achieved self-sufficiency in pulse production from the beginning of the green revolution until the era after liberalization. India achieved a record production of 17.1 million T in pulses production in 2014–2015 from an area of 23 million ha with a productivity of 753 kg ha^{-1} with an all-time high production of chickpea (8.25 million T), green gram (1.82 million T) and black gram (1.74 million T). Although the production of pulses has increased significantly during the last decade, continuing the rapid growth remains a major challenge for researchers, extension agencies, and policy makers. Nevertheless, the productivity of pulses in India (694 kg ha^{-1}) is low compared to most of the major pulse-producing countries. Pulses are the most suitable crops for growing in water-scarce areas. Pulses require less water, improve soil health, and adapt to the local microclimate for smart agriculture. Only 43 gallons of water (1 gallon = 3.785 L) are needed to produce 1 lb of pulses, while wheat, rice, and meat require 660 gallons of water, 1056 gallons of water, and 1142 gallons of water, respectively (https://krishijagran.com/featured/pulses-production-sustainability-and-its-role-in-human-nutrition/). Hence, pulse crops are considered smart crops in rainfed farming systems, which can contribute to addressing climate change in agriculture. Compared to resource-intensive crops such as wheat, soybean, rice, and maize, pulses are profitable crops because of their lower input requirements and higher market price. In addition, pulses improve soil health by fixing atmospheric nitrogen and adding humus content to the soil, thereby improving the biological, chemical, and physical properties of the soil (Nath et al. 2019; Borase et al. 2021). In rainfed farming systems, incorporating shorter-duration pulse crops into existing cropping cycles can increase total farm income. Therefore, there is a need for large-scale promotion and cultivation of pulses to increase income and nutritional security in rural and urban India. Methods and techniques that may be used to improve pulse production can be classified into vertical approach, horizontal approach, and policy intervention.

9.2 Pulse Production in India: The Need of the Hour

Food security in India has long been a major area of concern for agricultural scientists and planners. India is still largely vegetarian in its diet and relies heavily on vegetable sources to meet its daily protein requirement. India is bound to be the global leader in the production and consumption of pulses. Since, India is the major importer of pulses in the world, the production of pulse crops has remained stagnant between 11 and 14 million T over the past two decades. The gap between the demand and supply of pulses is widening. India still lags far behind in pulses production. About 20% of the total demand for pulses is met by imports. It also contributes to the maintenance of the production system through physical, chemical, and biological improvements of soil properties in the form of a rotational effect. Proper management practices are essential to harness high yields. Therefore, it is imperative to understand the constraints in pulse production to achieve high productivity and maintain soil health. The seed replacement rate is still (<30%) lower than that of cereals, especially wheat and rice. It is estimated that the total pulse requirement for India's growing population is 32.0 million T, which will rise to 1.69 billion by 2050. An annual growth rate of 2.2% is needed to reach this level. The demand for pulses is growing at the rate of 2.8% per annum. Although the challenges are varied, including climate change scenarios, and depleting land and water resources. Increasing the average productivity of pulses to more than 1200 kg ha^{-1} and bringing an additional area of about 35 million ha under pulses cultivation would be a concrete step in this direction. Increase in yield through the development of input-responsive varieties with multiple resistance to diseases and pests. In addition, short-duration varieties that fit well in different cropping systems and climate-resilient varieties of pulses are highly important in the vertical expansion of pulses in the country. Similarly, the development of new plant types for different agro-climatic conditions, and the development of photo-thermo-insensitive varieties in crops like black gram and green gram will help in expanding the areas of these crops in non-traditional areas of the country. To increase pulse production, emphasis should be given to the expansion of area under short-duration varieties, development of multiple disease/pest resistant varieties, use of micronutrients like zinc and sulphur, and the increase in area under winter pulse crops.

9.3 Growth and Instability in Area, Production, and Yield of Pulses in India

The area under pulses was recorded at 19.09 million ha during 1950–1951, which increased to 28.34 million ha during 2019–2020. The production of pulses increased at a slower date during 1950–1951 (8.41 million T) to 2000–2001 (11.08 million T). However, during the last two decades (2001–2021), there was a remarkable increase in pulses production in the country and the highest production was recorded in the year 2020–2021 (25.6 million T). The productivity of pulses in India has increased from 441 kg ha^{-1} in the year 1950–1951 to 789 kg ha^{-1} in the year 2012–2013 and further increased to 817 kg ha^{-1} in the year 2019–2020 (Directorate of Economics and Statistics 2021). The Tri-Annual Average End (TE) area for 2013–2014 and the share of different states in the production of pulses indicate that states like Madhya Pradesh, Maharashtra, Rajasthan, Uttar Pradesh, Andhra Pradesh, and Karnataka account for about 80% of the total area under pulses. The production pattern of different pulses was dominated by the states (Srivastava et al. 2010). The area under pulses in the country is almost constant, from 21.87 million ha during TE 1972–1973 to 24.42 million ha during TE 2013–2014, an increase of 11.6%. However, the production of pulses increased from 10.94 to 18.44 million T during the same period. The production of pulses in India showed a negative growth rate during the 1970s, which turned positive in the next decade and reached 3.35% per annum in the recent decade, mainly due to growth in productivity. Considering the rate of increase in production and productivity for the overall period, pulses can be considered slow-growing crops. Inadequate adoption of production technology, high price volatility, production risk, and low level of irrigation are important factors responsible for

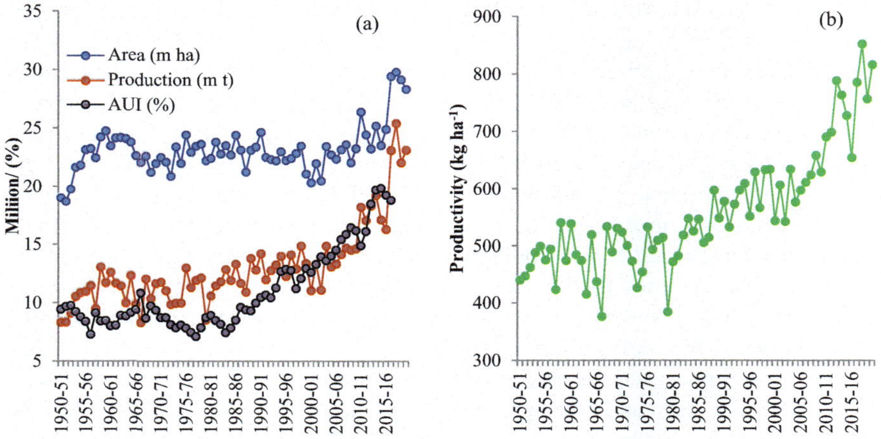

Fig. 9.2 The year wise trend in pulses area, production, area under irrigation (**a**) and productivity (**b**) during 1950–2020 (Directorate of Economics and Statistics 2021)

stagnation in the productivity of these crops (Ramasamy and Selvaraj 2002). The improvement in the productivity of pulses in India was mainly due to the adoption of improved varieties by farmers, the production of breeder seed, and the demonstration of pulse production technology and policy support (Gowda et al. 2013). The pulses sector has undergone significant changes during the period after the Green Revolution. The area under pulses has declined in North India, while it has expanded in Central and South India (Ali and Gupta 2012; Gowda et al. 2013). The regional area variation is more typical in the case of chickpeas. The production of chickpeas during TE 1972–1973 was 4.94 million T, which increased to 8.69 million T during TE 2013–2014, an increase of about 76%. Moreover, there is wide variation in the yield of chickpeas in different states, which varies from 558 kg ha^{-1} (Jammu and Kashmir) to 1281 kg ha^{-1} (Bihar). Although the area under chickpeas declined during the 1980s and 1990s and the area showed negative growth rates, the area and production of chickpeas in the country have increased rapidly in recent decades, with impressive increases in yield. The rate of growth in area and production of gram started in the 1990s. The production of chickpea has increased by 5.5% annually in the recent decade. The growth in the area under chickpeas remained steady for the overall period and the increase in production and yield was around 1% per annum, showing slow growth (Sharma and Brahm 2002; Srivastava et al. 2010). The area under chickpeas showed an increase during the 1950s, while the area during the 1960s to 1990s was less than in the 1950s. The decline in the chickpea area was mainly due to the replacement of cereals with higher-yielding varieties in Punjab, Haryana, Rajasthan, Uttar Pradesh, Bihar, and West Bengal (Tulsi 2015) (Fig. 9.2).

9.4 Nutritive Value of Pulses and Protein Dependence

Pulses are a good source of protein for most of the population in India. Protein malnutrition is prevalent among men, women, and children in India. Pulses account for 11% of the total protein intake in India (Reddy and Reddy 2010). In India, the frequency of consumption of pulses is much higher than that of any other source of protein, which indicates the importance of pulses in their daily food habits. About 89% of people consume pulses at least once a week, while only 35.4% consume fish or chicken/meat at least once a week. Pulses are consumed by the rich and the poor in almost the same frequency. The consumption of pulses increased from 0.15 kg $month^{-1}$ in 1973–1974 to 0.45 kg $month^{-1}$ in 2010–2020 in rural areas, which were very poor. It is necessary to increase the production of pulses to increase the balanced diet among the socially and economically backward classes while keeping it the cheapest source of protein. In addition, pulses provide healthy proteins, which are slow-digesting carbohydrates, and high in fibre compared to other protein-rich sources such as meat and meat products. In recent years, interest in pulses as a sustainable food source have grown notably for various reasons. They are low in fat, packed with nutrients, and rich sources of proteins and soluble fibre. The lack of gluten in pulses makes them suitable for gluten-free diets for individuals with celiac disease and gluten intolerance. During the Great Depression in the USA, beans were called poor man's meat and saved many lives thanks to their nutritional value and low price. The high nutritional value of pulses makes them ideal foods, specifically in regions where access to meat and dairy products are challenging for economic, distribution and marketing-related obstacles (Table 9.1).

9.5 The Impact of the Green Revolution on Pulses

The Green Revolution was initiated in India in the 1960s with high-yielding varieties of rice and wheat to increase food production and reduce hunger and poverty. As a result, the production of wheat and rice doubled after the Green Revolution; however, the production of other food crops such as pulses declined. The Green Revolution has led to the self-sufficiency of food grains in the country. However, it also destroyed the diverse gene pool available. The productivity of crops was increased by the use of fertilizers, pesticides, and groundwater resources. However,

Table 9.1 Nutritive value of pulses

Constituents	Magnitudes	Constituents	Magnitudes
Protein	>20%	Iron	7–10 mg 100 g^{-1}
Carbohydrate	55–60%	Vitamin C	10–15 mg 100 g^{-1}
Fat	>1.0%	Calcium	69–75 mg 100 g^{-1}
Fibre	3.2%	Calorific value	343
Phosphorus	300–500 mg 100 g^{-1}	Vitamin A	430–489 IU

Source: Singh et al. (2015)

the land became barren due to mismanagement and overuse of chemical fertilizers, pesticides, and lack of crop rotation, and groundwater loss became a common occurrence in agricultural areas.

Several high-yielding varieties (HYVs) were introduced during Green Revolution to increase agricultural productivity. These genetically improved varieties of wheat and rice were developed by the CIMMYT, Mexico, and IRRI, Philippines respectively. HYVs had 20% more grain than their previous varieties and were more responsive to nitrogen fertilizers. The incorporation of multiple traits and specific genes for short stature in HYVs. These efforts resulted in doubling of yield potential. Incorporation of genes responsible for photosensitivity in rice and wheat is possible throughout the year; thus, the Green Revolution included the use of HYVs of wheat and rice and the adoption of new agricultural practices involving the use of chemical fertilizers, insecticides, tractors, controlled water supply to crops, mechanical thresher, and pumps. The combination of these technologies was commonly referred to as "High Yielding Variety Technology (HYVT)". This technique was responsible for the growth rate of food grain production from 2.4% per annum before 1965 to 3.5% after 1965. The number of chemical fertilizers used was quite high after the advent of the Green Revolution, and the overuse of chemical fertilizers to achieve high yields causes physical and chemical degradation of soils by altering the natural microflora and increasing the alkalinity and salinity of the soil. The excessive use of groundwater for irrigation has depleted the water level in many parts of the country.

Beginning in the 1960s, the "Green Revolution" innovations brought major benefits to rice and wheat cultivation in India, but the benefits were unevenly distributed across regions and had an unintended but generally unintended impact on the pulses sector had a negative effect. The Green Revolution benefited rice and wheat farmers in irrigated areas in the North-West region of India, but failed to help farmers in other regions where farmers have to depend on rainfall. In addition, the expansion of rice and wheat cultivation had a detrimental effect on many other crops, including pulses, being largely replaced by acreage due to the easy availability of new, high-yielding varieties and access to irrigation. As they were displaced from their former traditional areas, pulses also experienced many changes, both in regional specialization and the adoption of technology. Overall, during the height of the Green Revolution (between 1960 and 1961 and 1980–1981), pulses production declined from 12.6 to 10.5 million metric T. In 2010 (Fig. 9.3), only 12% of the total irrigated area in India was sown under pulses while more than 60% of wheat and rice was sown.

9.6 Developmental Stages of Pulse Production in India

9.6.1 Pre- and Early Phases of the Green Revolution (1960–1970)

This decade was a period of stagnation in both the sectors devoted to pulses and their total production. There was no major breakthrough in pulses research and no government scheme was launched to support pulses production in the country.

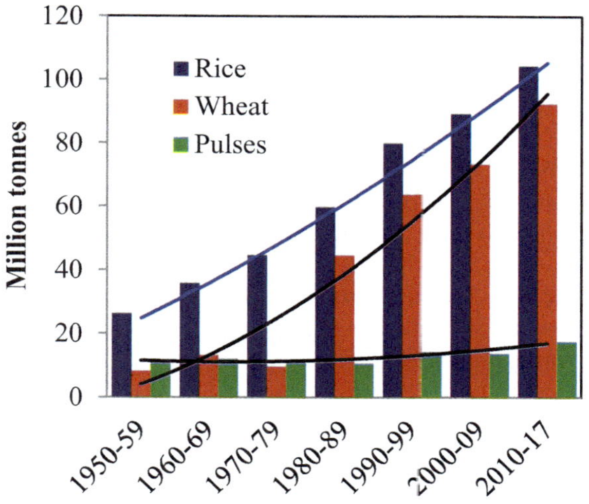

Fig. 9.3 The trend in the production of food crops in India from 1950 to 2017 (in million T). The period after beginning of Green revolution by introducing monohybrid crops in India showing increased production of rice and wheat. However, the production of pulses was almost stationary

9.6.2 Mature Phase of Green Revolution (1971–1990)

During this period, both the cultivation and production of pulses decreased in the northern and eastern regions of the country and increased in the southern and western regions. Several pulse development schemes were launched during this period, including Pulses Development Plan (Fourth Five Year Plan) (1969–1970 to 1973–1974); National Pulses Development Project (Seventh Five Year Plan) (1985–1986 to 1989–1990); Special food grain Production Program (1988–1989); Post-liberalization period (1991–2000). This period saw a further increase in the area and production of pulses in the southern and western states. Favourable terms of trade were achieved for agriculture relative to the industry, which may have influenced cropping patterns. Some short-duration and wilting-resistant varieties were developed during this period and Post-trade Spike Period (2001 and onwards). This period saw a third increase in the area and production of pulses in the southern and western states. The government has launched schemes for pulse development which include: Integrated Scheme of Oilseeds, Pulses, Palm Oil and Maize (ISOPOM) (2004); National Food Security Mission (NFSM) (2007–2008); Special plan to achieve more than 19 million T of pulses production by rainy season pulses (2012–2013).

These efforts may have increased the production of pulses to an extent and produced 18 million T in 2014–2015, but the imports of pulses also increased after 2001, which means that the increase in the production of pulses was not sufficient to meet the demand.

9.6.3 Comparative Performance: Pulses Versus Cereals

Table 9.2 summarizes the relative performance of pulses as compared with cereals between 1960 and 2010 in terms of area, production, and yields.

9.6.3.1 First Period: 1960–1970

During this period, the total pulse crop area in India declined by around 8% and, concomitantly, the area and production of both wheat and rice gained momentum.

9.6.3.2 Second Period: 1971–1990

Despite the implementation of the government's Pulses Development Scheme during this period (the Fourth Five-Year Plan), the total area, production, and yield of pulses increased by only 3.3%, 10.3%, and 6.7%, respectively. In comparison, for wheat, the increase in area, production, and yield was 48%, 171%, and 83%, respectively.

9.6.3.3 Third Period: 1991–2000

During the third period, the total pulse area declined, whereas the area under wheat and rice increased by 15% and 9%, respectively. The yields of both wheat and rice were also increased (Table 9.2).

9.6.3.4 Fourth Period: 2001 and Beyond

After 2001, there was a spike in the trade of pulses, with a 36% increase in imports, while at the same time chickpea, pigeon pea, wheat, and rice all showed a rise in production and yield. Consequently, there seems to be little evidence to support the idea of pulses imports crowding out domestic production. Importantly, during this period, for the first time, pulses' yield increased by more than 11%. The success of the Technology Mission on Oilseeds (TMO) induced the government to restructure

Table 9.2 Comparative performances of cereals and pulses, 1960–2010 (% change)

Crop	Particulars	1. Pre- and initial phase of Green Revolution (1960–1970)	2. Matured phase of Green Revolution (1971–1990)	3. Post liberalization period (1991–2000)	4. Post–trade spike period (2001–2010)
Pulses	Area	−8.02	3.31	−0.93	2.25
	Production	−0.87	10.34	2.15	14.08
	Yield	7.97	6.72	3.67	11.23
Wheat	Area	26.11	48.50	15.62	3.18
	Production	95.54	171.50	42.55	12.13
	Yield	55.13	83.00	23.33	8.73
Paddy/rice	Area	11.88	10.50	8.72	−1.53
	Production	33.77	70.60	28.55	10.34
	Yield	19.72	54.00	18.54	12.09

Source: Ministry of Agriculture, Directorate of Economics and Statistics, India 1950–2010 the program in 2004 by adding pulses and maize to create the Integrated Scheme of Oilseeds, Pulses, Oil Palm, and Maize (ISOPOM). Under ISOPOM, financial assistance was provided for the production and distribution of certified seeds, seed mini-kits, sprinkler sets, *Rhizobium* culture, PSB, gypsum/pyrite, plant protection chemicals, and biofertilizers. Demonstrations of integrated pest management (IPM) were organized on a large scale through the State Department of Agriculture. To address the shortage of quality seeds, a provision of credit to produce pulse seeds was also included in the scheme. In addition, in 2008 the Accelerated Pulses Production program was launched under the National Food Security Mission and implemented in 468 districts in 16 states.

9.7 Pulse-Growing Areas for Balancing Ecosystem

Among protein-rich foods, pulses have the lowest carbon and water traces. In addition, pulses improve soil health by naturally fixing atmospheric nitrogen in the soil; growing pulses reduces the need for the use of nitrogenous fertilizers. Thus, pulses provide valuable environmental services (Dudeja and Duhan 2005). The pulses are grown throughout the year in the country under diverse agro-climatic conditions. The pulses are important for their role in crop rotation and intercropping, as they help improve soil fertility by reducing soil pathogens and fixing nitrogen. Due to these factors, the yield of the post-pulses crop can increase by 20–40%. The low use of fertilizers, pesticides, and irrigation makes pulses an environmentally sustainable crop group. A policy option could be to assess the value of environmental services provided by pulses and to devise mechanisms to reward farmers or pulse-growing regions for these ecosystem services.

9.8 Production Trends and Geographical Distribution

Pulses are grown across the country, with the highest share being Madhya Pradesh (24%), Uttar Pradesh (16%), Maharashtra (14%), Andhra Pradesh (10%), Karnataka (7%) and followed by Rajasthan (6%) which accounts for about 77% of the total pulse production, while the remaining 23% is contributed by Gujarat, Chhattisgarh, Bihar, Orissa and Jharkhand. Among pulses, chickpea (45.1%) has a major share, followed by pigeon pea (15.7%), green gram (9.9%), blackgram (9.6%), and lentil (7.3%), which account for 87% of the total pulse production. The major pulse crops grown in India are chickpea, pigeon pea, green gram, black gram, lentils, peas and beans. They are important crops in terms of their contribution to the daily diet, human nutrition (protein), and their contribution to farmers' income and employment. Most importantly, all pulse crops improve soil fertility by stabilizing atmospheric nitrogen in the soil and helping to increase sustainability and soil productivity. Pulses are now grown in an area of 25–26 million ha, with an annual production of 24 million t. India accounts for 33% of the world's total area and 22% of the world production of pulses. India accounts for about 90% of global pigeon pea, 65% of chickpea, and 37% of lentils, accounting for 93%, 68%, and 32% of global production, respectively (FAOSTAT 2011). The shortage of supply to meet the increasing demand due to population growth has led to a steep rise in the prices of pulses. The net availability of pulses has come down from 70.1 g day person^{-1} in 1951 to 31 g day person^{-1} (Indian Council of Medical Research 65 g day person^{-1}) in 2008. Recently, under the National Food Security Mission (NFSM), high priority has been given to increase the production of pulses across the country to reduce the increasing imports, reduce protein malnutrition, and make pulses accessible to the common person. Most of the production of pulses has been gradually shifted from summer to winter and now the share of winter pulses has increased to about 61.0% of the total pulses production.

9.9 Constraints of Pulse Production in India

9.9.1 Extreme Weather Conditions

The poor soil and agro-climatic conditions of rainfed pulses reduce the length of the growing period and lead to the freezing of the early vegetative stage, which stagnate all biological activity for a longer period. The sudden temperature rise thereafter induces not only forced maturation but also invites several biotic stresses, such as diseases and insect pests (Ali et al. 2012; Reddy 2009; Singh and Singh 2008). Traditionally winter pulses sowing was delayed up to the last week of November and sometimes under extreme circumstances it goes up to the first fortnight of December. However, the optimum-sowing time for chickpeas and lentils is the first fortnight of October (Ramakrishna et al. 2000). Few winter legumes including lentils are also grown as a *paira* crop in eastern India, which helps in the timely planting of the crop even before, the paddy has been harvested (Singh and Singh 1995).

9.9.2 Unfavourable Weather Conditions

There has been a high degree of risk in the production of pulses. Presently more than 87% area of pulses is dependent on rainfall. The average rainfall of major pulse-producing states like Madhya Pradesh, Uttar Pradesh, Gujarat, and Maharashtra is about 1000 mm and the coefficient of variation of rainfall is 20–25%. Drought stress is a major cause of crop failure. Terminal drought and heat stress result in forced maturity with low yield levels. Drought stress alone can reduce seed yield by up to 50% in the tropics.

A huge jump in productivity can be achieved by implementing life-saving irrigation, especially in winter (*rabi*) pulses grown on residual moisture. The irrigated area under pulses has remained almost constant at 13% of the total area. The availability of sufficient soil moisture for crop growth depends on rainfall, water-holding capacity, and soil depth in rainfed areas. In South India, the water holding capacity of the soil often limits the grain yield to 50% under irrigation on

Alfisol. In contrast, on Vertisol soils, growth is reduced by 5–20% due to high water holding capacity. High evaporation during the *rabi* season in South India under drought poses a serious constraint to the gram yield. Other major problems are soil salinity and alkalinity, which are high in the Indo-Gangetic plains in semi-arid tropical and irrigated areas,

Grain yield is mainly affected by temperature. Freezing is an abiotic stress that limits the grain yield of pulse crops. All warm-weather pulses are sensitive to low temperatures but are generally not exposed to low temperatures. On the other hand, cold-season pulses (like chickpeas) are often subject to cold temperatures, especially in regions of northern India. However, there has not been much improvement in the development of chilled and frost-tolerant varieties.

Poor drainage or stagnation of water during the rainy season causes heavy damage to pigeon peas due to low plant conditions and incidence of *Phytophthora* blight disease especially in the states of Uttar Pradesh, Bihar, West Bengal, Chhattisgarh, Madhya Pradesh, and Jharkhand. Ridge planting is very effective in ensuring optimum plant stand and consequently high yield. Since most pulse crops are drought tolerant, much of the research effort has been focused on the genotype and associated production technology for dryland conditions.

There is a lack of germplasm adapted to high rainfall and irrigated conditions. Drought, water logging, temperature extremes, wind or hail, alkaline and saline soils, acidic soils, and various mineral nutrient deficiencies or toxicity are common abiotic stresses limiting pulse production in India. In chickpeas, pigeon peas, green gram, lentils, and black gram, drought is the main reason for the reduction in yield as they are grown in conditions of residual soil moisture or face soil moisture deficit during their breeding phase and end up with terminal drought.

To overcome such a situation, better management techniques to conserve soil moisture and maximize crop transpiration on soil evaporation can help reduce the impact of drought in pulses. In some situations, higher soil moisture induces more vegetative growth as well as living in chickpeas and lentils, as well as good drought resistance and production in lathyrus when sown on drought-prone top soils. Grain yields are likely to be highly affected by temperature extremes with recent changes in global temperature. Terminal drought stress or high temperature during maturation in chickpeas and lentils results in poor pod filling; Frost injury occurs due to low temperature.

Lathyrus and chickpeas can establish well in waterlogged soil even when sown in rice and can grow to maturity in waterlogged soil. Not all other pulses tolerate waterlogging conditions. Poor soil and agro-climatic conditions not only force the late sowing of legumes, but also reduce the length of the growing period and the need to sustain cold sores freezes all biological activities in the early vegetative stage that are long-term. A sudden temperature rise induces forced maturation as well as invites several biotic stresses, such as diseases and insect pests (Ali et al. 2012). Traditionally, the sowing of *rabi* pulses was delayed till the last week of November and sometime in extreme circumstances it goes up to the first fortnight of December, apparently due to the reasons already mentioned. However, some winter pulses including lentils are also grown as a *paira* crop. In eastern India, which helps in the early sowing of the crop, the paddy has been harvested. Not surprisingly, Indian farmers are still exposed to the uncertainties of monsoon. Therefore, Indian agriculture needs a fight against the monsoon.

9.9.3 Unfavourable Soil Conditions

In general, the pulse crop prefers neutral soil responses and is very sensitive to acidic, saline, and alkaline soils. Indian soils, in particular, the northwestern soils that have a high pH, in contrast to the eastern and northeastern parts, are known as acidic soils. Micronutrient deficiencies are evident to acute deficiency levels due to these soil conditions. Acute deficiency in relation to zinc, iron, boron, and molybdenum and secondary nutrients such as sulphur was observed in particular in conventional pulse growing (Singh et al. 2013). Most of the pulses in India are grown in low fertility, problematic soils, and unpredictable environmental conditions. India has experienced unprecedented population growth and India's population is estimated to be 1.69 billion in 2050. The increasing population is largely responsible for the fragmentation of land holdings resulting in low productivity of the land. In general, leguminous crops prefer neutral soil responses and are very sensitive to acidic, saline, and alkaline soil conditions. Sharp deficiencies with respect to secondary nutrients such as zinc, iron, boron and molybdenum, and sulphur are particularly observed in traditional pulse growing areas. Soil is an important resource and nutrient degradation in the soil is a serious problem, leading to the degradation of soil health. This is only due to deforestation and mismanaged agricultural practices. The increase in salinity and alkalinity is due to mismanagement and intensive use, which are other causes of loss of soil fertility. Micronutrient deficiencies also adversely affect the production of lentils, chickpeas and other pulses, which become more pronounced after the management of biotic stress. Experiments have shown that micronutrient deficiencies such as sulphur and zinc are very common in pulse growing areas. In traditional pulse-growing areas, *Rhizobium* spp. is present. There is effective nodulation by being present in the soil, but when introduced to new areas the host specific *Rhizobium* also needs to be introduced via inoculation also highly unscientific methods of irrigation cause further damage to the fertility level of the soil. As per the Agricultural Census 2010–2011, marginal and small land holdings (<2 ha) account for 85% of the total operational holdings and 44% of the total operated area. Small farm size precludes mechanization. This creates difficulties in the adoption of contemporary inputs, scientific land reclamation, water conservation, and plant protection measures.

9.9.4 Agronomic Limitations

An earlier study showed that the area under pulses is mostly predetermined, but as the irrigated area increases, pulses are shifted to rainfed areas and their area is replaced by cereals or some cash crops. In India, the irrigated area under pulses was only 12%, while under wheat and paddy; it was more than 60% of the total area (Reddy and Reddy 2010). Improper sowing time, low seed rate, faulty sowing method, inadequate irrigation, inadequate intercultural operation, and sowing below the utera without proper management is the major agronomic limitation (Ramakrishna et al. 2000; Reddy 2009). As a result of late sowing and early encounter with severe cold, the growth and development of winter pulses get stunted for a considerable period. Plants then have comparatively less time to complete their life cycle, which forces overall maturity (Ramakrishna et al. 2000). In eastern India, the normal sown lentil is a medium duration (130–150 days) crop, while under late sown conditions it is forced to complete its life cycle in 105 ± 5 days (Joshi 1998; Ramakrishna et al. 2000; Reddy 2009; Singh and Singh 2008; Singh et al. 2012). Generally, late-sown winter pulses chickpeas and lentils go through three distinct stages Phenological modifications at initial stages are bound to happen. Finally, these crops grow slowly during its early seedling stage due to the energy invested in the initial establishment (Singh et al. 2002). However, in the middle stage, very negligible growth and development is observed. This is a serious threat to the yield potential due to cold injuries. This stage is very important to create a source of channelizing energy in a later stage. In the last and most critical stage, pulses suffer heat injury, resulting in the onset of the reproductive phase, leading to imbalances in resources and inputs, biotic stress and forced maturation (Joshi 1998; Dixit et al. 2009; Reid et al. 2011; Singh and Bhatt 2013).

9.9.5 Huge Crop Loss Due to Insect Pests and Diseases

Pulses crops are affected by various pests and diseases during the growing stage and after the harvest. The estimated losses are in the range of 15–20% or more. Thus, India can increase the availability of pulses by 15–20% with appropriate crop protection research and development. It is an important way of bridging the gap between demand and supply. As a strategy to tackle this situation, there is a need to develop varieties with combined resistance to most common and major biotic and abiotic stress factors. Recent advances in genomics research have expanded the scope for the development of many resistant cultivars, which are to be exploited further.

9.9.6 Biotic Constraints

9.9.6.1 Occurrence of Insect Pests and Diseases

Pests, diseases, and weeds are the major biological factors limiting the productivity of pulse crops. Insects are one of the major determinants for achieving high production in agricultural crops. It was estimated that about 26% of the potential production is sheltered by consuming insects. In India, 30% of crop loss is recorded every year due to insect pests and diseases. The production deficit has increased over the years. Although legume crops are prone to many insect and seed-borne diseases, a major cause for concern is crop failure if it is not controlled. In addition, legume grains suffer heavy damage due to pests during storage. The perception and high incidence of disease and pests cause high losses resulting in low production and high protection costs. Secondly, there is limited availability of resistant/tolerant varieties to the farmers. The main reason for this could be a weak seed production programme. Incorporating insect resistance genes without compromising yield in field validation trials is still commercially viable. Presently, chemical pest control is the only option left for the farmers for effective control of pests and diseases. Although many plant protection chemicals have been developed over time and methodology, the use of pesticides in pulses is still very low. In general, farmers spray the chemical at the stage where the damage exceeds the economic threshold. This clearly shows that technological stagnation is primarily responsible for the backwardness of pulses across the country. More than 250 insect species are reported to affect pulses in India. Of these, about a dozen cause heavy damage to the crop. On an average 2–2.4 million T of pulses with a monetary value of about INR 6000 crore are lost annually due to the destruction of the insect pest complex. Among them, pod borer (*Helicoverpa armigera*) causes the most damage, followed by pod fly, wilting, and root rot. For example, pod borer (*Helicoverpa armigera*), *Fusarium* wilt, root rots, *Ascochita* blight, and *Botrytis* graymold are some of the major biological barriers to increasing chickpea productivity. Major barriers to productivity in pigeon pea are biotic stresses such as pod borer, pod fly, fusarium wilt, and sterility mosaic disease. Similarly, the major pests and diseases affecting lentil production in India are pod borers, aphids, cutworms, powdery mildew, rust and wilt. The abundance of legumes in N and P makes them attractive to pests and diseases. Recently several successful trials have been conducted to control pod borer using nuclear polyhedrosis virus (NPV), which has been found to be more effective in bringing about high and quick mortality. Another important pest affecting pulses are nematodes, of which root knot nematodes are important in terms of spread and damage to crop yields and have been effectively controlled by bio-agents. Pulses crops are prone to many insect and seed-borne diseases, crop destruction is a major cause of concern if its incidence is not controlled. *Fusarium* wilt is widespread in legume-growing regions. In addition, legume grains suffer heavy damage due to pests (bruchids) during storage. Legumes are a pest-free crop under normal conditions if proper crop rotation is followed. Although pod borers, aphids, and wilt are the major pest and disease pests (Singh et al. 2013) (Table 9.3).

9.9.7 Input Quality and Availability-Related Constraints

The timely availability of quality chemical fertilizers remains a major problem in many pulse-growing areas. The inadequate availability of gypsum or pyrite as a cheap source of sulphur continues to be a serious constraint in many states/territories. The nutrient requirement of legumes is much lower than that of cereals, primarily due to the level of biological nitrogen fixation, although legume crops generally respond favourably to higher doses of fertilizer nutrients than applied or

Table 9.3 Important diseases and insets pests of major pulses

Crop	Disease	Insect pests
Chickpea	*Fusarium* wilt, *Ascochyta* blight, *Botrytis* grey mould and stunt virus	Pod borer and cut worm
Pigeon pea	Sterility mosaic virus, *Fusarium* wilt, *Phytophthora* stem blight *Alternaria* leaf spot and powdery mildew	Pod borer and pod fly leaf tier
Blackgram/greengram	Yellow mosaic virus, *Cercosora* leaf spot, powdery mildew, leaf crinkle virus, and root rot	White fly, Jassid and pod borer
Lentil	Rust, wilt *Sclerotinia* blight, collar rot	Pod borer
Field pea	Powdery mildew, rust, downy mildew, wilt	Pod borer, stem borer, leaf minor

recommended. The availability of insecticides (including herbicides) has been comfortable in most states, but their quality in terms of effectiveness and eco-friendliness has been an issue, despite a well-designed regulatory mechanism. The sustainable use of fertilizers is one of the important reasons for low productivity and declining soil fertility. The average N:P:K ratio over the past two decades has been 7:3:1 as against the recommended 4:2:1 (https://silo.tips/download/indian-agriculture-challenges-and-prospects).

Except soybeans, most pulses are on the lowest priority for farmers to use the recommended amount of fertilizers. Average use of chemical fertilizers for pulses results in lower yields. For the growth and development of root nodules, phosphorus is essential and an average application of 40 kg P_2O_5/ha is recommended. With the withdrawal of subsidies on fertilizers, the decline in the use of non-nitrogen fertilizers has an adverse effect on yield. Evidence suggests that 100% of the nitrogen requirement for activating the nodulation process can be met through inoculation of efficient strains of *Rhizobium* when coupled with sound agricultural practices. However, studies show that the adoption of these biofertilizers is negligible. If this technology were as efficient, it would still have to be "pushed" by government agencies, as there would be substantial demand by this time. The continued use of *Rhizobium* inoculants in the long term appears to be difficult. Of course, strict quality control standards need to be implemented in the manufacture and sale of inoculants.

9.9.8 Varietal Constraints

The major constraints to the cultivation of pulses include the availability of desired quality and quantity of high-yielding varieties of pulses. Shortage of high-yielding varieties, low crop index, high susceptibility to diseases and pests, flower drop, short duration varieties, intermediate growth habits, poor response to inputs, and instability in performance are some of the varietal constraints which require immediate attention. High-yielding varieties need lots of water in addition to the support of fertilizers and pesticides, which becomes riskier due to the lack of assured irrigation facilities and erratic monsoon. The efficiency of inputs e.g. fertilizers, pesticides, and irrigation are primarily gritty by the quality of the seeds used. It is reported that seed quality is 20–25% of the productivity. Thus, the timely availability of quality seeds at reasonable prices to the farmers is essential for achieving high productivity and production. Several new high-yielding varieties were developed in the last two decades, but their performance is limited to a 10–20% higher yield than the local varieties. Because of inherited vulnerabilities, these varieties perform poorly at the field level, and a modest increase in yield does not appeal to farmers or cause any significant change in the national level of production. The need is to develop varieties with better yield gains and desirable characteristics, which are best suited for semi-arid climates. The ever-decreasing pattern of shrinking land holdings discourages farmers from growing medium to long-duration varieties, which occupy the land for 240–270 days. Medium and long-duration varieties do not allow farmers to grow cereal crops like wheat and paddy, which provide minimum cash income for the family and year-round food security.

9.9.9 Lack of Quality and Improved Seed Availability

Since the introduction of the Integrated Pulses Improvement Program in 1967, about 400 improved varieties of various pulse crops have been released for cultivation. However, at present only 124 varieties are being produced. Of these, a dozen are popular among farmers. The wide gap between the requirement of certified/quality seeds and their distribution in India is a major concern. The seed replacement ratio is very low (2–5%) while the required seed replacement ratio is 10% as both public and private agencies are not able to meet the quality seed requirement. Under NFSM, the responsibility of breeder and foundation seed production is entrusted to IIPR (Kanpur), while the production of certified seeds is given to National Seed Corporation and other state organizations.

9.9.10 Socio-Economic Constraints

Even though India is the world's largest producer of pulses, it still imports large quantities of pulses from the rest of the world. Farmers generally have limited access to inputs due to low purchasing power and access market to sell surplus produce pulses. Due to this situation, farmers give first priority to staple cereals and cash crops for input allocation and the second priority is given to pulses. In addition, there is a lack of policy support and post-harvest innovations related to pulse crops. The availability of quality seeds and other inputs of improved varieties is one of the major constraints in increasing the production of pulses. Pulses production is characterized by a very high degree of diversity in terms of both the number of crops and their spatial distribution under different agro-climatic conditions. Most of these crops are region specific in the sense that a state or a group of states is responsible for a large part of the area and production of a specific pulse crop.

Pulses like peas, lentils, Lathyrus, and even chickpeas show their regional distribution pattern. This diversity has many implications. Firstly, it places serious limitations on a single national policy for promoting pulses production in the country and regional crop-specific strategies for pulse development programmes. However, given the meagre resources available for pulse development, this diversified approach can mean spreading resources too little and in turn undermining the effort. This dilemma may partly explain the absence of any major emphasis on research on pulses, which in turn is partly responsible for their stagnation. The structure of pulse production is also characterized by the dominance of two crops, viz. chickpea and pigeon pea, which together account for more than half of the total pulse area in India. Therefore, if both these crops suffer from adverse climatic conditions, it reduces the production of pulses significantly.

9.9.11 Post-harvest Technology and Infrastructural Constraints

Post-harvest losses account for more than 10% of total pulses production, with the highest loss of 7.5% in storage followed by processing losses to the extent of about 2% including threshing and transportation. Damage by storage insects (5%) is more likely than wheat (2.5%), paddy (2%) and maize (3.5%) (Deshpande and Singh 2001). There is a need to popularize storage structures like metal storage bins. The processing capacity of pulse mills has recently increased from 65–66% in traditional pulse mills to 70–75%. Mini pulse mills should be popularized by the formation of pulse producer groups and processing unions to reduce post-harvest losses and increase rural employment. Rainfall during the maturation of summer and monsoon season pulses results in loss of yield and deterioration in the quality of cereals when farmers generally do not have a covered threshing floor. There is also a lack of awareness and means among farmers for the safe storage of grains/pulses. Many areas are accessible only during good weather. Storage facilities are either inadequate or inaccessible.

9.9.12 Credit, Marketing, and Policy Constraints

The farmers engaged in the cultivation of pulse crops are mostly small and marginal (Singh et al. 2015). Most of the farmers are present in areas with poor banking infrastructure. They have a poor resource base and lack of risk appetite ability. Therefore, they either do not have access to loans or become defaulters. Even the delivery of loans to such farmers is not hustle-free. A marketing network is lacking in remote areas. Procurement of produce by a dedicated Agency is virtually ineffective. However, there is a need to make the system of regulating the quality of inputs more effective in all states. Need to streamline the delivery of better technology, inputs, and credit through appropriate policy interventions. There is a need to give the benefit of crop insurance to pulse farmers.

9.9.12.1 Lack of Cash and Credit

Cash is a key element to enable small farmers to move from low input-low output to high input-high output agriculture (Reddy 2009). Nevertheless, these farmers have less access to credit because their asset base is less; the risk is less capacity and high-risk environments. Moreover, credit facilities for pulse crops from both formal and informal sources are limited due to volatile returns. This can be effectively dealt with by credit-linked insurance to pulse crops without any collateral security of recommended practices.

9.9.12.2 Marketing

The pulses markets are thin and fragmented due to scattered production and consumption across the states (Reddy et al. 2013). Farmers/rural traders sell their market surplus immediately after harvest, while some big traders/wholesalers trade in major markets and hoard pulses to take advantage of speculative profits. Due to this, the farmers are not getting the benefit of high price of pulses in the market. In recent years, market information and infrastructure have improved, and the gap between consumer price and producer price is narrowing, especially during the harvest season. Market prices are largely controlled by local buyers, regardless of the Minimum Support Price (MSP), and procurement of pulses by government agencies is very limited. The government has suggested procurement of pulses at MSP while local market trends suggest that the rates offered by local dealers are much lower than the MSP. Therefore, pulse growers have doubts about the effect of MSP on the actual market prices. There is a need to encourage investment in market infrastructure, warehouses, and market information systems in both public and private sectors through Public-Private Partnership (PPP) model and economic viability gap funding model. Presently, GOI has taken an initiative to increase MSP to promote pulse production in the country (Fig. 9.4), which has a great influence on farmers' interests in pulses production.

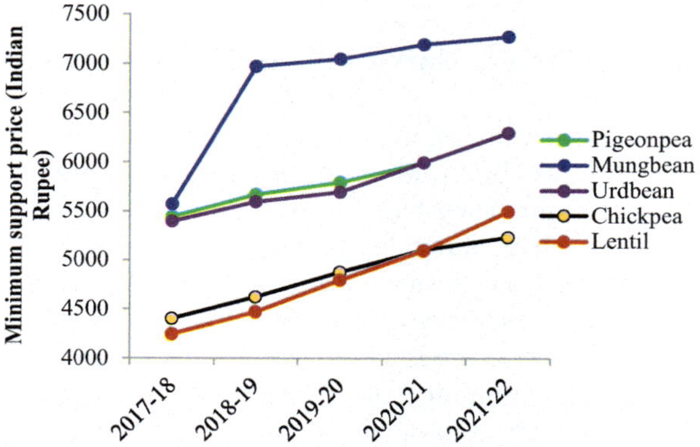

Fig. 9.4 Increase in minimum in support price in last 5 years (2017–2018 to 2021–2022) (Directorate of Economics and Statistics 2021)

9.9.12.3 Flaw in Policy Perspectives

Due to a number of economic and political compulsions, agricultural development strategies were concerned with the aim of achieving rapid growth in food production by concentrating resources and efforts at the level of better affluent areas and cultivators.

9.9.13 Farm Mechanization

Mechanization in agriculture is an essential input and increases productivity, and reduces human labour and the cost of cultivation. Agricultural electricity availability is to increase to 4.0 kW ha^{-1} by 2022 from the presently available 1.84 kW ha^{-1}, yet there are many parts of the country where agricultural operations are dependent on human labour and livestock. Mechanization of agricultural operations is one of the reasons for the success of the expansion of the area under chickpea in Andhra Pradesh, which needs to be replicated in Bihar. This can be further increased by developing suitable varieties for harvesting. There is a need to encourage agricultural mechanization activities such as harvesting and threshing in peak season through the distribution of subsidized agricultural machinery to tackle labour shortages and high wages. Different agro-climatic zones and soil types are suitable for pulses production. Mechanization is an essential tool for increasing productivity in many soils, which includes the adoption of deep ploughing, ridge making, combined harvesters, etc. The mechanization contributes to the timeliness of operations such as planting, line sowing, intercultural operations, reducing the cost of production, and improving resource (water, energy, and input) use efficiency. Keeping in view the smallholdings of farmers, custom hiring of implements or machines is the only practice to increase the reach of agricultural mechanization. In India, many pulses are harvested by hand, as the available varieties are not suitable for mechanical harvesting (Singh et al. 2019). In developed countries like Australia, Canada, and America, pulses like gram, lentils, etc. are harvested mechanically. With ever-increasing labour costs, manual harvesting has become an expensive field operation for many crops including pulses in India and farmers are increasingly opting for mechanical harvesting where it is possible. The availability of cultivars suitable for mechanical harvesting will decrease the cost of production and increase the cost of production to the farmers in the cultivation of pulses. The availability of cultivars suitable for mechanical harvesting will decrease the cost of production.

9.9.14 Post-harvest Handling of Grains for Reducing Losses

There is a need to promote mechanical threshing with the provision of incentives for the purchase of threshers. Procurement of pulses by government-authorized organizations will significantly reduce the storage requirement at the farmer level. Mini pulse mills should be popularized and promoted through various initiatives including incentives. The private sector should be involved to encourage the setting up of pulse mills in rural areas/districts having a large area under pulses on the lines of sugar mills. There is a need to involve private companies in the processing, packing, and marketing of pulses. Public sector procurement agencies are severely handicapped in funding and expertise in this area.

9.9.15 Subsistence Nature of Farming

Indian agriculture is classified based on its subsistence nature, i.e., most of the products are directly consumed by the producers, and the surplus, if any, is usually less. The reason for this is that most Indian farmers are out of use due to poor equipment and technology, and are not able to afford costly inputs. This result in low returns and meagre income, which means fewer savings and reinvestment. Indifferent behaviour towards the adoption of new technologies is also a major limiting factor for improving pulse productivity.

9.9.16 Awareness and Access to New Technology

Lack of awareness of farmers about improved varieties or availability of quality seeds as well as various new technologies is major constraint in the mass production of pulses. Television will be the most popular media for raising awareness; Trials and farmer fairs/field days would also be helpful. Subsidies need to be given for widespread adoption of the identified technology.

9.9.17 Physiological Limitations

The pulses being C3 plants inherently suffer from low yield potential and are a physically inefficient group of plants compared to cereals (C4 plants) such as sorghum and maize. However, the troubling future is that pulses have a lower Harvest Index (HI) than cereals. In pulses, it is only 15–20% as compared to 45–50% in the case of cereals like wheat and rice. Excessive vegetative growth results in a low HI but may result in the early splitting of dry matter into seeds and promote good growth of C4 plants. Low HI results from excessive vegetative growth but can be overcome by an early splitting of dry matter into seeds and by developing biotechnological and genomic tools to incorporate the good characteristics of C4 plants into C3 plants. In general, the rate of flower drop is high in pulses. In pigeon pea, more than 80% of the flowers produced in a plant are lost; yield can be increased significantly by reducing flower drop. This can be done either by breeding lines that maintain a large proportion of flowers that produce pods or through physical manipulations, such as the spray of hormones that reduce flower drop. Physiological studies at ICRISAT, India including the removal of pigeon pea flowers and young pods, have shown that plants compensate for the loss of flowers and young pods by setting pods from later formed owners, which otherwise would have dropped. This compensatory mechanism provides substantial plasticity for adaptation to intermittent adverse conditions such as moisture pressure or insect attack, which are common in warm rainfall regions of South India. The recent increase in yield levels in pigeon pea has been due to the release of longer-duration (annual) varieties, which make maximum use of assimilation to fill the available sink of a large number of flowers (Wani et al. 2002).

9.10 Realizing Potential Productivity

Indian agriculture is endowed with the second largest arable land area in the world with the largest number of different agro-climatic ones, almost round the year bright sunshine, 1120 mm of annual rainfall, numerous rivers across the country, 7000 km long coastline, exceptional biodiversity, and abundant labour force. Since pulses are always subjected to abiotic stresses leading to sub-optimal nutrient absorption, farmers use low doses of fertilizer nutrients, which are the major culprit in pulses production. In addition, nutrient use is imbalanced and rarely based on soil test values. However, the wide spectrum of agro-climatic conditions, favourable thermal regime or availability of temperature appropriation, and adequate rainfall almost throughout the year for cultivation point to the fact that there is immense untapped potential to improve the productivity of pulses and bring the additional area under pulses. There is overwhelming scientific evidence suggesting a vast gap between farmer's yield and front-line performance plot yield. There is a need for the promotion of high-yielding varieties along with a selection of suitable varieties for different agro-climatic conditions. Further use of disease and pest-resistant varieties is highly advocated under the current climate change scenario. Pulses are an important source of high-quality protein with cereal protein supplementation for India's substantial vegetarian population. Pulses can be produced with minimal use of resources and hence become less expensive than animal protein. Pulses meet their nitrogen requirements largely by fixing atmospheric nitrogen in their root nodules. Pulses can provide a sustainable solution in a rainfed area that is 67% net sown area contributes 44% of food grains and supports 40% of the population.

There are remarkable achievements in the development of high-yielding cultivars and pulse production technologies made by the dedicated efforts of the Indian Council of Agricultural Research (ICAR) through the Indian Institute of Pulses Research (IIPR) and various All India Coordinated Research Projects (AICRPs) on Pulses and a network of partners. Besides, the partnership of ICAR with international institutes, particularly International Crop Research for Semi-arid Tropics (ICRISAT) in chickpea (desi and kabuli) and pigeon pea; International Center for Agricultural Research in the Dry Areas (ICARDA) in lentil, chickpea (kabuli) and grass pea; and World Vegetable Center in mungbean and urdbean, have been very fruitful. Germplasm and breeding materials supplied by these institutes have contributed to the development of several widely grown cultivars across the country.

9.10.1 Vertical Approach to Increase Production

In the vertical approach, possible methods and techniques are used which help in getting more production without expanding the crop area, and here the focus depends on how to improve the productivity of the crop per unit area. The vertical approach can be used efficiently and effectively to increase pulse productivity.

1. Encourage promotion of sequential cropping, intercropping, and *utera* cultivation
   There are a good number of promising intercropping systems for pulses developed by Agricultural Research Stations. Farmers in rainfed states (Gujarat, Madhya Pradesh, Chhattisgarh, Maharashtra, Karnataka, and Andhra Pradesh) are familiar with some of them as they have been practicing them in traditional ways.

   Approaches to be adopted by rainfed states should include:
   (a) Identification of districts and promising intercropping systems and establishment of area coverage targets for each agro-climatic zone
   (b) Organizing intercropping demonstrations with farmer's active participation and comparing returns with a single cropping system
   (c) Ensuring the availability of seeds of improved pulse varieties recommended for intercropping
   (d) Demonstration of seeding equipment (animal-drawn and tractor-pulled) suitable for simultaneous planting of main and intercrop components
   (e) Pulses seed minikits can be given only to farmers opting for intercropping
   (f) KVKs at district level should be involved in farmers' training and field demonstration of production technology

9.10.1.1 Adoption of Promising Agronomic Practices

Agricultural practices that have a major impact on the productivity of pulses include tillage, crop geometry, plant population, planting method and timing, nutrient and water (rainwater and irrigation) management, seed treatment (with fungicides) and crop-specific bacterial culture, weed management, and plant protection. Crop-specific recommendations are developed by regional agricultural research stations based on applied and adaptive research findings generated in different agro-climatic zones (Ghosh et al. 2014).

9.10.1.2 Balanced Input Supply (Chemical Fertilizer, Bio-Fertilizer, and Plant Growth Regulator)

The widespread deficiency of zinc and sulphur in major pulse-producing states and boron deficiency in acidic soils of eastern and northeastern states has necessitated the use of sulphur containing fertilizers and zinc sulphate sources of zinc. For this, 20–40 kg ha^{-1} sulphur (through gypsum, SSP) at the time of sowing and zinc sulphate @ 25–50 kg ha^{-1} once in 2 years effectively removes the problem and helps in maximizing crop productivity (Singh et al. 2013).

9.10.2 Nutrient and Water Use Efficiency of Soil pH Improvement

The use of gypsum is limited in the western states and some parts of the eastern and southern states. The N requirement of pulses is much higher than that of cereals. However, most of the requirement is met through biological N-fixation. The growth and development of pulse crops are extensively controlled by Plant Growth Regulators (PGRs). To trigger vegetative growth, auxin plays an important role, especially under low ambient temperatures and cold rhizosphere regimes. However, the availability of the quality of *Rhizobium* inoculums is limiting. Phosphorus is becoming a limiting macronutrient that will affect the production of pulses. A common difficulty in reclaiming P from the soil is that it is not readily available to plants because P reacts with aluminium, iron, and calcium in the soil to form complexes (Hazra et al. 2018). These nutrients are essentially insoluble resulting in very little movement of P into the soil solution, and none of the compounds can be taken up directly by the roots. Phosphate solubilizing bacteria (strains from the genera of Pseudomonas, Bacillus, and *Rhizobium* are among the most powerful P solubilizers) as inoculants simultaneously increase P uptake by the plant and thus crop yields.

Therefore, fertilizers containing micronutrients readily available to small-holding farmers in remote areas will be of great help in increasing the productivity and production of pulses. Most pulse growing areas have low to moderate populations of native *Rhizobium* and seed inoculation with biofertilizer (*Rhizobium*) can increase pulse productivity by 10–12%. Lack of quality enhancement is also a major constraint in the use of biofertilizers. Vesicular-arbuscularmycorrhiza (VAM) Promises to improve the supply of Phosphate and micronutrients such as zinc in pulses. Phosphate-soluble bacteria (PSB) are another group of hydrotropic organisms that may be applicable in rainfed production systems, especially in soils with poor phosphorus availability (Venkatesh et al. 2019a, b). Combined use of cultures such as double inoculation of *Rhizobium* and VAM yields higher yields than *Rhizobium* alone.

9.10.3 Resource Conservation Technologies for Expansion of Irrigation

Supplemental irrigation with a limited amount of water, if applied to rainfed crops during critical stages can result in substantial improvement in yield and water productivity. In view of the good response for supplemental irrigation to pulse crops, the government should encourage policies to provide supplemental irrigation. The use of sprinkler irrigation has immense potential for saving water and expanding the area under irrigation. This method has gained popularity in many districts with limited water resources. Drip irrigation has also attracted the attention of policy makers, administrators, and social activists, as it has assumed social, economic, and ecological dimensions. Supplemental irrigation with limited amounts of water, if applied to rainfed crops during critical stages, can lead to substantial improvements in yield and water productivity. Given the good response to supplemental irrigation to pulse crops, the government should encourage policies to provide supplemental irrigation.

9.10.4 Use of Herbicides

Another production method where the cost of cultivation can be reduced substantially is to promote the use of post-emergence herbicides to control weeds by developing herbicide-tolerant varieties (Kumar et al. 2014; Nath et al. 2021). In general, pulses are sensitive to herbicides and at present manual weeding is the only option for weed control. Management of weeds in pulses is becoming costly and in some cases economically unviable due to the high labour cost involved in manual weeding. Herbicide-tolerant varieties offer an opportunity to control weeds through need-based applications of herbicides. Weed management through herbicides is not only economical but also facilitates zero-tillage or minimal tillage methods.

9.10.5 Seed Replacement/Multiplication Strategy

It is possible to increase the productivity in pulses by incorporating new HYV to achieve maximum yield. The major issue related to the promotion of quality seeds is the timely availability of quality seeds and the promotion of promising varieties in sufficient quantity to the farmers. The use of good quality/certified seeds in pulses is inadequate. Therefore, efforts should be made in this direction through various means sponsored by the government. Programs like the National Food Security Mission (NFSM), Integrated Scheme of Pulses, Seed Village Program, etc. These efforts have been successful to increase the seed replacement rate (SRR) of pulses to 22.5% by the year 2010–2011. Despite a long list of improved pulse varieties released for cultivation, their impact is yet to be fully realized by resource-poor farmers in many states in India. Small farmers' access to quality seeds of improved pulse varieties is hampered by both inadequate demand generation and limited supply.

The situation is further compounded by the unfavourable and inadequate policy support and regulatory framework, inadequate institutional and organizational setup, and deficiencies in production and supply infrastructure and the socio-economic status of the farmers. Several constraints limit the performance of seed systems in India, including limited access for smallholder farmers to seed-improved varieties; limited supply quality (breeder, foundation, and certified) seeds of farmer and market preferred varieties; lack of interaction between national seed production organizations and policy-making bodies. On the seed supply side, the pulses seed business generally does not attract large seed companies because the profit margin is low. More than 95% of lentil seeds (the leading global lentil producer) in India come from the informal sector (Materne and Reddy 2007). The situation concerning other pulses in India is similar. The seed replacement rate in chickpeas varies from 14% in

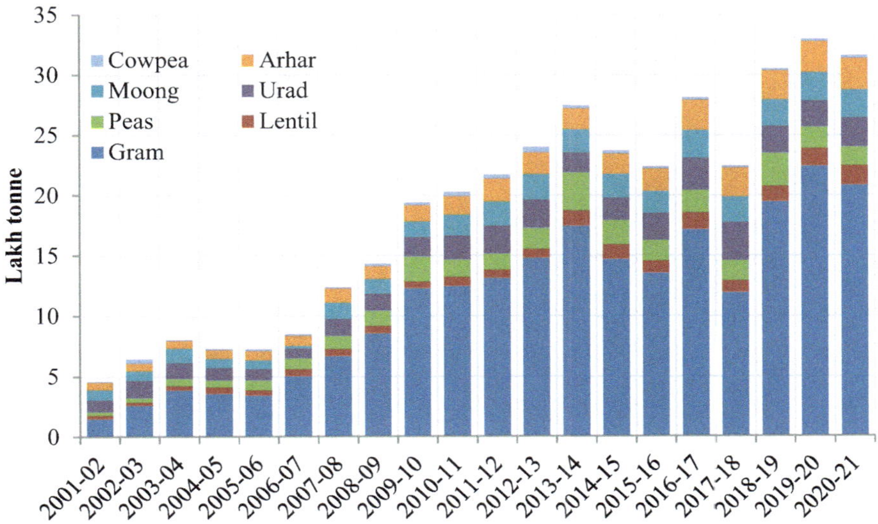

Fig. 9.5 Crop-wise distribution of certified/quality seeds of different pulses in India (Directorate of Economics and Statistics 2021)

soybean to 35% in India, thus indicating that most farmers still use their own saved seed (Fig. 9.5).

This situation is due to several factors including low seed multiplication rates of legumes and the reuse of grain from the previous crop as seed. Demand for specific varieties adapted to the often more narrow agroecology and consumer needs. Furthermore, when seed production occurs, it is often in high-potential areas, with seed reserves concentrated in areas with high population densities or areas with better infrastructure (i.e. remote, not stress-prone areas). As small and medium seed companies are emerging and gaining strength, they are also creating effective demand for pulses seed. However, their capacities are still limited by inadequate and discontinuous access to foundation seed, inadequate capital investment, and lack of appropriate marketing strategies including delivery systems targeting remote and small-scale farmers. Public and private partnerships would be the best approach to increase the availability of foundation seed needed for subsequent seed classes. In developing countries such as India, particularly for pulses, the formal seed sector is highly subsidized and evolving at different stages of development.

The informal seed sector is and will remain a major player in legumes. In recent times, development partners and researchers have realized the importance and importance of quality seeds in agriculture and several projects have been implemented or are in progress to improve the seed availability of improved farmer preferred varieties to the farmers. The key issue in resolving access to quality seed will be a thorough understanding and critical assessment of the existing seed sector (both formal and informal), their constraints and comparative advantage and complementarities status (Fig. 9.6).

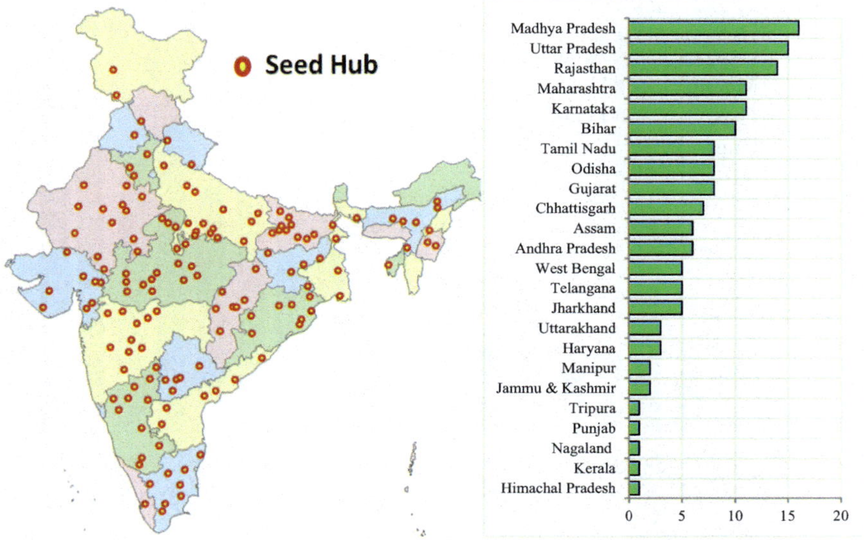

Fig. 9.6 State-wise distribution of 150 seed hubs for quality seed production in India (Source: ICAR-IIPR, https://iipr.icar.gov.in/seed_hub.html)

9.10.6 Efficient Pest Surveillance, Integrated Insect Pests, and Diseases Management

In the age of climate change, it is important that region-specific advisories be issued to guide pulse growers on pest control. There is a need to set up an effective pest monitoring mechanism at district level. Naturally, the pulse crop is attacked by more than one disease and pest at a time; Hence the need for multiple disease-resistant varieties. Recent developments in Integrated Pest Management (IPM) have given wide scope for cost-effective control of many pests and diseases. IPM is essentially a farmer's activity of using one or more management options to reduce pest populations below economic damage levels while ensuring the productivity and profitability of the entire agricultural system. A variety of chemical, biological and cultural methods have been found to reduce pest and disease damage. Properly planned cropping systems involving crop rotation or intercropping of non-host and host crops, different agronomic practices like the use of solar energy by summer ploughing preceding summer pulses are cost-effective components of IPM. Since IPM is knowledge-intensive, a systems approach involving various disciplines to evolve an Integrated Crop Management (ICM) should be the goal in the future.

9.10.7 Suitable Extension Approaches for the Adoption of Improved Pulse Production Techniques

Dissemination of better and better pulse production techniques should not only be region/agro-climatic region specific but also match with the resource base of the farming community. There is a need to look for an institution building, which collects the produce of the scattered farmers and connects them with businesses for better quality inputs and for efficient marketing of the produce. Similarly, the expansion strategy to be adopted should take into account the prevailing socio-economic condition of the farmers.

9.10.7.1 Short-Duration, High-Yielding Varieties

Matching the crop maturity period to the available soil moisture is a key strategy to survive drought stress. Therefore, the emphasis should be on crop improvement programs. Developing high-yielding, short-duration varieties that survive the ultimate drought is required. These shorter-duration varieties offer opportunities to incorporate a crop/variety into cropping systems with a narrow crop window or new production space. Introduction to high yield, short duration, *Fusarium* Wilt resistant varieties is a major breakthrough in improving pulse production under rainfed areas. Varieties for short seasons and hot climates with high adoption of improved varieties and production technologies are most suitable for south India. The major factors responsible for the significant increase in the area and production of chickpeas in central and southern India are:

Successful introduction of commercial farming through mechanized field operations and effective management of pod borer. Andhra Pradesh was once considered to be beyond the limits of chickpea cultivation, due to the hot and low season climate, but now has the highest average yield (1.4 T ha^{-1}) level in India. In Andhra Pradesh, more than 80% of the chickpea area is occupied by short-duration improved varieties (JG 11, KAK 2, JAKI 9218, and Vihar) (Gaur et al. 2012).

9.10.8 Improved Varieties with Drought Tolerance

Drought-tolerant varieties can provide a cost-effective long-term solution against the adverse effects of drought. The return on investment in breeding for drought tolerance is likely higher than other drought management strategies. Widespread dissemination of drought-tolerant material will provide sustenance to farmers who are more vulnerable to the shock of crop failure. Root traits, particularly root depth and root biomass, are known to play an important role in avoiding terminal drought through more efficient extraction of available soil moisture.

9.10.9 Horizontal Approach: Crop Diversification Under New Niches

Pulses have a wide variety of maturity periods, which enables their cultivation in multiple locations and in different production systems to increase production. Some examples are given below, but many more other crops and niches can be successfully exploited (Hazra and Bohra 2021). Efficient utilization of rice fallow lands and replacement of low productivity crops with pulses. The fallow land after the summer rice harvest that is estimated to be around 11.65 million ha can be potentially utilized for winter pulses (Kumar et al. 2020a, b, c). The rice fallow area is predominantly rainfed and is present over the states of Bihar, Madhya Pradesh, Chhattisgarh, Orissa, East Uttar Pradesh, West Bengal, and Jharkhand. About 25% of this area has the potential to support and strengthen post-rice winter pulses depending on soil type and depth. Thus, this 3–4 million ha of additional area can be brought under winter pulses. Assuming an average productivity of 600 kg ha^{-1}, this area can produce 1.8–2.4 million T of pulses. There is a need to encourage farmers through various incentives and area-specific expansion strategy for cultivation of pulses in the identified states. The necessary technical backup in terms of suitable short-lived varieties, nutritional supplements, and other agronomic practices must come from research organizations who has to conduct field demonstration on pulses in paddy fallow.

9.10.9.1 Chickpeas in Rice Fallows

The Gangetic plains of India, spanning four countries in South Asia, namely Bangladesh, India, Nepal, and Pakistan, is one of the most important agriculturally in the world. About 14.3 million ha rice area in IGP remains fallow during the winter season. These offer a huge potential for expansion of area under paddy fallow winter pulses such as chickpea, lentil, and Lathyrus (Kumar et al. 2020a, b). Lack of irrigation facilities, faster depletion of residual soil moisture, and poor accessibility of quality seeds are the major constraints to taking second crop pulses after rice in rainfed rice-fallow ecologies (Ghosh et al. 2016; Kumar et al. 2019a, b). Adoption of early maturing pulse crop cultivars and appropriate soil moisture conservation practices in rice fallow systems could be the strategic option for the sustainable intensification of rice fallows in India (Kumar et al. 2016; Hazra and Bohra 2021). Agricultural trials carried out by several state agricultural universities in five states of India (Chhattisgarh, Jharkhand, Orissa, West Bengal and East Madhya Pradesh) have shown extensively on agriculture that the short duration of gram and lentils after the harvest of rice can be grown successfully, and with a reasonably high yield level of 1–2.5 ha^{-1}.

9.10.9.2 Pigeon Pea in Rice-Wheat Cropping Systems

The rice-wheat cropping system is popular in the Indo-Gangetic plain of India and the system is currently facing several sustainability challenges including soil quality degradation, groundwater scarcity, declining factor productivity and environmental pollution (methane emission, residue burning etc.) (Dutta et al. 2022). Besides, the continued mono cropping of cereals has resulted in decreased soil fertility and increased incidence of pests and diseases, and posing a serious threat to the stability of the entire rice-wheat cropping system. Inclusion of pulses in the rice-wheat cropping system will go a long way in restoring soil fertility and reducing other related problems (Ghosh et al. 2020a, b; Nath et al. 2019; Hazra et al. 2020). Pigeon pea-wheat system could be Concerning the intensive energy, water, and input consumption in the rice-wheat production system in the northwestern plain regions, pigeon pea-wheat could be a diversification alternative for production sustainability (Singh et al. 2018).

9.10.9.3 Indo-Gangetic Plains and Rice Fallows

Pulses have been introduced in Indo-Gangetic plains with the development of short duration (150–170 days) varieties of pigeon pea like UPAS-120, Manak, AL-15, AL-201, etc. in irrigated areas of Punjab, Haryana, Delhi and western Uttar Pradesh under pigeon pea-wheat based cropping system. Similarly, short-duration (60–70 days) green gram varieties with synchronous maturity and resistance YMVs (PDM-11, PDM-54) offer good scope for their introduction as a catch crop in the rice-wheat system (Hazra et al. 2019). Pulse-based cropping systems are low-input, requiring less labour, water, pesticides, and fertilizers. The rice-pulse cropping system is also better than the rice-wheat cropping system. Overall, the pulse-based cropping system is more suitable for resource-poor farmers and water-scarce areas (Kumar et al. 2018). Therefore, policy options should be developed to include at least one pulse crop in the cropping systems to maximize the benefits of irrigated farming systems. However, these findings are applicable only under irrigated conditions. It should be noted that the scope for the introduction of pulse crops in rice-fallow (mostly unirrigated) crops needs to be tapped with supplementary irrigation, given the high profitability and scope for pulse crops as winter crops in cropping systems. The major future expansion of area under pulse crops may be in the rice fallow, where there are no other crops to compete with, although there are limits to the successful propagation of these crops in this system. Most of the farmers in South India, where large areas of paddy fallow are located, are not aware of the potential economic benefits of using fallow for the cultivation of pulses. In many cases, farmers get not only inadequate but also wrong information about the recommended pulse production technology. Governments should provide various incentives to increase the area under pulses in rice fallow land and train field workers and farmers to participate in demonstrations. About 5 lakh ha of winter pulses can be brought from upland rice, 4.5 lakh ha of bajra, and 3 lakh ha under barley, mustard, and wheat. Winter pulses like lentils and gram should be replaced by mustard, barley, and wheat. If possible, the collected rainwater should be used as life-saving irrigation for winter crop establishment.

9.10.10 Post-harvest Management

There is a need to promote mechanical threshing for post-harvest management of pulses by providing incentives for the purchase of threshers. Authorized organizations will significantly reduce the need for storage at the farmer level. Small pulse mills should be popularized and promoted. The private sector also encourages the setting up of 'Dal Mills' in rural areas/districts with a large area under pulses on the lines of sugar mills. Private companies should be concerned with the processing, packing, and marketing of pulses. Public sector procurement agencies are severely handicapped in funding and expertise in this area (Singh et al. 2013).

9.10.11 Expansion of Irrigation

Pulses are always grown in conditions of lack of moisture, which reduces the level of productivity. Scientific scheduling of irrigation, estimating the amount of water, and deployment of water-saving irrigation methods not only result in increased yield but also greater use of water and nutrients. The use of sprinkler irrigation has immense potential for saving irrigation water and expanding the area under irrigation. This method has gained popularity in many districts with limited water resources. Drip irrigation is the focus of policymakers, administrators, and social workers as it has assumed social, economic, and ecological dimensions. Fertigation holds promise for widely spaced crops such as pigeon pea. These devices can increase the irrigation area by 30–50% (Singh et al. 2013). Along with an expansion of irrigation facilities through ground water, there is scope for expansion of area irrigated under pulse crops, especially summer, rabi and spring season crops through canal irrigation system, as the yield response is high. In short, to increase the area and production of pulse crops, we need crop specific and region specific approach.

9.10.12 Farm Mechanization and Land Lease Market

One of the reasons for the success of expansion of the area under chickpeas in Andhra Pradesh is the development of varieties suitable for agricultural mechanization. Therefore, there is a need to encourage agricultural mechanization in peak-season activities like harvesting and threshing through the distribution of subsidized agricultural machinery to tackle labour shortage and high wage rates.

9.10.13 Pulse-Based Technologies

Already ICAR and ICRISAT, with the support of State and Central Governments, are involved in the development of short duration, photo-thermos-insensitive varieties for different agro-ecologies, development of hybrids in pigeon pea, efficient plant architecture in major pulse crops. Development of bio-intensive integrated pest management module, design of advanced machines to tackle labour shortage, production of breeder seeds of latest released varieties, and conducting frontline demonstrations in farmers' fields. R&D efforts under NFSM-Pulses and NARS need to be strengthened with a major emphasis on this.

- Replace cereal crops with high-yielding pulses in the prevailing rice-wheat cropping systems
- To encourage research and development on multi-cropping and additional maturation of pigeon peas for better cropping management
- Development of pigeon pea genotypes suitable for Rabi/Spring and Summer seasons
- Induction of short-duration varieties of pulses as catch crop;
- Introduction of spring green gram.
- Utilization of unused land and water in spring/summer season with high returns
- Using genomics and biotechnological tools for the development of multiple disease and insect repellents. Varieties to reduce yield loss of standing crop and increase yield
- Reducing storage losses and improving market knowledge and infrastructure.
- Technology dissemination and input delivery mechanisms for pulses were very weak. Research, extension, and coordination of farmers to encourage farmer-participatory research.
- Linking MSP to market prices can bridge the gap between demand and supply

There is a need for a holistic approach rather than a single approach method to increase the production and productivity of the pulse crop. Best Agricultural Science Practices (BAPs) and their various components have the potential to stand out under climate change conditions, all the components of pro-technology need to be adopted as a unit, and some of them need not be selected, which help to overcome many complications in soil health, poor response to threat and technology. These steps can be taken on a priority basis to improve the productivity of pulses, some of them are (1) encouraging rapid adoption of existing technology to reduce yield gap (2) seed replacement rate and quality production Institutional support for improvement. (3) Strengthening of life saving irrigation in pulse growing areas (4) Ensuring availability of critical inputs like seed fertilizers, pesticides (5) Gradual automation for pulses production (6) Public-private partnership for sustainable chain and to reduce post-harvest losses (7) Policy support for value chain of pulses.

9.11 Major Breakthroughs in Pulses

9.11.1 Hybrid Pigeon Pea Technology and Genomic Intervention

To break the yield barrier in pigeon peas, ICRISAT and partners have developed a cytoplasmic male-sterility (CMS) based hybrid breeding technique in pigeon peas. The CMS-based medium maturity hybrids, ICPH 2671 and ICPH 2740 produced 30–40% higher grain yield than the popular cultivars in farmers' fields in India. Mapping the genome of pigeon pea is a breakthrough that helps accelerate the development of improved varieties that can provide stable and high yield. A major breakthrough that helps accelerates the development of improved varieties that can provide stable and high yields (Saxena and Nadarajan 2010). Genome sequencing will be able to identify the structure and function of over 48,000 genes of pigeon pea. There are some unique genes that confer drought tolerance to pigeon pea. This characteristic can be transferred to identify high yielding varieties. This will also help in reducing the time taken for breeding new varieties from 6 to 10 years to about 3 years.

The production of pigeon pea can be increased through (1) wheat under irrigated conditions in Uttar Pradesh (UP), Haryana, Punjab, and northern parts in sequence with improved crop management (2) replacement of other dry land crops like cotton in low water availability states like Gujarat, Karnataka, Andhra Pradesh, Maharashtra, and Tamil Nadu, (3) the popularity of rabi pigeon pea in the states of Orissa, Gujarat, West Bengal, Bihar, and eastern Uttar Pradesh, and (4) growing area through inter-cropping of pigeon pea with soybean in Madhya Pradesh, Maharashtra, and Rajasthan; and with cotton, jowar, bajra, and groundnut in the states of Andhra Pradesh, Maharashtra, Karnataka, Gujarat, MP and UP, under which additional coverage is expected to at least 1 million ha by the end of the century, (5) Pest management of pod borer, *Fusarium* wilt, and sterility mosaic.

9.11.2 Success Story

9.11.2.1 Chickpea Revolution in South India

The introduction of the gram crop in non-traditional areas such as South Indian states is an example of technological and institutional success, which has the potential to be replicated in other crops. The area under chickpea is shifting from northern states to southern states. During the period 1991–1993 to 2006–2008, the highest increase in the productivity of gram has been recorded in Andhra Pradesh (124%), followed by Karnataka (63%), Maharashtra (52%), and Gujarat (40%), Still, there is scope for productivity enhancement in the states to increase production to meet growing demand at the national level (IIPR 2012). During the past four decades (1979–2019), chickpea production in central and southern India increased by 445% (from 1.27 to 6.95 million metric T) due to a 177% increase in area (2.42–6.71 million ha) and a 97% increase in yield (527–1036 kg/ha), while in the rest of the world, chickpea production increased 133% (1.85–4.31 million metric T) due to 49% increase in area (2.80–4.17 million ha) and 57% increase in yield (659–1033 kg/ha). Expansion of wheat area was crucial for India in ensuring food security and, fortunately, expansion in chickpea area in central and southern India fully compensated for the loss in chickpea area that occurred in Northern India. Several institutional and technological factors contributed to the expansion of the region in South India. These include the introduction of gram in black cotton soils, availability of rabi fallow land, adoption of short duration and high yielding varieties such as KAK-2, a Kabuli type with high market demand; and short duration and wilt-resistant varieties such as JG-11, stable yields, and prices, and well-developed land lease markets, which facilitated large-scale mechanization. This large-scale mechanization facilitated the consolidation of operational holdings, contracting major labour-demanding tasks such as harvesting and threshing that helped economies of scale in procuring inputs as well as production and marketing of output, to address the labour shortage.

Overall, even though the adoption of technology led to increased investment in gram cultivation, it helped reduce the cost of production due to a sharp increase in yield and profitability. Wide availability of highly subsidized cold storage godowns helped farmers to store chickpeas during the peak harvest season to overcome low market prices and to benefit from higher prices during the later period. The importance of successful government programs such as NFSM, subsidized seed distribution and mechanization, incentives for cold storage structures, and high MSP helped the gram revolution in South India.

9.11.2.2 Chickpea Revolution in Andhra Pradesh

The production of gram in Andhra Pradesh increased from 393 to 1375 kg ha^{-1} from 1987 to 2008, while the area increased from 52.2 to 542 thousand ha, because of which products in the same period increased from 19.9 to 730.7 thousand T. The annual compound growth rate of the crop is now 12.41% and the yield is 5.80% and has resulted in a massive 18.21% per year increase in production from 1987 to 2008.

9.12 Geographical Shifting of Pulse Cultivation

The traditional areas for pulse production have been switching to other crops as pulses have moved to non-traditional areas (ICAR-IIPR Vision 2025 2015). The pulses moved from northern to southern and from eastern to western zones, with central India becoming the hub for pulse farming. Andhra Pradesh, Karnataka, Madhya Pradesh, Maharashtra, and Rajasthan are emerging as the most promising states in pulse production. Madhya Pradesh dominates in chickpeas, and Maharashtra dominates in pigeon peas. Identifying the factors behind this transition is an important area for future research. Technology has played a crucial role in the shift in areas allocated to pulses from traditional to non-traditional areas. The states that do not have any highly commercialized or highly profitable crops seem to be the ones where the adoption of pulses has been easiest. For instance, the absence of any profitable star crop in the Telangana and Rayalaseema regions of Andhra Pradesh is facilitating chickpea cultivation so much, so that they have emerged as the major producer of chickpeas in India. In the same way, low-productive cotton in northern Karnataka and barley in Madhya Pradesh are being replaced by chickpeas. Reddy (2009) presents the case of chickpeas' rapid expansion in Andhra Pradesh as a suitable example to emulate for production growth in other pulse crops. The introduction of short-duration varieties of chickpeas, for example, has contributed to the increase in its area and production.

Proper soil and water conservation measures need to be taken to ensure that flooding does not occur to pigeon peas. At the same time, water logging could certainly play a role in diminishing the area planted, especially in deep black soil. Trends of regional specialization and geographic continuity are emerging regarding the area and production of pulses, which are visible at the regional and state level.

Pulse-producing states are rainfed conditions, absence of irrigation, and a general lack of alternative profitable crops. The pulses being cultivated predominantly in the marginal and rainfed regions under resource-starved conditions require a different approach to increase their area, production, and productivity. Indeed, research efforts have already shifted to develop varieties that can address the challenges of non-traditional areas.

In addition, research must be undertaken to find the potential of pulse production in the rainfed rice fallow systems spread across the states of Bihar, Chhattisgarh, Jharkhand, Madhya Pradesh, and Odisha. Post-rainy season fallows might comprise both the biggest challenge and the greatest opportunity for increasing pulse production. Alternative cropping patterns and the adoption of various technologies, coupled with seed availability, could increase pulse production in these marginal areas. Also, there is a need for further research to assess the feasibility and risks of policies increasing the MSP and expanding procurement in a deficit crop such as pulses.

9.13 Way Forward

Agriculture is an important sector of the Indian economy, contributing to about 15% of the national GDP and more importantly, almost half of India's population is wholly or significantly dependent on agriculture and allied activities for their livelihood. Nevertheless, agriculture in India faces many constraints and challenges, especially in the areas of food security, natural resource management, and the application of advanced agricultural technology to improve crop yield potential, better agricultural profitability, minimum adverse environmental impact, better soil Health and food self-confidence. Food security in India has long been a major area of concern for agricultural scientists and planners. At present, India produces over about 315 million T of food grains every year, an increase of four times since independence. However, increased efforts to produce more food have resulted in a tremendous shift in cropping systems towards cereal-grain-based cropping systems. This has marginalized pulses resulting in quantitative as well as qualitative degradation of the productive base, land, and agricultural resources. Pulses are a rich source of protein, vitamins, and minerals and popularly known as "poor man's meat" and "rich man's vegetable", contributing significantly to the nutritional security of the country. Pulses provide significant nutritional and health benefits and are known to reduce several non-communicable diseases such as colon cancer and heart disease. Pulses, supplemented with cereals, provide a perfect blend of vegetarian proteins of high biological value. India is the largest producer, importer, and consumer of pulses, and accounts for 25% of the global production from 35% of the global area under pulses. The productivity of pulses in India is less than half of the productivity levels in the USA and Canada, as pulses are mainly grown in rainfed conditions in areas of high rainfall variability. Persistent and increasing demand-supply gap is putting pressure on prices. The pulses are increasingly out of reach of the poor. Pulses production in India is caught in a vicious cycle Low and uncertain yield cycle, and poor returns per hectare resulted in the least preference for farmers to grow pulses. Inadequate adoption of production technology; high price volatility, production risk, and low level of irrigation are important factors responsible for stagnation in the productivity of these crops. The country will need 39 million T of total pulses by 2050, which would require pulses production to grow at an annual rate of 2.2%. The country will have to produce enough pulses to meet the growing requirement as well as remain competitive to protect domestic production. It is imperative to develop and adopt more efficient crop-production technologies with favourable policies and market support to encourage farmers to bring more area under pulses. To provide nutritional security to the poor population dependent on a vegetarian diet, making pulses affordable by promoting the domestic production of pulses is the best option. To increase the supply of pulses to the poor masses, the supply through Public Distribution System in the present scenario will not only distribute pulses to the poor at affordable prices and enhance nutritional security, but through stabilizing prices and promoting farmers. Lack of assured market is one of the major issues of poor performance of pulses. Government procurement will provide adequate marketing support to the producers. Alternative marketing arrangements through contract farming, a farmer-producer-company model needs to be promoted. Pulses can be grown on a wide range of soil and climatic conditions and play an important role in crop rotation, mixed and intercropping, *Rhizobium* bacteria is essentially maintaining soil fertility through biological nitrogen fixation (30–150 kg ha^{-1}). Pulses are a constant source of income and employment for small and marginal farmers along with the release of soil-bound phosphorus, and thus the stability of agricultural systems, and thus occupy a leading position in world agriculture. The United Nations declared 2016 the "International Year of Pulses" intending to increase the production and consumption of pulses by 10% by 2020 and to create awareness about the benefits of pulses using social media. Pulses are an essential part of the Indian diet for nutritional security and environmental sustainability. They are important food crops due to their high protein content (20–25%), carbohydrates (55–60%), calcium, and iron. Presently, we are in the midst of becoming self-reliant in pulses production, as we are the leading country in production, consumption, and import as well as in the whole world. By the end of 2050, we will be able to maintain our production. From a net exporter of pulses to a net importer if everything goes according to plan.

Since more than 80% of the area under pulses is in a stressed rainfed environment, quality seeds of improved varieties have emerged as one of the most important inputs for increasing the production of pulses in India. The per capita per day availability of pulses has declined from 61 to 32 g during 1951–2010 with the reduction in production; there has been a demand-supply imbalance resulting in rising import bills, unpredictable price hikes, and lower net profits than competing crops. The demand supply gap and shortfall in pulses have been attributed to several factors; the major ones are growing population, rising incomes of people, geographic change, sudden climate change, complex disease-pest syndrome, socioeconomic policies and input constraints. This coupled with other economic factors such as lack of an assured market, ineffective government procurement, lack of minimum support prices and trade liberalization make pulse cultivation less profitable than other crops. Stagnant pulse production in India has led to a steady decline in per capita pulse availability. Thus, pulse production in India has fluctuated widely leading to a steady decline in per capita availability over the past 20 years. To diversify our cropping systems with pulses to enhance pulse productivity and meet the national pulse requirement is the primary goal to achieve keeping in view of a high population growth rate of 1.44%. However, pulses are still an important component of the Indian agricultural economy, second only to cereals and oilseeds in terms of area, production, and economic value. India is the largest producer and consumer of pulses, accounting for about 25–28% of the global share (IIPR 2011) and 34% of food use. Thus, the challenge is for agricultural scientists, extension workers, planners, and the farming community to diversify their cropping systems with pulses to enhance and sustain pulses productivity and meet the national pulses requirement.

In view of this, India has to develop and adopt more efficient crop production technologies along with favourable policies to encourage farmers to bring more area under pulses. Advances in pulse production technologies have opened up a new hope of increasing the production and productivity of pulses in various irrigated and rainfed agricultural ecosystems of our country. We have to focus on bridging the yield gap in addition to area expansion and diversified production systems equipped with suitable pulse production technology already generated by research institutes to break the yield plateau through recent technological advancements. India needs about 32 million T of pulses by 2030 to feed an estimated population of about 1.68 billion. Global supply of pulses is limited, as India is the largest producer and consumer of pulses. Hence, India needs to produce the required quantity, but at the same time, it needs to remain competitive to protect indigenous pulse production. However, a concerted effort by farmers, researchers, development agencies, and the government is needed to ensure that India becomes self-reliant in pulses production in the next 5–10 years. The recent efforts and programs initiated by the Government are bearing fruit, and it is hoped that this momentum will continue and strengthen to make India self-reliant in pulses. The priority steps that can be taken to improve pulses productivity are: To encourage rapid adoption of advanced technologies to bridge the yield gap; Institutional support to promote seed replacement rate and quality production as well; Determination and strengthening of life-saving irrigation in pulse growing areas; To guarantee timely availability of critical inputs like seeds, fertilizers, pesticides, etc.; Gradual mechanization for production and storage of pulses; Public-private partnership to maintain the value chain and reduce post-harvest losses; Policy support for value addition and balancing chain for production and marketing of pulses.

In short, to increase the area and production of pulse crops, we need crop-specific and area-specific approaches, which should be adopted within the overall framework of the systems approach. Major thrust areas addressed are as follows

1. Use of high-yielding pulse varieties in place of cereal crops in the prevailing rice-wheat cropping systems
2. Induction of short-duration varieties of pulses as catch crop
3. Development of many diseases and pest-resistant varieties

4. Minimizing storage losses and improving market knowledge and infrastructure
5. Linking MSP to market prices can bridge the gap between demand and supply
6. Developing high nitrogen fixation varieties, which will play an important role in sustainable agriculture
7. Research, extension, and coordination of farmers to encourage farmer participatory research.

Farmers and extension functionaries should understand that there is an urgent need to incorporate better and more efficient management practices in intensive farming in a balanced manner to achieve higher pulses yield. Better quality earning maximum sustainable profitability. Finally, there is a need for a holistic approach to increase the production and productivity of the pulse crop rather than a single approach method. Majorly Good Agricultural Science Practices (GAP) with their various components for excellent production under the changing climate scenario require aggregation of all components of technologies.

The future thrust areas are development of varieties tolerant to drought, moisture stress, and high temperature; diversification of cereal-based cropping systems; management of pod borer, *Fusarium* wilt, *Ascochyta* blight; and emphasis on large-seeded kabuli chickpea.

References

Ali M, Gupta S (2012) Carrying capacity of Indian agriculture: pulse crops. Curr Sci 25:874–881

Ali M, Kumar N, Ghosh PK (2012) Milestones on agronomic research in pulses in India. Ind J Agron 57(3S):52–57

Borase DN, Murugeasn S, Nath CP, Hazra KK, Singh SS, Kumar N, Singh U, Praharaj CS (2021) Long-term impact of grain legumes and nutrient management practices on soil microbial activity and biochemical properties. Arch Agron Soil Sci 67(14):2015–2032

Deshpande SD, Singh G (2001) Long term storage structures in pulses. National Symposium on Pulses for Sustainable Agriculture and Nutritional Security (7–19 April). Indian Institute of Pulse Research, Kanpur

Directorate of Economics and Statistics (2021) Agriculture statistics at a glance. Department of Agriculture and Cooperation, Government of India, New Delhi

Dixit GP, Katiyar PK, Singh BB, Shivkumar (2009) Lentil varieties in India. AICRP on MULLaRP. Indian Institute of Pulses Research, Kanpur

Dudeja SS, Duhan JS (2005) Biological nitrogen fixation research in pulses with special reference to mungbean and urdbean. Ind J Puls Res 18(2):107

Dutta A, Trivedi A, Nath CP, Gupta DS, Hazra KK (2022) A comprehensive review on grain legumes as climate-smart crops: challenges and prospects. Environ Chall 17 100479

FAOSTAT (2011) Production. Crops. http://faostat.fao.org/site/567/default.aspx#ancor

FAOSTAT (2020) FAO Statistics. Food and Agriculture Organization of the United Nations. http://www.fao.org/faostat/en/

Gaur PM, Jukanti AK, Varshney RK (2012) Impact of genomic technologies on chickpea breeding strategies. Agronomy 2(3):199–221

Ghosh PK, Kumar N, Venkatesh MS, Hazra KK, Nadarajan N (2014) Resource conservation technology in pulses. Scientific Publishers, Jodhpur

Ghosh PK, Hazra KK, Nath CP, Das A, Acharya CL (2016) Scope, constraints and challenges of intensifying rice (Oryza sativa) fallows through pulses. Ind J Agron 61(4):122–128

Ghosh PK, Hazra KK, Venkatesh MS, Praharaj CS, Kumar N, Nath CP, Singh U, Singh SS (2020a) Grain legume inclusion in cereal–cereal rotation increased base crop productivity in the long run. Exp Agric 56(1):142–158

Ghosh PK, Nath CP, Hazra KK, Kumar P, Das A, Mandal KG (2020b) Sustainability concern in Indian agriculture: needs science-led innovation and structural reforms. Ind J Agron 65(2): 131–143

Gowda CLL, Srinivasan S, Gaur PM, Saxena KB (2013) Enhancing the productivity and production of pulses in India. In: Shetty PK, Ayyappan S, Swaminathan MS (eds) Climate change and sustainable food security. National Institute of Advanced Studies, Bengalaru and Indian Council of Agricultural Research, New Delhi, pp 145–159

Hazra KK, Bohra A (2021) Increasing relevance of pulse crops to sustainable intensification of Indian agriculture. Nat Acad Sci Lett 44:1–3

Hazra KK, Singh SS, Nath CP, Borase DN, Kumar N, Parihar AK, Swain DK (2018) Adaptation mechanisms of winter pulses through rhizospheric modification in mild-alkaline soil. Nat Acad Sci Lett 41(4):193–196

Hazra KK, Nath CP, Singh U, Praharaj CS, Kumar N, Singh SS, Singh NP (2019) Diversification of maize-wheat cropping system with legumes and integrated nutrient management increases soil aggregation and carbon sequestration. Geoderma 353:308–319

Hazra KK, Nath CP, Ghosh PK, Swain DK (2020) Inclusion of legumes in rice–wheat cropping system for enhancing carbon sequestration. In: Carbon management in tropical and sub-tropical terrestrial systems 2020. Springer, Singapore, pp 23–36

ICAR-Indian Institute of Pulses Research (2015) ICAR-IIPR vision 2025 Department of Agriculture, Research and Education, Ministry of Agriculture and Farmers Welfare. https://iipr.icar.gov.in/pdf/vision_250715.pdf

IIPR (2011) Vision 2030. Indian Institute of Pulses Research, Kanpur

IIPR (2012) AICRP on chickpea- Annual report. Indian Institute of Pulses Research, Kanpur

Joshi PK (1998) Performance of grain legumes in the Indo-Gangetic plain in residual effects of legumes in rice and wheat cropping systems of the Indo-Gangetic plain. In: Kumarrao JVDK, Johansen C, Rego TJ (eds) Legumes in rice and wheat cropping systems of the Indo-Gangetic plain. ICRISAT, Andhra Pradesh, pp 3–13

Joshi PK, Saxena R (2002) A profile of pulses production in India: facts, trends and opportunities. Ind J Agric Econ 57(3):326–339

Kumar N, Yadav A, Hazra KK, Yadav SL, Singh SS (2014) Prospects and potential of post-emergence herbicides for weed management in pulses. In: Pulses: challenges and opportunities under changing climatic scenario. ICAR-Indian Institute of Pulses Research, Kanpur, p 199

Kumar N, Hazra KK, Singh SS, Nadarajan N (2016) Constraints and prospects of growing pulses in rice fallows of India. Ind Farm 66(6):13–16

Kumar N, Hazra KK, Nath CP, Praharaj CS, Singh U (2018) Grain legumes for resource conservation and agricultural sustainability in South Asia. In: Legumes for soil health and sustainable management. Springer, Singapore, pp 77–107

Kumar R, Mishra JS, Rao KK, Bhatt BP, Hazra KK, Hans H, Mondal S (2019a) Sustainable intensification of rice fallows of Eastern India with suitable winter crop and appropriate crop establishment technique. Env Sci Pollut Res 26(28):29409–29423

Kumar R, Saurabh K, Kumawat N, Mishra JS, Hans H, Krishna B, Meena RS, Krishna Jha B, Upadhyay PK, Hazra KK (2019b) Conservation agriculture: perspectives on soil and environmental management in Indo-Gangetic plains of South Asia. In: Sustainable Management of Soil and Environment 2019. Springer, Singapore, pp 123–168

Kumar N, Singh SS, Ghosh PK, Hazra KK, Venkatesh MS, Praharaj CS, Singh MK, Kumar MS, Basu PS, Yadav A, Yadav SL (2020a) Improving chickpea productivity in rice-fallow of Indo-Gangetic Plain with soil moisture conservation and cultivar selection. J Food Legum 33(1): 28–35

Kumar N, Singh SS, Ghosh PK, Singh NP, Agrawal PK, Hazara KK, Praharaj CS, Yadav A, Yadav SL, Singh S (2020b) Issues and strategies for promotion of pulses in untapped rice-fallow in India: a review. J Food Legum 33(3):139–150

Kumar R, Mishra JS, Rao KK, Mondal S, Hazra KK, Choudhary JS, Hans H, Bhatt BP (2020c) Crop rotation and tillage management options for sustainable intensification of rice-fallow agro-ecosystem in eastern India. Sci Rep 10(1):1–5

Materne M, Reddy AA (2007) Commercial cultivation and profitability. In: Lentil. Springer, Dordrecht, pp 173–186

Nath CP, Hazra KK, Kumar N, Praharaj CS, Singh SS, Singh U, Singh NP (2019) Including grain legume in rice–wheat cropping system improves soil organic carbon pools over time. Eco Eng 129:144–153

Nath CP, Kumar N, Hazra KK, Praharaj CS, Singh SS, Dubey RP, Sharma AR (2021) Topramezone: a selective post-emergence herbicide in chickpea for higher weed control efficiency and crop productivity. Crop Prot 150:105814

Ramakrishna A, Gowda CLL, Johansen C (2000) Management factors affecting legumes production in the Indo Gangetic plain. In: Johansen C, Duxbury JM, Virmani SM, Gowda CLL (eds) Legumes in rice and wheat cropping systems of the Indo-Gangetic Plain-constraints and opportunities. ICRISAT, Andhra Pradesh, pp 156–165

Ramasamy C, Selvaraj KN (2002) Pulses, oilseeds and coarse cereals: why they are slow growth crops? Ind J Agric Econ 57(3):289–315

Reddy AA (2009) Pulses production technology: status and way forward. Econ Polit Wkly 44(52): 73–80

Reddy AA, Reddy GP (2010) Supply side constrains in production of pulses in India: a case study of lentil. Agric Econ Res Rev 23:129–136

Reddy A, Bantilan MC, Mohan G (2013) Pulses production scenario: policy and technological options. Policy brief. International Crops Institute for Semi-Arid Tropics, Hyderabad

Reid DE, Ferguson BJ, Hayashi S, Lin YH, Gresshoff PM (2011) Molecular mechanisms controlling legume autoregulation of nodulation. Ann Bot 108:789–795

Saxena KB, Nadarajan N (2010) Prospects of pigeonpea hybrids in Indian agriculture. Elec J Plant Breed 1(4):1107–1117

Shalendra KC, Gummagolmath PS, Sharma P, Patil SM (2013) Role of pulses in the food and nutritional security in India. J Food Legum 26(3–4):124–129

Sharma DK, Brahm P (2002) Analysis of growth and variability in area, production and productivity of pulses in India. Ind J Agric Econ 57(3):397–398

Sharma P, Dwivedi S, Singh D (2016) Global poverty, hunger, and malnutrition: a situational analysis. In: Biofortification of food crops. Springer, New Delhi, pp 19–30

Singh AK, Bhatt BP (2013) Effects of foliar application of zinc on growth and seed yield of late-sown lentil. Ind J Agric Sci 83(6):622–626

Singh KM, Singh RKP (1995) An economic analysis of lentil cultivation in N-E Alluvial Plains of Bihar. Econ Aff 40(3):157–163

Singh AK, Singh NP (2008) Yield and uptake of primary nutrients by large seeded varieties of lentil under varying seed rates in normal and late sown conditions. J Food Legum 20(2):187–189

Singh SK, Varma SC, Singh RP (2002) Integrated nutrient management in rice and its residual effect on lentil. Ind J Agric Res 36(4):286–289

Singh AK, Bhatt BP, Sundaram PK, Chndra N, Bharati RC, Patel SK (2012) Faba bean (*Vicia faba* L.) phenology and performance in response to its seed size class and planting depth. Int J Agric Stat Sci 8(1):97–109

Singh AK, Bhatt BP, Singh KM, Kumar A, Manibhushan, Kumar U, Chandra N, Bharati RC (2013) Dynamics of powdery mildew (Erysiphetrifolii) disease of lentil influenced by sulphur and zinc nutrition. Plant Pathol J 12(2):71–77

Singh AK, Singh SS, Prakash V, Kumar S, Dwivedi SK (2015) Pulses production in India: present status, sent status, bottleneck and way forward. J Agric Search 2(2):75–83

Singh U, Praharaj CS, Singh SS, Hazra KK, Kumar N (2018) Effect of crop establishment practices on the performance of component cultivars under pigeonpea (*Cajanus cajan*)–wheat (*Triticum aestivum*) cropping system in IGP. Indian J Agric Sci 88(5):691–697

Singh U, Gaur PM, Chaturvedi SK, Hazra KK, Singh G (2019) Changing plant architecture and density can increase chickpea productivity and facilitate for mechanical harvesting. Int J Plant Prod 13(3):193–202

Srivastava SK, Sivaramane N, Mathur VC (2010) Diagnosis of pulses performance of India. Agric Econ Res Rev 23:137–148

Tulsi L (2015) Pulses: need for production expansion. Econ Polit Wkly 50(35):133–136

Venkatesh MS, Hazra KK, Ghosh PK, Mishra JP (2019a) Integrated phosphorus management in maize-chickpea rotation in moderatelyalkaline Inceptisol in Kanpur, India: an agronomic and economic evaluation. Field Crops Res 233:21–32

Venkatesh MS, Hazra KK, Ghosh PK, Singh M (2019b) Integrated phosphorus management in maize–chickpea cropping system on alkaline Fluvisol. Nutr Cycl Agroecosyst 113(2):141–156

Wani SP, Sreedevi TK, Singh HP, Pathak P, Rego TJ (2002) Innovative farmer participatory integrated watershed management model: Adarsha Watershed, Kothapally, India, A Success Story!

Oilseeds and Oil Palm

10

R. K. Mathur, M. Sujatha, S. K. Bera, P. K. Rai, B. Kalayana Babu, K. Suresh, S. N. Sudhakara Babu, Mahaveer Sharma, Kirti Rani, B. C. Ajay, M. Kumar, H. K. Sharma, Pankaj Sharma, and V. V. Singh

Abstract

Oilseed self-sufficiency achieved during the Yellow Revolution in the early 1990s could not be maintained for more than a few years. Despite being the world's fifth-greatest producer of oilseed crops, India is also one of the world's largest importers of vegetable oils today. Towards becoming self-sufficient in edible oil', two new missions, i.e. (i) National Mission of Edible Oils–Oilseeds (NMEOOS), and (ii) National Mission on Edible Oils-Oil Palm (NMEOOP) have been launched by the Government of India with the aim to enhance the edible oilseeds production *vis-à-vis* edible oil availability in the country. Oilseed crops except for rapeseed and mustard, are mostly grown under sub-optimal rain-fed conditions, and as a result of unpredictable rainfall, the crops experience moisture stress at various phases of growth, resulting in low yields. Stringent strategies with an emphasis on quality improvement and value addition leveraging technologies with a bearing on employment through skill/entrepreneurship

R. K. Mathur (✉) · B. K. Babu · K. Suresh
ICAR-Indian Institute of Oil Palm Research, Pedavegi, Andhra Pradesh, India
e-mail: ravi.mathur@icar.gov.in

M. Sujatha · S. N. S. Babu
ICAR-Indian Institute of Oilseeds Research, Hyderabad, Telangana, India

S. K. Bera · K. Rani · B. C. Ajay
ICAR-Directorate of Groundnut Research, Junagadh, Gujarat, India

P. K. Rai
ICAR—Directorate of Rapeseed-Mustard Research, Bharatpur, Rajasthan, India

M. Sharma
ICAR-Indian Institute of Soybean Research, Indore, Madhya Pradesh, India

M. Kumar · H. K. Sharma · P. Sharma · V. V. Singh
College of Agriculture, Tharad, S.D. Agricultural University, Banaskantha, Gujarat, India

P. K. Ghosh et al. (eds.), *Trajectory of 75 years of Indian Agriculture after Independence*, https://doi.org/10.1007/978-981-19-7997-2_10

development will lead to improving the production and productivity of oilseeds. Development of next-generation technologies including dwarf and high oil-yielding hybrids, aid more productivity, and user-friendly harvesting mechanization will bring revolution in oilseed development in India.

Keywords

Oil seeds · Varieties · Hybrids · Yellow revolution · Oil content

10.1 Introduction

Within the section of field crops, oilseed crops are the second most important determinant of agricultural economics, trailing only cereals. Oilseed self-sufficiency achieved during the Yellow Revolution in the early 1990s could not be maintained for more than a few years. Despite being the world's fifth-greatest producer of oilseed crops, India is also one of the world's largest importers of vegetable oils today. In addition, there has been an increase in vegetable oil consumption in recent years, both for edible and industrial purposes. Majority of the oilseeds are cultivated under a rainfed ecosystem (70%). In general, the area under oilseeds has decreased, owing to their poorer profitability compared to competitive crops such as maize, cotton, chickpea, and others, in the current crop growing and selling conditions.

To meet the growing demand, India is importing edible oils of nearly 15 MT worth nearly rupees 70,000 crores at the cost of valuable foreign reserves. The per capita consumption of edible oils has grown by 2.48% annually with a CAGR of 2.17% from 15.8 kg/capita/year in 2012–2013 to 18.76 kg/capita/year in 2019–2020. If the trend continues, the per capita consumption in 2034–2035, with CAGR of 2.32%, will be 27.08 kg. The demand for edible oil was 25.63 MT in 2019–2020 against domestic production of 10.53 MT (7.03 MT from primary sources and 3.50 MT from secondary sources) necessitating the import of 15.10 MT to fill the demand and supply gap. Considering the low productivity of edible oilseeds (Groundnut, Soybean, Rapeseed-Mustard, Sesame, Sunflower, Safflower, and Niger) and predominantly rainfed cultivation in marginal land by marginal farmers in erratic and deficient rainfalls, the task set is abysmally difficult and ambitious. The situation is compounded further because of the recent declining trend in the area of major and minor edible oilseeds, increase in import of cheaper edible oil, non-availability of quality seeds, etc.

During 2021–2022, area under edible oilseed crops was 28.75 m ha with production and productivity of 37.10 MT and 1292 kg/ha, respectively. However, to boost domestic production, the anticipated edible oil production in the year 2024–2025 has been set to around 18 MT (13.50 MT from primary sources of seven edible crops and 4.50 million tonnes from secondary sources, viz. Coconut, Palm oil, Rice bran, cotton seed, and TBOs). Total demand for edible oil in 2024–2025 is estimated at around 28.00 million tonnes with an expected import of 10.00 million tonnes, assuming that present per capita consumption of 19.22 kg/year will remain unchanged. However, at the present rate of increased consumption of edible oil, the demand would be around 30.4 MT and there will be a need for import to the tune of12.4 MT.

Towards becoming 'self-sufficient in edible oil', a new mission i.e. National Mission on Edible Oils (NMEO) has been proposed to enhance the edible oilseeds production vis-à-vis availability in the country. The mission aims to enhance the production of seven edible oilseeds and boost edible oil from secondary sources to reduce the staggering burden of the import of edible oils. However, meticulous planning and concerted efforts of all stakeholders are required to achieve the mission. The details of such planning are discussed here in the context of improving the productivity of edible oilseeds and the area under cultivation.

10.2 Trends and Present Situation

In India, both annual and perennial edible oilseeds are in cultivation. Annual edible oilseeds include Groundnut, Rapeseed-Mustard, Soybean, Sunflower, Sesame, Niger, and Safflower, and perennials include Oil palm and Coconut. Besides, there are minor oil-producing species of forest and tree origin. Moreover, edible oils are also produced domestically from secondary sources like rice bran, cotton seed, corn, and other Tree Borne Oilseeds (TBOs). Presently, India produces edible oils in nearly 40% of its demand and the remaining 60% is imported. As per capita consumption of edible oils has grown by 2.48% annually with a CAGR of 2.17% from 15.8 kg/capita/year in 2012–2013 to18.76 kg/capita/year in 2019–2020, the demand and supply gap has widened over the years. The demand for edible oil was 25.63 MT in 2019–2020 against domestic production of 10.53 MT, necessitating the import of around 15.1 MT (palm oil: about 60%; Soybean oil: 25%; Sunflower: 12%; and rest others) costing around Rs. 69,000 crores to the national exchequer.

In 2021–2022, India produced 37.10 MT of edible oilseeds from 28.75 m ha area with a productivity of 1292 kg/ha and edible oil production of 11.20 MT. Among the edible oilseed crops, Soybean (35.63%), Groundnut (32.09%), Rapeseed-Mustard (28.96%) contributed 96.68% of the total edible oilseed production. Sesame, Sunflower, Safflower, and Niger contributed 3.32%. Contrary to seed production, the major contribution to domestic edible oil kitty comes from Rapeseed-Mustard (3.191 MT: 45.39%), Groundnut (1.769 MT: 25.17%), and Soybean (1.754 MT: 24.95%) amounting to 95.52%. The minor edible oilseeds (Sesame, Sunflower, Safflower, and Niger) contributed 4.48% of domestic production of 7.03 MT from primary sources during 2019–2020. Around 3.50 MT of edible oils comes from secondary sources (cottonseed oil, palm oil, corn oil, rice bran oil, coconut oil, and other TBOs). The major oilseed-producing states are Madhya Pradesh, Maharashtra, and Rajasthan (Soybean); Gujarat, Andhra Pradesh, Rajasthan, Tamil Nadu, and Karnataka (Groundnut); Rajasthan, Madhya Pradesh, Uttar Pradesh, West Bengal, and Haryana (Rapeseed-Mustard); Madhya Pradesh, Uttar Pradesh, Rajasthan, and Gujarat (Sesame); Karnataka (Sunflower); and Maharashtra and Karnataka (Safflower).

The overall yield gap in oilseeds is 65%. The maximum yield gap is in Sunflower (160%) among minor oilseeds and in major oilseeds (Groundnut, Soybean, and Rapeseed-Mustard), the yield gap ranged from 37% to 71%. Therefore, the production of edible oilseeds in India would substantially enhance the edible oil availability in India. There is a need to bridge the yield gap by technological interventions, without expanding the area under cultivation.

Sunflower, safflower, sesame, and niger crops are important edible oilseed crops with individual and combined strength to bridge the shortage and increase higher self-sufficiency in the edible oil requirement of the country. Though their per se contribution in acreage and production is less than 10% of the total oilseeds area and production, these crops have unique value and potential as hardy, wide adaptability, and high-quality oil, crop diversification, and secondary agriculture. Oil palm is the highest edible oil-yielding tree crop with a theoretical potential yield of 18 ton of oil/ha/year. It is capable of yielding 4–5 tonnes of palm oil and 0.4–0.5 tonnes of palm kernel oil.

10.3 Groundnut

Groundnut is an important food legume and oilseed crop of India, cultivated mainly by small and marginal farmers predominantly under rainfed conditions with minimal or no inputs. In India, the cultivation of groundnut is extended in almost all states and most of the districts. As per the latest statistics, it is grown in about 360 districts. The crop is grown mainly as a rainfed dry land crop on well-drained sandy soils in low (<750 mm) and medium (750–1000 mm) annual rainfall areas, often subject to the vagaries of the weather conditions. So far, only 29% of the groundnut area is under irrigation.

10.3.1 Shift in Area, Production, and Productivity

Between the decades of the 1960s and 1970s, there is practically little difference in productivity (700–800 kg ha^{-1}) and the increase in production was primarily due to the expansion in growing areas. But during the 1980s, particularly in 1988–1989 due to favourable seasons and the transfer of improved technologies through Technology Mission on Oilseeds, for the first time, the productivity crossed one ton (1132 kg ha^{-1}). However, there was a fluctuating trend and the average yield of the crop in 1990–2000 was 994 kg ha^{-1} and increased to1100 kg ha^{-1} in 2001–2010 and to 1459 kg ha^{-1} in 2007–2008 (Fig. 10.1). During the last decade (2010–2020) much attention was paid to increasing productivity. As a result, except for a bad season in 2012–2013, the productivity in India increased to around 9.3 million tonnes in 2017–2018. It further increased to 10.0 million tonnes of pods from about 4.5 million-hectare area in 2019–2020. The country is the second largest producer as well as consumer of groundnut in the world. Though its area and productivity are fluctuating due to abrupt fluctuations in the weather conditions, India has been

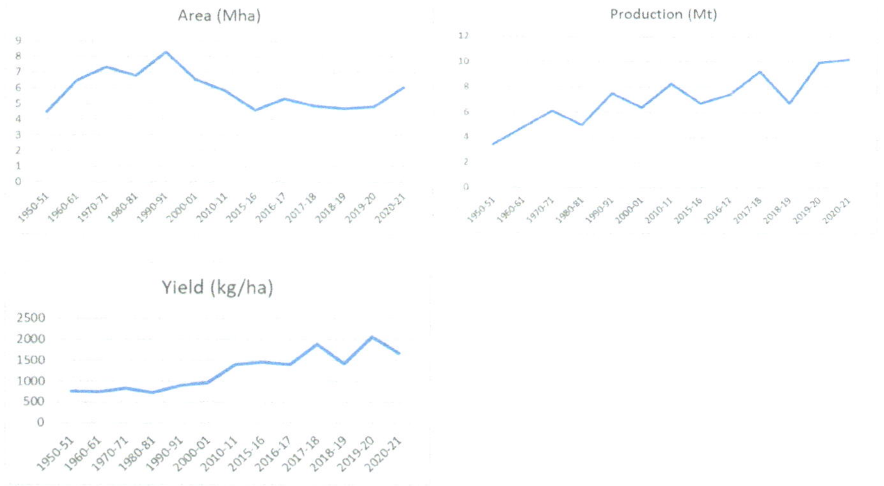

Fig. 10.1 Area, production, and yield statistics of groundnut in India from 1950 to 2021

Table 10.1 Major groundnut producing state of India year 2019–2020

(Area: million ha., Production: million tonne, Yield: kg/ha)					
State	Area	% to all India	Production	% to all India	Yield
Gujarat	1.69	34.54	4.64	45.99	2749
Rajasthan	0.74	15.11	1.62	16.04	2191
Tamil Nadu	0.35	7.09	0.98	9.74	2840
A.P.	0.66	13.52	0.85	8.39	1282
Karnataka	0.57	11.74	0.68	6.71	1180
Other states	0.88	17.99	1.32	13.06	1500
All India	4.89	100.00	10.10	100.00	2065

producing one of the best quality groundnuts in the world and has a lot of export demand due to its well-filled pods and flavour. Though in India groundnut is grown in almost all states, Gujarat (32% of total area and 39% of production), Andhra Pradesh (19% area and 10% production), Rajasthan (11% area and 14% of production), Tamil Nadu (7% area and 11% of production), Karnataka (11% area and 6% production), Madhya Pradesh (4% area and 5% of production), and Maharashtra (6% area and 4% of production) are the main groundnut growing states (Table 10.1).

The data for the last decade reveals that Gujarat, Rajasthan, and Tamil Nadu cover about 50% of the area producing about 65% of the groundnut of India and now the best belt for commercial production of peanuts. In India, as more than 70% of groundnut is grown during the rainy season (June to October), the productivity is relatively low and the production becomes rain-dependent. The average productivity of the post-rainy season (October to March, known as *Rabi* crop and February to May known as summer crop) is about 2000 kg ha^{-1}, much higher than rainy season productivity (1500 kg ha^{-1}), indicating more production potential during this season. In the rainy season, both short (less than 100 days duration) and long duration (110–140 days) groundnut varieties are grown based on weather, soil, and distribution of rainfall. While, in Post rainy season, only short-duration crop is grown.

10.3.2 Recently Released High-Yielding Groundnut Varieties

A total of 117 high-yielding varieties have been developed and recommended for commercial cultivation during the last 20 years. Among the varieties recommended for cultivation during post rainy season, J87 (4165 kg/ha, Dh86 (4020 kg/ha), TGGS341 (4000 kg/ha) and Greeshma (4000 kg/ha) are top recommended varieties for different zones. Among the varieties recommended for cultivation during the rainy season top five varieties are, Sneha (3820 kg/ha), RARS-T-2 (3734 kg/ha), Snigdha (3550 kg/ha), K 1812 (3514 kg/ha), and Pratap Mungphali (3388 kg/ha). The availability of quality seeds in peanuts is a major concern. Compared to other crops, the breeder seed requirement is very high in peanuts, due to high seed volume. Also, seed rate is very high (200 kg pods/ha), and around one million tonnes of peanut kernels are required to sow about 5 m ha area in a year. The notable varieties having high breeder seed demand are, Kadiri-6, GPBD-4, TG 37 A, Narayani, and Dharani over the years. However, in the rainy 2021 season, the highest DAC indent for breeder seed production was received for K 1812 (1080 q), GJG 32 (738 q), Dharani (611 q) GPBD-4 (450 q), and GJG-22 (310 q).

10.3.2.1 Rapeseed-Mustard

India is cultivating nine oilseed crops under the diverse agro-ecosystems, among which rapeseed-mustard stands second after soybean in terms of area (24%) and production (25%). The major rapeseed-mustard species cultivated in India include; *Brassica juncea, B. rapa, B. napus and B. carinata* (used for edible oil), and *B. oleracea and B. nigra*(for seed condiments). Among these, Indian mustard [*B. juncea* (L.) Czern], holding sizable contribution in terms of area and production of oilseeds, and edible oils. In the past, brown mustard (*B. rapa var. brown sarson*) was the dominant oilseed brassica crop in India; however, today, it is dominated by Indian mustard (*B. juncea*) (90% acreage). The different landraces of Brassica species have different ecological niches and the varieties were developed as per their agro-ecological suitability. Indian mustard (*B. juncea*) is predominantly cultivated in western to central parts of North India and in some non-traditional areas of southern India. Yellow mustard (*B. rapa var. yellow sarson*) and toria (*B. rapa var. toria*) are short-duration crops and cultivated in north-eastern India as catch crop. Taramira/rocket salad (*Eruca sativa*) is a drought-tolerant species grown in the drier parts of northwest India. Gobhi mustard (*B. napus*) and Ethopian mustard/Karan rai (*B. carinata*) are the new emerging oilseed crops having a limited area under cultivation in northern India (Jat et al. 2019). Indian mustard oil contains low amount of saturated fatty acids (palmitic acid and stearic acid) (8%) and a high amount of monounsaturated fatty acids (oleic, eicosenoic, and erucic acids) (70%), and polyunsaturated fatty acids linoleic and linolenic acids (22%). The mustard oil is high in omega-3, free from cholesterol and transfats, and has a very low N-6 to N-3 ratio compared to other oils. Indian mustard oil is blended with other vegetable oils (sunflower, soybean, corn, etc.) to improve its fatty acid profile. However, with the increasing awareness about health and wellness, the demand for mustard oil is increasing globally to be used for direct consumption and other value-added products (Kumar 2015; Jat et al. 2019).

10.3.3 Shift in Area, Production, and Productivity

A quantum jump in production of Rapeseed-mustard was recorded from a mere 0.76 million tonnes in 1950–1951 to 9.12 million tonnes in 2019–2020 (Fig. 10.2). Similarly, productivity levels increased from368 kg/ha in 1950–51 to 1345 kg/ha in the year 2019–2020 (Anonymous 2020). The per cent increase in total oilseeds area, production, and productivity in the country is +152, +548 and + 157% in2019–20, respectively, over the base period of 1950–1951. During the same period, the rapeseed-mustard area, production, and productivity increased by 227%, 1100%, and 265%, respectively. The compound annual growth rate (CAGR) in area, production, and yield levels of oilseed brassica is +1.71%, +3.61%, and + 1.87% during 2019–2020 over 1950–51, which is higher than total oilseeds (+1.33%, +2.7%, and + 1.36%), respectively. Rapeseed-mustard share in total oilseeds production is second (27%) after soybean (33%); however, stands first (23%)in edible oil production in the country (Fig. 10.3).

Despite significant progress achieved in oilseeds production in India, the increasing population and changing lifestyle of the people led to an increase in the import of

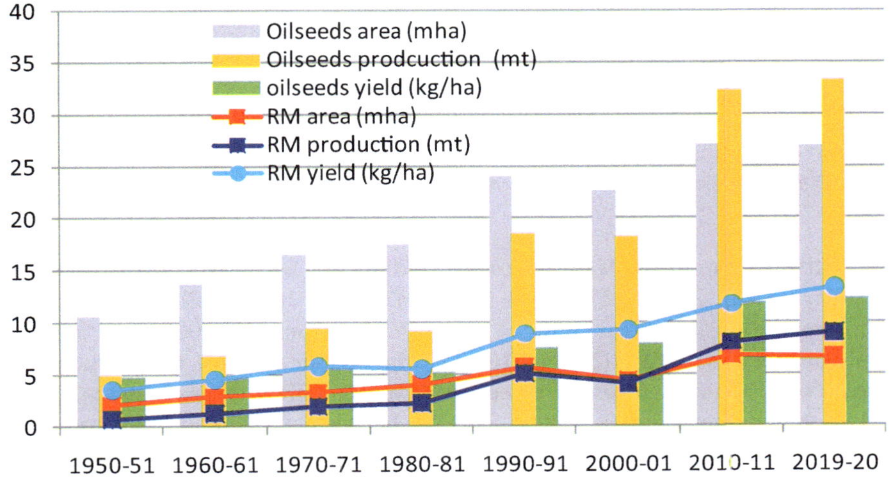

Fig. 10.2 Production trend in oilseed and rapeseed-mustard since 1950

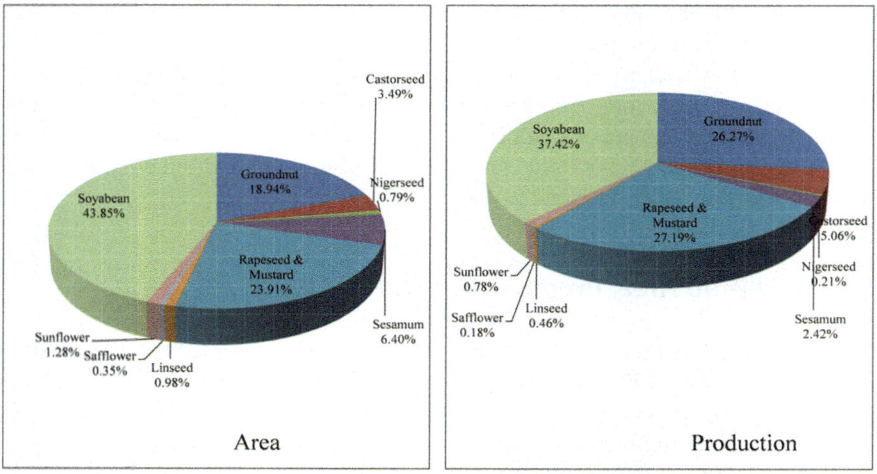

Fig. 10.3 Rapeseed-mustard share in total oilseed production in India (5-year average (2015–2016 to 2019–2020)

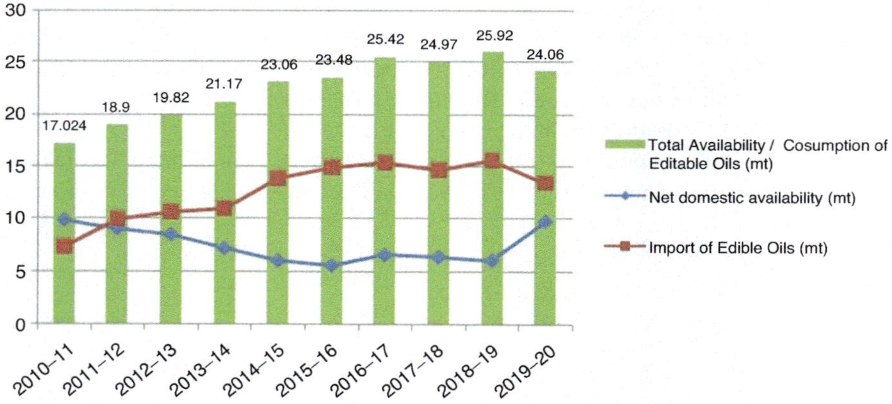

Fig. 10.4 Domestic production and import of edible oil in India (Source: Anonymous 2020)

vegetable oils up to 11.2% of the total world import, which is worth about 103.2 million US$ in 2016–2017 (Jat et al. 2019). During the last 20 years, the total consumption has increased (+148%) from 9.7million tonnes in 2000–2001 to 24.06 million tonnes in 2019–2020 (Fig. 10.4). During the same period, the production of edible oils from all sources increased (%) from 5.5 million tonnes in2000–2001 to 11.63 million tonnes in 2019–2020. This gap between consumption and production is just met out through import which has increased (+267%) from 4.2 million tonnes in 2000–2001 to 13.41 million tonnes in 2019–2020 (Jat et al. 2019; Anonymous 2020 Agril statistics at a glance, 2019–2020.

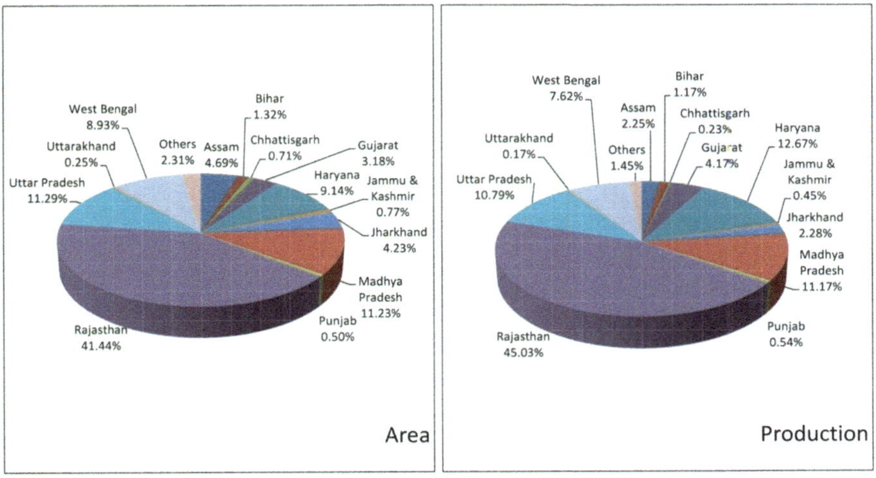

Fig. 10.5 Share of states in Rapeseed-mustard production (5-year average, 2015–2016 to 2019–2020) (Source: Anonymous 2020)

At present, vegetable oils hold a 50% share (in terms of value) in the country's total imports and drain a significant quantum of revenue to foreign countries. It is astern apprehension to the national economy while simultaneously fulfilling the domestic edible oils requirement. India mainly imports refined edible oil from other countries against the negligible export. The situation will be more challenging with rising consumption up to 2030 with the projected population and after that, may increase at a decreasing rate with a declining population growth rate. The limited and steadily declining natural resources (land and water), and competition from agriculture and nonagricultural sector, jeopardized the scope to increase the acreage under oilseeds. In such a situation, an increase in yield levels through translational scientific interventions is a conceivable alternative to increase production and edible oil availability. Rapeseed-mustard is grown all over India; however, it dominates in five states (Rajasthan, Madhya Pradesh, Haryana, Uttar Pradesh, and West Bengal) which contribute 82% area and 87% of production, with a maximum in Rajasthan (41% area and 45% production) (Fig. 10.5). Eco-regional variations are mainly noted due to biophysical fittingness of the crop.

10.3.4 Rapeseed-Mustard Growth in India

It is vital to look back through the CAGR per annum in the production of oilseeds as well as rapeseed-mustard during different periods of the pre-green revolution (1950–1951 to 1968–1969), green revolution (1969–1970 to 1990–1991) and post-green revolution (1991–1992 to 2019–2020). Govt. of India started Technology Mission on Oilseeds (TMO) in 1986 to enhance production, reduce imports and

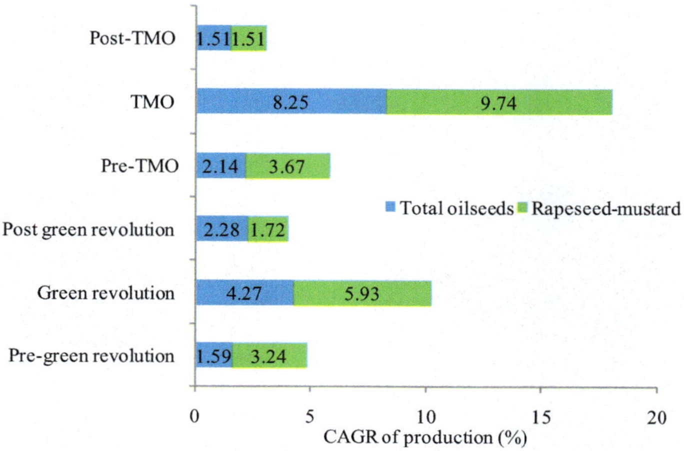

Fig. 10.6 CAGR of oilseed and rapeseed-mustard production during different periods

achieve self-sufficiency in edible oils in India. Thus breakthrough in the area, production, and yield of oilseeds as well as rapeseed-mustard came after the launch of TMO in the year 1986, by the central government to increase the production. Therefore, the period of pre-TMO (1966–1967 to 1985–1986), TMO (1986–1987 to 1995–1996), and post-TMO (1996–1997to 2019–2020)is also important to draw the future opportunities, constraints, and strategies. The period of TMO witnessed CAGR of 8.25% in oilseeds, and 9.74% in rapeseed-mustard production (Fig. 10.6). Similarly, during the green revolution period growth of oilseeds (4.27% CAGR) and rapeseed-mustard (5.93% CAGR) production was higher as compared to pre/post-green revolution period.

The growth was achieved mainly due to the development of new varieties, technological advancements in production and protection, and policy interventions during this period. In the era of globalization, during 1994–1995, the Government of India opened free trade under Open General License (OGL) and reduced the import duty on edible oils from 65 to 15% which resulted in heavy imports and drained money to other countries but also had the cascading effect on domestic oil prices, and impacted Indian oilseed industry with a considerable disincentive to the oilseed farmers. Though, during the TMO period, there was an effective implementation of minimum support price (MSP) that gave confidence to the farmers for minimum expected returns. Most of the time, the wholesale prices are much higher (Rs. 7450/q) than MSP (Rs. 5050/q) in 2021–2022.

10.3.5 Rapeseed-Mustard Breeding in India

The research work on the improvement of rapeseed mustard started at the beginning of the twentieth century at Pusa (Bihar), the then Bengal Presidency, through the collection and purification of land races. The initial scientific work for the varietal improvement of rapeseed-mustard in India started at Lyallpur (now Faisalabad in Pakistan), (NRCRM 2000), which led to the development of the Indian mustard variety RL 18 (Raya Layallpur 18), in 1937 and yellow sarson variety L1 through selection. Indian mustard strain RT 11 from Uttar Pradesh was released in 1936. During the course of about two decades (1947–1967), a number of high-yielding varieties of mustard (Laha 101, Varuna, Durgamani, Patan Mustard), toria (Abohar, BR 23, M 27, T 9, ITSA, T 36, DK 1), brown sarson (BSA, BSG, BSH 1, BS 2, BS 65, BS 70), yellow sarson (T 151, Patan sarson, YSPb 24, T 42), and taramira (ITSA) were developed (NRCRM 2007). The Indian Council of Agricultural Research (ICAR) in April 1967 launched the All India Coordinated Research Project on Oilseeds (AICORPO), including five major crops namely groundnut, rapeseed-mustard, sesame, linseed, and castor. However, the Project Coordinator (Rapeseed-Mustard) was established on January 28, 1981 at Haryana Agricultural University, Hisar, Haryana. In order to provide strong leadership to the country's rapeseed-mustard research activities, the ICAR established National Research Centre on Rapeseed-Mustard on October 20, 1993 at Bharatpur (Rajasthan), bringing under its umbrella the All India Coordinated Research Project on Rapeseed-Mustard with 19 research centres across the country. In February 2009, the National Research Centre on Rapeseed-Mustard was redesignated as Directorate of Rapeseed-Mustard Research (DRMR) (Chauhan et al. 2006; Chauhan et al. 2011).

Initially, Exotic germplasm was utilized in the Indian breeding programme to develop 21 varieties of rapeseed-mustard, especially for oil and seed meal quality characters. Two varieties of brown sarson and 10 varieties of gobhi sarson, 9 of Indian mustard and one in karan rai were developed. Until 1970, pure line and mass selection were the only breeding methods, employed to develop 26 varieties. Later on, hybridization was followed to develop varieties. A total of 221 varieties of rapeseed-mustard have been released during 1969–2021. All the varieties developed before the inception of AICRPO/ AICRP-R & M were selections from the local germplasm (Chauhan et al. 2006; Chauhan et al. 2011; Chauhan et al. 2020). Chauhan et al. (2011) reviewed the pedigree of mustard varieties and reported that most of the Indian mustard varieties are derived from a few common ancestors and a limited number of donors were utilized in the breeding programme, resulting in a narrow genetic base.

The first notified variety was ITSA of toria (*B. rapa*) in 1973. Bio 902 of Indian mustard, a somaclone was released in 1993. The list of trait-specific varieties is presented in Table 10.2. Indian mustard oil contains a high amount of erucic acid (40–57% of total fatty acids) which is not considered good for human health. Its seed meal contains high glucosinolate which is harmful to non-ruminant animals. Hence, efforts were made to develop varieties that contain low (<2%) erucic acid (Pusa Karishma, Pusa Mustard 21, Pusa Mustard 22, Pusa Mustard 24, RLC 1, RLC 2, RLC 3, and PM 32) and double low (low erucic acid and low glucosinolate 30 micro mole/g of defatted seed meal) varieties (PDZ 1 and PDZ 33). The varieties released during the last two decades (2001–2021) have been given in Table 10.2. The varieties have also been developed for various traits like tolerant to biotic (white

Table 10.2 Indian Mustard varieties recommended for specific conditions/specific traits (Source: Updated from Chauhan et al. 2011; Sharma et al. 2022)

Trait	Variety
White rust resistant	Basanti, JM 1, JM 2
Earliness	Kanti, Narendra Ageti Rai 4, Pusa Agranai, Pusa Mahak, Pusa Vijay, PM 25 (NPJ 112), PM 27 (EJ 17)
High oil content	Narendra Swarana Rai 8, NRCDR 02, Rohini
High temperature tolerant	Kanti, Pusa Agrani, RGN 13, Urvashi, NRCDR 02, PM 25, PM 27, CS 58, Pant Rai 18 (PR 2006–1), Pant Rai 19, Pusa Vijay, RH 0119, RH 406, RGN-229, RGN-236, RGN-298, PM 22
Intercropping	RH 30, RH 781, Vardan
Late sown	Ashirwad, RLM 619, Swarnajyoti, Vardan, Navgold (YRN 6), RGN 145, NRCHB 101, CS 56, PM 26 (NPJ 113), Radhika (DRMR 2017–15), Brijraj (DRMRIC 16–38)
Non-traditional areas	PusaAgrani, Pusa Jaikisan, Rajat, Gujarat mustard 2. TKM 1, Shatabdi
Rainfed	Geeta, GM 1, PBR 97, Pusa Bahar, Pusa Bold, RH 781, RH 819, RGN 48, Shivani, TM 2, TM 4, Vaibhav, RB 50, RH 406, RH 725, RGN 229, RVM 2, Pant Rai-20, RGN 298, Aravali mustard, DRMR 1165–40, DRMR 150–35, Birsa Bhabha Mustard 1 (BBM1)
Salinity tolerant	CS 52, CS 54, CS 56, CS 58, CS 60
Frost tolerant	RGN 13, RH 819, Swarnajyoti, RH 781, RGN 48
Low erucic acid (<2%)	Pusa Karishna, PM 21, PM 22, PM 24, ELM 079 (RLC1), RLC 2, RLC 3, Pusa mustard 32 (LES 54)
Double low	Pusa double zero mustard 1 (PDZ 1), Pusa double zero mustard 33 (erucic acid <2%, low glucosinolate<30 micromole/defatted oil cake)
Hybrid	NRCHB 506[a], DMH 1, Coral 432 (PAC 432), Coral 437 (PAC 437), Keshri (PRO 5111), RCH 1[a]

[a]Govt. sector hybrid

rust, Alternaria blight, powdery mildew) and abiotic stresses (drought, salinity, high temperature, and frost), and better quality for specific growing conditions. Various novel genetic stocks of oilseed brassica (with low erucic acid and low glucosinolates, high oil content, high oleic acid, and low linolenic acid, dwarfness, earliness, long main shoot, bold seed, yellow seed, tetra locular siliquae, white rust resistance, tolerance to high temperature and salinity, drought, high water use efficiency) were also registered with National Bureau of Plant Genetic Resources (NBPGR), New Delhi for the use in future breeding programmes. The major emphasis of the varietal improvement programme is now on genetic enhancement to widen the genetic basis for seed and oil yield with introgressing traits like, early, timely and late-sown conditions to cater to the need of diverse agroecological situations of the country, improvement of oil (low erucic acid) and seed meal (low glucosinolate), high omega oil quality, resistance/tolerance against biotic (white rust, Alternaria blight, Sclerotinia rot diseases, and aphid and painted bug insects) and abiotic stresses (drought, high temperature, frost, and salinity) (Jat et al. 2019) (Table 10.3).

Table 10.3 Recently released varieties of rapeseed-mustard developed in India (2011–2021) (Source: updated from Sharma et al. 2022)

Crop	Variety
B. juncea	Pusa Mustard 28 (NPJ 124), Coral 437 (PAC 437) (Hybrid), RLC 2 (ELM 123), RGN-229, RH 0406, Divya 33,Pant Rai-19 (PR 2006–1), PBR 378, RH 749, Raj Vijay Mustard 2 (JMWR 08–3), RRN-573, RGN-229, Pusa Mustard 29 (LET-36), Pusa Mustard 30 (LES-43),Giriraj (DRMRIJ 31), RGN-298, RGN-236,Gujarat Mustard-3 (GM-3),Albeli-1, Pant Rai-20, DRMR150–35, PBR 357, Gujarat Dantiwada Mustard 4, Gujarat Dantiwada Mustard 5, Raj Vijay Musatard-1, RLC-3, CS 1100-1-2-2-3 (CS 58), Pant Rai 21 (PRB2008–15), RH 0725, CS 2800-1-2-3-5-1 (CS 60), PDZ-1, Keshri (PRO 5111) (Hybrid), DRMR 1165–40, RH 761, TBM-204, Pusa Mustard 32 (LES 54), Radhika (DRMR 2017–15), Brijraj (DRMRIC 16–38), Azad Mahak [(KMR € 15–2)], RCH 1, Pusa Double Zero Mustard 33, TAM 108–1, PHR 126, Birsa Bhabha Mustard 1 (BBM1), Trombay Him Palam Mustard 1 (THPM-1)
B. carinata	PC 6 (BJC 1)
B. napus	GSC 7 (GSC 101), RSPN-25, ONK-1
*B. rapa*var. *Toria*	Sushree, TL17, Raj Vijay Toria 1, Pant Hill toria 1 (PT 2006–4), Pant Toria 508 (PTE 20008–2), Tapeshwari, Tripura Toria 1, RSPT-6, Raj Vijay Toria 3, (RTM 08–6), Jeuti(JT 90–1), Azad Chetna (TKM 14–2), Raj Vijay Toria 2 (RMT 08–2), AAU TS 38
*B. rapa*var. *Yellow sarson*	Pant Sweta (PYS 2007–10), Pant Girija (PYS 2012–6), Sachita (YSWB-2014/2), Anushka (YSWB-2011-10-1)
*B. rapa*var. *Brown sarson*	Shalimar Sarson 1, Shalimar Sarson-2 (KBS-49), KBS-3, HPBS-1, RSPT-6 (TCN 13–9),
Eruca sativa	Vallabh Taramira 1(PUT93–11), Vallabh Taramira 2 (PUT 93–1),Jwala Tara (RTM 1355), Jobner Tara (RTM 1351)

10.3.6 Future Projections, Challenges, and Strategies

At the current trends, 39.2 million tonnes of vegetable oil will be required to meet the demand of the 1.68 billion population in India by half of the twenty-first century due to changing lifestyle and dietary preferences of humans. Though the production of edible oils increased at a CAGR of 2.83% from 1950–1951 to 2019–2020. The maximum CAGR was witnessed during the green revolution period (oilseeds: 4.27%, Rapeseed-mustard: 5.93%) and TMO (oilseeds: 8.25%, Rapeseed-mustard: 9.74%). To fulfill the oilseed requirement, rapeseed-mustard has to contribute20.5 million tonnes of production (assuming a contribution of 25% in total oilseeds production) by the year 2050 which is a gigantic task for the scientists and policy makers. To meet out the projected demand of edible oils; area, production, and productivity of oilseed brassica should increase at a faster pace. At the present level of contribution of rapeseed-mustard (25%) in total oilseed stock, its area, production and yield have to increase to a tune of eight million ha, 20.51 million tons, and 2563 kg/ha by the year 2050 (Fig. 10.7) (Anonymous 2015; Jat et al. 2019).

The present level of the area, production and yield are 6.78 million ha, 9.12 million tonnes and 1345 kg/ha, respectively (2019–2020). To reach the level of

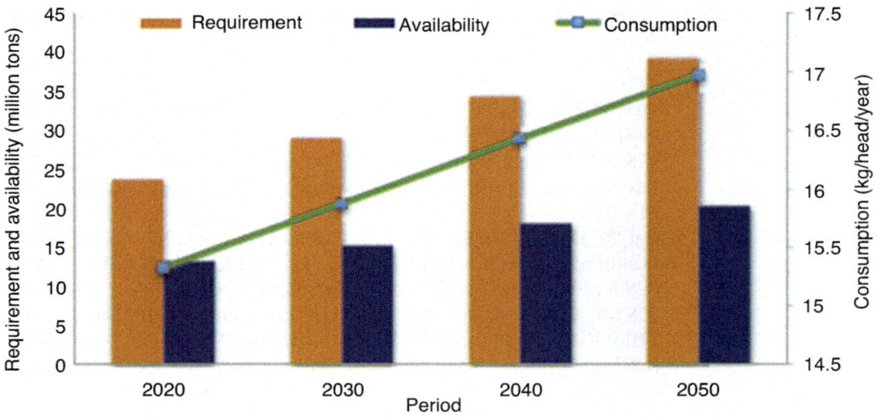

Fig. 10.7 Projections for demand, availability, and per caput consumption of edible oils up to year 2050 (Source: Jat et al. 2019)

projections in the year 2050, the country has to bring 1.22 million ha additional area under cultivation and double the productivity to produce 3-times more rapeseed-mustard at the present figures. Additional areas under cultivation for rapeseed-mustard may be explored in the rice-fallow system (North-eastern regions), and non-traditional areas (southern regions) provided with suitable technology and policy interventions. Under present climatic variability, fast depleting natural resources, limitations of food, and nutritional security, the realization of these targets will be unimaginable without path-breaking technological interventions (Jat et al. 2019). However, there is a need to integrate, frontier science research along with ongoing programs for yield improvement, resource use efficiency, diversified products, and uses, stakeholders involvement, and viable supply chain and marketing. Significant improvement has been made genetic enhancement of rapeseed-mustard, still, there is a need to develop varieties for diverse agro-ecological situations in traditional and non-traditional areas, input responsive, climate-resilient, tolerant to biotic and abiotic stresses, and of high oil content and quality. The use of pre-breeding along with advanced tools like GISH (Genomic In-Situ Hybridization), and FISH (Florescence In-Situ Hybridization) to harness the genes from wild and weedy relatives of rapeseed-mustard is required. Identification and development of high heterotic crosses, and exploitation of exotic germplasm to enhance the level of heterosis is the need of the hour. Effective utilization of genomic tools (transcriptomics, proteomics, metabolomics), gene editing, genetic engineering, maker assisted breeding is very much required for mining, tagging of the novel genes (for biotic, abiotic stress, quality traits, earliness, high photoperiodic responses), their functions and transfer into the agronomically superior genetic background is very much needed to break the yield barriers in rapeseed-mustard. Further, there is a need to freely exchange trait-specific germplasm with International organizations.

Application of proven technologies such as integrated nutrient management, precision agriculture, and resource conservation approaches for specific agro-ecology. Yield loss due to insect pests, diseases, and parasitic weeds (Orobanche) needs to be addressed through the development of resistance/tolerant varieties and holistic control measures to reduce the losses. Rapeseed-mustard productivity at the national level could be improved by bridging the yield gaps between farmer's fields and experimental plots through Farmer-Institution-Industry linkages. The development of double zero varieties has opened new avenues for mustard to be used in the development of value-added products. All the risk mitigation options like timely availability of inputs (quality seeds, fertilizers) and credit, high MSP and assured procurement, crop insurance, and linking farmers to market should be addressed for a better oilseed economy of our country.

10.4 Soybean

Soybean is one of the major economic crops in the world. The lower and middle yellow river of China is supposed to be the main centre of origin for soybean (Ruzhen 1989). It is believed that soybean was introduced to India from the north-eastern region of India and the Himalayan regions, specifically through the areas adjoining China (Tiwari 2006). The indigenous, black-seeded land race "Kalitur" has been the fountainhead for Indian soybean cultivation. It ruled soybean production for years, after which, yellow-seeded introductions, mostly from USDA took the lead (Tiwari 2006). Kalitur, for its seed longevity, wider adaptability, and climate-resilient, has been employed as a parent for hybridization to develop several varieties (Tiwari 2006). A systematic soybean breeding program has been started during the early 1900s at Pusa Agricultural Research Farm, Bihar. In 1943, selections have been made in 239 introductions at Kanpur. Lines such as T-1 and T-49 were recommended for plain areas, T-33 was recommended for hilly areas and T-2 was recommended for fodder purposes (Singh 2006). In 1947, Punjab Soy No. 1 has been developed in Punjab and recommended for its cultivation in Kangra valley. Likewise, in 1947, the Indian Agricultural Research Institute identified Monnetta as suitable for both grain and fodder purposes.

During the early years, despite its high protein and oil content, and nutritional qualities, several attempts have been made to popularize it in India. However, these efforts have been ineffective mainly due to a lack of knowledge on its cultivation, high-yielding varieties, market felicities, and awareness of its utility (Singh 2006). Owing to the stagnation of pulse production and the prevailing protein malnutrition problem, a collaborative effort has been made by GBPUAT, Pantnagar, JNKVV, Jabalpur, and the University of Illinois, USA for soybean popularization in India (Singh 2006). During 1965–66, soybean varieties from the southern USA have been evaluated at Pantnagar, and varieties like Bragg and Hardee have yielded about 3–4 t/ha with a maturity duration range of 110–130 days. This led to the establishment of AICRP on soybean in 1967, by ICAR (Singh 2006). Based on the multi-location trials, Bragg was released for its general cultivation across India.

The development of mega-varieties such as JS 335, JS 95–60, and JS 93–05 has been a milestone in boosting Indian soybean production. The variety JS 335 was released in 1994 and is known for its higher yield and wider adaptability across India. Early maturing varieties fit into existing cropping systems in central India. The varieties JS 93–05 and JS 95–60 were released in 2003 and 2006 respectively, and are early maturing (~90 days) and hold a major share in central Indian production (Agarwal et al. 2010). Few genetic stocks (MACS330, EC 390977, and EC 34101) with photoperiod insensitivity and early maturity traits have been registered with the Plant Germplasm Registration Committee (PGRC) of ICAR and are being used extensively in the development of high-yielding and early maturing varieties.

Soybean is a photo-sensitive short-day crop. Late sowing due to the delayed onset of monsoon will result in a reduction in biomass and yield. Exploitation of the long juvenility (LJ) trait aids in delayed flowering and biomass gain under short days. LJ trait has been identified in a germplasm accession AGS 25, and an association between delayed flowering and a functional allele *FT2a* has been established (Tripathi et al. 2021). Being a rainfed crop, and owing to erratic patterns in climate change, soybean has been challenged with several forms of biotic and abiotic stresses. A multiple stress tolerant variety, JS 97–52, was released in 2008. This has been reported to be resistant/tolerant to several forms of stresses like drought, water-logging, charcoal rot, YMV, colour rot, aerial blight, and different insect pests (Agarwal et al. 2010). This genotype has been extensively employed as a parent in varietal development programmes.

Varieties like Bragg were found susceptible to YMV and rust and epidemics of these diseases have been witnessed (Nene 1972; Singh et al. 1974). YMV has been the major diseases in soybean. Through larger-scale germplasm screening, genotypes-PI171443 (*G. max*) and *G. soja* (wild soybean) have been found to be resistant to YMV disease. Six introductions - PI200465, PI200466, PI200477, PI200490, PI200492, and PI224268 were found resistant to rust disease. These resistant donors have been extensively used as parents to develop high-yielding and disease-resistant varieties. YMV-resistant varieties such as Ankur, Alankar, PK 262, PK 327, PK 308, and PK 416 have been developed using resistant donors (Singh 2006). Recently, a YMV tolerant variety NRCSL 1 has been released for its cultivation across the eastern zone. Rani et al. 2017 identified a recessive gene governing YMV resistance in the donor accession PI171443. The gene was mapped on chromosome 6, flanked by two SSR markers GMAC7L and Satt322. Two duplicate dominant genes governing YMV resistance have been identified in *G. soja* (PI 393551). Linkage analysis revealed that the genes were tightly linked with SSR marker BARCSOYSSR_08_0867 on chromosome 8 and BARCSOYSSR_14_1416 and BARCSOYSSR_14_1417 on chromosome 14 (Rani et al. 2018).

Rust has been the major disease in the southern zone. Two germplasm lines EC 241778 and EC 241780 are rust-resistant. Employing these resistance sources, rust-resistant varieties like DSb 21, DSb 23, DSb 28–3 have been developed for their cultivation in the southern zone. In EC 241780, based on the sequence variations and functional annotations, three genes *Glyma18G51715, Glyma18G51741,* and

Glyma18G51765, encoding for NBS-LRR family protein, were identified as the most prominent candidate for *Rpp1* locus governing rust resistance (Ratnaparkhe et al. 2020). Anthracnose disease, especially in central India, has been a major economic disease in recent years. Through field screening under hot-spot conditions at ICAR-IISR, Indore, two germplasm lines, viz. EC 34372 and EC 457254, be anthracnose resistant (Nataraj et al. 2020). Further, genotypes NRC 128 and PS 1611 are durable resistance sources, when screened at multi-locations (Rajput et al. 2022).

Secondary agriculture plays a crucial role in enhancing the farmers' income through food-based industries. Despite being a nutrient high-rich crop, soybean carries the anti-nutritional factor-KTI (trypsin inhibitor) and beany flavour due to lipoxygenases, limiting its food use. In 2018, the first KTI-free variety- NRC 127 has been developed by ICAR-IISR, Indore. NRC 142, an early maturing variety free from KTI and Lipoxygenase 2 with multiple resistance to diseases and insect-pests was released for the Central and Southern Zone. In 2021, NRC 147 the first variety with 42 $\pm$ 5% oleic acid content was developed by ICAR-IISR, Indore for eastern and central zones.

A number of germplasm lines with quality traits such as high protein (G288 & G688), high oil content (AGS191, NRC7, G76), high oleic (IC210 & NRC106), low linolenic acid (VLS 59), Null KTI (NRC 101, 102), and null lipoxygenase (NRC 105) have been identified and have been registered with Plant Germplasm Registration Committee (PGRC) of ICAR.

10.4.1 Role of Soybean in Foreign Exchange

India ranks fifth in edible oil market in the world; nevertheless, 50% of its edible oil demand is met through import (Agarwal et al. 2013). On contrary, India has been the key player in the defatted oil cake (DOC) international market. Higher market price to the Indian farmers, than the MSP, is due to better Indian DOC price in the global market (Agarwal et al. 2013). During 2019, the Indian export share of soybean cake was 744,417 USD, with Iran being the leading importer (185,055 USD) followed by the USA (180,833 USD) (Table 10.1 and Fig. 10.1). EU28 aggregation accounts for about 10.7% (80,106 USD) of total Indian soybean cake export value, France being the leading importer (44,140 USD) among the European countries (UN Comtrade 2019; Directorate General of Commercial Intelligence and Statistics 2019).

About 80% of soybean grown across the globe is genetically modified and this is steadily increasing. Accordingly, the share of GM soybean and its meal is also increasing in the international market. On the other hand, compared to the rest of the world, demand for non-GM soybean is high in the EU with systematic segregation and identity preservation (IP) of non-GM soybean throughout the supply chain for non-GM preferring consumers. In 2013, the EU consumed about 28 million tonnes of soybean and derived products, of which 95% were imported. Soybean being the cheapest protein source has been in high demand in the EU mainly for animal feeding purposes, predominantly in the poultry and pork industry. About 11.3% of soy meal imported by 14 European countries (representing >90% of total EU soybean imports) is estimated to be segregated as non-GM. This represents 2.7 MT of soy meal with a market value of approximately 1.5 billion Euros. Hence, the market potential of non-GM soybean meals can hardly be considered. However, countries like Hungary and Sweden produce their compound animal feed almost exclusively through non-GM soybean. Germany being the largest producer of compound feed in EU is driven by the poultry sector to produce non-GM feed. About 50% of poultry feed is non-GM. In countries like the UK, the share of non-GM IP soybean in poultry compound feed is around 28%. In the EU, though the demand for non-GM IP soy meal is modest, a major concern is its premium which is about 20–30% of the non-segregated soy-meal accounting for 180 EUR per ton in markets such as the UK. Therefore, the premium for non-GM soy-meal paid globally by EU importers would be around hundreds of millions of Euros per annum which we cannot rule out. According to the operators in the supply chain, non-GM IP soybean will remain stable but this opinion can be withdrawn if the supply of non-GM soybean for the EU market becomes difficult. EU is highly dependent on Brazil for the supply of non-GM IP soybean and the future of the non-GM soybean market in EU depends on Brazil's ability to supply and the premium prices being paid. To combat high premium prices charged by Brazil, the EU has found new alternatives such as India where the quality of the production has increased in the last years. India as the potential alternative to the EU is mainly due to the harvest of soybean during September when Brazilian stocks usually decline resulting in a soybean price hike (Source: Tillie and Rodríguez-Cerezo 2015).

The USA is a leading GM soybean producer. Scientific reports confirmed that most of the soybean imported by Iran from the USA and Argentina was transgenic (Sarikhan 2006) and therefore, there is a huge market demand in Iran for safer GM soybean (Ghareyazie 2010). Hence, these two countries could be potential markets for GM soybean cake and GM soybean could be permitted in India for the same. On the other hand, to take the advantage of the premium price being paid by the EU for non-GM soybean cake, India may employ stringent non-GM soybean. But simultaneous cultivation of both GM and non-GM soybean may not be practically feasible since there is 50% of genetic contamination in soybean (Rissler 2004). Therefore, an exclusive economic impact analysis of GM and non-GM soybean cultivation in India has to be carried out to decide whether or not to allow GM soybean cultivation in India.

10.5 Sunflower

Sunflower (*Helianthus annuus* L.) was an addition to the edible oilseeds milieu post-independence that has established itself as an efficient oilseed crop in the country in the last five decades. The favourable features of the crop with short duration, wide adaptability to a wide range of seasons and soils, high seed multiplication rate, high-quality edible oil, suitability for both inter and sequence cropping, catch crop, non-branched plants easily adoptable to mechanized farming from seed to seed, support to the apiary and ecological farming, thalamus supporting dairy, etc., could

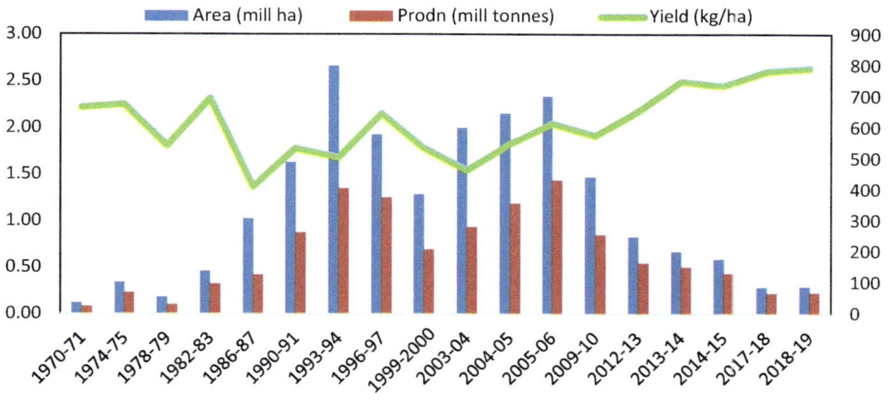

Fig. 10.8 Trends in the growth of sunflowers in India

achieve steep growth rate since its inception in seventies to reach 2.6 million ha in 1993–1994 by replacing the then less remunerative crops like millets and pulses in both *kharif* and *rabi* seasons. Since 2005–2006, the area has been continually declining due to low productivity leading to replacement with higher profitable competitive crops (Fig. 10.8). With this situation of the sharp decline in sunflower area, the private sector involvement in sunflower R & D has totally reduced and currently, the entire sunflower programme rests with public sector institutions.

During 2019–2020, sunflower was cultivated in 0.243 m ha with a production of 0.217 m tonnes and productivity of 891 kg/ha. The productivity is low due to the majority of rainfed cultivation and sub-optimal crop management. Karnataka continues to dominate sunflower cultivation in the country and accounts for more than 60% of the area but the productivity is lowest among major states cultivating sunflowers. Apart from Karnataka, Andhra Pradesh, and Maharashtra were the other major states cultivating sunflowers. There is a clear distinction between the crop expansion and productivity in the country from the 1990s, wherein the crop formed a promising niche in the north and east Indian states in Ingo-Gangetic Plains (IGP) of Punjab, Haryana, Uttar Pradesh, Bihar, West Bengal, and Odisha, cultivated in spring/*rabi*/ summer seasons as against its cultivation largely as rainfed in *kharif*, in marginal lands in the traditional area of the crop in peninsular India. The productivity of sunflowers in the spring season in the north and east India in the IGP region is the highest at about 2 ton/ha owing to its optimum agro-ecology of season, soil, and crop rotation effects.

From 2009 onwards, a very sharp decline in sunflower area and production (Fig. 10.9) is attributed to favourable market-driven profitability from competing for crops, viz. maize, cotton, pigeon pea, chickpea, and jowar in the same agro-ecologies of sunflower causing its replacement and losing area, in addition to adverse biotic, abiotic stresses, market, and non-market forces. Besides, sub-optimal nutrient management especially of inadequate and imbalanced nutrition, shifting sunflower cultivation to marginal lands with low organic manure application, improper soil and

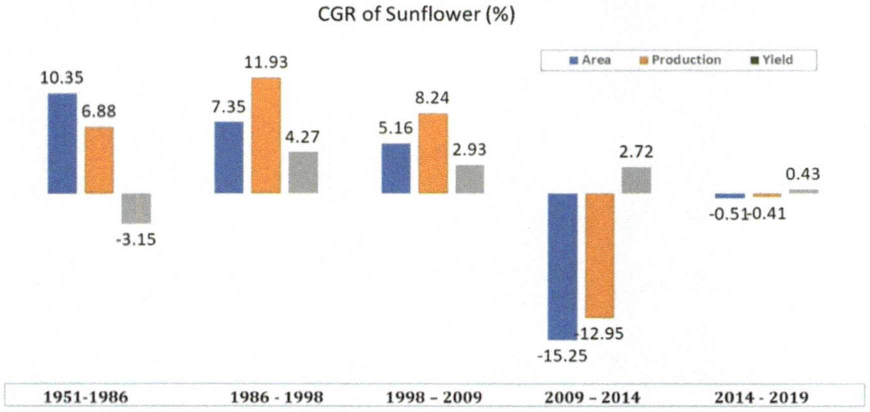

Fig. 10.9 Compound growth rates for area, production, and yield of sunflower in India

moisture conservation, improper plant population management, continuous cultivation without regard to rotation as well as non-adherence to recommended sowing window in the region, etc. have contributed to low yields and profits despite the availability of best management practices indicating a yield gap ranging from 53% to 120%. The low productivity despite the low acreage is a concern in general and especially in Karnataka and Maharashtra, the major states for sunflowers is attributed to the pushing of sunflowers to further low productive zones as a replacement to competing crops.

10.6 Safflower

Safflower (*Carthamus tinctorius* L.) is an important commercial oilseed crop. The crop is grown in about 60 countries around the world, the primary producers of which are India, USA, Mexico, Kazakhstan, Argentina, China, Ethiopia, and Australia. It is considered a minor crop with less than one million hectares area and around six lakh tonnes of production average in a year. It is a thistle-like herb belonging to the family Asteraceae or Compositae. It is one of humanity's oldest crops cultivated in India mainly for oil from the seeds and a reddish dye from the flowers. Nonetheless, it has mainly remained a minor crop grown on small plots for the growers' personal use.

Safflower is an important high-quality edible oilseed crop traditionally grown in black soils in *rabi* season in Maharashtra, Andhra Pradesh, and Karnataka contributing >90% of India's production of safflower. It is a hardy crop adapted to post rainy season under receding soil moisture conditions. The spines at the tip of leaves serve as natural protection against grazing and are also the deterrent for labour working in safflower fields, especially for harvesting and post-harvest operations with plants. Besides the main economic part of the seed for high-quality oil, the

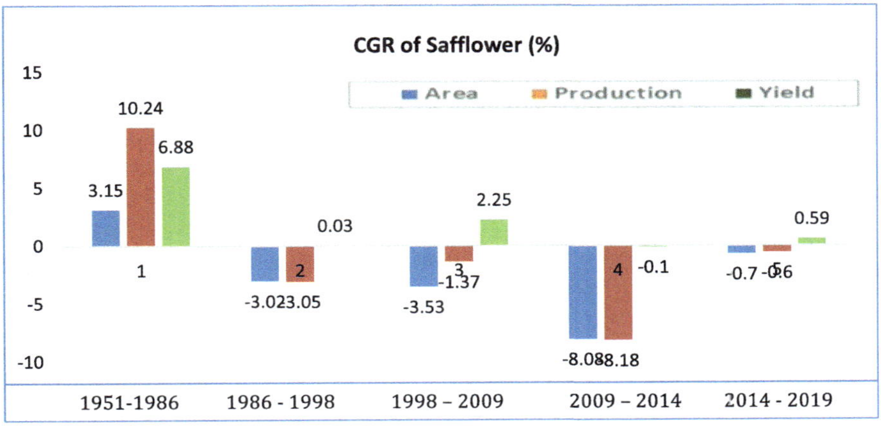

Fig. 10.10 Compound growth rates for area, production, and yield of safflower in India

flower petals post pollination serve as high-value herbal medicine against many human and veterinary ailments. The young seedlings at the thinning stage are used as a green leafy vegetables. The de-oiled cake is useful as poultry feed besides the protein hydrolysates is providing competitive alternatives to soy protein in the poultry industry.

The low productivity of crops is due to being grown under extreme situations of total rain-free conditions from sowing onwards and without any care of better management. The productivity of safflower varies directly with the *kharif* rainfall pattern and is decided by the soil profile moisture at sowing. With the availability of competing for profitable *rabi* crops of chickpea and *rabi* sorghum, the safflower area has drastically reduced. The period post-2000 till date witnessed rapid deceleration of the crop on the area front affecting production at the national level. Currently, it is grown in 40,000 ha with a production of 22,000 tonnes at a productivity of around 600 kg/ha. Recently the possibility of using a combine harvester of paddy in safflower is finding rejuvenation for the crop. The criticality of sowing time in October determines the yield. Aphid infestation with late sowing and Alternaria leaf spot with early sowing and intermittent rains are key limitations.

In India, the area under safflower has been steadily declining from about 1.0 M ha in 1986–1987 to 0.23 M ha in 2010–2011 to a mere 46,000 ha in 2018–2019 (Fig. 10.10). It is mostly cultivated in the states of Maharashtra and Karnataka and to a lesser extent in other states. In 2018–2019, Maharashtra and Karnataka accounted for about 93% of the area and production of safflower in the country. Yet, there has been a >50% decline in the area and production in these major states over the decade. The productivity of safflower is low at about 60% of the world as the crop is grown as rainfed and with poor agronomic management.

10.6.1 Shifts in Area, Production, and Productivity

India is the largest producer of safflower in the world producing approximately half of the world's annual safflower production followed by the USA. Safflower acreage and production around the world as well as in India is witnessing a steady decline and wide fluctuations over the last 2–3 decades. Safflower is cultivated in more than 20 countries on an average area of 9.55 lakh ha with a production of 8.09 lakh tonne and productivity of 846 kg/ha (2012–2013 to 2014–2015).

India is the largest producer of safflower (2.0 lakh tonnes) in the world with the highest acreage (4.3 lakh hectares) but with an average productivity of only 465 kg/ha. Poor crop management under input-starved conditions is the most important reason for such low per-hectare yields. It is mainly grown in Maharashtra, Karnataka, and parts of Andhra Pradesh, Madhya Pradesh, Orissa, Bihar, etc. Maharashtra and Karnataka are the two most important safflower-growing states accounting for 72% and 23% of the area and 63% and 35% of production, respectively. The area under safflower cultivation, which was about 787.3 thousand ha in 1983–1984 declined to about 676 thousand ha in the 1993–1994 and reached its minimum at 260.7 thousand ha in 2011–2012. It got further reduced, to 81 thousand ha during 2017–2018. The productivity witnessed a decline between 1983–1984 and 1993–1994 and showed some improvement and increased from 513 kg per ha in the early-1990s to 606 kg per ha in 2011–2012 and again declined to 556 kg/ha during 2017–2018 (DAC 2020). Over the last 11 years, the area under safflower has continuously decreased from 287.81 in 2009–2010 to 51.79 thousand hectares in 2019–2020. The production has been fluctuating from 178.82 million tonnes in the year 2009–2010, to the lowest of 24.64 thousand tonnes in the year 2018–2019. It improved to 43.67 thousand tonnes in the year 2019–2020. During the same period productivity varied from 621 kg/ha in the year 2009–2010 to a maximum of 843 kg/ha in 2019–2020.

10.7 Sesame

Sesame (*Sesamum indicum* L.) is one of the native crops of India and has many historical references in *Atharva Veda, Ashtadvaiji, Manusmriti,* and *Puranas* as *tila* indicating the presence of sesame in India since the pre-Christian era. It is considered the queen of oils for its high quality and wide uses and has equal demand as confectionery and food uses. Sesame is cultivated for its seeds which contain 38–54% oil of very high quality and 18–25% protein. Sesame is widely distributed across India in all seasons and soil types, but the varietal performance is specific to the region. India is the major producer of sesame in the world accounting for 27% of the world's area and 22% of its production. In India, sesame is grown in all the states and Uttar Pradesh, Madhya Pradesh, Rajasthan, Gujarat, West Bengal, Odisha, Maharashtra, Andhra Pradesh, and Tamil Nadu are the major states producing sesame.

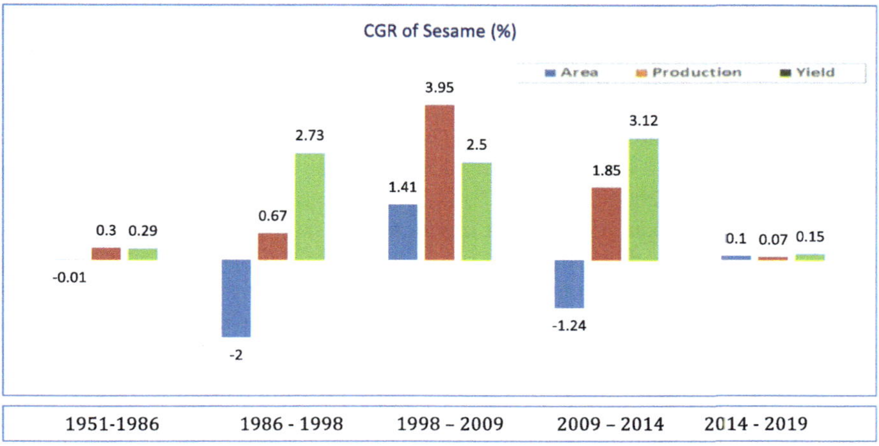

Fig. 10.11 Compound growth rates for Area, Production, and Yield of sesame in India

Among oilseeds, India earns a sizeable foreign exchange through the export of sesame (about 40% of production; Rs. 5000 crores annually) next only to groundnut. The export demand for Indian sesame is decreasing due to SPS restrictions especially on pesticide residues, despite being grown under low input conditions. Different sesame types especially seed coat colour are available, and so also the demand for specific regional preferences for specific sesame types in the country and globally. Several sesame varieties are developed especially of seed coat colour variation and are being grown widely. During 2019–2020, sesame was cultivated in an area of about 1.62 million hectares with a production of 0.75 million tonnes and productivity of 463 kg/ha (Fig. 10.11). Sesame productivity has increased by only 50% during the last 50 years. More than 85% production of sesame comes from West Bengal, Madhya Pradesh, Rajasthan, Uttar Pradesh, Gujarat, Andhra Pradesh, and Telangana, and West Bengal has the highest productivity of sesame at 951 kg/ha being grown as summer sesame.

Sesame growth rates were highest during 1998–2009 and later the area is decreasing (Fig. 10.11) and stabilizing at around 1.5 million ha. Nevertheless, huge scope exists for crop expansion across the country with higher productivity by bridging the yield gap and adopting best management practices.

In general, sesame has wider adaptability to different soils and seasons. Besides, due to the photo- and thermosensitivity of the crop, several varieties have developed that suit a particular region and season with tolerance to abiotic and biotic stresses. Among all oilseed crops, the demand for sesame is maintaining steady at around two million ha and increasing due to the global demand and utility both for whole seed and for high-quality oil. The growing opportunity for sesame as a competing crop under paddy fallow situations (residual fertility and moisture) across the country demands higher 'seed' of suitable varieties.

10.7.1 Shifts in Area, Production, and Productivity

In India sesame is grown practically in all the states. However, More than 85% production of sesame comes from West Bengal, Madhya Pradesh, Rajasthan, Uttar Pradesh, Gujarat, Andhra Pradesh, and Telangana. Sesamum production in India during the last decade averaged 17.84 lakh tonnes per year and has grown by about 19% between 2003–2004 and 2011–2012, while the area has increased by over 23% during the same period. The highest production of 8.93 lakh tonnes was recorded during 2010–2011, and the lowest was 4.41 lakh tonnes during 2002–2003. During the last 5 years highest production of 8.50 lakh hectares was recorded during 2015–2016, while the lowest 657 in the year 2019–2020. The productivity of sesame is low in India because the crop is mainly grown in *kharif*. It is significant to note that since 1965–1966, the productivity of sesame increased by 128% and the production increased by 54% despite a reduction of 32% in the area. In 2019–2020 maximum area was covered by Uttar Pradesh (356 thousand hectares) followed by Madhya Pradesh (315 thousand hectares) and Rajasthan (280 thousand hectares). The maximum production (134 thousand tonnes) was in West Bengal and the yield (933 kg/ha) was in Meghalaya. Since 1965–1966, the production of sesame increased by 192% in Madhya Pradesh and 4204% in West Bengal, area increased by 19% in Gujarat and 2186% in West Bengal, respectively, and the highest yield increase in Rajasthan in 2012–2013. Sesamum output and acreage, on the other hand, began to decline in the 1990s, with production fluctuating greatly over the last two decades. The area under sesamum witnessed a consistent decline during the last four decades. The area fell from 2.36 million ha during 1973–1974 to 2.32 in 1993–1994 and reached 1.65 million ha in 2001–2002, increasing marginally to 1.98 million ha in 2011–2012 and again declining to 1.42 million ha in 2018–2019 and again increased to 1.62 million ha in the year 2019–2020. However, production showed an increasing trend from 4.4 lakh tonnes in 1973–1974 to 6.76 lakh tonnes in 1993–1994 and a declining trend during the nineties while reaching 5.66 lakh tonnes in 2001–2002 and even up to 6.89 lakh tonnes in 2018–2019 and 6.57 lakh tonnes in the year 2019–2020. Despite a consistent decline in the area under sesamum in the country, production was increasing primarily due to yield improvements, even though the Indian yield is among the lowest in the world. The yield per ha witnessed a steady increase during the last four decades from less than 200 kg per ha in early 1970s to about 303 kg per ha in the 2010, 485 kg/ha in the year 2018–2019 and 405 kg/ha in the year 2019–2020 (Table 10.4).

10.8 Niger

Niger [*Guizotia abyssinica* L. Cass.] is a minor oilseed crop but is important in terms of its 32–40% content of quality oil with 18–24% protein in the seed. Niger is primarily grown on the denuded soils in the tribal pockets under input-starved conditions in India. Further it is the lifeline of tribal agriculture and economy. India ranks first in the area, production, and export of niger in the world. It has a

Table 10.4 Characteristic features of some promising varieties/hybrids of sesamum

Variety	Recommended states/ regions/situations	Oil content (%)	Average yield (kg/ha)	Specific features
TKG-306	Kharif, Madhya Pradesh	49–52	750–800	Duration 86–90 days, white seed, seed weight 2.8 g, resistant to *phytophthora* blight and moderately resistant to *Macrophomina*, *Cercospora*, powdery mildew, *Alternaria* leaf spot
SWB-32-10-1 (Savitri)	Rabi/summer, West Bengal	48–52	1200–1500 summer	Duration 84–88 days, light brown seed, tolerant to *Macrophomina*
Jawahar Til-12 (PKDS-12)	Summer, Madhya Pradesh	48–52	700–750	Duration 82–85 days, white seed, moderately resistant to *Macrophomina* stem/ root rot
Jawahar Til-14 (PKDS-8)	Summer, Madhya Pradesh	50–53	700–750	Duration 82–85 days
Gujarat Til-3	Kharif, Saurashtra region of Gujarat	48–52	750–800	Duration 84–88 days, white bold seed
RT-346	Kharif, Rajasthan Haryana, Panjab, Himachal Pradesh, Gujarat, utter Pradesh, Maharashtra	49–51	750–850	Duration 82–86 days, white seed, resistant to leaf curl, moderately resistant to Macrophomina, Alternaria, Cercospora
PKV-NT-11	Summer, Vidarbha region of Maharashtra	50–53	800–850	Duration 88–92 days, white seed, moderately resistant to root rot, bacterial blight
DSS-9	Zone I & II of Karnataka	48–50	600–650	Duration 85–90 days, white bold seed, early maturing
JLT-408	Assured rainfall zone of Khandesh and adjoining areas of Vidharba, Marathwada zones and rainfall zone of Khandesh and adjoining areas of Vidharba	51–53	700–800	Duration 80–85 days, white seed
RT-351	Rajasthan, Gujarat, UP, Maharashtra, Haryana, Punjab, HO, KK & J & K	48–50	700–800	Resistant to Macrophomia, leaf curl, Phillody, moderately resistant to Cercospora & capsule borer
TKG-308	Madhya Pradesh (*kharif*)	48–50	700–750	Moderately resistant to Macrophomia, Cercospora, bacteria leaf spot, leaf curl, tolerant to capsule borer

(continued)

Table 10.4 (continued)

Variety	Recommended states/ regions/situations	Oil content (%)	Average yield (kg/ha)	Specific features
Shubra	Odisha (kharif/summer)	48–52	800–900	Golden yellow bold seed, delayed shattering, synchronous maturity
Smarak	Odisha (kharif/summer)	46–50	800–900	Orisha (kharif, summer)
Gujarat til-4	Kharif, north Saurashtra zone IV of Gujarat	47–49	750–800	White seed, tolerant to leaf spot
DS-5	Kharif, summer Karnataka	49–51	600–700	White bold seed
Subhra (OSC-2007)	Odisha			
Smarak (OSC 560)	Odisha			

Source: https://oilseeds.dac.gov.in/Variety/Sesame.aspx

yield potential of 800–1000 kg/ha under optimum growing conditions. The states contributing primarily to the niger production of the country are Madhya Pradesh, Odisha, Maharashtra, Karnataka, and Chhattisgarh. Besides, the crop is also cultivated to some extent in hilly areas of Andhra Pradesh, Bihar/Jharkhand, Gujarat, Uttar Pradesh, Rajasthan, Tamil Nadu, West Bengal, Assam and Arunachal Pradesh in the North Eastern Hill region. It is a lesser-known edible oilseed crop with confined cultivation under hill slopes of Odisha, Andhra Pradesh, and Chhattisgarh with tribal communities on eroded shallow and marginal soils of coarse texture and low fertility. It is known to have many medicinal values and nutrition, is rich in P and fibre. The crop acreage is consistent at about 1.2 lakh ha in the country. The crop has export value as bird feed besides oil extraction and meal is high fibre and nutritious. On the other hand, despite its minor importance, India earns reasonable revenue through the export of niger as bird feed. During 2018–2019, India exported 13,370 tonnes of niger seed valued at Rs. 95.5 crores. Niger crop supports apiary and a source of wild honey of high value. As the crop has restricted cultivation with tribal communities, almost all the produce is considered organic and exploited.

In India, area under niger was a little less than 0.5 million ha in 1964–1965, reached to a high of 0.64 million ha in 1990s and has since then declined. The decline in area under niger cultivation in recent years (Fig. 10.12) can be ascribed to its lesser dependence for subsistence by tribal farmers for its primary product oil due to the ease and cheaper availability of edible oils even at remote hamlets, besides the need for embracing market driven remunerative crops. Currently (2019–2020) niger was grown in an area of 0.14 million ha with a production of 0.41 million t and productivity of 305 kg/ha.

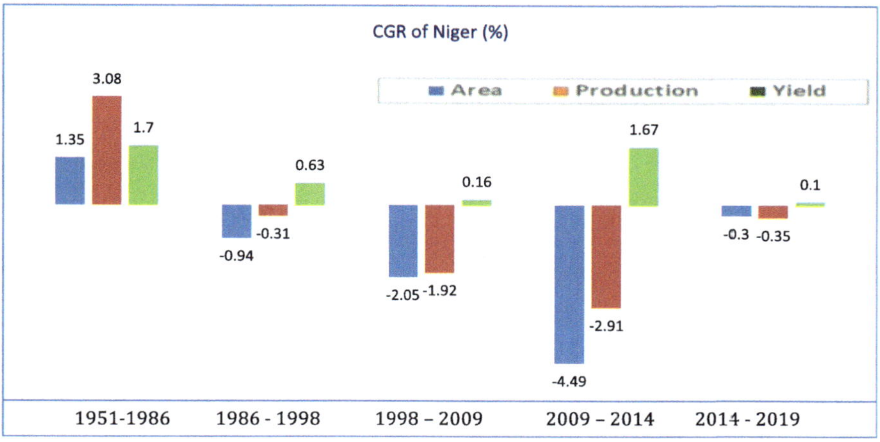

Fig. 10.12 Compound growth rates for area, production and yield of niger in India (Source: https://www.sopa.org/india-oilseeds-area-production-and-productivity/)

10.8.1 Shift in Area Production and Productivity

On average, the productivity of Niger is low, around 300–350 kg/ha in India. Niger production during 2012–2013 has increased by 7% and productivity by 88%, even after a reduction of 42% in the area over 1965–1966. Despite the maximum area (86.9 thousand hectares) and production (29.8 thousand tonnes) in Madhya Pradesh, Andhra Pradesh has the maximum seed yield of 750 kg/ha during 2012–2013. During last 11 years, the area under Niger seed has continuously decreased from 375.49 in 2009–2010 to 136.624 thousand hectare in 2019–2020. However, production has been fluctuating being 99.94 million tonnes in the year 2009–2010, to the lowest of 41.32 thousand tonnes in the year 2019–2020. During the same period productivity varied from 266 kg/ha in the year 2009–2010 to a maximum of 328 kg/ha in the year 2014–2015 and again declined to 303 kg/ha in the year 2019–2020 (Table 10.5).

10.9 Oil Palm

Oil palm (*Elaeis guineensis* Jacq.) is the highest edible oil-yielding tree crop which has a theoretical potential yield of 18 t of oil/ha/year. It is capable of yielding 4–5 tonnes of palm oil and 0.4–0.5 tonnes of palm kernel oil. Oil palm is a native of West Africa and is grown extensively in South-East Asian countries (Indonesia, Malaysia, Thailand, and Papua New Guinea), African countries (Cameroon, Ghana, Ivory Coast, Liberia, Nigeria, Republic of Congo, Sierra Leone, and Zaire) and South American countries (Brazil, British Guyana, Columbia, Costa Rica, Ecuador, Panama, Peru, and Venezuela). There are 17.00 million hectares planted with oil palm in the world with a total production of over 62.00 million tonnes of palm oil (2015),

Table 10.5 Characteristics features of some promising varieties of Niger

Variety	Oil content (%)	Average yield (kg/ha)	Recommended states/ regions/situations	Specific features
BNS-10 (Pooja-1)	36–38	650–700	All Niger growing states	Duration 95–100 days, shining black seed, resistant to pests, diseases.
KBN-1			Karnataka	Moderately resistant to *Alternaria* leaf spot
IGPN-2004-1, (Phule Karala-1)	39–41	650–700	Kharif season in MS and Karnataka (high rainfall areas of Maharashtra)	Duration 95–100 days, shining black seed, resistant to *Alternaria*, powdery mildew
Birsa Niger-3 (BNS-11)			Jharkhand, Chhattisgarh, MP, Odisha, Maharashtra, AP, WB, Kerala	Early maturing variety 85 days, drought tolerant
Utkal Niger-150	38–40	650–700	Orissa	Duration 105–110 days, black seed, tolerant to *Alternaria, Cercospora, Cuscuta*
DNS-4	39–41	500–600	Kharif, Karnataka	Shining black seed, robust growth habit

Source: https://oilseeds.dac.gov.in/Variety/Niger.aspx

Table 10.6 Current scenario of demand and supply of vegetable oils in India (Qty in LMT)

Year (Nov.–Oct.)	Domestic availability	Import	Availability/ consumption	% Self-sufficiency	% Share of imports	Per capita consumption (kg/yr)
2015–2016	86.30	148.50	234.80	36.8	63.2	19.00
2016–2017	100.99	153.17	254.16	39.7	60.3	18.90
2017–2018	103.80	145.92	249.72	41.6	58.4	19.50
2018–2019	103.52	155.70	259.22	40.0	60.0	19.80
2019–2020	106.55	134.16	240.71	44.3	55.7	18.00

Source: Annual Report 2020–2021, Department of Food and Public Distribution

which is expected to increase substantially due to growth in demand for oil by the food industry and use in bio diesel (Table 10.6). Though planted in 5.5% of the total world vegetable oil acreage, palm oil accounts for 32% of global supply of oils and fats. Malaysia, Indonesia, and Nigeria are the leading producers of palm oil in the world. Palm oil is the major source of dietary fat in Latin America, South-East Asia, China, parts of West Asia, and Africa. In India too, the highest consumption of edible oil is palm oil. This is probably the most economic source of edible fat available at present. It is a crop of the future and a source of health and nutrition, value addition, waste utilization, environment friendly, cogeneration, import substitution, and sustainability. The crop has the potential to contribute to agricultural diversification and economic development of rural communities by creating more job opportunities and provision of stable income. There is enormous scope for oil

Table 10.7 Import of palm oil vs other edible oils 2019–2020

Oil	Percentage
RBD Palmolein	3
Crude palm oil	51
Sunflower oil	19
Soybean oil	25
Rapeseed oil	0.13
Palm kernel oil	0.51
Others	1.12

Source: Annual Report 2020–2021, Department of Food and Public Distribution

palm development in India due to the availability of vast stretches of land under diversified agro-climatic conditions and with untapped underground water potential. In India oil palm is grown as small holders' irrigated crop unlike in other countries, viz. Malaysia and Indonesia where it is grown in very large areas under rainfed conditions (Table 10.7).

10.10 History of Oil Palm in India

In India, edible oil consumption has increased substantially over the years and the requirement is being met through imports. The quantum of import is likely to go up to meet the ever-increasing demand for edible oils, which necessitates increasing the production and productivity of vegetable oils in India. The Govt. of India and the National Agricultural Research System are giving adequate emphasis on improving the productivity of oilseed crops through the release of high-yielding varieties and the adoption of technology. During the last two decades, efforts have been made to introduce and exploit a number of new oil-bearing crops. However, only oil palm has shown promise for commercial cultivation under Indian conditions. Oil palm is the highest oil-yielding perennial crop and hence assumes significance for augmenting the indigenous availability of vegetable oils. Owing to its high productivity, palm oil is also one of the cheapest oils. Oil palm has the potential for production of 20–25 tonnes fresh fruit bunches per ha from eighth year of planting till almost 30 years, by adopting good management practices.

Oil palm was introduced in India at National Botanical Gardens, Kolkata during the year 1886. The Maharashtra Association for Cultivation of Sciences (MACS) later introduced African dura palms along canal bunds, home gardens, and, to some extent, in forest lands in Pune from 1947 to 1959. Large-scale planting of oil palm was launched from 1971 to 1984 in Kerala by Plantation Corporation of Kerala Ltd., (subsequently taken over by Oil Palm India Ltd.,) and Andaman Forest and Plantation Development Corporation in Andaman and Nicobar Islands from 1976 to 1985. Cultivation of Oil Palm in India by smallholders got initiated in 1986. The Department of Biotechnology, Govt. of India in collaboration with the state governments of

Andhra Pradesh, Karnataka and Maharashtra established three demonstration units of 1000 ha each during 1990–1991 to demonstrate the feasibility of growing oil palm under irrigated conditions to farmers and plantation companies. These plantations were raised with 'tenera' seeds of which 80% were imported from Costa Rica and the remaining 20% obtained from indigenous sources at Palode. In Andhra Pradesh and Karnataka, the demonstration was taken up on farmers' fields, while in Maharashtra it was taken up by the Development Corporation of Konkan Ltd. (DCKL) by obtaining the land on lease from farmers.

10.11 Indian Oil Palm Scenario

The Government of India has continuously been making efforts to increase the area and production of oil palm in order to increase the availability of edible oils in the country and to reduce the import burden. A comprehensive centrally sponsored scheme i.e. Oil Palm Development Programme (OPDP) was launched during 1991–1992 to promote oil palm cultivation on a massive scale, and the programme was brought under the purview of the Technology Mission on Oilseeds and Pulses (TMOP) by Department of Agriculture, Cooperation and farmers welfare (DAC & FW) during the VIII and IX five year plans. During the X and XI plans, support for oil palm cultivation was provided under the Integrated Scheme of Oilseeds, Pulses, Oil Palm and Maize (ISOPOM). A special programme on Oil Palm Area Expansion (OPAE) under RKVY was implemented from 2011 to 2012 till March 2014 to boost oil palm area expansion. During the XII plan, National Mission on Oilseeds and Oil Palm (NMOOP) launched under which Mini Mission-II (MM-II) has been dedicated to oil palm area expansion and productivity increases. Now the scheme is being implemented under National Mission of Edible Oils—Oil Palm (NMEO-OP). Oil palm cultivation and palm oil production in India is a tripartite one involving farmers-processors-government. DAC & FW has been implementing these programmes through the Department of Agriculture/Horticulture of the respective state governments. Under these programmes, financial assistance is being provided to the farmers at 85% cost of the planting material and 50% cost of the other components like maintenance cost of new plantations for four years, installation of drip-irrigation systems, diesel/electric pump-sets, bore-well/water harvesting structures/ponds, inputs for inter-cropping in oil palm (during gestation period), construction of vermi-compost units and purchasing of machinery and tools, etc.

The state governments of oil palm growing states have constituted Project Management Committee, Price Fixation Committee for looking after the oil palm development programme in the State. The present FFB price paid to farmers by the industry is based on the CACP recommended formula of 2012. This is regulated through the Oil Palm Acts of the State Governments. The States Governments further identified private entrepreneurs like M/s Godrej Agrovet Pvt. Ltd., M/s Ruchi Soya Industries, etc., for raising oil palm nurseries and establishment of processing mills in their respective States. The State Governments have allotted areas/ Mandals/Districts to the companies for the development of new plantations.

Accordingly, the companies have established nurseries in their allotted zone for developing seedlings of oil palm, which takes about 10–12 months. These companies purchase seed sprouts from indigenous seed gardens and also import sprouts from major oil palm-growing countries for the development of seedling nurseries in their respective States. These companies are also extending technical expertise to the farmers for development of oil palm plantation. After development of plantation, oil palm mills are also established by these companies and the FFB will purchased directly from famers under buy back policy of oil palm Act. Government of India has also provided financial support for the establishment of oil palm processing mill, especially in NE/LW areas/hilly states/regions. In the country, 24 oil palm processing mills have been established in different states having a capacity of 312 MT/hrs for crushing of oil palm FFB.

Oil palm has been established as a successful crop in a number of states in the country and productivity levels up to 6–8 tonnes oil per ha could be achieved. The fresh fruit bunch (FFB) yields obtained by progressive farmers of Andhra Pradesh and Karnataka, under optimum cultural and irrigated conditions, are between 20 and 25 tonnes of FFB/ha/year, i.e. 4–5 tonnes of oil per ha per annum from the fourth year onwards. The highest yield of 30–35 tonnes FFB/ha/year during the seventh year was also recorded in many plantations. A farmer could record 53.20 tonnes FFB per ha in Karnataka State, which indicates the potential of this crop under Indian conditions.

Though Oil Palm Development Programme in the country is progressing well, area coverage is not taking place as per the targets envisaged. The reasons are manifold, the most significant being the lack of assured price for FFBs and uncertainty in payment of FFB price since it is linked to the international price of CPO which is prone to major fluctuations. Further, there is no support from Govt. of India for the FFB price to farmers. However, the accelerated pace of oil palm cultivation is recorded when the support price and additional promotional activities were undertaken. As of 2020, various Expert Committees constituted by the Ministry of Agriculture, Government of India have identified a total of 27.99 lakh hectares in 22 states of the country as suitable for oil palm cultivation. So far, an area of 3.70 lakh ha has been covered under oil palm. In the areas where Oil Palm is grown, livelihood and social security have certainly improved. In India, oil palm is being grown in different agro-climatic regions deviating from the conventional requirements like temperature, rainfall, humidity, soil type, etc. Oil palm cultivation is being taken up in the agricultural land owned by farmers as irrigated (and rainfed in a few states) small holders' crop. Corporate plantations could not achieve desired results due to various reasons.

10.12 Oil Palm Research in India

With the present level of rapid increase in the per capita consumption of vegetable oils in the country, it is expected that the estimated population of 1620 million in 2050 may need 40.5 million tonnes of vegetable oils at the estimated per capita consumption of 25.00 kg. By 2050, if an area of 20 lakh ha is covered under oil palm, the country must be able to produce about 14 million tonnes of oil from oil palm. In order to tap the huge potential of this crop, Indian Council of Agricultural Research established ICAR-Indian Institute of Oil Palm Research (ICAR-IIOPR) at Pedavegi, Andhra Pradesh, India in 1995 with an aim to develop innovations and technologies for improving oil palm productivity and sustainability to address the challenges of producing more vegetable oil for the growing population. Since 1995, systematic efforts are being made at this Institute to cater to the research needs of the oil palm community in India and to increase the production and productivity of the crop. Though the present yield levels in plantations of progressive farmers are about 20–25 tonnes FFB/ha (on par with Indonesia and Malaysia), many plantations continue to record productivity levels of less than 10 tonnes per hectare. Thus, productivity improvement through better planting materials and efficient management of water, nutrients, and labour would be of crucial importance for achieving the targets. The salient research achievements of the Institute include Development of three oil palm hybrids with desirable characters (high FFB yield, high oil yield, and high sex ratio); Oil palm germplasm exchange under international collaboration with Malaysia; Development and supply of advanced parent materials to Karnataka, Mizoram and Andhra Pradesh for establishment of five new oil palm seed gardens in the country; Standardization of technology for oil palm hybrid seed production which was adopted at national level by public and private sector seed gardens; Standardization of best management practices for oil palm cultivation under irrigated conditions, first of its kind in the world; Development of oil palm cropping system models for doubling of farmers income; Conducting feasibility studies for identification of potential areas for oil palm cultivation in the country; Organizing training programmes on oil palm cultivation to officers and farmers from Oil Palm Development Programme implementing states.

10.13 Conclusions and Future Strategies

Oilseed crops are grown all over the world and are important crops because of their economic value. Oilseed crops are planted primarily for the production of edible oil. Oilseeds have recently gotten a lot of attention because of the growing demand for healthful vegetable oils, livestock feeds, medicines, biofuels, and other oleochemical industrial purposes. For the annual oilseeds and for oil palm in the country, solutions for increasing the productivity (and profitability) of oilseed-based production systems are being developed. As a result, to meet rising global demand, traditional breeding efforts must be combined with biotechnological technologies to increase oil yield per unit area. Another option for meeting this increased need is to expand the area where oilseeds are grown. Oilseed crops are mostly grown in rain-fed circumstances, and as a result of unpredictable rainfall, the crops experience moisture stress at various phases of growth, resulting in low yields. By extending irrigation facilities to oilseed, the yield of oilseed can be stabilized and greatly boosted. The oilseed interventions/strategies offered have been time-tested and are scale neutral, so they can be used to increase the productivity of oilseed-based agriculture systems with the requisite institutional support/handholding. The proposed strategies are categorized under three situations as follows:

- More seed production and distribution of newly released varieties.
- Developing low-cost technologies with a high impact on productivity resulting in higher income.
- Developing technologies with high impact that involve reasonable investment with a high return on investment (ROI), with emphasis on eco-friendliness, high input use efficiency, and.
- Strategies emphasizing quality improvement combined with value addition leveraging technologies, with a bearing on employment through skill/entrepreneurship development.
- Development of clonal seed gardens through mass multiplication of high-yielding elite palms.
- Development of oil palm seed gardens to meet the indigenous planting material requirement.
- Development of next-generation dwarf and high oil-yielding hybrids for improving productivity and ease in harvesting.
- Influence of climate change on regulatory processes involved in oil synthesis of oil palm mesocarp.
- Development of cropping system modules with medicinal and aromatic plants.
- Enhancing resource use efficiency through sensor based IOT applications in irrigation and fertigation.
- Development of suitable post-harvest technologies and value added products.
- Design and development of machineries for to reduce drudgery.
- Capacity building programmes for stakeholders on oilseed production technologies.

References

Agarwal DK, Husain SM, Ramteke R, Bhatia VS, Srivastava SK (2010) Soybean varieties of India. Madhya Pradesh, India, Directorate of Soybean Research, Indore

Agarwal DK, Billore SD, Sharma AN et al (2013) Soybean: introduction, improvement, and utilization in India—problems and prospects. Agric Res 2:293–300

Anonymous (2015) Vision 2050, ICAR-Directorate of RapeseedMustard Research. www.drmr.res.in/

Anonymous (2020) ICAR-Directorate of Rapeseed-Mustard Research, Bharatpur, India, pp 1–102

Chauhan JS, Singh KH, Kumar A (2006) Breeder seed production scenario of rapeseed-mustard varieties in India. NRCRM, Bharatpur, p 20

Chauhan JS, Choudhury PR, Pal S, Singh KH (2020) Analysis of seed chain and its implication in rapeseed-mustard (*brassica* spp.) production in India. J Oilseeds Res 37(2):71–84

Chauhan JS, Singh KH, Singh VV, Kumar S (2011) Hundred years of rapeseed-mustard breeding in India: accomplishments and future strategies. Indian J Agric Sci 81:1093–1109

Directorate General of Commercial Intelligence and Statistics. 2019. http://www.dgciskol.gov.in/

Ghareyazie B (2010) Food and environmental safety of biotech crops: Islamic perspective. In: International workshop for Islamic scholars on agribiotechnology: shariah compliance, pp 29–42

Jat RS, Singh VV, Sharma P, Rai PK (2019) Oilseed brassica in India: Demand, supply, policy perspective and future potential. OCL 26:8. https://doi.org/10.1051/ocl/2019005

Kumar S (2015) Edible oils quality and nutraceutical aspects. Stadium Press (India) Pvt. Ltd., New Delhi

Rajput LS, Nataraj V, Kumar S, Amrate PK, Jahagirdar S, Huilgol SN, Chakruno P, Singh A, Maranna S, Ratnaparkhe MB, Borah M, Singh KP, Gupta S, Khandekar N (2022) WAASB index revealed stable resistance sources for soybean anthracnose in India. J Agric Sci 159(9–10):710–720. https://doi.org/10.1017/S0021859622000016

Nataraj V, Shivakumar M, Kumawat G, Gupta S, Rajput LS, Kumar S, Sharma AN, Bhatia VS (2020) Genetic inheritance and identification of resistant germplasm sources for anthracnose resistance in soybean [*Glycine max* (L.) Merr.]. Genet Resour Crop Evol 67(6):1449–1456. https://doi.org/10.1007/s10722-020-00917-4(01)

Nene YL 1972. A survey of viral diseases of pulse crops in Uttar Pradesh. G B Pant University of Agriculture and Technology, Pantnagar, Uttaranchal, India. 191 pp.

NRCRM. 2000. *Rapeseed-mustard varieties of India*, 115pp + V

NRCRM (2007) Vision 2025. Perspective Plan National Research Centre on Rapeseed-Mustard, Bharatpur, pp 1–46

Rani A, Kumar V, Gili BS, Rathi P, Shukla S, Singh RK, Husain SM (2017) Linkage mapping of *Mungbean yellow mosaic India virus* (MYMIV) resistance gene in soybean. Breed Sci 67:95–100. https://doi.org/10.1270/jsbbs.16115

Rani A, Kumar V, Gill BS, Shukla S, Rathi P, Singh RK (2018) Mapping of duplicate dominant genes for *Mungbean yellow mosaic India virus* resistance in *Glycine soja*. Crop Sci 58:1–9. https://doi.org/10.2135/cropsci2017.11.0652

Ratnaparkhe MB, Marmat N, Kumawat G, Shivakumar M, Kamble V, Nataraj V, Ramesh SV, Deshmukh MP, Singh AK, Sonah H, Deshmukh R, Prasad M, Chand S, Gupta S (2020) Whole genome re-sequencing of soybean accession EC241780 providing genomic landscape of candidate genes involved in rust resistance. Curr Genomics 21(1):504–511. https://doi.org/10.2174/1389202921999200601142258

Ruzhen C (1989) Studies on the origin of cultivated soybean *(Glycine max(L.)* Mer.). Oil Crops China 1:1–6

Sarikhan S. 2006. GMO detection in foodstuffs imported into Iran's market. MsC thesis, University of Zanjan, Zanjan, Iran

Sharma HK, Arun K, Singh VV, Meena HS, Priyamedha BL, Meena PS, Rai PK (2022) Genetic Resources of Brassicas. In: Priyadarshan PM, Jain SM (eds) Cash Crops, pp 285–337. https://doi.org/10.1007/978-3-030-74926-2_1

Singh BB (2006) Success of soybean in India: the early challenges and pioneer promoters. Asian Agric History 10:41–47

Singh BB, Gupta SC, Singh BD (1974) Sources of field resistance to rust and yellow mosaic diseases of soybeans. Indian J Genet Plant Breed 34:400–404

Tillie P, Rodríguez-Cerezo E. 2015. Markets for non-genetically modified and identity-preserved soybean in the EU. JRC science and policy report. European Commission

Tiwari SP (2006) Plant introductions in soybean - achievements and opportunities. Indian J Plant Genet Resour 19(3):353–365

Tripathi R, Agrawal N, Kumawat G, Gupta S, Varghese P, Ratnaparkhe MB, Bhatia VS, Maranna S, Satpute GK, Chand S, Jain M (2021) QTL mapping for long juvenile trait in soybeanaccession AGS 25 identifies association between a functional al-lele of FT2a and delayed flowering. Euphytica 217:36. https://doi.org/10.1007/s10681-021-02775-2

UN Comtrade 2019. https://comtrade.un.org/

Horticultural Crops

11

J. Satisha

Abstract

The horticulture sector comprises several crops viz., fruits, vegetables, spices and medicinal crops, plantation crops, ornamental crops, mushrooms, tuber crops, etc. From pre-independence era till eighth 5-year plan no much importance was given to this sector as food security and poverty alleviation was the major issue. However, over the years the horticulture sector has emerged as one of the inseparable parts of agriculture providing a wide range of crop diversification to the farmers. Today, India has surpassed in the production of horticulture crops and agriculture crops. This sector also provides a lot of employment opportunities for sustaining many agro-industries which in-turn generates employment opportunities substantially. With an increase in the purchasing power of people, the cereal and pulses-based dietary style is switching towards vegetable and fruit-based dietary styles as horticultural crops provide nutritional security. To reach this stage of becoming the second largest producer of fruits and vegetables, the horticulture sector has faced several challenges and constraints. With the constant effort of government assistance by allocating funds to horticultural research in-terms of establishing horticultural research institutes and universities several new varieties and improved production practices have been developed which helped in achieving higher production of horticultural crops. In addition, horticulture is not only restricted to fruit and vegetable cultivation in rural areas but has stretched its wings to urban and semi-urban areas. The new modern concepts for landscape development are keen on using any kind of concrete or glass, turning them into real vertical gardens, being possible to overcome the

J. Satisha (✉)
Division of Fruit Crops, ICAR-Indian Institute of Horticultural Research, Bengaluru, Karnataka, India
e-mail: Satisha.j@icar.gov.in

P. K. Ghosh et al. (eds.), *Trajectory of 75 years of Indian Agriculture after Independence*, https://doi.org/10.1007/978-981-19-7997-2_11

development of the urban areas making a smooth transition for a healthy green urban environment. Because of the strong impetus for the development of the horticulture sector, there is tremendous scope for enhancing the productivity of Indian horticulture which is imperative to cater to the country's estimated demand of 650 million MT of fruits and vegetables by the year 2050.

Keywords

Horticulture · Policy issues · Challenges · Research contributions · Entrepreneurship—urban horticulture—sustainability

11.1 Introduction

Horticulture is one of the important sectors of agriculture in India, considering its potential with respect to increased income to farmers, providing both livelihood and nutritional security, and also its contribution towards a greater portion of foreign exchange to the country. Though at present the production of horticultural crops is surpassing the total agricultural crop production from the past several years, the importance or recognition given to horticultural crops is only from the seventh or eighth 5-year plan. This sunrise sector comprises a wide range of crops, viz., fruit crops, vegetable crops, tuber crops, flower crops, medicinal and plantation crops, mushrooms, potatoes, etc. Recent statistics from National Horticultural Board indicate that India is the second-largest producer of fruits and vegetables next to China. India contributes about 13% of global fruit production and about 21% of global vegetable production. The growth pattern of Horticulture is distinctly different from the agricultural sector. Though during the first few 5-year plans, the main focus was to achieve self-sufficiency in the production of food grains, over the years, horticulture was given more importance and has emerged as inevitable part of agriculture, offering varied choices to growers for diversification. Horticulture also provides greater scope for sustaining many agri-based industries that generate a lot of employment opportunities. Horticultural crops contribute to about 24.5% of the GDP from the total area of just 8% of the cultivable land in the country. The growth of horticulture sector in India from pre independence era passing through green revolution era till present is briefly discussed in this chapter.

11.2 Horticulture During the Pre-Independence Era

Before Independence, horticulture was just subsistence farming where it was either restricted to homestead or backyard gardens as for as fruits and vegetables were concerned. After practicing horticulture by Mughal emperors in the country, one of the English Baptist Missionary, Rev. William Carey founded Agri-Horticultural Society in India in 1820 with the objective of the promotion and development of agriculture in the country. Because of its establishment, a large variety of vegetables and fruits were introduced into the market to which Indian society was completely ignorant. Before the formation of society, the production of good quality vegetables was very scarce. Consequently, many new vegetables, herbs, fruits, etc. became easily procurable in the market. After the establishment of the society, Carey prepared an essay depicting the reasons for the establishment of an Agricultural Society in India. Some of the objectives proposed in the essay on supporting the formation of an agricultural and horticultural society in India are:

- Encouraging a better mode of cultivation for the improvement of the land.
- Evolving and using the best methods of crop rotation and land-cropping.
- Introduction of useful and new types of plants.
- Improvement in the implementation of husbandry practices.
- Improvement in animal husbandry.
- Inclusion of wastelands into a state of cultivation.

This society continued to help in introducing several better strains of vegetables, fruit trees, and flower crops. But, British rule had a very long impact on Indian gardens. They transformed the traditional Hindu and Mughal-style gardens into informal gardens. In addition, they also introduced a lot of fragrant trees, shrubs and bulbs. They also gave a lot of importance to colourful flowering annuals like asters, carnations, nasturtium, pansies, etc. which were introduced by the British. British also made a lot of ornamental and pleasure gardens in the form of parks and gardens around forts and monuments.

11.2.1 Post-Independence Era

The separation of horticulture from agriculture as a distinct activity is usually dated from the middle ages in Europe and most of the developed countries. But, in India, until the last two to three decades, horticulture was considered a part of agriculture. Despite that, the horticulture sector has made appreciable progress in the country with integrated efforts of researchers, farmers, decision-makers and other stakeholders. Development of horticulture was at a very low profile until the third 5-year plan and very little attention was received then onwards too. But, it is only after the eighth 5-year plan, investment in horticulture increased steadily resulting in the considerable strengthening of several horticultural development programmes in the country. As more than 30% of the population suffers from malnutrition, it was necessary for the country not only to provide food security but also access to a balanced nutritious diet. In this direction, horticultural crops like fruits, vegetables, medicinal plants and plantation crops play an important role, thereby enhancing access to quality food and opportunities for increased farm returns. The growth trend of horticulture is entirely different from that of the agriculture sector as a whole. During the 1960s when the country attained the green revolution, the main focus was to provide food security to growing population. Hence, the government policies concerning seeds, fertilizers, production technologies with respect to irrigation,

Table 11.1 Area and production of fruit crops

Year	Area ('000 Ha)	Production ('000 MT)	Increase over 1991–1992 (in percentage)	
			Area	Production
Fruit crop				
1991–1992	2870	28,630	–	–
1998–1999	5140	58,530	79.09	104.44
2005–2006	5509.6	58,740.3	91.97	105.17
Vegetable crops				
1991–1992	3729	44,042	–	–
1998–1999	5870	87,530	57.41	98.74
2005–2006	7164	109,050	92.12	147.60
Spices				
1991–1992	1618	1211	–	–
1998–1999	1751	1884	8.22	55.57
2005–2006	1731	2225	6.98	83.73

Database of National Horticulture Board (http://nhb.gov.in/); www.agricoop.nic.in; CMIE 2005

public investment, price supports and inputs facilitated India to achieve self- sufficiency. The government strategies focused on the development of high-yielding varieties, emphasized more on research and extension, irrigation and institutional credit, etc. These successful strategies in the 1970s made India not only self-sufficient in food grains but also emerged as a major exporter of food grains (rice and wheat). But this growth could not be sustained in the post-liberalization period, and as a result, the agriculture GDP growth declined from 3.3% in 1980–1995 to 2% in 1995–2002. The crop sub-sector growth rate declined from 3.3% in 1990–1991 to 0.61% 1996–1997 to 0.61% and further declined till 2003–2004 (Planning Commission XI Report). This stagnating/declining growth in the agriculture sector made a strong impetus to policy makers to search for an alternative mechanism to boost the growth of the agriculture sector. In this background, the diversification towards high-value crops was identified as one of the important components for the growth of the Indian agriculture sector (Chand 1999; Joshi et al. 2006).

Once, the country attained self-sufficiency with respect to food grains, the focus was to diversify agriculture towards other crops, viz. oil seeds, commercial crops like cotton, sugarcane and horticultural crops. Economic policies and reforms in the early 1990s further enhanced the speed of diversification in favour of horticultural crops. Once considerable growth was witnessed in horticulture crops mainly fruits and vegetables, more focus was again on diversification in the horticulture sector, viz. floriculture, medicinal and aromatic crops, mushrooms, post-harvest handling and processing, etc. (Jha et al. 2018). The area and production of horticulture during 1991–1992 to 2006–2007 is shown in Table 11.1. Thus, the diversification towards horticulture crops has made a strong impact to increase not only overall agricultural growth but also generated a lot of employment opportunities and increased farm income.

11.2.2 Development of Horticulture Sector—Plan Allocation

Due to the reasons mentioned in the above paragraphs about main focus of the country towards self-sufficiency in food grains, the budget allocation to the horticulture sector was very negligible in the post-independence era. The separate funding for horticulture has not been mentioned till fourth 5-year plans. In the fourth 5-year plan, the allocation was just 2.05 crores for the development of horticulture which is just 0.89% of the total plan allocation towards agriculture and allied sectors. The plan allocation for horticulture and allied activities is given in the Table 11.2.

There was a major boost in plan allocation for the development of the horticulture industry during the VIII plan. The allocation increased several times to 1000 crores during the VIII plan compared to mere 25 crore allocations during the VII plan which is in proportion terms from 0.23% to 4.45% of the total allocation for agriculture and allied sectors. Since then there were continuous increases in the percent allocation to Horticulture. The cabinet committee on economic affairs approved a centrally sponsored scheme for integrated development of horticulture for implementation during the XII plan with an outlay of INR 16840 crores, out of which INR 866 crores were contributed by state governments where the scheme was implemented.

First time in history, the horticulture section was established in the Division of Botany at IARI in 1954, and the focus on horticulture and research development was started from the IV plan onwards. During this period 2.05 crores were allocated for

Table 11.2 Plan allocation for agriculture and allied sectors and horticulture (in crores)

Plan period	Allocation for agriculture and allied sectors	Allocation for horticulture	Percentage of horticulture to agriculture
IV (1969–1974)	2320	2.05	0.85
V (1974–1979)	4865	7.62	0.16
VI (1980–1985)	5695	14.64	0.26
VII (1985–1990)	10,525	25	0.23
VIII (1992–1997)	22,467	1000	4.45
IX (1997–2002)	37,546	1453	3.87
X (2002–2007)	58,933	5025	8.53
XI (2007–2012)	136,381	15,800	11.59
XII (2012–2017)	363,273	16,840	4.63
Annual Plan (2012–13)	54,748	3062	5.60
Annual Plan (2013–2014)	18,781	4128	21.99
Annual Plan (2014–2015)	11,531	2922	25.34

Working Group on Horticulture, Plantation Crops and Organic Farming XI; Agricultural Statistics at a Glance 2013 and 2014, DES; Horticultural Statistics at a Glance 2015-(MIDH), DAC & FW the establishment of the Indian Institute of Horticultural Research at Hesaraghatta, Bangalore. Similarly, during the IX plan, a separate Division of Horticulture was created from the crops division of the Department of Agriculture and Cooperation of Ministry of Agriculture in 1981. The Horticulture Division was supported by National Horticulture Board, Coconut Development Board and National Bee Board to look after the overall development of Horticulture in India. The Department of Agriculture, Cooperation and Farmers Welfare (DAC & FW) of the Ministry of Agriculture and Farmers Welfare (MoA & FW) created a separate position of Horticulture Commissioner in 1985. Horticulture Development received major boost in the X plan where several schemes were launched for the development of horticulture. During this plan, several initiatives like Technological Mission for Integrated Development of Horticulture in North East Region, National Horticulture Mission, Micro-irrigation Mission Technology, Mission for Integrated Development of Horticulture in North East Region & Himalayan States, The Technology Mission for Integrated Development of Horticulture was launched in eight states of the north-east region in 2001–2002 to harness the potential of horticulture development.

11.2.3 Institutions and Policy Developments for Horticulture Development

To support the horticultural programs to increase production, organizational support is available from both Agriculture and Commerce ministries at the National level. Similarly at the state level, both Agriculture/Horticulture departments provide the necessary infrastructure. Many number of Directorates, Boards, Councils, Foundations and Authorities have been established under Central and State ministries to promote horticultural production for the both domestic and export market. Some of the infrastructures established for the holistic development of horticulture are listed in Table 11.3.

The Horticulture Division of IARI, New Delhi, State Agricultural Universities, National Institute of Postharvest Technology, National Institute of Agricultural Marketing, National Bank of Agriculture and Rural Development, etc. were focused to strengthen education, research development, marketing facilities, financial assistance, etc. Under the Indian Council of Agricultural Research, New Delhi a separate Division of Horticulture was established and vested with the responsibility of overseeing the overall accelerated development of horticulture from national perspective for improving nutritional, ecological and livelihood security. Under the Division of Horticultural Sciences, 10 Central Institutes, 6 Directorates, 7 NRCs, 13 AICRPs and 6 Network Projects/Outreach programmes are operating to cater to the needs of horticulture research, education and also extension services.

The National Horticulture Mission was established in 2004 with aim to double the production of horticultural crops, establishing convergence and synergy among various on-going and planned programs in the field of horticultural development and promoting the development and dissemination of technologies by blending traditional knowledge and new frontier knowledge.

Table 11.3 Institutes developed for holistic development of horticulture

Sl. No	Name	Year of establishment	Purpose
Ministry of Agriculture			
1	Horticulture Division	1980	Support and formulate policies aimed at accelerated growth of horticulture
2	National Horticulture Board, Gurgaon	1984	To promote integrated development of horticulture
3	Coconut Development Board	1981	Integrated development of coconut industry in the country through promoting production, processing, marketing and product diversification of coconut
4	Directorate of Cashewnut & Cocoa Development, Kochi	1966	Promotion of cashew & cocoa development by adoption of improved technology for production and making available quality planting material
5	Directorate of Arecanut & Spices Development, Calicut	1966	Yo formulate appropriate development schemes on spices, medicinal & aromatic plants and arecanut at the national level
6	National Committee on Use of Plastics in Agriculture (NCPA), New Delhi	1993	To implement and monitor the Centrally Sponsored Scheme on "Development of Horticulture through Plasticulture"
7	National Horticultural Research and Development Foundation, Nasik	1993	To cater to applied research and development of onion and garlic for promoting export
Ministry of Commerce			
8	Agricultural and Processed Food Products Export Development Authority (APEDA), New Delhi	1985	Export promotion and development of fruits and vegetables, processed products, etc.
9	Spices Board, Kochi	1987	It is an apex body for the export promotion of Indian spices and spice products
10	Tea Board, Kolkata	1954	Ensuring overall development of the Tea industry and trade
11	Coffee Board, Bangalore	1942	To facilitate the co-ordination of different segments viz: growers, processors, exporters, domestic traders, consumers and various associations representing the industry

(continued)

Table 11.3 (continued)

Sl. No	Name	Year of establishment	Purpose
12	Rubber Board, Kottayam	1947	Promoting the development of rubber industry, advise the Government of India on all matters related to rubber industry including the import and export of rubber
13	Cashew Export Promotion Council, Kochi	1955	Promotional activities related to exports

11.2.4 Horticultural Education

With increased awareness about horticulture, there was a need for specialized human resources in horticultural sciences. Accordingly, the first College of Horticulture was established in the year 1972 at Kerala Agricultural University (KAU), Kerala, to strengthen research, education and extension activities in horticulture. After that, faculties of horticulture were added to the majority of agricultural universities. Dr. Y.S. Parmar University of Horticulture and Forestry, Solan, Himachal Pradesh, was the first horticultural university, established in the country in 1985, which got great recognition not only in the country but in the whole of Asia for its mandate of imparting teaching, research and extension exclusively in horticulture and forestry sciences. At present, there are seven horticultural Universities exclusively for horticultural education in the country.

11.3 Horticulture Growth in the Last Two Decades (2001–2020)

The most significant change in the last two decades has been the use of technologies and private sector investment for production system management. The increase in area and production over the years from 2001 onwards till 2021 is shown in Table 11.4. The total production of horticultural crops during 2020–2021 is 326.5 million tonnes (mt), which was about 1.6% higher than the previous year and 8% higher than the previous 5 years' average (Anonymous 2017a, b). Within horticulture, production of vegetables was estimated at 181 mt in 2017–2018, about 1% higher than the year before, while that of fruits was estimated at 95 mt, 2% higher than the previous year with steady increase in area and production of both fruits and vegetables (Figs. 11.1 and 11.2).

However, the challenge before us is to reach the production of horticultural crops to 500 mt by 2050 to meet the rising demand for these crops due to changing lifestyles, enhanced purchase power, urbanization, food habits, awareness about medicinal plants, etc. under conditions of deteriorating production environment. Though there is an increase in per capita availability of fruits and vegetables, it is still below the recommended levels, and hence, nutritional and health security can be achieved through enhanced productivity. The impact of change in technologies like

Table 11.4 Area, production and productivity of horticulture crops in last two decades

Year	Area ('000 ha)	Production ('000 MT)	Productivity (MT/ha)
2001–2002	16,592	145,785	8.79
2002–2003	16,270	144,380	8.87
2003–2004	19,308	153,302	7.98
2004–2005	18,445	166,639	9.05
2005–2006	18,707	182,816	9.77
2006–2007	19,389	191,813	9.89
2007–2008	20,207	211,235	10.45
2008–2009	20,662	214,716	10.39
2009–2010	20,876	223,089	10.69
2010–2011	21,825	240,531	11.02
2011–2012	23,243	257,277	11.07
2012–2013	23,694	268,845	11.35
2013–2014	24,198	277,352	11.46
2014–2005	23,417	283,468	12.11
2016–2017	24,472	286,188	11.69
2017–2018	25,431	311,714	12.25
2018–2019	25,743	311,057	12.04
2019–2020	26,465	320,078	12.09
2020–2021	27,175	326,582	12.09
% Increase over 2001–2002	61.06	44.63	7.27

Horticultural Statistics at a glance 2017 and authors' personal compilation

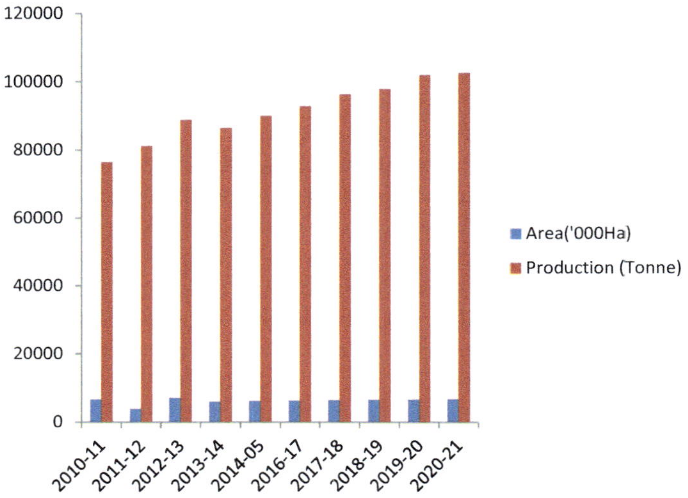

Fig. 11.1 (**a**) Trends in area and production of fruit crops from 2010–2011 to 2020–2021

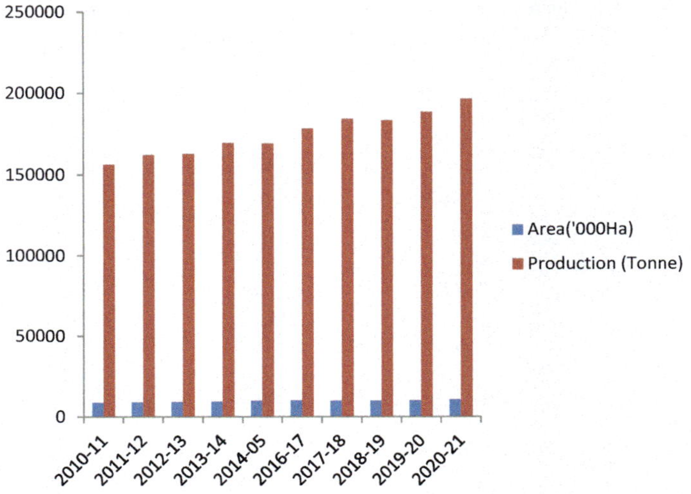

Fig. 11.2 (**a**) Trends in area and production of vegetable crops from 2010–2011 to 2020–2021

new cultivars, use of micro irrigation, improved quality seeds and plants and production system management is visible in increased production productivity, availability and export. Undoubtedly, the horticulture sector has moved dynamically despite numerous challenges and shortcomings and is in a crucial phase of development, needing initiatives for sustainable development. To achieve the targeted production of 451 million in the year 2022–2023, stipulated vertical growth will be required through the use of new cultivars, efficient water and nutrient management, effective plant health management coupled with strategies for reduced post-harvest losses and empowered human resources. This will require appropriate innovations and investment. Protected cultivation has shown yield enhancement up to 4 times, which is a potential technology to achieve vertical growth but would need investment and technological up gradation. Plant architectural engineering and management can mitigate the problem associated with seasonality in many crops, and the enhanced efficiency in water management, utilizing modern techniques, shall reduce water stress. Since horticulture provides variability and has the potential to adjust in different agro-climatic situation; technology-led development is inevitable, where in horticulture education to empower the youth with new knowledge becomes essential. Thus, it is essential to evolve technologies for enhancing productivity in horticultural crops through constant efforts in research, development and extension.

11.3.1 Recent Advances in Horticulture

Aiming to boost horticulture along with Indian agriculture, the government of India set a goal to double the farmers' income by 2022 as horticulture is a very important

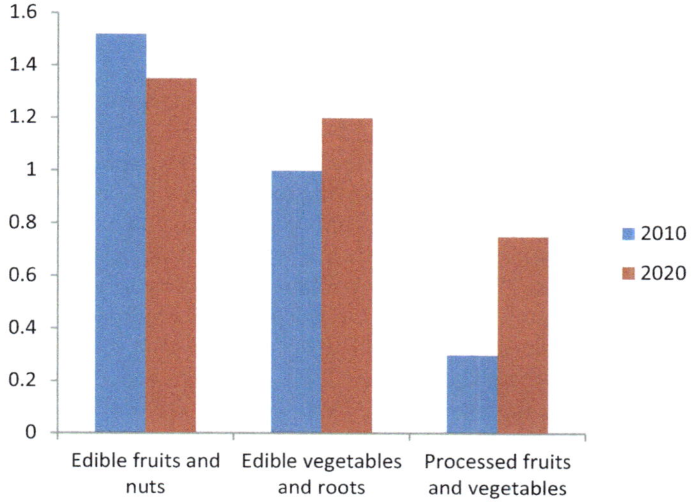

Fig. 11.3 India's export of fruits and vegetables (Source: UN Comtrade, figures in US$billion)

sector in Integrated Farming System in the country (Saikia and Bordoli 2014). The rise of farmers' income can be increased when risk is reduced, productivity goes up, cost of production comes down, post-harvest loss is minimized, and commodities are produced to get a remunerative price. Doubling farmers' income became a slogan for every agricultural institution and is formulated by bridging productivity gaps, employing the latest production technologies, processing technologies and marketing. Some of the advanced technologies in horticulture include high-density planting, urban and periubran horticulture, vertical gardening, micro greens, use of solar energy in horticulture, vegetable grafting, farm mechanization, etc. The goal set to double farmers' income by 2022–2023 is central to promoting farmers' welfare, reducing agrarian distress and bringing parity between the income of farmers and those working in non-agricultural professions. This can also be achieved through the use of the above-listed advanced technologies. Though India is the second largest producer of fruits and vegetables and with a steady increase in horticultural production over the years, it is still facing several constraints/challenges which need to be handled effectively (Prabhakar et al. 2014).

Despite leading in the production of fruits and vegetables, there is a minimum success with respect to their export. During 2010–2020, there has been an increase of US$2.8 million in the export of Indian horticultural products, with an export value of US$3.19 billion in 2020. The major export destinations are UAE, Netherlands, USA and Saudi Arabia. The export trend of fruits and vegetables is shown in Fig. 11.3

Some of the products which have outperformed the export performance of horticultural crops include onions, potatoes, chillies and fresh vegetables, grapes, etc. Various factors have been attributed to the export growth of these products like the initiative of Agri-export Zones (AEZs) and Grapenet, etc. Similarly, in fresh potatoes, exports increased from US$26.9 million in 2010 to US$ 70.7 million in 2020. Continuous efforts are being made to promote the export of agricultural products. With the help of APEDA, different certified agricultural products are being exported to other countries. In 2021, many fruits like chikoo, jackfruit, mango, litchi, jamun, grapes and dragon fruit have been exported, which has benefited the farmers. Now a special variety of bananas from the Jalgaon district of Maharashtra has been exported to Dubai with the help of APEDA. More recently, a GI certified mango from Malda district of West Bengal has been sent to Bahrain. Mangoes are processed by APEDA registered packhouse facility and then exported to various regions and countries including Middle East, European Union, Japan and South Korea. Indian floriculture industry comprises flowers such as roses, tuberose, gladiolus, anthurium, carnations, marigold, etc. The Indian floriculture market was worth INR 157 billion in 2018. The market is further projected to reach INR 472 billion by 2024, growing at a CAGR of 20.1% during 2019–2024. India's total export of floriculture was R 571.38 crore/81.94 USD million in 2018–2019. The major importing countries were USA, the Netherlands, the United Kingdom, Germany and United Arab Emirates. Though India's share in the global market is still 1% only, there is increasing acceptance of horticulture produce from the country. This has occurred due to concurrent developments in the areas of state-of-the-art cold chain infrastructure and quality assurance measures. Apart from large investments pumped in by the private sector, the public sector has also taken initiatives and with APEDA's assistance, several centres for perishable cargo and integrated post-handling facilities have been set up in the country. Capacity-building initiatives at the farmers, processors, and exporter's levels has also contributed towards this effort.

11.3.2 Challenges Faced by Indian Horticulture

11.3.2.1 Low Productivity and Non-availability of Quality Planting Material

One of the dissatisfaction facts of Indian horticulture sector is lower productivity of fruits, vegetables, ornamentals and medicinal plants though the productivity of few fruits and vegetable crops is highest in the world (Table 11.5). The prime reason for

Table 11.5 Productivity of important fruits and vegetables in India vis-a vis world

Crop	India (t/ha)	World (t/ha)
Banana	36	18.65
Grapes	26	8.83
Mango	6	7.23
Papaya	33	23.99
Pineapple	15	19.77
Brinjal	17.5	25
Cabbage	21.5	27.7
Cauliflower and broccoli	18.3	16.9
Okra	11.61	6.9
Onion	16.81	18.8
Potato	22.7	17.7
Tomato	19.1	32.8

low productivity are the non-availability of quality planting materials, dwindling natural resources, poor farmers, non-adoption of modern research findings etc. The challenge is to enhance productivity by increasing the factor productivity of all the production inputs and at the same time sustaining it by the adoption of good agricultural practices and precision farming principles.

11.3.2.2 Depleted and Degraded Production Environment

Most of the landholdings in Indian agriculture (>82%) are small to medium, and the contribution to GDP has declined drastically over the years. An estimate says that the average size of land holding by 2050 would be a meagre 0.30 ha. This will be further compounded by increasing urbanization, industrialization and other uses than farming. In this direction majority of the land held in horticulture is marginal and degraded lands which are unfit for production. Moisture, salinity and acidity stress, poor soil fertility, etc. are the characteristics of such marginal lands. In such lands, some of the challenges to be addressed are depleting organic matter and nutrients, erratic monsoon, depletion of groundwater resources, deteriorating water quality, etc. are major challenges to be addressed in the future.

11.3.2.3 Plant Health Management and Bio-Security Concerns

Plant health management by combating the effect of pests, diseases and weeds are the real challenge in improving the productivity of horticultural crops. In recent years, the advent of new pests and diseases with changing climate scenarios is emerging as major threat. The aftermath of indiscriminate and injudicious use of plant protection chemicals is the development of resistance to pests and diseases in addition to environmental pollution. The lack of foolproof quarantine measures also aggravates the introduction of new pests and diseases.

11.3.2.4 Post-Harvest Utilization

Substantiate proportion of the horticultural produce (from 5 to 39%) goes waste as post-harvest loss which is the main reason for not meeting the demand of fruits and vegetables. The overall loss for major fruits ranged from 6.7% to 15.88%. The overall loss was as high as 15.88% in guava, 10.39% in apple and 9.73% in fruit crops. For vegetables, the losses ranged from 4.58% to 12.44%. These losses occur at different stages of handling, transport, storage, processing and distribution. The biggest challenge in such situations is to find out the means to reduce the post-harvest losses of fruits and vegetables in addition to maintaining their quality during distribution and ensure their safety for consumption.

11.3.2.5 Changing Quality Consciousness and Global Competition

As socio-economic status of the countrymen is increasing so is their purchasing power of them. In recent years consumers are becoming more health and quality conscious, it is necessary to produce quality and safer horticultural produce. In addition, with the changing scenario of world trade, there is a requirement to produce horticultural commodities of international standards.

11.3.2.6 Climate Change and Horticulture

Global warming and climate change is the greatest concern of mankind in the twenty-first century. The established commercial varieties of fruits, vegetables and flowers will perform poorly in an unpredictable manner due to the aberration of climate. Commercial production of horticultural plants particularly grown under open field conditions will be severely affected. Due to high temperatures, the physiological disorders of horticultural crops will be more pronounced. Air pollution also significantly decreases the yield of several horticultural crops and increases the intensity of certain physiological disorders like black tip of mango. Hence there is a need to protect these valuable crops for sustainability against the climate change scenario. It was also revealed that changes in rainfall patterns can affect year-to-year differences in flowering, quality and productivity in tropical fruits. (Datta, 2013).

11.3.2.7 Inadequate Market Intelligence and Price Fluctuation

Lack of market infrastructure facilities for obtaining information on price fluctuation is one of the major constraints for realizing better prices for horticultural producers. As most of the horticulture farms are not well connected to the road as they are located on interior side, the producer cannot go to wholesale markets or long distant markets and has to depend on intermediaries to sell his produce. Therefore in the marketing of horticultural produce costs are involved for grading, packing, transport, loading and unloading, etc. In addition, middlemen also take some margins for them. These costs and margins determine the final price to be paid to the consumers. Hence, there is a need to develop good market intelligence to realize better prices for the growers.

These challenges demand a paradigm shift in research approach to harness the potential of modern science, innovative technology generation and its effective dissemination, enabling policy and investment support. Some of the critical areas which need to be looked at are genomics and molecular breeding, improved production technologies, diagnostic and vaccines, nanotechnology, secondary agriculture, farm mechanization, agri-incubators, technology dissemination and urban and peri-urban horticulture, etc. Multi-disciplinary and multi-institutional research should be of paramount importance.

11.4 Opportunities to Improve Production and Productivity in Horticultural Crops

11.4.1 Development of New Varieties by Integrating Conventional and Modern Breeding Techniques

From the perspective of evolving new varieties in horticultural crops, it is necessary to integrate both traditional and modern plant breeding techniques (Kumar, 2015). Fruit breeding comprises two parts, scion and rootstocks. The scion breeding objectives mostly include improving quality with respect to attractive colour and high yield, dwarfing for adoption to high-density planting, precocity and regular

Table 11.6 Varieties released for specific traits in some important horticultural crops

Crop	Traits	Varieties
Mango	Arka Udaya, Arka Suprabhath	High TSS
Guava	Arka Kiran	High lycopene
Papaya	Arka Prabhath	High carotene and TSS
Cauliflower	Pusa Betakesari	Beta carotene
Lemon	Pusa sarda	High TSS
Musk melon	Pusa madurima	High TSS
Carrot	Pusa Asita, Pusa Rudhira	Anthocyanin and lycopene
Radish	Pusa Jamuni and Pusa gulabi	Anthocyanin and lycopene
Red cabbage	Red Acre	Anthocyanin
Bitter Gourd	Pusa Aushadhi Pusa Vishesh	β carotene Ca and Fe
Tomato	Pusa upahar, Arka Vishesh	Vitamin C and Lycopene
Potato	Bhu Sona	β carotene
Black pepper	IISR Malabar Excel, Panniyur 2, Panniyur 9	>5% piperine
Turmeric	Sudarshana, Suvarna	High essential oil

bearing, seedlessness, soft seeds, improved nutraceutical properties, etc. However, the rootstocks are to be bred for imparting dwarfness to scions, effective uptake of nutrients, drought and salt tolerance, resistance to soil-borne pathogens, etc. Some of the trait-specific varieties released in various horticultural crops are shown in Table 11.6. Rootstock breeding is more difficult and time taking compared to scion breeding. Evaluation of rootstocks for desirable traits and stionic influence also requires more time.

Exploitation of wild species and incorporation of genes from wild species is one of the ways to improve the cultivated species. The main concern in this approach is most of these wild species of major fruit crops are in danger of extinction or some are grown as wild even today. However, it is very difficult to transfer these qualitative traits. Even if it is successful, the resultant hybrids may be sterile with low yield and of poor quality. Further mechanical isolation, chromosomal aberration and genetic and or cytoplasmic incompatibilities hinder hybridization between wild and cultivated species. In spite of these, wild species have been extensively used for breeding scions and rootstocks in citrus, banana, grape, papaya, blueberries and raspberries in other countries. Hence, such wild species of different crops are to be introduced or exploited and used in breeding programmes to incorporate some of the genes of interest (Chadha 2014). Some of the wild species used for incorporating some important traits in various horticultural crops are shown in Table 11.7

With the limitations in traditional breeding method, in the last two decades, the availability of many genomic resources like genome sequences, high-throughput analysis of gene expression, sufficient numbers of molecular markers, express sequence tags (ESTs) and high-density genetic maps has paved the way to the genetic engineering and molecular breeding of horticultural crops for improvement. The application of these modern biotechnological techniques can contribute

Table 11.7 Use of wild species in the improvement of horticultural crops

Crop	Species	Traits
Papaya	*Vasconcella cauliflora*	Papaya Ring Spot Virus tolerance
Strawberry	*Fragaria nilgirensis*	Peach aroma
Citrus	*Citropsis*	Less number of seeds
Mango	*Mangifera zeylanica* *Mangifera odorata*	Salt tolerance
Grapes	*Vitis rupestris*	Drought and Sodicity tolerance
Tomato	*Solanum hirsutum*	Whitefly and Fusarium wilt resistance
Brinjal	*Solanum stenotomum* *Solanum incanum*	Bacterial wilt resistance Shoot and fruit borer resistance
Okra	*Abelomoschus caillei* *Abelomoschus manihot*	YVMV resistance Fruit and Shoot borer resistance
Cucurbits	*Cucunis trigonus* *Cucumis hystrix*	Fruit fly resistance Downy mildew
Potato	*Solanum demissum* *Solanum vernei*	Late blight and leaf roll resistance Nematode resistance
Chilli	*Capsicum frutescence* *Capsicum chinense*	High Capsaicin Fruit rot resistance
Black pepper	*Piper calubrinum*	Phytophthora fruit rot
Rose	*Rosa damascene*	High flavour

efficiently to solving or reducing the problems faced by the horticulture industry. In the last decade, the emphasis on crop improvement using novel genomic tools have shifted toward the identification and functional analysis of miRNAs, one of the hottest research fields in plant sciences. Immediately after the publication of the first genome sequence of *Arabidopsis,* significant advancement was made in DNA sequencing technology in many horticultural crops. Among the fruit crops, the grapevine was the first crop to be sequenced. This was followed by other crops like apple, pear, papaya, strawberry, sweet orange, peach, eggplant, tomato, melons, cucumber, French beans and cassava and genome sequencing work is in progress in many other horticultural crops. Gene pyramiding in vegetable crops has paved the way for the development of many resistant varieties in different crops like tomatoes, peas, potatoes, beans, etc. Gene pyramiding entails the stacking of multiple genes thus leading to the simultaneous expression of more than one gene in a variety to develop desirable resistance expression. It is gaining importance as it improves the efficiency of plant breeding leading to the development of genetic stocks and the precise development of broad-spectrum resistance capabilities (Bhutia and Bhutia, 2017). Identification of major genes in a gene pool is through association genetics or association mapping. It involves searching for significant associations between changes in DNA sequence and changes in the phenotype of a trait in a large panel of unrelated species (Bharadwaj et al. 2014).

11.4.2 Improved Production Practices

One of the prime reasons for reduced productivity of horticultural crops in the country accounts for the non-adoption of advanced production technology practices. Some of the important practices to improve the productivity of horticultural crops are discussed below.

11.4.2.1 Production of Quality Seeds and Planting Material

As most of the horticultural crops especially fruit crops are propagated by vegetative means, they are produced in nurseries and sold to growers. Since they have a long gestation period, their full characteristics are shown after a long period. It is therefore essential to ensure that planting materials produced and supplied by the nurseries are of good quality, true to the type of said variety. State Governments ensure the supply of quality planting materials for fruit crops by enactment of the Fruit Plant Nursery (Regulation) Act and enforcement of its provisions through licensing of horticulture nurseries. However, all the States have not enacted Horticulture Nursery Acts. As per the information available in the report of the working group on Horticulture for the XI 5-year plan, at present only eight states have adopted Nursery Act and in nine states some system of registration/ monitoring exists for nurseries whereas in 13 states there is no nursery act at present. Though some states have enacted the Nursery Registration Act, the quality aspect of such vegetatively propagated planting materials is not ensured. Similarly, no mechanisms are available for ensuring the quality of tissue-cultured plants. Hence, there is a need to develop a mechanism to assure the quality of planting material. Recently, the Ministry of Agriculture & Farmers' Welfare finalized the draft Seed Bill 2019. The Bill aims to regulate the quality of seeds sold and facilitate the production and supply of these seeds to farmers. It aims to foster competition by amending the Seed Act, 1966 and Seed Rules, 1968. This bill also includes certification of horticultural nurseries for the supply of genuine planting material, guidelines for maintenance of nurseries and quality control to maintain true to the type of variety to be multiplied. National Horticulture Board, Gurgaon has launched the National Nursery Portal, which is a digital platform in the form of a Web Portal and a Mobile Application, which intends to create an online marketplace where Nurseries and Buyers can interact and sell or buy planting material. This portal facilitates availability of quality planting material and helps the buyers in procuring the planting material of different crops in nearby locations and to encourage nurseries to upgrade their infrastructure to get accreditation and expand their business.

11.4.2.2 Seed Village Concept for the Production of Hybrid Seeds in Vegetables

The seed village concept for the production of vegetable seeds is one of the paradigms in the horticulture production system. This not only helps in increasing the income of farmers but also helps in catering to the demand for seeds by the industry. One of the success stories in this concept has been documented at ICAR-IIHR, Bengaluru. Seeds of ICAR-IIHR bred hybrids and OP varieties of tomato, chilli, brinjal, onion, French bean and okra were produced under the seed village concept. Before the adoption of the Seed Village concept the average income generated by the cultivation of vegetables was INR 44,000 per acre. The intervention of the Seed Village Concept in these farmers' fields has resulted in an average net income of INR 85,000 per acre which amounts to a 94% increase as compared to commercial vegetable production. This concept might be one of the strategies for doubling the farmers' income.

11.4.2.3 High-Density Planting

The simple way of increasing the productivity of horticultural crops is to practice high-density planting. As it accommodates more plants per unit area, it ensures doubling the yield with good quality produce. High-density planting has been standardized for most the fruit crops like banana, pineapple, citrus, guava, apricot, peach, plum, mango, etc. Ultra-high-density planting and meadow orchards in guava are gaining popularity in recent years. In high-density orchards, tree size is regulated by employing dwarfing rootstocks or by use of growth retardants.

11.4.2.4 Use of Rootstocks

Employing rootstocks in most horticultural crops is an alternative strategy to combat major abiotic stresses and some soil-borne pests and diseases. It is an environmentally friendly and economically feasible approach to overcome some of the constraints related to abiotic and biotic stresses. A major breakthrough in the use of rootstocks in grapes has been achieved where the use of Dogridge rootstock could overcome the problems associated with soil salinity and water scarcity in addition to its effect on increasing the yield of most of the seedless grape varieties. In other crops also rootstocks have shown good potential in either increasing yield, altering the vigour of scion varieties or overcoming abiotic and stress tolerance. Some of the examples are Rangpur lime and Alemow in citrus for improving yield and resistance to phytophthora root rot, Malling and Malling Morton series for dwarfing and precocity in apple, *Psidium friedrichsthalianum* for wilt resistance in guava, Vellainkolumban, Nekkere, Olur, Kurukkan, and 13–1 in mango for dwarfing, salt tolerance, etc.; *Piper colubrinum* for quick wilt tolerance in pepper, *Rosa domesciana* in the rose; wild species of *Solanum* for wilt, nematode and flood tolerance in solanaceous vegetables, etc. Bacterial wilt disease caused by *Ralstonia solanacearum* is one of the most destructive diseases of economically important crops in Solanaceous vegetables which are a soil-borne pathogen. Grafting disease-resistant scions onto rootstocks with complementary disease resistance boosted the overall disease resistance compared to that possible with a single variety. Similarly, use of wild species of brinjal and cucurbits have been fully exploited for their tolerance to some of the biotic and abiotic stresses.

11.4.2.5 Integrated Soil, Nutrient and Water Management

Mulching, preparation of terraces, contour bunds, planting of a double row of herbaceous fruits crops across the slope and fruit trees on the contour line, use of drip irrigation system for high-value fruit crops and construction of water storage

Table 11.8 Comparison of traditional and drip irrigation in water saving and yield increase

	Traditional Irrigation		Drip irrigation			
Crop	Water	Yield	Water	Yield	% save in water	% increase in yield
Banana	1760	575	970	875	45	52
Grapes	532	264	278	32.5	48	23
Citrus	1660	100	640	150	61	50
Tomato	300	32	180	48	39	50
Brinjal	90	28	42	32	53	14
Chilli	100	42	42	3.1	62	44

Horticulture Science at Glance 2018

tank (farm ponds) for life-saving irrigation have been observed as proven technologies for soil and water management. Good Nutrient Management is an important method of growing crops without harming the environment or health. Integrated nutrient management considers nutrients from all sources and they should balance, but not exceed/fall short of the crop requirements. There are many agronomic practices that can influence and maximize the supply of nutrients and the grower should be aware of these and their potential to save money and replace the use of inorganic fertilizers. The application of inorganic fertilizers is one of the expensive inputs in horticulture. Indiscriminate use in horticultural crops like banana, grape, mango, papaya, cabbage, cauliflower, tomato, and ornamental crops has made the soil sick, polluted the groundwater and made it unsuitable for cultivation and consumption. Many of the nutrient mixtures to enhance fertilizer use efficiency have been developed by several institutes. Under the circumstances of indiscriminate use of inorganic fertilizers, the use of cost-effective and eco-friendly bio-fertilizers with suitable integration of organic manure will restore the soil health and keep the soil productive and sustainable. They are less expensive, eco-friendly and sustainable. The beneficial microbes in the soil, which are of greater significance to horticultural situations, are the biological nitrogen fixers, phosphate solubilizers and the mycorrhiza fungi. Many microbial formulations have been developed by several institutes which are used in integrated nutrient management programmes which improve nutrient availability to plants when applied to the soil. Efficient utilization of water through modern irrigation practices like drip irrigation, sprinkler irrigation and need-based irrigation based on the plant's physiological indices can save the irrigation water thereby increasing water use efficiency. Very clearly water use and increase in yield in some of the major fruits and vegetable crops have been demonstrated by following drip irrigation system (Table 11.8) Partial root drying technique and deficit irrigation technique are two important means of irrigation within drip irrigation which can double the water use efficiency. Great success has been achieved in many of the horticultural crops like grapes, papaya, and some vegetable growers are already practicing partial root drying techniques of irrigation.

11.4.2.6 Precision Farming

One of the principal features of post-green revolution agriculture is the site-specific management techniques that are beginning to be used for different production resources and the greater precision employed in research processes. Precision agriculture is gaining importance all over the world. This growth is being driven by the increased and intensified use of the many more sophisticated tools provided by information and communication technologies (Barrera 2011). Automated irrigation systems, sensor-based spraying technology, need-based application of irrigation based on the soil and plant sensors, mechanization of some farm operations, etc. are integrated into the component of precision farming and some encouraging results are observed in many crops. Thus precision farming is also can be one of the tools to maximize the input efficiency in terms of water, nutrients, labours, plant protection chemicals and maximizing the farmer's income. Precision Farming Development Centres (PFDC) have been established in 17 locations in the country which are mostly located in SAUs, ICAR institutes and IIT, Kharagpur. The PFDCs will be equipped further with the necessary hardware and software needed for generating information on precision farming techniques in the farmers' field. Recently the bid to promote precision farming in India got a major boost after the Union Ministry of Agricutlure and Farmer's Welfare issued guidelines to make drones more accessible to farmers and stakeholders in agricultural sectors. Indian precision agriculture market was valued at over $57 million in 2016 and is expected to grow at CAGR of over 10% to reach $99 million on account of increasing awareness related to the applications and precision agriculture and surging needs to ensure maximum yield from limited-sized farms.

11.4.2.7 Development of Horticulture-Based Cropping System

Suitable land use systems such as Agrihorti, Agri-horti-silvipastoral, mixed horti, horti-silvi-pastoral and multitierhorticulture system should be developed based on agroclimate zones, crop priority, topography and socio-economic factors. Production-related technologies can bring quick improvement in production and productivity of different regions. Short or medium-range programme on horticulture-based cropping systems; water management including micro-irrigation and fertigation, greenhouse cultivation of vegetables and flowers, integrated nutrient and pest management, environment pollution and pesticide residue problems have already received research attention. Further refinement of the technologies and their transfer will bring perceptible change/improvement in the production of different crop commodities.

11.4.2.8 Integrated Pest Management

In well-managed farming systems, crop losses to insects can often be kept to an acceptable minimum by utilizing resistant varieties, conserving natural enemies and managing crop nutrient levels to reduce insect reproduction. Recommended measures against diseases include the use of clean planting material, crop rotations to suppress pathogens, and eliminating infected host plants. Effective weed management entails timely manual weeding, minimized tillage and the use of surface residues. When necessary, lower-risk synthetic pesticides should be used for targeted control, in the right quantity and at the right time. Integrated pest management can be promoted through farmer field schools, local production of bio-control agents, strict pesticide regulations, and removal of pesticide subsidies.

11.4.3 Bio-Control Agents, Diagnostic Kits and Semiochemicals

Agrochemicals were used extensively to increase food production during the green revolution. The impact of such extensive use of agrochemicals resulted in environmental pollution, contamination of food and water and the development of resistance to chemical pesticides by several pest species. Minimizing the use of chemical pesticides and alternately use of bio-control agents (parasites, predators, etc.), vaccines and the development of diagnostic kits will help to counter the effect of the green revolution on the environment. Thus several research institutes in the country started to develop eco-friendly biological control methods for effective pest management. Efficient methods of mass multiplication of parasitoids, predators and pathogens against insect pests and antagonists against plant pathogens and plant parasitic nematodes have been developed. Similarly, BioControl technologies with pests and pathogens for bio Control of weeds have been developed. Spectacular results have also been obtained in the management of plant diseases using biocontrol agents by minimizing pesticide usage.

Some of the success stories in the use of biocontrol agents include mealy bug management in grapes and papaya using predators and parasitoids; pheromone trap for control of mango fruit flies and cue lure trap for cucumber fruit flies, etc.

Some of the alternate methods to control insect pests and nematodes were found to be more effective like the use of pheromone traps to control fruit flies in many fruits and vegetables, the use of entamophagus nematodes to control several pests. Some of the components in biopesticides include neem soap and pellets, pongamia soap, sealer cum healer (to control stem borers), *Beauveriabassiana and Metarrhizumanisopilae,* etc. Many of the bio-pesticide formulations have been commercialized which include *Paecilomyceslilacinus*—1% W. P.; *Verticilliumchlamydosporia*—1% W. P.; *Trichodermaharzianum*—1% W. P; *Pseudomonas fluorescens*—1% W. P.; *Trichodermaviride*—1.5% W. P., etc. These formulations are effective against many soil-borne pathogens and nematodes. Liquid and talc-based formulation of *Bacillus subtilis* for control of nematodes is also effective against nematodes in many horticultural crops.

Bacteria, viruses and MLOs are major diseases in many of the horticultural crops like papaya, grape, cucumber, chilli, tomato, beans, aster, etc. Identification of such organisms is the key step in controlling such diseases. Several diagnostic kits have been developed which not only help in identifying the viruses but also help in plant quarantine when any plant consignment reaches the country. In this context, many of the institutes are working in this direction to develop diagnostic kits. Some PCR-based kits for the identification of many plant viruses have also been developed and are being used in many horticultural crops.

Semiochemicals (behaviour modifying chemicals) are used by insects to interact with their environment for survival and reproduction. In other words, insects use chemical cues for identifying suitable mates (=pheromones) and host plants (= kairomones). This insect's reliance on chemical cues offers a number of opportunities for its management. In particular, insect pests of horticultural importance are evolving rapidly with changing climatic conditions, monocropping and constant selection pressure that is being exerted by intensive insecticidal applications. Recently, semiochemicals are increasingly being used as important components of integrated pest management (IPM) strategies for a number of insect pests worldwide. Push–pull strategies or stimuli-deterrent diversionary strategies (SDDS) (= behavioural manipulation methods) uses repellent/deterrent (push) and attractant/stimulant (pull) stimuli to direct the movement of pest or beneficial insects for sustainable pest management. However, their potential is less exploited in horticulture. This may be mainly due to a lack of thorough understanding of chemical-mediated processes of the intended pests. Therefore, the development of reliable, robust, and sustainable push–pull strategies requires a clear scientific understanding of behavioural and chemical ecology of pests, its interactions with hosts and non-specific and natural enemies in order to underpin key processes that can be exploited as weak links. The impetus to identify chemical cues that are involved in host location, oviposition site selection and mate location across the pests and host plants will help us to bring out viable behavioural cues that will aid in the formulation of end-to-end pest management programmes. Thus, to understand/manipulate various semiochemicals (pheromones, allelochemicals-kairomones, allomones and synnomones) and maximize their usefulness in IPM, collaborations between entomologists and chemists are essential (Anonymous 2017a, b).

11.4.4 Secondary Agriculture

Secondary agriculture is a ray of hope for the Indian economy because of its immense untapped potential. This has a large scope in horticulture compared to agriculture as there is wide scope for the development of value-added products from fruits, vegetables, flowers and medicinal crops (Chengappa 2013). The impetus for the development of secondary agriculture has been firstly due to consumers' demand for value-added goods like ready-to-eat, ready-to-serve, convenience food, functional food, and nutraceuticals in both domestic and international markets.

In the current scenario of Indian agricultural production and post-production, secondary agriculture is a more appropriate action for minimizing post-harvest losses, increasing the income of farmers and generating employment and export growth by encouraging processing and value addition. The opportunities for promoting secondary agriculture would be increased demand for the processed products, assured remunerative prices for the processed products coupled with the constant availability of raw materials and workforce. But the limitations of secondary agriculture include inadequate infrastructure, high upfront capital investment, poor marketing linkages, lack of electricity supply, maintenance of quality standards, etc. Processing activity not only increases the value of raw material by several times but also brings about desirable socio-economic benefits such as increased employment opportunities and improvement in income and lifestyle.

Thus looking into the opportunities and limitations for secondary agriculture, to promote secondary agriculture in India, there is a need for the establishment of agri incubation centres, encouraging cooperative and contract farming, development and dissemination of cost-effective processing technologies, integration technologies for different products to operate the enterprises for a substantial period, establishing synergy between departments and convergence of institutional support and development of integrated cold chain infrastructure and not but least educating or creating awareness among farmers about secondary agriculture.

Value-added process products, especially fruits and vegetables improve the national economy by reducing post-harvest losses at one end and on the other end, provide quality nourishment to millions of consumers at affordable costs. Present-day processed value-added innovative horticultural products include pre-cut minimally processed vegetables, fruit juice concentrates, semi-concentrates, improved dehydrated fruits and vegetables, freeze-dried fruits and vegetables. Many of the institutes working on horticultural crops have standardized the technology to extend the shelf life of such crops and standardized the protocol for storage such as modified atmosphere storage, control atmosphere storage, shrink wrapping, etc. Value addition through product development is one of the main mandate in most of the institutes in which hundreds of processed products like osmo-dehydrated products, fruit-based beverages, squashes, culinary pastes and purees, lactic acid fermentation of vegetables and protocols for minimally processed foods has been standardized. India, being a leading producer of fruits and vegetables, must play to its strength and leverage this agricultural bounty by adding value to its produce. While the government has taken some positive steps like priority sector lending to food processing companies, FDI in the food processing sector and establishing mega food parks, there is still room for improvement. Value addition to horticultural products has numerous advantages for all stakeholders. Besides enhancing the shelf life of the product & reducing food wastage, value addition leads to higher monetary returns for the food & beverage industry. This will also mean higher incomes for farmers as exports and profits in the sector increase. The current level of processing in the horticulture sector stands at a mere 2% of the total output produced, but the sector presents an immense opportunity to increase the processing levels and boost investment. While food wastage at farm-gate and various supply chain stages was reported at about INR 630 billion in 2018–2019, the food processing sector is reported to grow at an Average Annual Growth rate of about 7.4% between the year 2014 and 2018. The sector constitutes 8.83% and 10.66% of gross value added in the manufacturing and agriculture sector, respectively. As per the latest Annual Survey of Industries (2016–2017), the horticulture sector, especially food processing has accounted for 15.95% of the total number of factories, employment to 11.36% of the total workforce, 14.09% of the output, and 16.78% of the operational factories. However, India has less than a 1% share in the global market in the export of value added. During 2018, there were about 427 registered food processing units in India and about 7917 cold storage facilities for storing agricultural produce. NHM/MIDH has sanctioned about 1507 cold storage projects with a capacity of 6,815,383 MT.

11.5 Farm Mechanization

According to the 2011 census, 263 million people (54.6%) are engaged in the agriculture sector, which is likely to decline to 190 million (33%) by 2020 (Agricultural Situation in India 2017). These figures show that during important seasons such as sowing and harvesting there will be a decrease in the workforce and it will harm production. Thus, the additional demand for energy for various agricultural works needs to be completed through the medium and for this, the agricultural mechanization sector needs to grow rapidly. Farm mechanization is one of the prime aspects for the development of the agricultural sector in a sustainable mode which helps in increasing production through agricultural works, reducing the deficit, reducing the cost of various farm operations through better management of costly inputs, increasing the productivity of natural resources and it also helps in reducing the constraints related to various agricultural activities. Agricultural mechanization is an important input to agriculture for performing timely farm operations; reducing the cost of operation; maximizing the utilization efficiency of costly inputs (seeds, fertilizer, plant protection chemicals, water and agricultural machinery); improving the quality of produce; reducing drudgery in farm operations; improving the productivity of land and labour and for improving the dignity of labour. The strategy for mechanization in different regions will be different depending on the conditions and resources of that region.

Agriculture engineering division in most of the research institutes is involved in developing many farm machineries to mechanize most of the farm activities. In horticulture sector, machines related to nursery are needs of the day as nursery involves major manpower in terms of potting mixture preparation and pot filling, seed sowing, etc. In this direction many of the machines have been developed and commercialized which include media, mixing, portray filling, seed drilling, etc. for raising flower and vegetable seedlings. Similarly, motorized pot-filling machines have been developed for filling polythene bags in fruit nurseries. Power and computer-operated mist chamber and poly houses have been established which increases the efficiency of inputs and reduces the requirement of a manual labour force. Complete mechanization has been achieved in some of the crops like onion right from seed sowing, raised bund formers, transplanters, inter-row cultivators, harvesters, de-toppers, dryers and graders. Management of pests and diseases is a very labour-intensive aspect of horticulture crops. Tractor-operated hydraulic platform for spraying, electrostatic sprayers are being used by many fruit growers reducing manual labour with increased spray efficiency. Harvesting machines for some of the fruit crops like mango, guava, sapota, lime, apple and other temperate fruits have been developed.

11.6 Agri Incubators

Agricultural entrepreneurship is one of the important aspects which play a critical role in the development pathway as it creates jobs, drives shapes and innovation and brings competition which ultimately improves productivity. It acts as a catalyst for economic growth and national competitiveness. The growth of entrepreneurship in the horticulture sector is very much the need of the hour. However, there is a need for a mechanism that can facilitate agribusiness ventures. Business incubators have shown that they can be an effective model that can be considered not only for technology transfer and commercialization but also for achieving growth of the venture by offering the required business environment. Agribusiness incubators are relatively new to the arena but have seen considerable success, which can be a compelling reason to be viewed as an alternate and effective extension model for the growth and prosperity of the agricultural sector and rural sector (Sharma et al. 2014).

The initiatives proposed by the ICAR in the Plan period such as National Agricultural Entrepreneurship Project, Student Rural Entrepreneurship and Awareness Development Yojana (Student READY), Attracting Rural Youth in Agriculture (ARYA), and Farmer FIRST, all focus on the grassroots level and in promoting entrepreneurship and technology development. While technology commercialization is important, the incubators bring in the much-needed perspective of market-oriented research to the research institutions. The incubators will provide an opportunity for farmers and rural youth to proceed beyond just farming and engage in more meaningful mechanisms of economic contribution through the creation of ventures and employment opportunities (Philroy et al. 2014).

National Agriculture Innovative Project (NAIP) has funded to establish Business Planning and Development (BPD) units in several ICAR research institutes and agricultural universities across the country. This helped in promoting numerous technologies from these institutes, most of which were kept on the shelf, and starting up ventures based on these technologies. This has helped in commercializing 331 technologies among 518 licensees through the BPD units, which has brought in additional revenue to the NARS and much higher revenue than before BPD units were conceptualized.

11.7 Technology Dissemination

Innovation of new technologies and their effective dissemination are the key drivers in attaining food security in the country in addition to providing farmers a competitive edge over the traditional method of farming which will improve their standard of living. To achieve this, farmers should have easy access to state of art technologies, their methods of production, marketing, etc. This can be possible only when extension workers go beyond technology dissemination by facilitating broader changes supportive of evolving rural livelihoods. KVK under the umbrella of the National Agricultural Research and Education System are the true carriers of such frontline technologies and impart knowledge and critical input support for the farmers (Anonymous 2015).

Dissemination of technologies through a group of farmers is an effective means as it can reach more farmers in a single shot. In this direction, the initiative started as the collectivization of producers, especially small and marginal farmers, into farmers' producer organizations (FPO) has emerged as one of the most effective pathways to address the many challenges of agriculture but most importantly, improved access to investments, technology and inputs and markets. Department of Agriculture and Cooperation, Ministry of Agriculture, Govt. of India identified farmer producer organization registered under the special provisions of the Companies Act, 1956 as the most appropriate institutional form (NABARD 2015). These organizations mobilize farmers and build their capacity to collectively leverage their production and marketing strength. There are FPOs formulated for different crops which can help in the dissemination of the latest technologies related to that crop right from sowing to harvesting, processing and marketing. Many of the states in the country documented many success stories with respect to FPOs. Some of the success stories through FPOs in the country include the establishment of coconut drying unit, production of dehydrated tomato powder, banana fibre extraction, grading and marketing of grapes, marketing of vegetable crops, mushroom production technology, etc.

Information and communication technologies (ICT) are the most natural allies to facilitate the outreach of the Agricultural Extension System in the country (Singh et al. 2015). Despite large, well-educated, well-trained and well-organized agricultural extension manpower, around 60% of farmers in the country still remain un-reached, not served by any extension agency or functionary. Of the 40%, who have some access to Agricultural Information, the major sources of this information are Radio and Television. The telephone has just started to make its presence felt in this scenario. Internet supporting Information-Kiosks are also serving the farming community, in many parts of the country. Hence ICTs are highly relevant for Agricultural Extension scientists, researchers, functionaries and organizations.

Through the use of ICT in agriculture many success stories have been documented. Some success stories in this context are the management of grape diseases through weather-based forecasting system, disease monitoring and management of apple scab through weather forecasting, e-auction of flowers, use of Krishi portals for effective technology dissemination, development of Decision Support System, mobile-based apps, etc.

11.8 Urban and Peri-Urban Horticulture

More than 40% of the Indian population live in urban areas by 2030, and this may grow up to 65% by 2050. Hence, horticulture in urban and peri-urban areas becomes most important to meet the need for fruits, vegetables, flowers and other components of horticulture (Sumangla et al. 2013). Looking into this, the Government of India emphasized vegetable and flower production in such areas and many working groups have been constituted. As an outcome of many deliberations, it has been advocated to give attention not only to fruit and vegetable production but also to environmental services and health care. Roof and terrace gardening is finding now a place to get fresh fruits, vegetables and flowers, and also for environmental services. Vegetable production has expanded in and around cities as an informal activity practiced by poor and landless city dwellers. The broad diversity of horticultural crop species allows year-round production, improved employment and income. Growers have realized that intensive horticulture can be practiced on small plots, making efficient use of limited water and land resources. In addition, due to their short cycle vegetables provide quick to emergency needs for food. Leafy vegetables provide a quick return to meet a family's daily cash requirements for purchasing food. Because they are produced in close vicinity of consumers post-harvest losses can be reduced significantly. Mushroom production is one of the promising activities in urban and peri-urban areas. Appropriate management of water, land, air and city wastes helps in improving hydrological functioning through soil and water conservations and microclimate and avoids costs of disposal of recycled urban wastes. Identifying suitable avenue trees and planting them in urban areas can reduce the air pollution level.

The lack of vegetation in urbanized areas, as a result of human establishments, directly affects the quality of life, from a physical and aesthetical point of view. The construction of vertical gardens is recommended both in interiors and especially in the exterior of buildings. By applying these technologies, any kind of area can be used at its maximum capacity. Even if the price of constructing and maintaining vertical gardens is higher than a classical landscape it's compensated by the environmental benefits, raising the vegetation surfaces, with an impact on reducing the pollution effect. The new modern concepts for landscape development are keen on using any kind of concrete or glass, turning them into real vertical gardens, being possible to overcome the development of the urban areas making a smooth transition for a healthy green urban environment. Vertical Gardening is a special kind of urban gardening suitable for small spaces, particularly for decorating walls and roofs in various styles. This is an alternative method for gardening by expanding the scope of growing plants in a vertical space. Intensive urbanization has left hardly any horizontal space for outdoor gardens. Green walls are not only spectacularly beautiful but also helpful in enlivening the ambiance. Green walls can absorb heated gas in the air and lower both indoor and outdoor temperatures, providing healthier indoor air quality as well as a more beautiful space.

11.9 Conclusions and Way Forward

Horticulture in India was a subsistence farming prior to independence and has emerged today as one of the important components in Indian agriculture contributing about 30% of India's GDP and providing about 37.1% of the total export of agricultural commodities. The historical growth rate of horticultural crop production accounts for 2.7% per year which may be attributed to improved production technologies, favouring government policies to encourage horticultural growth in terms of establishment of research infrastructures, increased budget allocation to the horticultural sector, establishment of processing and package industries, cold storage, etc. Still, several constraints are being faced by the horticulture industry with respect to global climate change, post-harvest losses, insufficient trained manpower, etc. Hence, there is a strong need to strengthen the research on horticultural crops to develop demand-driven technology by improved variety, pest management, etc., in both public and private sector institutes. But with constant efforts, this sector can be developed as an organizational industry and has to be managed by stakeholders with farmers as entrepreneurs. Keeping in view the great potential of horticulture in doubling farmer's income, the Government of India has allocated INR 2250 crores for horticulture development for 2021–2022 for 'Mission for Integrated Development of Horticulture' (MIDH), a centrally sponsored scheme. The Ministry is implementing MIDH with effect from 2014–2015, for realizing the potential of the horticulture sector covering fruits, vegetables, root and tuber crops, mushrooms, spices, flowers, aromatic plants, coconut, cashew and cocoa. Government intervention in the horticulture sector has led to a situation wherein horticulture production has surpassed agriculture production in the country. During the year 2019–2020, the country recorded its highest-ever horticulture production of 320.77 million tonnes from an area of 25.66 million hectares. As per the first Advance Estimates for 2020–2021 the total horticulture production in the country is 326.58 lakh MT from an area of 27.17 lakh ha.

In spite of achieving higher production over agricultural production, this sector is still facing several challenges with respect to huge post-harvest losses and gaps in post-harvest management and supply chain infrastructure. There is tremendous scope for enhancing the productivity of Indian horticulture which is imperative to cater to the country's estimated demand of 650 million MT of fruits and vegetables by the year 2050. Some of the new initiatives like focus on planting material production, cluster development programme, credit push through Agri Infra Fund and formation and promotion of FPOs are the right steps in this direction. Thus, there are options of opportunities and challenges, which will need attention. The issues to be addressed are: Innovations in technologies through institutional support as well as the import of knowledge and technology backed by the development through skills. Development strategies should be for cluster approach linked with post-harvest management and marketing, quality seeds and planting material, precision farming and smart horticulture, environmentally controlled horticulture, efficient management of nutrients and water, and enhanced ICT use to add efficiency to input management including use of Block Chain Technology and artificial intelligence and knowledge transfer.

References

Agricultural Situation in India, 2017. VOL. LXXIV August, 2017 No. 5 pp: 60. (Ed: P.C. Bodh)

Anonymous (2015) KVKs impact on dissemination of improved practices and technologies. In report of National Institute of Labour Economics Research and Development (NILERD), pp 302. http://iamrindia.gov.in/writereaddata/UploadFile/KVK0Final0Report020150December022_1648.pdf

Anonymous (2017a) Horticultural Statistics at a glance 2017. Government of India, Ministry of Agriculture and farmers welfare, Ministry of Agriculture and cooperation and farmers welfare, horticulture statistics division, pp 481. https://nhb.gov.in/statistics/Publication/Horticulture%20At%20a%20Glance%202017%20for%20net%20uplod%20(2).pdf

Anonymous (2017b) ICAR-IIHR, News Letter, 2017, vol 38(2), pp 3. https://iihr.res.in/sites/default/files/October%20to%20December%20Newsletter%202017.pdf

Barrera A (2011) New realities, new paradigms: the new agricultural revolution. In: Communicate (Technological Innovations), pp 10–21

Bharadwaj C, Goswami AK, Kumar N (2014) Molecular mapping for important traits in fruit crops. In: Jaiprakash GAK, Singh SK (eds) e Book: Recent Advances in breeding approaches and varietal development of perennial horticultural crops

Bhutia, ND, Bhutia KL, 2017. Pyramiding of disease resistance genes in vegetables. Biotech Articles https://www.biotecharticles.com/Agriculture-Article/Pyramiding-of-Disease-Resistance-Genes-in-Vegetable-Crops-4181.html

Chadha KL (2014) Revisiting breeding priorities and strategies in perennial horticultural crops of India. In: Jaiprakash GAK, Singh SK (eds) e book: recent advances in breeding approaches and varietal development of perennial horticultural crops

Chand R (1999) Agriculture diversification in India. Mittal Publications, New Delhi

Chengappa PG (2013) Secondary agriculture: a driver for growth of primary agriculture in India. Indian J AgriEcon 68:1–19

Datta S (2013) Impact of climate change in Indian horticulture – a review. Int J Sci, Environ Technol 2:661–671. https://doi.org/10.2139/ssrn.2570710. https://www.researchgate.net/publication/273242368

Jha GK, Suresh A, Punera B, Supriya P (2018) Growth of horticulture sector in India – trends and prospects. Indian J Agric Sci 89(2):314–321

Joshi PK, Birthal PS, Minot N (2006) Sources of agricultural growth in India: role of diversification towards high value crops. In: MTID discussion paper no. 98, IFPRJ, food policy, Washington D.C., pp 20–23

Kumar N (2015) Breeding of horticultural crops: principles and practices (2nd Edn). New India Publishing Agency, p 219

NABARD Report, 2015. Farmer producer organizations - frequently asked questions (FAQs)' Farm Sector Policy Department & Farm SectorDevelopment Department, NABARD Mumbai

Philroy J, Karuppanchetty SM, Nancy D, Bhubesh Kumar G, Aravazhi R (2014) Agribusiness incubation transforming Indian agriculture. In: A business incubation approach of NARS – BPDs. ICARISAT, Hyderabad, p 65

Prabhakar et al (2014) Vision 2050. ICAR-Indian Institute of Horticultural Research, Benglauru, p 21

Saikia A, Bordoloi R (2014) Strategic Interventions on Horticultural Sector for Doubling Farmers Income in Assam by 2022–2023. Acta Sci Agric 3.1:56–58

Sharma KK, Karuppanchetty SM, Aravazhi S (2014) Developing entrepreneurs through an Agribussiness incubator at ICRISAT. A report published by ICARISAT, Hyderabad, India

Singh, K.M, Abhay Kumar and Singh, R.K.P.,2015. Role of Information and Communication Technologies in Indian Agriculture: An Overview, MPRA Paper, University Library of Munich, Germany

Sumangla HP, Malhotra SK, Chowdappa P (2013) Urban and peri-urban horticulture-a perspective. In: Confederation of Horticulture Associations of India, New Delhi, p 94. https://pib.gov.in/PressReleaseIframePage.aspx?PRID=1717447

Livestock Sector 12

S. K. Mahanta and Shilpi Kerketta

Abstract

Livestock is an integral part of the socio-economic fabric of rural India since ancient times, being a source of livelihood and a provider of draft power, manure and fuel. Sustained increases in income and urbanization have led to rapid growth in demand for animal food products, and now the livestock sector is under pressure to produce more and more. Over the last few decades, livestock production sector grew faster than crop sector as a whole and made a significant contribution to overall agricultural growth, which is considered to be a vital factor in poverty reduction in most developing countries including India. India's livestock sector is one of the leading in the world with a holding of 11.6% of the world livestock population as per the nineteenth livestock census report, and performing well with respect to production, value addition and export of milk, meat and other products. The gross value added (GVA) of livestock sector was about INR 962,682 crore at current prices during the financial year 2019–2020, which was about 28.36% of agricultural and allied sector GVA and 5.21% of total GVA. This chapter deals with the historical progression of Indian livestock development, programmes/activities initiated under three major eras, pre-independence and independence era, green revolution era and post-green revolution era, and issues that need to be resolved in the future for improved livestock production.

S. K. Mahanta (✉) · S. Kerketta
School of Animal and Fishery Sciences, ICAR-Indian Agricultural Research Institute, Gauriakarma, Hazaribagh, Jharkhand, India

P. K. Ghosh et al. (eds.), *Trajectory of 75 years of Indian Agriculture after Independence*, https://doi.org/10.1007/978-981-19-7997-2_12

Keywords

Livestock sector · Growth and development · National livestock mission · Future needs · Policy issues

Livestock rearing along with crop farming is an integral part of human life since the beginning of civilization. It has enormously contributed to the food basket and draft animal power, besides maintaining ecological balance. Due to the conducive climate and topography, livestock rearing has played a prominent socio-economic role in India. It has a significant role in generating gainful employment in the rural sector, particularly among the resource-poor farmers and women, besides supplying low-cost and quality food to millions of people. Indeed, livestock production and crop farming are inherently linked, each being dependent on the other, and both are vital for overall food security. It supports crop farming in the form of critical inputs, contributes to the health and nutrition of the household, supplements incomes, offers employment opportunities, and acts as a dependable 'bank on hooves' in times of need. Thus it acts as a supplementary and complementary activity to crop farming in India (Anonymous 2019).

To know the dynamics of the process of development inherent in a given socio-economic scenario, it is necessary to place the present situation in the appropriate historical context. A historical progression of livestock rearing in India is expected to help us understand the process of development through which the present situation has emerged. The historical progression can be divided broadly into three major eras, namely the pre-independence and independence era, the green revolution era and post-green revolution era.

12.1 Pre-Independence and Independence Era

A reasonably stable structure of livestock and crop farming in India was destroyed by the colonial exploitation during the imperial British rule during the eighteenth and nineteenth century. In the nineteenth century, the twin processes led to the destruction of important Indian hand industries, mainly textiles, and the conversion of the Indian agrarian economy into a source of raw materials for the workshops in Britain. This general phenomenon also resulted in destruction of traditional communal grazing lands and adversely affected the Indian milk economy. Indeed, Britishers changed the use of land which was traditionally meant for the cultivation of grains and fodder crops. Hundreds of acres of land were used for cash crops such as cotton and jute for the workshops in Britain. During this period, there was a proportionate increase in the acreage devoted to cash crops at the expense of food grains (Habib 1963). Division of lands with the division of households also started. Dependence on agriculture and on smaller and smaller pieces of divided land kept increasing. Intensive cropping for grains essential for living increased. Gradually by the beginning of twentieth century, the situation was such that most of the villages lacked suitable grazing areas and even there were no forest grazing grounds for livestock. Fodder crops represented only 4% of the cropped area.

The change in the scenario of livestock farming in India was started primarily as a consequence of the demands of British military troops. Voelkar (1893) a consulting chemist to the Royal Agricultural Society of England recommended the establishment of dairy farms for the supply of milk to troops and government institutions in 1893. An imperial dairy expert was appointed in 1920, whose responsibilities included evaluating the status of the sector and outlining feasible growth plans for the dairy industry. Gradually the move of commercializing dairy industry in India caught speed and various reports from the Royal Commission on Agriculture, Colonel Olver, an animal husbandry expert of the Imperial Council of Agriculture and the Wright Report were placed in this direction.

12.1.1 Recognition of Milk as a Marketable Commodity

Finally at the beginning of twentieth century, milk recognized as marketable commodity. Otherwise, milk was traditionally produced and consumed for immediate needs. The primary impetus for this change came from two distinct demand centres, viz., the British troops stationed in India and the growing urban centres like Bombay. Reports available on the existing conditions revealed that despite the fact that 90% of the human population and 95% of the bovine population lived in rural areas, practically no milk was assembled there for sale. Milk is a perishable commodity and requires quick, regular and special transport. A lack of adequate facilities for such a means of transport that milk produced in the interior could not be economically used for meeting urban demand. Even though fluid milk fetched a higher price in the market, rural producers were forced to convert it into less perishable, but also less profitable products. In the 1930s only 31.2% of the total milk produced was used as liquid milk and 68.8% was converted into various products (Wright 1937). The first systematic attempts to link rural production with the urban market were made in the 1940s, primarily by establishing collection centres, transport facilities and distribution link between consumers in Bombay and producers in rural areas north of Bombay.

The military farms gave requisition for land, bought the best young stocks and obtained abundant fodder supplies from adjacent villages. According to Pepperall Report, the prices they paid for the fodder frequently induced the villagers to sell a major proportion of their fodder crops and thereby deprived their own stock (Anonymous 1946). Thus the fodder economy started to be commercialized. The area under fodder crops increased by 51% from 1915 to 1937 (Table 12.1). However, the situation was markedly different in the case of fodder crops as the area under other crops such as sorghum and gram declined by 6% and 12%, respectively. However, this increase in the acreage under fodder, grown mainly as a commercial crop, was not able to compensate for the destruction of communal grazing lands, since an increasing fodder shortage was observed in this period. By 1944–1945, the fodder shortage became an acute problem (Table 12.2). This problem was further

Table 12.1 Area (million acres) under fodder crops[a] in India (1915–1942) (Burns 1944)

Year	Area
1915–1916	7.0
1918–1919	7.0
1936–1937	10.6
Average (1915–1942)	9.1

[a]Fodder crops included oats, fodder sorghum, lucerne, guinea grass, etc.

Table 12.2 Demand and supply (million tons) of feedstuffs in the year 1944–1945 (Anonymous 1945)

Feedstuffs	Demand	Supply
Green fodder	226.50	78.10
Straw	127.88	87.20
Oil seed cakes and seeds	30.87	3.40
Cereal by-products	20.85	1.47

aggravated by the export of oilseeds and oilcakes. As a result, the price of milk was increased sharply. From 1939 to 1942, the average increase in the prices of milk was 50% and in 1947, the prices further increased by 230% compared to that in 1939, in the major cities (Anonymous 1949).

12.1.2 Initiation of Crossbreeding in Cows

In 1935 out of the total of 24,596,116 adult cattle, 49% were cows and out of total 5,709,614 adult buffaloes, 82% were females. The all-India average lactation yield of cows was 188 kg in 1937, whereas in the case of buffaloes it was 500 kg. However, buffaloes' contribution to the total milk produced was 54.4%. But the number of buffaloes was much smaller than that of cows. Thus buffaloes were more important than cows as milch animals during this period. Despite this predominance of buffalo as the milch animals in India, the first half of twentieth century was an era in which efforts to improve indigenous cows through crossbreeding were started. Since the Britishers were used to cow's milk and products, made from cow's milk such as cheese. Some degree of mechanization in agriculture was also started by the second quarter of the twentieth century. The use of oil engines, mechanized transport, power-driven oil mills, had slowly started replacing bullock as a source of draft power. As a result, the milch values of indigenous cows became comparatively more important than the draft power of indigenous bullocks. Moreover, it was also technologically possible to improve cows by cross-breeding because good European milch breeds of cows were available, while good buffalo breeds were not available.

The Imperial Department of Agriculture and Provincial Agricultural Department attempted cross-breeding Indian cattle with European breeds before the First World War. The experiment was found successful in raising the milk yield of indigenous cows but it was observed that the susceptibility of the crossbreds to diseases prevalent In India increased as compared to the pure Indian breed. Cross-breeding became an accepted practice in military dairy farms. The military farms established a technological base from which further activities related to cross-breeding by artificial insemination (AI) came up. However, a systematic and thorough study of AI with reference to Indian conditions started at the Imperial Veterinary Research Institute via an Indian Council of Agricultural Research–sponsored scheme only in 1942. This detailed study on the collection, preservation and transport of semen to rural areas and the technique of AI revealed that it was possible to introduce AI in India, Consequently, after the establishment of major institutes at Calcutta, Patna, Bangalore, etc. in 1945, it was recommended that State Governments may set up AI centres in selected areas having high cattle density and adequate transport facilities.

12.1.3 Planned Livestock Development

The Government played a crucial role in the planned development of agriculture in general and livestock/dairy development in particular, since independence. On the one hand, it ensured the cheap and easy availability of essential- commodities for the consumers at large; on the other hand, it also managed that the producers get their fair returns for milk and other agricultural products. Accordingly, the supportive prices were declared by the government so that no one is a loser. Indeed, in the first post-independence phase of planned livestock development, the major importance was firstly given to improving the indigenous breeds through selective breeding and secondly to organize the milch herds located in metropolitan towns on more hygienic and scientific lines. During the early 1950s, the Government adopted a policy of discouraging cross-breeding of cows and encouraging selective breeding of indigenous milch breeds. Various military and Government farms were instructed to discontinue cross-breeding of cows. Simultaneously, programmes to maintain quality bulls and castration of 'undesirable sire stocks' were launched under the Key Village Centre Schemes. This was just opposite to cross-breeding of cows in the immediate post-independence era and perhaps due to the close linkage between cross-breeding and the needs of British troops during the last phase of colonial rule.

12.2 Green Revolution Era

When the British left India in 1947, India continued to be haunted by memories of the Bengal Famine, the world's worst recorded food disaster that occurred in 1943 under British rule. It was, therefore, natural that food security was one of the main items on free India's agenda. This awareness led, on one hand, to the Green Revolution (GR) in India, and on the other, legislative measures to ensure that businessmen will never again be able to hoard food for reasons of profit. The GR, spreading over the period from 1967–1968 to 1977–1978, changed India's status from a food-deficient country to one of the world's leading agricultural nations. Until 1967 the government mainly gave attention to expanding the farming areas. But the growth of the population was much faster than food production. This called for immediate and drastic action to increase production. The action came in the form of the GR. The GR resulted in a record grain output of 131 million tonnes in 1978–1979. Yield per unit of farmland enhanced by more than 30% between 1947 (when India gained political independence) and 1979. The crop area under high-yielding varieties of wheat and rice increased considerably. It also created plenty of jobs not only for agricultural workers but also for industrial workers through the creation of related facilities such as factories and power stations.

During the GR era, the growth and development of the livestock sector were also unique. By mid-1960s it became clear that the strategy of encouraging selective breeding of indigenous milch breeds was not yielding the desired results and per capita availability of milk was continuously declining. By this time, the Anand experiment showed that it was feasible to link rural producers to urban markets. The cross-breeding technology was then possible to be taken to the doorsteps of milk producers because of the developments, in the field of cryogenics, particularly the manufacture and transport of liquid nitrogen. Another event of crucial importance was the excess milk production in the European Economic Community (EEC) countries and consequently their anxious search for markets for milk, milk products and dairy equipments. As a result of all these factors, a new strategy was adopted in the early 1970s. This strategy had two major elements- (a) linking the rural producers with urban consumers by creating an integrated infrastructure for the collection, storage, processing, transport and distribution of milk, and (b) importing the semen of proven western dairy breeds and establishing a delivery system for carrying out artificial insemination in the villages, A major investment component for this strategy was to be financed by loans and donations from the EEC countries in form of milk powder. Besides this, the increase in livestock production was also governed by several schemes run during this period as described below.

12.2.1 Operation Flood

One of the world's largest rural development programme, operation flood was launched in 1970. It helped dairy farmers direct their own development, placing control of the resources they create in their own hands. A national milk grid linked milk producers throughout the country with consumers in over 700 towns and cities, reducing seasonal and regional price variations while ensuring that the producer gets fair prices in a transparent way on a regular basis. The main foundation of operation flood was village milk producers' cooperatives, which procured milk and provided inputs and services, making modern management and technology available to members. Operation flood's main goals were increase in milk production (a flood of milk), augmenting rural incomes and providing reasonable prices for consumers. Indeed from the outset, operation flood was conceived and implemented as much more than a dairy programme. Rather, dairying was seen as an instrument of development, generating employment and regular incomes for millions of rural people. It was implemented in three phases.

In phase I (1970–1980), it was financed by the sale of skimmed milk powder and butter oil gifted by the European Union then EEC through the World Food Programme. NDDB planned the programme and negotiated the details of EEC assistance. During its first phase, it linked 18 of India's premier milk sheds with consumers in India's four major metropolitan cities namely Delhi, Mumbai, Kolkata and Chennai. In later phases, the number of milk sheds increased from 18 to 136 and 290 urban markets expanded the outlets for milk. Thus a self-sustaining system of 43,000 village cooperatives covering 4.25 million milk producers came up. Domestic milk powder production increased from 22,000 tonnes in the pre-project year to 140,000 tonnes, In this way EEC gifts and World Bank loans helped to promote self-reliance. Direct marketing of milk by producers' cooperatives was increased by several million litres a day. Later dairy cooperatives were again expanded and strengthened with the infrastructure required to procure and market increasing volumes of milk. Veterinary first-aid health care services, feed and artificial insemination services for cooperative members were extended, along with intensified member education. Milksheds even reached 173 with the numbers of women members and Women's Dairy Cooperative Societies increasing significantly.

12.2.2 All India Coordinated Research Project (AICRP) on Cattle

This project was initiated in 1968 with the aim to improve the production performance of cattle maintained under farm and field conditions. Under this project, Field Progeny Testing (FPT) was initiated to bring about genetic improvement in cross-bred cattle at farmers' herds in collaboration with the SAUs, State Government, and Non-Government organizations. This project contributed a lot in enhancing the productive capacity of milch cattle in the country and presented a remarkable share of the cattle population too.

12.2.3 All India Coordinated Research Project (AICRP) on Buffalo

This project started in 1970 with the objective to improve the production potential of buffaloes through the assessment of genetic merit of sires. It was initiated with the selection of bull based on progeny selection and its distribution in the field to evaluate the performance at several State and Central Government farms. The project at various institutes contributed in increasing the elite buffalo breeds and their production performance.

12.2.4 All India Coordinated Research Project (AICRP) on Sheep and Goat

This programme was initiated in the year 1971, which led to steady growth of wool and hair by around 2% during the 1970s and 4% in the 1980s and 1990s. But in the later phase growth rate showed a decreasing trend that remained below 1% which could be due to liberalization of the import of wool at a low rate of 15% of duty leading to an adverse impact on domestic output.

12.2.5 Special Livestock Breeding Programme

The programme was launched in 1975–1976 with an objective to enhance the production of various livestock products like milk, eggs, wool, etc. and to provide employment opportunities and supplement income to the weaker sections of the rural poor. The programme includes crossbred heifer rearing along with the setting up of other livestock production units. Simultaneously the scheme also aimed at supporting the landless agricultural labourers and marginal and small farmers in improving the quality of crossbred heifer calves. Thus the programme assisted in improving the performance of female calves from a young age to fully exploit their utility when matured.

12.2.6 Programme on Fattening of Male Buffalo Calves

Meat production showed tremendous growth during the 1960s as compared to 1950s. However, its growth remained passive during the 1970s, when total livestock output witnessed a substantial acceleration in growth. Increment in meat production growth was observed around 1980–1981 as a result of a phenomenal increase in buffalo meat which was supported through this programme and taken up under the aegis of Hind Agro Industries Limited.

12.3 Post-Green Revolution Era

The green revolution had a massive impact on the social and economic development of people whose consequence was observed during post green revolution era. The modification of agriculture practices such as intensification over the years led to overall disturbance in delicate agro-ecosystem. Although it resulted in three times more production of cereal crops with a slight increase in cultivated land (30%). Additionally, there were significant impacts on poverty reduction and lower food prices. But the combination of high production cost with low economic returns from agricultural practices are directly affecting the livelihood of farmers. The green revolution made India a country of self-sufficiency in cereal grains, thereby storing surplus unused grains as livestock feed. It has also a huge impact on the feeding behaviour of humans, as millets once cultivated as a major cereal crop to alleviate malnutrition now is replaced by rice as a staple food, making millets as a fodder crop. The Food and Agriculture Organization (FAO) has recorded that over the years 1961–2017, there are a decrease in the production of millet and an increase in the production of rice; thus, rice became the staple diet of the country. Ultimately the GR increased the availability of fodders and grains for sustaining more livestock and enhanced their productivity.

12.3.1 Growth Trend in Livestock Population

Over the years, the need for livestock products has increased significantly being attributed to rapid urbanization, changing lifestyles, rising income levels, increased nutritional awareness among people, etc. Fortunately with the ever-increasing food needs of a growing human population, there is also an increase in almost all the livestock species across the world including India to satisfy the growing demands. As per the nineteenth Livestock Census (BAHS 2014) India's livestock sector is one of the largest in the world with a holding of 11.6% of the world livestock population which consists of buffaloes (57.83%), cattle (15.06%), sheep (7.14%), goats (17.93%), camel (2.18%), equine (1.3%) and pigs (1.2%). Indeed, India's position in the world in terms of livestock population is quite impressive, being first in the case of cattle and buffalo, second for goats and third for sheep, and tenth in the case of camel population.

There was a remarkable growth in livestock population (million heads) from 1951 (292.80) to 2019 (535.82). The twentieth livestock census released by the Department of Animal Husbandry and Dairying, Government of India showed that the livestock population increased by 4.6%, from 512.06 million in 2012 to 535.82 million in 2019 with an annual growth rate of 0.66%. This increment was led by a sharp increase in a number of small ruminants like sheep and goats, which was nearly 95% of the total livestock increase. Based on the species-wise livestock population during this period (Table 12.3), it was observed that the number of cattle gradually increased from 1951 (155.30 million) to 1992 (204.58 million), but it significantly decreased to 190.90 million in 2012. But the notable fact was that number of cattle increased in 2019 with an annual percentage growth rate of 0.12%. Among cattle, there was a marked change from indigenous cattle to crossbreds. Crossbred cows had grown at a much faster rate than indigenous stock. According to the livestock census 2019, there was a 6% decline in a total number of indigenous cattle over the previous census despite the launch of government schemes like the Gokul Mission for the improvement of indigenous breeds. However, the pace of decline of the indigenous cattle population during 2012–2019 was much lesser as compared to 2007–2012 which was about 9%. On the contrary, the population of total exotic or crossbreed cattle increased by 26.9% in 2019 as compared to the previous census of 2012. The decline in numbers was due to people's abandoning indigenous cattle. The population of buffalo increased from 43.40 million in 1951 to 109.85 million in 2019. This trend clearly showed that now-a-days people are rearing more buffaloes than cattle and other livestock due to globalization. For meat purposes, buffaloes have more commercial value than others in world trade. The annual percentage growth rate of buffalo significantly declined to 0.15 from 2012 to 2019 which was 1.45 from 1992 to 2012. A striking feature of the Indian livestock sector was that number of sheep and goats consistently increased.

Table 12.3 Trends in livestock population (millions head) from 1951 to 2019 (Mondal and Mishra 2019)

Species	1951	1972	AGR (%)	1992	AGR (%)	2012	AGR (%)	2019	AGR (%)
Cattle	155.30	178.30	0.71	204.58	0.74	190.90	−0.33	192.52	0.12
Buffalo	43.40	57.40	1.54	84.21	2.34	108.70	1.45	109.85	0.15
Sheep	39.10	40.00	0.11	50.78	1.35	65.07	1.41	74.26	2.02
Goat	47.20	67.50	2.04	115.28	3.54	135.17	0.86	148.88	1.45
Horse	150	0.90	−1.90	0.82	−0.44	0.63	−1.16	0.34	−6.58
Camel	0.60	1.10	3.97	1.03	−0.32	0.40	−3.06	0.25	−5.36
Pig	4.40	6.90	2.71	12.79	4.27	10.29	−0.98	9.06	−1.71
Mule	0.06	0.08	1.59	0.19	6.88	0.20	0.26	0.08	−8.57
Donkey	1.30	1.00	−1.10	0.97	−0.15	0.32	−3.35	0.12	−8.93
Yak	–	0.04	–	0.06	2.50	0.08	1.67	0.06	−3.57
Total	292.80	353.60	0.99	470.86	1.66	512.06	0.44	535.82	0.66

AGR annual growth rate

According to the latest livestock census of 2019, sheep and goats accounted for 13.87 and 27.80% of total livestock in the country. During the last inter-census period number of sheep increased from 65.07 million in 2012 to 74.26 million in 2019 with a sharp annual percentage growth rate of 2.02%. While during the same time frame the number of goats increased from 135.17 to 148.88 million at a 1.48% annual percentage growth rate which was quite significant. On the contrary, the numbers of other livestock like donkeys, camels, pigs, horses and mules declined during this period.

12.3.2 Growth Trend in Milk and Meat Production

Presently livestock is one of the fastest-growing agricultural subsectors in developing countries including India. The global market for milk and meat is growing fast which is driven by population growth, rapid urbanization and increasing incomes in developing countries. This is an opportunity for India to improve its participation in the global market. India is the world's top milk-producing nation since 1998 and it accounts for 22% of the world's total milk production according to the UN's Food and Agriculture Organization (FAO). The country improved its milk production from 17 million tons only in 1951 to 187.7 million tons in 2019 (Table 12.4). The growth in milk production remained sluggish for more than two decades after independence (around 1% per annum) whereas the growth of the population was closer to 2% (BAHS 2014). The substantial improvement in milk production was achieved with the launch of a nationwide dairy development programme (operation flood) in the year 1970. During the last inter-census period, the production of milk increased from 127.9 million tonnes in 2012 to 187.7 million tons in 2019 at a 5.63% compound annual growth rate which was quite impressive.

Unlike milk, meat production in India mainly falls under the unorganized sector. India is the sixth largest producer of meat in the world. According to the livestock census report, the growth rate of meat production was sluggish up to 2001–2002 in India. But in 2002–2003, first time it increased significantly from 1.9 million tonnes to 2.1 million tonnes. But the highest compound annual growth appeared during 2007–2008 which was 73.91%. But it failed to maintain its growth rate afterwards.

Table 12.4 Trends in production (million tonnes) and growth of milk and meat from 1951 to 2019 (Mondal and Mishra 2019)

	Milk		Meat	
Year	Production	CAGR (%)	Production	CAGR (%)
1951	17.0	–	–	–
1972	23.2	1.49	–	–
1992	55.7	4.48	–	–
2012	127.9	4.24	5.5	–
2019	187.7	5.63	8.1	5.69

CAGR compound annual growth rate

However, it was noticeable that the production of meat increased in our country from 5.5 million tons in 2012 to 8.1 million tons in 2019 with a 5.69% compound annual growth rate. During the year 2020, it further increased to 8.60 million tons with a positive growth rate of 5.98%. Meat from Indian animals contains less fat, and while the present international trend favours low-fat meat, thus Indian meat industry has a promising future in the global market.

Besides continuation of earlier programmes on livestock development, several new measures/schemes (Anonymous 2021a) were initiated during post green revolution era by the Government to increase the productivity of livestock, which led to increase in milk production significantly to 198.40 million tons with an annual growth rate of 5.68% in 2020 and India continued to be the largest producer of milk in the world. The gross value added (GVA) of the livestock sector reached about INR 962,682 crore at current prices during the financial year 2019–2020, which was about 28.36% of agricultural and allied sector GVA and 5.21% of total GVA (Anonymous 2021b).

12.3.3 National Livestock Mission (NLM)

This scheme was started in the financial year 2014–2015 and revised from time to time. It aims towards employment generation, entrepreneurship development, increase in per-animal productivity and thus targeting increased production of meat, milk, and wool under the umbrella scheme development programme. The surplus production will assist in the export earnings after meeting the domestic demands. National livestock mission plans to develop entrepreneurs to create forward and backward linkage for the produce available at the unorganized sector and to link with the organized sector. It has three sub-missions. Sub-mission on breed development of livestock aims at entrepreneurship development and breed improvement in sheep, goat and piggery by providing the incentivization to the individual, FPOs, FCOs JLGs, SHGs, etc. for entrepreneurship development and also to the state government for breed improvement infrastructure. While sub-mission on feed and fodder development aims towards strengthening of fodder seed chain to improve the availability of certified fodder seed required for fodder production and encouraging entrepreneurs to establish units of fodder block/hay bailing/silage making through incentivization. The last sub-mission on innovation and extension aims to incentivize the institutes, universities and organizations carrying out research and development related to sheep, goat, pig and feed and fodder sector, extension activities, livestock insurance and innovation.

12.3.4 Rashtriya Gokul Mission (RGM)

It was implemented for the development and conservation of indigenous bovine breeds in December 2014. This mission is very important for the betterment of resource-poor as more than 80% low producing indigenous animals are with small and marginal farmers and landless labourers. It is important for enhancing milk production and productivity of bovines to meet the growing demand for milk and making dairying more remunerative to the rural farmers of the country. It is leading to the multiplication of elite animals of indigenous breeds and increased availability of indigenous stock. It is proposed to be continued under the umbrella scheme Rashtriya Pashudhan Vikas Yojna. The RGM is expected to enhance productivity and benefit the programme, percolating to all cattle and buffaloes of India, especially with small and marginal farmers. It will also benefit women in particular since over 70% of the work involved in livestock farming is performed by women.

12.3.5 Promoting Sex-Sorted Semen

With the mechanization of crop farming, the utility of male cattle and buffaloes has been reduced. Livestock farmers are not willing to maintain bullocks/male buffaloes for agriculture or any other draft work. Hence, male calves born at farmer's houses have become a liability. Farmers often let the male calves loose which are resulting into increase in the stray animal population. Only female calves can be produced (with more than 90% accuracy) by use of technology like sex-sorted semen in an artificial insemination programme. Extensive use of this technology is also expected to increase the number of female animals, thereby improving the income of farmers through the sale of female animals or by the sale of milk.

12.3.6 Implementing National Digital Livestock Mission (Livestock)

Livestock-related activities and transactions including health and breeding services, sale and purchase, etc. should take place in entirely digital mode based on unique animal ID Pashu Aadhaar which is currently being assigned through ear tagging to all large and small animals throughout the country. A complete open-source tech stack is expected to enable inputs of all reporting including disease and outbreak reporting by veterinarians and field-level workers and service providers through a user-end digital interface. Farmers can access their own data, make service requests, and access the latest technical and business information through an updated version of the farmer-facing app e-GOPALA or through a connected national level four-digit call centre number. Since all farmers will be linked, direct benefit transfers from all central or state-level schemes, and e-vouchers giving the power to choose a service provider, will also be possible through this mechanism. Even dairy processors, other private companies, app developers, and researchers can access the database through established data-sharing standards, and product traceability regulations can be designed and enforced on the basis of this database. Indeed, not only the ease of working and accountability of service providers is increased manifold, and farmers are fully empowered, but also the entire economy around livestock is multiplied through this database.

12.3.7 National Animal Disease Control Programme (NCDP)

India's livestock wealth is quite impressive, but the prevalence of animal diseases is a serious impediment to the growth of the livestock sector. Losses are humongous and often beyond estimation due to diseases like foot and mouth disease (FMD), brucellosi, etc. It is because of FMD that there is not only a reduction in milk production and trade in livestock products but also there is infertility and a reduction in the quality of hides and skins of the animals, including their draught power. Thus FMD has a direct negative impact on the trade of milk and other livestock products. Similarly, brucellosis is another important disease of livestock resulting in huge financial losses and has an adverse impact on human health, as it has zoonotic potential. Farm workers and livestock owners are always at risk of contracting as well as spreading this disease. Hence, control of Brucellosis will have a double impact, both in human and livestock health, besides rich economic gains to the animal owners/farmers. Hence, it is crucial to control FMD by vaccination of all cattle, buffaloes, goats, sheep and pigs and brucellosis by vaccination of all female bovine calves (4–8 months old) in the country. This will not only make animals healthy but will also result in better productivity and acceptability of our animal products the world over. Besides, efforts in this direction will lead to contribute towards improving farmers' income.

12.3.8 Livestock Health and Disease Control Scheme

This scheme aims to improve the animal health sector by way of the implementation of prophylactic vaccination programmes against various diseases of livestock, capacity building, disease surveillance and strengthening of veterinary infrastructure. It is visualized that implementation of the scheme will ultimately lead to prevention and control, subsequently eradicating diseases, increased access to veterinary services, higher productivity from animals, boosting up of trade in livestock, in livestock products and improving the socio-economic status of livestock farmers.

12.4 Future Projections and Needs

Earlier smallholder livestock production was widespread in the country, but recently it has witnessed a gradual transformation to a semi-commercial or commercial mode. However, the requirements for age-old traditional production system and the current as well as future production systems are not similar and the country needs to be equipped with effective technological backstopping and efficient input delivery systems besides facilitating favourable market and marketing networks. Indeed, there are number of critical gaps, which needs to be bridged for improved livestock production, like resource-driven livestock production than demand-driven, mindset of people to manage animals under a low/negligible input system, the presence of huge non-descript livestock population, low genetic improvement in organized and unorganized rural herds, poor productivity of native animals, inadequate availability of superior germplasm, inadequate feed and fodder supply, indiscriminate breeding of animals in field conditions, inadequate availability of vaccines, cold-chain and other health measures, unorganized marketing of animal products and problems associated with the diffusion of new technologies. So we need to put our concerted efforts on following issues (Srivastava 2018).

12.4.1 Recording Performances of Animals

At present we do not have an effective mechanism to identify the animals and performance recording, although small-scale level progeny testing programs are in use at a few places. The elite germplasm of all breeds needs to be identified and subjected to performance recording under systematic breeding programmes so that more numbers of elite germplasm become available and male offspring from these elite females can be exploited for future artificial breeding purposes.

12.4.2 Working on Indigenous/Native Animals

In our country, we have a huge number of indigenous animals who are capable of sustaining productivity under even very low inputs and also known for their drought tolerance and disease resistance. But there has been a change in the utilization pattern of these genetic resources, now they are facing stiffer competition for their survival, which warrants urgent measures to be taken for their conservation. Indigenous breeds like Gir, Sahiwal, Tharparkar, Red Sindhi, etc. are potential milk yielders, producing 1200–3000 L or even more milk per lactation. These breeds need to be promoted and used for the grading-up of non-descript cattle in different regions of the country for improving their milk productivity. Besides the regular programs for conservation and improvement such as indigenous cattle, we have several *gaushalas* which need to be considered for the development and conservation of indigenous breeds in their respective regions. Similarly, we have other species of indigenous animals like buffalo, sheep, goat, pig, etc. which are highly suitable and adapted to Indian conditions also need to be considered for conservation and genetic improvement.

12.4.3 Augmenting Availability of Quality Feed Resources

We need to have an effective plan for augmenting the availability of quality feed resources in the country, as there is a shortage of feeds and fodders for livestock, leading to an inadequate supply of nutrients especially protein, energy and minerals. The regional and seasonal deficits of fodders and feeds are more important than the national deficit as it is not economical to transport them over long distances. In order to meet the nutritional requirements of animals, there is a need to increase the bioavailability of nutrients from feeds and fodders using chemical, biological and biotechnological techniques. Indeed, it is required to improve the productivity of the land and also to meet the feed and fodder requirements from the limited area available for this purpose. We also need to look at newer feed resources and evaluate them for livestock feeding and find out how much these can be exploited in bridging the gap between supply and demand of the nutrients.

12.4.4 Prevention and Control of Animal Diseases

It is an important issue that needs cooperation from many stakeholders in the livestock production chain. Particularly at the farm level, better prevention and control of diseases lead to much higher output. We need to develop a strong and reliable epidemiological status of economically and zoonotically important diseases in different regions of the country to identify high, moderate and low prevalence areas and to formulate control or eradication strategies. It is true that vaccination is the most capable way to control many diseases, but it is not 100% effective in preventing the diseases like brucellosis. Thus, vaccination coupled with good husbandry practices is essentially needed to control the disease. Exact diagnosis of diseases is not possible under field conditions due to the inadequate availability of veterinary laboratories/health centres. Thus the diagnosis is very often made based on the symptoms only. We need to keep pace with the contemporary developments in the technical improvements of the conventional vaccines to make them more useful and also continue with research and development efforts for newer generation vaccines in the long term.

12.4.5 Creation of Special Animal Product Zones (SAPZ)

Zone-specific planning is the need of the hour for any livestock development program. Intensive livestock development programs requiring germplasm improvement through selection and upgrading need to be put in place along with region-specific production packages. In order to promote the export of livestock products, we need to set-up Special Animal Product Zones (SAPZs). SAPZs should be identified keeping in view the basic infrastructure and other requirements including the availability of land and irrigation facilities for the production of fodders, healthcare and marketing facilities in the region. These SAPZs need to be linked with an export-based processing plants, the compound feed industry, regulated animal market, veterinary polyclinics, semen bank, etc. in the cooperative or private sector. The farmers in SAPZs need to be offered incentives to set-up medium scale (20–100 animals) or large-scale (more than 100 animals) commercial livestock farms depending upon the purpose (milk, meat, etc). The farmers also need to be assured of the minimum price for animal products and an added premium for quality.

12.4.6 Policy Issues

We have a huge population of non-descript livestock that need to be characterized and the homogenous populations deserve the status of breeds to be recognized as a new breed. Once they are recognized properly, suitable policies need to be evolved to improve their productivity while restricting the growth of low producing population. Exclusively by facilitating the large-scale commercial livestock production, it may not be possible to obtain inclusive growth, however, boosting the smallholder farmers as a whole and commercial livestock production in identified areas will put livestock as an instrument of inclusive economic development. To achieve this, we need to have a stringent mechanism/legislation in place that protects the interests of the large smallholder population, while facilitating scaling up commercial livestock production. Livestock producers also need to be brought under an organized umbrella by establishing livestock cooperatives/groups/breed societies etc. to encourage the concept of 'quality assured production' and incentivizing in the form of better prices for the quality producers. Although it requires establishing quality test centres at various levels. Indeed, a check system needs to be established for tracing 'from table to farm' for the prevention of the sale of substandard livestock products as it is expected affects adversely the consumer health.

12.5 Conclusions

Livestock has been an important source of livelihood for resource-poor farmers over the years. The gross value added (GVA) of the livestock sector was about 28.36% of agricultural and allied sector GVA and 5.21% of total GVA during the financial year 2019–2020. Among the different livestock components, a major share of value comes from milk followed by meat. It has been observed that in India the total number of livestock is increasing day by day at a steady rate. However, this increment is largely driven by a sharp increase in the number of sheep and goats, which is nearly 95% of the total livestock increase. Among the livestock, there has been a marked shift from indigenous cattle to crossbreds. Based on the livestock census in 2019, there was a 6% decline in the total number of indigenous cattle and a 26.9% rise in total exotic or crossbred cattle over the previous census. Similarly, there was also a decline in number of other livestock like donkeys, camels, pigs, horses and mules. India possesses huge livestock wealth, but the productivity of livestock is pretty low due to many constraints like the dominance of animal population with local low-yielding breeds, shortage of feed and fodder, inappropriate management and disease control measures, etc. Proper attention need to be given to the above-mentioned constraints along with policy/input support from the government and other stakeholders. The poor productivity of Indian livestock can be improved by adopting suitable breeding strategies like cross-breeding, upgrading and selection based on the agro-climatic conditions, farmers' resource availability and preferences. Similarly, the necessary supporting infrastructure with effective input delivery systems and viable processing enterprises need to be encouraged for the better contribution of livestock in the Indian economy in years to come.

References

Anonymous (1945) Report of the milk sub-committee appointed by the policy committee on agriculture, poultry and fishery. Ministry of Agriculture, Government of India, New Delhi

Anonymous (1946) The dairy industry of India: report on an investigation with recommendations, also known as the Pepperall report. Manager of Publications, Delhi, pp 1–17

Anonymous (1949) Marketing of milk in the Indian union. Directorate of Marketing and Inspection, Ministry of Agriculture, Government of India, New Delhi, pp 13–14

Anonymous (2019) Annual Report (2018–19). Department of Animal Husbandry, Dairying and Fisheries, Ministry of Agriculture & Farmers Welfare, Government of India

Anonymous (2021a) Schemes/programmes. Department Animal Husbandry and Dairying. Ministry of Fisheries, Animal Husbandry and Dairying, Government of India, New Delhi. https://dahd.nic.in/schemes-programmes. Accessed 16 Dec 2021

Anonymous (2021b) Annual report (2020–21). Department Animal Husbandry and Dairying. Ministry of Fisheries, Animal Husbandry and Dairying, Government of India, New Delhi. https://dahd.nic.in/schemes-programmes. Accessed 18 Dec 2021

BAHS (2014) Basic animal husbandry statistics. Department of Animal Husbandry, Dairying and Fisheries, Krishi Bhawan, New Delhi, pp 1–156

Burns W (1944) Technological possibilities of agricultural development in India. Government Printing Press, Punjab, pp 1–20

Habib I (1963) The agrarian system of Mughal India (1556–1707). Asia Publishing, New Delhi, pp 1–53

Mondal S, Mishra AP (2019) Dynamics and performance of livestock and poultry sector in India: a spatio-temporal analysis. Natl Geogr J India 65(4):389–402

Srivastava AK (2018) Strategy paper: livestock development in India. Trust for Advancement in Agricultural Sciences, New Delhi, pp 1–20

Voelkar JA (1893) Report on the improvement on Indian agriculture. Printed by Eyre and Spottiswoode, London, pp 1–216

Wright NC (1937) Report on the development of cattle and dairy industries in India. Manager of Publications, Delhi

Fisheries and Aquaculture 13

Lopamudra Sahoo, Bijay Kumar Behera, Debabrata Panda, Janmejay Parhi, Chandan Debnath, Sumanta K. Mallik, and Subal K. Roul

Abstract

India, a country with rich aquatic diversity endowed with vast aquatic resources has made tremendous progress in the field of fisheries and aquaculture. At the national level, the GDP contribution of Fisheries to Agriculture is 7.28%. At the global level, India holds the second rank in aquaculture. But looking back there are many Individual, Institutional and Governmental efforts in achieving present day's success. At the individual level, the contribution of Dr. Hiralal Chaudhuri for the development of artificial spawning is a major landmark. Development of induced breeding followed by composite fish culture during the 1980s revolutionized the sector, today which we term as Blue Revolution. At Institutional Level various research Institutes were established like Central Inland

L. Sahoo (✉)
ICAR Research Complex (RC) for NEHR (North Eastern Hill Region), Agartala, Tripura, India
e-mail: Lopamudra.Sahoo@icar.gov.in

B. K. Behera
ICAR-CIFRI (Central Inland Fisheries Research Institute), Barrackpore, West Bengal, India

D. Panda
ICAR-CIFA (Central Institute of Freshwater Aquaculture), Bhubaneswar, Odisha, India

J. Parhi
College of Fisheries, (Central Agricultural University), Agartala, Tripura, India

C. Debnath
ICAR Research Complex for NEH Region, Umiam, Meghalaya, India

S. K. Mallik
ICAR-DCFR (Directorate of Coldwater Fisheries Research), Nainital, Uttarakhand, India

S. K. Roul
ICAR-CMFRI (Central Marine Fisheries Research Institute), Digha, West Bengal, India

P. K. Ghosh et al. (eds.), *Trajectory of 75 years of Indian Agriculture after Independence*, https://doi.org/10.1007/978-981-19-7997-2_13

Fisheries Research Institute (CIFRI), Central Marine Fisheries Research Institute (CMFRI), Central Institute of Freshwater Aquaculture (CIFA), Central Institute of Brackishwater Aquaculture (CIBA), Central Institute of Fisheries Education (CIFE), Central Institute of Fisheries Technology (CIFT), Directorate of Coldwater Fisheries Research (DCFR), National Bureau of Fish Genetic Research (NBFGR) to develop cutting edge aquaculture technologies. Noticeable technologies that need to be highlighted are Captive breeding of high-value freshwater, brackishwater and marine species, Development of improved variety of fish species, Enclosure culture, Disease management, Feed Management, Cataloguing and conservation of novel species, etc. A lot has been achieved, but there are several challenges that also need to be addressed like climate change, the emergence of new aquatic diseases, fulfilling the feeding requirement, the establishment of value chains and domestic market linkages, development of motivated entrepreneurs. Besides, there is also a need for the development of information and communication technology and precision farming in this field. The whole idea in developing this sector further will be sustainable development of fishers and fisheries for a better tomorrow in the form of economic growth, nutritional fulfilment with a healthy environment.

Keywords

Fisheries · Aquaculture · Pre-independence · Post-independence · Growth and development · Blue revolution

13.1 Introduction

India, a blessed country with rich and diverse natural resources, is endowed with a wide plethora of aquatic and fisheries resources ranging from deep seas to lakes, ponds, and rivers more than 10% of the global biodiversity in terms of fish and shellfish species. India is bested with marine fisheries resources that are spread along the country's vast coastline with Exclusive Economic Zone (EEZ) of 2.02 million square km and a continental shelf of 0.53 million sq. km. The inland resources are in the form of reservoirs (31.5 lakh hectares), floodplain lakes (8.12 lakh hectares), brackish water (12.4 lakh hectares), rivers and canals (1.95 lakh km), ponds and tanks (24.1 lakh hectares), saline/alkaline affected areas (12 lakh hectares), etc. Fish production in India has shown tremendous growth that can be understood by the increase of production from 0.75 million MT during 1950–1951 to the current production of 14.1 million MT (Figs. 13.1 and 13.2). Apart from this, Fisheries and aquaculture-based activities serve the livelihood of 15 million people. On average around 50 different types of Fish and shellfish are being exported to 70 nations. In terms of quantity fish and processed products contribute to 14 lakh tonnes and in terms of value, they contribute INR 46,000 crores to agricultural export of the nation. Fisheries sector contributed around Rs. 2100 billion to the country's GVA. This accounted for 7% of the agricultural GVA and 1.25% of the

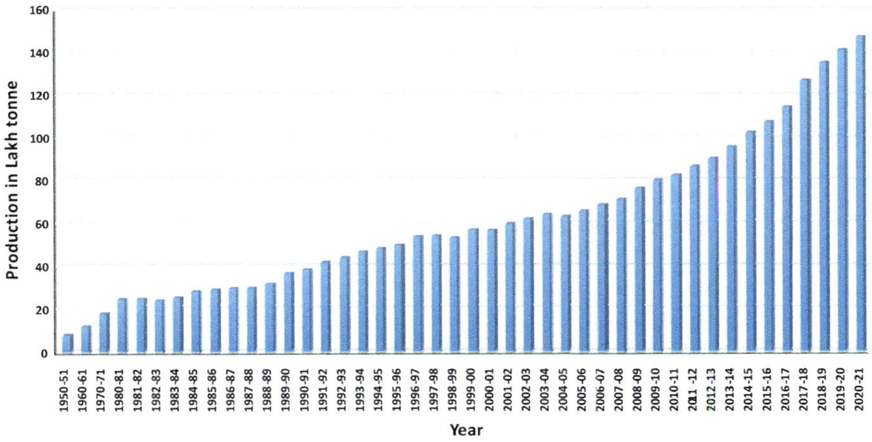

Fig. 13.1 Fish production scenario of India. (Handbook on Fisheries Statistics 2000, GOI)

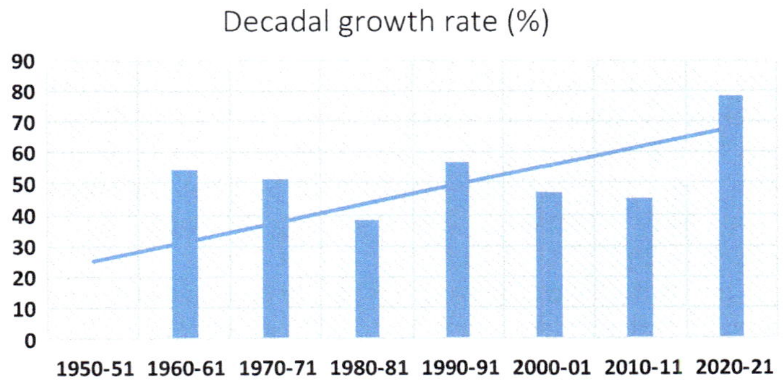

Fig. 13.2 Decadal growth rate in fisheries production. (Data from Handbook on Fisheries Statistics 2014, GOI)

total GVA in the year 2018–2019. This sector with a growth rate of 10% recorded the highest among the other agricultural sectors.

Fisheries and aquaculture are sunshine sectors because of their faster growth than the crop and livestock sectors (Kumar et al. 2006). But everything was not as rosy as it appears now. Indian Fisheries and Aquaculture have continuously evolved and developed with many stakeholders and professionals' continuous and untiring efforts. To know it, we must travel to the past and dig out some of the significant contributions of the sector.

13.2 Status of Fisheries in Pre-Independence and Independence Era

The recorded observation of fish occurrence in India can be traced back to 3 millennium BC. Evidence of fish imprints being used for human consumption has been found in Mohenjodaro and Harappa of Indus Valley Civilizations (2500 BC–1500 BC). In a book compiled by Mansoltara in 1127 AD, King Someswara and King Vikramaditya VI were the first to record the common sport fishes of India and they had further classified them into freshwater and marine fishes.

The foundation for modern scientific research in fisheries dates back to the British era when various Naturalists, Taxonomists, Zoologists and Botanists recorded the fishes and other aquatic organisms from India and catalogued them. A veterinary surgeon by profession Sir Francis Day contributed significantly to catalogue of the fishes of India through his research articles and monographs. Indian Fisheries Act is a stepping stone in our history that conferred upon the rights of developing fisheries to the different provinces in their respective states.

Traditionally aquaculture has been in practice for self-consumption and local trade in the eastern part of the country in small homestead ponds. During the nineteenth century, advances were made over the traditional homestead aquaculture with the breeding of carps in controlled conditions with the simulation of river-like conditions. Pisciculture was popularized in the state of Tamil Nadu followed by other states like Uttar Pradesh, West Bengal, etc. with the establishment of Fisheries Departments. The Indian fisheries sector was characterized as primitive, ignorant, unorganized, ill-equipped and caste-based by the National Planning Committee in 1946 (Kurien 1985), suggesting a social rigidity that would prevent economic efficiency (Karnad 2017).

Owing to this status, much attention was paid to the planned development of this sector through 5-year plans and the establishment of technical research organizations. As a result of this, in 1947, two major research stations were set up under the Ministry of Food and agriculture: The Central Inland Fisheries Research stations (CIFRS) in Kolkata, West Bengal and Central Marine Fisheries Research Stations (CMFRS) in Mandapam, Tamil Nadu (Silas 2003).

Traditionally spawns were collected from rivers during the breeding season (Fig. 13.3). The artificial spawning of minnow *Esomus danricus* through pituitary injection was a remarkable achievement by Prof. Hiralal Choudhuri in the year 1955. This was followed by the captive spawning of other culturable species like Rohu, Bata, Mrigal and many other cultivable species during 1957 (Chaudhuri and Alikunhi 1957). This was the first-ever aquaculture breakthrough, which impeded the subsequent Blue Revolution. The Indians were the first to develop hypophysation technique for artificial captive breeding of Indian Major Carps, way ahead of the Chinese, who created the same method for Chinese carps in 1958.

Subsequently, after mastering the art of captive breeding of fish, another significant achievement was using synthetic artificial inducers in fish breeding. During the 1970s and 1980s, due to the paucity of carp pituitary extracts, research was targeted to use Human Chorionic Gonadotropin (HCG) for breeding. Various synthetic

Fig. 13.3 Collection of riverine fish seed using a *gamcha*, a rectangular mosquito netting cloth. (*Pic. Courtesy: CIFRI, Barrackpore from web*)

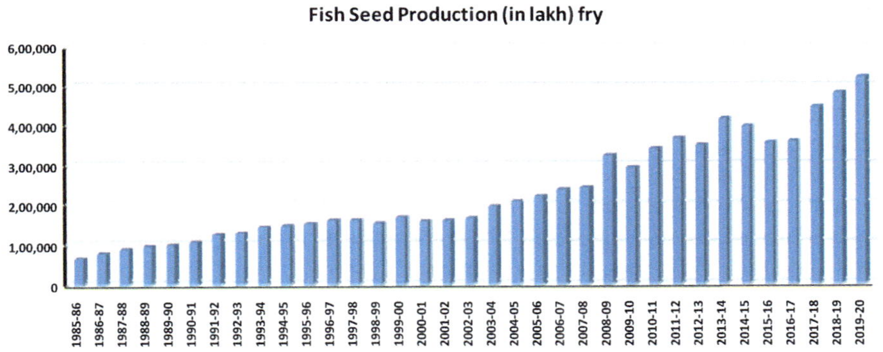

Fig. 13.4 Fish seed production scenario of the country. (Data from Handbook on Fisheries Statistics 2020, GOI)

hormones are available in the market with different combinations and trade names like ovaprim, ovatide, Wova-FH, etc. Because of these artificial hormones, fish seed production was boosted by from 63,000 lakh fry in the 1980s to over 5 lakh fry in the present day (Fig. 13.4). Subsequently, because of these technological advancements, riverine seed collection and bundh breeding became obsolete.

Scientific carp farming was established in India with the launch of an All India Coordinated Research Project (AICRP) on 'Composite Culture of Indian and Exotic Fishes' by the CIFRI in 1971 where as demonstrations as high as 8–10 tonnes/ha/year was obtained. This AICRP was operated in 12 centres across the country until 1984. It was followed by the launch of three more AICRPs on 'Spawn Prospecting',

'Air-breathing Fish Culture' and 'Brackish water Fish Culture' (FAO 2003). Because of all these efforts, farmers were convinced to take up carp seed production and culture as a business enterprise in different parts of the country like Andhra Pradesh, Punjab, Haryana, Maharashtra, etc. leading to a blue revolution in the country. Govt launched the formal Blue Revolution in India during the seventh 5-Year Plan (1985–1990) on the occasion of the Fish Farmers Development Agency (FFDA) sponsorship.

13.3 The Blue Revolution Era

13.3.1 Technologies in Freshwater Aquaculture

13.3.1.1 Composite Fish Culture

Hypophysation and Captive breeding laid the foundation of modern scientific aquaculture in the country, The third significant breakthrough in bringing about the blue revolution in the country was composite fish culture, i.e. the culture of compatible combinations of Indian and Chinese carps, i.e. culturing of fish based on their biology, habitat, feeding and compatibility (Alikunhi et al. 1971). From a meagre production of 600 kg/ha/year traditionally, it has been made possible to obtain a show of 15,000 tonnes/ha/year through composite fish culture with scientific management practices.

Till 2000, marine fish production dominated India's total fish production. However, due to improved management practices adopted through scientific backing, inland fisheries and aquaculture have taken a U-turn and presently contribute almost 90% of total fish production from a contribution of 46% in the 1980s.

India holds the second rank in aquaculture at the global level. Freshwater aquaculture production registered a 10 times growth from 0.37 million tonnes in 1980 to 9.0 million tonnes in the present time. Currently, the growth rate is hovering over 6% and contributing to more than 95% of the total aquaculture production. The main species in freshwater aquaculture are *Labeo rohita*, *Catla* and *Cirrhinus mrigal*. They are contributing to around 75% of the total production. The rest 25% is coming from Chinese carp and catfish.

It is estimated that only about 40% of the available area of 2.36 million hectares of ponds and tanks has been put to use and immense scope for expansion of the area exists under freshwater aquaculture (Handbook of Fisheries and Aquaculture 2013). Due to different schemes under the blue revolution, the national average productivity has risen from 600 kg/ha/year to 3000 kg/ha/year in a span from 1974 to the present times.

Technology of Intensive carp culture developed by ICAR-CIFA has the capacity to produce 10–15 tonnes/ha/year with supplementary feeding, aeration and biofertilization with Azolla (Tripathi et al. 2000). There are many progressive fish farmers who are achieving a higher production from 8 to 12 tonnes/ha/year (Handbook of Fisheries and Aquaculture 2013).

Technologies of induced carp breeding and composite fish culture in captive conditions converted the otherwise traditional sector into a lucrative, viable commercial enterprise. Breeding technology has been standardized for around 50 species of food and ornamental category (Raizada et al. 2019). Another technology developed by ICAR-CIFA called Multiple Spawning has the potential to produce 2–3 times more spawn compared to traditional breeding practices (Gupta et al. 1995). Much advancement was made in modernizing the traditional hatchery system. The eco-hatchery and FRP carp Hatchery could provide an impetus to fish seed production in the country.

13.3.1.2 Freshwater Prawn Culture

Macrobrachium rosenbergii, *M. malcolmsonii* and *M. gangeticum* are three freshwater prawn species brought into the culture system because of high consumer demand. The production of freshwater prawns is 10,000 tonnes during the present time. But it was 35,000 tonnes during 2002–2003. Andhra Pradesh was the chief contributor to the sector with a 73% contribution from an area of 38,820 ha (MPEDA). Because of the introduction of exotic Vannamei, production has declined in present times. But this scenario can be improved through popularization in landlocked parts of the country with standardization of seed production in artificial seawater.

13.3.1.3 Cold Water Aquaculture

Recognizing the necessity of cold water fisheries, ICAR-CIFRI started a cold water unit in 1963 to assess and utilize the fisheries resources available in this sector. Subsequently, it was given the status of NRC on Coldwater Fisheries in 1987 and further upgraded to the status of Directorate on Coldwater Fisheries Research in 2008. The institute has mapped and documented 258 species, 76 genera and 21 families of the Coldwater fish resource. This resource accounts for 17% of the total fish fauna of the country. Out of these, 203 species have been recorded from the Himalayas and 91 from the Deccan Plateau Total fish production from hill states is only 75,000 MT, accounting for 1.5% of the country's inland production. Food fishes of cold waters are mahaseer and schizothoracids belonging to the indigenous species and trouts among the exotic varieties. Schizothoracids and Mahseers support capture and sport fisheries, whereas aquaculture is limited mainly to carp culture in mid-hill cold waters and rainbow trout in higher altitudes.

13.3.1.4 Flood Plains and Wetlands

There are around 31.5 lakh hectares of small, big and medium reservoirs in the country. ICAR-CIFRI has developed various technologies for improving the productivity of these reservoirs. Notable among these are fish seed ranching and cage culture. Due to these technologies, productivity could be increased to 500, 200 and 100 kg/ha/year from the small, medium and large reservoirs, respectively. Floodplains/wetlands are productive ecosystems that can be effectively utilized for aquaculture. Enclosure fisheries like cage and pen culture developed by ICAR-CIFRI has resulted in increasing the productivity from 200 to 400 kg/ha/year.

13.3.1.5 Scientific Interventions in Culture Practices

Diversification is an essential practice for the long-term sustainability of the sector. It was also realized in due course of time, that aquaculture is being dominated by a few Indian major carps and Chinese carps. To address this and to make the whole system more sustainable, species diversification with various endemic species with local and regional demand like *L. gonius, L. bata, Cirrihinus cirrhosa, Puntius sarana, L. dussumieri, P. jerdoni, Labeo calbasu, C. reba, L. fimbriatus* were also introduced into composite fish culture system (Jena et al. 2020).

Catfish breeding and culture were considered a national priority. Under that, ICAR-CIFA standardized several catfish species' breeding and culture techniques like Magur (*Clarias batrachus*), Singhi (*Heteropneustes fossilis*), Murrels and *Anabas testudineus,* etc. A production of 3–5 tonnes of Magur could be demonstrated in small shallow ponds in a duration of 6 months.

Monoculture and Polyculture of Pabda has been standardized and popularized in eastern and North eastern parts of the country.

Many exotic species were introduced to the country because of their fast growth. Notable among them are *Pangassius, Pangassius sutchi. Pangasius sutchi*, was introduced to India in 1995–1996 from Thailand to the state of West Bengal. This fish has the potential to grow up to 1.5 kg in a year with a production capacity of 20 tonnes/ha/year (Belton et al. 2017). The Government of India has stipulated strict guidelines for regulating the introduction of *P. sutchi* in the country. The guidelines suggested keeping the upper limit of production to 20 tonnes/ha/year.

Another exotic fish that Govt has permitted is *Oreochromis niloticus* in 2012 with a prescribed set of guidelines for the propagation of the species. As per the guidelines, farming of monosex male/sterile is permitted.

The scientific basis of culture techniques were standardized for nursery, rearing and grow out ponds for spawn, fry and fingerling rearing respectively. Of the total farmers in the country 90% are small and marginal. Besides, there are farmers with high investment capacities. So technologies for all categories of farmers were developed and classified as Low, Medium and High input Technologies. In high input technology, also called high-tech aquaculture, there are two primary components that lead to higher production: Water exchange and aeration. Production levels of over 10–15 tonnes/ha/year have been possible through this system.

Then again, based on the production cycle, composite fish culture technology was classified into single stocking, single harvesting and multiple stocking multiple harvesting. In this system, marketable size fishes are harvested from ponds at regular intervals with periodical restocking.

Many scientific interventions have been done in Integrated fish farming. This integration has been most successful with livestock and horticulture with fish production of 3–55 tonnes/ha (Ayyappan and Jena 2003). This system is a very sustainable system as the by-products/wastes are utilized as principal inputs thereby making the farming practice highly remunerative and farmers friendly.

Traditional practices of Paddy and Fish culture were improvised based on location. Production of 500 kg fish and around 3 tonnes of paddy can be achieved from a hectare of land.

In the high altitude raceway culture has been standardized for commercially important species like rainbow Trout by ICAR-DCFR.

13.3.1.6 Technologies in Genetics, Nutrition and Health Management

Developing improved rohu (Jayanti) through selective breeding with a record of 17% higher growth response per generation after 11 generations is a significant achievement from ICAR-CIFA. Jayanti Rohu has recorded a 55% higher growth in farmers' fields in various parts of the country. Similarly improved Catla and freshwater prawn have recorded 2% and 10% higher growth, respectively.

Various types of fish feed like CIFABROOD™, CIFRI CageGrow, Vanamiplus, Varsha, Varna have been developed for freshwater, brackish water and marine water species. Private entrepreneurs have played a major role in the popularization of these feeds.

In the 1980s, Epizootic ulcerative syndrome, a transboundary disease caused huge economic losses to the farmers of the country. CIFAX™ as a prophylactic and therapeutic measure was developed by ICAR-CIFA to EUS. A National Surveillance Program for Aquatic Animal disease initiated through research Institutions along with NBFGR and NFDB is in place for aquaculture disease monitoring in India.

India is home to 10% of the global ichthyofaunal biodiversity. The database of ICAR-NBFGR records around 2650 fin fishes, which includes around 290 exotic species. (NBFGR). Out of around 2300 indigenous species, 880 fishes are found in freshwaters, 115 in brackish waters and 1350 fishes are found in marine waters, belonging to 39 orders, 225 families and 852 genera. The biodiversity-rich areas such as North East India and Western Ghats have been explored through various network programmes. The recent discovery of more than 39 new species warrants more intensive exploratory surveys to map further unknown resources (NBFGR).

13.3.1.7 Technologies in Brackishwater Aquaculture and Mariculture

Brackishwater farming in India is an age-old traditional system confined mainly to bheries (artificial impoundments in coastal wetlands) of West Bengal and Pokkali (Salt Resistant deep water paddy) fields along the Kerala coast. The fisheries in these systems involved trapping the seeds of fish and shrimps and culturing them without providing any external input. The first scientific attempt at developing various mariculture technologies in India was during the 1960s by ICAR-CMFRI (CMFRI). Initially, the focus was on marine molluscan species like Oyster, mussel, clam, etc., and the main species that were brought to culture practices were *Pinctada fucata, Pinctada maragaritifera* and *Crassostrea madrasensis* (Mojada et al. 2021).

In the early 1970s, the major was focused on the development of coastal aquaculture and mariculture mainly because of dwindling capture fisheries resources and export pressure. Consequently, important Penaeid species were bred in captivity at Narakkal farm of CMFRI like *Penaeus monodon, P. japonicus, P. semisulcatus,* etc. (Ponnaiah 2011).

As such the production was of subsistence level. ICAR launched AICRP on "Brackishwater Fish farming" in 1973 (CIBA). Under this project, several technologies of fish and shrimp farming were developed. The brackishwater aquaculture in India got its fillip with the demonstration by the DBT-sponsored project on semi-intensive shrimp farming at Nellore, Andhra Pradesh, by TASPARC, during 1989–1993 (CIBA). To provide further impetus to brackishwater aquaculture, ICAR-CIBA, a part of CIFRI, was established as a separate institute in 1985.

In the year 1997, ICAR-CIBA developed the technology for the breeding of *Lates calcarifer* throughout the year (Arasu et al. 2009). Then in the year 2015, the same institute developed the captive breeding technology of *Chanos* followed by the successful artificial breeding of *Mugil cephalus* in 2016. The culture technologies were also standardized for the above 3 species. Besides captive breeding of *Etroplus suratensis*, *Lutjanus argentomaculatus* and *Tenuilosa ilisha* were also standardised.

Another significant achievement is the promotion of Saline water aquaculture to transform wasteland into wetlands. Under this 13,000 ha have been covered with a production of 4331 tonnes with a productivity of 6 tonnes/ha. Many potential candidate species like *Etroplus, Chanos,* etc. have proven to be commercially suitable in inland saline areas. Production potential ranging from 0.5 to 3 tonnes/ha/year has been demonstrated from such waters (Handbook of Fisheries and Aquaculture 2013).

ICAR-CMFRI developed seed production and culture techniques for oysters and mussels in 1970s. Besides it has also developed induced breeding technology for Cobia and Silver pompano in 2010 and 2011 followed by seed production technology *for Epinephelus coioides* in 2013. Simultaneously culture practices of various fin fish species like Pompano, Cobia, Grouper etc. were standardized. Apart from this culture of crustaceans, oysters, mussels, clams, and seaweed culture was also practised. In the year 1972, CMFRI initiated a pearl culture program and successfully developed the technology for pearl production in Indian pearl oysters. The Mandapam Research Centre of CMFRI has developed the technology for the culture of agar-yielding algae Gracillaria and Gelidiella that led to a rise of entrepreneurship for producing agar in various areas of Tamil Nadu.

Biofloc Technology that operates on Zero water Wastage Concept has been in vogue in the present days. In a technology developed by ICAR-CIBA, 40–60 kg of finfish like Tilapia and 5 kg of shellfish like shrimp can be produced from a biofloc unit of 1 tonne capacity.

Integrated Multitrophic aquaculture or IMTA is a technology that utilizes different species at various trophic levels so that nutrient loads of fed species can be used as input for extractive species. ICAR-CMFRI has developed IMTA for Cobia with sea weed *Kappaphycus alvarezzi*.

The current mariculture production is to the tune of 0.01 million tonnes. For the first time in the country, Open Sea aquaculture was initiated by ICAR-CMFRI in 2005 by establishing the open sea floating cage in Vishakapattanam. Further refining technologies and adoption have led to the rapid spread of cage mariculture.

As a result of all these concerted efforts, brackishwater and mariculture gained momentum in our country. With only 13% of the area utilized under brackishwater aquaculture, the production is 0.7 million MT with a productivity of 4 tonnes/ha. In 1990s, MPEDA established shrimp hatcheries in Andhra Pradesh and Odisha that

Table 13.1 Technologies in freshwater aquaculture

Sl. no	Technology	Year
1	Induced breeding of *Esomus danricus* using pituitary gland extract	1955
2	Induced breeding of *Cirrhinus reba,* followed by three Indian major carps	1957
3	Induced breeding of Chinese carps	1962
4	Development of composite fish culture technology	1970
5	Breeding of Magur	1982
6	Intensive culture technology	1994
7	Selectively bred Jayanti Rohu was released	1996
8	CIFAX developed against EUS	1998
9	Diversified species were promoted, different culture techniques were developed	1999–2009
10	CIFABROOD™, a diet for broodstock was developed	2009

led to immense growth in the production. The most significant contribution was from Tiger shrimp *P. monodon*. A major havoc in the rapid growth of the shrimp hatchery was the white spot syndrome virus (WSSV) in 1994 when the industry experienced heavy losses. The time was to bring in alternate species to keep the industry running. *L. vannamei*, an exotic shrimp species fitted to the void with Government's strict control and monitoring measures. This species has the potential of 10–12 tonnes/ha in 3–4 months duration. The total shrimp production of the country is 8 lakh tonnes from an area of 1.7 lakh hectare. Ninety percent of the contribution comes from Vannamei. CIBA has played a crucial role in developing diagnostic kits for white–spot syndrome virus (WSSV), the first of its kind in India, and supported the formulation of best management practices (BMPs) for Indian shrimp culture to motivate and support the farmers to continue the production of Tiger shrimp along with vannamei production.

Shrimp export was the main reason for boost in India's export growth. Farmed shrimp production grew from 20 MT in 1947 to 7.47 lakh MT in 2020, thereby contributing to the major export share of INR 46,662 crores fisheries export earnings.

The major technological advancements that lead to blue revolution in the country have been summarized in Tables 13.1 and 13.2.

13.3.1.8 Major Schemes by the Government for the Promotion of the Sector

The first Blue Revolution was launched during the seventh 5-Year Plan (1985–1990) under the sponsorship of the Fish Farmers Development Agency (FFDA). Later during the eighth 5-year plan (1992–1997), the intensive Marine Fisheries Program was launched. Subsequently, it was decided by the Ministry of Agriculture and Farmers Welfare along with the Department of Animal Husbandry, Dairying and Fisheries to restructure this scheme along with all other ongoing plans under a single umbrella "Blue Revolution". The Blue Revolution in India and the Fish Farmers

Table 13.2 Technologies in brackishwater and mariculture

Sl. no	Technology	Year
1	Pearl culture programme initiated	1972
2	Culture techniques of agar yielding seaweed Gracillaria and Gelideiella was standardized	1974
3	Captive breeding and larval rearing of Penaeid species	1975
4	Culture techniques of mussel and oyster	1977
5	AICRP on brackishwater aquaculture	1987
6	Seed production techniques of *Lates calcarifer*	1997
7	Open sea aquaculture through cages	2005
8	Seed production of cobia	2010
9	Seed production of silver pompano	2011
10	Seed production of *Epinephelus cooides*	2013
11	Seed production and culture of *Chanos chanos*	2015
12	Seed production and culture of *Mugil cephalus*	2016
13	Biofloc technology for shrimp	2016
14	Integrated multitrophic aquaculture with fed and extractive species	2018

Development Agency (FFDA) improved the aquaculture and fisheries sector with the introduction of new techniques of rearing, marketing, exporting, and breeding.

Some of the significant outcomes of the first Blue Revolution in India are mentioned below:

1. From a meagre production of 60 thousand tonnes, Fisheries production reached 4.7 MT that included 1.6 Mt. from freshwater. India was recorded to achieve an average annual growth of 14.8% as compared to the global average percentage of 7.5 in the production of fish and fish products.
2. In the export scenario, the Fishery sector recorded a growth rate of 6–10%, thereby becoming the largest contributor amongst other agriculture sectors.
3. India became the world's second-largest producer of fish with exports worth more than INR 47,000 crores.
4. Globally India ranked second in aquaculture.
5. The fisheries and aquaculture production contributed 1% and 5% to India's GDP and Agricultural GDP, respectively.

13.3.1.9 Blue Revolution 2.0/Neel Kranti Mission

The focus of the Blue Revolution 2.0 is on development and management of fisheries. This covers inland fisheries, aquaculture, marine fisheries, including deep sea fishing, mariculture and all activities undertaken by the National Fisheries Development Board. The Government of India in December 2014 launched "Blue Revolution" Mission with a central outlay of INR 3000 crores. Under this scheme, it was aimed for a sustained annual growth rate of 6–8% in fish production. In these 5 years, main focus was on private investment and entrepreneurship development with capacity building and development of the value chain. The scheme was implemented from 2015–2016 to 2019–2020. The government under the MGNREGA started to develop the farm ponds, where pisciculture would take place.

The salient features of BR scheme and achievements made are given below:

- Different GOI projects to be converged and linked with Sagarmala project like MNREGA scheme, NRLM and RKVY. Increasing the yield as well as productivity in culture and capture fisheries from both inland and marine waters.
- Motivating the socially and economically backward sections particularly schedule castes and tribes to take up fishing-based activities. The Blue Revolution Scheme also encourages entrepreneurship development, private investment, Public-Private Partnership (PPP), and better leveraging of institutional finance.

13.3.1.10 Pradhan Mantri Matsya Sampada Yojana

Government of India launched its flagship scheme "Pradhan Mantri Matsya Sampada Yojana" as a COVID-19 relief package under Self Reliant India on 10 September 2020. The main focus was to bring about Blue Revolution in the country through responsible and sustainable development in Fisheries Sector. The total outlay of this scheme is INR 20,050.00 crores. The main features of this scheme are as follows:

- Enhancement of Production and Productivity.
- Infrastructure and Post-harvest management.
- Fisheries Management and Regulatory Framework.

The goals of PMMSY is additional fish production by 70 lakh metric tonnes, enhancing productivity to 5 tonnes/ha, increasing the export value to 1 lakh crores from 46 thousand crores and generating 55 lakh additional employment with an objective of doubling farmers income by the financial year 2025.

13.3.1.11 The Way Forward

The number of undernourished people in India is expected to rise to 70 million by 2050. To suffice the need for food and nutritional security of the underprivileged and to cater to the needs of individuals with increased living standards, fish and fishery products are going to play a vital role. With increased anthropogenic pressure and an ever-developing and expanding economy, resources like land, water, energy, labour, etc. are likely to deplete and diminish in the coming days. The average size of operational holdings has declined progressively from 2.28 ha in 1970–1971 to 1.55 ha in 1990–1991 and 1.23 ha in 2005–2006. Compared to the present situation, the per capita land holdings will be much smaller by 2050 (ICAR-CIFA). To meet the demand of the country, a contribution of 17.5 million metric tonnes/year needs to be contributed from fisheries. The per capita water availability was 5500 cubic metres in 1955, and it has reduced to 1850 cubic meters in 2008 and would be further reduced to 1250 cubic meters by 2050. This warrants an urgent need to make aquaculture activity sustainable on a long-term basis. Hence utilization of diverse water resources like seasonal ponds, irrigation water, open wells, etc. and reusing and recycling waste through recirculation, utilization of rainwater, etc. will be both a challenge and an opportunity. Also, value addition and processing need to be addressed in a larger dimension. The idea is to increase productivity by obtaining more values per unit of water used.

Aquaculture through feeding constitutes about 20% of total aquaculture production, and to increase the production three times by 2050, the fed fish culture practices are to be raised at least four times the present level and that is going to pose a major challenge. Since 1995, aquafeed production has grown globally at an average rate of about 10.7%/year. The output of formulated feed increased from 7.7 million metric tonnes in 1995 to over 35 million metric tonnes in 2010, and it is expected to reach around 70 million metric tonnes in 2020. The total volume of feed sold in 2010 was 60,000 tonnes of pelleted feed and 3,72,000 tonnes of extruded floating feed. Currently, the aquaculture sector uses nearly 20% of total available concentrated feed and by 2050, about 23 million metric tonnes of feed would be required considering the three-fold increase in aquaculture production (CIFA). The fish feed resources have multiple competitors like the dairy and poultry industries. The feed resources used are generally the by-products of agro-processing industries. The availability of these products depends on agriculture which is heavily dependent on different climatic variables like rainfall, temperature, etc. The last decade witnessed a 3–4 times increase in the price of fish feed ingredients. Therefore, the future will need to identify, analyse and utilize alternate ingredients in fish feed.

The disease will continue to be a menace in the growth of aquaculture in the country owing to increased globalization and trade, Intensification of framing practices, transboundary and continental movement of different stages of fish, the introduction of novel species, etc. Diseases account for 10–15% losses in aquaculture, which will continue to rise in the coming future. For disease management, Better Management Practices (BMPs) and Regulatory framework needs to be in place for both existing and emerging pathogens. In the shrimp industry, the development of specific-pathogen-free (SPF) broodstock and strict biosecurity protocols need to be strengthened. There are various novel biotechnological tools like Loop-Mediated Isothermal Amplification (LAMP), Bead Array, Microarray, Multiplex test, etc. for the exclusion of pathogens in aquaculture, which needs to be further refined and standardised. Bacteriophage therapies are already under and have proved useful in the prophylaxis and treatment of bacterial infections in aquaculture. Immunostimulants like lipopolysaccharide and vitamin C have proved effective in controlling diseases (Sahoo et al. 2016, 2017, 2020). Bioactive compounds from sea and plants could be explored to find novel molecules to treat diseases. The focus must be on making available low-cost, readily available farmer-friendly diagnostics, eco-friendly therapeutics, common cost vaccines, etc.

In India, there are basically two categories of farmers: Farmers with more extensive land holdings and farmers with small land holdings of less than 2 ha. This is again less in NE India where the average land holding is 0.54 ha. These farmers constitute 90% of the total farmers. Hence Research Organizations will have to develop technologies which should be feasible enough for upscaling and downscaling as per the need. These farmers will also need assistance in a package of practice for input use efficiency, value and marketing channels, etc.

East India is a hot spot of biodiversity, but even then, the significant species cultured are the Indian Major carps and Chinese carps. Diversification is essential for sustainability in future aquaculture practices. *Osteobrama belangeri* is a high-demand fish in Manipur, whose breeding and culture are practiced, but productivity enhancement can be taken up. Similarly, *Ompak bimaculatus* is preferred in Tripura and Assam. The of an economically viable and sustainable package of practice for the species in mono and polyculture systems is much more essential. There are a few cultivable species in the cold water region; *Labeo dyocheilus, L. dero, Neolissocheilus hexagonolepis,* etc. whose breeding, feed, culture practice need to be refined to introduce into the culture system. *Bangana devdevi,* a minor carp, has good potential as food fish. The composite culture of *Bangana devdevi* (locally known as Khabak in Manipur) and *Osteobrama belangeri* (Pengba) can be standardized in Manipur.

Other minor endemic carps such as *Semiplotus semiplotus* for the state of Arunachal Pradesh, Assam and Sikkim have more significant potential. The breeding and culture practice for the species can be taken up. Genetic selection of *Neolissocheilus hexagonolepis, Ompak bimaculatus, Osteobrama belangeri* and other minor carp species having demand and culture potential can be taken up.

The disease component may be a part of the selection programme. Since the genetic characterization of chocolate mahseer populations has already been done, selection should be the next step.

Integration with agriculture and horticulture is within reach of farmers as these traditional crops do not require much expertise, but the focus should be on the use of improved varieties. The same is with the rice cum paddy culture system, i.e. in Apatani (Arunachal) and Manipur where a rich resource such as water is available in plenty. The presently available fish stock for this purpose has lower productivity may be due to low genetic worth.

Total fish production from hill states accounts for 1.5% of the total production. But there is tremendous potential for hill aquaculture. Rainbow trout, a high-value species whose current production is just 700 tonnes, can be increased to ten times through technological improvement, commercialization and popularization.

Brackishwater aquaculture currently utilizes only 14% of the available resources, pointing out that it can significantly increase aquaculture production. Policymakers worldwide are of the consensus that in comparison to other sectors, growth originating in food production sectors can improve the livelihoods of small and marginal farmers. In this way, the Indian brackishwater aquaculture sector has tremendous potential and a prominent role to play with 857 million rural populations with access to brackishwater resources. Engaging them for the transition towards blue socio-economic growth, commonly called "Blue Growth", is envisaged for 2050 (ICAR-CIBA).

India has enormous mariculture potential with 2.2 million sq. km of Exclusive Economic Zone (EEZ), 0.5 million sq. km of the continental shelf, 8129 km of coastline, and 1.2 million ha of brackishwater and 20 million ha for marine farming.

The country is still nascent in world marine and coastal fish production, with just about 3.08% contribution in 2018 (FAO 2020).

India needs to produce about 18 million tonnes of fish by 2030 as compared to 14.1 million tonnes produced through capture and culture. The additional fish production has to come through aquaculture. As stated in the National Policy on Marine Fisheries, 2017 (NMF), an assessment of the exploited fish stocks in EEZ indicated overcapacity in the territorial waters and noted that further increase in capture fisheries has limited scope. Hence steps for the promotion and further development in the mariculture sector are the only options for meeting the fish demand in the coming years. This is attainable, as the projected mariculture production based on the area available is four to eight million tonnes against the current output of fewer than 0.01 million tonnes.

Aquaculture and conservation can go hand in hand, as in the case of endangered white sturgeon, which has been successfully conserved through aquaculture. Hence the development of region-specific aquaculture and conservation practices can benefit aquaculture diversification and play a crucial essential role in the conservation of that species. Hence Research should be targeted to focus on conservation through aquaculture.

India has rich and diverse ichthyo-faunal biodiversity. The current genetic research in India is limited to only a few species. Hence capacity and infrastructure to undertake -scale genomic analysis is the need of the hour, focusing on comparative genomics and phylogeny study. Traits are essential for launching any breeding programme. But traits are not easy to manifest in phenotyping as they change according to environmental conditions. Tagging the polymorphism at the gene level and using them to map the QTLs through marker-assisted selection (MAS) can be done. This will enable efficient and precision breeding for improved and desired traits. Through this, genetic progress in breeding could increase by more than 50%.

Precision farming is another important aspect that needs to be focussed to optimize field-level requirements and management regarding farming practices, input requirements, environmental protection, and saving labour and energy to derive maximum outputs. Mechanizations and other work need to be further scaled up. Climate resilient Aquaculture technologies need to be popularized like Integrated Multitrophic Aquaculture (IMTA), Recirculating Aquaculture Systems (RAS), Raceways, Floating cages, Penculture, etc. Popularization of ICT-based innovative tools like sensors, automated solutions for water quality, feeders, aerators, etc. can be improvised and popularized for further improvement in the aquaculture sector.

As the purchasing capacity of the domestic consumer is increasing each day, supply chains should be strengthened to cater to the need of the domestic consumer instead of relying heavily on global markets. This is particularly relevant to shrimp, prawns and other highly valued fishes. Supply chain systems to collect farm produce and deliver it to human consumption are in high demand. Developing domestic fish markets with these networks will be an excellent opportunity to generate employment and business.

13.4 Conclusions

Fish production in India has registered an annual growth rate of more than 7%. India's total estimated fisheries potential is 22.31 million metric tonnes, out of which the marine fisheries potential is 5.31 million metric tonnes and inland fisheries potential is 17 million metric tonnes. It is also important to mention that 71% of marine fisheries potential and 58% of inland fisheries potential has been harnessed (2017–2018). The three Indian Major carps *Labeo rohita, Catla catla and Cirrhinus mrigala,* with a production of more than 1.8 million tonnes, contribute the significant chunk of production, followed by the Chinese carps: silver carp, grass carp and common carp (FAO 2003). Somewhere, in 2014 the culture fisheries surpassed capture fisheries globally as a source of seafood for human consumption (Tacon and Metian 2018). Fish production in the country has shown continuous and sustained increments since independence. Only after the Indian independence, has fisheries and agriculture, been recognized as an essential sector. There are lots that have been achieved, but there are several challenges that also need to be addressed like climate change, emergence of new aquatic diseases, fulfilling the feed requirement, establishment of value chains and domestic market linkages, development of motivated entrepreneurs. Besides there is also need for development of information and communication technology, precision farming in this field. The whole idea in developing this sector further will be sustainable development of fishers and fisheries for a better tomorrow in the form of economic growth, nutritional fulfilment with a healthy environment.

References

Alikunhi KH, Sukumaran KK, Parameswaran S (1971) Studies on composite fish culture: production by composite combinations of Indian and Chinese carps. J Ind Fish Assoc 1(1):26–57

Arasu ART, Kailasam M, Sundaray JK (2009) Asian seabass, fish seed production and culture. CIBA Special Publication 42, Central Institute of Brackishwater Aquaculture, Chennai, p 158

Ayyappan S, Jena JK (2003) Grow-out production of carps in India. J Appl Aquac 13(3/4):251–282

Belton B, Padiyar A, Ravibabu G, Gopal Rao K (2017) Boom and bust in Andhra Pradesh: development and transformation in India's domestic aquaculture value chain. Aquaculture 470:196–206

Chaudhuri H, Alikunhi KH (1957) Observations on the spawning in Indian carps by hormone injection. Curr Sci 26(12):381–382

FAO (Food and Agriculture Organization of the United Nations) (2003) Fisheries statistics. FAO, Rome

FAO (Food and Agriculture Organization of the United Nations) (2020) Fisheries statistics. FAO, Rome

GOI (2000) Hand book on fisheries statistics. Ministry of Agriculture, Government of India, New Delhi

GOI (2014) Hand book on fisheries statistics. Ministry of Agriculture, Government of India, New Delhi

GOI (2020) Hand book on fisheries statistics. Ministry of Agriculture, Government of India, New Delhi

Gupta SD, Rath SC, Dasgupta S, Tripathi SD (1995) A first report on quadruple spawning of *Catla catla* (ham). Vet Arhiv 65(5):143–148

ICAR (2013) Handbook of fisheries and aquaculture. Indian Council of Agriculture Research publication (ICAR), New Delhi

Jena JK, Das PC, Panigrahi A (2020) Local self-reliance in aquaculture and fisheries: Gandhian models. In: Pathak H, Pal S, Mohapatra T (eds) Mahatma Gandhi's vision of agriculture: achievements of ICAR. ICAR, New Delhi, pp 48–59

D Karnad (2017) Locating effective commons and community in Maharashtra State's fisheries, India. PhD thesis, Rutgers University

Kumar P, Dey MM, Paraguas FJ (2006) Fish supply projections by production environments and species types in India. Agric Econ Res Rev 19(2):327–351

Kurien J (1985) Technical assistance projects and socio-economic change: Norwegian intervention in Keralas fisheries development. Econ Polit Wkly 20(25 and 26):A70–A88

Mojada SK, Divu D, Johnson B, Joseph I, Gopalakrishnan A (2021) Mariculture advancements in India: towards a new epoch. Infofish International, Selangor. www.infofish.org

Ponnaiah AG (2011) Shrimp farming. In: Ayyappan S, Moza U, Gopalakrishnan A, Meenakumari B, Jena JK, Pandey AK (eds) Handbook of fisheries and aquaculture. Indian Council of Agricultural Research, New Delhi, pp 548–560

Raizada S, Pravin P, Kumar AT, Jena JK (eds) (2019) ICAR technologies: breeding and seed production of finfishes and shellfishes. Indian Council of Agricultural Research, New Delhi, p 66

Sahoo L, Choudhury TG, Debnath C, Parhi J, Datta M, Purusothamman CS, Paniprasad K (2016) Immunostimulatory effect of vitamin C on hematobiochemical parameters of *Labeo bata* (Hamilton, 1822). Fish Technol 53:59–63

Sahoo L, Parhi J, Debnath C, Prasad KP (2017) Effect of feeding lipopolysaccharide as an immunostimulant on immune response and immune gene expression of *Labeo bata*. Vet Immunol Immunopathol 188:48–58

Sahoo L, Debnath C, Parhi J, Choudhury J, Choudhury TG, Pani Prasad K, Kandpal BK (2020) Molecular characterization and immunostimulant-induced expression analysis of type I interferon gene in *Labeo bata* (ham). Aquac Rep 18:100490

Silas E (2003) History and development of fisheries research in India. J Bombay Nat Hist Soc 100(2 and 3):502–520

Tacon AG, Metian M (2018) Food matters: fish, income, and food supply—a comparative analysis. Rev Fish Sci Aquac 26(1):15–28

Tripathi SD, Aravindakshan PK, Ayappan S, Jena JK, Muduli HK, Chandra S, Pani KC (2000) New dimension in intensive carp polyculture in India. J Aquacult Trop 15(2):119–128

Commercial Crops (Jute, Cotton and Sugarcane)

14

Gouranga Kar, D. Blaise, T. K. Srivastava, Chandan Sourav Kar, P. Verma, A. R. Reddy, and Pushpa Singh

Abstract

Jute, cotton and sugarcane are the principal commercial crops in India contributing. While India is the largest producer of jute and cotton; but is the second-largest producer of sugarcane in the world. Research on jute, cotton and sugarcane has long history of research in pre independence era, but more emphasis was given after post-independence period coinciding with the green revolution of India during 1960–1970s. There was a total shift from white jute (*Corchorus capsularis*) to tossa (*C. olitorius*) in jute; from the *desi* (*G. herbaceum*) to the American cotton (*G. hirsutum*) in cotton and hybrid varieties in sugarcane in this period due to high yield and better fibre quality in jute and cotton and higher sugar content and disease resistance in sugarcane. Major path-breaking achievements in jute consist of the introduction of resistance to premature flowering was a *tossa* jute, the development of microbial retting consortium "CRIJAF SONA" comprising pectinolytic bacteria and genome sequencing, while in cotton it was the development of the inter-specific and intraspecific hybrids and the *Bt* cotton hybrids and varieties. Interspecific hybridization through nobilization and improved methods of planting were major achievements in sugarcane. With technological advancement, jute productivity increased from 1.04 to 2.56 t/ha

G. Kar (✉)
Central Research Institute on Jute and Allied Fibre, Barrackpore, West Bengal, India

D. Blaise (✉) · P. Verma · A. R. Reddy
ICAR-Central Institute for Cotton Research, Nagpur, India

T. K. Srivastava (✉) · P. Singh
ICAR-IISR, Lucknow, India

C. S. Kar
ICAR-CRIJAF, Kolkata, India

P. K. Ghosh et al. (eds.), *Trajectory of 75 years of Indian Agriculture after Independence*, https://doi.org/10.1007/978-981-19-7997-2_14

since independence to now. India could take the credit for developing the world's first cotton hybrid (H-4) in 1970. In cotton, productivity which was stagnant at about 300 kg/ha prior to the introduction of *Bt* cotton, now jumped to more than 500 kg/ha. Sugarcane yield also almost doubled from 3.21 to 7.65 t/ha during the corresponding period. There is huge scope in all three commercial crops like diversified use of jute and cotton and alternative source of energy like biofuel or bioenergy from jute and sugarcane. Proper exploration of the untapped potential of this group of crops can not only increase farmers' income but also contribute to the environment and ecology of intensive cropping system.

Keywords

Commercial crops · Fibre crops · Bt cotton · Sugar crops · Breeding program · Cropping systems · Mechanization · Processing · Value addition · Breakthrough research

14.1 Introduction

Jute is the second most important natural fibre crop of India after cotton and the principal jute species cultivated are tossa (*Corchorus olitorius*), covering 94% area, and white jute (*Corchorus capsularis*) which is restricted to 6% area. India is also the largest producer of raw jute and jute products in the world. India with a production of 2.1 million tonnes and Bangladesh with 1.6 million tonnes together contribute 93% share of global jute production. Presently, jute is cultivated in around 7.0 lakh ha with national average productivity of around 2.50 t/ha. The jute sector also supports the livelihood of around four million farm families, provides direct employment to 3.7 lakh workers, and generates employment of ten million man-days/annum.

Cotton plays a significant role in the Indian economy as the textile industry is based on cotton. India is now the largest producer of cotton and is one of the leading exporters of cotton yarn around the world. The Indian textile industry contributes around 5% to the country's gross domestic product (GDP), 14% to industrial production, and 11% to total export earnings (Directorate General of Commercial Intelligence and Statistics). Cotton is cultivated on ~13.0 M ha area in the country producing ~37 million bales with a share of about 36–38% of the total global cotton acreage and a quarter of the World's cotton production.

Sugarcane (*Saccharum officinarum* L.) is the most important sugar-producing crop in the world contributing about 80% to total sugar production. It contributes about 0.69% to the national GDP of India and supports 7.5 million livelihoods of farm families and is the only source of raw material for all 533 sugar mills. The annual sugar production of the sugar industry is 30 million tonnes and India is the second-largest producer of sugarcane and sugar in the World after Brazil, contributing about 20 and 16.66% of total annual global sugarcane (1900 million tonnes) and sugar production (180 million tonnes).

In this chapter, we have covered the domestication and production of jute, cotton and sugarcane in three different phases, (1) Pre-independence era, (2) Post-independence era, (3) Green Revolution era and (4) Post-Green Revolution era.

14.2 Pre-Independence Era

14.2.1 Jute

British East India Company, which first exported 100 tonnes jute consignment in 1793, was the first jute trader, mainly during the start of the twentieth century to trade with Dundee's Jute Industry in Scotland and enjoyed a monopoly in this trade. The modified power-driven flax spinner by Dundee spinners during the 1830s played important role in increasing the production and export of raw jute from India. The first jute mill was established at Rishra in 1855 on the River Hooghly near Calcutta and subsequently shifted to power-driven weaving machinery and continued for the next three decades. Coarse bags were the earliest goods woven of jute in Dundee and after experience gained finer fabrics called burlap, or hessian (in India) were produced.

14.2.2 Cotton

The varietal improvement programme in cotton started as early as 1904 and was further strengthened in 1923 with the constitution of the Indian Central Cotton Committee (ICCC). India produced nearly 5.9 million bales from an 8.1 million hectares area during1938–1939 (Fig. 14.1). The production received a setback from 1945 to 1946 because of the greater focus given to food production during the Second World War (Agarwal 2007). Further setback to production was following the division of India during the partition of India and Pakistan.

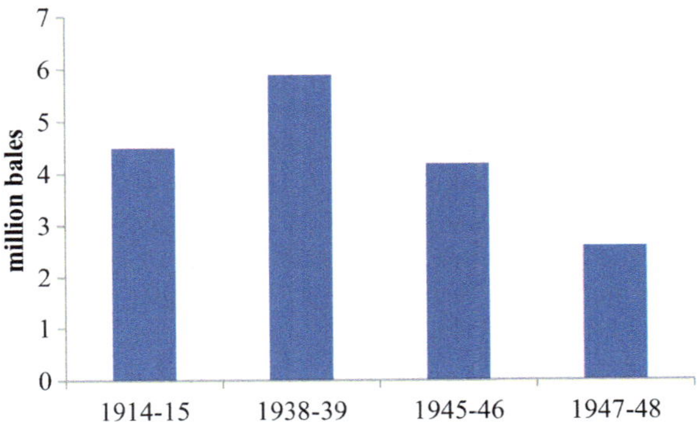

Fig. 14.1 Cotton production (million bales) in India during the pre-independence and at the time of partition following independence

14.2.3 Sugarcane

The cultivation of sugarcane in India finds a mention in Vedic literature and is considered to be in vogue since 800–1000 BC. Initial varieties were low in sugar as well as juice because of their higher fibre content. For the development of high-yielding and high-sugar varieties of sugarcane the ICAR—Sugarcane Breeding Institute, Coimbatore, was established as early as 1912. The imposition of tariffs on imported sugar in the year 1932 brought a major revolution to sugarcane cultivation which established India as one of the important sugar-producing countries and the area under the crop began to expand with an increasing trend. Consequently, a Central Sugarcane Committee was formed in the year 1944 under the ministry of agriculture to solve problems associated with sugarcane cultivation. Sugarcane research and development in eastern India comprising eastern UP, Bihar, Odisha, West Bengal and Assam holds a special significance as the region witnessed the very beginning of the modern sugar industry in the nation. The first ever modern sugar factory was established on the banks of the Hugli River in West Bengal in the early eighteenth century. The cultivation of sugarcane in its modern form in sub-tropical India emanates from the replacement of indigo cultivation with sugarcane farming during the famous 'Champaran movement' of Mahatma Gandhi.

14.3 Independence and Post-Independence Era (1947–1967)

14.3.1 Jute

During the post-independence era, most of the jute-growing areas went to Pakistan (Now Bangladesh) and only 0.57 million ha areas remained in the country with merely a total production of 3.31 million bales (1 bale = 180 kg) and productivity of 1.04 t/ha which was insufficient to cater the need of the jute industries in the country. Consequently, the increasing demand for jute fibre as raw material by the jute mills was satisfied mainly through an increase in the acreage of the crop (from 0.57 to 1.08 million ha) and a relatively less increase in productivity (1.04–1.13 t/ha) till the 1980s. Thereafter, the area of raw jute declined from 1.02 to 0.68 million ha though the productivity was more than double (2.54 t/ha). The trend in the area, production, and productivity of raw jute on a decadal basis is depicted in Table 14.1.

Two very popular white jute varieties (JRC 212 and JRC 321) and one tossa jute variety JRO 632 with high yield and early maturity were developed through selection and released in 1954 for growing in low-lying areas. From 1961–1962, till 1968, India surpassed the production of 50 lakh bales mark (0.9 million tonnes) which was more than sufficient for the requirement of the mills. Although there was considerable fluctuation in productivity during this period, in the latter part (i.e., 1968) it touched nearly 1.4 t/ha.

Table 14.1 Area, production and yield of raw jute in India since 1950

Year	Area (million ha)	Production (million bales)	Yield (t/ha)
1950–1951	0.57	3.31	1.04
1960–1961	0.90	5.26	1.05
1970–1971	1.08	6.19	1.03
1980–1981	1.30	8.16	1.13
1990–1991	1.02	9.23	1.63
2000–2001	1.02	10.56	1.87
2010–2011	0.77	10.00	2.33
2019–2020[a]	0.68[a]	9.50[a]	2.53[a]

Note: 1 bale = 180 kg
Source: Directorate of Economics and Statistics, DAC, Govt. of India
[a] *First advance estimates; Source: DES, MoA & FW*

14.3.2 Cotton

The partition in August 1947, affected the cotton industry severely because the bulk of the irrigated area (approx. 40% of the total area) under cotton was ceded to Pakistan. On the other hand, most of the cotton mills remained in India. Therefore, the production in India plummeted to 2.3 million bales in 1948 (Fig. 14.1). Most of the cotton produced was of short and medium staple fibre length. During the early part of the twentieth century, the country had to import huge quantities of cotton in the range of 0.8–0.9 million bales (1 bale = 170 kg lint) per annum (https://cotcorp.org.in/). However, with concerted efforts initiated by the Government under the aegis of the Indian Central Cotton Committee (ICCC), the area and production of cotton increased to 7.8 M ha and 5.3 million bales, respectively by 1966–1967 (Barik and Gautam 2009).

14.3.3 Sugarcane

Sugarcane is an asexually propagated crop but for breeding, true seeds provide the basic material. The fertility of sugarcane seed was first reported in Barbados by J.W. Parris in 1858. Interspecific hybridization and sugarcane breeding work in India started in 1912 with the establishment of the Sugarcane Breeding Station at Coimbatore. Co 205 hybrid developed by crossing noble cane (*Vellai*) with *S. spontaneum, was* released in 1918. It became very successful in the major sugarcane-growing tract of subtropical north India. The introduction of POJ varieties in the breeding programme resulted in the development of many good sugarcane varieties like Co 213, Co 244, Co 312, and Co 313 which were very successful from 1920 to 1940. For tropical India, another Java variety POJ 2878 was used as a parent for the development of wonder variety Co 419 with very good yield and sugar content (Nair 2011). The trend of area, production and yield in sugarcane during the last 70 years has been depicted in Fig. 14.2.

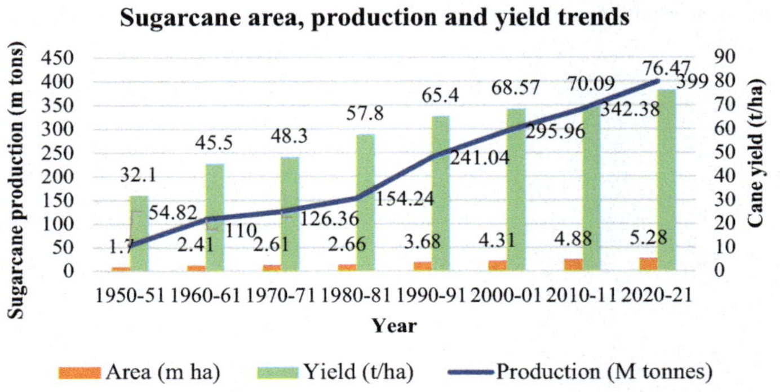

Fig. 14.2 A trend of area, production and cane yield during the last 70 years

Presently the main aim of sugarcane breeding is to develop varieties having high cane yield and sugar content accompanied by traits like resistance to red rot, wilt and smut diseases. Other significant attributes are resistance to abiotic stresses like waterlogging, soil salinity, and extreme high and low temperatures. Good ratooning ability is also a required characteristic for a variety to be released.

14.4 Green Revolution (GR) Era

14.4.1 Position of Jute, Cotton and Sugarcane in GR Era

14.4.1.1 Jute

As a result of stiff competition from cheaper synthetic fibre, export markets of jute goods started shrinking during the late sixties and early seventies which resulted in shrinkage of jute area and production. This ultimately reduced the area to 5 lakh ha and production of 29 lakh bales in 1969. All India Coordinated Research Project on Jute and Allied Fibres (AICRP on JAF) was established in 1967 with its headquarters at JARI, Barrackpore, West Bengal. The AICRPJAF project was modified to the All India Network Project on Jute and Allied Fibres (AINPJAF) in the year 2004. Initially (1969–1970) AINPJAF started with five centres, which has presently expanded to nine (09) regular centres. The production of jute has remained stable in the range of 10–12 million bales during the last decade. Although an area under cultivation has witnessed a decline of 15.5%, the productivity has increased during the same period.

14.4.1.2 Cotton

After independence, cotton breeders concentrated their efforts on developing improved cotton varieties of the Upland cotton or 'American' cotton 'tetraploids' ($2n = 52$) to replace diploid desi cotton ($2n = 26$). Varietal improvement got momentum with the inception of the All India Coordinated Cotton Improvement

Project (AICCIP), at Coimbatore in April 1967. Indian cotton breeders utilized wild species of cotton as novel gene sources to introgress traits like fibre quality, and tolerance to biotic and abiotic stresses into cultivated species. Since the inception of AICCIP, more than 375 non-Bt cotton varieties and hybrids have been released so far in different states of India (Singh and Kairon 2001). This resulted in great improvements in production and productivity (Fig. 14.3).

During this period, cotton area increased from 48 lakh ha to 76 lakh ha with an annual growth rate of 0.89% per annum. But the yield showed a higher growth rate of 1.26% per annum due to which production increased to 45 lakh bales with an annual growth of 2.15% (Table 14.2).

Within decades, there was a total shift from the *desi* to the American cotton (Table 14.3). By the end of the twentieth century, much more rapid changes were witnessed in the species composition. At the time of Independence, more than 97% of the area was under desi cotton.

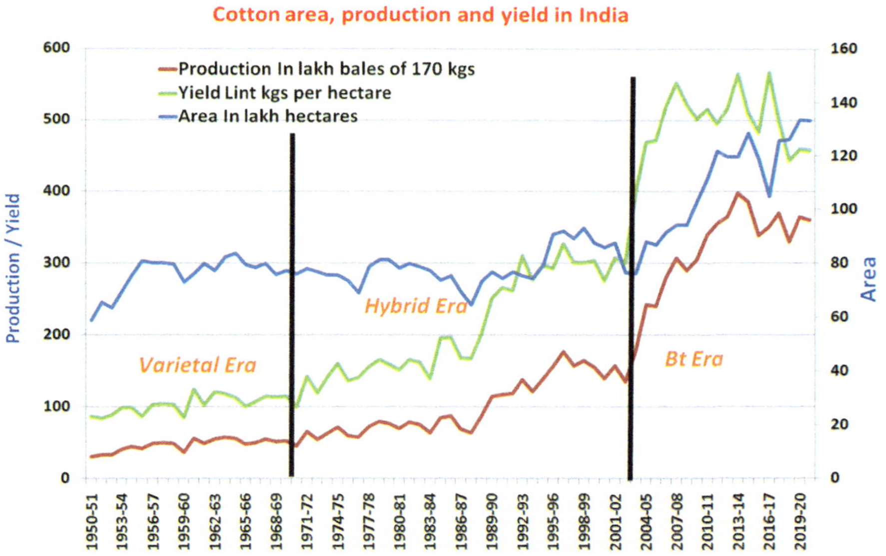

Fig. 14.3 Area production and productivity of cotton in India over the years

Table 14.2 Compound growth rate (percent per annum) of cotton area, production and yield during different periods

Period	Area	Production	Yield
Varietal era (1947–1970)	0.89	2.15	1.26
Hybrid era (1970–2002)	0.43	3.73	3.28
Bt-cotton era (2002–2020)	3.06	3.96	0.87
2002–2010	4.45	10.24	5.54
2010–2021	1.22	−0.12	−1.35

Table 14.3 Change in cotton species composition in India

Species	% of total cotton area						
	1947	1970	1980	1990	2000	2007	2020
G. arboretum	65	30	20	30	17	4	~2
G. hirsutum	3	53	54	48	69	90	97
G. herbaceum	32	17	14	12	11	5	Negligible
G. barbadense	–	–	11	10	3	1	0.9
	100	100	100	100	100	100	100

The area under *desi* cotton (*G. herbaceum)* is now restricted to the "Wagad" area in Gujarat. The Indian *desi* cotton lacked parameters like high strength and length, a high spinning ability which made the pressing need of growing the American cotton all over the country.

During this period, two major changes in agronomy were noticed (1) line sowing replacing the broadcast method of sowing and (2) application of fertilizers. Gradually with increasing cases of pest incidence, there was an increase in the usage of pesticides.

14.4.1.3 Sugarcane

The breeding strategy was oriented towards inter-varietal crosses among popular and promising varieties. This helped in the identification and release of many promising varieties in the early 1950s and 1960s. Identification of distinct sugarcane cultivation zones has been done under AICRP on Sugarcane and varieties are identified and released for specific zones. Thus, the concept of in situ selection was introduced. In order to facilitate sugarcane breeding and selection in different parts of the country, a "National Hybridization Garden" for sugarcane was established by the Indian Council of Agricultural Research at SBI, Coimbatore in 1971 (Sinha et al. 2016) (Table 14.4).

14.4.2 Biodiversity, Varietal Development and Breeding Programme

14.4.2.1 Jute

Sudan Green (SG), exotic germplasm from Sudan, Africa, having premature flowering resistance was identified and hybridized with JRO 632 and JRO 620 and pedigree selection, which resulted into varieties like JRO 524, JRO 7835 and JRO 878. JRO 524 (Navin) replaced 90% *capsularis* jute rapidly due to its early maturity (120 days) and high yield (34–36 q fibre/ha). Like Sudan Green, Tanganyika-1, an exotic strain was utilized in the hybridization programme and JRO 8432 (Shakti) (IC 15901 × Tanganyika-1) was developed and released in 1999. In 2002, JRO 128 (Surya) was released for quality fibre (fineness 2.7 tex) with a yield potential of 35–4.0 t/ha. S 19 (Subala) was released in 2005 with a high yield (3.5–4.0 t/ha) and better-quality fibre (fineness 2.7 tex, strength 25.95 g/tex with less lignin content) along with fine fibre *capsularis* variety JRC 80, having potential for textile blending.

Table 14.4 Sugarcane growing agro-climatic zones of India

Sl. no.	Name of zone	Area coverage	Particulars
Subtropical			
1.	North western zone	Uttar Pradesh (western and central), Punjab, Haryana, and parts of Rajasthan	Extreme moisture deficit during summer. Short elongation phase and less growth due to low winter temperature
2.	North central zone	Eastern U.P., Bihar and West Bengal	Flood results in inundated crops for long periods
3.	North eastern zone	Assam and other northeastern states	Sugarcane grown in patches and area under cultivation is very less
Tropical			
4.	East coast zone	Coastal Odisha, Coastal Andhra Pradesh, Coastal Tamil Nadu	Very high productivity, most suitable climatic conditions for sugarcane growth
5.	Peninsular zone	Maharashtra, Gujarat, Karnataka, Kerala, Madhya Pradesh and parts of Andhra Pradesh and Tamil Nadu	Longer duration. (*Adsali* crop of 18 months), most favourable for sugar accumulation and long sun-shine hours, low temperature at night adds to high sugar recovery

Source: Project coordinator (sugarcane), AICRP, Lucknow

Another variety, JRO 204 (Suren) was released in 2007 which outperformed both JRO 524 and JRO 8432 with a yield potential of 36–40 q fibre/ha. JBO 1 (Sudhangshu), JRO-2407 (2.30 tex) and JROM-1 (2.57 tex) were found to have better fibre fineness. *Capsularis* variety JRC 532 was released which covered 90% of the *capsularis* area. CO 58 and JBO 1 were released in 2010 for late sown condition. JBC 5 (1.45 tex), JRCM 2 (1.25 tex) and *olitorius* variety JROM 1 (2.57 tex) were released in 2013, having better fibre fineness, and being more suitable for numerous textile and non-textile diversified applications. JRO 2407 (2.30 tex), JRC 9057 (1.31 tex) and KJC 7 (1.30 tex) with better quality and suitable for lowland situations were released in 2016. An *olitorius* variety JROB 2, having high biomass-yielding potential (>60 t/ha) was released in 2021.

14.4.2.2 Cotton

The world's first cotton hybrid (H-4) was developed in 1970 by Dr. C T Patel and became popular and was cultivated on a large scale by farmers. The release of the first inter-specific hybrid *Varalaxmi* by the UAS, Dharwad followed. The success of these two hybrids resulted in the development of several hybrids by the different state agricultural universities and the prominent ones are listed in Table 14.5.

The cotton area did not show a significant increase during this period. Cotton productivity during this period increased to 301 kg lint/ha from 101 kg/ha with an annual growth of 3.28% while production registered a growth of 3.73% per annum

Table 14.5 Popular hybrids in India

Name of State	Tetraploid cotton	Diploid cotton
North Zone		
Punjab	FHH 209, LHH 144	DDH 11, Moti, FMDH 3, FMDH 8, FMDH 9
Haryana	Dhanlaxmi, Om shankar	AAH 1, CICR 2, AAH 32
Rajasthan	Maru vikas	RAJDH 9
Central Zone		
Madhya Pradesh	JKHy 1, JKHy 2, JKHY 11	–
Gujarat	H 4, H 6, H 8, H 10	DH 7, DH9,
Maharashtra	PKV Hy 2, NHH 44, NHH 250, Savithri, RHH 195, NHH 302, CICR HH 1	AKDH 7, AKDH 5, PhA 46
South Zone		
Andhra Pradesh	LAHH 1, LAHH 4, NHB 80, Bunny, Mallika	–
Telangana	LAHH 1, LAHH 4, NHB 80, Bunny, Mallika	–
Karnataka	Varalaxmi, DCH 32, DHB 105, DHH 11, RAHH 455, Bunny, Mallika	DDH 2
Tamil Nadu	Savita, TCHB 213, Surya, Shruthi, TSHH 0629, Suguna, Bunny, Mallika	–

by increasing from 4.5 million bales in the previous period to 13.6 million bales by the end of this period (Table 14.2).

14.4.2.3 Sugarcane

Sugarcane is an asexually propagated crop for cultivation, however, for breeding purposes, true seeds provide the basic material. In general, six different approaches have been recognized as far as sugarcane breeding is concerned. This includes poly cross-breeding, nobilization breeding, wide crossing, close breeding, and inbreeding. The sugarcane varieties at present in commercial cultivation are the results of spectacular advances in breeding by interspecific hybridization and they continue to provide varietal improvement.

Interspecific hybridization and sugarcane breeding work in India started in 1912 with the establishment of the Sugarcane Breeding Station at Coimbatore. Barber crossed noble cane (*Vellai*) with *S. spontaneum*. The resulting F_1 hybrid was named Co 205 and it was released in 1918. With the success of interspecific hybridization, tri-specific hybridization was tried using clones of *Saretha, Kansar,* etc. The introduction of POJ varieties in the breeding programme resulted in the development of many good sugarcane varieties like Co 213, Co 244, Co 312, and Co 313 which were very successful from 1920 to 1940. For tropical India, another Java variety POJ 2878 played an important role as a parent for the development of the wonderful variety Co 419 with very good yield and sugar content.

Presently the main aim of sugarcane breeding is to develop varieties having high cane yield and sugar content accompanied by traits like resistance to red rot, wilt, and smut diseases. Other significant attributes are resistance to abiotic stresses like waterlogging, soil salinity, and extreme high and low temperatures. Good ratooning ability is also a required characteristic for a variety to be released.

14.5 Post-Green Revolution Era

14.5.1 Biotechnological Approach

14.5.1.1 Jute

The development of molecular markers and genomics technologies during the twenty-first century has triggered the augmentation of these technologies in traditional plant breeding. Though a number of molecular markers, particularly *SSRs* have been developed in jute, the application of these technologies is limited due to several bottlenecks, such as delayed developments in marker and genomics technologies, low genetic diversity, incomplete genetic maps, low marker coverage in genetic maps, difficulty in trait-marker linkage establishment, low power of QTL detection and preponderance of dominance variation for the quantitative traits related to fibre yield and component characters. Despite these biological and developmental roadblocks, significant achievement has been made in marker discovery and genetic map construction.

Genome size has been determined by various authors, e.g., *C capsularis* and *C. olitorius* are ~ 280 Mb and ~ 324 Mb, respectively (Sarkar et al. 2011), *C. capsularis* ~ 274 Mb (Akashi et al. 2012); *C. capsularis* and *C. olitorius* ~ 336 Mb and 361 Mb (Zhang et al. 2021), respectively. In general, *C. capsularis* contains a smaller genome compared to *C. olitorius*.

A preliminary genetic map of *C. olitorius* was developed by Das et al. (2012) and Topdar et al. (2013) using SSR markers but a map of *C. capsularis* was developed using RAPD, ISSR and SRAP markers (Chen et al. 2014). Subsequently, high-density genetic map of *C. olitorius* and *C. capsularis* was developed (Kundu et al. 2015; Tao et al. 2017). Yang et al. (2019) developed a *C. olitorius* map containing 4839 SNP markers over a length of 1375.41 cM. A few QTLs for fibre yield and component characters have been mapped on these maps, such as QTLs for plant height, stem diameter, node number, fibre yield, wood yield, green biomass and root weight (Sarkar et al. 2016). Three fibre quality-associated traits, namely fibre strength, fibre fineness and histological fibre content were also mapped by Kundu et al. (2015). A total of 16 QTLs for salt tolerance were identified in *C. olitorius* (Yang et al. 2019). In addition, a number of mapping populations using multiple parents are being developed to reduce linkage drag. A multi-parent advanced generation inter-cross (MAGIC) population involving 20 parental lines of diverse geographical origin has been developed (Sarkar et al. 2016), which shows significant variability for fibre yield, plant height, base diameter and green biomass. While the genetic markers developed in jute have shown good potential for population structure and diversity analyses, jute breeding is yet to gain the benefits of genomic selection approaches.

14.5.1.2 Cotton

Cotton is the first genetically engineered crop that was approved for commercial cultivation in the country in 2002. The crop was genetically modified in order to overcome the problem of serious pest infestation, especially the American bollworm (*Helicoverpa armigera*). Of the six events which are being commercialized in India, only Mon531 and Mon15985 events that harbour full-length Cry toxin was able to capture the cotton seed market. Modified cry1Ac gene developed and used in India is a truncated version of Cry toxins having three domains of activated toxins. The transgenic developed with the truncated cry1Ac gene has resulted in 2–2.5 fold higher protein expression than Mon531 event containing full-length Cry toxin. It also posed 100% mortality to the American bollworm (Singh et al. 2016). However, work of exploring new Cry protein/toxins for the pink bollworm has not been exercised much in India.

Domain swapping strategy is another recent expansion to transgenics development, which has been used to generate Cry1AcF, wherein Domains I and II of Cry1Ac have been swapped with domain III of Cry1F. Cry1AcF efficacy has been validated through transgenic development in crops such as chickpea, castor, groundnut and pigeonpea against lepidopteran insects (Rathinam et al. 2019). Further, the deletion of few amino acids of N-terminal end of Cry2A with replacement and introduction of Lysine and Proline has shown increased toxicity up to 6.6-fold against three lepidopteron insect pests. Transgenic events of cotton generated by transformation of chimeric cry2AX1 gene, a synthetic codon optimized gene designed from cry2Aa and cry2Ac genes of *Bacillus thuringiensis* exhibited 43.33–86.66% mortality of *H. armigera* (Dhivya et al. 2015, 2016).

14.5.2 Multiple Stress-Tolerant Varieties

14.5.2.1 Biotic Stress-Resistant Varieties

Over 40 insect-pest species infect jute (~30% loss from stem rot diseases *Macrophomina phaseolina*). However, only a few of them are major insect-pests of jute, namely, yellow mite (*Polyphagotersonemus latus*), hairy caterpillar (*Spilosoma obliqua*), stem weevil (*Apion corchori*), indigo caterpillar (*Spodoptera exiqua*) and jute semilooper (*Anomis sabulifera*). Most of the resistance sources have come from the indigenous and exotic germplasm of jute. The cultivar 'JRO-204' exhibits moderate resistance to the yellow mite and stem rot. The pathogens are in soil-borne/seed borne/airborne and infect over 150 species. Despite many attempts, the genetics of resistance to stem rot is not well understood. Among, wild species *C. aestuans*, exhibits good resistance against hairy caterpillars and 'WCIN-136-1' (INGR21036) exhibits high resistance against stem rot. Interspecific hybridization between *C. olitorius* and *C. aestuans* has resulted in the development of advanced breeding lines showing resistance to stem rot (Mandal et al. 2021). Another wild species, *C. fascicularis* has resistance against indigo caterpillars.

14.5.2.2 Abiotic Stress-Resistant Varieties

Jute yields are seriously impacted by abiotic stresses (such as waterlogging, drought and salinity). Among two cultivated species, *C. capsularis* L. is a drought-sensitive species, and C. *olitorius* L. is a drought-tolerant species. The jute crop faces drought stress in the early growth phase and waterlogging stress during the later growth stage. Early flowering of jute specifically in *C. olitorius* jute is the abiotic stress-induced trait. Low night temperatures, cloudy sky and short day length coupled with drought often induce early flowering in jute, which is detrimental to fibre quality. Early flowering of jute is always followed by stem bifurcation at the top and branching which impact fibre grade. Therefore, it is necessary to develop jute varieties that are tolerant to changing environmental conditions via molecular breeding strategies. Gene and QTL mapping involved in jute drought stress is not reported to date. Drought tolerance in jute plants is carried out mainly on the evaluation of drought-resistance germplasm and morphological, physiological, and biochemical changes during drought response and transcriptome sequencing (Yang et al. 2017; Kabir et al. 2021).

14.6 Improved Management Practices

14.6.1 Production Technologies

14.6.1.1 Jute

Cropping Systems

The promising location-specific jute and allied fibre-based crop sequences for both rainfed and irrigated conditions have been identified which had certainly strengthened the stability of production systems. Under rainfed conditions some recommended jute-based crop sequences are—jute-lentil, jute-blackgram, jute-toria for north and south Bengal; jute-blackgram-toria, jute-blackgram-wheat, jute-toria, jute-rice-toria for Assam; jute-mustard-cowpea for Orissa, Bihar and Uttar Pradesh, respectively. Under irrigated conditions, the recommended sequences are jute-rice-potato, jute-rice-vegetables, and jute-rice-mustard for north and south Bengal; jute-rice-potato/wheat/pea for Assam; jute-rice-lentil, jute-rice-wheat for Bihar, Orissa and West Bengal, respectively (Mitra et al. 2006).

Resource Use Efficiency

Resource use efficiency in terms of land use efficiency (LUE), water use efficiency (WUE), energy use efficiency (EUE), production efficiency (PE) and economic efficiency (EE) were varied in different cropping systems under nutrient and crop residue management practices (Kumar et al. 2016). Jute-rice-baby corn-jute (leafy vegetable) cropping system recorded the highest system productivity (19.24 t/ha) followed by jute-rice-garden pea (8.86 t/ha), water use efficiency (34.86 kg/m^3), production efficiency (65.9 kg/ha/day), and economic efficiency (Rs. 724/ha/day) followed by jute-rice-garden pea recorded those parameters were (8.94 t/ha), 27.01 kg/m^3, 30.31 kg/ha/day, and Rs.346/ha/day, respectively. Assuming jute fibre yields 3.0 t/ha and water consumed per ha of jute 5371 cum, to produce 1 kg jute fibre, about 1.98 cum or 1980 L water is used.

Intercropping

Large-scale adoption of jute and green gram/mung intercropping (1:1) resulted in a higher total return from the intercropping system (27–33 q jute fibre and 5–6 q mung/ha) and also adds 2 ton/ha of pulse threshing waste (nitrogen 2.35%) which is the source of organic matter for improving soil health. Jute-green gram intercropping smothers dicot weeds and *Cyperus rotundus* by 54% reducing 95% sunlight light entry below the canopy and eliminating second weeding (30 man-days/ha) in jute crop (Ghorai et al. 2010). Suitable early jute varieties resistant to pre-mature flowering, short-duration green gram varieties ('TMB -37', 'Virat', 'Pant mung 5', maturity in 52–60 days, resistant to Mung bean Yellow Mosaic Virus) made this system successful.

Integrated Nutrient Management (INM)

The results of long-term fertilizer trials at ICAR-CRIJAF had revealed that fibre crop of jute on average, adds 3.80 t/ha of organic C along with an adequate amount of macro and micronutrients to the soil thereby, enriching the fertility status and reducing the requirement of the inorganic fertilizer of the cropping system, in the long run. The nutrient requirement of jute was estimated at 2.06–3.14 kg N, 1.50–1.66 kg P_2O_5, 5.18–7.97 kg K_2O, 4.70–4.99 kg CaO, 1.04–2.15 kg MgO, 36.7–78.4 g Fe, 11.2–25.1 g Mn, 1.8–2.6 g Cu and 13.9–21.4 g Zn for each quintal of dry fibre production. The average removal of nutrients by a jute crop yielding 20–25 q fibre/ha has been estimated at 84–111 kg N, 37–64 kg P_2O_5 and 177–197 kg K_2O, 84–124 kg Ca, 25–29 kg Mg and 9–13 kg S/ha. Research have shown that the average requirement of N and K per quintal of fibre production is relatively lower for *olitorius* than for *capsularis* variety as *olitorius* is more efficient in utilizing nutrients from soil and fertilizer.

Water Management

Jute, commonly grown as a rainfed crop, requires about 500 mm of water for its growth and development but is also grown under irrigation on a small scale in some parts of Bihar, Uttar Pradesh and coastal Andhra Pradesh. Irrigation application has recorded an increase of 20.8% fibre yield of jute over no irrigation. With the onset and peak of the monsoon, excess rainfall may cause waterlogging in the soil. *Olitorius* jute cannot thrive in standing water. It is more drought-resistant and is therefore grown on lighter soils. *Capsularis* jute, on the contrary, can grow even in standing water, especially towards the later part of its growth. Studies indicate that the application of irrigations at 50% soil water availability at 60 cm soil depth is beneficial to the crop and involves three irrigations during the pre-monsoon season for both *capsularis* and *olitorius* jute. The fibre yield was found to be the best when the available soil water regime was maintained at a 100–50% level. This required three irrigations and water use was 528 mm (rainfall received was 295 mm).

Conservation Agriculture (CA)

Jute has a much higher CO_2 absorption rate which is creating new avenues in the era of ecological concern. Jute can sequester a significant amount of carbon in its vegetative stage and reports confirm that 1 ha of jute plant can absorb about 15 tonnes of CO_2 from the atmosphere and liberate about 11 MT of O_2 in 120 days (IJSG 2013). Field experiments conducted at ICAR-Central Research Institute for Jute and Allied Fibres, Kolkata, with zero tillage (ZT), zero tillage + residue (ZT + R) along with conventional tillage (CT) (control) under the most predominant cropping systems, i.e. jute–rice–wheat/lentil/mustard systems revealed that minimum soil disturbances coupled with residue retention optimized soil properties and provided better soil environment for plant growth. Soil organic carbon (SOC) was positively correlated with mean weight diameter (MWD) (0.83) and available N (0.68) but negatively correlated with BD (−0.74). Evaluation of soil quality index (SQI) showed better values in jute–rice–lentil (range 0.42–0.62) under zero tillage with residue (range 0.45–0.62) compared to other systems, depicting significant cumulative effect of lentil crop along with zero tillage (ZT) and residue incorporation on soil quality.

Mechanization

Due to poor farm mechanization in inter-culture operations, fibre extraction and retting of jute, the cost of cultivation in raw jute is quite high. In 2007, ICAR-CRIJAF developed two portable farm implements "CRIJAF Nail Weeder" and "CRIJAF Multi Row Seed Drill" to promote farm mechanization in jute agriculture. CRIJAF Herbicide Applicator was developed and commercialized in 2008. The development of the "CRIJAF Multi Row Seed Drill" and its improved type "Jute Seeder" had significantly contributed to replacing broadcasting by line-sowing to a large extent. Both the implements reduce seed rate by 50%, reduce labour requirement for manual weeding and thinning by 55–60 man-days/ha compared to broadcasted jute (110–115 man-days/ha), and increase fibre yield by 15–20%. The low-cost and lightweight manual weeders "CRIJAF Nail Weeder" and "Single Wheel Jute Weeder" developed by ICAR-CRIJAF give effective control of the composite weed flora (80–85%) at the very early stage (5 days after emergence) in jute and other upland crops.

Improved Retting

ICAR-CRIJAF had made a landmark achievement of developing a microbial retting consortium "CRIJAF SONA" comprising three pectinolytic bacterial strains of *B. pumilus* (*Bacillus pumilus* IMAU80221; *Bacillus pumilus* GVC 11 and *Bacillus pumilus* SYBC-W) having xylanase activity without any cellulase activity which facilitates the jute retting process by reducing the retting time by 6–7 days than traditional retting (14–19 days), increases fibre productivity by 8–10% and improving fibre quality, particularly in stagnant water retting by 1–2 grades with an additional net income of INR 12,000–15,000/ha over conventional retting. The use of "CRIJAF SONA" had also reduced the retting water requirement of jute by 75%.

Energy and Economics

The energy consumption calculation showed that total energy consumption was 10,904 MJ/ha in jute (Singh et al. 2018). The total cost of cultivation for 1-ha jute crop excluding land rent has been worked out during 2020–2021 to be INR 1.12 lakh. With an average fibre yield of 28.5q the gross return is INR 1.56 lakh per ha with the sale price of jute fibre at INR 5100/q and jute stick at INR 200/q which results in net return of INR 4400 per ha.

14.6.1.2 Cotton

Conservation Tillage

Deep ploughing followed by land levelling and planting was the common practice in the irrigated north as well as in the rainfed tracts of central India. With the development of planters, farmers in north India began to adopt rotary tillage with relatively minimum soil disturbance and planting with cup-type seed drills. But better plant stand was obtained using inclined plate planters and facilitated timely planting on a large-scale opening with the possibility of high-density planting (Venugopalan et al. 2013).

Cropping Systems

In north India, a single crop of long-duration cotton was the common practice. But with the development of the short-duration cultivars, it was transformed into a double-crop. Timely sowing of wheat after cotton became possible with irrigation facilities and conservation practices. In central India, strip cropping of cotton + pigeonpea was the most common practice (Blaise et al. 2005). The adoption of high-density planting systems in the American and *desi* cotton varieties improved cotton productivity (Blaise et al. 2021a).

Weed Control

Until the hybrid era, farmers used animal power and manual weeding with traditional implements but significant changes in weed control methodology happened due to the availability of effective selective herbicides (Blaise 2011) and the use of high clearance tractors (Punjab and Haryana). In central and south India, bullock-operated blade hoes are still a popular method. Alternative methods such as the use of cover crops with allelopathic effects (Blaise et al. 2020), stale seed-bed technique, and use of mulch (Nalayini et al. 2009; Blaise et al. 2020) were also identified.

Fertilizer Use

Soil fertility depletion is now a major cause of concern for cotton farming. On average, fertilizer use was 93 kg/ha in 2000–2001 which increased to more than 240 kg/ha in 2017–2018. Soil fertility issues became a growing concern for want of application of organic manures and increasing micronutrient deficiencies (Srinivasa Rao et al. 2012). Bioconversion of cotton stalks into valuable bio-enriched compost at the on-farm level can bring down the dependency on inorganic fertilizers for sustainable cotton production apart from the safe disposal of cotton wastes thus protecting the environment (Velmourougane et al. 2022). Physical and biological soil health can be improved by integrating legumes into cotton-based cropping systems (Blaise et al. 2021b, c).

Irrigation

Water application efficiencies are as low as 70% in surface-flood irrigation when compared to 90% with the pressurized irrigation system. Furthermore, fertigation through drip has now become popular and the farmers are invariably using the fertigation technique wherever the drip system is being operated.

14.6.1.3 Sugarcane

Germination

Soaking of setts in water for 6 h before planting enhanced germination and also resulted in higher cane yield. Horizontal placement of setts is better than vertical placement and continuous laying of setts along the furrow requires about 40,000 three-bud setts/ha ensuring optimum growth and yield. The adverse effect of low temperature on germination was also made to overcome using growth regulators like gibberalic acid, inter cropping and inoculation of sugarcane setts with *Gluconacetobacter diazotrophicus*.

Seed-Cane, Planting and Tillering

Sugarcane is asexually propagated through pieces of its stalk called setts. The use of stalk pieces with two or three buds instead of whole cane as planting material resulted in a higher yield in comparison with one bud or whole cane. To upgrade seed multiplication ratio of 1:10 with these three bud setts, Spaced Transplanting (STP) technique was devised and was found to be successful in increasing the seed multiplication ratio to 1:40. Inter-row spacing was found crucial in determining the number of millable cane and row spacing of 90, 75 and 60 cm for autumn, spring, and late planting of sugarcane was found better to ensure ample and early synchronous tillering. Tillers in sugarcane are considered the base for the final commercial yield. The number of millable canes directly contributes to 40% of the yield followed by length (27%), girth (3%) and weight (30%) of the cane stalk in subtropics (Yadav 1993). Row spacing using double, triple and paired row (staggering the rows at 70: 20, 150:30 and 120:30 cm) were also studied for significant yield advantage. Wider planting (<120 cm row distance) is being advocated for mechanized cultivation, but for small farmers spacing up to 90 cm is optimum. The trench method of planting is better than the flat method of planting. *Vertical Planting* of sprouted three-bud setts in 45 cm deep and 20 cm wide trench gave 30 t/ha more cane yield than the flat method. *The ring pit* (90 cm diameter and 30 cm depth) *method and its modification*

(75 cm × 30 cm) of planting was another breakthrough to give 100% more yield as compared to the conventional flat planting.

Intercropping

In the initial stage (130–135 days) of sugarcane growth, short-duration pulses, oilseeds, vegetables or cereals are advocated. In spring planted sugarcane; green gram, cowpea, black gram, soybean, dhaincha, and in the autumn-planted cane Gram, lentil, potato, and peas mustard are suitable for intercropping. In the spring, ratoon cane, guar and green gram can be intercropped while in the autumn ratoon, berseem, peas and gram are suitable. Sugarcane + green gram intercropping system recorded the highest cane and cane equivalent yields. An innovative wheat-sugarcane intercropping system under Furrow Irrigated Raised Bed (FIRB) was developed for sandy loam and loamy soil, by sowing wheat in November on raised beds and sugarcane in the furrows in February, resulting in 35–40% more yield. Potato companion cropping with wheat after every four rows have also been suggested. Intercropping cowpea, green gram, and black gram in the spring crop during the early growth stage (60 days after planting) reduces harmful weed effects. Intercropping of high-density early bulking forage crops (Egyptian clover/Indian clover) was found to increase bud sprouting of stubbles. Intercropping potato with sugarcane ratoon crop has also been reported for a favourable increase in sprouting and thus the number of tillers and yield.

Nutrient Management

Sugarcane as a long-duration crop removes a substantial amount of essential nutrients from the soil. As nitrogen is more crucial for sugarcane tillering and yield, the blanket recommendation is 112–300 kg N/ha based on the soil, climate, and duration of the crop. Sugarcane roots absorb both the nitrate and ammonium ions alike and various fertilizer sources of nitrogen were found to have similar effects. Trash mulching is also reported for its importance in increasing N uptake and apparent recovery of N. Sugarcane is less responsive to phosphorus and potassium application. Among macro and micronutrients, Sulphur (S) was reported critical, especially in coarse-textured low organic matter soils, and application of 20–40 kg S/ha is recommended. Crop residues and sugar factory by-products were used and the application of sulphitation press mud along with *Gluconacetobacter diazotrophicus* was compared to that the recommended dose of fertilizer (150: 60:60 kg/ha NPK) in a plant-ratoon system. The use of sugarcane trash (3.5 t/ha) as mulch, was found important in sequestrating carbon in the soil and hence balancing CO_2 emission. A package of practices for organic sugarcane cultivation was perfected in the Institute as nutrient management for organic sugarcane cultivation under different crop rotations evinced profitable cane yield with sulphitation press mud or FYM at 20 t/ha.

Water Management

Sugarcane being a long-duration crop requires 20 ML of irrigation water/ha for its cultivation. Plants and ratoon differ in their efficiency to use water. The sugarcane yield is higher when more water is made available to the crop during the tillering phase. Irrigation during the first, second and third orders of tillering ensured the highest cane yield. The furrow method of irrigation was found better in comparison to the flood method. Skipping of alternate furrows would save 36% of irrigation water and enhance water use efficiency by 64% along with improved yield. Scheduling irrigation using IW/CPE ratio is a better approach to realizing higher cane yield with limited use of water. Among current high wastewater-efficient ligation methods like sprinkler and drip reported, a yield of 158 t/ha using the drip method of irrigation. To combat waterlogging conditions, early planting, trench and raised bed method of planting, enhancing seed rate to compensate the tiller mortality, spray application of urea, propping with earthing up, etc. have proved efficient. Drought or moisture stress particularly during the initial stages causes a 30–70% loss in sugarcane yield. Soaking of setts in saturated limewater, application of FYM, foliar spray of KCl and urea (2.5% each during the stress period), trash mulching and pitplanting proved effective.

Ratoon Management

Ratooning reduces cost and also ensures high sugar recovery at the beginning of the crushing season. In India sugarcane is ratooned at least once after the plant crop. Ratooning results in substantially reduced yield compared to that of plant crops. Application of K (66 kg/ha) along with irrigation water, 1 month before harvesting of sugarcane plant crop increased sprouting, number of millable canes, and cane yield of subsequent ratoon. Besides this, the application of CCC at 8 kg/ha and Ethrel at 500 ppm have been found useful for stubble sprouting. Agro-techniques for multi-ratooning viz., trash mulching, gap filling, phorate application (15 kg/ha), stubble shaving, and 20% extra seed in the plant crop enhanced ratoon yield to the tune of 34, 38, 43 and 77% over the control in the first, second, third and fourth ratoons, respectively. Inoculation of *Trichoderma* with trash mulch increased the cane yield by 12.8% over no trash removal and no *Trichoderma* application. Trash mulching and gap-filling contributed 21–28 and 10–29% increases in yield, respectively. Ratoon Management Device (RMD) which involves earthing up from both sides of the cane stubbles leading to protection of subterranean buds from cold injury has been reported to enhance the survival and sprouting of ratoon buds by providing a congenial rhizospheric environment.

14.6.2 Integrated Pest Management (IPM)

14.6.2.1 Jute

Climate change is affecting the weed infestation in jute and influencing the dynamics of crop–weed competition. The weed survey conducted in jute growing states revealed that during the last few years, several weeds like *Trianthema* sp., *Portulaca* sp., *Portulaca microphylla, Euphorbia microphylla, Digera arvensis, Alternanthera sessilis* has become invasive in jute causing considerable yield loss of the crop. Jute crop suffers from heavy weed infestation during the early stage of their growth, which significantly reduces the fibre yield and increases cultivation cost. The magnitude of yield loss ranged between 52 and 55% in *C. capsularis* and 59–75% in *C. olitorius* (Sarkar et al. 2005). The growth of weeds in the jute field showed a sigmoidal pattern and maximum dry biomass weight of grassy weeds was observed at 87 days in *C. olitorius* and 112 days in *C. capsularis* jute (Sarkar and Bhattacharya 2005). The critical period of crop-weed competition in jute was found to be between 15 and 60 days after sowing (Gogoi and Kalita 1992). Manual weeding in jute is costly incurring around 35% of the total cost of cultivation (Saraswat 1980) and the energy requirement for weeding is quite high (543 MJ/ha) (Borkar et al. 1999). Maximum control of weeds was achieved with two manual weeding at 3 and 5 weeks after sowing (Guha 1999) or with hoeing at 15 DAS and two hand weeding at 3 and 5 weeks after sowing(Prusty et al. 1988). Mixed cropping with Red amaranth (seeding at 10–30 kg/ha) resulted in 22–25% more weed suppression than two hand weedings (Ghorai 2007). Butachlor 50 EC or 5 g at 1–1.5 kg a.i./ha (rainfed or irrigation) or Pretilachlor 50 EC at 0.5–0.9 kg a.i./ha (45–48 h after sowing with irrigation) were found to be effective pre-emergence herbicide for weed control in jute. Post-emergence spray of quizalofop ethyl 5% EC (40–50 g a.i./ ha) + adjuvant at 1 mL/L of water at 15–21 days after emergence (DAE) controlled the grassy weeds very effectively (Ghorai et al. 2004).

Similarly, pest and disease incidence are getting significantly influenced by a change in climate, viz. elevated temperature, high humidity and erratic rainfall and also with changes in input use patterns (Satpathy et al. 2016). Lepidopteran insects including semi looper (*Anomis sabulifera* Guenee) and hairy caterpillar (*Spilarctia obliqua* Walker), are the major ones resulting 20–48.5% fibre losses in jute. In the last decade, the indigo caterpillar (*Spodoptera exigua* Hubner.) has become an important pest in jute. The reports of gram pod borer (*Helicoverpa armigera*), bacterial leaf spot, little leaf and bunchy top in jute also indicate the emergence of new pests and diseases in jute with changing climatic conditions. Stem rot caused by *Macrophomina phaseolina* is the major disease in jute and spread of the disease is increasing with increased rainfall and humidity particularly in the jute growing areas suffering from impeded drainage condition.

14.6.2.2 Cotton

In the pre-Bt era, bollworms (spotted bollworm—*Earias vitella* and *E. insulana*; pink bollworm—*Pectinophora gossypiella*) were the main constraints in the cotton crop. After the introduction of *Bt* cotton, sucking pests like aphids, leafhoppers, thrips and whiteflies (virus vectors causing leaf curl) became a threat. The most important strategies are the growing of tolerant *Bt* hybrids, timely sowing, and avoiding crop extension. However, reports on resistance breakdown to cry toxins in pink bollworm and its recent outbreaks in the last decade have resulted in serious loss to cotton growers in India.

The Bollgard II (BGII) hybrids offered effective control against the American cotton bollworm and the spotted bollworm (Kranthi et al. 2005). Pink bollworm has now evolved resistance to Cry1Ac + Cry2Ab in BGII (Kranthi 2015). A spray of any synthetic pyrethroid such as cypermethrin, fenvalerate or organophosphates such as quinalphos or profenophos, is now recommended for controlling the pink bollworm whenever the economic threshold level (ETL) level is exceeded. Excessive use of insecticides for the control of sucking pests and bollworms has led to a situation where some minor pests are being reported to emerge as important pests (Blaise and Kranthi 2019).

14.6.2.3 Sugarcane

The extent of loss to cane yield due to weeds varies from 10% to total failure depending on the composition and diversity of weed flora. Manual hoeing performed at 30, 60 and 90 days after planting suppressed the weed population and dry matter accumulation most effectively and resulted in the highest number of millable canes and cane yield. Pre-emergence spray of sulfentrazone at 720 g/ha has been found effective in controlling C. *rotundus*. Pre-emergence application of atrazine at 2 kg/ha followed by 2,4-D spray at 1.0 kg/ha 60 days after planting is found to be most effective in controlling the weeds below the economic threshold level. For integrated weed management (IWM), the best treatment for controlling weeds in sugarcane was atrazine (2.0 kg/ha) applied immediately after planting + manual hoeing 45 days after planting. Periodical hoeing (30, 60, 90 DAP) or hoeing with herbicides gives the best weed control and thereby increases cane and sugar yields.

Sugarcane is vulnerable to several insects-pests and diseases and these problems are more serious in sub-tropics. The major sugarcane insect-pests are moths, borers, scales, white-grubs, termites, white-flies, black bugs and mealy bugs. The mechanical (removal of egg masses or adults, and their destruction and rogueing of infested plants), cultural (light earthing-up, avoiding late planting, de-trashing of cane stalks) and biological methods, as well as the use of resistant varieties, are some of the means of controlling pests. Pesticide formulations of chlorinated hydrocarbons, organo-phosphates and carbamates have effectively been used in reducing damage to the crop caused by a variety of insect-pests. But the moths, borers and scale insect have defied chemical control. Red rot, smut, rust (fungi), grassy-shoot disease (mycoplasma), and ratoon-stunting disease (bacteria) are the important diseases of sugarcane, transmitted through seed-cuttings (seed cane). They are, to a great extent, responsible for lowering cane yield, its quality or both. The raising of seed nurseries from heat-treated canes offers the best scope for eliminating primary disease infection of ratoon stunting and grassy shoot diseases.

14.6.3 Drivers Affecting Productivity

14.6.3.1 Jute

The major factor affecting the productivity of jute and allied fibres are: (1) Improved high-yielding variety, (2) Effective weed management technology, (3) Mechanization of intercultural operations, (4) IPM and INP protocols and (5) Efficient harvesting and retting technology.

14.6.3.2 Cotton

The 'telephone method' used to lend support to the branches by farmers using telephone wires and shift from the broadcast method of sowing to the square or rectangular method recorded. In the irrigated regions, a shift in the planting on the furrow irrigated raised bed system became popular and was gradually replacing the flatbed method. Hybrids were found to be more responsive to fertilizer application and the adoption of phosphate and potash fertilizers was popularised (Kairon and Venugopalan 1999). With the decontrol of fertilizers, the phosphate and potash fertilizers increased resulting in an imbalanced fertilizer supply, lopsided in favour of nitrogenous fertilizers. The shift in the cotton species from the *desi* to the American cotton; gradually lead to an increase in productivity and ultimately production.

14.6.3.3 Sugarcane

Ratoon Management Device (RMD) which involves earthing up from both sides of the cane stubbles leading to protection of subterranean buds from cold injury have reported to enhance the survival and sprouting of ratoon buds by providing a congenial rhizospheric environment.

14.6.4 Value Addition

14.6.4.1 Jute

Apart from traditional uses of hessian and sacks, jute has tremendous potential for diversified uses like paper pulp, biofuel, jute as a leafy vegetable, blended textiles and technical textiles.

Paper Pulp: Jute can serve as an excellent alternate natural source for paper pulp. The average biomass production of *C. capsularis* and *C. olitorius* ranges between 48 and 55 t/ha. ICAR-CRIJAF has developed a tossa jute variety 'JROB-2' for this purpose.

Biofuel: At present, only 5.5–6% of the green biomass of jute is extracted as commercial fibre, the remaining huge amount of biomass (55–60 t/ha) can be successfully converted to bioenergy, particularly to biofuel due to less lignin content (Satya and Maiti 2013).

Vegetable Crop: Traditionally consumed as a vegetable in India and South East Asia for their nutritional values, jute leaves contain a high amount of antioxidants.

Technical Textile and Composites: Jute fibre is being used for many "technical textiles" like Agrotech, Buildtech, Clothtec, Geotec, etc. which are extensively used for manufacturing of soil savers, wall hangings, soft luggage fabrics, etc. Fibre composite is an important area where jute fibre is used in combination with synthetic or other natural fibres. Some of the commercial uses of these composites are false ceilings, panels, doors, automobile interiors and many other innovative products.

14.6.4.2 Cotton

Coloured Cotton: Naturally coloured cotton has a history of more than 5000 years in India. The colour of the natural cotton fibre varies widely from dark tan, brown, khaki, grey and green. The coloured cotton fibres are not suitable for machine spinning due to their short-staple and low strength, thus their future dwindled in comparison with the white cotton. Moreover, the neglect of the coloured cotton stocks owing to the shrink in demand, remained the poor yielders. However, the short supply of chemical dyes during the Second World War spurred the interest in naturally coloured cotton (Kranthi 2016). Gradually, India reinitiated the efforts to explore the potential of cultivating the coloured cotton. These efforts resulted in the development of improved varieties with three basic colours in naturally coloured cotton brown, grey and green in *G. arboreum* and *G. hirsutum* background. Some of the recent naturally coloured varieties of *G. hirsutum* are; MSH-53 (Vaidehi 95) Dark brown, JCC 1 (KC 94–2)- Bright Almond brown coloured cotton, NDLH 1, HC 2, DDB 12: Dharwad dark brown, DMB 225-Dharwad Medium Brown and DGC 78 Dharwad green, whereas AC 2 Brown, DDCC 1-Brown and DGC 78 Green lint belonged to *G. arboreum*.

Surgical Cotton: It is a specialty product that is used for medical purposes. The best cotton cultivars are those that have very short fibre lengths and high absorbency with low wax content (Blaise et al. 2021a, 2022). At present, the combernoil waste that remains in the textile and ginning industry is commonly used because of its low price. Further, the linters (small fibres attached to the seed) that remain after ginning had high-end uses such as the making currency notes and high-quality paper.

The cottonseed that is left after the seed cotton is ginned has multiple uses such as (1) extraction of vegetable oil (Prasad and Blaise 2020); (2) cotton seed meal left after oil extraction is used as animal feed and (3) protein-rich cotton seed can be used as a protein supplement just like the soybean.

14.6.4.3 Sugarcane

Sugarcane, a C4 photosynthetic large stature perennial grass, has one of the most efficient photosynthetic mechanisms among commercial crops since it allows it to fix almost 2–3% of radiant solar energy and transforms it into chemical energy that is usable as a food and fuel source. The high photosynthetic capability also allows it to show a high coefficient of CO_2 fixation, comparable to the moderate climate zone woods thus contributing to the decrease of the greenhouse effect. A hectare of sugarcane produces about 100 tonnes of green matter every year, which is more than twice the agricultural yield of most other commercial crops. A crop of 100 tonnes leaves around 60–65 tonnes of dry matter (8–10 tonnes of trash (dry leaves) in the field), while after extraction of sucrose, nearly 50–60 tonnes of bagasse is generated in sugar factory zones. This dry matter when burned has the potential of producing 4000 kcal/kg (7200 Btu/lbs). The total dry matter content thus has a fuel equivalent of about 10–20 tonnes of oil and efficient use of the energy potential of sugar cane can result in approximately 1 tonne of oil equivalent for every tonne of sugar produced. Bagasse is generally used as a raw material for heat energy to run the sugar mills and also for co-generating electricity, whereas trash except for its use as mulch is yet to be exploited for its fuel potential. As a metabolic energy carrier for animal feeding, each cultivated hectare delivers 75,000 million calories each year, equivalent to more than eight times the yield of other fodder (60–120 kWh of electric power per tonne of cane).

From cane harvest and processing, it is possible to obtain more than eight products and by-products, which are potential raw materials for the extractive, chemical, biochemical and fuel industries, leading to the production of more than 50 commercial products. Practically all products and by-products obtained from sugar cane have the potential to serve as substrates for liquid or solid-state fermentation processes and via the usage of the available second- and third-generation technologies, a significant number of production processes and products could be developed. Sugar cane thus has a significant advantage as a renewable raw material, in the production of basic chemicals, with a yield not equalled by any other plant. Since the energy-delivering capacity of sugarcane is equivalent to five times that used by the crop, the energy produced by cane plants and variations among different varieties or germplasm make it an extremely important renewable and sustainable energy bio-resource. All the above factors taken together and the possibilities offered by further genetic improvement turns sugar cane into an ideal energy crop for the next century. Along with annual sugar production of 35 million tonnes the sugar industry produces 333 crore litres of ethanol used for blending with petrol to run the automobiles. Using sugarcane bagasse as fuel the industry also generates 8000 MW of electricity of which 6500 MW is supplied to the grid as surplus. Considering the targeted 20% ethanol blending of petrol by 2030 AD under enhanced emphasis on the utilization of renewable sources of energy and its ever-increasing demand, there exists vast scope for transformation of sugarcane into energy cane.

Export-Import of Jute, Cotton and Sugarcane

The export of jute goods fetches a significant amount of foreign exchange and the total export of jute goods had increased from INR 2092 crores INR in 2010–2011 to INR 2427.6 crore INR in 2019–2020 (Source: IJMA, Directorate General of Commercial Intelligence and Statistics, Kolkata). India has exported 46.05 lakh bales (0.85 million metric tons) of cotton valued INR 6123 crores during 2018–2019 (DGCIS, Kolkata). Exports of sugar in 2020–2021 stood at seven million tonnes and 5.96 million tonnes in 2019–2020 respectively. India's sugar exports increased to $4.6 billion (about INR 35,000 crore) in 2021–2022 from $1.17 billion (about Rs 9000 crore) in 2013–2014.

14.6.5 Breakthrough Research

14.6.5.1 Jute

Early Flowering Resistance

Introduction of resistance to premature flowering was a path-breaking achievement in *tossa* jute. Although *tossa* jute had higher productivity, it did not fit well into the cropping system as farmers had to sow the crop almost 1 month later than white jute. Still, during the 1970s the area under jute cultivation was dominated by *capsularis* varieties *(capsularis: olitorius = 75:* 25). 'Sudan Green' (SG), exotic germplasm from Sudan, Africa, having premature flowering resistance was identified and hybridized with 'JRO-632' and 'JRO-620' followed by pedigree selection, which resulted into varieties like 'JRO-524', 'JRO-7835' and 'JRO-878' during the 1970s at ICAR-CRIJAF, Barrackpore. Among these 'JRO-524' (Navin), began to replace *capsularis* varieties rapidly. Being released in 1977, it became popular among farmers during the early 1990s, and still maintains dominance.

Genomics

ICAR-CRIJAF has recently measured and reported that two *Corchorus* species' genome sizes are about 300% smaller than that of their reported estimates and for *C. capsularis* and *C. olitorius* are ~280 Mb and ~324 Mb, respectively (Sarkar et al. 2011). Scientists from ICAR- CRIJAF in collaboration with ICAR-National Research Centre on Plant Biotechnology (NRCPB) successfully accomplished whole genome sequencing of *C. olitorious* cv. 'JRO-524' (Navin) with a draft genome size of 377.3 Mbp.

Microbial Retting Consortium

ICAR-CRIJAF had made a landmark achievement of developing a microbial retting consortium "CRIJAF SONA" comprising of three pectinolytic bacterial strains of *B. pumilus* (*Bacillus pumilus* IMAU80221; *Bacillus pumilus* GVC 11 and *Bacillus pumilus* SYBC-W) having xylanase activity without any cellulase activity which facilitates the jute retting process by reducing the retting time by 6–7 days than traditional retting (14–19 days), increases fibre productivity by 8–10% and improving fibre quality, particularly in stagnant water retting by 1–2 grades with an additional net income of INR 12,000–15,000/ha over conventional retting.

14.6.5.2 Cotton

First among the technological changes that swept cotton farming over the past 20 years was the development of the *Bt* cotton hybrids. Improved agronomic management practices such as high-density planting, water management, new selective herbicides and insecticides, and partial mechanization, brought in major changes in production technologies.

Bt Cotton

Genetically modified cotton, popularly known as '*Bt* cotton' in India was first developed by US multinational company Monsanto in 1996 and later by several companies including JK Seeds, Metahelix in our country (Table 14.6).

The adoption of *Bt* cotton in 2002 brought about drastic changes in cotton cultivation in the country. *Bt* cotton has replaced the area under varieties of *desi* and upland cotton and also non-*Bt* hybrids. The area under *Bt* hybrids increased to an extent of 94% in 2020 (Fig. 14.4).

Table 14.6 Transgenic cotton events approved for commercial cultivation in India

S. no.	Event	Developer	Year of approval	Name of the gene
1	MON 531	Mahyco/monsanto	2002	Cry 1 Ac
2	MON 15985	Mahyco/monsanto	2006	Cry 1 Ac + cry2AB
3	Event-1	JK agri genetics	2006	Cry 1Ac
4	GFM event	Nath seeds	2006	Cry 1Ab + cry1Ac
5	BNLA 106 Dharwad event	ICAR-CICR, UAS, Dharwad	2008	Cry1ac
6	Event 9124	Metahelix	2009	Cry 1 C

Source: Karihaloo and Kumar (2009)

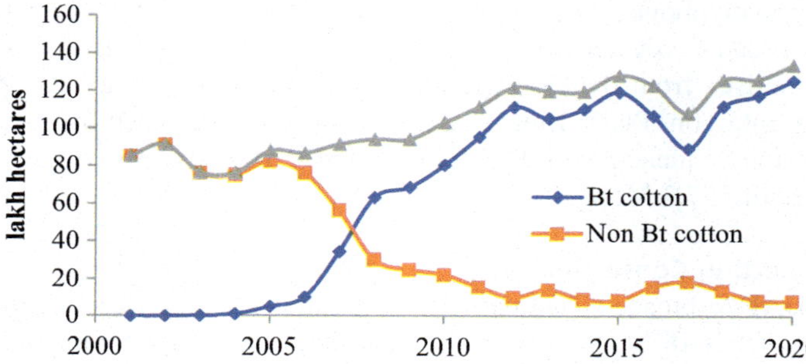

Fig. 14.4 Area under non-Bt cotton and Bt cotton hybrids after the introduction of *Bt* cotton in 2002

To date more than 2000 *Bt* cotton hybrids containing various approved events have been approved for commercial cultivation in India. Twelve *Bt* cotton varieties containing Cry 1Ac gene have been released commercially for cultivation in India by the public sector institutions (Table 14.7).

The lint yield per hectare which was stagnant at about 300 kg/ha prior to the introduction of *Bt* cotton, jumped to 472 kg in the year 2005–2006 and reached 566 kg/ha in 2013–2014 (Fig. 14.4). The growth of the cotton area, production and productivity was spectacular during this period, especially the first half of this period. Cotton production reached 39.8 million bales in 2013–2014. Similarly, the cotton area reached 12.8 M ha during 2014–2015. However, after 2013–2014, the lint yield is stagnant at around 500 kg lint/ha. Further, the *Bt* hybrids were found to be susceptible to most of the sucking pests. Consequently, there is an interest in desi varieties having superior fibre quality and high yield potential. Probably, this may prove as the turning point for the stronger comeback of desi varieties. However, with the *Bt* hybrids ruling the roost, the species composition is totally lopsided with more than 97% area under the American cotton and the rest under the Extra Long Staple and *desi* cotton. This has created a short fall in the short-staple category of cotton.

Table 14.7 List of *Bt* cotton varieties released for commercial cultivation in India

S. no.	Name of the Bt variety	Developer	Year	Zone/state	Condition
1.	ICAR-CICR Bt 6 (RS2013)	ICAR-CICR	2017	Haryana	Irrigated
2.	ICAR-CICR Bt 14 (CPT2)	ICAR-CICR	2017	Maharashtra	Rainfed
3.	ICAR-CICR Bt 9 (SRI1)	ICAR-CICR	2017	Maharashtra	Rainfed
4.	ICAR-CICR GJHV 374 Bt	ICAR-CICR	2017	Maharashtra	Rainfed
5.	ICAR-CICR PKV 081 Bt	ICAR-CICR	2017	Maharashtra	Rainfed
6.	ICAR-CICR Rajat Bt	ICAR-CICR	2017	Maharashtra	Rainfed
7.	ICAR-CICR Suraj Bt	ICAR-CICR	2017	Maharashtra	Rainfed
8.	ICAR-CICR 16 Bt	ICAR-CICR	2020	Central zone	Irrigated
9.	ICAR-CICR 23 Bt	ICAR-CICR	2020	South zone	Rainfed
10.	ICAR-CICR 21 Bt	ICAR-CICR	2021	Central zone	Rainfed
11.	ICAR-CICR 25 Bt	ICAR-CICR	2021	South zone	Rainfed
12.	PAU Bt 3	PAU	2020	North	Irrigated

Source: Singh et al. (2021)

14.6.5.3 Sugarcane

Interspecific hybridization through mobilization and improved method of planting (trench method of planting) to be better over the flat method of planting. A new method of planting, IISR 8626 was found superior over others. Vertical Planting of sprouted 3 bud setts in a 45 cm deep and 20 cm wide trench gave 30 t/ha more cane yield than the flat method. The ring pit method of planting was another breakthrough in the list of planting methods. This planting method produced cane with 0.1 unit higher pol percent than flat planting mainly due to uniformly grown mother shoots at harvest. The system has also been reported to give 100% more yield as compared to the conventional flat planting in the subtropics.

14.6.6 Future Strategies

14.6.6.1 Jute and Allied Fibre Crops

The challenges in scientific development are very complex and diverse for jute and allied fibre crops. The narrow genetic base of cultivated jute and allied fibre crops, less harvest index, high cost of cultivation, unavailability of low-cost mechanization, lack of simple and economical retting technology and vulnerability to biotic and abiotic stresses are the major constraints for jute farmers. Development of varieties and plant types for multiple stresses and diversified use are the challenges that need to be addressed. However, there are tremendous opportunities that may be exploited through advanced scientific concepts, and trained manpower with adequate infrastructural backup.

- Improvement of jute and allied fibre crops to suit the changing environment/ requirement.
- Value addition for diversified use and increasing profitability.

- Application of biotechnology/genetic engineering tools for improvement of jute and allied fibre crops.
- Controlling increased incidence of diseases and pests in jute and allied fibre crops.
- Integrated weed control measures in jute and allied fibres.
- Mechanization of jute and allied fibre crops cultivation in view of shortageof labour and cost-effectiveness.
- Retting of jute and allied fibres with low volume of water and/or with waste water.
- Effective dissemination of jute and allied fibre production technologies.

14.6.6.2 Cotton

The National Gene Bank of the Central Institute for Cotton Research, Nagpur has more than 50 genetic stocks of coloured cotton which were indigenously collected or were imported from Mexico, Egypt, Peru, Israel, the Soviet and USA. The future of naturally coloured cotton is vibrant as it is a pollution-free, eco-friendly, energy-efficient, cost-effective, nontoxic, novel viable textile material that in the near future has the potential to be explored for a wide range of colours with different shades, uniformity and stability. Presently, the country has approximately one-third share in the global cotton area and one-fourth in cotton production. A stable and good price is one major reason for the increase in the cotton area. During the past few years, cotton is commanding remunerative prices in the market. Further, a favourable Minimum Support Price and effective procurement policy encouraged farmers to prefer cotton to other crops which was evidenced by the increase in cotton area. During the last two seasons (2019–2020 and 2020–2021), though there was COVID-19 pandemic, cotton farmers were able to market their products due to the MSP operations by the Cotton Corporation of India (Mayee et al. 2020). During the current season (2021–2022) cotton prices touched a new height of INR 10,000/quintal. There is an apprehension that the cotton area further increases by 15–20% during the next season.

14.6.6.3 Sugarcane

- Precision application technology of all inputs for enhancing resource use efficiency in sugarcane agriculture.
- Techniques for high-density shoot management during the initial stages of crop growth to promote healthy tillers are needed.
- Concerted efforts to reduce excessive tiller mortality during the grand growth phase are desirable.
- The yield gap in ratoon is still a concern; hence research on rhizosphere modulation and population management is required.
- Studies on the effects of climate change on sugarcane productivity and adaptation strategies are requisite.
- Green House Gas (GHG) emission quantification and balancing under the sugarcane system is the need of the hour.
- Addressing labour scarcity through viable mechanization and befitting agronomy needs to be devised.

14.7 Conclusions

Jute occupies a unique position in the national economy in terms of its contribution to employment (about five million) in agriculture and industry. It is a natural bast fibre-yielding crop with considerable commercial, environmental and socio-economic importance having diversified end-users and is mostly grown in tropical and south-east Asian countries. India is the major producer of jute goods in the world, contributing about 60% of global production. Compared to the Fifties, productivity has been enhanced by almost 3 times which is presently 2.68 t/ha. Although the area under jute is declining or constant (~8 lakh ha) since the nineties, the production is stable over decades due to improved varieties and a package of practices. Apart from traditional uses such as hessian and sacking, various value-added diversified products opened up the export market which gained to earn more than INR 2500 crores per annum. The demand for natural geo-textiles is increasing day-by-day. The jute sector needs radical changes in many other aspects beyond the traditional use of fibre, i.e. pharmaceutical, nutraceutical, diversified commercial uses and exploiting the environmental benefit as an effective carbon sequester in the backdrop of climate change and environmental stresses created by the industrialization and urbanization.

Cotton is the most important fibre crop and sustains the livelihood of more than 6 million farm households and contributes to more than a fourth of the global production. The introduction of the Upland cotton varieties into the country resulted in a significant increase in production in the late 1960s and early 1970s. The major breakthrough occurred when the world's first hybrid was developed in 1970 that lead to other major changes in the production technology such as square planting, telephone method, etc. and consequently yield levels of 300 kg lint/ha were achieved. The development of the *Bt* hybrids and its approval for commercial cultivation in 2002 brought about nearly a doubling in the yield as well as a significant decrease in the usage of pesticides. Cotton is a multipurpose crop, (1) which caters to the oil pool in the country and is presently the third major vegetable oil after rapeseed/mustard and soybean, (2) is important in surgical cotton manufacture, (3) cotton seed meal can be used as an animal feed and (4) in the making of currency notes and high-quality paper. Major changes are needed to meet future requirements in light of climate change such as the development of short-duration compact varieties that are amenable to machine harvest.

Sugarcane is the most important sweetener-producing crop and is grown worldwide under tropical and sub-tropical climatic conditions. Besides sugar, it is used as raw material for the production of jaggery (gur), khandsari, bio-ethanol and various other distillery products. Indian farmers grow sugarcane as a cash crop which grows well in medium to deep soils with adequate irrigation and drainage. Cultivars developed in India as inter-specific hybrids by crossing *S. officinarum* with *S. barberi* and *S. spontaneum* respond positively to agronomic management and high input levels. Besides sugar, the crop holds great scope as an 'energy crop' by providing raw material for electricity and bio-fuel generation that entails the further expansion of the area under the crop and the need for the development of varieties with suitable production and protection technologies for its altered utilization. The increasing cost of production and growing scarcity of labour makes the mechanization driven sugarcane cultivation an immediate necessity.

References

Agarwal OP (2007) Cotton economy in India. Cotaap Research Foundation, Mumbai. https://Wcrc.Confex.Com/Wcrc/2007/Techprogram/P1780.Htm 6/7

Akashi R, Fancy N, Tanmoy AM, Khan H (2012) Estimation of genome size of jute *Corchorus capsularis* (L.) var. CVL-1 using flow cytometry. Plant Tissue Cult Biotechnol 22(1):83–86

Barik A, Gautam HC (2009) Revolution in Indian cotton. Directorate of Cotton Development, Ministry of Agriculture, Department of Agriculture & Cooperation, Government of India, Mumbai

Blaise D (2011) Tillage and green manure effects on Bt transgenic cotton (*Gossypium hirsutum* L.) hybrid grown on rainfed vertisols of central India. Soil Tillage Res 114:86–96

Blaise D, Kranthi KR (2019) Cotton in India. In: Chauhan BS, Jabran K (eds) Cotton production. Wiley, Hoboken, pp 193–215

Blaise D, Bonde AN, Chaudhary RS (2005) Nutrient uptake and balance of cotton + pigeonpea strip intercropping on rainfed vertisols of central India. Nutr Cycl Agroecosyst 73:135–145

Blaise D, Manikandan A, Verma P, Nalayini P, Chakraborty M, Kranthi KR (2020) Allelopathic intercrops and its mulch as an integrated weed management strategy for rainfed Bt-transgenic cotton hybrids. Crop Prot 135:105214

Blaise D, Kranthi KR, Ravindran CD, Thalal K (2021a) High plant density can improve productivity of Asiatic cotton (*Gossypium arboreum* L.). Arch Agron Soil Sci 67:607–620. https://doi.org/10.1080/03650340.2020.1741553

Blaise D, Manikandan A, Desouza ND, Bhargavi B, Somasundaram J (2021b) Intercropping and mulching in rain dependent cotton can improve soil structure and reduce erosion. Environ Adv 4:100068

Blaise D, Velmourougane K, Santosh S, Manikandan A (2021c) Intercrop mulch affects soil biology and microbial diversity in rainfed transgenic *Bt* cotton hybrids. Sci Total Environ 794:148787

Blaise D, Kranthi KR, Saxena S, Venugopalan MV, Mohan P (2022) Productivity and fibre traits of diverse surgical cotton (Gossypium arboreum L.) cultivars under rainfed conditions. Indian J Agric Sci 92:300–304

Borkar UN, Palit P, Singh DP (1999) Energy productivity of jute and sisal. In: Palit P, Pathak S, Singh DP (eds) Jute and allied fibres-agriculture and processing. Golden Jubilee Publication, CRIJAF, Barrackpore, pp 129–131

Chen Y, Zhang L, Qi J, Chen H, Tao A, Xu J, Lin L, Fan P (2014) Genetic linkage map construction for white jute (*Corchorus capsularis* L.) using SRAP, ISSR and RAPD markers. Plant Breed 133(6):777–781

Das M, Banerjee S, Dhariwal R, Vyas S, Mir RR, Topdar N, Kundu A, Khurana JP, Tyagi AK, Sarkar D, Sinha MK (2012) Development of SSR markers and construction of a linkage map in jute. J Genet 91(1):21–31

Dhivya K, Natarajan B, Varatharajalu U, Duraialagaraja S (2015) Expression and functional analysis of Cry2AX1 gene in insect resistant T3 progeny cotton plants. Biosci Trends 8(18): 5055–5060

Dhivya K, Satish S, Balakrishnan N, Udayasuriyan V, Sudhakar D (2016) Genetic engineering of cotton with a novel cry2AX1 gene to impart insect resistance against *Helicoverpa armigera*. Crop Breed Appl Biotechnol 16:205–212

Ghorai AK (2007) Weed management of jute (*Corchorus olitorius*) by soil solarisation. Indian J Agric Sci 77(6):390–392

Ghorai AK, Chakraborty AK, Pandit NC, Mondal RK, Biswas CR (2004) Grass weed control in jute by targa super (quizalofop-ethyl 5% EC). Pestology 28:31–34

Ghorai AK, Chowdhury H, Dey RK (2010) Intercropping of jute with pulses: a novel approach to smother purple nut sedge. JAF News 8(2):32–33

Gogoi AK, Kalita H (1992) Crop-weed competition in capsularis jute. Indian J Weed Sci 24(3 and 4):74–76

Guha B (1999) Cultural weed management in tossa jute (*Corchorus olitorius*). Ann Agric Res 20(4):406–410

IJSG (2013) Jute matters, International Jute Study Group, Dhaka, Bangladesh, 1(5)

Kabir SMT, Hossain MS, Bashar KK, Honi U, Ahmed B, Emdad EM, Alam MM, Haque MS, Islam MS (2021) Genome-wide identification and expression profiling of AP2/ERF superfamily genes under stress conditions in dark jute (*Corchorus olitorius* L.). Ind Crop Prod 166:113469. https://doi.org/10.1016/j.indcrop.2021.113469

Kairon MS, Venugopalan MV (1999) Nutrient management research in cotton—achievements under all India coordinated cotton improvement project. Fertil News 44(4):137–144

Karihaloo JL, Kumar PA (2009) Bt cotton in India-a status report, 2nd edn. Asia Pacific Consortium on Agricultural Biotechnology, New Delhi, p 56

Kranthi KR (2015) Pink bollworm strikes Bt-cotton. Cotton Statistics and News 35:1–5. http://www.cicr.org.in/pdf/Kranthi_art/Pinkbollworm.pdf

Kranthi KR (2016) How colourful is the future of naturally coloured cotton? Cotton Stat News 2014:1–4

Kranthi KR, Naidu S, Dhawad CS, Tatwawadi A, Mate K, Patil E, Bharose AA, Behere GT, Wadaskar RM, Kranthi S (2005) Temporal and intra-plant variability of cry1Ac expression in Bt cotton and its influence on the survival of the cotton bollworm, *Helicoverpa armigera* (Noctuidae: Lepidoptera). Curr Sci 89:291–297

Kumar M, Ghorai AK, Mitra S, Majumdar B, Kundu DK (2016) Productivity and resource use efficiency of different jute based cropping systems under nutrient and crop residue management practices. J AgriSearch 3(2):76–81

Kundu A, Chakraborty A, Mandal NA, Das D, Karmakar PG, Singh NK, Sarkar D (2015) A restriction-site-associated DNA (RAD) linkage map, comparative genomics and identification of QTL for histological fibre content coincident with those for retted bast fibre yield and its major components in jute (*Corchorus olitorius* L., Malvaceae sl). Mol Breed 35(1):1–17

Mandal K, Kumar AA, Ghosh D (2021) Development of screening techniques to evaluate response of jute (*Corchorus olitorius*) to *Macrophomina phaseolina*. Australas Plant Pathol 50:621–630

Mayee CD, Choudhary B, Blaise D, Patil PG (2020) COVID-19 impact on Indian cotton. ICAC Recorder 39(2):60–64

Mitra S, Maiti SN, Sarkar S (2006) Recommendations for jute and allied fibre crops: an endeavour of all India network project. CRIJAF, Barrackpore, Kolkata, p 28

Nair NV (2011) Sugarcane varietal development programmes in India: an overview. Sugar Tech 13(4):275–280

Nalayini P, Anandham R, Sankarnarayanan K, Rajendran TP (2009) Polyethylene mulching for enhancing crop productivity and water use efficiency in cotton (*Gossypium hirsutum*) and maize (*Zea mays*) cropping system. Indian J Agron 54:409–414

Prasad R, Blaise D (2020) Low gossypol containing cottonseed: not only a fibre but also a food crop. Natl Acad Sci Lett 43(7):599–602. https://doi.org/10.1007/s40009-020-00931-1

Prusty JC, Behera B, Parida AK, Lenka D (1988) Effect of herbicides on fertilizer use economy in jute-rice cropping system. Indian J Agron 33(3):315–316

Rathinam M, Kesiraju K, Singh S, Thimmegowda V, Rai V, Pattanayak D, Sreevathsa R (2019) Molecular interaction-based exploration of the broad-spectrum efficacy of a *Bacillus thuringiensis* insecticidal chimeric protein, Cry1AcF. Toxins 11:143. https://doi.org/10.3390/toxins11030143

Saraswat VN (1980) Ecology of weeds of jute fields in India. Int J Pest Manag 26(1):45–50

Sarkar S, Bhattacharya SP (2005) Economics of different weed management methods in both the species of jute. J Crop Weed 1(1):57–60

Sarkar S, Bhattacharjee AK, Mitra S (2005) Weed management in jute by Trifluralin (48%) EC in the early jute-weed competition phase. J Crop Weed 2(1):30–33

Sarkar D, Kundu A, Saha A, Mondal NA, Sinha MK, Mahapatra BS (2011) First nuclear DNA amounts in diploid (2 n= 2 x= 14) *Corchorus* spp. by flow cytometry: genome sizes in the cultivated jute species (*C. capsularis* L. and *C. olitorius* L.) are~ 300% smaller than the reported estimate of 1100–1350 Mb. Caryologia 64(2):147–153

Sarkar D, Satya P, Mandal NA, Das D, Karmakar PG, Singh NK (2016) Jute genomics: emerging resources and tools for molecular breeding. In: Ramawat KG, Ahuja MR (eds) Fiber plants. Springer, Cham, pp 155–200

Satpathy S, Gotyal BS, Selvaraj K (2016) First report of cotton mealy bug *Phenacoccus solenopsis* Tinsley on cultivated jute (*Corchorus olitorius* L.) in India. Entomol Gen 36:55–61

Satya P, Maiti R (2013) Bast and leaf fibre crops: kenaf, hemp, jute, agave, etc. In: Singh BP (ed) Biofuel crops: production, physiology and genetics. CABI, Boston, MA, p 292

Singh P, Kairon MS (2001) Cotton varieties and hybrids. CICR technical bulletin no. 13. ICAR-Central Institute for Cotton Research, Nagpur, MH

Singh AK, Paritosh K, Kant U, Burma PK, Pental D (2016) High expression of cry1Ac protein in cotton (*Gossypium hirsutum*) by combining independent transgenic events that target the protein to cytoplasm and plastids. PLoS One 11(7):e0158603

Singh AK, Kumar M, Mitra S (2018) Carbon footprint and energy use in jute and allied fibre production. Indian J Agric Sci 88(8):1305–1311

Singh SB, Venugopalan MV, Santosh HB, Raghavendra KP, Balasubramani G (2021) Bt varieties for increasing cotton yields under rainfed ecosystem. Cotton Innovate 6(1):1–3

Sinha OK, Kumar R, Chaudary SK (2016) Forty five years of AICRP on sugarcane all India coordinated research project on sugarcane. ICAR-Indian Institute of Sugarcane Research, Lucknow, p 77

Srinivasa Rao C, Blaise D, Venugoapalan MV, Patel KP, Biradar DP, Aladakatti YR, Marimuthu S, Buttar GS, Brar MS, Ratna Kumari S, Reddy VC (2012) Soil fertility management strategies for maximizing cotton production of India. Indian J Fertil 8(12):80–95

Tao A, Huang L, Wu G, Afshar RK, Qi J, Xu J, Fang P, Lin L, Zhang L, Lin P (2017) High-density genetic map construction and QTLs identification for plant height in white jute (*Corchorus capsularis* L.) using specific locus amplified fragment (SLAF) sequencing. BMC Genomics 18(1):1–12

Topdar N, Kundu A, Sinha MK, Sarkar D, Das M, Kar CS BS, Satya P, Balyan HS, Mahapatra BS, Gupta PK (2013) A complete genetic linkage map and QTL analyses for bast fibre quality traits, yield and yield components in jute (*Corchorus olitorius* L.). Cytol Genet 47(3):129–137

Velmourougane K, Manikandan A, Blaise D, Mageshwaran V (2022) Cotton stalk compost as a substitution to farmyard manure along with mineral fertilizers and microbials enhanced Bt cotton productivity and fibre quality in rainfed vertisols. Waste Biomass Valor 13:2847. https://doi.org/10.1007/s12649-022-01689-x

Venugopalan MV, Kranthi KR, Blaise D, Lakde S, Sankaranarayana K (2013) High density planting system in cotton—the Brazil experience and Indian initiatives. Cotton Res J 5(2): 172–185

Yadav RL (1993) Agronomy of sugarcane, principle and practices, 1st edn. NTEK Computer System, Lucknow

Yang Z, Yan A, Lu R, Dai Z, Tang Q, Cheng C, Xu Y, Su J (2017) De novo transcriptome sequencing of two cultivated jute species under salinity stress. PLoS One 12(10):e0185863

Yang Z, Yang Y, Dai Z, Xie D, Tang Q, Cheng C, Xu Y, Liu C, Deng C, Chen J, Su J (2019) Construction of a high-resolution genetic map and identification of quantitative trait loci for salt tolerance in jute (*Corchorus* spp.). BMC Plant Biol 19(1):1–9

Zhang L, Ma X, Zhang X, Xu Y, Ibrahim AK, Yao J et al (2021) Reference genomes of the two cultivated jute species. Plant Biotechnol J 19:2235

Part III

Natural Resource Management to Attain Food and Environment Security

Soil and Plant Nutrition 15

T. J. Purakayastha, Debarati Bhaduri, Dhiraj Kumar, Rajendra Yadav, and Ankita Trivedi

Abstract

The green revolution which was particularly the interplay of high-yield varieties, irrigation facilities, high inputs of fertilizers, energy, etc. especially in wheat and rice crops had led to a dramatic increase in food grain production in the country. India achieves self-sufficiency in total food grain production. That was no doubt warranted as the condition in the country was precarious before the green revolution and also just after independence, with lots of imports from the western world. However, agricultural intensification over the years added to the degradation of our fragile ecosystems. Loss of fertility of the soil, soil alkalinity, acidity, salinization, water logging, erosion, water pollution, air pollution, heavy metal loading in agricultural fields, a decline in the water table, micronutrient deficiency, incidence of diseases, climate change, etc. is few of the challenges which the present day agriculture is facing in the aftermaths of the green revolution. Fertilizers and pesticides have been used indiscriminately which led to the pollution of soil, air and water. Even indigenous cultivars are slowly being lost in tandem with mono-cropping systems. Several resource conservation tools and innovative resource conservation management strategies have been developed that may serve very beneficial in terms of improving soil fertility, and grain

T. J. Purakayastha (✉) · A. Trivedi
Division of Soil Science and Agricultural Chemistry, ICAR-Indian Agricultural Research Institute, New Delhi, India

D. Bhaduri
ICAR-National Rice Research Institute, Cuttack, Odisha, India

D. Kumar
ICAR-Indian Institute of Soil Science, Bhopal, Madhya Pradesh, India

R. Yadav
Agriculture University, Kota, Rajasthan, India

P. K. Ghosh et al. (eds.), *Trajectory of 75 years of Indian Agriculture after Independence*, https://doi.org/10.1007/978-981-19-7997-2_15

production and also taking care of environmental concerns. It is high time that we should adopt and promote those strategies among the farmers and other stakeholders to have a win-win situation under the present scenario.

Keywords

Environment · Food grains · Green revolution · Natural resources · Technologies

15.1 Introduction

India stands second globally, for having the largest agricultural land, covering 20 agro-ecological regions and (159.7 M ha) of arable land. Rural households' dependency on agriculture is around 60%. In order to examine and report the situation of Indian agriculture and its rural economy, the Royal Commission of Agriculture was set up in 1926. The Commission made elaborate recommendations for the improvement of Indian agriculture and in 1928, submitted its report. It evaluated agriculture in erstwhile British India encompassing Pakistan, Myanmar and Bangladesh. The country was having (91.85 M ha) of a net sown area with a 151 million of cattle population. The cereals and pulses were cultivated in about 75% of the area. Fruits and vegetables occupy around 2.5% of the area and non-food crops and oil seeds cover 20% (Report of Committee on Doubling Farmers' Income 2018). Agricultural production increased from 50 to 308 Mt during the last seven decades. After the partition of Myanmar in 1937 accompanied by Indian independence, India became food deficient and the situation was very precarious, with the import of food from western countries and thus rises in food prices along with the increasing population. It was indeed difficult to feed the mammoth population.

Post-colonial period was associated with a higher level of agricultural production than during colonial period. The agricultural output increased by 2.7% annually between 1949–1950 and 1973–1974. This figure was marginally higher than the population growth rate during that time. However, in the first half of the century, the agricultural growth rate was just 0.8% per annum (Bhalla 1979; Dantwala 1986). Among the total food grain production, the contribution of rice was nearly 50% which has declined by 0.09% annually, while the increase in population was 0.67% annually. Between 1911 and 1941, the decline in per capita food grain availability was 26% (Dantwala 1979).

Prior to green revolution the major crops grown were millet, rice, sorghum, wheat, maize and barley (Hall 1964; USDA 1963), and production of millet and rice were even higher than maize, wheat and barley production combined. Meanwhile, millet production has gone down and the commonly consumed and known crops in every household soon became fodder crops in a few decades following the green revolution. Further, we have nearly lost 1 lakh indigenous varieties of rice after the 1970s which took thousands of years to evolve and the traditionally grown rice varieties which were consumed before the green revolution shortly become nonexistent (The Hindu 2012). The loss of traditional varieties was mainly owed to the adoption of subsidized high-yielding varieties and due emphasis on monoculture.

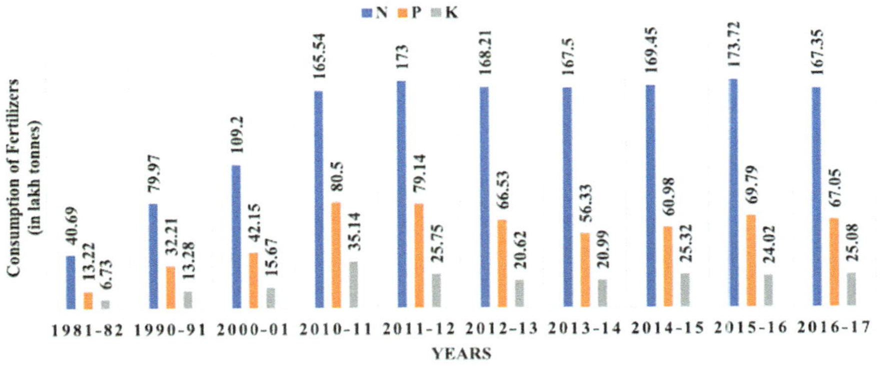

Fig. 15.1 Fertilizer consumption post-green revolution period (Source: DoF and DAC & FW 2017)

With an increase in the production of food grains, fertilizer consumption also increased over the years and mainly consumption of nitrogenous fertilizers escalated during post nutrient-based subsidy scheme in 2010 that led to skewed consumption of fertilizers (Fig. 15.1). The number of essential nutrients also got deficient in our soils with time, starting from N in the 1950s to nine nutrients (N, P, K, B, S, Zn, Cu, Mn and Fe) in 2005–2006. Widespread deficiency of micronutrients not alone but multiple nutrient deficiencies led to a risk of deficiency in humans as well due to the intricately linked food chain. Approximately, 20 Mt of three major nutrients are removed by growing crops annually (Tandon 1992), but the subsequent addition of nutrients through both inorganic and organic fertilizers falls short of this harvest. Further, as per an estimate, the gap that exists among removal and additions of nutrients is around 8–10 Mt of N, P_2O_5 and K_2O every year for the last 5 decades (Tandon 2004).

In the early 1960s, the severity was due mainly to a shortage of food grain and low productivity so there was a risk of famine. However, in present times, we are struggling for attaining sustainability, concerns with green issues, climate change, etc.

15.1.1 Pre-independence and Independence Era

Research on soils and soil properties started in India way back in 1806 during the British colonial period by Buchanan, who coined the term 'Laterite' (a local terminology resembling brick making) for the red-coloured and honey-combed like structure in earth surface lying in Malabar coast of India. Much later, in the year 1892, J.W. Leather started the research on soil science and is often popularly known as the 'Father of Indian Soil Science and Agricultural Chemistry'. He initiated research in the field of soil science at Pusa (Bihar), which was later transformed into Rajendra Agricultural University in 1970. With Dr. Leather's effort, the first long-term manure experiment for studying soil productivity was started in India;

firstly, in Kanpur (Uttar Pradesh) in 1885; thereafter at Pusa (Bihar) in 1905, and Coimbatore (Tamil Nadu) in 1908 following the Rothamsted model. Dr. Leather also demarcated four major groups of soils in India: Alluvial soils (across the Indo-Gangetic plain), black cotton soils (Deccan plateau) and red and lateritic soils (Peninsular and eastern India) (Abrol and Nambiar 1997).

With the establishment of the Royal commission (1929), the scenario of management and productivity of Indian soils changed and it helped to make subsequent efforts to improve the fertility levels of different soils in India. The commission reported that red and black soils were deficient in nitrogen, phosphorus and humus, but potash and lime were sufficient.

In India, soil research in the 1930s focused mainly on soil colloids and led to the concept of determining the role of Al^{3+} in the availability of nutrients and mainly phosphates. The physical chemists got involved in studying soil colloids' relationship to soil fertility. During this time, soil fertility research was initiated by using organic manures and fertilizers. Much emphasis was given to the use of manure, composting, and green manure and composting by Bangalore method was also developed during that phase (Randhawa 1979). Three permanent manorial trials were initiated at Jalandhar (1934), Shahajanpur (1935) and Padegaon (1939) in the subsequent years.

After World War II, the insufficiency of food grains was a major challenge and during that time FAO (Food and Agricultural Organization) was established considering the urgency. Indian agriculture was at a bad phase when we got Independence. India had just passed through the major Bengal famine (during 1942–1943) and was struggling with dangerously very low per capita food availability of 417 grams per day during 1946. Priority to agriculture was thus realized at that time and reflected in various agricultural transformations, like land reform, irrigation facilities, production of major inputs, particularly seeds and fertilizers emphasized and greater support for research and extension was given (Swaminathan 2006).

Across the diverse agro-climatic zones of the country, Indian agriculture was predominantly dependent on the monsoon. Due to the lack of adequate artificial irrigation, rainfall represented the most determinant factor in tropical agriculture. From the published literature of the pre-independence and independence era, it can be found that India's farming was backward and most of the farmers were resource-poor. Most of the farmers practised subsistence farming and cultivated crops with limited land and financial capacity. Lack of technical knowledge and financial ability restricted the farmers from using costly agricultural inputs and relevant agricultural machinery. Soil fertility was maintained by most of the farmers either by fallowing the fields regularly or through the application of refuse from household and animal sources as manures (Saha 2013).

International Rice Research Institute, Philippines and Rockfellar foundation surveyed Indian soil types during this era and concluded that improvement of soil management is essentially required to bridge the yield gap of major cereals like rice, wheat and maize. Both teams recommended that appropriate fertilizer application for the particular soil type is required. Based on this recommendation, the Indian government signed an agreement with the US government that allowed the export of fertilizers. That was the time when the very first project started in India, which leads to the initiation of the largest fertilizer demonstration program in the world to popularize the usefulness of commercial fertilizers and farmers were supplied with free fertilizers. Indian scientists were also involved in the collection of fertilizer use statistics data under various soil and climatic conditions in India. Extensive research by researchers on soil fertility and fertilizers during 1954–1956 was undertaken and emphasized to conduct national-level soil surveys and prepare soil maps. Twenty-four soil-testing service centres throughout India were also opened during that phase, with the purpose to get soil samples from farmers for testing and giving specific recommendations regarding the type and dose of fertilizers to be used by the farmers on different crops.

Proceedings of the eleventh meeting of the crops and soils wing of the board of agriculture and animal husbandry in India give us an idea about various soil research conducted before and during the 1950s. The use of radio-active isotopes in soil science had already started. The utility of both organic manures and chemical fertilizers was realized for the improvement of soil fertility and crop productivity. Recommendations were made for importing and subsidizing chemical fertilizer. The importance of rapid soil testing for evaluating soil fertility status was realized. Experimentation on the reclamation of saline and alkaline soils with various soil amendments was initiated. The use of shelter belts and wind breaks for controlling soil erosion was emphasized. Ghosh and Dutta (1960) initiated preliminary studies on soil fungi isolation from five localities of Cuttack and Orissa. Nitrogen fixation by blue-green algae in Indian rice was exhaustively studied by De (1939). Regur soils or black cotton soils of India were characterized and studied by Simonson (1954). Clay mineral identification of the black and red soil of Hyderabad was done by Nagelschmidt et al. (1940) and it was concluded that both the soils are derived from the same or very similar parent rock, a coarsely crystalline granite or gneiss. Most of the fertilizer experiments mainly focused on the response of nitrogenous and phosphatic fertilizers (Abichandani 1959). By that time, the yield response from nitrogenous fertilizer was found overwhelming and the need for phosphatic fertilizer was also realized at some localities; however, the response from potassic fertilizer was not much encouraging.

The fertilizer association of India (FAI) was established in 1955 the goal to promote the best use of balanced fertilizers for enhancing production and productivity, and ensuring the availability of adequate domestic supply of fertilizers with a liaison to industry. The FAI maintained a connection with different state governments, government of India, and other relevant research bodies and played a prominent role in serving Indian agriculture.

15.1.2 Post-green Revolution Era

The pre-green revolution period is roughly considered from 1949–1950 and 1964–1965, and the post-green revolution period from 1967–1968 and 1977–1978. Pre-green revolution era of agriculture production was dependent on area expansion. But for achieving self-sufficiency in food grain production, a new agricultural revolution known as the "green revolution" started during the mid-1960s. High-yielding varieties, chemical fertilizers, pesticides, improved mechanization and irrigation brought about the much-aspired green revolution that brought a significant shift in Indian agriculture. But several criticisms of the green revolution at different forums and reports had happened, mostly by environmentalists and critics of globalization. The intensification of agriculture caused massive environmental concerns, viz. desertification, soil degradation in terms of salinity, alkalinity, sodicity, erosion, waterlogging, heavy metal pollution, a decline in the water table, eutrophication, etc. that became evident in the 1980s.

The green revolution is also partially blamed for deteriorating soil health. Soil health is an attribute of chemical, physical and biological processes constantly declining and is often cited as one of the prime reasons for stagnating or declining crop yields and low input use efficiency. Imbalanced fertilization caused the multinutrient deficiency. The abuse of nitrogenous fertilizers, in part, contributed by government policies in fertilizer pricing and subsidization, and indiscriminate use of pesticides has caused major problems of pollution in rural India. Intensification of crop production, especially cereals, has eroded agricultural biodiversity not only of plants but also animals, insects and soil bacteria. Faulty irrigation also resulted in soil salinity and alkalinity.

During past decades, interest in soil fertility improvement and its maintenance had increased visibly due to mounting pressures on our natural resources. Increasing population, conversion of arable land into non-agricultural use, and land degradation coupled with climate change had saddened the situation. The challenge is very big for India which supports 18% of the population globally on 2.4% of the land resource, because per capita land availability is continuously decreasing from 0.34 ha (1951–1952) to 0.12 ha (2018). Further, the total food grain production which was 50 (Mt) in 1950–1951 had increased to 308 Mt during 2020–2021. The fertilizer use intensity had jumped from mere 70,000 tonnes during 1950–1951 to 133.44 kg ha^{-1} during 2019–2020. Similarly, the net area that is under irrigation increased from 21 M ha (1951–1952) to 68.4 M ha (2019–2020).

15.2 Dwindling Resource Base in Post-green Revolution Era

The green revolution of the mid-1960s no doubt brought about the country's food security and transformed the image of our nation from ship-to-mouth to self-sufficiency and further to food exporting nation. However, misuse of agricultural technologies and the greed for obtaining more led to what we called second-generation problems in the aftermath of the green revolution. The dwindling natural resource base had challenged the long-term sustainability of agriculture. The major effects in the recent past on soil health concerning different aspects is as follows:

15.2.1 Soil Organic Matter

The soil organic matter (SOM) stands as the focal point of soil health management. As SOM is the source of all the essential nutrients, a storehouse of microorganisms and captures carbon, thus improving the health of soils. Thus the restoration of soil organic carbon (SOC) is very crucial in tropical soils. SOC is a sensitive indicator that alters due to the impact of land use change, biomass alteration, global warming, pollution, deforestation, etc. From terrestrial ecosystems to the atmosphere, transfer of 1–2 Pg C yr^{-1} takes place due to land use change of which only the decomposition of SOC contributed 15–17% of carbon to this (Houghton and Hackler 1994). It is estimated that Indian soils have a total SOC pool of 21 Pg to 30 cm depth and about 63 Pg to 150 cm depth, respectively. The SOC pool contributes 2.2% of the world pool in 1 m depth and 2.6% to 2 m depth in Indian soils (Lal 2004). But the soils mainly of arid and semi-arid regions extended mostly to central and southern Indian states need immediate attention. The level of carbon is continuously declining in our soils. Proper management of soils is crucial for maintaining and or increasing the soil's carbon content. The SOC and SIC have an inverse relationship with depth. The role of SIC is immense in dry climates for controlling soil carbon dynamics (Bhattacharyya et al. 2000). The climatic adversity plays important role in low rainfall and high-temperature regions for accumulation of carbon as carbonates in soils, which drastically reduce the water infiltration, increase the soil pH, sodicity, etc. that in turn hampers the successful establishment of crops. Thus, SOM management through appropriate technological interventions and practices is the need of the hour.

15.2.2 Soil Degradation

Prior to the green revolution, the increase in food grain production was achieved through the expansion of the cultivated area. But after the green revolution, in order to achieve food grain sufficiency, the introduction of high-yielding varieties, use of chemicals and fertilizers, irrigation, improvement in farm implementation, etc. were done. India has succeeded in attaining food sufficiency, but in the aftermath, it created problems related to desertification, salinization, alkalization, water logging, deforestation, increase in the ground water table, etc. By the 1980s, these environmental problems became quite crucial and increasingly becoming prominent through time (Kumar and Pasricha 1999; Singh 1997). Around 147 M ha of land is suffering from some or other kind of degradation in India. Inappropriate agricultural practices led to widespread land degradation problems in the country, which hampers directly the food, nutritional and livelihood security of the farmers, especially small landholders. Erosion is the main contributor towards soil erosion either through water or wind that results in loss of top soil or often leads to water logging that creates soil salinization. The most severe form of erosion is water erosion which results in topsoil loss or terrain deformation. Based on soil loss data, (Dhruvanarayan and Ram 1983) in their first approximation, computed that annual total soil loss occurs @ 5.3 billion tonnes throughout the country which leads to mean soil erosion of 16.4 t ha^{-1} yr^{-1}. Of the total soil erosion, 61% got transferred from one place to another, 29% of the eroded soil was lost permanently to the sea and the rest 10% was deposited in reservoirs. The water-borne soil erosion leads to annual crop loss of cereals (13.4 Mt), pulses and oilseeds equal to 162 billion US $ (Sharda et al. 2010).

Thus, in order to combat land degradation, work on innovative soil and water conservation technologies should be adopted. Emphasis on watershed-based development, reclamation of problem soils, vegetative barriers to arrest soil loss, grassland and agroforestry development and promotion, etc. should be needed.

15.2.3 Micronutrients

The nutrient deficiency in soil reflects in the nutrient content of crops, which in turn affects the deficiency in humans and animals. Due to excessive use of high-analysis fertilizers, neglect of micronutrient application, exhaustive cropping systems, high-yielding varieties, little use of organic manures, and advancement in technological interventions for analysis, not only single but multiple micronutrients got deficient in our soils. The study undertaken by the All India Coordinated Research Project on Micro and Secondary Nutrients and Pollutant Elements in Soils and Plants showed that in 615 districts of 28 states, it was observed that widespread and variable deficiency of sulphur and micronutrients observed in soils of different states. More than 50% of soils in 86, 101, and 131 districts were found to be deficient in available B, S, and Zn out of 615 studied districts. More than 25% of soils in 41, 83, and 5 districts face deficiencies of available Mn, Fe, and Cu (Shukla et al. 2021a). The deficiency of micronutrients became a restriction to the stability, productivity, and sustainability of soils (Bell and Dell 2008). At both regional and global levels, the micronutrient deficiency affects more than two billion people (De-Regil et al. 2013; FAO 2015; Bailey et al. 2015). Meanwhile, the availability of the micronutrients in soils and their plant uptake depends upon several factors, such as soil pH, the content of organic matter, soil type, clay fractions, chelation, rhizosphere environment, external application, micronutrient efficient genotypes, root activity, etc. Thus, the fluctuation of micronutrients in the soil is a global phenomenon. Continuous application of high-analysis fertilizers in mono-cropping systems led to over-exploitation of micronutrients. Thus utmost care should be given to improving the micronutrient status of soils, which otherwise improves the nutrition of the human and animal continuum.

15.2.4 Low Nutrient and Water Use Efficiency

During 2018–2019, the total consumption of NPK fertilizer was 27.23 Mt which is likely to touch 48.0 Mt by 2050. Continuous soil fertility decline, low factor productivity, very less nutrient use efficiency (NUE), and poor environmental quality is a matter of serious concern that is creating obstacles to attaining sustainability in Indian agriculture. The use efficiency of nutrients is low in the soils of India. It varies from 30–50% relative to nitrogen (N), 15–20% for phosphorus (P) and 60–70% for potassium (K), 8–10% for sulphur (S) and 1–2% in case of micronutrients. The N that is not utilized by plants goes easily for polluting the ground water by way of leaching, to nearby rivers, volatilization, denitrification, etc. A considerable amount of P and K also got lost due to fixation, or loss due to erosion.

Water and nutrients are two very important resources for plants and animals that are needed for proper function and wellbeing. As per an estimate, around 83% of total freshwater is consumed in agriculture followed by 12% in the industrial sector and 5% in domestic use. Although the share of the latter two sectors is likely to shoot in near future owing to an increase in demand, population, economic growth, urbanization, income, etc. Fossil fuels are fast declining in the world, thus the prices of fertilizers are expected to increase in the future. Therefore, instead of increasing the production of fertilizers, the focus should be on increasing their use efficiency. With time, the consumption of fertilizers mainly the primary one had increased exponentially, but the use efficiency of nutrients was remarkably low due to their various losses on application or conversion to or cycling to some recalcitrant pools in the soil. It was estimated that if not properly managed 70–80% of N addition may be lost from rain-fed agro-ecosystems and 60–70% from irrigated agro-ecosystems (Roberts 2008; Ladha et al. 2005). The use efficiency of N in Indian conditions is day by day declining as depicted in Table 15.1. The decline in per capita water availability, water quality deterioration, and others, will surely affect the water resources to be used sustainably that make it quite difficult for attaining the target of 345 Mt of food grain production by 2030 and 494 Mt by 2050. However, this can be achieved by improving crop productivity under irrigated conditions from the current 2.3 t ha^{-1} to 4.0 t ha^{-1} and in the rainfed area from 1.0 t ha^{-1} to 1.5 t ha^{-1} (Rattan et al. 2014). This can only be achieved provided water use efficiency (WUE) and NUE are enhanced. Unregulated irrigation practices lead to the loss of about 70% of applied water as runoff, leaching, and other avoidable losses that in turn increase the cost and drastically reduces the WUE and NUE. Thus, these losses need to be checked and if possible avoided (Thompson 2012).

Table 15.1 Nitrogen use efficiency and its time trend in different countries

Country	Year	Nitrogen use efficiency (kg grain kg^{-1} N)	Change (%)
USA	1980	42	–
	2000	57	+36
UK	1981–1985	26	–
	2001–2002	4	+23
Japan	1985	57	–
	2001	75	+32
India	1970	60	–
	2004	20	−60

Source: Patil (2009)

15.2.5 Biodiversity Loss

In relation to crops, India is one of the 12 mega-diversity centres of the globe. Before the green revolution in 1960s, India harbours more than 100,000 rice varieties (Prasad 2016), comprising a stunning diversity w.r.t. nutrition, taste, pest and disease resistance, and adaptability to a range of conditions in the age of climate change. The quest for the adoption of high-yield varieties (HYVs) and hybrids led to a loss of much of the needed biodiversity. The genetic bases of HYVs were very narrow and the farmers were left with no other options rather to sown one type of seed every year. All this resulted in a loss of several thousand of indigenous species that were grown locally and the systems in agriculture that have evolved owing to the basis of the accumulated wealth of knowledge over centuries (Vandana 1993). The native oilseed crops like mustard, sesame, etc. and the pulse crops like moong, tur, etc. were not cultivated further in large areas as was done earlier. The traditional grown and consumed crops like millets and other coarse grains were very well adapted to grow in arid and semi-arid regions with minimum water requirements. However, farmers were obliged to move to only rice and wheat owing to the unavailability of high-yielding seeds of millets (Srivastava et al. 2020). The old system that considers farmers mere recipients of the technology not as a participant-led to increased dependency on relatively few plant varieties. Further, increased industrialization and similar trend become the key players of what we called "genetic erosion".

15.2.6 Heavy Metal Pollution

The persistent use of agrochemicals together with waste water for irrigation purposes resulted in soil quality deterioration due mainly to heavy metal accumulation which is non-degradable and thus persists for longer periods. The metal contamination of soils not only negatively impacted the growth and yield of plants but also leads to food chain bioaccumulation and hampers the soil's biological activity. Due to the direct consumption of food crops grown on contaminated soils, heavy metal soil pollution becomes a great concern for human health (Ling et al. 2007; McLaughlin et al. 2000). Agricultural activities also lead to the potential level of toxic metal addition in soil and different activities from irrigation through waste water in peri-urban areas to fertilizer and pesticides application causes heavy metal load (Table 15.2).

Continuous application of large quantities of inorganic fertilizers leads to heavy metal accumulation in intensively cultivated agricultural soils (Khan et al. 2008) and thus alters soil health. For example, phosphatic fertilizers add cadmium, mercury, arsenic, lead, and other potentially toxic elements to the soil that are considered to be carcinogenic (Pierzynski et al. 2000). Nearly 20% of the pesticide spray goes waste as runoff in the soil that either leaches down and pollutes the ground water or surface water or vaporizes along with the soil water (McLaughlin et al. 2000). The application of waste water for irrigation on long term basis often leads to heavy metal accumulation in soils. Singh et al. (2010) assessed the impact of heavy metals like

Table 15.2 Heavy metal contamination of soils due to agricultural sources (mg kg^{-1})

Element	Nitrogenous fertilizers	Phosphate fertilizers	Manures	Pesticides	Sewage sludge
Arsenic	2–120	2–1200	3–150	22–60	2–26
Copper	1–15	1–300	2–60	12–50	50–3300
Chromium	3–19	66–245	5.2–55	–	20–40,600
Cadmium	0.05–8.5	0.1–170	0.3–0.8	–	2–1500
Nickel	7–38	7–38	7.8–30	–	16–5300
Lead	2–1450	7–225	6.6–15	60	50–3000
Mercury	0.3–3	0.01–1.2	0.09–26	0.8–42	0.1–55
Zinc	1–42	50–1450	15–250	1.3–25	700–49,000

Source: Pendias and Pendias (2000)

Cd, Pb, Ni Cu, Zn, and Cr on human health through the consumption of vegetables and cereals grown locally and observed that the level of Pb, Cd, and Ni concentrations were quite higher than the permissible limits of WHO/FAO standards in all the studied crops. Such sort of heavy metal load leads to soil health degradation and contaminates the vegetables grown on that soil through the food chain (Rattan et al. 2002). Solid sludge application for improving soil fertility and the use of sewage water for irrigation has become a common phenomenon in majority of the developing countries (Hashimoto et al. 2009).

Sewage-sludge also contains heavy metals like Cu, Pb, Cr, Hg, Ni, Cd, and Zn in toxic levels. In suburban areas, farmers use untreated sewage to grow crops owing to its higher nutrient concentration, but its long-term use is a matter of serious concern (Saha et al. 2010). The heavy metal contamination of soil not only alters the soil pH, porosity, colour, and biological properties but also impacts the surface and groundwater leading to the deterioration of soil and water quality (Musilova et al. 2016). Heavy metal accumulation in soil on a long-term basis leads to its translocation to vegetables and crops through root uptake. Soil fertility can be reduced with toxic metal concentration and leads to the accumulation of these metals in food stuff that cause potential health hazards to consumers (Khan et al. 2008). One of the reasons for increased neurological disorders nowadays is due to consumption of heavy metal-contaminated food and vegetables (Duruibe et al. 2007).

15.2.7 Ground Water Contamination

In India, the success of the green revolution has also been attributed due to the assured groundwater supply. But, unabated pumping of water associated with climatic variability, had drastically reduced the recharge of groundwater. It created the problem of salinization, water logging, pollution, the decline in the water table, etc. Intensive extraction of ground water during the last four decades in order to fulfil the agricultural demand led to a dramatic decline in the level of groundwater in north-west India so much so that, the region is referred as the largest hotspot of groundwater depletion in the world (Long et al. 2016). Often excess fertilization causes soil, water, and air pollution. In coarse-textured soils, the situation becomes very pathetic, whereby excessive use of N fertilizer is generally recommended 25% higher than the loamy soils. High nitrate level in groundwater in intensively cultivated rice-wheat crop rotation was reported by Bajwa (1993). Improvement in water use efficiency, best-suited crops as per the water requirement, and strict regulations could arrest the over-exploitation of groundwater.

15.2.8 Greenhouse Gas Emissions

The agriculture sector is often cited as a major source of Greenhouse Gases (GHGs) and contributes around 24% of total global anthropogenic GHGs emissions (IPCC 2014). Mineral fertilizers as a major production input have contributed as a potent greenhouse gas emitter in Indian agriculture. In India, agricultural production accounts for 18% of total GHG emissions thus being a significant emitter of GHGs (INCCA 2010). Livestock and the rice cultivation are considered as the major contributor of total GHGs emissions with national average of 45.54 kg CO_2.e kg^{-1} mutton meat, 2.4 kg CO_2.e kg^{-1} milk, and 5.65 kg CO_2.e kg^{-1} rice. While, the production of non-rice cereals, fruits, and vegetables together contributes less GHGs <1 kg CO_2.e kg^{-1} product. This suggests that a shift in the dietary pattern towards the consumption of animal-based protein could increase greatly the GHGs emissions from Indian agriculture (Vetter et al. 2017). The Government of India had not committed to reducing GHGs emissions by any fixed proportion but had announced voluntarily that by 2005 to 2020, it would reduce its GHGs by 20–25% of gross domestic product (Pahuja et al. 2014; MoEF 2010). In the recently held UN climate conference (COP26) in November 2021, India committed to become a "carbon neutral" country through a net zero goal by 2070 and appealing to world leaders to make "lifestyle change".

Efficient management of input fertilizer applications and their subsequent uptake by crops would remarkably reduce the loss. Efficient management of N leads to the reduction of N_2O emission (Bhatia et al. 2012) and also through the use of nitrification inhibitors. Several nitrification inhibitors found as very effective and efficient in reducing the emission of N_2O, meanwhile, the costs are a little bit high. Thus, the use of natural nitrifications inhibitor and slow-release N fertilizers like neem-coated urea can serve as an environment-friendly technology. The lower rate of nitrification in zero-tilled soils over conventionally-tilled soils from semi-arid climates led to lower N_2O emissions (Rochette 2008). Further, the direct-seeded rice (DSR) does not require puddling and seeds could be sown easily in tilled or non-tilled soils which can save water and labour and result in the reduction of CH_4 emissions (Pathak et al. 2012). However, few natural inhibitors or slow-release fertilizer materials like neem oil-coated urea (NOCU) are cost-effective and environment-friendly technology. Puddling of rice contributes 24% of total agricultural emissions (3.37 Mt) in India, while, about (0.14 Mt) N_2O is emitted from nitrogenous fertilizer application in wheat, rice, and other crops (Bhatia et al. 2013). Thus the resource conservation strategy like conservation agriculture, smart farming, precision agriculture, zero tillage, DSR, and others may be promoted and adopted for reducing emissions from soil and side by side improving the SOC status of soils.

15.2.9 Imbalance Fertilization

The imbalance ratio of N:P:K is a major concern for Indian agriculture. While the general recommended ratio for cereals is 4:2:1 but is around 6.8:2.7:1 (2018–2019). Excessive nitrogenous fertilizers induce initial vegetative growth, which makes the crops susceptible to pests, diseases, lodging, and poor floral induction finally delaying maturity thereby leading to a reduction in yield. The consumption of fertilizers has grown almost 245 times in the last five decades. During 1950, the fertilizer use per hectare was 0.55 kg ha^{-1} but in 2017–2018, this changed to 134.07 kg ha^{-1} in India. The green revolution of the early 1960s and the increased intensification of agriculture are considered the major causes behind this growth determinant. Further, another estimate by Tandon (2004) suggested that since the past 50 years still the gap between total nutrients applied and removal is about (8–10 Mt) of N + P_2O_5 + K_2O per year. In reality, 20–80% of the nutrients in the fertilizer materials are susceptible to loss which contaminates the environment in different forms or temporarily accumulates in the soil about different complex reactions that hamper the immediate availability. Thus, this disparity can only be minimized by the efficient use of fertilizers.

15.2.10 Declining Factor Productivity

At the country level, the partial factor productivity, i.e., the ratio of grain yield and per unit of particular nutrient input is fast declining over the years. The response of cereal crops has declined considerably i.e., from 13.4 in 1970 to 3.7 kg grain kg^{-1} applied nutrient in 2005. This has really become a matter of serious concern and an issue that needs to be addressed on priority (Tewatia 2012). The probable reasons for the declining factor productivity in the country might be due to excessive fertilizer application, injudicious or imbalance of fertilization, a decline in soil fertility, soil degradation, inappropriate timing and rate of fertilizer application, intensive monocropping, antagonistic reactions between some plant nutrients, lack of soil testing facilities, unawareness among the farmers regarding balanced nutrition, fertilizer subsidy, environmental degradation, low water efficiency, etc.

15.3 Lack of Awareness and Linkages

Agricultural growth and development should not be done at the environmental cost and human health degradation. In order to eradicate malnutrition and starvation, food access in both quantity and quality terms, its accessibility and approachability at all-time should be done. Apart from the major cereal crops, production and access of fruits, vegetables, millets, pulses and oilseeds needs attention. Promotion of nutrient rich crops, diversification of cropping systems, and securing the indigenous varieties of crops are the need of the hour. The focus should be given towards small and marginal farmers, with livelihood security through making agriculture remunerative. Farmers have to be given adequate space in decision-making as a participant in the process rather than just a recipient of agricultural innovation. There is an urgent need for different agencies to come along and chalk out strategies for achieving sustainable food production, which is accessible to everyone, affordable, nutritious, and also environmentally sustainable.

15.4 Technologies for Sustainable Soil Health Management

Detailed retrospection of green revolution-style farming questioned the sustainability of soil health. So, the lessons learnt from this were thoroughly analysed and researches were conducted to decipher appropriate management practices to maintain or restore soil health. The technologies developed after the green revolution era are briefly discussed below:

15.4.1 Integrated Nutrient Management (INM)

The INM involves the integration of both inorganic fertilizers and organic manures in order to improve and maintain the fertility of soil or health and to sustain crop productivity. Incorporation of organic sources, i.e. farmyard manure (FYM), vermicompost, poultry manure, enriched compost, press mud, biofertilizers, phosphate solubilizing bacteria (PSB), vesicular arbuscular mycorrhiza (VAM), and azotobacter along with chemical fertilizers can effectively increase the nutrient availability in soil, improves physical condition and its organic carbon status. A proper balance of inorganic fertilizer and organic manures is important not only for increasing yield but also for sustaining soil health.

15.4.2 Micronutrient

Indian soils are poor in micronutrient content as the soils are under continuous mining due to intensive agriculture for a long time without micronutrient fertilizer application. Increased micronutrients demand by high-yielding cultivars along with the adoption of intensive cropping, use of low micronutrient content high-analysis fertilizers, a decline in the use of organic manures and residues of crops, crop cultivation on soils deficient in micronutrient reserves and other factors like natural and anthropogenic affects adversely the micronutrients phytoavailability and thus aggravated the situation (Takkar and Shukla 2015). On average 36.5, 12.8, 7.1, 4.2

and 23.2% of soils are deficient in Zn, Fe, Mn, Cu and B, respectively (Shukla et al. 2021b).

Micronutrient management should be based on the demand and supply of micronutrients in soil-plant system, which differs with crops, soil types, severity of deficiency, source, method, time, rates and frequency of application. Results emanating from a large number of field studies indicated that 2.5–10 kg Zn ha^{-1} as $ZnSO_4.7H_2O$ proved most effective in mitigating its deficiency and sustaining high soil productivity in the majority of crops grown on diverse Zn-deficient soils. As compared to top dressing, Zn application through broadcast or its band placement proved to be superior. Similarly, other Zn application methods like foliar application with 0.5% to 2.0% $ZnSO_4.7H_2O$ solution, side dressing, or soaking and coating of seeds with Zn solution, etc. In upland crops, Fe chlorosis is observed mainly in sorghum, rice, sugarcane, groundnut and chickpea under highly calcareous soils, or soils with limited aeration, having high P and bicarbonate content and low Fe. Ferrous sulphate (19–20.5% Fe) is the major source used for managing Fe deficiency in the country. However, Fe-EDDHA (10.0% Fe), Fe-EDTA (9–12% Fe), biotite, pyrite and organic manures (FYM 0.15% Fe), poultry and pig manure (0.16% Fe), sewage sludge also been used as Fe source to correct its deficiency in crops. By and large foliar application of $FeSO_4$ at 10–12 kg ha^{-1} or 50–150 kg ha^{-1} soil application alleviated Fe deficiency in the majority of crops (Takkar et al. 1989).

Severe Mn deficiency is difficult to manage with soil application due to soil-applied Mn oxidation, especially in high-pH soils. Foliar application of $MnSO_4.H_2O$ is an immediate measure to overcome Mn deficiency in wheat though it needs annual application. Copper deficiency in Indian soils is very less. Boron (B) deficiency is an important nutritional problem limiting crop production in acid and calcareous soils. To correct the deficiency of B, soil application of borax ($Na_2B_4O_7.10H_2O$, fertilizer grade, 10.5% B) is commonly employed. Boric acid (17% B), solubor (19% B) are mostly used as foliar sprays.

15.4.3 Biofertilizer and Its Application

Biofertilizers, referred to as microbial inoculants, are a group of microbial cultures which are artificially multiplied and can improve the fertility of soil as well as crop productivity. Biofertilizers supplement chemical fertilizers and 20–180 kg N ha^{-1} in soil. Bio-fertilizers improve soil fertility, physical characteristics of soil, tilth, and crop productivity. They also suppress various disease-causing pathogens in soil besides promoting plant growth by releasing growth-stimulating substances. Different groups of biofertilizers are enlisted in Table 15.3.

Plant growth-promoting bacteria colonize the rhizosphere of plants and facilitate the growth of plants, through various mechanisms, e.g., by altering the selectivity for Na^+, K^+, Ca^{2+} resulting in higher K^+/Na^+ ratios, production of osmolytes, IAA aided root surface enhancement, increased ACC deaminase activity resulting in reducing stress ethylene level. Different PGPRs that promote salt tolerance in various crops (Singh et al. 2017) have been presented in Table 15.4.

Table 15.3 Different groups of biofertilizers

S. No.	Groups	Examples
N fixing biofertilizers		
1.	Free-living	*Azotobacter, Clostridium, Anabaena, Nostoc*
2.	Symbiotic	*Rhizobium, Anabaena azollae, Frankia*
3.	Associative symbiotic	*Azospirillum*
P solubilizing biofertilizers		
1.	Bacteria	*Bacillus megaterium var. phosphaticum*
		Bacillus circulans, Pseudomonas striata
2.	Fungi	*Penicillium* sp., *Aspergillus awamori*
P mobilizing biofertilizers		
1.	Arbuscular mycorrhiza	*Glomus* sp., *Acaulospora* sp., *Gigaspora* sp.
		Sclerocystis sp. and *Scutellospora* sp.
2.	Ectomycorrhiza	*Laccaria* sp., *Boletus* sp., *Pisolithus* sp., *Amanita* sp.
3.	Orchid mycorrhiza	*Rhizoctonia solani*
Biofertilizers for micronutrients		
1.	Silicate and Zinc solubilizers	*Bacillus* sp.
Plant-growth-promoting rhizobacteria		
1.	Pseudomonas	*Pseudomonas fluorescens*

Table 15.4 PGPR-induced salt tolerance in crops

Salt-tolerant PGPR	Crop species	Reference
Azospirillum brasilense	Pea (*Pisum sativum*)	Dardanelli et al. (2008)
Pseudomonas syringae, Enterobacter aerogenes, Pseudomonas fluorescens,	Maize (*Zea mays*)	Nadeem et al. (2007)
Pseudomonas fluorescens	Ground nut (*Arachis hypogaea*)	Saravanakumar and Samiyappan (2007)
Azospirrilum	Lettuce (*Lactuca sativa*)	Barassi et al. (2006)
Achromobacter piechaudii	Tomato (*Lycopersicon esculentum*)	Mayak et al. (2004)

Source: Singh et al. (2017)

Broadly, there are three methods of biofertilizer application. Seed treatment is the most general mode of application where the seed is first coated with *Rhizobium* or *Azotobacter*, followed by inoculating with P solubilising biofertilizer. Root dipping method of biofertilizer application is basically applicable for *Azospirillum* where paddy or vegetable plant roots were dipped in *Azospirillum* mixture half an hour before sowing. Soil application of biofertilizer is also done. Phosphate solubilizing bacteria (PSB) is applied in soil. Rock phosphate, cow dung, and PSB mixture are used in between rows or during levelling of the field through soil application.

However, there are various constraints which limit the applicability of biofertilizers in agriculture. Lack of good quality strain, non-availability of storage facility, field conditions like extremely high or low pH, temperature, nutrients deficiency not only influence the response of inoculants but also limit their benefits.

15.4.4 Customized Fertilizer

Customized fertilizers are one kind of multi-nutrient carrier that were manufactured through a specific process of granulation, being site-specific, can satisfy the crop needs in a particular situation, and are comprised of macro and micro nutrients through both inorganic and organic sources. Through customized fertilizer production, site-specific nutrient management is being focussed to maximize fertilizer use efficiency in an effective manner. It is a combination of nutrients as per crop requirement and soil testing and may be a formulation of primary, secondary as well as micro-nutrients.

15.4.5 Enhancing Fertilizer Use Efficiency

Application of fertilizer in the soil is associated with several forms of losses, thus not all of the nutrients were absorbed by the plants. Thus, only a fraction of the nutrient (s) is utilized by the crops. Fertilizer use was directly linked with increasing agricultural productivity during and post green revolution era. But, incomplete use of applied fertilizer by crops due to low fertilizer use efficiency caused leaching or runoff loss of nutrients, especially nitrogen (N) and phosphorous (P) thereby raising environmental concerns. So, the challenge is to keep agricultural productivity increasing with the least harm to the environment because of fertilizer. Fertilizer use efficiency may be improved by integrated application of inorganic fertilizer with organic amendments, site-specific nutrient management, split application of nitrogenous fertilizer, choosing the correct source of fertilizer and applying the right amount at right time using the right methodology, application of fertilizer based on soil test as well as crop response, by adopting appropriate agronomic management practices, making nano-based formulations of fertilizers, inoculating microorganisms to solubilize native soil nutrients, etc.

15.4.6 Fertigation

Along with fertilizer, water was also considered one of the most crucial resources for increasing productivity. Post-green revolution, irrational use of water incurred in the past that resulted in numerous problems including water scarcity, soil salinity, and alkalinity were realized and alternate water management strategies were explored. Among those, micro irrigation using a drip system emerged as the most efficient water management strategy, where fertilizer mixing along with the water was also a convenient and effective way of fertilizer application. Fertigation is a process of fertilizer application along with irrigation. In order to maximize yield, and improve water and nutrient use efficiency, fertigation is the answer whereby both the water and nutrients were delivered to crops through a drip irrigation system. The fertigation provides essential elements to the active root zone directly, thereby minimizing the loss of expensive nutrients, which in turn improves the quality of produce and productivity of crops. It makes economical use of water thus saving time as well as money. Thus through fertigation, the idea of uniform application of water and nutrients, saving of resources, yield enhancement, quality of food and environmental gains may be achieved.

15.4.7 Soil Health Card (SHC)

Management of soils through Soil health card (SHC) based recommendations is needed to have management of resources in an efficient manner and for sustainable crop productivity (Bhaduri et al. 2020). Soil health index (SHI) developed by integrating sensitive chemical, physical and biological parameters of soil could serve as a yardstick for assessing the overall health of the soil. The soil health card thus developed had information about the farmer, the land (altitude, longitude, and latitude), soil type, and various soil health indicators, along with soil health index and appropriate recommendation practices on fertilizers, manures, and reclamation measures like gypsum application, leaching of salts for salt-affected soils (Fig. 15.2) (Purakayastha et al. 2019). Online assessment of soil health is also an important area of research and development. In this direction, Purakayastha et al. (2016) developed an online conceptual framework-based decision support system (DSS) for soil health assessment for alluvial and hill soils of India (Fig. 15.3) (http://ssacdss.iari.res.in/home.php). The DSS has 10–11 chemical, 2–4 physical, and 3–5 biological parameters for the above two soils. By inserting the values of analysed data in respective listed soil health indicator parameters, the in-built software computes the SHI value. This type of DSS may be useful for researchers, soil testing laboratories, farmers, agricultural departments, policy makers, etc. for judging health. The major challenge is to integrate this DSS with fertilizer and manure recommendations and amelioration measures if any are required.

Meanwhile, several other constraints need to be addressed on priority to reap the benefits of SHC as mentioned below:

- The campaigns on explaining the content of SHC to the farmers, how to use recommended doses, benefits of SHC for reducing the cost on excess fertilizers and enhancing profitability need to be done on regular basis.
- SHC extension services to be strengthened for better advisories.
- Sampling should be practised as per the field variability and in a grid fashion in front of the farmers to build confidence. The general grid size for the irrigated system is 10 ha and the rainfed is 2.5 ha, but it may be changed depending on the site variability.

National Innovations in Climate Resilient Agriculture
ICAR-Indian Agricultural Research Institute
New Delhi

Soil Health Card

Name Mr. Tulsi Ram

Address Village Mumtajpur
Block Patuadi
District Gurgaon
Haryana

Location E 76°46'54.7"
N 28°17'2.5"
Altitude 228 m

Soil type Alluvial soil (Inceptisol)
Sandy loam texture

Crop Rice, wheat

Soil health: Alkaline with moderate salinity and good biological and physical properties and moderate soil health

Amelioration: Gypsum application (5t ha^{-1}) followed by leaching.

Manure application:
5 ton FYM ha^{-1} yr^{-1}

Fertilizer application:
Rice: 120 kg N, 60 kg P_2O_5 and 45 kg K_2O ha^{-1}
Wheat: 120 kg N, 60 kg P_2O_5 and 45 kg K_2O ha^{-1}

Soil health indicator	Value	Rating
Chemical indicator		
pH	9.3	High
Electrical conductivity (dS m^{-1})	0.7	Medium
Soil organic carbon (g kg^{-1})	7.0	Medium
Labile carbon (mg kg^{-1})	363	Medium
Available nitrogen (mg kg^{-1})	158	Medium
Available phosphorus (mg kg^{-1})	6.2	Medium
Available potassium (mg kg^{-1})	160	High
Available iron (mg kg^{-1})	8.7	Medium
Available manganese (mg kg^{-1})	2.8	Medium
Available zinc (mg kg^{-1})	1.4	Medium
Available copper (mg kg^{-1})	1.3	Medium
Physical indicator		
Bulk density (Mg m^{-3})	1.3	Medium
Water holding capacity (%)	22	Medium
Soil aggregate stability (%)	72	High
Mean weight diameter	1.5	High
Biological indicator		
Microbial biomass carbon (mg kg^{-1})	342	High
Dehydrogenase activity (ug TPF g^{-1} 24 h^{-1})	122	Medium
Potentially mineralizable nitrogen (mg kg^{-1})	74	High
Soil health index	0.7/1.0	Medium

Contact: Dr. Tapan Jyoti Purakayastha, Principal Scientist, Division of Soil Science and Agricultural Chemistry, ICAR-Indian Agricultural Research Institute, New Delhi 110012, E-mail: tpurakayastha@gmail.com, Mob. No. 9868091129.

Fig. 15.2 Sample of soil health card developed for farmers (Purakayastha et al. 2019)

- Soil testing facilities should be made better and the results should be given expediously.
- Human resources should be developed for sampling, testing, analysis, and preparation of results to improve the quality of services.
- Results should be given before the sowing of crops so that the farmers can practice as per the recommendations.
- Generation of soil fertility maps and display in villages/blocks for wider publication and awareness of soil health issues.
- Government agencies should make available organic amendments and fertilizers at reasonable prices in villages.

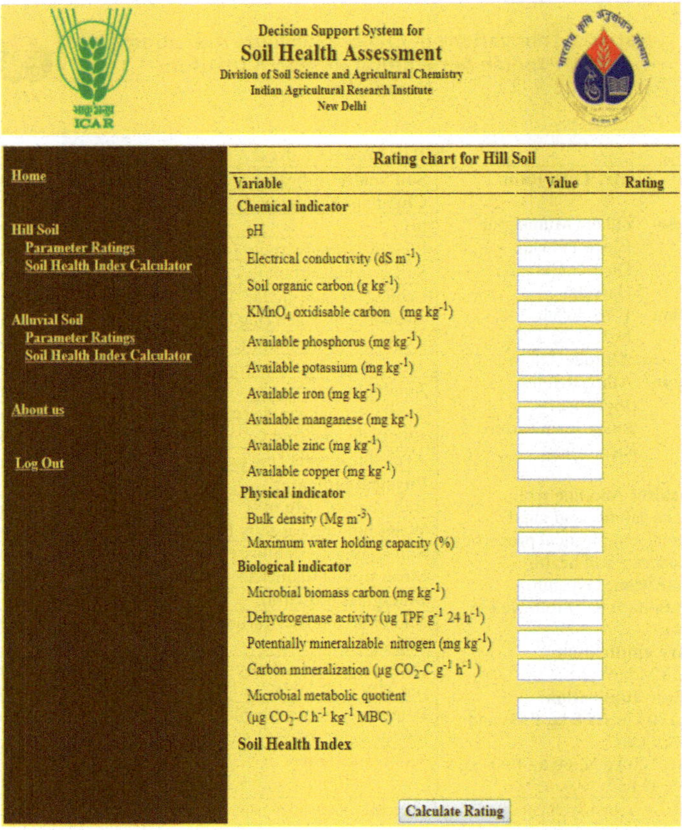

Fig. 15.3 IARI web portal developed for on-line assessment of soil health (http://ssacdss.iari.res.in/home.php)

Government of India has taken a big leap in initiating a new program called Soil Health Card (SHC) Scheme in February 2015 to promote balanced fertilization to protect soil health and for sustainable agriculture. The SHC is a soil fertility status report of a specific field and integrates important soil properties that affect the crop productivity. It is just like a physician's prescription where the health status of soil is reflected and recommendation is provided to the grower accordingly. Details in a SHC (Fig. 15.4) includes 12 important parameters such as pH, EC, organic carbon, macronutrients like N, P and K, secondary nutrients like sulphur and micro-nutrients such as zinc, iron, manganese, copper and boron. After proper examination of soil, the SHC is provided to the farmers once in 3 years, based on which farmers decide the type of crops to be taken and management practices to be followed for enhanced income.

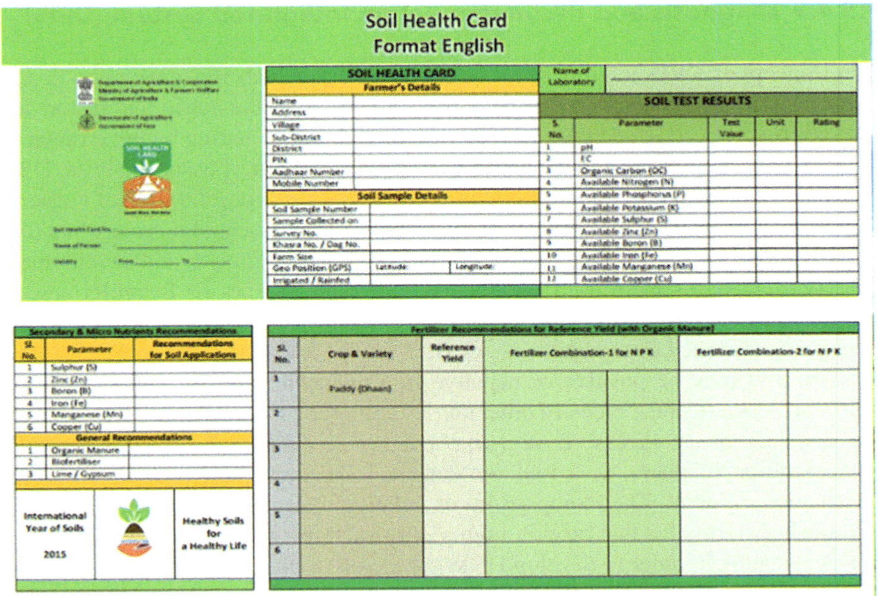
Soil Health Card
Format English

SOIL HEALTH CARD	
Farmer's Details	
Name	
Address	
Village	
Sub-District	
District	
PIN	
Aadhaar Number	
Mobile Number	
Soil Sample Details	
Soil Sample Number	
Sample Collected on	
Survey No.	
Khasra No. / Dag No.	
Farm Size	
Geo Position (GPS)	Latitude: Longitude:
Irrigated / Rainfed	

Name of Laboratory

SOIL TEST RESULTS

S. No.	Parameter	Test Value	Unit	Rating
1	pH			
2	EC			
3	Organic Carbon (OC)			
4	Available Nitrogen (N)			
5	Available Phosphorus (P)			
6	Available Potassium (K)			
7	Available Sulphur (S)			
8	Available Zinc (Zn)			
9	Available Boron (B)			
10	Available Iron (Fe)			
11	Available Manganese (Mn)			
12	Available Copper (Cu)			

Secondary & Micro Nutrients Recommendations

Sl. No.	Parameter	Recommendations for Soil Applications
1	Sulphur (S)	
2	Zinc (Zn)	
3	Boron (B)	
4	Iron (Fe)	
5	Manganese (Mn)	
6	Copper (Cu)	
General Recommendations		
1	Organic Manure	
2	Biofertiliser	
3	Lime / Gypsum	

International Year of Soils 2015

Healthy Soils for a Healthy Life

Fertilizer Recommendations for Reference Yield (with Organic Manure)

Sl. No.	Crop & Variety	Reference Yield	Fertilizer Combination-1 for N P K		Fertilizer Combination-2 for N P K	
1	Paddy (Dhaan)					
2						
3						
4						
5						
6						

Fig. 15.4 Format of a typical soil health card

15.4.8 Soil Test Crop Response (STCR)

Soil test involves chemical analysis, interpretation, evaluation and further recommendations based on the test results and other considerations. STCR studies take into consideration the targeted yield in order to develop relationship between crop yields and soil test estimates and inputs of fertilizers. Thus the concept of target yield strikes a balance among fertilization of soils and crops. At present times, with increasing prices, demand of fertilizer consumption is increasing alarmingly, besides depletion of soil fertility status thus necessitates adoption of integrated application of inorganic and organic sources of nutrients for sustaining soil health and crop production. STCR developed algorithms of leaf colour chart, SPAD and fields count CM 1000 meter values at three critical crop growth stages in rice-wheat system with yield; also developed prescription equation for fertilizer for even secondary nutrient like (sulphur). For organic farming system, soil testing protocols including microbiologically exploited organic phosphorus-pools characterization and quantification also been developed.

Methods of STCR involve (a) gradient experiment and (b) test crop experiment. However, the conditions for targeted yield equations are: (1) used for similar soils of particular agro-eco region (2) for particular crop in an area, the maximum yield may not exceed 75–80% of highest yield achieved. (3) N recommendations through fertilizer should be similar as the general recommended dose of the crop in that area.

15.4.9 Diagnosis and Recommendation Integrated System (DRIS)

The concept of DRIS interprets leaf or plant analysis which was developed first by Beaufils (1973). This comprehensive system identifies all the nutritional factors that limits crop production and thus increases the probability of getting higher crop yields by improved fertilizer recommendations. For interpretation of tissue analysis, DRIS considers nutrient ratios rather than individual or absolute concentration of nutrients. There are already a set of optimum nutrient ratios of the essential nutrients like (N/P or N/K or K/P) within a plant that takes cares of its proper growth and development. The nutrient ratios were used for obtaining nutrient indexes or Beaufills nutrient Indexes (BNI). DRIS utilizes the nutrient balancing concept for detection of any nutrients deficiency or toxicity.

The BNI may be positive or negative in a plant and the concentration of any nutrients may affect the index of other nutrients, thereby making either the nutrient to be deficient or in excess. It is a mathematical technique that applies nutrient concentration in plants as a tool for detecting the most limiting nutrient in the production system. The comparison of relative balance of nutrients with the established norms for that crop in high yield situations was done for evaluation of nutrient status. In order to develop DRIS for a crop or plant, the given requirements should be met if possible.

1. The factors that certainly affects the crop yield must be well defined.
2. The yield and these factors relationship must be explained.
3. The norms for calibration must be established.

For particular conditions the recommendations based on accurate and judicious use of norms must be refined continually.

15.4.10 Nutrient Cycling

Nutrient cycling refers to their movement within and in between different biotic or abiotic entities in which the nutrients occur in global environment. These nutrients can either be mined from geogenic sources, converted from organic sources to available forms or taken up from the atmosphere through different mechanisms, thereby being taken up by the plants and again their return to soil or the atmospheric continuum. Soil microbial diversity plays an important role for nutrient cycling through release of many compounds, solubilisation, change of structures, other mechanisms that regulates the fluxes of nutrients. The distribution of nutrients in various living or dead pools, their abundance in different compartments is specific to a particular ecosystem. For example, in terrestrial ecosystems, either nutrients greatly confined in living biomass (tropical rainforests) or found in SOM or humus (tundra ecosystems) (Lavell and Spain 2001).

Nutrient cycling happens in all ecosystems that dictate the proper functioning of the system though with different rates. The largest flux of nutrients takes place through conversion or decomposition of organic matter, and is very difficult to monitor as major portion immediately incorporated in the biomass of microbes. In intensively cultivated system, the nutrient fluxes greatly been affected due to management practices like tillage, application of external inputs like nutrients, water, amendments, nitrogen fixing crops etc. and the inherent fertility of soil greatly supplemented through external inputs. In aquatic ecosystems, eutrophication occurs due to non-deliberate release or leaching of nutrient applied to nearby agricultural fields or industrial release. The soil fertility can also be altered and severely affected due to anthropogenic activities like mining of nutrients through intensive cultivation, acidification due to fertilizer application, heavy metal pollution due to waste water and sewage application, waterlogging, salinization, soil erosion, crop residue burning etc.

The ecosystem maintains a stoichiometry, where there exists a balance between nutrients. Living organisms tend to contain a relatively constant proportion of elements, especially C, N, and P. In the freshwater microalgae, the contents of the elements expressed as molar stoichiometric ratios are C (125), N (19), and P (1) (Reynolds 1990); the same order of magnitude as the Redfield ratio (106:16:1) proposed for suspended particulate matter in oceans (Redfield 1958). For land plants, the mean proportion is about 200:13:1 (although quite higher for woody plants) because the C content is derived by the need for structural tissue. If the nutrients get imbalanced then the organisms have different mechanisms to drive out excess nutrients or capture nutrients if found limiting. However, any change in nutrient cycling includes situations of excess nutrients, through eutrophication or deficiency of nutrients or some specific situations in oceans.

15.4.11 Composting Processes and Techniques

Composting refers to decomposition or rotting of organic materials by microbes through natural process in a controlled situation. Besides being a source of plant nutrients, compost improves the overall health of soil. Due to the above benefits of compost, soil becomes (1) resilient against any type of stress like drought, salinity, sodicity, toxicity diseases etc., (2) it helps the crops to take up nutrients in right proportion, (3) active nutrient cycling in soil due to their vigorous microbial activity.

Types of Composting Composting is basically of aerobic and anaerobic types. Anaerobic composting involves anaerobic microbes for organic matter decomposition into intermediate products like methane, hydrogen sulphide, organic acids and other compounds. It takes place under absence or limited supply of oxygen. The anaerobic composting is a low temperature technique of composting, thus the pathogens and weed seeds remain intact. Moreover, this process takes more time than aerobic method. The merits of anaerobic composting like less work involves less nutrients lost often being offset by the limitations. Aerobic composting occurs under ample supply of oxygen, and the aerobic microorganisms' breaks down organic materials into CO_2, ammonia, water, heat, stable organic end products and humus etc. Though, more nutrients get lost in aerobic method of composting, but it is considered as more efficient and useful for agricultural production. Salient features of few selective aerobic techniques of small scale are given below (Table 15.5).

15.4.12 Organic Manure

Manures are generally animal or plant-origin wastes that are used as a prime source of nutrients for plants. On decomposition, the manures release nutrients, often composed of almost all essential nutrients but in limited amounts. They are basically of two types, bulky and concentrated organic manure based on nutrient concentration. Manures improve the overall condition of the soil that in turn helps in the proper growth and development of plants. It often acts as mulch that maintains the soil temperature and checks the evaporation loss of moisture. Due to the good extent of microbial activity, the nutrients at lower depths were made available to the plants.

Apart from several factors, a decline in the SOC status of the Indian soils is an important cause of declining soil health and crop productivity (FAO 2005). Manures generally provide nutrients very slowly and thus make them available for longer periods for optimum growth and development of crops. If soil health is improved, then the response of the crops against the added nutrient also gets improved. The most commonly used manure is farm yard manure (FYM) which is comprised of nearly all of the essential nutrients required by crops. Similarly, green manures have got a lot of potential for the addition of organic matter in soil apart from its source as an N fixer. They have fast decomposability and 40–45 days grown green manure crop may supply an estimated 100–125 kg of N which is equivalent to the N requirement of any cereal crop. Several management practices of crops (viz. cover crops, inter-cropping with legumes and residue retention, appropriate crop production systems/cropping systems, tree-based green leaf manuring, biofertilizers use, integrated nutrient management), etc. were considered as effective for improvement of organic matter status in soils that leads to soil health improvement and crop yield sustainability (Indoria et al. 2016, 2017; Srinivasarao et al. 2017). Further, the inclusion of energy crops, dung, and crop residues as agricultural feed stocks could reduce GHGs emissions provided substitution of fossil fuels is done by these stocks (Smith et al. 2008).

Indian organic source contributes annually about 5 Mt of available N, P and K and is further expected to reach 7.75 Mt by 2025 (FAO 2005). Use of crop residues, green manures, legumes, off-farm organic waste, and management practices that improve soil and crop found to help in C sequestration through various ways (Lampkin 1999).

The increasing interest for use of organic sources of nutrients and soil amendments is because they serve as a source of C which helps in better soil health and mitigates climate change. Some of the potential alternative organic sources availability in our country are presented in (Table 15.6). However, due to the non-availability of organic amendments (FYM), lower nutrient use efficiency, declining soil fertility, and factor productivity, amendments/relevant organic sources

Table 15.5 Small scale aerobic composting and their salient features (Misra et al. 2003)

Method	Salient features					Duration
	Substrate size reduction	Turnings at intervals of (days)	Added aeration provision	Microbial inoculation	Supporting microbial nutrition	
Indore pit		+15, +30, +60		Inoculum from old pit		4 months
Indore heap	Shredded	+42, +84				4 months
Chinese pit		+30, +60, +75			Superphosphate	3 months
Chinese high temperature compost	Shredded	+15	Aeration holes in heap through bamboo poles/ maize stalks		Superphosphate	2 months
Ecuador on-farm composting		+21	Lattice of old branches/ poles at heap base			2–3 months in summer; 5–6 months in winter
Berkley rapid composting	Shredded to small size	Daily or alternate day turning				2 weeks with daily turning and 3 weeks with alternate day turning
North Dakota State University hot composting	Shredded	+3 or +4	4–5 holes punched in Centre of pile		0.12 kg N per 90 cm dry matter	4–6 weeks
EM-based quick composting		+14, +21		EM	Molasses	4–5 weeks
IBS rapid composting	Shredded	+7, +14, then every 2 weeks	Raised platform ground/ perforated bamboo trunks	*Trichoderma* sp.		3–7 weeks.

Table 15.6 Potential alternative organic sources availability in India

Alternative organic sources	Total availability/ yr	References
Crop residues	500–550 million tonnes (Mt)	NAAS (2012)
Municipal biosolid	48 Mt	Pappu et al. (2007)
Rice husk	20 Mt	Sengupta (2002)
Sugarcane bagasse	90 Mt	Sengupta (2002)
Groundnut shell	11 Mt	Sengupta (2002)
Sugarcane press mud	9.0 Mt	Chanakya et al. (2006)
Poultry manure	6.25–8 Mt	The Hindu (2009)
Coir pith	7.5 Mt	Vijaya et al. (2008)
Food/fruit processing industries	4.5 Mt	Chanakya et al. (2006)
Distillation waste of plant material after the essential oil extraction	2–3 t	Singh et al. (2013)
Seri waste	5000 tonne	Gunathilagaraj and Ravignanam (1996)
Willow dust	30,000 tonnes	Chanakya et al. (2006)
Area under green manuring crop	About 7 M ha	FAO (2005)

Source: Indoria et al. (2018)

are quite necessary for soil fertility enhancement and productivity, the food production should not be hampered for increasing population on a sustainable basis.

15.4.13 Green Manures

Green manure is composed of un-decomposed green plant material that is obtained either through in-situ green manure by growing green manure crops or by collecting the leaves and twigs of manure crops and then application in the field. The green manure crops that were grown in the field belong to the leguminous family; their growth could be enhanced by the application of phosphatic fertilizers. After attaining sufficient growth they were applied to the fields. The most common green manure crops are Dhaincha (*Sesbania aculeata)*, Sunhemp (*Crotalaria juncea*), Cluster beans (*Cyamopsis tetragonoloba*), *Sesbania rostrata* and Pillipesara (*Phaseolus trilobus*). Apart from green manuring crops grown in agricultural fields, the forest plantations' leaves, twigs, and green manure crops grown on bunds, wastelands, etc. were also utilized for the purpose. Important species used for green leaf manuring are Glyricidia (*Glyricidia sepium, Glyricidia aculata*), Subabul (*Leucaena leucocephala*), Neem (*Azadirachta indica*), Calotropis (*Calotropis gigantean*), Wild Indigo (*Baptisia australis*), Mahua (*Madhuca longifolia*), Karanj (*Millettia pinnata)*, Avise (*Linum usitatissimum*), and other herbs or shrubs.

15.4.14 Conservation Agriculture

The practice of conservation agriculture (CA) technology worked on the basic principle of soil disturbance at a minimum level, crop diversification or rotation, and surface crop residue retention. Crop residue retention and tillage greatly affects soil health and aids in enhancing/improving soil organic carbon stock. Reduction in labour and fuel cost, retention of soil moisture is considered basic factors behind the adoption of CA (Thomas et al. 2007; Llewellyn et al. 2012). The adoption of CA in India is underway with efforts to refine and develop for different suited conditions, although there remain several constraints along its adoption in a full way. It includes lack of suitable seeders, especially for small farmers, crop residue diversion for other purposes, technical know-how, mindset of the growers about the tillage practices, etc. There are more payoffs than tradeoffs in CA technology that needs to be understood by both growers and promoters. CA improves the soil's biological activity, better soil aggregation, water movement, and a high level of protection of soil against physical alteration. The soil erosion reduced drastically, and with improved agricultural inputs movements, the biodegradation of pesticides enhanced. The CA adoption has been done extensively in the irrigated belts of Western Indo-Gangetic plains. The major input in CA is no-till seed cum fertilizer drill and interventions include alternate crops than rice-wheat rotation, raised bed plantings, laser levelling, crop residue management, etc. The underground, as well as surface water pollution, also gets less hampered through CA along with the check on GHGs emissions. At different locations of Indo-Gangetic plains, zero tillage led to saving of land preparation costs by INR 2500 (41.7$) ha^{-1} and diesel consumption reduction by 50–60 L ha^{-1} (Sharma et al. 2005a). Zero tillage under wheat allows timely sowing of seeds, nutrient use efficiency improvement, saving of water, uniform drilling, and increases by about 20% of yield.

15.4.15 Biochar

Biochar is a pyrolysis product of plant biomass, dark in colour, and recalcitrant in nature which helps to enrich the carbon stock and nutrient balance in soil (Lehmann et al. 2006, 2008; Majumder et al. 2019). It is composed of oxygen-containing aromatic and functional groups and is a plant and animal residue product from biomass (Tan et al. 2015). The proper use of biomass residues for the preparation of amendments or nutrients is a noble way to manage soil fertility and improve soil health. Biochar is having negative charges on the internal surface that help to enhance the cation exchange capacity to attract metal ions from the soil solution.

(Mukherjee et al. 2011) and also reduce the metals and organic contaminants in soil effectively (Beesley et al. 2010). Its use in agricultural soils has the capacity to enhance carbon sequestration rates, reduce farm waste and enhance soil quality. It serves as a slow-release source of carbon and or nutrients (Crombie et al. 2013; Spokas et al. 2010) due to its recalcitrance nature. The mineralization of carbon to CO_2 occurs slowly thereby improving the soil organic matter (SOM) (Cheng et al.

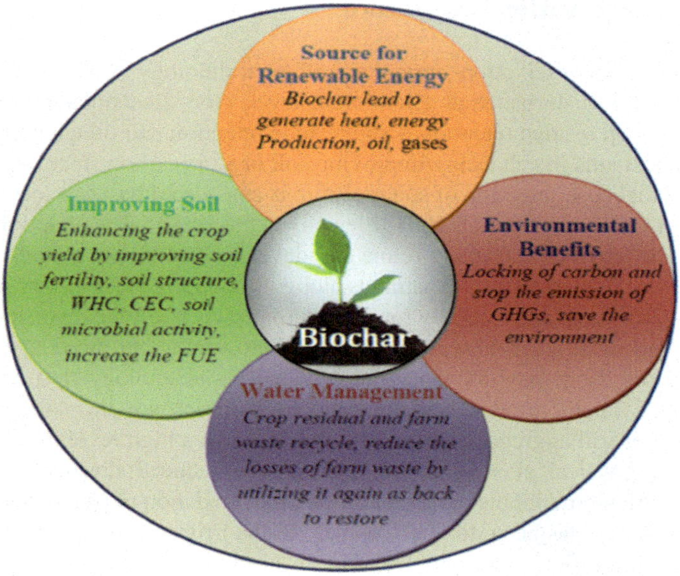

Fig. 15.5 Sustainable management of the agriculture sector through Biochar as a potential component. (Source: Jatav et al. 2021)

2006; Masiello 2004). For sustainable management of agriculture, biochar may serve as a potential source for overall soil quality improvement (Fig. 15.5).

Biochar when added to soil it alters soil health (Paz-Ferreiro and Fu 2013). In recent times, a lot of interest arises in biochar application in soils as a means of C sequestration while improving soil health simultaneously (Laird 2008; Lehmann et al. 2006). Biochar research in India is very new and very limited literature is available on this aspect. Purakayastha et al. (2015) reported that the physical, chemical, morphological, and spectral properties of kiln-produced biochar are greatly influenced by the type of feedstock. Maize stover biochar was reported to have a higher content of N and P and thus had good potential to enhance soil fertility if used as amendments. Meanwhile, the maize stover biochar is inherently alkaline in nature and can act as liming material for the reclamation of acid soil. Besides this, biochar having high structural stability due to the presence of aromatic C=C stretching and other surface functional groups, might possess greater potential for long-term C sequestration. However, rice straw biochar has more labile C, thus its addition led to a sudden increase in biological characteristics of the soil and thus may be beneficial from a soil health point of view. The wheat straw was found to have a higher content of K and thus can be used in K-deficient soils or crops requiring high K. At Tirunelveli, Tamil Nadu, India, a plot scale biochar application on agricultural soils was evaluated to investigate biochar potential for improvement in fertility and moisture status of soil (Mankasingh et al. 2011). The feedstock for biochar should be used sustainably from all the available sources like cassia stems, palm leaves, rice husk, saw dust, etc. Different biomass feedstock of biochar was found to have a C content of more than 20% and comprise both macro and micronutrients. Results suggested cassia biochar @ 6.6 metric t ha^{-1} was enough to initiate the process of C sequestration that is being shown by a decrease in bulk density and an increase in OM status of the soil. Additionally, biochar enhances SOC, organic matter, pH, EC, cation exchange capacity, and water holding capacity. Biochar can act as a potential site for microorganisms to act and survive. The porous structure of biochar creates new habitat for soil microorganisms (Egamberdieva et al. 2018; Gorovtsov et al. 2020), larger pores colonized rapidly but do not provide shelter to soil micro fauna (Noyce et al. 2016). It markedly influences the enzymatic activities by adsorption of microbes on its particles (Foster et al. 2018). The applications of biochar improve soil fertility and may address the issue of climate change.

15.4.16 Soil Amendments

Soil amendments include both inorganic and organic matter that has been added to the soil for improvement in soil texture, aeration, drainage or water retention, etc. Clay soil requires amendments for the improvement of texture, aeration, and drainage, while sandy or rocky soils also improve textural characters and water retention characteristics on the application of amendments. Soil amendments are from various sources. The inorganic and organic amendments are listed in Table 15.7.

15.4.17 Precision Agriculture

Precision agriculture is an information and technology-based agricultural management system that in turn analyses and takes care of variability in the field in order to conduct experimentation through the right time, right place, and right way for gaining optimum sustainability, profitability, and taking care of resources. Nowaday's sensor-based technologies are in the way of the improvement of water as well as nutrient use efficiency in a timely and precisely applying them when the crop needs it. It saves a lot of money, time, and electricity and enhances profitability. The drip and sprinkler systems were reported to enhance the application rate effectively in comparison to the surface irrigation system. The site-specific management of the 4 R approach for having the application of inputs at right time, the right amount, the right source, the right place allows for fine-tuning crop management systems. Precision agriculture works in tandem with proper consideration of the environmental situation. The farmers can prepare themselves accordingly for any sort of farm management practices. It drastically reduces the labour cost, improves fertility, uses the efficacy of inputs, saves precious resources, reduces environmental damage, enhances livelihood and skilled manpower generation, etc. If properly managed and implemented, it can fully revolutionize modern farm management practices. However, still, large-scale precision farming adoption is at its nascent stage owing to a lack of technical know-how, lower risk-bearing ability by farmers, social, economic,

Table 15.7 Common used inorganic and organic amendments

Amendment	Function/purpose
Inorganic	
Sand and/or profile soil conditioner	For improvement of aeration and drainage
Lime or Sulphur	To increase or decrease pH
Vermiculite, a natural balsamic mineral and perlite made from heated amorphous volcanic glass	Increase aeration; improve texture and helps in water retention
Organic	
Manures	For addition of nutrients
Compost	Improve the soil texture and aeration, helps in drainage and water retention, and also release nutrients for plants
Mulch	Helps in the retention of soil moisture
Humus	It helps in soil stabilization, and adds nutrients in high proportion thus rectifying deficiencies and better soil health
Peat Moss	Improves the soil's ability to retain more moisture and is best suited for sandy or rocky soils. Also used for stabilization of clay soils
Wood products like shavings and wood chips	To improve soil, but may create a nitrogen deficiency

and demographic conditions, and small land holdings that make it difficult for financial gains from currently available precision farming technologies, etc.

15.4.18 Carbon Sequestration Options

The potential of soil/terrestrial C sequestration is immense in India has so vast land area and ecosystem diversity. Out of 329 (M ha) of total land area, 297 M ha of land comprised 162 M ha (arable land), 69 M ha (forest and woodlands), 8 M ha (permanent crops), 11 M ha (permanent pasture), and 58 M ha (other land uses) respectively (Lal 2004). The soil organic carbon (SOC) pool is estimated to be 21 Pg to 30 cm in depth and 63 Pg to 150 cm in depth. While, the soil inorganic carbon (SIC) pool is estimated to be 196 Pg to 1 m depth. As compared to uncultivated soils (15–20 g kg^{-1}) SOC content, most of the cultivated soils have less than 5 g kg^{-1} of SOC (Lal 2004). Degraded soil restoration and recommended management practices (RMPs) adoption on agricultural and forestry land are the important strategies for soil C sequestration. The potential of soil C sequestration for Indian soils is 39–49 (44 ± 5) Tg C yr^{-1} that comprises 7–10 Tg C yr^{-1} (degraded soils and ecosystems restoration), 5 to 7 Tg C yr^{-1} (erosion control), 6–7 Tg C yr^{-1} (RMPs adoption on agricultural fields), and 22–26 Tg C yr^{-1} (from secondary carbonates). The degraded soils and ecosystems restoration, control of erosion, and conversion of marginal soils under agricultural practice to restorative land use are crucial options for SOC sequestration (Aggarwal et al. 1997; Kaur et al. 2000, 2002a, b; Moench 1991;

Singh and Singh 1996). The degraded soil restoration including eroded soils and salt-affected soils may lead to the enhancement of biomass and thus helps in the improvement of SOC pools. The SOC pool enhanced substantially from 10 to 30–45 Mg ha^{-1} after 8 years of establishment of tree species (i.e., mesquite) in a sodic soil (Garg 1998). Salt-affected soils restoration led to a substantial increment of SOC pool (Bhojvaid and Timmer 1998). A similar kind of potential exists if the restoration of extensive wastelands throughout India will be done (Gupta and Rao 1994). A lot of literature exists concerning SOC dynamics in Indian soils (Mohan-Rao and Shantaram 1978). Utilization of crop residues and manures in the soil to improve soil biodiversity, mainly the earthworm activity, is an important strategy for enhancing both SOC and soil quality (Bhadauria and Ramakrishnan 1996; Lavelle et al. 1998). The long-term impact of integrated nutrient management (INM) including NPK + Manuring @8–10 Mg ha^{-1} yr^{-1} on SOC content in surface depth was done across several ecoregions of India (Swarup 1998). A low rate of SOC sequestration rate of 15–120 kg C ha^{-1} yr^{-1} was observed for NPK + Manuring over that of control while considering the soil bulk density of 1.4 Mg m^{-3} and plow depth of 20 cm. The low rates of soil C sequestration are due mainly to high soil temperature, low soil water content, and high oxidation rate. Continuous adoption of 100% NPK + FYM in maize-wheat-cowpea rotation in semi-arid sub-tropical India might sequester nearly 1.83 Tg C yr^{-1} that corresponds to about 1% of fossil fuel emissions by India (Purakayastha et al. 2008).

Increased content of soil organic matter and biological activity, improved soil structure with the maintenance of soil aggregates, and reduced oxidation of soil organic matter. Similarly, crop rotations through the diversification of crops affect soil health through the activity of roots and the chemical composition of residues of different crops added (Srinivasrao et al. 2013). For the sustainability of soil productivity in intensive agriculture, it is desirable that the cropping system should be changed after a few years but unfortunately, modern agriculture has ignored this concept, especially in India. Most of the cropping systems are cereal-cereal and very little area is under cereal-legume. The Indo-Gangetic Plain which is considered the food basket of India needs to be explored by introducing legume crops into the system. It may be exploited by introducing summer moong in between wheat-rice or by replacing rice with rainy season pulse e.g., pigeon pea or wheat by chickpea/pea. Several studies proved that legume-based systems improved SOC in a short span of 5 years (Singh and Singh 1996). Data obtained in the soybean-wheat cropping system under LTFE resulted in an improvement of organic carbon from 0.57% to 0.99% (Dwivedi et al. 2007) at Jabalpur.

Climate is an important natural factor that decides carbon sequestration in soil. An increase in mean annual temperature reduces soil organic carbon content in humid regions, while in temperate climates, the soils are several times richer in carbon than in warmer climates (Swarup and Singh 2009). High rainfall coupled with low temperature is very conducive for SOC accumulation while high temperature with low rainfall decreases it.

The amount of carbon needed to maintain soil organic carbon equilibrium is very much important in terms of the cost of carbon resources and GHGs emissions into the atmosphere. The amount of carbon added to soil varied from 2264 to 6360 kg ha^{-1} yr^{-1} in the soybean wheat system in Jabalpur, Madhya Pradesh, and this amount varied from 220 to 6576 kg ha^{-1} yr^{-1} in the sorghum wheat system in Akola, Maharashtra (Singh 2016). Long-term fertilizer experiments proved that increase in SOC was almost similar to the application of 5–15 t ha^{-1} of FYM each year. Instead of applying manure in large quantities in to a field and then return back to the same field after 3 years should be our management strategy. At Pantnagar, the minimum amount of carbon required to maintain 1.5% carbon, there is a requirement of 1337 kg C ha^{-1} yr^{-1}.

15.4.19 Soil Quality Assessment and Potential Indicators in Indian Soil

Researchers in the world and in India used the "Soil Management Assessment Framework (SMAF)" for soil health indices development (Andrews et al. 2004; Bhaduri and Purakayastha 2014; Masto et al. 2007, 2008a, b; Purakayastha et al. 2019). Presently the soils were analysed for selective chemical parameters like pH, EC, SOC, available N, P and K on a routine basis in different soil testing laboratories but this only provides a glimpse of soil status and ignores greatly the physical and biological properties that otherwise of very high significance and affects numerous soil functions like (i.e., nutrient cycling, physical support and stability, biodiversity and habitat for organisms) (Purakayastha et al. 2019). By using the principal component analysis (PCA) tool, SOC, available Zn, available Mn, available K, aggregate stability, microbial biomass carbon, and potentially mineralizable nitrogen were identified as the most sensitive seven soil health indicators in alluvial soils of northwest India. Various sensitive soil quality indicators had been identified in the different cropping systems, soil types, and climate in India (Table 15.8). All these results indicate that there occurs little commonality among the screened indicators even for similar production systems raised at different geographical sites.

The soil health index (SHI) was developed for a wide range of treatments, cropping systems under long-term fertility trials across India by several investigators (Bhaduri and Purakayastha 2014; Bhaduri et al. 2014; Basak et al. 2016a, b, c; Chaudhury et al. 2005; Kundu 2014; Mandal et al. 2005; Mohanty et al. 2007; Masto et al. 2007, 2008a, b; Sharma et al. 2005b, 2008). Several of them observed higher SHI under balanced fertilization than imbalanced ones. Further, SHI was generally higher with organics than without organics/FYM. Recently, while assessing SHI for soils in farmers' fields, Basak (2011) as well as Biswas (2011) not only identified the master indicators for assessing the SHI but also simultaneously determined their critical values for maintaining good soil health and higher crop yields in rice growing belts of Alfisols, Entisols and Inceptisols of West Bengal. Critical limits of soil quality indicators and soil quality indices are extremely important for yield sustainability. In this direction, Biswas et al. (2017) established the critical threshold limits of soil quality indicators and soil quality indices in three rice-growing soils in

Table 15.8 Few important indicators identified for different soil types and cropping systems

Centre	Soil type	Cropping system	Indicators identified
AAU	Clay loam	Rice-rice	Available K, Zn, OC, % BS and dehydrogenase activity
ANGRAU	Sandy loam	Groundnut-redgram	MBC, pH, available P, and K, WHC, and Zn
BHU	Sandy loam	Rice-lentil	OC, available P, Ca, Mg
CRIDA	Sandy	Sorghum-castor	HC, available N, P, K, and S, and MBC
CRIJAF	Sandy loam	Jute-rice-wheat	Available P and K, MBC, MWD, and OC
CRRI	Sandy clay loam	Rice-rice	DHA, available K, and OC
OUAT	Silty clay loam	Rice-field pea	MBC, OC, mineralizable N, and alkaline phosphatase activity
IARI	Loam	Maize-wheat	Available N, OC, alkaline phosphatase activity, MBC, BD, and available Zn
CRIDA	Sandy loam	Sorghum-mung bean	Available N, available Zn, available Cu, MBC, MWD, and HC

Source: Mandal et al. (2005), Masto et al. (2007), and Sharma et al. (2008)

India. For the rice-rice cropping system under Inceptisols, the upper and lower critical limits of the indicators were worked out as available Zn (1.7 and 1.2 mg kg^{-1}), bulk density (1.2 and 1.6 Mg m^{-3}), β-glucosidase activity (68 and 18 μg p-nitrophenol g^{-1} soil h^{-1}) and urease activity (64 and 24 μg NH4 g^{-1} soil h^{-1}), dehydrogenase activity in Entisols were (93 and 12 μg TPF g^{-1} soil 24 h^{-1}), aggregate stability (66 and 11%), total organic C (11.6 and 10.7 g kg^{-1}) and pHw (5.7 and 5.3) and oxidizable organic C in Alfisols were (7.8 and 5.0 g kg^{-1}), β-glucosidase activity (51 and 15 μg p-nitrophenol g^{-1} soil h^{-1}), aggregate stability (52 and 19%) and mineralizable C (273 and 173 μg C g^{-1} soil), respectively (Biswas et al. 2017). For key soil quality indicators and quality indices, both upper and lower critical limits were established such as Inceptisols (0.85 and 0.56), Alfisols (0.37 and 0.56), and Entisols (0.23 and 0.65). This limit needs to be periodically analysed and judged for enhancing/maintaining soil quality and sustainability in yield and thus optimum management practices under rice-rice cropping systems in sub-tropical India should be employed.

15.4.20 Nanofertilizers

Nanoparticles (NPs) are defined as small particles that are either natural or engineered with a range of size 1–100 nm, and compared to their counterparts, express different physical and chemical characteristics. They have a long-term impact on agriculture and the environment as their usage greatly enhances efficacy, and uptake by plants, reducing wastage, improving yield and other parameters, and reducing the cost of cultivation. NPs have a unique set of characteristics that includes size, large surface area, zeta potential, functionality on surface, crystallinity, hydrophobicity, and hydrophilicity—that allows it to use as targeted controlled release kinetics of NPs for use in smart delivery systems (Solanki et al. 2015).

The NPs slowly release the nutrients and thus favouring premature losses (Chhowalla 2017). The NPs hold the nutrients by adhering through thin nono-film or nano-emulsion that in turn ensures a strong hold of nutrients on plant surfaces due to higher tension of the surface of the nano-coating (Lavicoli et al. 2017). Recent advancement in the field of nanotechnology ensures the use of nano-zeolites (Jahangirian et al. 2020) and nano-clays (Mandal et al. 2019). The NPs zeolites have a particular network of pores that releases the nutrients slowly (Jahangirian et al. 2020). Still, lot of improvement and adaptation, implementation through numerous field trials, and its efficacy w.r.t., environmental and other parameters need to be done. The opportunities and challenges pertaining to nano-biotechnology have been presented in (Fig. 15.6).

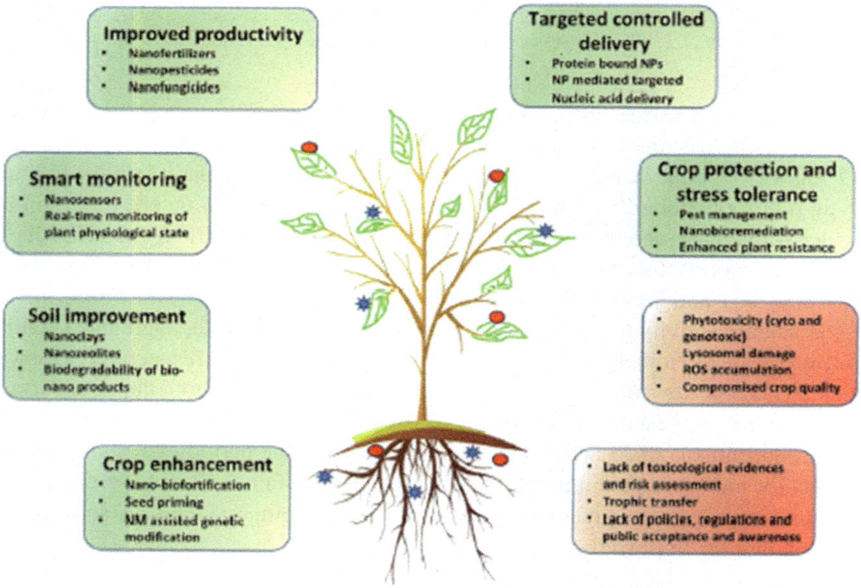

Fig. 15.6 Opportunities in (green) and challenges in (red) for nanobiotechnology in agriculture (Source: Chugh et al. 2021)

15.4.21 Microorganisms in Soil Fertility and Crop Production

Microorganisms in soil are very crucial as nearly every chemical transformation occurring in the soil is possible due to the active participation of soil microorganisms.

1. Soil fertility maintenance which involves the cycling of nutrients mainly C and N is dictated by the involvement of soil microorganisms and thus helps in the proper growth and development of plants.
2. The soil microbes like mycorrhizal fungi plays important role in making the mineral nutrients available to plants, especially phosphorus, which is relatively immobile in soil.
3. Other soil microorganisms can enhance the available nutrients in the soil. For example, nitrogen-fixing bacteria could transform nitrogen present in soil/atmosphere in gaseous form into soluble forms which are further taken up by the plants.
4. Several microbes are thought to have beneficial effects on soil fertility improvement thus; they are being utilized as biofertilizers and presently used as microbial cultures in agriculture.
5. Similarly, some other soil microorganisms are found to produce compounds (such as vitamins and plant hormones) that can improve plant health and contribute to higher crop yield. These are termed "phytostimulators" and are currently employed for possible uses as microbial inoculants for crop yield improvement. (i.e., strain of Azospirillum a bacterial inoculant used as wheat root colonizer).
6. Apart from these beneficial soil microorganisms, some other soil microbes are considered pathogenic to plants and thus might cause considerable damage to the crops. However, certain native soil microorganisms are antagonistic to these pathogens and thus prevent the infection of crop plants.
7. Antagonism against plant pathogens generally involves competition for nutrients and thus led to the production of some inhibitory compounds i.e., secondary metabolites (antimicrobial metabolites and antibiotics) and extracellular enzymes.
8. Some soil microorganisms produce compounds that stimulate natural defence mechanisms of the plants and improve their resistance to pathogens. These are collectively termed "biopesticides" and they are an emerging and important alternative (i.e. biological control) in comparison to chemical pesticides for the protection of crops against specific pathogens and pests.
9. Mycorrhizal fungi colonize plants' root systems and thus aid in nutrient uptake, thereby improving plant growth and overall health.

15.4.22 Biofertilizers

Biofertilizers (BFs) are generally cost-effective, eco-friendly, non-toxic, and easy-to-apply material that can enhance crop yields and the biodiversity of the soil. The major biofertilizers practised for enhancing the nutrient supply, crop yield, and quality includes Azotobacter, Azospirillum, Rhizobium, N fixers, Phosphate solubilizers and Vesicular Arbuscular Mycorrhizae (VAM). BFs act as a good substitute for chemical fertilizers (Deepali and Gangwar 2010; Thomas and Singh 2019). These bio-inoculants colonize the rhizosphere, or inside the plants it promotes growth when applied to the seeds, plant surface, or in soil (Raghuwanshi 2012).

Biofertilizers are of various types as follows:

- **Nitrogen-fixing:** These help in the fixation of atmospheric N and made it available to plants. **Example: Free Living**—*Clostridium, Rhodospirillum, Anabaena, Azotobacter, Nostoc, Klebsiella, Aulosira Bejerinkia, Desulfovibrio*, etc.
  Symbiotic—Frankia, Rhizobium, Trichodesmium, etc.
  Associative symbiotic—Enterobacter, Azospirillum spp., *Acetobacter diazotrophicus, Alcaligenes, Azoarcus* spp., *Herbaspirillum* spp.
- **Phosphorus solubilizing:** They help in the solubilization of insoluble P to soluble forms through secretion of acids, lowers down the soil pH in order to dissolve the bound phosphates.
  Example: Bacteria—*Pseudomonas striata, Bacillus circulans, Microccocus, Agrobacterium, Penicillium* spp., *B subtilis.*
  Fungi: *Aspergillus awamori, Penicillium* spp., and *Trichoderma.*
- **Phosphorus mobilizing:** They transfer the P from the soil to the cortex of the roots and are a category of broad-spectrum biofertilizers.
- **Mycorrhiza:** *Arbuscular mycorrhiza*, Acaulospora spp., Glomus spp., Scutellospora spp., Gigaspora spp.
- **Potassium solubilizing:** They solubilize the potassium and help in metal ions removal.
  Bacteria: *Bacillus mucilaginosus, Pseudomonas fluorescens, B. edaphicus.*
  Fungi: *Aspergillus niger.*
- **Potassium mobilizing:** They mobilize potassium that are in inaccessible forms in soil.
  Bacteria—*Bacillus* sp.
  Fungi—*Aspergillus niger.*
- **Micronutrient Oxidizing**:
  Sulphur oxidizing: *Thiobacillus* spp. They solubilize zinc by proton, acidification, chelated ligands, and by oxidoreductase systems.
- **Zinc solubilizing:** *Mycorrhiza, Bacillus* sp., *Pseudomonas* sp.
- **Plant growth promoting:** They release hormones to improve nutrient availability, help in root growth, and improve yield. Examples: *Rhizobacteria, Pseudomonas fluorescens, Pseudomonas* spp., *Erwinia, Agrobacterium, Enterobacter, Arthrobacter, Rhizobium, Bacillus.*

There are still a lot of constraints for promotion and application of biofertilizers that include: Lack of regulation and standards for BFs production, unavailability of desired strains, standardization of dose of BFs for particular crops and soil, soil conditions like acidity, alkalinity, salinity, water logging, etc., proper labelling of BFs, promotion, and awareness of BFs, a climatic abnormality that hampers its application and efficacy, types of equipment required for proper culture, production and maintenance of BFs, lack of funding for its promotion and large scale production, etc.

15.4.23 Waste Management

Rapid urbanization, changing food habits, lifestyles, industrialization, the density of population, etc. have created a problem for the management of municipal solid waste (MSW) in India. Per capita waste generation increased exponentially at 0.26 kg day^{-1} to 0.85 kg day^{-1} (CPCB India 2018). In India, the daily generation of MSW is nearly 143,449 metric tonnes (MT), out of which nearly 111,000 MT used be collected, and about 35,602 MT treated (Kumar et al. 2017). In landfill sites, about 80–90% of municipal waste used to disposed off without proper treatment and that causes air, water, and soil pollution (Ahluwalia and Patel 2018; Joshi and Ahmed 2016). Due to migration and natural resources depletion, the municipalities are going through overburden of waste at the environmental and socio-economic prospects (Gerdes and Gunsilius 2010). According to the reports of United nations Environmental Programme (UNEP), the world is producing 300 Mt of plastic waste of which 9% is recycled, 14% is collected for purpose of recycling and the rest goes to oceans annually (United Nation Environment Programme 2019), this is a very serious concern. Further, the city waste and hospitals generate a lot of hazardous wastes leading to breathing complications and premature deaths (Joshi and Ahmed 2016; Mohan 2019). In India, the high-income groups utilize more packaged products, which leads to excess volumes of paper, plastics, metals, glass, and textiles as compared to low-income groups (Sridevi et al. 2012). MSW might also comprise paints, medicines, batteries, e-waste, pesticides, etc. The majority of the municipalities cannot handle such a huge amount of waste generated from the unorganized sector due to the non-availability of suitable machinery and tools (Annepu 2012).

Agricultural waste comprised of agro-waste containing cattle waste, animal carcasses, food products, residues, crop waste, bagasse, drops, pruning, weeds, wastes of pesticides, etc. Around 998 Mt of agricultural waste is generated yearly (Agamuthu 2009). The reasons for less focus on agricultural waste management and waste recycling is due to social taboo, non-sorting of waste at source, unavailability of residue and labour, the nutritional value of crop residues is generally low, intensive recycling leads to loss of nutrients, not enough organic sources to meet the crop demands on wider level (Choudhary 2018), unorganized informal sector involved in the collection, grading and recycling of wastes, poor implementation and government policies.

15.5 Future Strategy and Way Forward

Post green revolution period is associated with achieving self-sufficiency in food grain production but also witnessed a dramatic decline in environmental conditions coupled with declining soil fertility, partial factor productivity, multi-nutrient deficiencies, low use efficiency, salinization, water logging, fluctuating water table associated with climate change had aggravated the situation. Over the years, the organic matter addition has also declined, owing mainly due to conversion towards alternative purposes, additionally, low use efficiency had challenged the sustainability and productivity of agricultural systems greatly. Supplementation of plant nutrients in a balanced dose is quite necessary that should match the crop needs.

A few points, as discussed below, need to be resolved to address the in-time solutions related to soil fertility and health management in a sustainable manner for the longer run:

- Detailed characterization of the "locked-up nutrient pool" in soil and their release dynamics in the soil through different management interventions.
- Micronutrient profiling of important food crops to develop biofortified micronutrient-rich cultivars.
- Balanced fertilization especially for potassium with other nutrients is to be mandatory to check the mining of potassium from potassium-bearing clay minerals.
- Integrated nutrient management strategies need to be refined in tandem with locally available organic resources to supplement the costly chemical fertilizer, synchronize the nutrient requirement of the crop, and less greenhouse gas emissions.
- Biofertilizers need to be popularised among farmers through a fertilizer distribution system with a proper storage facility at the right time and right quantity.
- Soil testing and fertilizer recommendation for different crops and cropping systems need to be revisited and focus must be given to the development of multinutrient extractants for the evaluation of soil fertility.
- Research on nutrient dynamics under conservation tillage systems to optimize the nutrient management strategies under such conditions.
- Modelling the dynamic processes governing nutrient release (and fixation) in soil for plant uptake, with due care being taken of the corresponding soil buffering power, thereby leading to a better understanding of the basic processes involved.
- Nutrient dynamics in rain-fed agro-ecosystems, an overlapping field, where the crop yield potentials are far from being realized. Indeed rainfed agriculture by far received relatively scanty research attention, where the potential for increasing agricultural productivity lies chiefly.
- Water and nutrients are so intrinsically related that these cannot be and should not be managed independently of one another. Crop, climate, and soil-specific schedules for irrigation and nutrients need to be developed at the regional level. Management of inputs especially nutrients and water has the potential to critically

reduce input costs, maintain soil health, close yield gaps, and make agriculture production systems more sustainable.

- Agricultural practices that reduce or offset greenhouse gas emissions can improve soil productivity, and water quality and increase farmers' income.
- The major task for researchers are to work on finding commonality in soil quality indicators for different soil types, production systems, etc. so that easy and inexpensive methods are developed for those common master indicators for their inclusion in routine analysis in soil testing laboratory for assessment of soil health.
- Again, studies assessing the health of soils to meet other ecosystem services have hardly been done in India, although its role and capability for providing water security (NO_3^{-1}pollution), waste recycling, biodiversity, and environmental protection and climate change abatement are in urgent demand for assessment. A modest beginning can be made with the existing long-term experiments where along with biological productivity, biological diversity, carbon budgeting, crop quality, etc. may be included as goal variables.
- Increase the research focus on the identification of best management practices that enhance the carbon sequestration potential of soils under different agroecological regions of the country.
- Research and development support for innovative technologies to accelerate total C sequestration in cropping systems.
- Develop and promote regional and state-specific approaches to enhance soil C stocks.
- Policy on minimum tillage and conservation agriculture is needed to be developed for major soil groups of India and promoted to adopt.

There is a need for new technologies, greener innovations, and smooth policy support to achieve sustainability and to combat new challenges emerging in the arena of natural resource management. Due to ever-increasing climate variability, global warming, soil degradation, distorted market policy, etc., agriculture in coming years is very challenging. The goal should not be only to increase the yield for having food security but proper environmental gains, nutrition, quality, reducing harvest losses, recycling of waste, etc. is highly needed in present times. Borlaug once shed light on changing diets, reducing waste, crop intensification, and expanding aquaculture. We have a lot of technologies with us to fight against hunger and malnutrition.

References

Abichandani CT (1959) Present position of fertilizer experiments in rice in India. Proc Natl Acad Sci India Sect B Biol Sci 49(5):319–330

Abrol IP, Nambiar KKM (1997) Fertility management of Indian soils-a historical perspective. Adv Geoecol 29:293–310

Agamuthu P (2009) Challenges and opportunities in agro-waste management: an Asian perspective. Inaugural meeting of First Regional 3R Forum in Asia 11–12 November, Tokyo

Aggarwal RK, Kumar P, Power JF (1997) Use of crop residue and manure to conserve water and enhance nutrient availability and pearl millet yields in an arid tropical region. Soil Tillage Res 41:43–51

Ahluwalia IJ, Patel U (2018) Working paper no. 356 solid waste management in India: an assessment of resource recovery and environmental impact. Indian Council for Research on International Economic Relations (Issue 356). https://icrier.org/pdf/Working_Paper_356.pdf

Andrews SS, Karlen DL, Cambardella CA (2004) The soil management assessment framework: a quantitative soil quality evaluation method. Soil Sci Soc Am J 68:1945–1962

Annepu RK (2012) Sustainable solid waste management in India. MS Dissertation, pp 1–189. https://doi.org/10.1007/978-981-4451-73-4

Bailey RL, West KP Jr, Black RE (2015) The epidemiology of global micronutrient deficiencies. Ann Nutr Metab 66(2):22–33

Bajwa GS (1993) Nitrate pollution of groundwater under different systems of land management in Punjab. In: Proceedings of the agricultural science congress. National Academy of Agricultural Sciences, New Delhi, pp 223–230

Barassi CA, Ayrault G, Creus CM, Sueldo RJ, Sobrero MT (2006) Seed inoculation with Azospirillum mitigates NaCl effects on lettuce. Sci Hortic 109:8–14

Basak N (2011) Assessment of soil health under rice-based cropping systems. PhD thesis, Bidhan Chandra Krishi Viswavidyalaya, West Bengal, India, pp 1–221

Basak N, Datta A, Mitran T, Singha Roy S, Saha BN, Biswas S, Mandal B (2016a) Assessing soil quality indices for sub-tropical rice-based cropping systems in India. Soil Res 54:20–29

Basak N, Datta A, Biswas S, Mitran T, Mandal B (2016b) Organic amendment influences soil quality in farmers' field under rice-based cropping systems in Indo-Gangetic Plains of India. J Indian Soc Soil Sci 64(2):138–147

Basak N, Datta A, Mitran T, Mandal B, Mani PK (2016c) Impact of organic and mineral inputs onto soil biological and metabolic activities under a long-term rice-wheat cropping system in subtropical Indian Inceptisols. J Environ Biol 37:83–89

Beaufils ER (1973) Diagnosis and recommendation integrated system (DRIS) (No Folleto 5743)

Beesley L, Moreno-Jiménez E, Gomez-Eyles JL (2010) Effects of biochar and greenwaste compost amendments on mobility, bioavailability and toxicity of inorganic and organic contaminants in a multi-element polluted soil. Environ Pollut 158:2282–2287

Bell RW, Dell B (2008) Micronutrients for sustainable food, feed, fibre and bioenergy production. International Fertilizer Industry Association, Paris, p 175

Bhadauria T, Ramakrishnan PS (1996) Role of earthworms in nitrogen cycling during the cropping phase of shifting agriculture (Jhum) in Northeast India. Biol Fertil Soils 22:350–354

Bhaduri D, Purakayastha TJ (2014) Long-term tillage, water and nutrient management in rice–wheat cropping system: assessment and response of soil quality. Soil Tillage Res 144:83–95

Bhaduri D, Purakayastha TJ, Patra AK, Chakraborty D (2014) Evaluating soil quality under a long-term integrated tillage–water–nutrient experiment with intensive rice–wheat rotation in a semi-arid Inceptisol, India. Environ Monit Assess 186:2535–2547

Bhaduri D, Purakayastha TJ, Pramanik P, Saha A, Verma BC, Shahid M, Saha D (2020) Chapter: 14 – Indexing methods of soil quality in agro-ecosystems: an overview of Indian soils and beyond. In: Soil analysis: recent trends and applications. Springer, Singapore, pp 255–284. https://doi.org/10.1007/978-981-15-2039-6_14

Bhalla GS (1979) Transfer of technology and agricultural development in India. Econ Polit Weekly 14(51/52):A130–A142. JSTOR, http://www.jstor.org/stable/4368252

Bhatia A, Pathak H, Jain N, Singh PK, Tomer R (2012) Greenhouse gas mitigation in rice–wheat system with leaf color chart-based urea application. Environ Monit Assess 184(5):3095–3107

Bhatia A, Jain N, Pathak H (2013) Methane and nitrous oxide emissions from Indian rice paddies, agricultural soils and crop residue burning. Greenh Gas Sci Technol 3:196–211

Bhattacharyya T, Pal DK, Velayutham M, Chandran P, Mandal C (2000) Total carbon stock in Indian soils: issues, priorities and management. Special Publication of the International Seminar on Land Resource Management for Food, Employment and Environmental Security (ICLRM), New Delhi, 8–13 Nov, pp 1–46

Bhojvaid PP, Timmer VR (1998) Soil dynamics in an age sequence of *Prosopis juliflora* planted for sodic soil restoration in India. For Ecol Manag 106(2–3):181–193

Biswas S (2011) Assessment of biological health of soil for rice-rice cropping systems. PhD thesis, Bidhan Chandra Krishi Viswavidyalaya, West Bengal, India, pp 1–207

Biswas S, Hazra GC, Purakayastha TJ, Saha N, Mitran T, Roy SS, Mandal B (2017) Establishment of critical limits of indicators and indices of soil quality in rice-rice cropping systems under different soil orders. Geoderma 292:34–48

Chanakya HN, Ramachandra TV, Vijayachamundeeswari M (2006) Anaerobic digestion and reuse of digested products of selected components of urban solid waste. In: Technical Report of Centre for Ecological Sciences and Centre for Sustainable Technologies, Technical Report No 114, Indian Institute of Science, Bengaluru, pp 1–109

Chaudhury J, Mandal UK, Sharma KL, Ghosh H, Mandal B (2005) Assessing soil quality under long-term rice-based cropping system. Commun Soil Sci Plant Anal 36:1141–1161

Cheng CH, Lehmann J, Thies JE, Burton SD, Engelhard M (2006) Oxidation of black carbon by biotic and abiotic processes. Org Geochem 37:1477–1488

Chhowalla M (2017) Slow release nanofertilizers for bumper crops. ACS Cent Sci 3:156–157

Choudhary S (2018) Chapter—7 modern concept of farm waste management. https://doi.org/10.22271/ed.book01.a07

Chugh G, Siddique KHM, Solaiman ZM (2021) Nano biotechnology for agriculture: smart technology for combating nutrient deficiencies with nanotoxicity challenges. Sustainability 13:1781. https://doi.org/10.3390/su13041781

CPCB India (2018) MSW—generation and composition CPCB 2018. https://cpcb.nic.in/uploads/MSW/Waste_generation_Composition.pdf

Crombie K, Mašek O, Sohi S, Brownsort P, Cross A (2013) The effect of pyrolysis conditions on biochar stability as determined by three methods. GCB Bioenergy 5:122–131

Dantwala ML (1979) Agriculture policy in India since independence. In: Shah CH (ed) Agriculture development of India: poverty and problems. Orient Longman, New Delhi

Dantwala ML (1986) Technology growth and equity. In: Meller JW, Desai G (eds) Agricultural change and rural poverty. International Food Policy Research Center, Washington, DC

Dardanelli MS, Fernández de Córdoba FJ, Espuny MR, Rodríguez Carvajal MA, Soria Díaz ME, Gil Serrano AM et al (2008) Effect of *Azospirillum brasilense* co-inoculated with *Rhizobium* on *Phaseolus vulgaris* flavonoids and Nod factor production under salt stress. Soil Biol Biochem 40:2713–2721

De PK (1939) The role of blue-green algae in nitrogen fixation in rice-fields. Proc R Soc Lond B Biol Sci 127(846):121–139

Deepali GK, Gangwar K (2010) Biofertilizers: an ecofriendly way to replace chemical fertilizers. Available online: https://www.krishisewa.com/articles/organic-agriculture/237-biofertilizer-substitute-chemical.html

Department of Fertilizers and Department of Agriculture, Cooperation & Farmers Welfare (DAC & FW), India (2017). http://agricoop.nic.in/sites/default/files/Krishi%20AR%202017-18-1%20for%20web.pdf

De-Regil LM, Suchdev PS, Vist GE, Walleser S, Peña-Rosas JP (2013) Home fortification of foods with multiple micronutrient powders for health and nutrition in children under two years of age. Evid Based Child Health 8(1):112–201

Dhruvanarayan VVN, Ram B (1983) Estimation of soil erosion in India. J Irrig Drain Eng 109:419–434

Duruibe J, Ogwuegbu MOC, Egwurugwu J (2007) Heavy metal pollution and human biotoxic effects. Int J Phys Sci 2(5):112–118

Dwivedi AK, Singh M, Kauraw DL, Wanjari RH, Chauhan SS (2007) Research bulletin on impact of fertilizer and manure use for three decades on crop productivity and sustainability and soil quality under Soybean-Wheat system on a Vertisol in Central India. IISS (ICAR), Bhopal, pp 224–235

Egamberdieva D, Hua M, Reckling M, Wirth S, Bellingrath-Kimura DS (2018) Potential effects of biochar-based microbial inoculants in agriculture. Environ Sustain 1:19–24

FAO (2005) Fertilizer use by crop in India, Land and Plant Nutrition Management Service, Land and Water Development Division, Food and Agriculture Organization of the United Nations, Rome, pp 1–45

FAO (2015) The state of food insecurity in the world 2015. Meeting the 2015 international hunger targets: taking stock of uneven progress. FAO, Rome

Foster E, Fogle E, Cotrufo M (2018) Sorption to biochar impacts-glucosidase and phosphatase enzyme activities. Agriculture 8:158

Garg VK (1998) Interaction of tree crops with a sodic soil environment: potential for rehabilitation of degraded environments. Land Degrad Dev 9:81–89

Gerdes P, Gunsilius E (2010) The waste experts: enabling conditions for informal sector integration in solid waste management. www.gtz.de/recycling-partnerships

Ghosh GR, Dutta BG (1960) Soil fungi from Orissa (India)—I. Mycologia 52(6):915–918

Gorovtsov AV, Minkina TM, Mandzhieva SS, Perelomov LV, Soja G, Zamulina IV, Rajput VD, Sushkova SN, Mohan D, Yao J (2020) The mechanisms of biochar interactions with microorganisms in soil. Environ Geochem Health 42:2495–2518

Gunathilagaraj K, Ravignanam T (1996) Vermicomposting of sericultural wastes. Madras Agric J 83:455–457

Gupta RK, Rao DLN (1994) Potential of wasteland for sequestering carbon by reforestation. Curr Sci 66:378–380

Hall WF (1964) Agriculture in India. Regional Analysis Division, Economic Research Service, United States Department of Agriculture, p 13. https://archive.org/details/agricultureinind64hall

Hashimoto Y, Matsufuru H, Takaoka M, Tanida H, Sato T (2009) Impacts of chemical amendment and plant growth on lead speciation and enzyme activities in a shooting range soil: an X-ray absorption fine structure investigation. J Environ Qual 38(4):1420–1428

Houghton RA, Hackler (1994) The net flux of carbon from deforestation and degradation in south and Southeast Asia. In: Dale V (ed) Effects of land-use change on atmospheric CO_2 concentrations: South and Southeast Asia as a case study. Springer, New York, pp 304–327

INCCA (2010) Indian network for climate change assessment. Ministry of Environment and Forests, Government of India: Greenhouse Gas Emissions

Indoria AK, Sharma KL, Reddy KS, Srinivasarao C (2016) Role of soil physical properties in soil health management and crop productivity in rainfed systems–II. Management technologies and crop productivity. Curr Sci 110:320–328

Indoria AK, Sharma KL, Reddy KS, Srinivasarao C (2017) Role of soil physical properties in soil health management and crop productivity in rainfed systems–I: soil physical constraints and scope. Curr Sci 112:2405–2414

Indoria AK, Sharma KL, Reddy KS, Srinivasarao C, Srinivas K, Balloli SS, Osman M, Pratibha G, Raju NS (2018) Alternative sources of soil organic amendments for sustaining soil health and crop productivity in India – impacts, potential availability, constraints and future strategies. Curr Sci 115(11):2052–2062

IPCC (2014) Summary for policymakers. In: Climate Change 2014, mitigation of climate change. Contribution of working group III to the fifth assessment report. Intergovernmental panel on climate change. Cambridge University Press, Cambridge, UK and New York, NY, p 8

Jahangirian H, Rafiee-Moghaddam R, Jahangirian N, Nikpey B, Jahangirian S, Bassous N, Saleh B, Kalantari K, Webster TJ (2020) Green synthesis of zeolite/Fe_2O_3 nanocomposites: toxicity and cell proliferation assays and application as a smart iron nanofertilizer. Int J Nanomedicine 15: 1005

Jatav HS, Rajput VD, Minkina T, Singh SK, Chejara S, Gorovtsov A, Barakhov A, Bauer T, Sushkova S, Mandzhieva S, Burachevskaya M, Kalinitchenko VP (2021) Sustainable approach

and safe use of biochar and its possible consequences. Sustainability 13:10362. https://doi.org/10.3390/su131810362
Joshi R, Ahmed S (2016) Status and challenges of municipal solid waste management in India: a review. Cogent Environ Sci 2(1):1–18. https://doi.org/10.1080/23311843.2016.1139434
Kaur B, Gupta SR, Singh G (2000) Soil carbon, microbial activity and nitrogen availability in agroforestry systems on moderately alkaline soils in northern India. Appl Soil Ecol 15:283–294
Kaur B, Gupta SR, Singh G (2002a) Bioamelioration of a sodic soil by silvopastoral systems in southwestern India. Agrofor Syst 54:13–20
Kaur B, Gupta SR, Singh G (2002b) Carbon storage and nitrogen cycling in silvopastoral systems on a sodic soil in northwestern India. Agrofor Syst 54:21–29
Khan S, Cao Q, Zheng YM, Huang YZ, Zhu YG (2008) Health risks of heavy metals in contaminated soils and food crops irrigated with wastewater in Beijing, China. Environ Pollut 152(3):686–692
Kumar R, Pasricha NS (1999) Land use and land cover changes in the Indus plains of Punjab in the post Green Revolution period 1965–1995. In: Proceedings of the international seminar on historical perspectives of land-use/land-cover change in South Asia for the study of global change, April 11–13, 1999. NPL, New Delhi
Kumar S, Smith SR, Fowler G, Velis C, Kumar SJ, Arya S, Rena KR, Cheeseman C (2017) Challenges and opportunities associated with waste management in India. R Soc Open Sci 4(3). https://doi.org/10.1098/rsos.160764
Kundu S (2014) Assessment of quality and resilience of soils in diverse agro-ecosystems. NAIP sub-project C-2060. Indian Institute of Soil Science, Bhopal, pp 1–245
Ladha JK, Pathak H, Krupnik TJ, Six J, van Kessel C (2005) Efficiency of fertilizer nitrogen in cereal production: retrospect and prospects. Adv Agron 87:85–156
Laird DA (2008) The charcoal vision: a win–win–win scenario for simultaneously producing bioenergy, permanently sequestering carbon, while improving soil and water quality. Agron J 100(1):178–181
Lal R (2004) Soil carbon sequestration in India. Clim Chang 65:277–296
Lampkin NH (1999) Organic farming in the European Union: over-view, policies and perspectives. In: Organic farming in the European Union: overview, policies and perspectives for the 21st century, proceedings of a joint EU and Austrian Conference, Avalon Foundation and Eurotech Management, Vienna, Baden, 27–28 May 1999, pp 23–30
Lavell P, Spain AV (2001) Soil Ecology. Springer
Lavelle P, Barois I, Blanchart E et al (1998) Earthworms as a resource in tropical agroecosystems. Nat Resour 34:26–41
Lavicoli I, Leso V, Beezhold DH, Shvedova AA (2017) Nanotechnology in agriculture: opportunities, toxicological implications, and occupational risks. Toxicol Appl Pharmacol 329:96–111
Lehmann J, Gaunt J, Rondon M (2006) Bio-char sequestration in terrestrial ecosystems – a review. Mitig Adapt Strateg Glob Chang 11:403–427. https://doi.org/10.1007/s11027-005-9006-5
Lehmann J, Skjemstad J, Sohi S, Carter J, Barson M, Falloon P, Coleman K, Woodbury P, Krull E (2008) Australian climate–carbon cycle feedback reduced by soil black carbon. Nat Geosci 1: 832–835
Ling W, Shen Q, Gao Y, Gu X, Yang Z (2007) Use of bentonite to control the release of copper from contaminated soils. Aust J Soil Res 45(8):618–623
Llewellyn RS, D'Emden FH, Kuehne G (2012) Extensive use of no-tillage in grain-growing regions of Australia. Field Crops Res 132:204–212
Long D, Chen X, Scanlon BR, Wada Y, Hong Y et al (2016) Have GRACE satellites overestimated groundwater depletion in the Northwest India aquifer? Sci Rep 6:24398
Majumder S, Neogi S, Dutta T, Powel MA, Banik P (2019) The impact of biochar on soil carbon sequestration: meta-analytical approach to evaluating environmental and economic advantages. J Environ Manag 250:109466

Mandal B, Ghoshal SK, Ghosh S, Saha S, Majumdar D, Talukdar NC, Ghosh TJ, Balaguravaiah D, Vijay SBM, Singh AP, Raha P, Das DP, Sharma KL, Mandal UK, Kusuma GJ, Chaudhury J, Ghosh H, Samantaray RN, Mishra AK, Rout KK, Behera BB, Rout B (2005) Assessing soil quality for a few long term experiments – an Indian initiative. In: Proceedings international conference on soil, water and environmental quality-issues and challenges, New Delhi, pp 1–25

Mandal N, Datta SC, Manjaiah KM, Dwivedi BS, Kumar R, Aggarwal P (2019) Evaluation of zincated nanoclay polymer composite in releasing Zn and P and effect on soil enzyme activities in a wheat rhizosphere. Eur J Soil Sci 70:1164–1182

Mankasingh U, Poon-Chung C, Ragnarsdottir V (2011) Biochar application in a tropical, agricultural region: a plot scale study in Tamil Nadu, India. Appl Geochem 26:S218–S221

Masiello C (2004) New directions in black carbon organic geochemistry. Mar Chem 92:201–213

Masto RE, Chhonkar PK, Singh D, Patra AK (2007) Soil quality response to long-term nutrient and crop management on a semi-arid Inceptisol. Agric Ecosyst Environ 118:130–142

Masto RE, Chhonka PK, Singh D, Patra AK (2008a) Alternative soil quality indices for evaluating the effect of intensive cropping, fertilisation and manuring for 31 years in the semi-arid soils of India. Environ Monit Assess 136:419–435

Masto RE, Chhonkar PK, Purakayastha TJ, Patra AK, Singh D (2008b) Soil quality indices for evaluation of long term land use and soil management practices in semi-arid sub-tropical India. Land Degrad Dev 19:1–14

Mayak S, Tirosh T, Glick BR (2004) Plant growth-promoting bacteria confer resistance in tomato plants to salt stress. Plant Physiol Biochem 42:565–572. https://doi.org/10.1016/j.plaphy.2004.05.009

McLaughlin MJ, Hamon RE, McLaren RG, Speir TW, Rogers SL (2000) Review: a bioavailability-based rationale for controlling metal and metalloid contamination of agricultural land in Australia and New Zealand. Aust J Soil Res 38(6):1037–1086

Misra RV, Roy RN, Hiraoka H (2003) On-farm composting methods. In: FAO Land and water discussion Paper 2. Publishing Management Service Information Division FAO, Viale delle Terme di Caracalla, 00100 Rome, Italy. ISSN 1729-0554

MoEF (2010) India: taking on climate change post-Copenhagen domestic actions. Ministry of Environment and Forests, Government of India

Moench M (1991) Soil erosion under a successional agroforestry sequence – a case study from Idukki district, Kerala, India. Agrofor Syst 15:31–50

Mohan V (2019) Topic: 91% premature deaths due to air pollution in low and middle-income countries: UNEP 1. The Economics Times (2019, May 10) https://economictimes.indiatimes.com/news/environment/pollution/91-premature-deaths-due-to-air-pollution-in-low-and-middle-income-countries-unep/articleshow/69264417.cms

Mohan-Rao N, Shantaram MV (1978) Select bibliography on soil organic matter research in India. Agricola, p 155

Mohanty M, Painul DK, Mishra AK, Ghosh PK (2007) Soil quality effects of tillage and residue under rice–wheat cropping on a vertisol in India. Soil Tillage Res 92:243–250

Mukherjee A, Zimmerman A, Harris W (2011) Surface chemistry variations among a series of laboratory-produced biochars. Geoderma 163:247–255

Musilova J, Arvay J, Vollmannova A, Toth T, Tomas J (2016) Environmental contamination by heavy metals in region with previous mining activity. Bull Environ Contam Toxicol 97:569–557

NAAS (2012) Management of crop residues in the context of conservation agriculture. Policy Paper No. 58, National Academy of Agricultural Sciences, New Delhi, pp 1–12

Nadeem SM, Zahir ZA, Naveed M, Arshad M (2007) Preliminary investigations on inducing salt tolerance in maize through inoculation with rhizobacteria containing ACC deaminase activity. Can J Microbiol 53:1141–1149. https://doi.org/10.1139/w07-081

Nagelschmidt G, Desai AD, Muir A (1940) The minerals in the clay fractions of a black cotton soil and a red earth from Hyderabad, Deccan state, India. J Agric Sci 30(4):639–653

Noyce GL, Winsborough C, Fulthorpe R, Basiliko N (2016) The microbiomes and metagenomes of forest biochars. Sci Rep 6:26425
Pahuja N, Pandey N, Mandal K, Bandyopadhyay C (2014) GHG mitigation in India: an overview of the current policy landscape. Working Paper, World Resource Institute, Washington, DC
Pappu A, Saxena M, Asolekar SR (2007) Solid wastes generation in India and their recycling potential in building materials. Build Environ 42:2311–2320
Pathak H, Sankhyan S, Dubey DS, Bhatia A, Jain N (2012) Dry direct-seeding of rice for mitigating greenhouse gas emission: field experimentation and simulation. Paddy Water Environ 11(1–4): 593–601
Patil VC (2009) Precision nutrient management: a review. Indian J Agron 54(2):113–119
Paz-Ferreiro J, Fu S (2013) Biological indices for soil quality evaluation: perspectives and limitations. Land Degrad Dev 27(1):14–25. https://doi.org/10.1002/ldr.2262
Pendias AK, Pendias H (2000) Trace elements in soils and plants. CRC Press, pp 10–11
Pierzynski GM, Sims JT, Vance GF (2000) Soils and environmental quality, 2nd edn. CRC Press, London
Prasad SC (2016) Innovating at the margins: the system of rice intensification in India and transformative social innovation. Ecol Soc 21(4):7. https://doi.org/10.5751/ES-08718-210407
Purakayastha TJ, Rudrappa L, Singh D, Swarup A, Bhadraray S (2008) Long- term impact of fertilizers on soil organic carbon pools and sequestration rates in maize- wheat-cowpea cropping system. Geoderma 144:370–378
Purakayastha TJ, Savita K, Pathak H (2015) Characterisation, stability and microbial effects of four biochars produced from crop residues. Geoderma 239-240:293–303
Purakayastha TJ, Kamble K, Pathak H, Kumari S, Singh A (2016) Decision support system for online assessment of soil health. http://ssacdss.iari.res.in/home.php
Purakayastha TJ, Pathak H, Kumari S, Biswas S, Chakrabarty B, Padaria RN, Kamble K, Pandeya M, Sasmal S, Singh A (2019) Soil health card development for efficient soil management in Haryana, India. Soil Tillage Res 191:294–305
Raghuwanshi R (2012) Opportunities and challenges to sustainable agriculture in India. Nebio 3: 78–86
Randhawa MS (1979) History of the Indian Council of Agricultural Research, 1929–1979, ICAR, New Delhi
Rattan RK, Datta SP, Chandra S, Saharan N (2002) Heavy metals and environmental quality: Indian scenario. Fertilizer News 47:21–40
Rattan RK, Biswas P, Ranjan D (2014) Efficient water management for sustainable agriculture. J Indian Soc Soil Sci 29:1–86
Redfield AC (1958) The biological control of chemical factors in the environment. Am Sci 46(3): 205–221
Report of the Committee on Doubling Farmers' Income (2018) Input management for resource use efficiency and total factor productivity, vol VII. Document prepared by the Committee on Doubling Farmers' Income, Department of Agriculture, Cooperation and Farmers' Welfare, Ministry of Agriculture and Farmers' Welfare
Reynolds CS (1990) Temporal scales of variability in pelagic environments and the response of phytoplankton. Freshw Biol 23(1):25–53
Roberts TL (2008) Improving nutrient use efficiency. Turk J Agric 32:177–182
Rochette PH (2008) No-till only increases N_2O emissions in poorly aerated soils. Soil Tillage Res 101:97–100
Saha M (2013) The state, scientists, and staple crops: agricultural modernization in pre-green revolution India. Agric Hist 87(2):201–223
Saha JK, Panwar N, Srivastava A, Biswas AK, Kundu S, Subba Rao A (2010) Chemical, biochemical, and biological impact of untreated domestic sewage water use on vertisol and its consequences on wheat (*Triticum aestivum*) productivity. Environ Monit Assess 161:403–412

Saravanakumar D, Samiyappan R (2007) ACC deaminase from *Pseudomonas fluorescens* mediated saline resistance in groundnut (*Arachis hypogea*) plants. J Appl Microbiol 102:1283–1292. https://doi.org/10.1111/j.1365-2672.2006.03179.x

Sengupta J (2002) Recycling of agro-industrial wastes for manufacturing of building materials and components in India: an overview. Civ Eng Constr Rev 15:23–33

Sharda VN, Dogra P, Prakash C (2010) Assessment of production losses due to water erosion in rainfed areas of India. J Soil Water Conserv 65:79–91

Sharma AR, Singh R, Dhyani SK (2005a) Conservation tillage and mulching for optimizing productivity in maize-wheat cropping system in the outer western Himalaya region – a review. Indian J Soil Conserv 33(1):35–41

Sharma KL, Mandal UK, Srinivas K, Vittal KPR, Mandal B, Grace JK (2005b) Long-term soil management effects on crop yields and soil quality in a dryland Alfisol. Soil Tillage Res 83: 246–259

Sharma KL, Grace JK, Mandal UK, Gajbhiye PN, Srinivas K, Korwar GR, Bindu VH, Ramesh V, Ramachandran K, Yadav SK (2008) Evaluation of long-term soil management practices using key indicators and soil quality indices in a semi-arid tropical Alfisol. Aust J Soil Res 46:368–377

Shukla AK, Behera SK, Prakash C, Tripathi A, Patra AK, Dwivedi BS, Trivedi V, Rao S, Chaudhari SK, Das S, Singh AK (2021a) Deficiency of phyto-available sulphur, zinc, boron, iron, copper and manganese in soils of India. Sci Rep 11(1):19760. https://doi.org/10.1038/s41598-021-99040-2

Shukla AK, Behera SK, Prakash C, Patra AK, Rao CS, Chaudhari SK, Das S, Singh AK, Green A (2021b) Assessing multi-micronutrients deficiency in agricultural soils of India. Sustainability 13(16):9136

Simonson RW (1954) The regur soils of India and their utilization. Soil Sci Soc Am J 18(2): 199–203

Singh RB (1997) Impact of agriculture and land use/cover change on soil and water resources and ecosystem sustainability: a case study of Punjab. In: Epps R (ed) Sustainability of rural system in the context of global change. University of New England, Armidale

Singh M (2016) Impact of nutrient management and changing climate on productivity and carbon sequestration: a case study from long-term fertilizer experiment. J Indian Soc Soil Sci 64:S67–S76

Singh S, Singh JS (1996) Water-stable aggregates and associated organic matter in forest, savanna, and cropland soils of a seasonally dry tropical region, India. Biol Fertil Soils 22:76–82

Singh A, Sharma RK, Agrawal M, Marshall FM (2010) Health risk assessment of heavy metals via dietary intake of foodstuffs from the wastewater irrigated site of a dry tropical area of India. Food Chem Toxicol 48(2):611–619. https://doi.org/10.1016/j.fct.2009.11.041. Epub 2009 Nov 24. PMID: 19941927

Singh R, Singh R, Soni SK, Singh SP, Chauhan UK, Kalra A (2013) Vermicomposting from biodegraded distillation waste improves soil properties and essential oil yield of *Pogostemon* cablin (patchouli) Benth. Appl Soil Ecol 70:48–56

Singh SR, Joshi D, Tripathi N, Singh P, Srivastava TK (2017) Plant-growth promoting bacteria: an emerging tool for sustainable crop production under salt stress. In: Arora S, Singh AK, Singh YP (eds) Bioremediation of salt affected soils: an Indian perspective. Springer, pp 101–131

Smith P et al (2008) Greenhouse gas mitigation in agriculture. Philos Trans R Soc B Biol Sci 363: 789–813

Solanki P, Bhargava A, Chhipa H, Jain N, Panwar J (2015) Nano-fertilizers and their smart delivery system. In: Nanotechnologies in food and agriculture. Springer, Cham, pp 81–101

Spokas KA, Baker JM, Reicosky DC (2010) Ethylene: potential key for biochar amendment impacts. Plant Soil 333:443–452

Sridevi V, Musalaiah M, Ch MVV, Lakshmi A, Kesavarao L (2012) A review on integrated solid waste management. Int J Eng Adv Technol 2(5):1491–1499

Srinivasarao C, Indoria AK, Sharma KL (2017) Effective management practices for improving soil organic matter for increasing crop productivity in rainfed agroecology of India. Curr Sci 112: 1497–1504

Srinivasrao C, Gopinath KA, Govindarajan V, Sikka A (2013) Use of biochar for soil health management and greenhouse gas mitigation in India: potential and constraints. NICRA Bulletin 1/2013 Central Research Institute for Dryland Agriculture, Hyderabad

Srivastava P, Balhara M, Giri B (2020) Soil health in India: past history and future perspective. In: Giri B, Varma A (eds) Soil health. Springer, New Delhi, Noida, pp 1–19

Swaminathan MS (2006) An evergreen revolution. Crop Sci 46(5):2293–2303

Swarup A (1998) Emerging soil fertility management issues for sustainable crop productivity in irrigated systems. In: Swarup A, Reddy DD, Prasad RN (eds) Long-term soil fertility management through integrated plant nutrient supply. Indian Institute of Soil Science, Bhopal, pp 54–68

Swarup A, Singh M (2009) Soil carbon dynamics under intensive cropping system. J Indian Soc Soil Sci 57(4):469–476

Takkar PN, Shukla AK (2015) Management of soil fertility: micronutrients. In: Pathak H, Sanyal SK, Takkar PN (eds) State of Indian agriculture–soil. NAAS, New Delhi, pp 121–152

Takkar PN, Chhibba IM, Mehta SK (1989) Twenty years of coordinated research on micronutrients in soils and plants, 1967–1987, Indian Institute of Soil Science, Bhopal, India

Tan X, Liu Y, Zeng G, Wang X, Hu X, Gu Y, Yang Z (2015) Application of biochar for the removal of pollutants from aqueous solutions. Chemosphere 125:70–85

Tandon HLS (1992) Assessment of soil nutrient depletion. In: Proceedings of the FADINAP regional seminar on fertilization and the environment, Chiangmai

Tandon HLS (2004) Fertilizers in Indian agriculture—from 20th to 21st century. FDCO, New Delhi, p 240

Tewatia RK (2012) Developments in fertilizer consumption in India. Indian J Agron 57(3):116–122

The Hindu (2009) Benefits from poultry manure – no chicken feed. The Hindu, Chennai, 22 Oct 2009

The Hindu (2012) From 1,10,000 varieties of rice to only 6,000 now. The Hindu. http://www.thehindu.com/news/national/karnataka/from-110000-varieties-of-rice-to-only 6000-now/article3284453.ece

Thomas L, Singh I (2019) Microbial biofertilizers: types and applications. In: Biofertilizers for sustainable agriculture and environment. Springer, Berlin/Heidelberg, pp 1–19

Thomas GA, Titmarsh GW, Freebairn DM, Radford BJ (2007) No-tillage and conservation farming practices in grain growing areas of Queensland—a review of 40 years of development. Aust J Exp Agric 47:887–898

Thompson H (2012) Food science deserves a place at the table –US agricultural research chief aims to raise the profile of farming and nutrition science. Nature, July 12

United Nation Environment Programme (2019) Plastic recycling: an underperforming sector ripe for a remake. https://www.unenvironment.org/news-and-stories/story/plastic-recycling-underperforming-sector-ripe remake

United States Department of Agriculture (USDA) (1963) Economic research service, regional analysis division. The 1964 Far East, Communist China, Oceania agricultural situation: supplement no 4 to the 1964 world agricultural situation, p 49–50. https://archive.org/details/1964fareastcommu74unit

Vandana S (1993) The violence of the green revolution: third world agriculture, ecology and politics, 2nd edn. Zed Books Ltd, London

Vetter SH, Sapkota TB, Hillier J, Stirling CM, Macdiarmid JI, Aleksandrowicz L, Green R, Joy EJM, Dangour AA, Smith P (2017) Greenhouse gas emissions from agricultural food production to supply Indian diets: implications for climate change mitigation. Agric Ecosyst Environ 237:234–241. https://doi.org/10.1016/j.agee.2016.12.024

Vijaya D, Padmadevi SN, Vasandha S, Meerabhai RS, Chellapandi P (2008) Effect of vermicomposted coir pith on the growth of *Andrographis paniculata*. JOS 3:51–56

16 Water

Gouranga Kar

Abstract

Water forms the backbone for all future endeavors to achieve food security for ever-increasing population of our country. In the present context, upscaling agricultural economic growth to more than 4% annually is the main challenge. Taking water technologies for better water management from lab to land is a formidable task to be addressed. Modernization/automation of irrigation systems, precision irrigation, land reforms, corporate farming, cooperative farming, water and energy pricing, crop insurance, institutional mechanism for better governance, and water rights are some of the key issues for better water management in agriculture. The projected food requirement demands a pronounced role for research, development, and training in the water and agriculture sector. It is evident that the water availability for agriculture is declining, and to enhance agricultural production, more water is needed. Therefore, concerted and holistic efforts are required in increasing the overall water use efficiency at system level which would be achieved through various measures like timely execution of projects, minimizing the losses, better operational efficiency through stakeholders' participation, implementation of on-farm water management technologies, conjunctive use of water, and changes in irrigation policy. The need of the day is to economize water in agriculture, bring more area under irrigation, and increase the yield per unit area and unit quantum of water through multiple use of water, conjunctive use, recycling of poor quality/waste water, pressurized irrigation system, crop diversification, and other water-saving agricultural methods.

G. Kar (✉)
Central Research Institute on Jute and Allied Fibre, Barrackpore, West Bengal, India

P. K. Ghosh et al. (eds.), *Trajectory of 75 years of Indian Agriculture after Independence*, https://doi.org/10.1007/978-981-19-7997-2_16

Keywords

Water resources · Agricultural water management · Irrigation use efficiency · Water productivity

16.1 Water Resource Scenario of India

The main water resource of India consists of the precipitation on the Indian territory which is estimated to be around 4000 km^3/year and transboundary flows received in rivers and aquifers from the upper riparian countries which are around 500 km^3. Out of the total precipitation including snowfall, the availability from surface water and replenishable groundwater is estimated as 1869 km^3. Due to various constraints of topography, uneven distribution of resource over space and time, it has been estimated that only about 1123 km^3 including 690 km^3 from surface water and 433 km^3 from groundwater resources can be put to beneficial use. The water resources of the country at a glance are given in Table 16.1. Water resources of major river basins of the country are given in Table 16.2. Indian precipitation has high spatial and temporal variability. Precipitation over a large part of India is concentrated in the monsoon season during June to September/October. Precipitation varies from 100 mm in the western parts of Rajasthan to over 11,000 mm at Cherrapunji in Meghalaya. Water resources of major river basins of the country are given in Table 16.2.

The spectacular growth in food production in India (from merely 50.8 million tons in 1950–1951 to an estimated production of more than 300 million tons in 2021–2022) is mainly attributed to three major factors, namely, development of high-yielding varieties, expansion in irrigated area, and fertilizer use. Though agriculture sector is the largest consumer of water (82.8%), the requirement of water from other sectors like domestic and industrial needs is increasing due to escalating population, urbanization, and industrialization (Table 16.3).

The Ministry of Water Resources (MoWR) has projected that in 2025 the total demand of water will be 1093 billion cubic meter (BCM), the bulk of which

Table 16.1 Water resources of India

Estimated annual precipitation (including snowfall)	4000 km^3
Runoff received from upper riparian countries	500 km^3
Average annual natural flow in rivers and aquifers	1869 km^3
Estimated utilizable water	1123 km^3
(a) Surface	690 km^3
(b) Ground	433 km^3
Water demand utilization (for year 2000)	634 km^3
(a) Domestic	42 km^3
(b) Irrigation	541 km^3
(c) Industry, energy, and others	51 km^3

Source: Water Resources at a Glance. Central Water Commission (2021), GoI

Table 16.2 Water resources of major river basins of the country

S. no.	River basin	Catchment area (km^2)	Average water resource potential (BCM)	Utilizable surface water resources (BCM)
1	Indus (up to border)	317,708	45.53	46
2	(a) Ganga	838,803	509.52	250
	(b) Brahmaputra	193,252	527.28	24
	(c) Barak and others	86,335	86.67	–
3	Godavari	312,150	117.74	76.3
4	Krishna	259,439	89.04	58
5	Cauvery	85,167	27.67	19
6	Subarnarekha	26,804	15.05	6.8
7	Brahmani and Baitami	53,902	35.65	18.3
8	Mahanadi	144,905	73	50
9	Pennar	54,905	11.02	6.9
10	Mahi	39,566	14.96	3.1
11	Sabarmati	31,901	12.96	1.9
12	Narmada	96,659.79	58.21	34.5
13	Tapi	65,805.80	26.24	14.5
14	West-flowing rivers from Tapi to Tadri	58,360	118.35	11.9
15	West-flowing rivers from Tadri to Kannyakumari	54,231	119.06	24.3
16	East-flowing rivers between Mahanadi and Pennar	82,073	26.41	13.1
17	East-flowing rivers between Pennar and Kanyakumari	101,657	26.74	16.5
18	West-flowing rivers of Kutch and Saurashtra including Luni	192,112	26.93	15
19	Area of inland drainage in Rajasthan	144,835.90	Negi.	N.A
20	Minor river draining into Myanmar (Burma) and Bangladesh	31,382	31.17	N.A.
	Total	3,271,953	1999.2	690.1

Source: Water Resources at a Glance. Central Water Commission (2021), GoI

(910 BCM) will be consumed by irrigation and the country will face the stiff competition for water from different sectors (Table 16.4). The National Commission on Integrated Water Resources (NCIWRD) projected much lower water demand of 784–843 BCM under different scenarios than MoWR. Further, the average consumption per person per year for all users is around 680 m^3, but it is projected to increase. There is continuous decline in per capita water resources of the country from 5176 m^3 in 1951 to 4732 m^3 in 1955, 2209 m^3 in 1991, 1820 m^3 in 2001, and

Table 16.3 Water requirement or demand estimation by uses (in BCM) projections as per standing subcommittee of MoWR, RD, and GR

Sector	2025	2050
Irrigation	910	1072
Domestic	73	102
Industrial	23	63
Energy	15	130
Other	72	80
Total	1093	1447

Source: Standing Subcommittee of MoWR, RD, and GR Report-Basin Planning Directorate, CWC, XI Plan Document; Report of the Standing Subcommittee on "Assessment of Availability and Requirement of Water for Diverse Uses 2000"

Table 16.4 Sector-wise projected demand of water in 2025 (billion cubic meter)

		National Commission for Integrated Water Resources Development (NCIWRD)	
Sector	Ministry of water resources	Low	High
Irrigation	910	561	611
Drinking water	73	55	62
Industry	23	67	67
Energy	15	31	33
Other	72	70	70
Total	1093	784	843

Source: Water Resources at a Glance. Central Water Commission (2021)

Table 16.5 Categorization of river basins by water stress

Water availability (m^3/year/capita)	Category
≥1700 both in 2025 and 2050	Safe
≥1700 both in 2025 but <1700 in 2050	Moderately stressed
<1700 both in 2025 and 2050	Critically stressed
<1700 in 2025 and <1000 in 2050	Moderately scarce
<1000 in 2025 and <1000 in 2050	Critically scarce

Source: Water Resources Sector Report, NITI, DMEO, NITI Aayog 2021, GoI

1703.6 m^3 in 2005. The per capita water availability has been reported to be 1656 m^3 in 2007 implying that the entire country is now classified as "water-stressed" (Table 16.5). Indian population is expected to rise to around 1640 million by 2050. As a result, by 2050, the per capita water availability will further decline to 1140 m^3/year. For adequate living standards as in western and industrialized countries, a renewable water supply of at least 2000 m^3 per person per year is necessary (Postel 1992). A country will be water-stressed if water availability is only 1000–2000 m^3 per person per year, while the country is called "water-scarce" if the value comes below 500 m^3 per person per year (Bouwer 2000). With the decline in per capita water availability to 1140 m^3/person/year by 2050, the country will enter into the "water-stressed" category (Kumar and Kar 2013).

16.2 Plan-Wise Gross and Net Irrigated Area

An overlook of different 5-year plan (FYPs) depicts that the net sown area in first FYP was 125.95 Mha which has increased to 139.6 Mha in fourth FYP and is almost constant thereafter (Table 16.6). There is a consistent growth (0.58% per year) in gross sown area from 140 Mha in first FYP to 189.84 Mha in tenth FYP. The increase in gross sown area was a result of improvement in cropping intensity from 111.17% in first FYP to 136.05% in tenth FYP. The major reason for increase in cropping intensity was due to assured supply of irrigation water through different irrigation projects. This is reflected through significant increment in the net as well gross irrigated area of the country with the growth of 2.08% and 2.52% per annum, respectively, during 1950–2007. Over different FYPs, increase in net irrigated area and gross irrigated area led to higher share in gross and net sown areas. Reports indicate that around 42% of gross as well as net sown area was irrigated. Cropping intensity, irrigation intensity, share of net, and gross irrigated area witnessed increasing trend in all regions of the country over different FYPs reflecting an improvement in status of agriculture. State-wise irrigation potential created in major, medium, and minor irrigation is depicted in Table 16.7.

16.3 Water Productivity

Water productivity (WP, expressed in kg of dry matter per m^3 of water) can be defined in different ways (Molden et al. 2001). The numerator may refer to dry matter (DM), e.g., total DM or yield DM. Among different states, crop water productivity varies from 1.01 kg/m^3, the highest being in Punjab to 0.21 kg/m^3, the lowest in Odisha (Table 16.8). These differences are mainly due to varying cropping and land-use patterns, yield levels, and consumptive water use (CWU). Punjab, Haryana, and Uttar Pradesh (UP) in the Indo-Gangetic Basin (IGB) have the highest water productivity. Although irrigation contributes a major part of CWU in AP and TN, its contribution is low in West Bengal and Kerala. Maharashtra, Madhya Pradesh (MP), Karnataka, and Gujarat, with a mixture of cropping patterns (more than 50% of the area under maize and other coarse cereals and pulses), have lower WP. In Maharashtra and Karnataka, irrigation covers only 15 and 23% of area, respectively, and contributes 17 and 28% of the CWU. Irrigation in MP and Gujarat covers 29% of the grain area but contributes 52 and 41% of the CWU. Orissa, Chhattisgarh, and Jharkhand have the lowest water productivities and share 12.8% of the total CWU but contribute only 6.3% of the grain production. Rainfed rice dominates cropping patterns in these states.

Table 16.6 Details of plan-wise position of irrigation potential created and utilized (in Mha)

Plan		Potential created					Potential utilized				
		Major and medium	Minor			Total	Major and medium	Minor			Total
			S.W.	G.W.	Total			S.W.	G.W.	Total	
Up to 1951 (pre-plan)	Cumulative	9.7	6.4	6.5	12.9	22.6	9.7	6.4	6.5	12.9	22.6
I plan (1951–1956)	During	2.5	0.03	1.13	1.16	3.66	1.28	0.03	1.13	1.16	2.44
	Cumulative	12.2	6.43	7.63	14.06	26.26	10.98	6.43	7.63	14.06	25.04
II plan (1956–1961)	During	2.13	0.02	0.67	0.69	2.82	2.07	0.02	0.67	0.69	2.76
	Cumulative	14.33	6.45	8.3	14.75	29.08	13.05	6.45	8.3	14.75	27.8
III plan (1961–1966)	During	2.24	0.03	2.22	2.25	4.49	2.12	3.03	2.22	2.25	4.37
	Cumulative	16.57	6.48	10.52	17	33.57	15.17	6.48	10.52	17	32.17
Annual plans (1966–1969)	During	1.53	0.02	1.98	2	3.53	1.58	0.02	1.98	2	3.58
	Cumulative	18.1	6.5	12.5	19	37.1	16.75	6.5	12.5	19	35.75
IV plan (1969–1974)	During	2.6	0.5	4	4.5	7.1	1.64	0.5	4	4.5	6.14
	Cumulative	20.7	7	16.5	23.5	44.2	18.39	7	16.5	23.5	41.89
V plan (1974–1978)	During	4.02	0.5	3.3	3.8	7.82	2.7	0.5	3.3	3.8	6.5
	Cumulative	24.72	7.5	19.8	27.3	52.02	21.09	7.5	19.8	27.3	48.39
Annual plans (1978–1980)	During	1.89	0.5	2.2	2.7	4–59	1.48	0.5	2.2	2.7	4.18
	Cumulative	26.61	8	22	30	56.61	22.57	8	22	30	52.57
VI plan (1980–1985)	During	1.09	1.7	5.82	7.52	8.61	0.93	1.01	4.24	5.25	6.18
	Cumulative	27.7	9.7	27.82	37.52	65.22	23.5	9.01	26.24	35.25	58.75
VII plan (1985–1990)	During	2.22	1.29	7.8	9.09	11.31	1.9	0.96	6.91	7.87	9.77
	Cumulative	29.92	10.9	35.62	46.52	76.44	25.4	9.97	33.15	43.12	68.52
Annual plans (1990–1992)	During	0.82	0.47	3.27	3.74	4.56	0.85	0.32	3.1	3.42	4.27
	Cumulative	30.74	11.46	38.89	50.35	81.09	26.25	10.29	36.25	46.54	72.79
VIII plan (1992–1997)	During	2.21	1.05	1.91	2.96	5.17	2.13	0.78	1.45	2.23	4.36
	Cumulative	32.95	12.51	40.8	53.31	86.26	28.38	11.07	37.7	48.77	77.15

IX plan (1997–2002)	During	4.1	1.09	2.5	3.59	7.69	2.57	0.37	0.85	1.22	3.79
	Cumulative	37.05	13.6	43.3	56.9	93.95	30.95	11.44	38.55	49.99	80.94
X plan (2002–2007)	During	4.59	N.A.	N.A.	3.2	7.79	2.73	N.A.	N.A.	1.49	4.22
	Cumulative	41.64	N.A.	N.A.	60.1	101.74	33.68	N.A.	N.A.	51.48	85.16
XI plan (2007–2012)	During	6.34	N.A.	N.A.	5.45	11.79	1.33	N.A.	N.A.	1.43	2.76
	Cumulative	47.97	N.A.	N.A.	65.56	113.53	35.01	N.A.	N.A.	52.91	87.92

S.W. surface water, *G.W.* groundwater.

Source: Water Resources at a Glance. Central Water Commission (2021), GoI

Table 16.7 Irrigation potential created in major, medium, and minor irrigation

					IPC up to XI plan		
	State	UIP of MMI projects	UIP of minor projects	Total UIP	MMI	Minor	Total
S. no.	A	B	C	D	E	F	G = E + F
1.	Andhra Pradesh	5000.00	6260.00	11,260.00	4803.73	3340.550	8144.28
	Telangana						
2.	Arunachal Pradesh	0.00	168.00	168.00	1.20	132.248	133.448
3.	Assam	970.00	1900.00	2870.00	455.96	1016.820	1472.783
4.	Bihar	5223.50	5663.50	10,887.00	3054.46	5924.780	8979.240
5.	Chhattisgarh	1146.93	571.00	1717.93	1269.32	842.295	2111.610
6.	Goa	62.00	54.00	116.00	55.55	25.927	81.478
7.	Gujarat	3000.00	3103.00	6103.00	3679.09	2071.970	5751.060
8.	Haryana	3000.00	1512.00	4512.00	2206.29	1637.670	3843.960
9.	Himachal Pradesh	50.00	303.00	353.00	30.45	186.217	216.667
10.	Jammu Kashmir	250.00	1183.50	1433.50	325.61	745.661	1071.270
11.	Jharkhand	1276.50	1108.00	2384.50	530.71	534.200	1064.905
12.	Karnataka	2500.00	3474.00	5974.00	2965.83	1704.170	4670.00
13.	Kerala	1000.00	1679.00	2679.00	715.69	763.650	1479.340
14.	Madhya Pradesh	4853.07	11,361.00	16,214.07	2506.43	2534.340	5040.772
15.	Maharashtra	4100.00	4852.00	8952.00	4128.71	3185.600	7314.310
16.	Manipur	135.00	469.00	604.00	158.50	120.690	279.190
17.	Meghalaya	20.00	148.00	168.00	–	77.770	77.770
18.	Mizoram	0.00	70.00	70.00	–	51.740	51.740
19.	Nagaland	10.00	75.00	85.00	–	124.510	124.510
20.	Odisha	3600.00	5203.00	8803.00	2147.36	1887.430	4034.790
21.	Punjab	3000.00	2967.00	5967.00	2684.39	3497.710	6182.100
22.	Rajasthan	2750.00	2378.00	5128.00	3167.13	2487.760	5654.890

23.	Sikkim	20.00	50.00	70.00	–	42.740	42.740
24.	Tamil Nadu	1500.00	4032.00	5532.00	1578.27	2331.990	3910.260
25.	Tripura	100.00	181.00	281.00	29.25	161.863	191.113
26.	Uttar Pradesh	12,154.00	17,481.00	29,635.00	9288.09	25,320.130	34,608.220
27.	Uttarakhand	346.00	518.00	864.00	288.98	585.347	874.327
28.	West Bengal	2300.00	4618.00	6918.00	1901.41	4159.680	6061.090
	Union Territories	98.00	46.00	144.00	0.00	61.935	61.935
	All India total	58,465	81,428	139,893	47,972.41	65,557.393	113,529.798

UIP ultimate irrigation potential, *MMI* major and medium irrigation
Source: Water Resources at a Glance. Central Water Commission (2021), GoI

Table 16.8 Variation in water productivity among various Indian states

Sl. no.	States	Total (irrigated + rainfed) Area (Mha)	Production (M Mt)	Yield (t/ha)	CWU (mm)	WP (kg/m^3)
1.	Uttar Pradesh	20.3	43.4	2.13	351	0.61
2.	Madhya Pradesh	11.2	11.1	0.99	278	0.36
3.	West Bengal	6.6	15.2	2.31	447	0.52
4.	Bihar	7.1	12.1	1.71	373	0.46
5.	Rajasthan	11.7	11.7	1.00	220	0.46
6.	Punjab	6.3	25.5	4.07	404	1.01
7.	Haryana	4.3	13.4	3.13	363	0.86
8.	Uttaranchal	1.0	1.7	1.75	298	0.59
9.	J&K	0.9	1.2	1.38	271	0.51
10.	Himachal Pradesh	0.8	1.5	1.78	245	0.73
	India	123	205.4	1.66	344	0.48

Source: Amarasinghe and Sharma (2008)

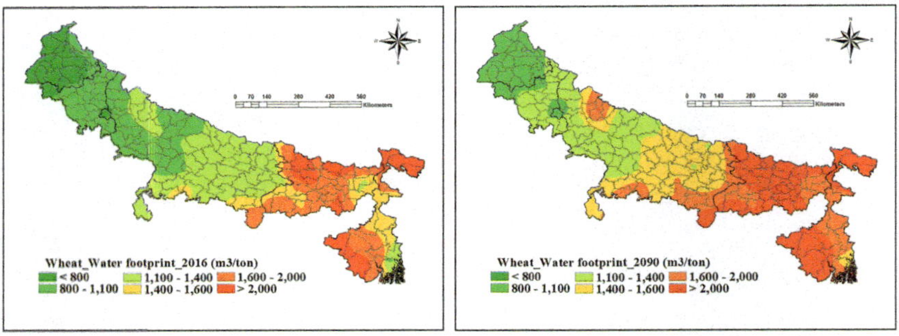

Fig. 16.1 Impact of climate change (RCP 4.5) on water footprints of wheat in Indo-Gangetic region. (Source: Kar et al. 2019)

16.4 Water Footprint

The increased temperature leads to increase in potential evapotranspiration which will result in more water demand for irrigation and ultimately lowering of groundwater table at some places. It has been estimated that the water footprint in wheat under Indo-Gangetic plains is 800–1700 m^3/ton during 2016, which will be around 1100–2000 m^3/ton in 2090 (Fig. 16.1). Decrease in rice yield by 20% due to water shortage has been predicted in India. There will be short-term increased availability of water in the rivers and their tributaries due to melting of glaciers in the Himalayas.

16.5 Suitable Agro-techniques and Water Management Practices to Enhance Water Productivity, Water Use, and Irrigation Efficiency

Strategies to enhance water use efficiency and irrigation efficiency include improved irrigation system management to provide more reliable water supply to farmers through storage and improved operation of reservoirs, better distribution of water with improved control structures, as well as more responsive management. More reliable water supply allows farmers to invest in better on-farm water management such as better land leveling, zero tillage, or pressurized irrigation. Improved management usually requires improved institutions as well as improved technologies. Irrigation efficiency mainly depends on water loss in conveyance system including distribution system and in the field. Loss of water in conveyance system is mainly due to seepage and evaporation. Evaporation loss takes place from the exposed water surface.

Some of the strategies needed to improve water and irrigation use efficiency under rainfed and irrigated agriculture are discussed below:

16.6 Rainwater Harvesting and Management

Rainfed farming supports the livelihood of millions of farmers and households even after realizing the complete irrigation potential. Under the changed climate scenario for effectiveness of rainwater harvesting, emphasis should be given on the following issues:

- Best designs for different rainfall zones and soil types.
- Optimum size of farm pond given the catchment area available under different farming situations.
- How can the capital cost of farm ponds be reduced through convergence of other developmental programs?
- What are the innovations in checking evaporative losses, cost-effective sealants, and water-lifting devices for conveyance?
- What are the best options in terms of crop choices to realize the best returns from stored water?
- How to resolve the issue of sharing water in case of smallholders where catchment and command area belong to different farmers?
- What are the on-site and off-site benefits including environmental payoffs due to rainwater harvesting?
- What are the indigenous techniques of rainwater harvesting which can help farm pond technology become more cost-effective?
- Potential of water harvesting through farm ponds in adaptation and mitigation of climate change.

16.7 Watershed Plus

The current watershed technologies should be further upgraded to diversify and enhance productivity, income, employment, and environmental service under the changed climate scenario. This should include value addition, new variety commodities, horticulture, livestock rearing, and aquaculture with greater emphasis on high and low rainfall regions. Soil and water conservation, efficient inputs, practices, seeds, enterprising by landless, equity, and convergence are the uppermost consideration of technology generation. Market- and agro-industry-driven employment generation, empowerment of local communities, panchayats, and women should be essential ingredients of all inclusive development process.

Convergence of diverse resources of various ministries like MGNREGA, eco-afforestation, million wells, and watershed allocation should generate critical mass and adequate synergies. Private-public-social capital linkage should target internalization of wastelands of private, panchayats, and public departments. Recharging of aquifers in forest and gully lands by runoff from agricultural fields, wasteland, urban areas, metaled roads, flyover, etc. is essential. Contamination of groundwater recharging through shafts, abandoned bore wells, open wells, etc. is a larger technological challenge. This technological demand presumes latest art of modeling contamination, bioremediation, and convergence of nontraditional sectors, departments, agencies, and resources.

16.8 Multiple Use of Water for Realizing Higher Water Productivity

Per capita availability of land, water, and other natural resources is already low and declining further. Multiple use of water through agricultural diversification which has several components like poultry, duckery, fisheries, apiculture, etc. along with diversified crop production or value-added products has proved to be an effective means in mitigating drought and eliminating poverty as it provides high income, regular employment, and balanced and quality food, even with less water. As the agricultural diversification comprises of several farm enterprises in addition to diversified crops, it provides opportunity to farmers to have a "basket of complementary options" for reducing the weather risks involved in single commodity-based agriculture. Water productivity can be improved by five to seven times by integrated planning through multiple use of water. In pond-based integrated farming, culturing of fish can be done in dug-out pond, and vegetables and horticulture can be grown on raised embankment. To enhance the productivity and profitability from harvested water, it has to be utilized for both consumptive and nonconsumptive purposes. Freshwater Indian major fish species (Catla, Rohu, Mrigal) of 3–4 units size at 10,000/ha can be released into the pond during August. The stored water of the pond can be utilized for providing supplemental irrigations to grow *rabi* crops like groundnut, sesamum, mustard, sunflower, and vegetables (cucumber, watermelon, okra, and ridge gourd). The vegetables like tomato, cauliflower, brinjal, and bottle gourd can be grown on the pond bund with the help of harvested water. Short-duration fruit crops like papaya and banana can be grown on the pond bund. After adoption of pond-based farming system, farmers' net returns were found to be higher by three to four times than that of earlier in a study at Balasore, Odisha.

16.9 Crop Diversification with In Situ Rainwater Management

Crop diversification with less water-requiring crops through intercropping/mixed cropping/sequential cropping will be useful for climate change mitigation. By introducing legumes in rice-wheat cropping system, natural resources can be conserved, and soil fertility can be increased. Crop diversification in rainfed rice field was implemented in Odisha on large scale with maize (cv. Navjyot), pigeon pea (cv. UPAS-120), groundnut (cv. Smriti), black gram (cv. T9), and cowpea (cv. Pusa Kamal), and their productivity and water use efficiency were compared with that of sole rice (cv. Vandana). Partial substitution of rice through rice-based intercropping, viz., rice + pigeon pea (4:1), rice + groundnut (4:1), and rice + black gram (4:1), was also practiced with proper weed management. Complete substitutions of rice through legume-based intercropping, viz., groundnut + pigeon pea (4:1), groundnut + green gram (4:1), groundnut + black gram (4:1), groundnut + cowpea (4:1), were also implemented. Study revealed that adoption of crop diversification enhanced water productivity and water use efficiency by two to three times (Kar et al. 2003). Crop diversification (sole cropping/intercropping) with maize, groundnut, black gram, green gram, cowpea, pigeon pea, etc. ensured higher net economic return in rainfed upland rice field even in rainfall deficit years when net return from sole rice was nil or negative.

16.10 Irrigation Scheduling Based on Scientific Approach

Irrigation scheduling is an approach of deciding the quantity and timing of irrigation. It helps the irrigator to decide on when and how much water to apply for minimizing crop yields and efficiency of water use. The basic objective of irrigation scheduling is to make available the correct amount of water for the biological processes of plants at appropriate time by applying water needed to replenish the soil moisture to the desired level. A phenological-based irrigation scheduling in maize was followed as per description in Table 16.9, and water productivities were compared. A comparison of crop water productivity (CWP) between treatments receiving irrigation at flowering and milk ripening and grain filling stages and not receiving irrigations at these stages with the same amount of irrigation (I_4 and I_5) showed that water was more efficiently utilized when irrigation was not skipped at flowering and milk ripening and grain-filling stages (Table 16.10). With the same amount of irrigation (300 mm) in I_4 and I_5, less crop yield was obtained in I_4 because irrigation was skipped at flowering stage of the crop under this treatment. Better water utilization efficiency and higher CWP in treatment I_5 were obtained which might be associated

Table 16.9 Irrigation treatment for field experiments

Stages	Description	Irrigation (mm) treatments					
		I_1	I_2	I_3	I_4	I_5	I_6
Stage 0	Period of germination of seed in the soil	X	X	X	X	X	X
Stage 1	Emergence of coleoptile from the soil and seedling growth up to three leaves unfolded	X	X	X	60	60	60
Stage 2	Stem elongation (1): internodes below fifth, sixth, and seventh leaves have begun to elongate	60	60	60	X	X	X
Stage 3	Stem elongation (2): 8–11 leaves unfolded, stem elongation rapidly, internodes below fifth and sixth leaves fully elongated	X	X	X	60	60	60
Stage 4	Stem elongation (3): 12–15 or more leaves unfolded, stem still elongates, emergence of tassel from the whorl	60	X	60	60	60	60
Stage 5	Flowering (start of pollen shedding, 50% pollen shedding, 50% silking, end of flowering)	X	60	60	X	60	60
Stage 6	Water ripe stage of caryopsis, start of silk drying	X	60	60	60	60	60
Stage 7	Milk ripening stage (milk to solid conversion of endosperm, but whole kernel content is still milky liquid)	X	X	X	60	X	60
Stage 8	Dry ripe stage (kernel is no longer milky, reached physiological maturity)	X	X	X	X	X	X
Stage 9	Ripeness	X	X	X	X	X	X
	Total irrigation during crop growth (mm)	120	180	240	300	300	360

I_1, I_2, I_3, I_4, I_5, and I_6 are irrigation treatments, *X* no irrigations were applied
Source: Kar and Kumar (2015)

with adequate water applied during flowering stage. This result implies that the crop growth stage at which deficit irrigations are imposed on the crop is also a determining factor to achieve higher CWP.

16.11 Pressurized Irrigation System

Conventional flood method of irrigation has low water use efficiency due to loss of water during conveyance and distribution. During the last few decades in Indian agriculture, pressurized irrigation systems are being promoted, which include both drip and sprinkler methods of irrigation. Since these methods need water under pressure, they are classified as pressurized irrigation systems. Pressurized irrigation system is a well-established efficient method for saving water and increasing water

Table 16.10 Water productivity of maize under different irrigation and nitrogen levels (pooled data of 2 years)

Irrigation treatments	GY (kg/ha)	NR (Rs./ha)	IWA (mm)	ER (mm) + SPC (mm)	SCWU (mm)	WP_{CWU} (kg/m^3)	WP_{NR} (Rs./m^3)
I. Irrigation treatments							
I_1	^{E}2129	11,661	120	131	251	^{D}0.849	^{D}4.65
I_2	^{D}2490	14,910	180	127	307	^{D}0.811	^{C}4.86
I_3	^{C}3553	21,977	240	118	358	^{B}0.993	^{A}6.14
I_4	^{B}3785	24,065	300	112	412	^{C}0.913	^{B}5.84
I_5	^{A}4534	25,806	300	105	405	^{A}1.120	^{A}6.37
I_6	^{A}4675	27,075	360	99	459	^{B}1.019	^{B}5.90
II. Nitrogen treatments							
N_1	^{D}2295	13,155	300	114.8	365	^{D}0.629	^{C}3.61
N_2	^{C}2714	20,526	300	116.8	366	^{C}0.742	^{B}5.61
N_3	^{B}3319	19,871	300	115.6	366	^{B}0.908	^{B}5.44
N_4	^{A}4385	29,465	300	114.5	365	^{A}1.203	^{A}8.08
N_5	^{A}4525	25,725	300	115.5	365	^{A}1.239	^{A}7.04

GY grain yield, *NR* net returns, *IWA* irrigation water applied, *ER* effective rainfall, *SPC* soil profile contribution, *SCWU* seasonal crop water productivity, *WPcwu* water productivity in terms of crop water use, WP_{NR} water productivity in terms of net return in rupees, WP_{NR} water productivity in terms of net return

I_1, I_2, I_3, I_4, I_5, and I_6 are irrigation treatments (in this table), and N_1, N_2, N_3, N_4, and N_5 are nitrogen treatments (nitrogen at 30, 60, 90, 120, and 150 kg N/ha)

Table 16.11 Irrigation efficiencies in pressurized irrigation system

Factors	Sprinkler irrigation system	Drip irrigation system	Surface irrigation system
Overall irrigation efficiency (%)	50–60	80–90	30–35
Application efficiency (%)	70–80	90	60–70
Water savings (%)	30	60–70	NA

Source: Kumar and Kar (2013)

use efficiency as compared to the conventional surface method of irrigation (Table 16.11).

16.12 Drip Irrigation

Drip irrigation is an efficient method of micro-irrigation since it applies water directly into the crop root zone to meet the crop water requirement. It permits application of pesticides, fertilizers, and other soluble chemicals through fertigation.

The system applies water at low rate and under pressure to keep the soil moisture within the desired range for plant growth. The system has overall application efficiency around 90% as compared to 25–30% for surface irrigation.

16.13 Sprinkler Irrigation System

The role of the sprinkler irrigation system has been universally acknowledged in achieving high water use efficiency, in improving crop productivity and quality of produce, and in saving of irrigation water and labor costs. The sprinkler irrigation system simulates the water application as that of rainfall. The overall efficiency of sprinkler method is as high as 65% as compared to 25–30% of surface method or irrigation. The system has an advantage for the close growing crops in supplying the required quantity of water. This saves the irrigation water and helps in uniform application of water in the field. To adopt sprinkler system, a farmer has to incur an average cost of Rs. 15,000/hectare. Classification of this irrigation system has been done based on arrangement of spraying water or on the basis of their portability. In India, the rotating head portable systems are most commonly being used by the farmers.

16.14 Improving Water Use Efficiency of Canal Command

Considerable amount of water is lost in canal command through evaporation, seepage, and percolation. In no irrigation project in India, the total losses in the canal distribution system and field have been less than 50% of the head discharge. A review of 90 irrigation projects of the world indicated generally low irrigation efficiencies, with only 20–40% of water diverted from the reservoir being effectively used by the crop, while in India the irrigation efficiency is around 10–20%. The losses of irrigation water are unlined canal distribution system in North India (Table 16.12). In any irrigation project, the total losses in the canal distribution system and field have not been less than 50% of head discharge.

Table 16.12 Losses of irrigation water in unlined canal distribution systems and in the field

	% of supplies at canal head		
Source of loss	Seepage	Evaporation	Total
Main canals and branches	13.6	3.4	17.0
Distributaries (10% of supply at distributor)	6.4	1.6	8.0
Field water courses (27% of supply at outlet head)	16.0	4.0	20.0
Losses from field during application (30% of supply reaching the head)	13.2	3.3	16.5
Total	49.2	12.3	61.5

Source: Kumar and Kar (2013)

Studies conducted in Nagarjuna Sagar Project have revealed that only 40–60% of water released from the reservoir reaches the field and only 20–40% is used by the crops. The main components of water losses in an irrigation system are (a) water losses in storage (10–20%), (b) water losses in conveyance systems (25–40%), (c) water losses in operation (10–30%), and (d) water losses in application (45–70%). To increase the overall irrigation efficiency of 40–50% for a project, improved irrigation and drainage practices should be adopted based on scientific land water management principles. The methods to reduce conveyance losses include lining the bed and sides of the canal, eradication of weeds, and reduction of wastage from escapes and tail ends. The use of plastic films as lining material also has tremendous scope in India.

16.15 Water-Saving Technologies in Canal Command

Irrigated agriculture can be made vibrant by increasing farm productivity in the country through scientific and efficient management of canal systems, and water production can be enhanced through adoption of water-saving technologies. Rotational irrigation is often recommended to irrigate a large area with limited water supply and to ensure better equity among water users. Rotational irrigation is the application of required amount of water to fields at regular intervals. The field may often be without standing water between irrigations, but ideally the soil does not dry enough for moisture stress to develop. A major advantage of rotational irrigation is possibly the more effective use of rainfall. Plant height, tiller number, leaf area index, dry matter production, and grain yield generally decrease as the irrigation interval increases from 4 to 10 days (Table 16.13). The results of experiments conducted by WALAMTARI on rice in Andhra Pradesh under Sriram Sagar Project in sandy loam soils are given in Table 16.14.

16.16 Reuse of Irrigation Water

About 30% of water applied to the field is lost as surface or subsurface flow reaches drains. The water lost through the drains can be stored at suitable sites and pumped back for irrigation by giving due consideration of water quality-related issues.

Table 16.13 Effects of rotational irrigation in IR 20 rice variety

Irrigation interval (days)	Plant height (cm)	Tillers (no. of hill)	LAI	Total dry matter (g/hill)	Yield (t/ha)
4	102	18	6.5	53	7.2
6	97	18	54	47	7.1
8	98	17	5.2	49	6.8
10	92	12	4.2	37	5.6

Source: Reddy and Reddy (2009)

Table 16.14 Savings in water due to rotational irrigation for rice crop

Irrigation interval (days)	Quantity of water applied (mm)	Yield of rice (kg/ha)	Reduction of yield (%)	Water saved (%)
1	1717	6324	–	–
2	1117	6056	4.42	34
3	1027	5924	6.75	40
4	913	5720	10.55	46
5	877	5646	12.00	48

Source: Reddy and Reddy (2009)

16.17 Better Operation Planning

Irrigation system operation is the process of releasing, conveying, and diverting water in the canal systems to ensure predetermined flows at prescribed times for specified duration at all designated points of delivery. The operation plan considers the water available in the reservoir at the beginning of each crop season and spells out the starting date of release of water; the mode of supplies, i.e., whether intermittent and continuous; the detailed schedule of releases; and the closing date of release of water. The operation plan is prepared with active participation and involvement of farmer's representatives so that it may be implemented without any resistance from the farmers. This will reduce wastage and better utilization of irrigation water will be ensured.

16.18 Systematic Canal Operation

The object of systematic canal operation is to ensure equitable distribution of available flows at the heads of all offtakes on a distributory. A schedule of canal operation has to be prepared in advance. For instance, a distributory may be divided into seven reaches each covering one or more offtakes whose total discharge is about a seventh of the designed discharge of the distributory. One-seventh of the total discharge can be saved on each day by closing the offtake in each reach of 24 hours in succession in a week. This water can be made available to tailenders and thus it will help saving water.

16.19 Response to Rainfall in Canal Operations

Canal operations should respond meticulously to rainfall in the command area for efficient use of storage water in the reservoir. The demand of crops for water is met in the wet season by both rainfall in the command area and water stored in the reservoir. Water requirement of crops is worked out using one of the empirical methods suggested by FAO, Rome. Commonly used method is Penman-Monteith method. While assessing the water requirements of crops during the wet season for preparing the operation plan, effective rainfall worked out from the data of average rainfall over a period during the past 20 years is deducted from the water requirements of crops worked out by the Penman-Monteith method or any other empirical method. But in real-time situations, the pattern and amount of rainfall vary influencing the scheduled deliveries. When rainfall is in excess, the scheduled irrigations are skipped, and when rainfall is deficient, unscheduled deliveries have to be made. This will ensure maximum utilization of rainfall as well as protecting the crops from water stress or from excess water. Irrigation deliveries are based mainly on the status of water balance in the cropped fields. The water balance in paddy fields is on the basis of ponding depth, while for light-irrigated crops, it is on the basis of water storage in the root zone. The allowable depletion of soil water storage is limited by readily available soil moisture capacity. Response to rainfall for paddy depends on adopted ponding depth, while for light-irrigated crops, response is related to the depletion level of readily available soil water from the root zone. The mode of irrigation supplies differ during various stages of crop growth and development, as water requirements of crops vary at vegetative and reproductive stages.

16.20 Precision Land Leveling

Declining irrigation water availability, sustainable crop production to meet increasing food demand necessitates adoption of advanced technologies for better water management. Land leveling at farmers' field enables efficient water utilization of scarce water resources through elimination of unnecessary depression and elevated contours. Laser leveling is a laser-guided precision leveling technique used to achieve very fine leveled land with desired grade on agriculture field. It is becoming popular in almost entire state of Punjab. Precision land leveling must be treated as a promising technology for improving crop yields, enhancing input use efficiency, and ensuring long-term sustainability of the resources based on intensively cultivated areas. Keeping in view the benefits of laser leveling, a study was conducted to evaluate the performance laser leveler at farmers' field and to estimate the savings in water resources at different level of adoption. Water saving in different crops at different level of laser leveler is presented in Table 16.15.

16.21 Groundwater Recharge

Groundwater played a major role in achieving green revolution and contributes to 60% of the gross irrigated area of the country. Overexploitation of groundwater has reached critical level in Haryana, Punjab, Rajasthan, and Tamil Nadu. The Punjab-Haryana region could lose its production potential in a few decades if current pattern of groundwater extraction and pollution, soil salinization, and rice wheat cropping system persists. Necessary steps must be taken immediately for groundwater

Table 16.15 Water saving in different crops at different level of laser leveler

Crop	Water saved (%)	Water required (cm)	Water saved (cm)	Area (ha)	Water saved (ha-m) at different level of irrigation				
					10	25	50	75	100
Maize	27.1	35	9.5	152,560	1447	3618	7235	10,853	14,470
Wheat	26.0	35	9.1	3,469,520	31,573	78,932	157,863	236,795	315,726
Cotton	27.25	20	5.5	541,060	2949	7372	14,744	22,116	29,488
Paddy	26.33	160	42.1	2,625,204	110,595	276,486	552,973	829,459	1,105,946
Total	26.25	81.3	21.3	6,788,344	146,563	366,408	732,816	1,099,224	1,465,631

Source: Annual Report (2009)

recharge in these areas. Rainwater harvesting has considerable potential for recharging depleted groundwater aquifers meeting irrigation demand. If water cannot be stored above the ground, it should be stored underground via artificial groundwater recharge, considering that 98% of Earth's freshwater resources are stored in underground and there is enough space for more storage.

16.22 Conjunctive Use of Surface Water and Groundwater

Conjunctive use of multiple water resources should be practiced to mitigate the effect of the shortage in canal water supplies subjected to variation in depth round the year, increase the availability of existing water supplies, alleviate the problems of high water table and salinity, and facilitate the utilization of saline groundwaters which, otherwise, cannot be used without appropriate dilution. Strengthening of knowledge based on geology and aquifer characteristics, hydrology of surface, and groundwater facilities is required to develop appropriate conjunctive use system.

16.23 Management of Poor Quality Groundwater

Groundwater in arid regions is largely saline and is sodic in nature under semiarid areas. The use of poor-quality water for irrigation deteriorates the soil health owing to salinity, sodicity, and toxic effects. In addition to reduced productivity, it deteriorates the quality of produce and also limits the choice of cultivable crops. In saline groundwater areas, where canal water availability is limited, conjunctive use of canal water and saline groundwater is recommended. Dilution or mining of available poor-quality water with good water should be done in such proportion that resultant EC is acceptable for the range of crops to be grown in a given area.

Some of the management options to manage the saline groundwater are selection of semi-tolerant crops, salinity-tolerant cultivars, proper selection of crop sequences, avoiding saline water use during initial growth stages, addition of farm yard manure and organic matter, application of additional dose of nitrogen, etc. Some important points to be noted for management of conjunctive use practices are analysis of saline water to evaluate its use potential, selection of crops/varieties, presowing irrigation with good quality water so that germination and seedling emergence are not affected, adequate leaching of salts, change of cropped area, and adoption of improved cultural and nutrient management practices.

Management technology options for use of sodic groundwater include selection of crops, conjunctive use of canal water, river water management and leaching strategies to maintain a high level of soil moisture and low level of salts, exchangeable sodium in the rhizosphere, use of land management practices to increase the uniformity of water distribution, infiltration and salt leaching besides the optimal use of chemical amendments like agricultural grade gypsum and acidic pyrites at proper time, and mode of their application with judicious use of organic materials and chemical fertilizers.

16.24 Mitigating Groundwater Pollution

The choice of appropriate cropping systems and management practices helps in minimizing nitrate leaching besides improving N-use efficiency. Legume intercropping in cereals grown with wider row spacing reduces nitrate leaching. Parallel multiple cropping (a system of growing two dissimilar growth habit crops with minimum competition) of sugarcane and black gram and that of pigeon pea and maize resulted in low NO_3-N content in soil profile as compared to sole cropping.

As a crop management strategy for minimizing NO_3-N leaching, delaying large N applications until the crop can utilize it, avoiding irrigation when large amounts of NO_3-N are present in the root zone and split application, etc. have also been recommended. There is a specific need to create awareness among people about the presence of fluoride in groundwater, its action on body tissues, and available remedial technologies. Various artificial recharge techniques including Aquifer Storage Recovery (ASR) technique may be applied to improve the water quality by dilution. The ASR technique is a viable or cost-effective option for storing large volumes of freshwater. Looking at the success achieved through ASR in many countries, there is a case to plan trial of this technology at suitable locations in the country. Besides this, the concept of mixing two different aquifers/sources of water can also be tried to overcome the fluoride problem (Bajwa et al. 1993).

Investigations of geochemistry are very essential to understand the occurrence and mobilization of arsenic in the aquifer system. Thus, there is urgency for an integrated study on geological, hydrological, and geochemical characterization of the multilevel aquifer system to predict the origin, occurrence, and mobility of arsenic in groundwater. There is a priority to arrive at suitable and economically viable alternatives to maintain sustainable drinking water supply. The traditional system of using surface runoff water can be augmented by system for reducing dependence groundwater extraction.

16.25 Farmers' Participation for Managing Irrigation Water

The effective water management is critically linked with the performance of local-level water institutions. The innovation of Water User Associations (WUA)/Pani Panchayat institution has diverted larger flow of information as well as initiative from government irrigation departments toward it. The transfer of irrigation management responsibility from the government irrigation authority to local body depends on allocative and investment decisions by the farmers' group/organization. There has to be a paradigm shift giving the irrigators (farmers) real decision-making power in managing the irrigation system as a whole system. Watershed management is by nature beyond the work of individuals, and thus collective effort by all farmers concerned is required for successful management. Replication of successful samples and community-led success stories is likely to lead to most sustainable results. The objectives of farmers' groups cannot be realized overnight and it takes time. Therefore, it is of paramount importance to keep it functional and effective for a long time.

It should become a part of the tradition of the village over time, as is the case with the already existing traditional village organizations.

16.26 Water Management in Waterlogged Areas

The saucer-shaped land forms, high rainfall (average 1500 mm) due to southwest monsoon (June–September), poor drainage condition, and slow disposal of accumulated water in the plains to the ocean make the coastal region susceptible to waterlogging and flood-prone, and the area remains submerged for about 5–-6 months (July–November) under water depths varying from 0.75 to 2.0 m. Alternatively, the winter and summer rainfall (November–May) is meager and erratic. As a result, after December, the land becomes dry, and available soil moisture in the land is not sufficient to meet evapotranspiration loss of any crops. Thus, coastal waterlogged areas are subjected to receive both extreme events. In one season, the area is underproductive due to excess water; in other season, agriculture is not possible because of the lack of soil moisture. Due to constant waterlogging of 0.5- to 2.0-m depths, the normal rice fails to grow in seasonal flood0prone areas. Improved deep waterlogging rice cultivars like 'Hangseswari' and 'Saraswati' were introduced with improved sowing methods (line sowing with 20 cm distance, row to row, and fertilizer dose of 20:20:20) in the seasonal waterlogged areas (Kar et al. 2010).These varieties can produce up to 2.8 t/ha grain yield in *kharif* season in deep waterlogged situation if flood commences after second week of August.

The suitable water harvesting structure depending upon the maximum depth of waterlogging was designed and constructed by taking 25–30% of the total field in representative waterlogging areas of Puri District. Waterlogging-tolerant rice varieties were grown surrounding the structure during *kharif* season. The water inside the structure was utilized for fish rearing, and the bund was utilized to grow vegetable short-duration fruit and vegetable crops initially and to grow firewood plants at later stages. The harvest was also utilized to grow high-yielding medium duration rice (cv. 'Lalat,' 'Konark') and vegetable crops during *rabi* season. Through this approach, farm returns enhanced up to about 1.0 lakh/hectare from waterlogged area which remained unproductive earlier. The cropping intensity of farmers was increased between 150 and 250%. Water chestnut and medicinal plant (*Acorus calamus*) were introduced in seasonal flood-prone areas, and package of practices for their cultivation was standardized. From water chestnut and calamus cultivars, Rs. 19,000/ha and Rs. 35,000/ha net returns were obtained, respectively.

16.27 Indigenous Technical Knowledge (ITK) for Water Resource Management

Our agriculture has a lot of inherited sustainable practices passed from one generation to another generation. The present and future generations must be aware of the ancient technologies practiced by our ancestors to build future research strategy.

With the success of Green Revolution, India has progressed in every field of agriculture and allied studies. But due to the indiscriminate use of chemicals, intensive cropping system with intensive input use on a long-term basis has been a major constraint to maintain sustainability. Our agroecosystem is threatened due to irrational use of chemicals in the form of pesticide, fertilizer, and herbicide. Soil health is getting deteriorated, water and air sources are getting polluted. and there is an increasing erosion of plant and animal genetic resources. Therefore, sustainability is the need of the hour.

The indigenous technical knowledge (ITK) provides an insight into the sustainable agriculture, since these innovations have been passed from one generation to another as a family technology. Various ITKs have been followed by our ancestors, but unfortunately these small local systems are dying out. Chadwick et al. (1998) explored ITK of water harvesting and artificial recharge and found that the most important strategy is with the Maldharis. The local people have traditionally managed to safeguard their livelihood through rainwater harvesting. The techniques developed by them harvest sporadic floodwater, which ensures that their drinking needs are fulfilled even during water-scarce years. Their extensive knowledge of water harvesting based on the local ecosystem and of the complex water harvesting system subsequently developed by them is based on their knowledge, culture, and experience. Though the area is flat, some depressions, known locally as tanks or jells, of various sizes exist and act as storage areas for flood water. The water lasts several months before drying up.

The local nomads found that after infiltration of water in the soil, the freshwater gets stored at shallow depths in a layer "floating" above salty groundwater. The dug shallow wells in this zone are called "virda" reaching down as far as the freshwater. Each virda is used until the water begins to get salty at which point the virda is filled with grasses and silt from the next virda and left to replenish. Other methods include the development of a series of tanks sunk parallel with the slope. These progressively fill during the rainy season with each feeding to one larger deeper tank. Similarly, sprinkling of thin slurry from cow dung over drought-affected paddy so as to increase the water retention properties of the soil is also a common practice. In the field of local Aus rice, at an early stage, the use of a hand hoe followed by laddering is undertaken. This process thins out unwanted seedlings and weeds and aids mulching.

16.28 Government Initiative/Policies for Higher Irrigation and Water Use Efficiency

Government has given considerable importance to the development of command area under canals. Earlier during 1950–1951, the canal-irrigated area was 8.3 million ha which is now 17 million ha. Wells and tube wells accounted for 29% total irrigated area in 1951, and they had a share of 64% of the total irrigated area in 2010–2011. It is worth mentioning that the Government of India has taken many initiatives in the past. Between 1991 and 2007, India invested (approx.) ~ USD 4000 million in public canal systems. Yet the canal-irrigated area decreased by 38 lakh hectares during that period, as infrastructure is old, water supply is unreliable, and further there are no incentives.

Similarly, even after 50–90% subsidy for the micro-irrigation, it covers less than 5% of India's cultivated area. The government schemes have succeeded in some states although faltered in others. Electric-powered groundwater exploitation has thus emerged as a unique confluence of physical, policy, and political factors that have trapped many states in a vicious spiral of declining groundwater, deteriorating water quality, stagnant crop productivity, deteriorating power service delivery, and poor financial health of power generation companies. Most state governments provide subsidized or free electricity to farmers which results in overuse of water and declining groundwater tables. It has been reported that Indian farmers use two to four times more water to produce a unit of major food crop than in China or Brazil. Of this maximum, 45% is shared by tube wells followed by canals and wells. In this context, the Indian government has tried to inculcate new policies and schemes to improve agricultural productivity while simultaneously increasing water use efficiency. The Indian government introduces schemes as commendable effort to increase irrigated area. One example is the launching of (approx.) ~ USD 7.5 billion "Pradhan Mantri Krishi Sinchai Yojana (PMKSY)." This scheme provides a sound framework for the expansion and effective water use in irrigation. This scheme can be enhanced by restoring Mahatma Gandhi National Rural Employment Guarantee Act.

16.29 Water Rights and Water Pricing

Water rights: In India, water rights are related with the ownership of land. That essentially means that the landowners have rights to extract water through wells on their lands. Also, they are encouraged to collect rainwater on their land.

Water pricing: Pricing water and water-related services adequately can boost people to invest more in water-related infrastructure and value watershed services rather than wasting them. In most states, there is no payment of water fees or any other charge. Even in many states, electricity is provided free to pump water if water is to be used for irrigation purposes. The distorted water pricing is resulting in overexploitation of the natural resources which may have long-term implication such as salination, thus rendering good agricultural land unfit for growing crops and presence of heavy metals. The state governments avoid withdrawing these provisions as farmers may consider high water pricing as depriving of their entitlement, which could in turn lead to conflict and may also result in increase in food prices.

16.30 Water Resources of North Eastern Hill (NEH) Region

The northeastern region of India comprising eight states, viz., Meghalaya, Nagaland, Sikkim, Tripura, Assam, Manipur, Mizoram, and Arunachal Pradesh, with 39 million population has a total geographical area of 262,180 km^2 (nearly 8% of the total area of India). The per capita availability in Brahmaputra Basin is more than 14,000 m^3 compared to only about 120 m^3 in Kutch area of Gujarat. The total annual water resources of the country have been estimated at 1953 km^3. The utilizable resource has been estimated as 690 km^3 surface and 450 km^3 groundwater equalizing a total of 1140 km^3. The annual demand for irrigation water in India and Northeast region during the year 2020 was 460 and 20 km^3 and for 2025 estimated at 810 and 32 km^3, respectively (Satapathy and Sharma 2006). The average annual precipitation in the country is about 4000 km^3, a part of it goes as runoff, a part of it lost as evapotranspiration, and the remaining goes to recharge groundwater. The NE region has a surface water and groundwater potential of 1487.2 and 25.3 km^3 (Table 16.16), respectively (Borthakur et al. 1989). Out of irrigation potential of 36,810 km^2, only 5120 km^2 or 15.3% of the potential has been exploited in NE region (Satpathy and Sharma, 2006). Out of total irrigated area, 69.8% is in Assam alone. Manipur has 25.5% of the net sown area, compared to 13.8% in Meghalaya. Present cropping intensity of 135% of the region would go up to 200% if full irrigation potential is utilized.

16.31 Pradhan Mantri Krishi Sinchayee Yojana (PMKSY)

Pradhan Mantri Krishi Sinchayee Yojana (PMKSY) was launched during the year 2015–2016, with an aim to enhance physical access of water on farm and expand cultivable area under assured irrigation, improve on-farm water use efficiency, introduce sustainable water conservation practices, etc. PMKSY has various components, viz., Accelerated Irrigation Benefits Programme (AIBP);

Table 16.16 Inland water resource potential in NE states (000 ha) and utilizable groundwater resource for irrigation (m ha m)

State	River length and canals (km)	Reservoirs	Tanks/ ponds	Lakes	Groundwater resource (m ha m)
Arunachal Pradesh	2500	0.38	1.00	2.50	0.1223
Assam	4850	36.04	16.47	83.84	1.8421
Manipur	3360	0.10	5.00	4.00	0.2681
Meghalaya	5600	1.17	1.82	0.21	0.1042
Mizoram	1748	–	1.60	–	–
Nagaland	1600	9.50	4.94	0.21	0.0615
Tripura	1200	4.50	9.60	0.50	0.2135
Total	19,468	51.69	40.44	91.26	

Source: Satapathy and Sharma (2006)

PMKSY-Har Khet Ko Pani (HKKP) including Command Area Development and Water Management (CADWM), Surface-Minor Irrigation (SMI), and Repair, Renovation and Restoration (RRR) of Water Bodies [implemented by DoWR, RD & GR, MoJS]; PMKSY-Per Drop More Crop (PDMC) [implemented by the Department of Agriculture, Cooperation and Farmers' Welfare]; and PMKSY-Watershed Development Component(WDC) [implemented by the Department of Land Resources]. The PMKSY-PDMC focuses on water use efficiency at farm level through precision/ micro-irrigation. Surface Minor Irrigation (SMI) schemes with irrigation potential less than 2000 ha were included under Accelerated Irrigation Benefits Programme (AIBP) for providing central assistance (CA) since 1999–2000 for Special Category States. Subsequently, the Scheme extended to area covering DPAP, tribal, DDP, flood-prone, left-wing extremist, and Kalahandi Balangir Koraput (KBK) region of Odisha. The scheme of SMI is now a part of PMKSY (HKKP).

16.32 Salient Features of the National Water Policy (2012)

The National Water Policy (2012) has been formulated for planning, development, and management of water resources. Some of the salient points regarding National Water Policy (2012) are given below:

- Planning, development, and management of water resources need to be governed by common integrated perspective considering local, regional, state, and national context, having an environmentally sound basis, and keeping in view the human, social, and economic needs.
- Principle of equity and social justice must inform use and allocation of water. Water needs to be managed as a common pool community resource held by the state under public trust doctrine to achieve food security, support livelihood, and ensure equitable and sustainable development for all. Water is essential for sustenance of the ecosystem, and therefore, minimum ecological needs should be given due consideration.
- Safe water for drinking and sanitation should be considered as preemptive needs, followed by high-priority allocation for other basic domestic needs (including needs of animals), achieving food security, and supporting sustenance agriculture and minimum ecosystem needs.
- Available water, after meeting the above needs, should be allocated in a manner to promote its conservation and efficient use. Given the limits on enhancing the availability of utilizable water resources and increased variability in supplies due to climate change, meeting the future needs will depend more on demand management, and hence, this needs to be given priority, especially through (a) evolving an agricultural system which economizes on water use and maximizes value from water and (b) bringing in maximum efficiency the use of water and avoiding wastages.
- Water quality and quantity are interlinked and need to be managed in an integrated manner, consistent with broader environmental management

approaches inter alia including the use of economic incentives and penalties to reduce pollution and wastage.

- The impact of climate change on water resource availability must be factored into water management-related decisions. Water-using activities need to be regulated keeping in mind the local geoclimatic and hydrological situation.
- There is a need to evolve a National Framework Law as an umbrella statement of general principles governing the exercise of legislative and/or executive (or devolved) powers by the center, the states, and the local governing bodies. This should lead the way for essential legislation on water governance in every State of the Union and devolution of necessary authority to the lower tiers of government to deal with the local water situation.
- There is a need for comprehensive legislation for optimum development of interstate rivers and river valleys to facilitate interstate coordination ensuring scientific planning of land and water resources taking basin/subbasin as a unit with unified perspectives of water in all its forms (including precipitation, soil moisture, groundwater, and surface water) and ensuring holistic and balanced development of both catchment and command areas.
- Such legislation needs, inter alia, to deal with and enable the establishment of basin authorities, comprising party states, with appropriate powers to plan, manage, and regulate the utilization of water resource in the basins.
- The center, the states, and the local bodies (governance institutions) must ensure access to a minimum quantity of potable water for essential health and hygiene to all its citizens, available within easy reach of the household.

16.33 Future Irrigation Water Resources

The ultimate irrigation potential is estimated to be 139 Mha. For meeting the future food grain demand of 494 million tons in 2050, it is required to bring 146 Mha under irrigation which is quite high compared to the present gross irrigated area of 89.55 Mha. Thus, it is difficult to achieve the requirement of 146 Mha of irrigated area considering the area that would continue to remain rainfed. Alternatively, it is essential to increase the crop productivity from present 2.3 t/ha to 4 t/ha under irrigated conditions and from less than 1 t/ha to 1.5 t/ha for rainfed area. Most of the irrigation projects are operating at a dismally low efficiency of 35%. Besides, there is an increase in waterlogged and saline areas due to erratic rainfall distribution. With increasing pressure on land and water, immediate action is needed to increase irrigation efficiency to 60% for surface water and 75–80% for groundwater. It has been predicted that with almost no increase in the net sown area in the near future, the cropping intensity needs to be increased from 132% to 145% by 2025 to step up the gross cultivated area from 194 Mha to 210 Mha.

Likewise, about 80 Mha would continue to remain as rainfed, and very less area is expected to switch from rainfed to irrigated due to erratic rain distribution. Management of declining water resources in a sustainable manner for enhancing their productivity needs to be addressed in the agricultural water management research and development of the country. The need of the day is to economize water in agriculture, bring more area under irrigation, and increase the yield per unit area and unit quantum of water through multiple use of water, conjunctive use, recycling of poor quality/waste water, pressurized irrigation system, crop diversification, and other water-saving agricultural methods.

16.34 Conclusions

Water forms the backbone for all the future endeavors to achieve the vision of food security. In the present context, upscaling agricultural economic growth to more than 4% annually is the main challenge. Taking water technologies for better water management from lab to land is a formidable task to be addressed. Modernization/automation of irrigation systems, precision irrigation, land reforms, corporate farming, cooperative farming, water and energy pricing, crop insurance, institutional mechanism for better governance, and water rights are some of the key issues for better water management in agriculture. The projected food requirement demands a pronounced role for research, development, and training in the water and agriculture sector. It is evident that the water availability for agriculture is declining and to enhance agricultural production, more water is needed. Therefore, concerted and holistic efforts are required in increasing the overall water use efficiency at system level which would be achieved through various measures like timely execution of projects, minimizing the losses, better operational efficiency through stakeholders' participation, implementation of on-farm water management technologies, conjunctive use of water, and changes in irrigation policy. Simultaneously, the efforts of research and development institutions are required in the development of water management technologies, suitable database development, economic studies of various irrigation systems, policy guidelines for on-farm water management, and adoption of participatory irrigation management. Serious efforts from developmental agencies as well as research institutes are required to develop a suitable water perspective plan for various regions in the country for its implementation.

References

Amarasinghe UA, Sharma B (2008) Spatial variation of crop water productivity of india and potential for improvements. Draft report prepared for the IWMI-CPWF project 'Strategic Analysis of India's River Linking Project'

Annual Report (2009) All India coordinated research project on water management

Bajwa MS, Singh B, Singh P (1993) Nitrate pollution of groundwater under different systems of land management in the Punjab. In: Proceedings of first agricultural science congress. National Academy of Agricultural Sciences, New Delhi, pp 223–230

Borthakur HP, Bora CK, Phukan U, Bhattacharya BK, Bowmie BC, Khound HP (1989) Agro-climatic regional planning of the eastern himalayan region—profile and strategy. Assam Agricultural University, Jorhat

Bouwer H (2000) Integrated water management: emerging issues and challenges. Agric Water Manag 45:217–228

Chadwick M, Soussan J, Mallick D, Alam S (1998) Understanding indigenous knowledge: its role and potential in water resource management in Bangladesh. LEEDs and BCAS, pp 13–15

Kar G, Kumar A (2015) Effect of phenology based irrigation scheduling and nitrogen on light interception, water productivity and energy balance of maize (*Zea mays* L.). J Indian Soc Soil Sci 63(1):39–52

Kar G, Singh R, Verma HN (2003) Crop diversification in rainfed rice area of eastern India to increase productivity and rainwater use efficiency. Research Bulletin 13/2003. Water Technology Centre for Eastern Region, Bhubaneswar, p 51

Kar G, Sahoo N, Ashwani K (2010) Deep water rice and pond based farming for higher productivity of waterlogged areas. Research Bulletin No 48. Water Technology Centre for Eastern Region, Bhubaneswar, pp 1–32

Kar G, Mohanty S, Anand PSB, Panda DK, Raviraj A, Rank RD, Singh PK (2019) National innovations for climate resilient agriculture, annual report, 2018–19. Indian Institute of Water Management, Bhubaneswar, pp 14–18

Kumar A, Kar G (2013) Sustainable agricultural water management strategies for higher water use efficiency. J Sustainable Planet 4(1):47–60

Molden D, Murray-Rust H, Sakthivadivel R, Makin I (2001) A water productivity framework for understanding and action. Workshop on Water productivity. Wadduwa, Sri Lanka, 12–13 November, 2001

Postel S (1992) Last oasis: facing water scarcity. Worldwatch Institute, Washington DC

Reddy GHS, Reddy TY (2009) Efficient use of irrigation water. Kalyani Publishers, New Delhi

Satapathy KK, Sharma UC (2006) Water and water resources management. In: Sharma UC, Datta M, Samra JS (eds) Soils & their management in North East India. ICAR Research Complex for NEH Region, Umiam

Water Resources at a Glance. Central Water Commission (2021) GoI, Central Water Commission, Sewa Bhawan, R.K. Puram, New Delhi

Land

17

Suresh Kumar, Dibyendu Sarkar, Krishnendu Das, B. Mondal, H. Biswas, Rajesh Bishnoi, D. C. Sahoo, G. W. Barla, Surender Kumar, V. Ramamurthy, and B. S. Dwivedi

Abstract

Since independence, India has made much progress in agriculture. Indian agriculture, which grew at the rate of about 1.0% per annum during the 50 years before independence, has grown at the rate of about 2.6% annum in the post-independence era. The green revolution (GR) provided India a greater genetic diversity and institutional capacity to produce more. It stimulated infrastructural and rural development, increased prosperity of villages and improved quality of life. During the period 1950–1951 to 2016–2017, the area of other cultivable land, barren and uncultivable land, culturable wasteland, land not available for cultivation and fallow lands showed a steady decline. There was a greater use of such land for agricultural and non-agricultural uses. The area under permanent pasture and other grazing lands also decreased. Fertilizer-responsive varieties/

S. Kumar · R. Bishnoi · D. C. Sahoo · G. W. Barla · S. Kumar
ICAR-Indian Institute of Soil and Water Conservation, Research Centre, Koraput, Odisha, India

D. Sarkar (✉)
Department of Agricultural Chemistry and Soil Science, Bidhan Chandra Krishi Viswavidyalaya, Mohanpur, West Bengal, India

K. Das
ICAR-National Bureau of Soil Survey and Land Use Planning, Kolkata Regional Center, Kolkata, India

B. Mondal
ICAR-National Rice Research Institute, Cuttack, Odisha, India

H. Biswas
ICAR-National Bureau of Soil Survey and Land Use Planning, Nagpur, Maharashtra, India

V. Ramamurthy · B. S. Dwivedi
ICAR-National Bureau of Soil Survey and Land Use Planning, Bangalore Regional Centre, Bangalore, India

P. K. Ghosh et al. (eds.), *Trajectory of 75 years of Indian Agriculture after Independence*, https://doi.org/10.1007/978-981-19-7997-2_17

hybrids were introduced, and as more irrigation facilities were created, acreage under low water requiring traditional crops paved the way to water-intensive cropping. This transformation through GR also showed its side effects like regional imbalance, social inequality and problems of soil degradation. Since land resources are finite, suitable measures are required to reclaim and rehabilitate degraded lands and wastelands. The target 15.3 of the Sustainable Development Goals (SDGs) deals with land degradation neutrality (LDN) and aims to combat desertification, restoring degraded land and soil—including land affected by desertification, droughts and floods to achieve a land degradation-neutral world. India, being a signatory to the SDGs, has voluntarily taken the onus to establish a national framework containing processes and mechanisms for achieving the land restoration target of 26 Mha by 2030. In this chapter, we briefly discussed (1) pre- and post-independence status in land resources, land uses, land resource development, land degradation and measures of land rehabilitation; (2) land use changes, status and causes of land degradation, measures to mitigate land degradation, cost/economics of land restoration, national programmes/policies of soil and water conservation during pre- and post-GR era; and (3) LDN for achieving food security and environmental sustainability in India.

Keywords

Land resources · Land use patterns · Land degradation · Green revolution · Food security · Soil and water conservation programmes · Land degradation neutrality

17.1 Introduction

Agriculture plays a vital role in the economic development of India. Besides providing food to the nation, agriculture provides employment to people, contributes to market of industrial goods and earns foreign exchange. Since independence, progress in Indian agriculture has been accelerated. The growth of Indian agriculture was at the rate of about 1.0% per annum during the 50 years before independence and at the rate of about 2.6% per annum in the post-independence era. Area expansion was the main source of growth during the 1950s and 1960s; after that, the contribution of land area under agricultural production has declined over time, and increase in productivity became the main source of agricultural growth through green revolution (GR). The GR helped India to attain self-sufficiency in food and fibre requirements of the over-increasing population of the country. Agricultural production of India increased from 50 Mt to over 300 Mt since the 1950s. All these developments in Indian agriculture are contributed by a series of steps initiated by the Indian Government. Land reforms, inauguration of the Commission for Agricultural Costs and Prices to ensure remunerative prices to producers, new agricultural technologies, investment in research and extension services, provision of credit facilities and improvement of rural infrastructure are some of these steps. However, this had further consequences of loss of plant biodiversity and increased environmental pollution. Land degradation caused by inappropriate agricultural practices has direct consequences on the food and livelihood security of farmers. The causes of land degradation are both natural and anthropogenic. Natural causes are earthquakes, landslides, volcanic eruptions, tsunamis, droughts, avalanches, floods, tornadoes and wildfires. Anthropogenic causes of land degradation are land clearing and deforestation, inappropriate agricultural practices, improper management of industrial wastes, careless forest management, overgrazing, surface mining, urban sprawl and industrial development. Land degradation caused by inappropriate agricultural practices has direct consequences on the food and livelihood security of farmers.

17.2 Food and Environmental Security During Pre-independence and Independence Era

The country witnessed 14 famines between eleventh and seventeenth centuries (Bhatia 1985). Nevertheless, these famines affected a particular region due to the scarcity in rainfall or other climatic hazards, but after 1860, famines started to spread to different regions of the country with overall scarcity of food grains. Incidence of such shortages also appears to have augmented, and 20 famines occurred during the period between 1860 and 1909. During 1880, the Famine Commission was constituted, which perceived that every province in British India (including Burma) had excess of food grains, to the tune of 5.16 million tons (Mt) (Bhatia 1970). In fact, about 1 million tons of rice and other grains were also annually exported from India. However, the condition has altered severely during the time of the Second World War, and the infamous Bengal famine occurred during 1943.

17.3 Pre-independence Land Resources

Land resource development and conservation programme during pre- and post-independent India has been advanced through several phases. During the Mughal period, there were different forms and arrangement of land revenue collection and management of land resources. Tenants had customary rights over the land they cultivated as long as land revenue and taxes were paid to the Government. The task of collecting taxes and revenue was assigned to a category of people termed *zamindars* (Bhaumik 1993). The *zamindars*, to whom they vended the landownership, customarily selected a series of middlemen for income collection. The increasing number of such middlemen implied that there was remarkable raise in revenue collected from the tenants, and the incapability in paying the extended amount led to their expulsions from the right to cultivate, exploitation and declining agrarian production (Bhaumik 1993). During independence, the agrarian system was considered as opportunistic, rent-seeking middlemen, differential land revenue and proprietorship through different provinces, few land owners held/owned a bigger portion of the land, a large number of tenant farmers, numerous among them had uncertain contract, and unfair production relations (Appu 1996).

17.4 Land Reforms During Post-independence Period

The uniqueness of farming in India is linked with the specified goal to achieve financial progress along with the social equity for assuming few extensive steps for land alterations or conservation in the post-independence period. These changes, earlier from the pre-independence situations, involved were (a) eradication of intermediaries, (b) ceiling on holding size, (c) legislations for tenants, (d) cooperative farming, (e) elimination of involuntary labour and (f) amalgamation of holdings. During the 1960s, many states had passed regulations imposing ceilings on holding of land. The ceilings specified varied from state to state and with types of land and crop.

17.5 Land Use Pattern in the Pre-GR, GR and Post-GR

During the last century, India has endured notable changes in land cover and land use including deforestation, shifting of crop area and urbanization (Roy et al. 2015). The basic factor in agriculture is land. More than 50% of the total geographical area (TGA) is used for cultivation of crops, which made India as the leading nation for production of farm produce globally (Teluguntla et al. 2015). In India, land classification was largely dependent on agriculture, and up to 1950, the land was broadly classified into five categories: (1) area under forests, (2) area not available for cultivation, (3) uncultivated lands including current fallows, (4) area under current fallows and (5) net area sown. Later, it was realized that such a classification did not give a clear picture of the actual area under different categories of land use required for agricultural planning. Hence, a reclassification was adopted from March 1950. Then land was classified under nine categories: (1) forests, (2) barren and uncultivable lands, (3) land put to non-agricultural uses, (4) cultivable wastes, (5) permanent pastures and other grazing lands, (6) miscellaneous tree crops and groves not included in the net area sown, (7) current fallows, (8) other fallows and (9) net sown area.

The total cropped area in the country has increased from 131.9 Mha in 1950–1951 to about 200.2 Mha in 2016–2017. The total arable land and land under permanent crops are quite high in India (51.2%) as compared to China (10.4%), Japan (13.0%) and Mexico (11.7%). The areas of other cultivated lands, barren and uncultivable land, culturable wastelands, land other than cultivated excluding fallow lands and pasture and grazing lands have been showing a steady decline during the GR period. This indicates a greater use of those lands for agricultural use. Barren and uncultivated lands have paved the way for the increase in area under forest and area under cultivation. On the other hand, the area under fallow lands has been increased from 19.3 to 26.4 Mha during the period between

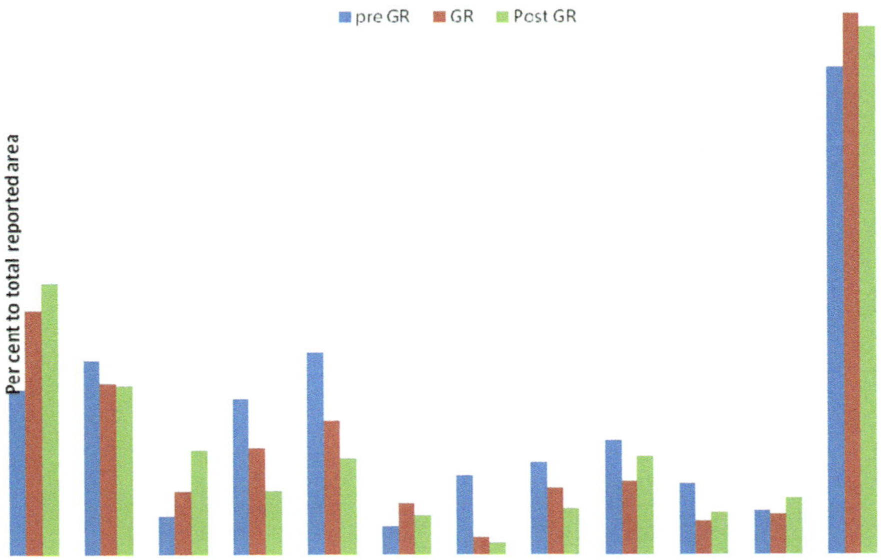

Fig. 17.1 Pre- to post-GR and changes under main land use categories

1970 and 2017 (Fig. 17.1). The area under current fallows steadily increased by nearly 4.31 Mha from 1970–1971 to 2016–2017, and the net decrease was nearly 1.1 Mha from 1980–1981 to 2016–2017 (Table 17.1). Timely correction measures to rejuvenate these fallow lands contributed for the decreasing trend under current fallows.

Forests play a vital role in keeping the ecological balance by protecting land, conserving biological diversity and recycling air and water components for the sustenance of life. The Government Forest report of 1989 has indicated that India possesses only about 2.0% of the world's forest to cater to the needs of about 17.0% of the world's human population besides the rich livestock population. Of the 64.2 Mha (19.5%) of forestland, only 36.1 Mha was good natural forest with a crown cover of at least 40%, accounting for 11.5%. The present per capita availability of forestland is 0.1 ha, while the minimum requirement of forestland is 0.5 ha.

Simultaneously, the increase of non-agricultural area was also observed from 9.4 Mha in 1950–1951 to 27.8 Mha in 2016–2017. However, it is not clear as to what extent and what quality of the land has been converted for non-agricultural uses like urbanization and industrialization. But it is clear that inadvertent expansion of urban areas, including small towns, led to alteration of primary farmland for utilization under non-agricultural purposes. The area under permanent pastures and grazing lands decreased from 13.3 to 10.34 Mha during the GR and post-GR period. The poor quality of the forest cover and pastures not only led to accelerated environmental degradation but also created severe deficit in fodder and fuel. At present, the country is facing a nearly 40% deficit in livestock feed. The problem became more

Table 17.1 Land use classification in India during different decades including pre-green revolution (GR), GR and post-GR periods

	Pre-GR			GR		Post-GR		
Land unit\year	1950–1951	1960–1961	1970–1971	1980–1981	1990–1991	2000–2001	2010–2011	2016–2017
Geographical area	328.7	328.7	328.7	328.7	328.7	328.7	328.73	328.73
Reporting area for land utilisation statistics	284.3	298.5	303.8	304.2	304.9	305.2	307.48	308.32
Forest	40.5 (14.2)	54.1 (18.1)	63.8 (21)	67.5 (22.2)	67.8 (22.2)	69.84 (22.88)	71.59 (23.28)	72.02 (23.36)
Not available for cultivation								
Area under non-agricultural uses	9.4 (3.3)	14.8 (5)	16.5 (5.4)	19.6 (6.4)	21.1 (6.9)	23.75 (7.78)	26.4 (8.59)	27.84 (8.8)
Barren and uncultivable land	38.2 (13.4)	35.9 (12)	28.1 (9.3)	20 (6.6)	19.4 (6.4)	17.48 (5.73)	17.18 (5.59)	16.99 (5.52)
Total	47.5	50.8	44.6	39.6	40.5	41.2	43.58	44.82
Other uncultivated lands excluding fallow land								
Permanent pasture and other grazing lands	6.7 (2.4)	14 (4.7)	13.3 (4.4)	12 (3.9)	11.4 (3.7)	10.66 (3.49)	10.3 (3.35)	10.34 (3.33)
Land under miscellaneous tree crops and groves not incl. in net area sown	19.8 (7)	4.5 (1.5)	4.4 (1.4)	3.6 (1.2)	3.8 (1.3)	3.44 (1.13)	3.2 (1.04)	3.12 (1.00)
Culturable wasteland	22.9 (8.1)	19.2 (6.4)	17.5 (5.8)	16.7 (5.5)	15 (4.9)	13.63 (4.47)	12.65 (4.11)	12.24 (3.99)
Total	49.5	37.6	35.1	32.3	30.2	27.7	26.15	25.70
Fallow lands								
Fallow lands other than current fallows	17.5 (6.1)	11.2 (3.8)	8.7 (2.9)	9.7 (3.2)	9.7 (3.2)	10.27 (3.36)	10.32 (3.36)	11.27 (3.68)
Current fallows	10.7 (3.8)	11.6 (3.9)	10.6 (3.5)	14.8 (4.9)	13.7 (4.5)	14.78 (4.84)	14.28 (4.64)	15.09 (5.01)
Total	28.1	22.8	19.3	24.6	23.4	25.0	24.60	26.36

Agricultural lands								
Net area sown	118.8 (41.8)	133.2 (44.6)	140.9 (46.4)	140.3 (46.1)	142.9 (46.9)	141.34 (46.31)	141.56 (46.04)	139.42 (45.33)
Total cropped area (gross cropped area)	131.9 (46.39)	152.8 (51.19)	165.8 (54.58)	172.6 (56.74)	185.7 (60.91)	185.34 (60.73)	197.68 (64.29)	200.2 (64.93)
Area sown more than once	13.2 (4.64)	19.6 (6.57)	24.9 (8.2)	34.6 (11.37)	42.9 (14.07)	44 (14.42)	56.12 (18.25)	60.79 (19.72)
Net irrigated area	20.9 (7.35)	24.7 (8.27)	31.1 (10.24)	38.7 (12.72)	48 (15.74)	55.2 (18.09)	63.67 (20.71)	68.65 (22.27)
Gross irrigated area	22.6 (7.95)	28 (9.38)	38.2 (12.57)	49.8 (16.37)	63.2 (20.73)	76.19 (24.96)	88.94 (28.93)	98.15 (31.83)
Cropping intensity	111.1	114.7	117.7	123.1	130	131.1	139.64	143.6

Source: Directorate of Economics and Statistics, DAC & FW and Agricultural Statistics at a Glance severe in the GR period than in the pre-GR period as traditional varieties were serving a dual purpose providing grain and fodder than present improved varieties.

The net sown area increased from 118.8 to 139.42 Mha (4.53%) during 1950–1951 to 2016–2017, which can be attributed to expansion of farming in inferior lands comprising culturable wastes and land covered by bushes, shrubs and grasses. Similarly, area under current fallow has been increased by 1.21%. Area under permanent fallow has remained stagnant at nearly 10 Mha during last several decades. The area under cultivation and forest during the pre-GR increased by around 10.1% at the expense of the area under non-agricultural uses mainly from barren and uncultivated and culturable wastelands and fallow lands (Table 17.2).

17.6 Soil Erosion and Conservation in Ancient India

Soil loss is not a new phenomenon; possibly it is as old as the creation of the universe itself (Cogo and Levien 2002). Though a portion of the soil loss is attributed to inevitable natural hazards, there is great evidence that it is caused by anthropogenic events also in terms of swelling, breach of jungles, deforestation, overgrazing of grasslands and expansion of farmland (Blaikie 1985).

During the British period, the lower Indo-Gangetic Plains had a swampy environment enclosed with dense forest, which inhibited the farming activities. Slowly, agrarian habitations in clean forest lands started for many reasons. Though deformation of natural order from the initial and ancient times has been performed by mankind, the intensity and resulting effects have been enhanced many-folds during the colonial times. During 1880–1950, about 20 Mha of forest area was transformed to farmland. The main emphasis of the British Government was to augment the revenue from export of timbre and development of infrastructure (Gadgil and Guha 1992). Construction of road, mining activities in hill areas and transportation have resulted in severe landslide problems. Many regions like large semi-arid tract and the foothills of the Shivaliks and Aravalis are exposed to severe sheet and gully erosion. The Chambal and Yamuna ravines were incessantly moving into the fertile Gangetic Plains. Farmers cultivated much of these erosion-prone areas, leading to further degradation. In Punjab, a large area remained trapped under *Cho* (torrents) beds in Hoshiarpur District (FAO 1985). Also, under 'grow more campaigns' during the period 1920–1950, even hilly regions like Nilgiris in the south were brought under cultivation, which augmented erosion risks in already fragile areas.

After independence, adequate emphasis was given to control erosion, both soil and water, across the diverse agro-climatic regions of the country. Farmers generally appreciate and support soil and water conservation activities carried out by the Government and other agencies. But they have remained aloof from adopting such activities for effective control of soil and water erosion due to the fact that they are resource-poor and their land holdings are fragmented. Rich people intensified and competed for living space, water and fertile as well as marginal lands. Further, transformations of economy through energy generation, building large dams, over-mining and over-pumping of water largely are the key factors now causing

Table 17.2 Programmes and policies for soil and water conservation in India

Year	Programmes/policies	Feature/aim
1900	The Punjab Land Preservation (*Chos*) Act, 1900, which also known as Punjab Land Preservation Act, 1900	Recognized for soil erosion problem
1920	Ravine reclamation project	Reclamation of ravines in United Province (Uttar Pradesh)
1928	Royal Commission on Agriculture	Emphasized the importance of moisture conservation in dryland region
1923	Establishment of Dry Farming Research Station at Manjari	First systematic approach to tackle the problems of dry farming areas on small plots at Manjari, Pune
1930	Dryland farming research scheme and setting up of the Indian Council of Agricultural Research (ICAR)	To develop soil water conservation techniques for better farming in semi-arid lands at Manjari, Pune
1938	Dry farming scheme	Stress on contour bunding
1945	Report of the Famine Inquiry Commission	Soil water conservation was adopted as a part of famine relief measures
1948	Damodar Valley Corporation Act	To control flood, generate electric power for distribution and provide water for irrigation and other purposes
1953	Establishment of Central Soil Conservation Board	As per the recommendation of the First FYP, the Central Soil Conservation Board was set up to organize a national soil conservation programme
1954–1956	Central Soil and Water Conservation Research and Training Institute (CSWCRTI), Dehradun	To conduct studies on run-off and soil loss
1961	River valley schemes	For conservation, control and optimum utilization of water resources
1962	Soil conservation in the catchments of river valley projects (RVPs)	To control premature siltation of reservoirs
1973–1974	Drought Prone Areas Programme (DPAP)	To minimize unfavourable effects of drought on crop production and livestock
1977–1978	Desert Development Programme (DDP)	To minimize the adverse effect of drought and control desertification through rejuvenation of natural resource base
1980	Kandi Watershed and Area Development Project	A World Bank-funded programme for integrated development strategy to rehabilitate the sub-Himalayan Shiwalik area
1980–1981	Integrated watershed management in the catchment of flood-prone rivers (FPR)	Enhance the productivity and tackle menace of floods
1983	Starting 47 model watersheds (operational research projects)	For soil and water conservation in drylands

(continued)

Table 17.2 (continued)

Year	Programmes/policies	Feature/aim
1983	Himalayan Watershed Management Project	A World Bank-funded programme to increase and stabilize crop and forage yields and production of fuel wood and timbre in rainfed farming areas
1983	Rainfed Area Watershed Development Project	A World Bank-funded programme to minimize the degradation of Himalayan ecosystems caused by depletion of forest cover, overgrazing and bad land use and decrease erosion and flooding in Gangetic Plains
1984	The World Bank-Assisted Comprehensive Watershed Project, Kabbalnala	To conserve soil and moisture to improve the overall productivity
1984–1985	Participative Integrated Development of Watersheds (PIDOW)	To integrate forestry programmes, animal husbandry practices and credit programmes with soil and water management at the watershed level
1985	Established the National Wasteland Development Board	To tackle the problem of degradation of lands and restoration of ecology and to meet the growing demands of fuel wood and fodder at the national level
1985	National Land Use and Conservation Board	Formulate a national policy plan for conservation, management and development of land resources
1985–1986	Reclamation and Development of Alkali and Acid Soil (RADAS)	Reclamation of problematic soil
1988	National Land Use Policy	To devise an effective administrative procedure for regulating land use to prevent further deterioration of land resources
1989	Integrated Afforestation and Eco-Development Scheme (IAEPS)	To restore and regenerate the ecological balance of degraded forests on a watershed basis following participatory approach
1990	Integrated Watershed Development (Hills) Project	A World Bank-funded programme to improve the productive potential of the project area using evolving watershed treatment technologies and community participatory approaches
1990	Integrated Watershed Development Project–Plains	A World Bank-funded programme to stabilize selected watersheds through land treatments, emphasizing soil and moisture conservation, and by introducing more sustainable land management systems
1989–1990	Integrated Wasteland Development Project (IWDP) Scheme	To regenerate degraded non-forest land through silvopasture and soil and water conservation on the village and micro-watershed scale

(continued)

Table 17.2 (continued)

Year	Programmes/policies	Feature/aim
1990–1991	National Watershed Development Programme in Rainfed Areas (NWDPRA)	Promote sustainable natural resource management by holistic and sustainable development of rainfed areas through watershed development (WSD) approach
1992	National Conservation Strategy and Policy Statement on Environment and Development	To tackle the problem of sustainable land use management system
1992	Indo-German Watershed Development Programme	To rehabilitate micro-watersheds for the purpose of regeneration of natural resources and sustainable livelihoods, using a participatory approach
1992	Constitution (74th Amendment) Act, 1992	Regulation of land use and urban planning brought under the domain of urban self-governing bodies
1992	Policy Statement for Abatement of Pollution	Advocate the use of mixed policy instruments in the form of legislation, regulation and fiscal incentives
1994	Guidelines for watershed development	Provide common guidelines for WSD
1994–1995	Watershed Development Project in Shifting Cultivation Areas (WDPSCA)	To protect and develop the hill slopes of *jhum* areas through different soil and water conservation measures on watershed
1995	Karnataka Watershed Development (KAWAD) Project	For ensuring livelihood security in the drought-prone and degraded areas of Karnataka through the restoration of degraded habitats
1999	The Department of Wasteland Development was renamed as the Department of Land Resources	To formulate integrated land resource management policies and implementation of land development programmes
1999	Integrated Watershed Development (Hills II) Project	A World Bank-funded programme to improve productive potential using evolving watershed treatment technologies and community participatory approaches
1999–2000	Watershed Development Fund (WDF) in NABARD	Provide financial support to scale up successful participatory WSD projects in 100 priority districts; promote a more unified strategy to WSD
2001	Common Guidelines for Watershed Development (revised)	Update the 1994 WSD guidelines to have a more participatory and project-specific focus with greater flexibility in implementation. Applicable to IWDP, DPAP, DDP and other programmes notified by the GOI

(continued)

Table 17.2 (continued)

Year	Programmes/policies	Feature/aim
2001	Karnataka Watershed Development Project	A World Bank-funded programme to improve productive potential and strengthen community/institutional arrangements for natural resource management
2002	National Afforestation Programme	For ecological restoration of degraded forests and to develop forest resources with peoples' participation, with focus on improvement in livelihoods of the forest-fringe communities, especially the poor
2002	Sujala Watershed Programme	A World Bank-funded programme to increase the productive potential of the watersheds by involving the communities in the process through building appropriate people's institutions and capacitating them to plan, implement and manage their resources to achieve more sustainable development
2003	Hariyali Guidelines	Integrate community institutions more meaningfully in DPP, DPAP and IWDP and simplify procedures
2004	Uttarakhand (initially Uttaranchal) Decentralized Watershed Development Project	A World Bank-funded programme to improve the productive potential of natural resources and increase incomes of rural inhabitants through socially inclusive, institutionally and environmentally sustainable approaches
2005	Himachal Pradesh Mid-Himalayan Watershed Development Project	A World Bank-funded programme to improve productive potential and increase incomes of rural inhabitants through socially inclusive, institutionally and environmentally sustainable approaches
2005	Mahatma Gandhi National Rural Employment Guarantee Scheme (MGNREGS)	Enhance livelihood security in rural areas by providing at least 100 days of guaranteed wage employment a year to every household whose adult members volunteer to do unskilled manual work (e.g. soil and water conservation, afforestation and land development)
2006	Parthasarathy Committee Report	The Parthasarathy Committee was established as a technical committee to evaluate the DPAP, DDP and IWDP. In 2006, the Committee released a report that served as a review of India's watershed programme. The Committee's report serves as the basis of the Neeranchal Guidelines and the NRAA

(continued)

Table 17.2 (continued)

Year	Programmes/policies	Feature/aim
2006	National Rainfed Area Authority	Sustainable and holistic development of rainfed areas and to create common guidelines for all WSD schemes under the different ministries for the development of rainfed farming systems
2008	Common Guidelines for Watershed Development (Neeranchal) released	Promote a fresh framework to guide all WSD projects in all departments and ministries
2009	Sustainable Land, Water and Biodiversity Conservation and Management for Improved Livelihoods in Uttarakhand Watershed Sector	A World Bank-funded programme to improve the productive potential of natural resources and increase incomes of rural inhabitants in selected watersheds through socially inclusive, institutionally and environmentally sustainable approaches
2009–2010	Integrated Watershed Management Programme (IWMP)	IWMP was launched in 2009–2010 with the objective of bringing various programmes such as the IWDP, DDP and DDAP under one common integrated programme
2011	Revised Edition of the Common Guidelines for Watershed Development released	Provide amendments to the 2008 guidelines based on clarifications and suggestions from concerned ministries, departments, state government and NGOs
2013–2014	Karnataka Watershed Development II (Sujala-III)	A World Bank-funded programme to demonstrate more effective watershed management through greater integration of programmes related to rainfed agriculture, innovative and science-based approaches and strengthened institutions and capacities
2012	AF-HP Mid-Himalayan Watershed Development Project	A World Bank-funded programme to reverse the process of degradation of the natural resource base and improve the productive potential of natural resources and incomes of the rural households
2013	Revisions added to 2008 Common Guidelines (known as *Neeranchal* Guidelines)	Added new features to the 2008 Common Guidelines to ensure momentum to the IWMP while strengthening its innovative feature
2015	Operational Guidelines of PMKSY	IWMP is implemented as WSD Component of PMKSY (WDC-PMKSY)

(continued)

Table 17.2 (continued)

Year	Programmes/policies	Feature/aim
2016–2017	Neeranchal – World Bank-Assisted National Watershed Management Project	A World Bank-funded programme to support WDC-PMKSY through technical assistance to improve incremental conservation outcomes and agricultural yields for communities in selected sites and adoption of more effective processes and technologies into the broader PMKSY in participating nine states: Andhra Pradesh, Chhattisgarh, Gujarat, Jharkhand, Madhya Pradesh, Maharashtra, Odisha, Rajasthan and Telangana
2020	Integrated Project for Source Sustainability and Climate Resilient Rain-fed Agriculture in Himachal Pradesh	A World Bank-funded programme to improve upstream watershed management and increase agricultural water productivity in selected *Gram Panchayats* in Himachal Pradesh
2021	Rejuvenating Watersheds for Agricultural Resilience through Innovative Development (REWARD)	A World Bank-funded programme to strengthen capacities of national and state institutions to adopt improved watershed management for increasing farmers' resilience and support value chains in selected watersheds of participating Andhra Pradesh, Karnataka and Odisha

degradation of land. Thus, safeguarding and maintaining a balance with the nature are a great challenge in a country like India with a fast-expanding market economy (Rangarajan 2015).

17.7 Status and Sources of Land Degradation in India

Assessment of the status and extent of land degradation in India was first undertaken by the National Commission on Agriculture, which, based on secondary data, placed the extent of degradation at 148 Mha (NCA 1976), followed by the estimates of 175 Mha by the Ministry of Agriculture (MoA 1978). Salt-affected soils and waterlogged areas were pegged at 7 and 6 Mha, respectively (Bali 1985). The National Wasteland Development Board (NWDB 1987) estimated an area of 123 Mha under wastelands. Using the GLASOD methodology outlined by Oldeman (1988), the National Bureau of Soil Survey and Land Use Planning (ICAR-NBSS & LUP) reported 187 Mha (Sehgal and Abrol 1994) to be degraded under various processes. Assessment of degradation by other agencies can be referred in Gautam and Narayan (1988), who opined those differences in estimations of degradation are due to the use of varying definitions of land degradation, data sources, classification systems, methodologies and scales. In order to do away with the varying estimation figures, a consortia of NRM Institutes of the ICAR including NBSS & LUP, Nagpur; IISWC, Dehradun; CSSRI, Karnal; and CAZRI, Jodhpur, in association with NRSC, Hyderabad, agreed upon a harmonized national land degradation figure of 120 Mha (NAAS 2010).

Water erosion has been the most serious land degradation problem in India, resulting in loss of surface soil and terrain deformation. The average soil erosion rate is ~16.4 tons/ha/year, resulting in an annual total soil loss of 5.3 billion tons throughout the country (Dhruvanarayan and Ram 1983). Approximately 29% of the total eroded soil is permanently lost to the sea, while 61% is transferred from one place to another, and the remaining 10% is deposited in reservoirs.

17.7.1 Overgrazing, Deforestation and Poor Forest Management

Loss of vegetation occurs due to cutting beyond the silviculturally permissible limit, unsustainable fuel wood and fodder extraction, encroachment by agriculture into forest lands, forest fires and overgrazing. High livestock density, particularly in arid regions, causes overgrazing, which results in decreased infiltration and accelerated run-off and soil erosion. Due to overgrazing, soil loss is 5 to 41 and 3 to 18 times greater than normal at the mesoscale and macroscale, respectively (Sharma 1997).

17.7.2 Urbanization, Industrialization and Mining

An increase in urbanization, industrialization and infrastructure development is continuously dwindling land of agriculture, forestry, grass and pasture and unused lands with wild vegetation. Opencast mining is also an issue, because it disturbs the physical, chemical and biological features of the soil and alters the socio-economic behaviour of a region. Negative effects of mining are water scarcity due to lowering of water table, soil contamination, part or total loss of flora and fauna, air and water pollution and acid drainage. Overburden removal from mine area results in significant loss of vegetation (Sahu and Dash 2011). Overburden removal is done by blasting or using excavators, which generates large volume of waste (soil, debris and other materials).

17.7.3 Natural and Social Causes of Land Degradation

Natural causes of land degradation are earthquakes, tsunamis, droughts, avalanches, landslides, volcanic eruptions, floods, tornadoes and wildfires. In addition to these social causes of soil degradation are land shortage, decline in per capita availability of land, economic pressure on land, land tenancy, poverty and population increase.

17.7.4 Land Shortage, Fragmentation and Poor Economy

In India, small land holdings are very common. About 80% of farmers' holdings are ≤2 ha, accounting for >50% of agricultural output. Size of land holding declined from 2.3 ha to 1.3 ha from 1970 to 2000 with per capita availability of land of 0.32 ha in 2001 (Mythili 2013). Small land holdings lead to severe pressures on land resources. Because of such pressure, labour, land and capital resources limit the use of green manuring or soil conservation measures. Land shortage coupled with poverty leads to non-sustainable land management practices and land degradation. There are two other direct causes of land degradation, i.e. improper crop rotations and unbalanced fertilizer use (FAO 2010).

17.7.5 Agricultural Activities Causing Land Degradation in India

Land degradation caused by inappropriate agriculture has negative impact on the food and livelihood security of farmers. The C-sustainability index was high in 1960, which is indicative of the minimum usage of inputs prior to the onset of the GR, and the C-sustainability index decreased because of greater C-based inputs, in which a linear relationship exists between C inputs and C outputs (Bhattacharyya et al. 2015). Agricultural practices result to land degradation in a number of ways depending on land use, crops grown and management practices. The common causes of land degradation by agriculture include cultivation in fragile deserts and marginal sloping lands without any conservation measures, land clearing through clear cutting and deforestation, agricultural depletion of soil nutrients through poor farming practices, overgrazing, excessive irrigation, overdrafting, urban sprawl and commercial development and land pollution including industrial waste disposal to arable lands.

17.7.5.1 Imbalanced Fertilizer Use

The imbalanced NPK consumption ratio of 6.2:4:1 in 1990–1991 has widened to 7: 2.7:1 in 2000–2001 and 5:2:1 in 2009–2010 as compared to the ideal ratio of 4:2:1. Excess nitrate leaches down to groundwater due to heavy N fertilization. As food grain production increased with time, deficiency of the number of elements in Indian soils increased from one (N) in the 1950s to nine (N, P, K, S, B, Cu, Fe, Mn and Zn) in 2005–2006. Although the use of fertilizers has increased several folds, the overall consumption continues to be low in most parts of the country. Widespread Zn deficiency, followed by S, Fe, Cu, Mn and B, is common throughout the country. Every year, ~20 Mt of the three major nutrients is removed by growing crops (Tandon 1992), but the corresponding addition through inorganic fertilizers and organic manures falls short of this harvest. Soil loss through erosion is another reason for soil fertility depletion, accounting for an annual loss of ~8 Mt of plant nutrients through 5.3 billion tons of soil loss (Prasad and Biswas 2000).

17.7.5.2 Excessive Tillage and Use of Machinery

Excessive tillage coupled with the use of heavy machinery for land preparation and harvesting causes a multitude of soil and environmental problems. Puddling of soil for rice degrades soil physical properties and has negative impacts on soil biology (Hobbs et al. 2008). Poor physical soil health leads to poor crop establishment. Excessive tillage for land preparation, indiscriminate irrigation and excessive fertilizer applications are the main sources of greenhouse gas (GHG) emission from agricultural systems.

17.7.5.3 Crop Residue Burning

The NBSS & LUP reported that nearly 3.7 Mha suffer from nutrient loss and/or depletion of soil organic matter (SOM). Burning of crop residues or its disposal is a pervasive problem, which contributes to SOM loss. According to the Ministry of New and Renewable Energy (MNRE 2009), ~500 Mt of crop residues are generated every year, and ~125 Mt is burned. Crop residue generation is the highest in Uttar Pradesh (60 Mt) followed by Punjab (51 Mt) and Maharashtra (46 Mt). Among different crops, cereals generate 352 Mt of residues followed by fibre crops (66 Mt), oilseeds (29 Mt), pulses (13 Mt) and sugarcane (12 Mt). Rice (34%) and wheat (22%) are the dominant cereals contributing to crop residue generation (NAAS 2012).

17.7.5.4 Poor Water Management

Improper use of canal irrigation has resulted to soil degradation problems like waterlogging and salinization. Expansion of canal irrigation (i.e. Indira Gandhi Nahar Project) has been associated with widespread waterlogging and salinity problems in areas, such as in the Indo-Gangetic Plains (IGP). A large part of arid, semi-arid and sub-humid areas has become barren due to the development of saline-sodic soils because of poor irrigation and drainage management.

17.7.5.5 Inefficient Cropping

Improper crop rotation coupled with lack of proper soil and water conservation measures is important reason contributing to soil erosion in cultivated lands. Cultivation of marginal lands on steep slopes, in shallow or sandy soils with laterite crusts and in arid or semi-arid regions has resulted in land degradation. Cultivation of marginal land with low SOM with unsuitable cropping systems has been the major cause of accelerated wind and water erosion. Wind erosion is a serious problem in arid and semi-arid areas, in coastal areas with sandy soils and in the cold desert areas of Leh in the extreme north of India.

17.7.5.6 Overuse of Pesticides and Environmental Pollution

Indiscriminate use of pesticides together with sewage sludge and composted municipal wastes leads to contamination of soil and water with toxic substances and heavy metals. Some commercial fertilizers also contain significant amount of heavy metals. Long-term use of those fertilizers exerts undesirable effects on the environment.

Indiscriminate use of agrochemicals, such as fertilizers and pesticides, is often the cause for land degradation.

17.8 Region-Wise Land Degradation in India

Major causes and extents of soil degradation in India are presented region-wise, and the regions included are (1) hilly areas, (2) trans- and upper-Indo-Gangetic Plains (IGP), (3) middle and lower Indo-Gangetic Plains and coastal areas, (4) dryland and desert areas, (5) southern peninsular India and (6) central India.

17.8.1 Hilly Areas

The Indian Himalayas occupy an area of 53.7 Mha (16.4% of the TGA of the country), which consists of two distinct sub-regions, viz. eastern Himalayas region or northeastern hills (NEH) and western Himalayas region or northwestern hills (NWH). Annual rainfall in the NEH region is much higher (2800–12,000 mm) than the NWH region (350–3000 mm). Other than soil erosion by water, mass erosion and landslide/landslips cause soil loss. Human-induced intensification of landsliding has been caused by vegetation clearing, construction of roads and buildings, mining and building of hydropower projects.

The major cause of land degradation in hilly areas is soil erosion by water. The Himalayas have steep slopes, fragile geology and intense storms, all of which accelerate soil erosion. Erosion rates are high for the Shiwalik hills (~80 tons/ha/year) and shifting cultivation areas (~40 tons/ha/year). Nearly 39% of the area has a potential erosion rate of >40 tons/ha/year. The severity of water erosion is more in the NEH region (22.3% of TGA) than in the NWH region (12.6% of TGA). Rivers and torrents (seasonal streams with flash flow during monsoon) in the Himalayan region also cause land degradation. A recent example is that of the flood-induced deluge occurred in the Kedarnath Valley of Uttarakhand State during mid-June 2013, which took a toll of thousands of lives and destroyed property including agricultural lands.

Shifting cultivation, locally named as *jhum* cultivation, is a traditional form of land use system in the NEH. On average, 3.9 Mha of land is under shifting cultivation every year (CSWCR & TI 2011). In this system, crops are cultivated on steep slopes. Land is cleared by cutting the forest or bush to stump level, leaving cut materials to dry and eventually burn to prepare the land ready for sowing before the onset of rains. Deforestation coupled with shifting cultivation has resulted in loss of even >200 tons of soil/ha/year due to erosion. Soil erosion in catchment area results in siltation of reservoirs and streams, leading to frequent floods on the plain/low-lying areas. Removal of top soil leads to loss of fertility, shallow soil depth, which is not easily built up. Besides, reduction of *jhum* cycle to 3–5 years from the earlier 10–15 years or more due to excess demographic pressure has been causing further land degradation as there is less time left for restoration of soil fertility. On average, ~10 tons of biomass per ha is burnt annually in such cultivation practices leading to air pollution due to the release of large quantity of CO_2 into the atmosphere.

17.8.2 Trans- and Upper-Indo-Gangetic Plains (IGP)

The trans-Gangetic Plains consists of Punjab, Haryana, Delhi and union territory Chandigarh and parts of Uttar Pradesh (UP). The upper IGP consists of parts of UP and Delhi. In this region, the major causes of land degradation are water erosion; residue burning; SOM and nutrient depletion; soil physical, chemical and biological degradation caused by inappropriate crop rotation; faulty irrigation and water management; overuse of pesticides; reduced use of organic matter; and poor crop cycle planning. Erosion by water is the major process of soil degradation in this region too.

Excessive tillage using heavy machinery and formation of new beds for vegetable cultivation cause soil physical degradation. Crusting and compaction associated with a decline in soil structure is a common form of physical soil degradation. Compaction of soil at the plough layer is more under rice-wheat systems, where puddling is done for rice followed by several tiller and disc harrow passes for wheat cultivation. It was also noticed that puddling increased soil bulk density in the sub-surface layer (15–30 cm) in rice-based systems (Aggarwal et al. 1995). When high-intensity raindrops strike bare soil surfaces, soil aggregates are broken down and clog soil pores, causing more run-off and topsoil loss. All these practices with little or no addition of organic manure aggravate the problem of soil physical degradation due to continuous reduction in SOM.

Soil chemical degradation is mainly in the form of development of acidity and salinity and decrease in SOM and available nutrients including N. Other chemical processes include leaching, decrease in cation retention capacity and fertility depletion. In sodic soils, soil physical degradation (either by crusting or by hardpan formation) rather than soil chemical degradation affects crop production. Many irrigated areas are prone to salinity and sodicity problems, when water evaporates leaving on the land surface. Soil salinity and sodicity also interfere with nutrient management during crop cultivation. Salinity and alkalinity problems are much more aggravated in low-precipitation areas with multiple sources of irrigation, improper cropping pattern and poor drainage that led to accumulation of salt in the root zone. The expansion of canal irrigation has been associated with waterlogging and salinity problems in several areas. Use efficiency of N fertilizers for crops is very low due to loss by volatilization in saline and alkaline soils and by leaching and run-off in waterlogged areas. Rates of both mineralization and immobilization of N in soils decreased considerably at higher levels of soil sodicity (Bandyopadhyay and Rao 2001). Zinc availability was also inhibited in saline soils used for rice production due to reduction of sulphate (SO_4^{-2}) to sulphides and subsequent precipitation of Zn as Zn-S. Zinc deficiency is also widespread in alkaline soils, where it and other micronutrients like Fe, Mn and Cu are precipitated as hydroxides and carbonates.

17.8.3 Middle and Lower IGP and Coastal Areas

The middle and lower IGP consists of parts of Uttar Pradesh and West Bengal and entire Bihar. The area is characterized by subtropical monsoon climate with smaller and more diversified farm holdings. The coastal zones of east and west India occupy an area of about 10.8 Mha (Velayutham et al. 1998). In the eastern IGP, drainage and flood water management are the major problems. Some major rice-growing soils of this region suffer from waterlogging. The waterlogged alluviums in eastern India have water stagnation above ground for about 6 months each year. The adverse physical conditions allow only one anaerobic paddy crop with a very low yield potential of <1.0 tons/ha.

Some agricultural practices such as continuous cropping with limited supply of organic amendments, removal of crop residues and excessive tillage cause loss of soil functioning. Mandal et al. (2007) observed significant loss of soil organic C (7–33%) from this region under long-term rice-based cropping systems without addition of any fertilizer inputs. Saha et al. (2000) also reported similar decline in SOM under rice-wheat-jute cropping. Nutrient depletion is also a concern in the middle and lower IGP. Growing two rice crops in a rice-jute-rice system exhausts more nutrients than a single crop of wheat in the traditional jute-wheat system. Negative P balance has been reported from the lower IGP (Biswas et al. 2006). In the case of rice-wheat and rice-rice-rapeseed systems, available P declined, despite a slightly positive P balance. The negative annual K balance in many rice-based cropping systems in West Bengal ranges from 39 to 179 kg/ha/year, and there is a declining trend in available soil K.

Approximately 3.5 Mha in West Bengal adjoining Bhagirathi River are contaminated with As. Contamination of groundwater with As concentrations exceeding the permissible limit of 50 μg L^{-1} has also been detected in Bihar (Acharyya and Shah 2004). The primary cause of the problem is excessive withdrawal of groundwater for summer paddy rice. This causes oxygenated decomposition of pyritic sediments, which contain high amounts of As and are deposited in the Bengal deltaic plains. The sediments upon oxidation release sulphuric acid that solubilizes the As. The solution moves down to aquifers, polluting the groundwater. Irrigation with As-rich water has resulted in considerable accumulation of As in soils. Summer (*Boro*) rice irrigated primarily with groundwater contained more As than *Kharif* rice. Edible parts of leafy and underground vegetables (e.g. spinach, fenugreek and radish) had much greater As concentrations than vegetables with edible fruit parts (e.g. brinjal, beans, okra and tomato).

Presently, ~23% of the coastline along India's mainland is affected by erosion (Kumar et al. 2006). A tsunami can generate fast moving, large onshore currents that move objects far inland, forcing collapse of structures, eroding beaches of sand and stripping coastal vegetation. It is also capable of inundating coastal lands. In December 2004, India was the third most affected country by tsunamis with ~4000 ha of land devastated (DiMaRF 2005).

17.8.4 Dryland and Desert Areas

About 25% of India's TGA is desert, and as much as 69% of the country's area is classified as 'dryland'. The dryland and desert areas consist of parts of Gujarat, Andhra Pradesh and the whole of Rajasthan. In the desert region, land degradation is caused by high wind speed, increased frequency of high wind events, less rainfall, high rainfall variability with many drought spells each year, soils with low resistance to wind erosion, little or no natural vegetation, overgrazing, little or no residue recycling, crop residue burning, imbalanced plant nutrition, termite infestation and deep tillage (Srinivasarao et al. 2013).

Little crop residue recycling, excessive tillage and long non-crop periods are important reasons why SOM and soil N are being depleted. Many dryland crop residues are burned regularly, particularly of cotton, pigeon pea, castor, chilli and maize. Farmers realize the importance of crop residue recycling, but decomposition of these hardy residues is considered a barrier for subsequent crop production, and therefore the residues are burned (Srinivasarao et al. 2013).

Wind erosion is a serious problem leading to loss of fertile topsoil, especially for Camborthids and Solorthids, which are major soils with sandy, loamy sand or sandy loam textures. The coarse-textured soils hold little moisture and nutrients. Soil salinity is another common problem associated with extreme aridity and poor-quality groundwater. Causes of desertification include change in frequency and amount of rainfall, reduction in vegetal cover, poor agricultural management practices, cultivation on marginal lands, over-exploitation of natural resources and excessive grazing.

17.8.5 Southern Peninsular Region

Water erosion is the major problem in this region. This is mainly due to high-intensity rainfall, deforestation, overgrazing and poor land use practices. Soil loss is severe to extremely severe in most of the hilly areas of Madhya Pradesh, Chhattisgarh and Western Ghats. Waterlogging is a major problem in Kerala, as well as in Andhra Pradesh, Karnataka and Tamil Nadu. The main factors causing waterlogging are lack of natural drainage, upheaval in the river bed or drainage channel, indiscriminate cultivation in the drainage channel bed, interception of natural drainage due to construction of roads and embankments and discharge of surplus canal water into channels with inadequate capacity.

Temporal calculation of soil nutrient showed that there was a decrease in N fertility index in Kerala (from 2.11 to 1.66) and Karnataka (2.33 to 2.05) and an increase in Tamil Nadu (1.1 in 1967 to 1.34 in 1997). Soil organic C content in forest soils under teak (*Tectona grandis*) and sal (*Shorea robusta*) with low management was twice that under cropped land. Soils in the peninsular plateau are also characterized by low soil organic C (SOC) concentration (0.4%) due to high temperature. As a result, this region contributes only 17% of the total SOC stock of the country despite occupying about one-third of the land area (Bhattacharyya et al. 2000).

About 25% of sand supplied for construction work in Bangalore is extracted from coarse surface soils by washing (Hegde et al. 2008). Sand mining causes severe ecological imbalance by the loss of surface soil along with plant nutrients, decline in crop yield, siltation of tanks, exploitation of groundwater and increased soil erosion. About 43.8 million m^3 of soil is used per year for sand extraction near Bangalore, which uses about 48,180 million litres of water.

17.8.6 Central India

The central zone consists of three states, namely, Madhya Pradesh, Chhattisgarh and Maharashtra, and covers 23% of the TGA of the country. Black soil or vertisols are prevalent in this zone. The black soils have high moisture holding capacity (150–250 mm/m), but water is not all available to plants because the water is held very strongly by the smectite clays. These soils are very sticky when wet and hard when dry. These soils have low permeability as well. The major problems to crop production are poor workability and drainage, high water erosion and loss of soil fertility due to residue burning.

Prevalence of montmorillonitic clays in vertisols makes them difficult to work. Under drying, they crack; ploughing becomes very difficult and produces very large clods. Under wetting, they are sticky; tillage is difficult because of poor trafficability. Thus, farmers get a narrow window of time for crop sowing, so some vertisols remain fallow even during the rainy season. Poor drainage condition of vertisols is one of the main constraints of low production. With the initial showers of rain, particularly when the soil is dry, the rate of infiltration is high because of cracks and granular structure. But as soon as the soil profile becomes filled, swelling of the soil decreases infiltration rates and affects internal drainage. Therefore, without proper land care and crop management during the rainy season, vertisols are prone to huge soil and water loss. Vertisols generally have low soil organic C and N, P, S and Zn contents. Low and imbalanced use of fertilizers result in micronutrient deficiencies. Since most of the residues are either removed from the field or burnt, nutrient recycling in soil is badly affected. The calcareous nature of these soils also affects nutrient availability to crops.

17.9 Cost of Land Degradation in India

In India, the annual cost of land degradation was assessed in terms of production loss, replacement cost analysis and TEV approaches. However, loss of productivity is widely used for assessing the impact of land degradation in India (Mythili 2003). Recently, TERI (1998) estimated that the economic losses from land degradation and change of land use in 2014–2015 stood at 2.54% of India's GDP and land degradation accounts for 82% of losses.

17.10 Mitigation of Land Degradation

Soil conservation measures like contour ploughing, bunding and use of strips and terraces are useful for decreasing soil erosion and slow run-off water. Mechanical measures like physical barriers, such as embankments and wind breaks, or vegetation cover (and use of vegetative buffer strips and geo-textiles) and soil husbandry are important measures to control soil erosion (Srinivasarao et al. 2014). Besides, conservation agriculture (CA), agroforestry, integrated nutrient management (INM) and crop diversification also conserve soil and water.

Mechanical soil and water conservation practices are necessary for reducing soil erosion, conserving rainfall within the slope and safe disposal of excess run-off from top to foot hills. The structures often used in case of extreme soil degradation are (1) bunding, used in agriculture to collect surface run-off, increase water infiltration and prevent soil erosion; (2) graded bunds, constructed in medium to high rainfall areas of ~600 mm/year; (3) contour bunds, mechanical or vegetative barrier created across the slope; (4) bench terrace and half-moon terrace, adopted where soil depth is >1.0 m; (5) grassed waterways, channels laid out preferably on natural drainage lines in the watershed; and (6) water harvesting ponds, dug-out embankment type of water harvesting structure used for creating seasonal and perennial water harvesting structure at the bottom of a micro-watershed for irrigation and fish farming. In vertisols, graded broad bed and furrow system of land configuration improves surface drainage and allows better water infiltration. It also facilitates drainage of excess water through grassed waterways.

For river bank erosion control, bioengineering treatments like spurs, retaining walls and earthen embankments may be used combined with suitable vegetation such as giant cane (*Arundo donax*), five-leaf chaste trees (*Vitex negundo*), morning glory (*Ipomoea* sp.), bamboo (*Bambusa vulgaris*), Napier grass (*Pennisetum purpureum*) and munja (*Saccharum munja*) (CSWCR & TI 2011).

Agronomical practices like use of cover crops, mixed cropping/inter-cropping/strip cropping, crop rotation, green manuring and mulching are vital practices associated with INM. When crops like maize, sorghum and castor are cultivated along with legumes like groundnut, green gram, black gram, soybean and cowpea in inter-row spaces, sufficient cover on the ground is ensured and erosion hazards decreased. Poor infiltration rate is one of the major constrains of black soils. In black soils, deep tillage (subsoiling with chisel plough) can improve soil water storage by better infiltration and minimizing water stress.

Integrated watershed management, which involves soil and water conservation coupled with suitable crop management, is another good strategy for mitigating soil erosion. This strategy involves construction of check dams along with gullies, bench terracing, contour bunding, land levelling and planting of grasses. An operational research project on watershed management at Fakot by the CSWCR & TI, Dehradun, during 1975–1986 is a successful example of participatory integrated watershed management (Joy et al. 2005). A case study in Netranahalli Watershed (Karnataka) in the Southern Peninsular India highlighted the importance of involvement of communities for conservation of soil and water and their sustainable management.

Liming is the most preferred practice for managing of acid soils. Lime raises soil pH, thereby increasing the availability of plant nutrients and reducing toxicity of Fe and Al. A recommendation of low dose of lime (i.e. one-tenth to one-fifth of lime requirement) applied along with fertilizers in furrows at the time of sowing is found to be effective (Sharma and Sarkar 2005; Bhat et al. 2010).

Management of saline soils involves tillage, irrigation with good quality water and leaching of excess salts. Inversion tillage has an edge in decreasing potential soluble salt accumulation in the root zone over zero tillage. However, deep tillage may bring more salts to the soil surface and root zone. Salinity-tolerant crops also support the formation of stable soil aggregates and improvement of soil tilth. Rice is a potential crop for reclamation of saline/alkali soils. Salt-affected soils are reclaimed by leaching followed by application of green manures. Gypsum is the major amendment used for amelioration of alkali soils. Other amendments are phosphogypsum and acid formers like pyrites, sulphuric acid, aluminium sulphate and sulphur. The treated field should be kept submerged with good quality water to facilitate reaction and subsequent leaching.

Reclamation of acid sulphate soils follows the approaches like the following: (1) pyrite and soil acidity can be removed by leaching after drying and aeration; and (2) pyrite oxidation can be checked or stopped and existing acidity inactivated by maintaining a high water table, with or without (3) additional liming and fertilization with P, though liming may often be uneconomical in practical use. For coastal acid sulphate soils of Sundarbans, application of lime, superphosphate and rock phosphate has been found useful (Bandyopadhyay 1989).

Mitigation of As contamination could be achieved by replacing *Boro* (summer season) rice requiring more groundwater with summer legumes and pulses, decreased irrigation coupled with addition of zinc sulphate, greater use of organic/green manures that moderate As toxicity in soils and plants and phytoremediation employing hyper-accumulating plants like brake fern (*Pteris vittata*) and water hyacinth (*Eichornia crassipes*). Blue-green algae can also decontaminate paddy soils through accumulation of As in its biomass and subsequent removal.

Domestic and municipal wastes, sludges, pesticides and industrial wastes need to be used with utmost caution to avoid the possibility of pollution of soil. Mined land can be better reclaimed by back filling and spreading topsoil over the surface. Reclaimed land after mining can be used for planting trees. Sen and Oosterbaan (1992) presented a practical working method on integrated water management for Sundarbans through surface gravity-induced drainage during the rainy season (through land shaping)-cum-excess rainwater storage for irrigation during dry season. Treated distillery effluent as a waste by-product can be used for irrigation nearby distillery industries, and this technique has a significant influence on mitigating land degradation. The SOC of the surface (0–15 cm) soil and aggregate stability are improved with application of those effluents.

In the hills, most of the upper slope is covered with horticultural tree plantation using half-moon terraces and contour bunds, and the remaining lower section is used for cultivation of field crops by forming bench terraces. The following crops may be grown: (1) fruit trees in half-moon terraces on contour, (2) pineapple in two rows planted closer together in contour bunds, (3) vegetables (bean, cowpea, guar or cluster bean, pea) and cover crops like sweet potato in the interspaces of the contour and (4) ginger and turmeric grown in the interspace area of contours.

Vegetative barriers can decrease run-off by 18–21% and soil loss by 23–68% on slopes varying from 2 to 8% in the Himalayan region. Vegetative barriers of Guinea grass, Khuskhus and Bhabar were effective (after 3–4 years) in reducing soil loss by 6–8 tons/ha/year and run-off by 33–38%. Maize and wheat yield was increased by ~32 and 10%, respectively, due to moisture conservation in the hilly region (Ghosh 2009). Pigeon pea, due to its extensive canopy cover (95–98%), was effective in reducing run-off (28–29%) and soil loss (2.1 to 2.6 tons/ha) in finger millet/kodo millet-lentil cropping systems.

The presence of trees improved the soil properties by accumulating more organic C in soil. Trees in an agroforestry serve two important purposes: (1) the root system holds soil in place and reduces susceptibility to erosion, and (2) plant stems decrease the flow velocity of run-off and enhance sedimentation. Legume-based agroforestry has the capacity to support biological N fixation to enhance subsequent soil N availability and improve soil fertility and crop yields (Ramesh et al. 2010). Cultivation of trees and/or crops in salt-affected soils is known as biosaline agroforestry. In India, many tolerant species for saline soils have been tried since long, like *Prosopis juliflora*, *Salvadora persica*, *S. oleoides*, *Tamarix ericoides*, *T. troupii*, and *Salsola baryosma* successful on sites with ECe > 35 dS/m; *Tamarix articulata*, *Acacia farnesiana*, and *Parkinsonia aculeata* on sites with moderate salinity (ECe 25–35 dS/m); *Casuarina* (*glauca*, *obesa*, *equisetifolia*), *Acacia tortilis*, *A. nilotica*, *Callistemon lanceolata*, *Pongamia pinnata*, *Eucalyptus camaldulensis*, and *Albizia lebbeck* on sites with moderate salinity (ECe 15–25 dS/m); and trees like *Casuarina cunninghamiana*, *Eucalyptus tereticornis*, *Acacia catechu*, *A. ampliceps*, *A. eburnea*, *A. leucocephala*, and *Dalbergia sissoo* on sites with lower salinity (ECe 10–15 dS/m).

Conservation tillage improves soil structure and labile C pools in surface soil for different cropping systems. Introduction of a legume crop in the sequence improved C retention in surface soils with only short-term adoption of zero or minimum tillage. Recently, controlled traffic using permanent tram lines is included in the farming practices under CA in south Asia. In the IGP, planting on bed with either CT or ZT generally saves irrigation water and labour requirements without sacrificing crop yield. Despite many benefits of CA practices, the adoption rate in India is very low. Farmers prefer to practice CA partially (i.e. transplanted rice in the *Kharif* season followed by ZT wheat with residue retention) in the *Rabi* season. The reasons behind this are (1) poor germination and low crop productivity under direct seeded rice, because puddling and waterlogged condition help to reduce soil pH in alkaline soils and thus improve soil chemical health during the rice-growing season, (2) availability of rain and irrigation water for raising a good rice crop under puddled condition, and (3) less care is needed for transplanted rice in puddled soil compared with direct seeding. Some farmers even grow CT maize/jowar in the *Kharif* season (for better weed control, aeration and reduction in surface compactness/crusting) and raise wheat under CA in the *Rabi* season. Looking at these facts and due to constant efforts by several institutions, some farmers have started adopting full or complete CA (i.e. direct seeded rice followed by ZT wheat).

In drylands, the major constraints to the use of CA include insufficient amounts of residues due to water shortage and degraded nature of soil resource, competing uses of crop residues, resource-poor smallholder farmers and lack of in-depth research (Jat et al. 2012). Even then, CA holds considerable promise in the arid region, because it can control soil erosion by wind and water and reduce compaction and crusting. Due to limited production of biomass, competing uses of crop residues and shortage of firewood, farmers often find it hard to use crop residues to cover soil surface in dryland ecosystems, where only a single crop is grown in a year. With CA (soil cover with crop residues), it is sometimes possible to grow a second crop with residual soil moisture in the profile. It is better to use the chopped biomass of semi-hard woody perennial plants instead of crop residues to cover the soil surface (NAAS 2012).

17.11 Programmes for Soil and Water Conservation in India

Conservation of soil and water, which are the most important natural resources for sustainable crop production, has for long been a prime objective of the land policy in India. During the pre-independence era, *Zamindari* and *Ryotwari* systems prevailed for collecting land revenue or tax from the land users. Therefore, after independence, land reforms were brought in to abolish these exploitative systems. The evolution of the soil and water conservation programmes/policies can be broadly categorized into three phases, viz. pre-independence, post-independence and watershed development period. During the pre-independence period, soil and water conservation activities used to be a part of programmes aimed for improving dryland farming and drought mitigations. In the post-independence phase, many programmes were formulated having major focus on soil and water conservation along with creation of different institutions and policies. During the third phase, most of the soil and water conservation programmes were formulated and implemented having watershed development as their basis. The concept of watershed devolvement was adopted in multipurpose Damodar Valley Development Project. In 1970, it was adopted by the Soil Conservation Department and became a part of the national approach to improve agricultural production and alleviate poverty in rainfed regions since then (Gray and Srinidhi 2013). The Central Government formulated programmes like the Drought Prone Areas Programme (DPAP) in 1973–1974 and the Desert Development Programme (DDP) in 1977–1978 to support drought proofing in ecologically sensitive areas of the country. The brief chronology of various programmes and policies for conserving soil and water is given in Table 17.2.

17.12 Success Stories of Land Management: *Fallow Land Turns into Income Booster by Tapping Every Drop of Water—A Success Story*

Eval Village is situated at the northwestern part of Santalpur Taluka of Patan District in Gujarat having average rainfall of 400 mm, and its soil type is sandy loam. Farmers were engaged in agriculture during *Kharif* season (rainfed), and during *Rabi* season, agricultural fields were fallow due to the lack of irrigation water. The main crops growing during *Kharif* season were rainfed jowar and bajra. During *Rabi* and summer season, farmers were engaged in day-to-day labour-related works. A group of farmers in village were motivated by state officials for adopting soil and water conservation measures for enhancing the potential of land. With the help of IWMP, an earthen bund was constructed for rain water harvesting in the year 2009–2010 with the average water storage capacity of nearly 75,000 cum. After the earthen bund construction, the cropping pattern of the village was entirely changed. Farmers of this group are growing cumin and castor in the area. In terms of economic benefits, the total gross income of the group from both the crops in the subsequent 1 year was Rs. 16.20 lakhs; the net income was Rs. 9.00 lakhs by considering the average market rate of castor; the cumin seed was Rs. 650 per 20 kg and Rs. 3000 per 20 kg, respectively, during 2010–2011; and the average costs of production for cumin and castor are nearly Rs. 35,000/ha and 20,000/ha, respectively (GoI 2021). This earthen bund structure also helped in providing water for supplemental irrigation to meet water requirement of the crop and also provided residual moisture for a crop in the same field, thus supporting the food and nutrition security of the farmers. In the first year itself, the net income of the group of farmers itself exceeded the cost of construction of earthen bund. Other benefits were the significant rise in the water levels of the wells in the Eval and adjacent villages. This small intervention has directly increased the farmers' income potential manifolds just by increasing the irrigated area. With the increase in income level and cropping intensity, the farmers are planning to improve varieties of seed and follow water-efficient technology like drip and sprinkler system for better productivity.

17.13 Land Degradation Neutrality: Talk of the Day for Attaining Sustainability

Target 15.3 of the SDGs deals with land degradation neutrality (LDN) and aims at combating desertification; restoring degraded land and soil including land affected by desertification, droughts and floods; and striving to achieve a land degradation-neutral world by the year 2030. Land degradation neutrality is a state whereby the amount and quality of land resources, necessary to support ecosystem functions and services and enhance food security, remain stable or increase within specified temporal and spatial scales and ecosystems. India, being a signatory to the SDGs, has voluntarily taken the onus at the Conference of Parties (COP14) of the UNCCD in September 2019 to establish a national framework containing processes and mechanisms for achieving the land restoration target of 26 Mha by 2030. Initiatives of the Central Government such as zero-budget natural farming, doubling farmers' income and national afforestation programme, *Pradhan Mantri Krishi Sinchayee Yojana* (more crop per drop), *Swacch Bharat Yojana*, increased use of biofertilizers and reduced use of pesticides, creating a carbon sink of around 2.5–3.0 billion tons and soil health card scheme have been indicated in the New Delhi Declaration for achieving LDN (GoI 2019).

A land degradation-neutral (or zero net degradation) situation is basically one where the rate or extent of degradation (loss) is balanced by sustainable land management practices (gains). The range of activities pursued to achieve LDN fall under one of the three aspects, viz. avoid (or protect), reduce and reverse (Table 17.3). These aspects must be continuously monitored for assessing the impact of the LDN activities and obtaining feedback for midterm corrections.

17.14 Way Forward and Conclusions

India is facing widespread problems of land degradation such as soil erosion, declining soil fertility and disturbed hydrological ecological functions; hence, the focus should be on restoration of degraded lands. To achieve the 26 Mha target of restoring degraded land by 2030, efforts must be made to increase the adoption rate of science-led technologies and approaches integrating different dimensions like topographical, ecological, economic, biophysical, social and cultural aspects. For sustainable management, there is a need to improve market access and land tenure securities for smallholders including tribal farmers; strengthening of extension services with adequate financial and technical support is required for the adoption of sustainable land management measures. Farmer's awareness about the sustainable land management practices and their significance towards crop patterns and systems may be enhanced by organizing demonstrations, field days, farmer field schools, *Kisan Mela*, etc. India's commitment to restore 26 Mha of degraded land by 2030 may be achieved by enabling improved legal and policy conditions across sectors and ensuring the involvement of all the stakeholders like community leaders at local level, producers, governments at different levels and scientists and technicians in policy development. Productive potential of land maintained in the long term will require implementation of adaptive sustainable land management practices that seeks to monitor outcomes and also learn from previous ample experiences and emerging new knowledge. On the basis of these, management practices should be modified for maintaining the productive potential of land.

Table 17.3 Achieving land degradation neutrality

	Land degradation neutrality activity	
Aspect	What	How
Avoid/ protect	Conservation of biodiversity, protection of forests and wildlife	Adoption of suitable policy frameworks (e.g. National Water Policy, National Land Use Policy, National Forest Policy, National Biodiversity Action Plan, etc.) for sustainable land management, creating community awareness, bringing about behavioural changes in society, institution building
Reduce	Reduce agricultural emissions and environmental pollution	Reduce enteric fermentation
		Improve rice ecosystem management
		Improve manure/compost management
		Increase nitrogen use efficiency
		Minimize liquid discharge from industries, waste water treatment, say no to single use plastics
	Increase crop production while maintaining soil quality	Crop rotation, Site-specific nutrient (Soil Health Card) and pest management
		Advocating the use of compost, biofertilizers, organic pesticides
		Promote zero-budget natural farming
		Judicious water management – micro-irrigation, more crop per drop
	Reduce soil erosion	Watershed-based land management (PMKSY), prioritization to rainfed areas and fragile ecosystems
		Maintaining adequate and appropriate soil cover
		Adopting in situ soil moisture conservation practices in arable (e.g. mulches, bunds, ridges and furrows) and nonarable (e.g. trenches) land
		Drainage line treatment to arrest scouring and sedimentation
		Water harvesting, groundwater management
Restore	Land reclamation and decontamination	Amelioration of problem soils (e.g. liming, drainage, gypsum application), bio-drainage, phytoremediation, mine spoil stabilization
	Carbon-positive budgeting in agricultural land	Minimize C losses through leaching, erosion and decomposition
		Maximize C gains through composting, crop residue management, root biomass
	Increase vegetative cover	Afforestation/reforestation, adopting agroforestry systems, pasture development
	Land use planning	Based on capability and resilience of soils and its suitability of different uses, viz. agriculture, horticulture, grassland, forestry
Monitor	Appraisal of planning, implementation and achievements	Effective use of geospatial technology, being proactive in the implementation of plans, obtaining regular feedback for midterm corrections

References

Acharyya SK, Shah BA (2004) Risk of arsenic contamination in groundwater affecting the Ganga alluvial plain. Environ Health Perspect 112:19–20

Aggarwal GC, Sidhu AS, Sekhon NK, Sandhu KS, Sur HS (1995) Puddling and N management effects on crop response in a rice–wheat cropping system. Soil Tillage Res 36:129–139

Appu PS (1996) Land reforms in India. Vikas Publishing House, New Delhi

Bali JS (1985) Problems of gullied and ravinous area - policy, programmes and progress. In: Proceedings National Seminar on Soil Conservation and Watershed Management. New Delhi, September 17–18

Bandyopadhyay AK (1989) Effect of lime, superphosphate, powdered oystershell, rock phosphate and submergence on soil properties and crop growth in coastal saline acid sulphate soils of sundarbans. In: Proceedings of the international symposium on rice production on acid soil of the tropics

Bandyopadhyay BK, Rao DLN (2001) Integrated nutrient management in saline soils. J Indian Soc Coastal Agric Res 19:35–58

Bhat JA, Kundu MC, Hazra GC, Santra GH, Mandal B (2010) Rehabilitating acid soils for increasing crop productivity through low-cost liming material. Sci Total Environ 408:4346–4353

Bhatia BM (1970) India's food problem and policy since independence. Somaiya Publications, the University of Michigan, Ann Arbor

Bhatia BM (1985) Famines in India: a study in some aspects of the economic history of India with special reference to food problem. Konark Publishers, New Delhi, p 251

Bhattacharyya T, Pal DK, Mandal C, Velayutham M (2000) Organic carbon stock in Indian soils and their geographical distribution. Curr Sci 79:655–660

Bhattacharyya R, Ghosh BN, Mishra PK, Mandal B, Rao CS, Sarkar D, Das K, Anil KS, Lalitha M, Hati KM, Franzluebbers AJ (2015) Soil degradation in India: challenges and potential solutions. Sustainability 7:3528–3570

Bhaumik SK (1993) Tenancy relations and agrarian development: a study of West Bengal. Sage Publications, New Delhi

Biswas B, Ghosh DC, Dasgupta MK, Trivedi N, Timsina J, Dobermann A (2006) Integrated assessment of cropping systems in the Eastern Indo-Gangetic plain. Field Crops Res 99:35–47

Blaikie P (1985) The political economy of soil erosion in developing countries. Longman, London, p 188

Cogo NP, Levien R (2002) Erosion and productivity, human life. In: Lal R (ed) Encyclopedia of soil science. CRC Press, Boca Raton, pp 428–431

CSWCR & TI (2011) Vision, 2030; vision document of the central soil and water conservation research and training institute. Allied Publisher, Dehradun, pp 1–46

Dhruvanarayan VVN, Ram B (1983) Estimation of soil erosion in India. J Irrig Drain Eng 109:419–434

DiMaRF (2005) Disaster management and research foundation; Government of India, Kerala

FAO (1985) Soil Conservation and Management in Developing Countries. FAO Soils Bulletin No. 33, Rome, p 208

FAO (2010) "Climate-smart" agriculture. Policies, practices and financing for food security, adaptation and mitigation. Rome, p 41

Gadgil M, Guha R (1992) This fissured land: an ecological history of India. Oxford University Press, New Delhi

Gautam NC, Narayan LRA (1988) Wastelands in India. Pink Publishing House, Mathura, p 96

Ghosh BN (2009) Vegetative barriers for erosion control in western Himalayan region. CSWCR & TI, Dehradun, pp1-8.

GoI (2019) PM's address at the 14th conference of parties of the un convention to combat desertification. https://pib.gov.in/PressReleasePage.aspx?PRID=1584534

GoI (2021) View best practices/innovations/case studies. Department of land Resources, Ministry of rural Development, Government of India. https://iwmpmis.nic.in/mainPage.jsp?requestAction=UserUploadDetailsReportofSLNA
Gray E, Srinidhi A (2013) Watershed development in India: economic valuation and adaptation considerations. World Resources Institute, Washington
Hegde R, Kumar R, Kumar A, Srinivas R (2008) Sand extraction from agricultural fields around Bangalore: ecological disaster or an economic boon. Curr Sci 95:243–247
Hobbs P, Sayre K, Gupta R (2008) The role of conservation agriculture in sustainable agriculture. Philos Trans R Soc Lond Ser B 363:543–555
Jat RA, Wani SP, Sahrawat KL (2012) Conservation agriculture in the semi-arid tropics: prospects and problems. Adv Agron 117:191–273
Joy KJ, Parnjpe S, Shah A, Badigar S, Lele S (2005) Scaling up of watershed development projects in India: learning from the first generation projects. In: Proceedings of the fourth IWMI-tata annual partners meet. International Water Management Institute, Anand, pp 133–134
Kumar VS, Pathak KC, Pednekar P, Raju NSN, Gowthaman R (2006) Coastal processes along the Indian coastline. Curr Sci 91:530–535
Mandal B, Majumder B, Bandopadhyay PK, Hazra GC, Gangopadhyay A, Samantaroy RN, Misra AK, Chowdhuri J, Saha MN, Kundu S (2007) The potential of cropping systems and soil amendments for carbon sequestration in soils under long-term experiments in subtropical India. Global Change Biol 13:357–369
MNRE (2009) Annual report of the Ministry of New and Renewable Energy. Government of India, New Delhi
MoA (1978) Draft report on state of land degradation. Government of India, New Delhi
Mythili G (2003) Land degradation due to agriculture. Envisage Newslett Environ Econ Node Madras Sch Econ 2:1–7
Mythili M (2013) Intensive agriculture and its impact on land degradation. http://coe.mse.ac.in/pdfs/coebreifs/Mythili.pdf
NAAS (2010) Degraded and wastelands of India-status and spatial distribution. New Delhi, ICAR, p 158
NAAS (2012) Management of crop residues in the context of conservation agriculture. Policy Paper No. 58. ICAR, New Delhi, p 12
NCA (1976) Report of the National Commission on Agriculture, part V, IX and abridged. Ministry of Agriculture and Irrigation, New Delhi
NWDB (1987) Description, classification, identification and monitoring of wastelands. New Delhi, National Wastelands Development Board, Ministry of Environment and Forests. Government of India
Oldeman LR (ed) (1988) Global assessment of soil degradation (GLASOD). Guidelines for general assessment of status of human-induced soil degradation. ISRIC, Wageningen, p 12
Prasad RN, Biswas PP (2000) Soil resources of India. In: Singh GB, Sharma BR (eds) 50 years of natural resource management. ICAR, New Delhi
Ramesh P, Panwar NR, Singh AB, Ramana S, Yadav SK, Shrivastava R (2010) Status of organic farming in India. Curr Sci 98:1190–1194
Rangarajan M (2015) Nature and nation: essay on environmental history. Ashoka University History Series
Roy PS et al (2015) Development of decadal (1985–1995–2005) land use and land cover database for India. Remote Sens 7:2401–2430
Saha MN, Saha AR, Mandal BC, Roy PK (2000) Effect of long-term Jute-R-W cropping system on crop yield and soil fertility. In: Long-term soil fertility experiments in R-W cropping systems. R-W consortium Paper Series 6, New Delhi, pp 94–104
Sahu HB, Dash S (2011) Land degradation due to mining in India and its mitigation measures. In: Proceedings of the second international conference on environmental science and technology, 26–28 February
Sehgal J, Abrol IP (1994) Soil degradation in India: status and impact. Oxford & IBH, New Delhi

Sen HS, Oosterbaan RJ (1992) Research on water management and control in the Sunderbans, India. Annual Report. ILRI, Wageningen

Sharma KD (1997) Assessing the impact of overgrazing on soil erosion in arid regions at a range of spatial scales. Human impact on erosion and sedimentation. In: Proceedings of the international symposium of the fifth scientific assembly of the international association of hydrological sciences. April–3 May 1997, Rabat, Morocco, pp 119–123

Sharma PD, Sarkar AK (2005) Managing acid soils for enhancing productivity. Technical bulletin, NRM division. ICAR, New Delhi, p 23

Srinivasarao CH, Venkateswarlu B, Lal R, Singh AK, Kundu S (2013) Sustainable management of soils of dry land ecosystems for enhancing agronomic productivity and sequestering carbon. Adv Agron 121:253–329

Srinivasarao CH, Venkateswarlu B, Lal R (2014) Long-term manuring and fertilizer effects on depletion of soil organic stocks under Pearl millet-cluster bean-castor rotation in Western India. Land Degrad Dev 25:173–183

Tandon HLS (1992) Assessment of soil nutrient depletion. In: Proceedings of the FADINAP regional seminar on fertilization and the environment. 7–11 September 1992, Chiangmai, Thailand

Teluguntla PG et al (2015) Global cropland area database (GCAD) derived from remote sensing in support of food security in the twenty-first century: current achievements and future possibilities. In: Thenkabail PS (ed) Remote sensing handbook. Volume II, Land resources monitoring, modeling, and mapping with remote sensing. CRC Press, Boca Raton, p 849

TERI (1998) Green India 2047: looking back to think ahead. Tata Energy Research Institute, New Delhi

Velayutham M et al (1998) Soil resources and their potentials in coastal areas of India. In: Proceedings of the national seminar on frontiers of research and its application in coastal agriculture. Gujarat Agricultural University, Anand

Farm Mechanization 18

Gopal U. Shinde, Sandip Mandal, P. K. Ghosh, Sushant Bhalerao, Omkar Kakade, Jyothi Motapalukula, and Anup Das

Abstract

Mechanization of agriculture plays a key role in improving production and productivity. There is a positive correlation between the level of mechanization and productivity. The farm power availability in Indian farm has reached to 2.84 kW/ha. The level of farm mechanization in India is about 55%. It is further expected to increase due to scarcity of farm labor and the need to increase farm productivity. The development of agricultural automation activities and digital technology promotion in India has become a priority of national policy agenda with the digital technology intervention being expected to enhance the productivity by 5–10%. The digital technology implementation in India has challenges to establish an infrastructural setup and development of technical skills in human resource which will be exercised with overall attention toward operational time optimization, precision of farming operations, human fatigue minimization, and efficient outcome-oriented production activities. This chapter discuss about the past activities in farm mechanization with an achievement during

G. U. Shinde (✉)
NAHEP-CAAST, DFSRDA Centre, Vasantrao Naik Marathwada Krishi Vidyapeeth, Parbhani, Maharashtra, India
e-mail: gushindevnmkv@vnmkv.org.in

S. Mandal
ICAR-Central Institute of Agricultural Engineering, Bhopal, India

P. K. Ghosh
National Institute of Biotic Stress Management, Raipur, Chhattisgarh, India

S. Bhalerao · O. Kakade · J. Motapalukula
NAHEP-CAAST, Vasantrao Naik Marathwada Krishi Vidyapeeth, Parbhani, Maharashtra, India

A. Das
ICAR Research Complex for Eastern Region, Patna, Bihar, India

P. K. Ghosh et al. (eds.), *Trajectory of 75 years of Indian Agriculture after Independence*, https://doi.org/10.1007/978-981-19-7997-2_18

pre-independence, Green Revolution era, post-Green Revolution era present status, and strategies for future development.

Keywords

Mechanization · Agricultural automation · Machinery · Pre-independence · Green revolution

18.1 Introduction

Farm mechanization is essential to replace human or animal drudgery. It is also required to increase productivity by bringing timeliness in farm operations and to increase cropping intensity for better management of high-priced inputs and reduction in the difficulties associated with various agricultural activities. When farm operations are carried out using tools, equipment, or machinery, it minimizes energy consumption. Farm mechanization increases productivity of crops as well as of the land. It also improves the way of life by eliminating the drudgery operations associated with human work in farming, increases output benefits toward the rural economic cycle by improving the economy of farmers, and increases access to input supply chains and integration in modern food systems. Moreover, it produces numerous and renewed business opportunities and achieves better livelihoods for families. Hence, mechanization can have significant contribution to the sustainable development of global food systems. Farm mechanization involves the use of engines, tractors, threshers, motors, pumps, harvesters, small tillers, and so on. In pre-independence era, agriculture in India was done by human and animal muscle power using mostly wooden tools. This was slowly shifted to small equipment powered by animals. The post-independence era witnessed the import of tractor and machinery. Gradually, Indian farms shifted to the adoption of more advanced machines and power sources (Foster and Rosenzweig 2010; Singh et al. 2014; Singh and Singh 2021).

The total agricultural land in India is 159.7 million hectare which is second in the world after the United States. The total irrigated area is 82.6 million hectares. However, the average farm size is less than 2 ha. The farm power before independence was negligible about 0.1 to 0.3 kW/ha, but after independence, it has increased to 2.84 kW/ha as reported in 2022. The country achieved a record production of about 314 million tons in 2021–2022 compared to only about 50 million tons during the 1950s. This clearly indicates there is a positive impact of farm mechanization in Indian agriculture (Kale 2015).

India faced severe food scarcity during the 1950s and was dependent on imports to feed its people. To mitigate this problem, “Green Revolution” took place in the 1960s, where several new practices like improved high-yielding hybrid seeds, chemical fertilizers, development and expansion of irrigation systems, and education to farmers were practiced. Technological breakthroughs like improved water system and increased mechanization in agricultural operations helped to shape the “Green Revolution.” As there was success of the “Green Revolution,” cereal imports in

India were generally negligible in the 1970s, except massive crop failures in 1979–1980 led to renewed imports of 2.3 million tons of food grains in 1981–1982. By the end of the twentieth century, India had achieved self-sufficiency by producing enough food, for example, 212 million tons of food grains in the year 2001–2002. In countries like India where scarcity of capital persists and supply of labor is abundant, partial mechanization will be the most suitable remedy. The factors which are creating hurdles against the mechanization of agriculture include small and fragmented holding, poverty, and ignorance among the farmers (Nikhade and Gunaki 2020).

The rapid expansion of mechanically powered equipment over the last six decades has had a significant impact on crop productivity. Mobile farm power sources (tractors, draught animals, power tillers, combine harvesters) and stationary farm power sources (diesel engines and electric pumps) had both contributed significantly to the increase in food grain productivity in India. The proportion of agricultural workers to total workers fell from 59.1% in 1991 to 54.6% in 2011 and 39.4% in 2021. The population of draught animals decreased from 78.42 million in 1971–1972 to 38.74 million in 2018–2019, resulting in a decrease in the combined share of agricultural workers and draught animals in total farm power availability from 60.8% in 1971–1972 to less than 10% in 2019–2020. As the population of working tractors increased to 7.5 million, the net sown area per tractor decreased from 487 ha $tractor^{-1}$ in 1975–1976 to 18 ha $tractor^{-1}$ in 2019–2020. For timely and precise field operations, the average farm power availability target by 2030 is 4.0 kW ha^{-1}. According to the current level of farm mechanization assessed for major cereals, pulses, oil seeds, millets, and cash crops, seedbed preparation is highly mechanized (more than 70%), while harvesting is the least mechanized (less than 32%). While about 65% of wheat field is mechanically sown, only about 20% and 30% of sugarcane and rice fields are mechanized, respectively. Wheat and rice harvesting and threshing are 60% mechanized. In recent years, laser levelers, zero-tillage equipment, seeder, and inclined plate planters have become very popular (Kishtwaria and Rana 2012; Kumar and Kutumbale 2019).

The development of agricultural mechanization in India is both fascinating and remarkable in many ways. Over the last six decades, the country has progressed from a situation in which it faced severe food shortages to one in which it is now an exporter of many food commodities and a major exporter of other industrial products, including agricultural tractors. This has been accomplished despite a more than threefold increase in population and a negligible increase in arable land area. Approximately two-thirds of the population live in rural areas, with approximately 50% still relying on agriculture for a living. While it is primarily an agrarian economy, agriculture now accounts for 14% of the economy, down from 56% in 1950. The manufacturing and service sectors currently account for 27% and 59% of the economy, respectively (Lipton 2006; Pathak et al. 2022).

Agriculture of the country is very diverse, owing to the country's varied soil and climatic conditions. Climate is full of "extremes"; temperatures range from arctic cold to equatorial hot, and rainfall ranges from extreme aridity with less than 100 mm in Western India's Thar Desert to the world's maximum "rainfall" of

11,200 mm at Mawsynram in the Meghalaya. The available rainfall varies greatly in both space and time. Despite a significant increase in irrigated area, approximately 65% remains without assured irrigation, and agricultural productivity in rain-fed areas is low (Mehta et al. 2012; Murali and Balakrishnan 2012; Bala Parameswara 2016).

With approximately 0.8 million tractors sold annually, India has emerged as the world's largest tractor producer. Farm mechanization, dominated by tractor-powered improved implements and machinery, has improved timeliness and precision in field operations; saved labor, inputs, and operation costs; and resulted in a 20% increase in yield. Threshing of major crops is now completely mechanized, and primary processing of food grains is mechanized in 68.2% of cases. India now manufactures a wide range of agricultural implements that are designed in the country's large manufacturing industry. Machines, equipment, and tools designed and developed by ICAR are also mass produced by the country's industries (Gulati and Juneja 2020).

These developments over the last few decades have contributed to the country's overall farm mechanization of crop production reaching about 50%. Precision agriculture, which makes use of artificial intelligence, machine learning, and robotics technologies, is an emerging area for automated farm mechanization and postharvest operations. This will improve precision in farming while addressing the issue of human labor scarcity.

18.2 Pre-independence and Pre-Green Revolution Era

The first-degree program in agricultural engineering was established at Allahabad Agriculture Institute in 1942; this marked the first formal training and education in the field of agricultural engineering/mechanization in India. Several tractors and bulldozers were imported in the mid-1940s, and the Central Tractor Organization and State Tractor Organizations were formed. The first Indian Institute of Technology (IIT) was established in Kharagpur in 1951, and its agricultural engineering program began in 1952, indicating its significance alongside other engineering disciplines. Tractors were not manufactured in India during this time period, and all tractors were imported (Pathak et al. 2022).

Tractor use increased from 8000 in 1950 to 20,000 in 1955 and 37,000 units by 1960 (Fig. 18.1). These were primarily used on large government and private farms. Prior to 1960, draught animals were used for most farm operations and transportation. From 1961 to 1970, India advanced to the second stage of agricultural mechanization, with tractor production beginning in 1961 with the production of 880 tractors by Eicher Tractors Ltd. During this decade, five companies were granted licenses to manufacture tractors, and production of power tillers began in 1965. This was the most important period in agricultural engineering education. The first College of Agricultural Engineering was established in 1962 at Pant Nagar with the assistance of the University of Illinois, followed by six more colleges and two degree-granting divisions/departments at institutes affiliated with the Indian Council of Agricultural Research. During this time, large farmers with land holdings of more

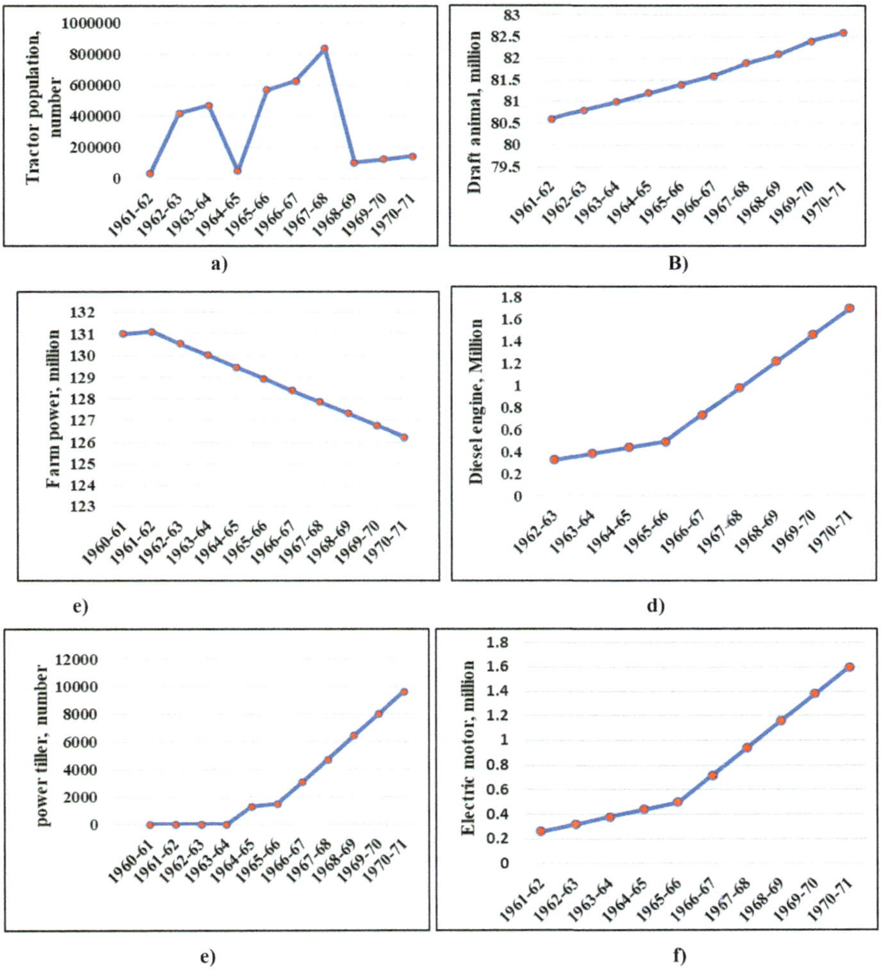

Fig. 18.1 Population of various farm power sources in the country (year-wise): (**a**) tractor population in number, (**b**) draft animal in number, (**c**) farm power in million, (**d**) diesel engine in million, (**e**) power tiller in number, and (**f**) electric motor in million

than 10 ha owned 96% of the tractors. The Green Revolution began in the late 1960s, and a large quantity of high-yielding variety of wheat and rice seeds was distributed to farmers following extensive research trials. Another feature of this time period was the development of irrigation facilities using water-lifting devices. Traditional water-lifting devices (Persian wheels) could only provide two to three irrigations per cropping cycle for wheat, whereas high-yielding varieties required six to eight irrigations. Large- and medium-sized farmers purchased diesel engines to power irrigation pumps. During this decade, rural electrification to power irrigation pumps expanded (Dixit and Bhardwaj 1990).

18.2.1 Equipment Used in Pre-independence Era

18.2.1.1 Land Preparation Equipment

(a) Wooden plow (Nagar) is used for plow the land, introduced in 1797 by Charles Newbold. (b) Kirloskar iron mold plow was introduced by Mr. Laxman Rao Kirloskar in 1903. (c) Iron mold plow for two furrow plow was invented by Keeneland in 1928 (Fig. 18.2).

18.2.1.2 Land Cultivation Equipment

Animal-drawn cultivators Doofan (two rows) and Tifan (three rows) are used for cultivating the land or for intercultural operation and also used for sowing the seed (Fig. 18.2).

18.2.1.3 Harrowing Equipment

Animal-drawn harrow or blade harrow (Bakhar or Guntaka) is generally used in clay soil for preparation of seedbeds and covering seed after sowing which is also used to control the weeds. Sickle is used for harvesting the crop and cutting other vegetation (Fig. 18.2).

18.2.1.4 Threshing

It is performed manually (Fig. 18.2).

Farm tools and equipment used in pre-independence era

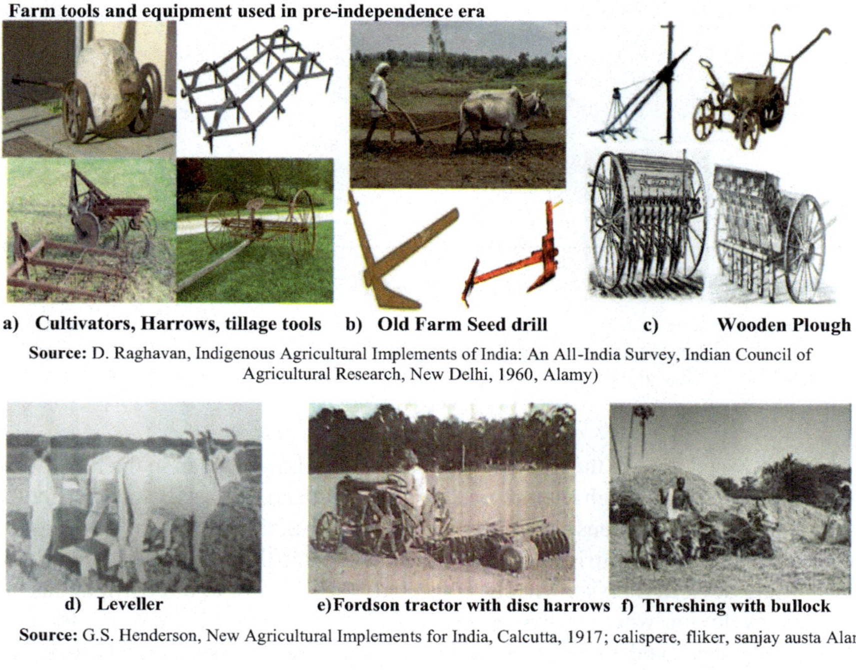

a) Cultivators, Harrows, tillage tools b) Old Farm Seed drill c) Wooden Plough

Source: D. Raghavan, Indigenous Agricultural Implements of India: An All-India Survey, Indian Council of Agricultural Research, New Delhi, 1960, Alamy)

d) Leveller e)Fordson tractor with disc harrows f) Threshing with bullock

Source: G.S. Henderson, New Agricultural Implements for India, Calcutta, 1917; calispere, fliker, sanjay austa Alamy

Fig. 18.2 Equipment used in pre-independence era

18.3 Green Revolution Era

In India, the new agricultural strategy began in 1966 with the Kharif crop. The period from the mid-1960s to the 1980s can be considered the second phase of Indian agriculture. During this time, the new agricultural strategy, also known as the Green Revolution strategy, was successfully implemented. The "new agricultural strategy" was launched as a pilot project in seven districts, known as the Intensive Agriculture District Programme. The Indian Council of Agricultural Research (ICAR), formerly known as the Imperial Council of Research, was re-established in 1965. It is now one of the world's largest national agricultural systems, with 101 ICAR institutes and 71 agricultural universities. Through its research and development during the Green Revolution, this research institution became the key to all subsequent development in farming practices. Since 1951, these studies have increased production of food grains by 5.4 times, horticulture crops by 10.1 times, fish by 15.2 times, milk by 9.7 times, and eggs by 48.1 times (Singh and Singh 2021).

The development of basic infrastructure became necessary during the wave of the Green Revolution. The Command Area Development Programme began in 1974–1975, during the fifth 5-year plan, with the launch of the irrigation project for various sizes of farmland. Diversification from food grains to nonfood grains such as poultry, fisheries, vegetables, and fruits increased in the late 1980s, accelerating agricultural GDP growth from 1980 to 1990. The area under primary food crops was 137.10 million ha in 1985–1986 and 138.61 million ha in 1990–1991, while nonfood crops were 34.53 million ha in 1985–1986 and 40.68 million ha in 1990–1991. On 12 July 1982, the National Bank for Agriculture and Rural Development (NABARD) was established on the recommendation of the Shivaraman Committee to provide a rural credit facility, seen in a new era for agriculture in India. The Agricultural Price Commission 1965, later renamed the CACP (Commission for Agricultural Costs and Prices), was the institution in charge of recommending the minimum support price based on the variable input price index. In terms of infrastructure and institutional transition, the years 1980–1990 had a significant impact on agricultural production and productivity.

To meet the demand of high farm power in the Green Revolution era, there was double-fold increase in tractors and other farm machinery sources and a gradual decrease in animate power sources (Fig. 18.3). However, there was an increase in human power due to the increase in population as well as involvement of more people in farming.

18.4 Post-independence Progress in Agricultural Mechanization

The four decades from 1961 to 2021 are very important in terms of mechanization because this is when India made significant progress in agricultural mechanization. Between 1961 and 2021, the number of power tillers increased from 9500 and to 100,000 and tractors from 146000 to 2.6 million, while the number of draught

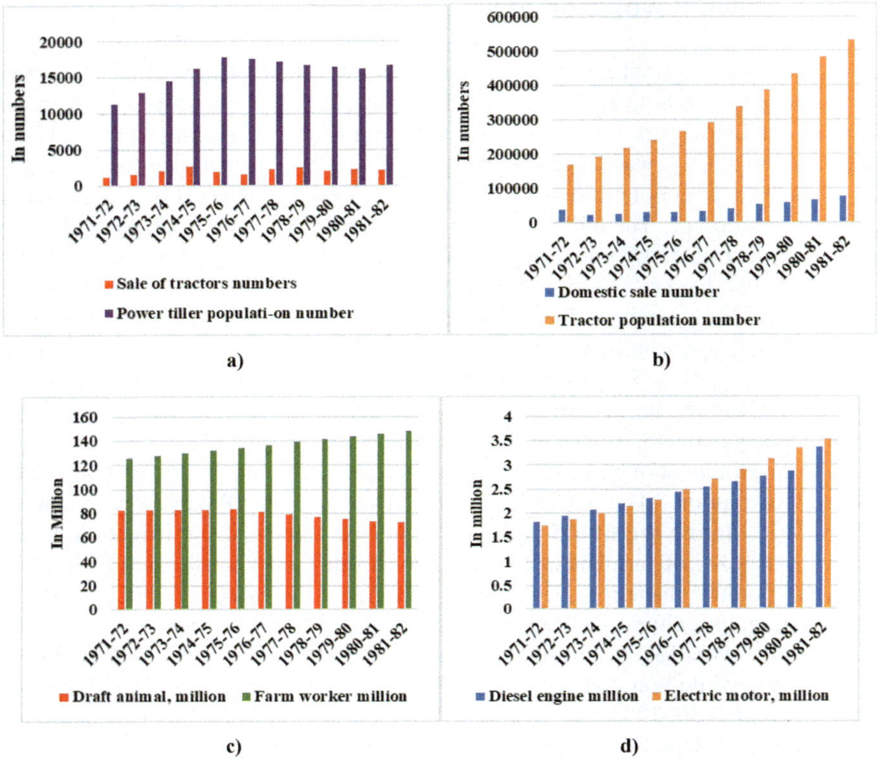

Fig. 18.3 Year-wise population of various farm power sources in the country: (**a**) sales of tractor number and power tiller population number, (**b**) domestic sale number and tractor population number, (**c**) draft animal and farm worker in million, (**d**) diesel engine and electric motor in million

animals in use decreased from 82.6 to 60.3 million, demonstrating a significant increase and shift in farm power sources in India (Fig. 18.4). The first decade (1971–1980) of these four decades was a period of growth of agriculture in India, as well as increased demand for mechanization services. Six new tractor manufacturing plants were established, while three existing plants were closed. Six new units were also licensed to manufacture power tillers, while two older units were closed. As banks opened branches in rural areas, rural credit became a reality. Farmers' access to credit increased, and the tractor market expanded rapidly. A concerted effort was launched to provide farmers with incentives through appropriate price support mechanisms. Minimum support prices for food grains and sugarcane had been declared, ensuring farm profitability. Rural electrification increased significantly, resulting in increased power availability to agriculture, particularly for

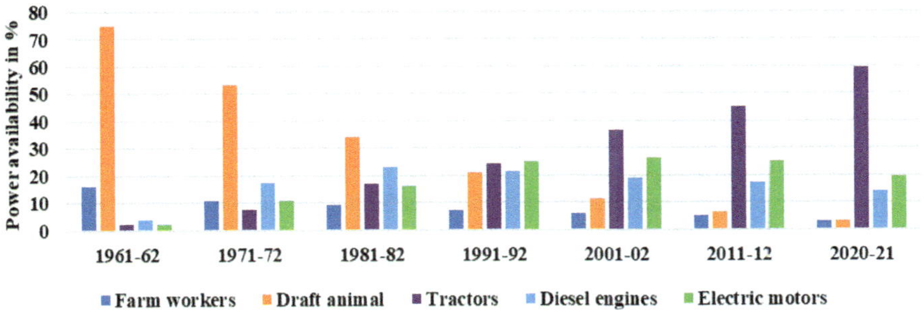

Fig. 18.4 Population of various farm power sources (million) in the country

irrigation. During this decade, a new College of Agricultural Engineering was also established. Farmers installed electric motor and diesel engine-driven irrigation pumps and purchased threshers to handle the increased volumes of produce as irrigation pump and thresher manufacturing expanded (Pathak et al. 2022).

During this time, custom hiring of threshers and pump sets increased because many farmers could not afford to purchase these machines. For tillage and transportation, custom hiring of tractor implement systems and tractor trailer units grew rapidly. In fact, custom hiring accounted for roughly 60% of the tractors' annual usage. Draught animals, on the other hand, remained the primary source of power, with a peak population of 83.4 million in 1975. The second decade (1981–1990) saw a quantum leap in food production as well as a rapid increase in agricultural mechanization. Four new tractor manufacturing units were established, but four older units were closed. One new power tiller manufacturing unit started but four older units were closed. The Indian government made a concerted effort to popularize tractors and make them more affordable to farmers. There was a significant increase in agricultural engineering education capacity, with the establishment of seven new agricultural engineering colleges. The ICAR established a separate agricultural engineering division in charge of coordinating R & D projects related to agricultural mechanization and postharvest processing at its institutes and state agricultural universities. To meet the demand for custom work, medium farmers and small entrepreneur farmers purchased tractors. Tractor-powered threshers have grown in popularity. The minimum support price was raised on an annual basis, and grain bulk storage facilities were significantly expanded.

During the 1980s and 1990s, the use of power tillers nearly doubled. Agricultural mechanization had spread widely by the third decade (1991–2000), and policy decisions were made to allow mechanization to grow on its own merits and strength. The requirement for a license to manufacture tractors was abolished in 1992, and two new tractor units began production during this time. Eight new Colleges of Agricultural Engineering were established in the field of education. Custom hiring of machinery and implements became extremely popular. The majority of farmers hired threshers (powered by tractors via PTO) from entrepreneur operators who were not necessarily farmers. The use of custom-hired combine harvesters has grown in popularity.

During the fourth decade (2001–2010), India's agricultural situation improved noticeably, resulting in a significant increase in agricultural machinery manufacturing. John Deere, New Holland, and SAME are three major international manufacturers that have established plants in India. Because of mergers, Mahindra & Mahindra (M & M) and Tractors and Farm Equipment (TAFE) have grown into massive conglomerates with global operations. Three new Colleges of Agricultural Engineering were established, bringing the total number of colleges to 30 as of 2010. The annual admission capacity at the bachelor's level had reached 1200 places, 300 at the master's level, and 100 at the doctoral level. Between 2008 and 2010, the minimum support price for farm produce increased significantly. High food prices in India and around the world at the time heightened interest in energy and water-saving technologies. In North India, zero-till drilling of wheat after rice was becoming popular, owing to cost and time savings. Custom-hired combine harvesters have grown in popularity. The number of draught animals, on the other hand, was rapidly decreasing (Mani et al. 2008; Pathak et al. 2022).

Figure 18.5 represents the overall changes in the farm power sources from post-independence era. Animal and human power sources reduced drastically to whwaas mechanical and electrical power sources increased many fold. Very recently renewable source like solar pumps and electric-operated prime movers are gradually increasing owing to the rising fuel price.

18.4.1 Equipment Used in Post-independence Era

Reversible mold board plow, disc plow, subsoiler and chisel plow, and rotary plow are used for plowing the land. Tine cultivator, disc cultivator, and rotary cultivator can be used for seed bed preparation and for sowing with seeding attachment. Spike tooth harrow and disc harrow, which is used for pulverizing, smoothen the soil in seed bed preparation or control the weeds. Tractor-drawn seed drill and seed cum fertilizer drill are used for sowing the seeds. Automatic planter for tomato, semiautomatic planter for sugarcane. Power thresher operated by a prime mover such as electric motor, engine, and tractor is used for threshing. Tractor mounted combine harvester is used for harvesting the crop (Fig. 18.6).

18.5 Modern Trends of Farm Mechanization in India

During the last decade, the use of crop residues and by-products for animal feed, organic manure production, and energy generation has taken the center stage. Technological interventions and infrastructure development have reduced

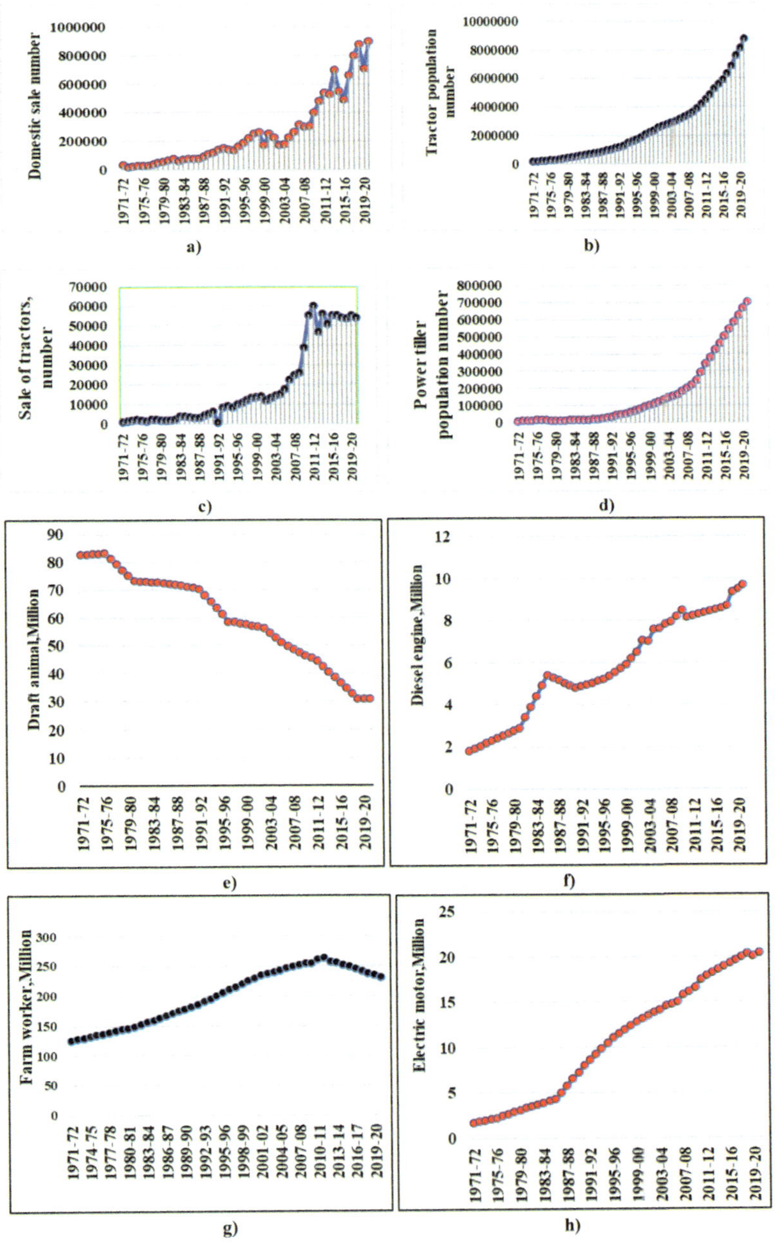

Fig. 18.5 Year-wise population of farm power sources: (**a**) domestic sale in number, (**b**) tractor population, (**c**) sale of tractors, (**d**) power tiller population, (**e**) draft animal in million, (**f**) diesel engine, (**g**) farm worker in million, (**h**) electric motor in million

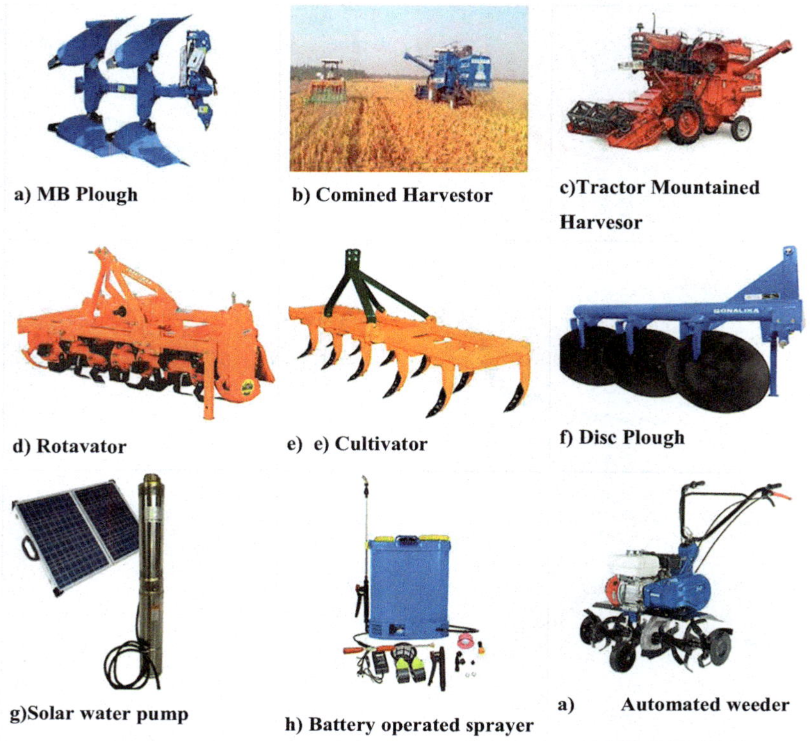

Fig. 18.6 Equipment used in post-independence era

postharvest losses by 2% (from 2005–2006 to 2013–2014), saving approximately 30 million tons of food annually. Precision machinery, equipment, and instruments are increasingly in demand. A wide range of new technologies are being developed, including automated hardware and software, autonomous ground vehicles, drones, GPS guidance, robotics, sensors and telemetric, clean and green energy, sensor-based harvesting, storage, smart packaging, and nondestructive evaluation of food safety and quality. With advances in smart technology and digitalization, the future of agricultural machinery for food production and processing is highly promising. The large size of machines and their high operating costs are major barriers to small farms transitioning from conventional farming to smart farming.

The research is now focused on the development of affordable and dependable smart farm machines for small farmers. Future technology must be portable and plug-and-play in order to be successful and sustainable (Fig. 18.7). Advanced mechanization systems will play a significant role in future agriculture sustenance

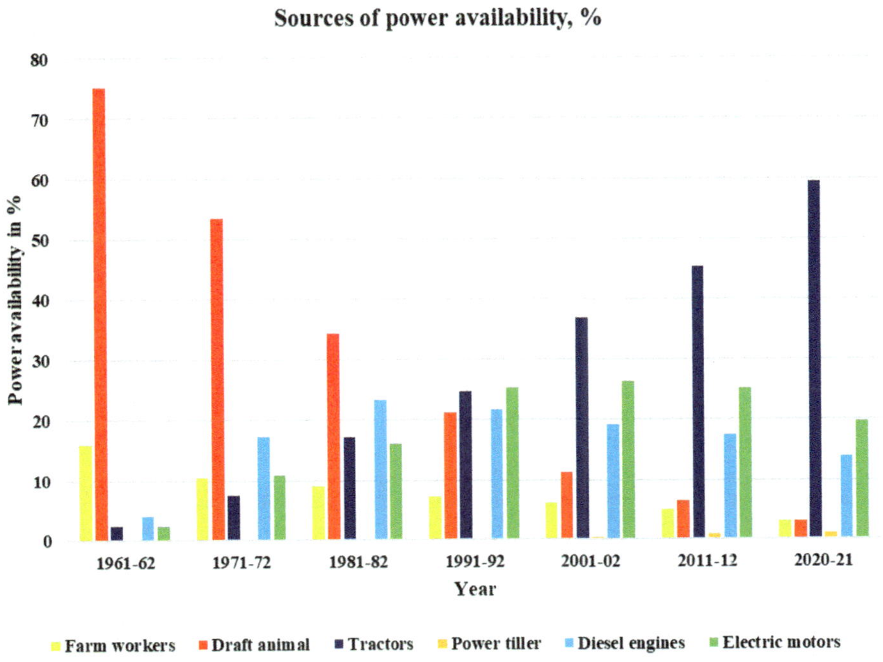

Fig. 18.7 Share of different sources of power availability (%) in the country. (Sources: Singh et al. 2014)

and postharvest processing technologies, ensuring not only food security but also nutritional security. Tables 18.1 and 18.2 presents the decadal changes in stats of farm mechanization and farm power sources in Indian agriculture.

18.6 Future Scope

18.6.1 Artificial Intelligence (AI) in Farming

Influence of automation in agriculture increased over the world since the twentieth century. There are a lot of applications of AI in farming for farmer to do more with less effort or negligible efforts. Artificial intelligence is the most disruptive technology in agriculture service as computing system interacts with different environments to maximize productivity. There are three types of robotic farming in digital technology: (1) agri-bots, (2) agri-drone, and (3) agri-AGV (automated guided vehicle) and the automation and digital solutions can be obtained by CAD/CAM/CAE soft tool, Mechatronics and sensors applications in designing and formulating by

Table 18.1 Progress of mechanization and power consumption in India

Pre-green revolution era (before 1965)	Green revolution era (1965–1975)	Post-green revolution era (1975–1990)	Post-economic reform period (1990–2000)	Digital agriculture period (2000 onward)
Farming with traditional method	HYVs, fertilizer, irrigation, chemical inputs	Use of more scientific method/ machinery/ implements/ precision	Agricultural sub-sector growth was tremendous	Use of sensors, automation switches, automatic operations, Wi-Fi, robots, drones, GIS/GPS, AGVs
Farm power availability: 0.27 kW/ha	Farm power availability was about 0.47 kW/ha	Farm power availability was about 0.48 kW/ha	Farm power availability was about 2.02 kW/ha	Farm power availability was about 2.84 kW/ha
Share of animate power sources: 98%	Share of animate power: 62%	Share of animate power: 21.7%	Share of animate power: 11.8%	Share of animate power decreased to 13%
Low productivity of food grain: 0.58 t/ha	The productivity of food grain: 0.95 t/ha	The productivity of food grain: 1.184 t/ha	The productivity of food grain: 2.11 t/ha	The productivity of food grain: 3.14 t/ha
Enhanced production through increase in cultivated area by natural resources and traditional methods	Improved production/ productivity through adoption of HYVs, fertilizer, irrigation, and chemical inputs	Improved production/ productivity through adoption of upgraded farm machine/ implements/ precision agricultural inputs	Liberalization, privatization, and globalization encourage to export more, increasing productivity	Enhanced productivity with intelligent operation handling system and human comforts with quality production

machine learning, Deep Learning, deep machine learning and Bigdata software technologies to develop an excellent logic of farm operations in machine or equipment. (ref: https://nahep.vnmkv.org.in). The future scope with participation of interdisciplinary research field experts such as a) Climate based Decision knowledge support (CDKS) b) Seedling Seed Processing Nursery automation (SSPN),c)Smart Portable Machinery(SPM) d)Food Processing Automation (FPA) and e)

Table 18.2 Comparison of farm machinery in pre- and post-independence era. (Sources: Jagbishwar 2021)

Operation	Pre-independence	Post-independence	Present status in the twenty-first century
Primary tillage	Wooden plow (Naggar), Kirloskar iron mold plow, iron mold plow	Reversible mold board plow, disc plow, subsoiler, chisel plow, rotary plow	Rotary plows and tined cultivators are mostly used tillage implements now
Secondary tillage	Doo fan (two rows), Tifan (three rows), blade harrow (Bakhar, Guntaka)	Tine cultivator, disc cultivator, rotary cultivator, tractor blade harrow	Vertical rotor rotary cultivator, self-propelled cultivators
Weeding	Khurpi, Vila (sickle)	Handle wheel hoe, peg weeder, Cono weeder	Power weeder, automatic weeder, robots
Harvesting	Sickle (plain, serrated)	Combine harvester, tractor mounted	Automated combine harvester, crop-specific harvesters
Threshing	Manually threshing or animal trampling	Power thresher, pedal-operated paddy thresher, Olpad thresher	
Irrigation	Rehat (Persian wheel)	Centrifugal pumps, piston pump, rotary pump	Sensor-operated automatic irrigation system (sprinkler, drip)
Spraying	brooms	Knapsack sprayer, engine power sprayer, airplane sprayer, spraying by drone	Automatic sprayers, drones, selective sprayers

Instrumentation Cell to develop a think-tank for agricultural automation system design (Fig. 18.8).

18.6.1.1 Agri-bots

Artificial intelligence has been directly applied in agri-bots; there are different robots that fall in this agri-bot. The following operation works have been done by agri-bots: (1) harvesting of fruit, (2) cotton picking, (3) packaging of fruit, (4) packaging of flower, (5) sorting of fruit, and (6) automatic sprayer (Fig. 18.9).

An agbot, also called an agri-bot, is an autonomous robot used in farming to help improve efficiency and reduce reliance on manual labor. Future farms are expected to be tilled, sown, tended, and harvested solely by fleets of cooperating autonomous robots called swarm robots that will weed, fertilize, and control pests and diseases, all the while collecting valuable data:

(a) A tomato-harvesting robot from Suzhou Botian Automation Technology Co., Ltd (Jiangsu, Suzhou, China).
(b) The Energid robotic citrus-picking system (Bedford, MA).
(c) An apple-harvesting robot with custom-built manipulator mounted on top of a modified crawler mobile robot.

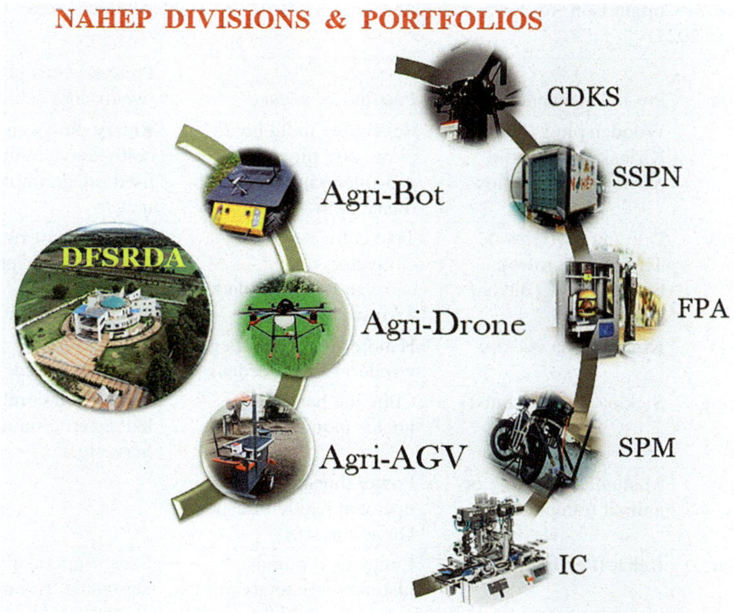

Fig. 18.8 Divisions and Portfolios for future digital farming solution activities as per NAHEP-ICAR Project

(d) A vacuum mechanism robot for apple picking from Abundant Robots (Hayward, CA, USA).
(e) An apple-catching prototype robot developed at the Washington State University. Most of these projects have used eye-in-hand look-and-move configuration in their visual servo control.
(f) A linear actuator robotic system for apple picking developed by FFRobotics (Gesher HaEts 12, Israel).
(g) The Dogtooth strawberry robot (Great Shelford, Cambridge, UK).
(h) A robotic harvester, dubbed the "vegebot," has been trained to identify and harvest iceberg lettuce, a crop which has so far resisted automation (University of Cambridge).
(i) Hortibot, a semiautonomous robot, is a navigational platform designed to have different agricultural tools fitted to it to either mechanically remove weeds or precision-spray them with herbicide.
(j) It was introduced by the Sweeper EU H2020 project consortium (www.sweeper-robot.eu) on July 4, 2018. It is an assembly of an autonomous mobile platform with Fanuc LRMate 200iD robot manipulator (Fanuc America Corporation, Rochester Hills, MI) holding an end effector and catching device for fruit harvesting.

Fig. 18.9 Different types of agri-bots

(k) A common approach in fruit detection and counting is by using a single viewpoint, as in the case of a cucumber-harvesting robot or multiple viewpoints with additional sensing.
(l) An outdoor robot based on a commercial Bosch Indigo lawn mower platform and Kinova robotic arm for automatic bush trimming and rose pruning.
(m) Spray robot developed by Holland Green Machine for smart chemical application in greenhouses.
(n) The AgBot II is an innovative agricultural robot prototype fully designed and fabricated by QUT researchers and engineers with significant co-funding from the Queensland Government. AgBot II forms part of a new generation of crop and weed management machinery, intended to work in autonomous groups across both broad acre and horticultural crop management applications. The robot's cameras, sensors, software, and other electronics enable it to navigate through a field, apply fertilizer, detect and classify weeds, and kill weeds either mechanically or chemically, providing a tool for farmers to help reduce operational costs and efficiency losses.
(o) A robotic platform equipped with drive belt operating based on the FroboMind software that can be connected to different modules and implements for automated and semiautomated mechanical weed control, precision seeding, furrow opening, and cleanings.

18.6.1.2 Agri-drone (Fig. 18.10)

An agricultural drone is an unmanned aerial vehicle used in agriculture operations, mostly in yield optimization and in monitoring crop growth and crop production. Agricultural drones provide information on crop growth stages, crop health, and soil variations. Artificial intelligence also applies in agri-drone. Spectral cameras and sensors are able to view and capture very broad or narrow bands within the electromagnetic spectrum. Combining these spectral sensors with drone technology, a new visual perspective of agriculture can be achieved.

Agri-drones are used to spray pesticides on crops and to detect pests and diseases. These are also used in the agriculture sector in Rajasthan, and an action plan has been developed for their multipurpose use in spraying farm chemicals and water-soluble fertilizers on crops. This is very precise for respective working. Types of drones are fixed-wing drone, LTA and tethered system drones, and rotary-wing drones (vnmkv.org.in).

(a) Surveying drone, an agriculture survey drone, is an unmanned aerial device equipped with sensors and digital imaging capabilities to monitor crop growth remotely. Agricultural drones provide a bird's-eye view of the farms, thus revealing issues related to irrigation, soil, and pest and fungal infestations. (b) Apple-picking drone: As the vehicle moves, the tethered drones use their arms and articulated grippers to pick ripe apples from the trees. Once picked, the apples are deposited on a tarp featuring QR codes. When deposited on the QR codes, the apples are fed into nearby crates. (c) Agri-drones can be used to spray chemicals as they have reservoirs, which can be filled with fertilizers and pesticides for spraying on crops

a). Surveying drone b). Fruit picking drone c). Spraying drone

Fig. 18.10 Different types of agri-drones

in a very little time, as compared to traditional methods. Thus, drone technology can usher in a new era for precision agriculture.

18.6.1.3 Agri-AGV (Automated Guided Vehicle) (Fig. 18.11)

Research and development in robotic harvesting date back to the 1980s, with Japan, the Netherlands, and the USA as the pioneer countries. The first studies used simple monochrome cameras for fruit detection inside the canopy. Other than the visible light RGB cameras and the ultrasonic radar sensors that are commonly used for object detection due to their affordable cost, advances in the sensing and imaging technology have led to the employment of sophisticated devices such as infrared, thermal, hyperspectral cameras, LiDAR, or combination of multi-sensors that are adopted with novel vision-based techniques for extracting spatial information from the images for fruit detection, recognition, localization, and tracking. A common approach in fruit detection and counting is by using a single viewpoint, as in the case of a cucumber-harvesting robot, or multiple viewpoints with additional sensing from one or multiple vision sensors that are not located on the robot.

(a) A little robot named Wall-Ye is trying to get involved in the process from the ground up by helping out in vineyards in France. (b) QUT has developed a prototype robotic capsicum (sweet pepper) harvester nicknamed "Harvey," combining robotic vision and automation expertise to benefit agricultural producers. The project, which ran from 2013 to 2017, was led by QUT researchers and engineers with significant co-funding from the Queensland Government Department of Agriculture and Fisheries. (c)Vine agent, a robot equipped with advanced sensors and artificial intelligence to monitor the field for plant's health assessment, was developed at the Universitat Politècnica de València. (d) Autonomous off-road mobile robot to

Fig. 18.11 AGVs used for different operations in agriculture

optimize vineyard performance, VinBot is an all-terrain autonomous mobile robot (Summit XL) with a set of sensors capable of capturing and analyzing vineyard images and 3D data by means of cloud computing applications, to determine the yield of vineyards and to share information with the wine growers. VinBot responds to a need to boost the quality of European wines by implementing precision viticulture (PV) to estimate the yield (amount of fruit per square meter of vine area: kg/m^2). Winegrowers need to be able to estimate yield accurately to perform yearly canopy management techniques and harvest vineyard areas sequentially, according to the optimal ripeness of the grape in each area, which improves wine quality. (e) GRAPE (Ground Robot for vineyard Monitoring and ProtEction) is an Echord++ experiment on agricultural and food robotics. The project is developing the onboard intelligence, ranging from autonomous navigation algorithms to perception and manipulation. The system is able to navigate autonomously over different types of terrain in vineyards. (f) An integrated multipurpose farming robotic platform for row crop weed control developed by interdisciplinary teams which is also capable of creating detailed map of the field (Redmond et al. 2018).

18.7 Future Strategies and Conclusions

The country achieved a record production of about 314 million tons in 2021–2022 compared to only about 50 million tons during the 1950s. As the population of working tractors increased to 7.5 million, the net sown area per tractor decreased from 487 ha tractor^{-1} in 1975–1976 to 18 ha tractor^{-1} in 2019–2020. For timely and precise field operations, the average farm power availability target by 2030 is 4.0 kW

ha^{-1}. The influence of automation in agriculture increased over the world since the twentieth century. No farm machinery research/development project should be initiated without conducting a market survey to access the client needs and perceptions. Greater industry-institution collaboration by undertaking joint research projects and use of reverse engineering would be helpful for speedy development and commercialization of new equipment. Computer-aided design (CAD) must be used for optimum design, cost reduction, and reliability. All R & D organization must have a CAD facility with latest design packages. Train R & D engineers to develop proficiency in computer-aided design. R & D engineers must ensure compatibility of their design with BIS/ISO standards, norms, and practices. Standardization of critical component to ensure quality, durability, and interchangeability is essential. Upgrade manufacturing technology to upgrade quality and reduce the cost. It is understood that a proposal is foot to establish a farm mechanization institute under the auspices of the Ministry of Agriculture and Cooperation. This institute will intensify research on different aspects of farm mechanization including techno-socioeconomic aspect with a view to develop a long-range farm mechanization policy. A draft agricultural mechanization policy has already been evolved, and it awaits approval of the government. Since bulk of tractor and farm machinery manufacturers are located in the northern states of India, it might be desirable to locate such an apex institution in the Punjab, as this state in spite of being one of the most mechanized states in the country has just one ICAR institute, whereas her neighboring states have two to three ICAR central institutes.

References

Bala Parameswara B (2016) Abolition of Zamindari system and its impact on agriculture. Imp J Interdiscip Res 2(4):9–14

Dixit VK, Bhardwaj JL (1990) The impact of tractorisation on farm employment in Raipur district of Madhya Pradesh. Agric Situat India 45(4):233–237

Foster A, Rosenzweig MR (2010) Barriers to farm profitability in India: mechanization, scale and credit markets. In: Conference agriculture for development-revisited. University of California at Berkeley, Berkeley, pp 1–2

Gulati A, Juneja R (2020) Farm mechanization in Indian agriculture with focus on tractors. ZEF-Discussion Papers on Development Policy, p 297

Kale VKN (2015) Sub mission on agricultural mechanization (SMAM) –a new initiative of Govt. of India. Agricultural Machinery Manufacturers' Meet (AMMM) –2015. Held at Hotel Le Meridien, Coimbatore during July 17–18

Kaur S (2017) Different types of agricultural implements and water lifting devices used in British India. Int Res J Manag Sci Technol 8(5). ISSN 2250-1959

Kishtwaria J, Rana A (2012) Intervention of gender friendly land preparation technologies for drudgery reduction of hill farm women. Work 41(1):4342–4348

Kumar V, Kutumbale V (2019) Perspectives of farm mechanization in India–a review paper. J Gujarat Res Soc 21(x):0374–8588

Lipton M (2006) Can small farmers survive, prosper, or be the key channel to cut mass poverty? eJADE 3:58

Mani I, Panwar JS, Adlakha SK (2008) Custom-hiring of agricultural machine - a major drift. In: Proceedings of the 42nd ISAE convention held at CIAE, Bhopal, 1–3 February 2008

Mehta M, Gandhi S, Dilbaghi M (2012) Intervention of drudgery reducing technologies in agriculture and impact evaluation. Work 41(1):5003–5500
Murali P, Balakrishnan R (2012) Labour scarcity and selective mechanisation of sugarcane agriculture in Tamil Nadu, India. Sugar Tech 14(3):223–228
Nikhade SR, Gunaki AS (2020) Agricultural mechanization in India. Economics 8(5):2320–28820
Pathak H, Mishra J, Mohapatra T (2022) Indian agriculture after Independence. Indian Council of Agricultural Research, New Delhi, p 426
Sahay J (2021) Elements of agricultural engineering. Sri Hari Publication
Singh SP, Singh S (2021) Farm power availability and its perspective in Indian agriculture. RASSA J Sci Soc 3(2):114–126
Singh S, Singh RS, Singh SP (2014) Farm power availability on Indian farms. Agric Eng Today 38 (4):44–52

Forestry and Agroforestry 19

Sunil Kumar, Inder Dev, and Naresh Kumar

Abstract

With the global population projected to exceed about nine billion people by 2050, the agricultural output must expand by an estimated 60% to meet the global food requirement. Forests are critical for the survival of forest dwellers, which include many indigenous peoples, and they help in delivering water to agricultural lands and protect the catchments. Forests, trees, and agroforestry systems contribute to food security and nutrition in many ways. Globally, forests cover nearly 4.06 billion ha land which is around 31% of the total land area. Globally, forestry sector has employed 12.5 million people, out of which India accounted for 6.23 million. India's forests are home to hundreds of millions of people, and forests are the main source of subsistence for them. Forests provide livelihood support to them in the form of minor forest produce, water, grazing grounds, and habitat for shifting cultivation. Globally, forestry sector has employed 12.5 million people, out of which India accounted for 6.23 million. The total forest cover of India in 2001 was 20.55% of the total geographical area. India's forest cover is consistently on the rise. India State of Forest Report (ISFR) 2019 showed an increase of 5188 km^2 of forest and tree cover across the country compared to the ISFR 2017. Presently, nearly 24.5% of India's total land area is under forest and tree cover and still far from the eventual target of 33%, which India has committed to raise to by 2030.

S. Kumar (✉)
ICAR-Indian Grassland and Fodder Research Institute, Jhansi, Uttar Pradesh, India

ICAR-Indian Institute of Farming System Research, Modipuram, Uttar Pradesh, India

I. Dev (✉) · N. Kumar
ICAR-Central Agroforestry Research Institute, Jhansi, Uttar Pradesh, India

P. K. Ghosh et al. (eds.), *Trajectory of 75 years of Indian Agriculture after Independence*, https://doi.org/10.1007/978-981-19-7997-2_19

Keywords

Forests · Forest diversity · Forest policy · Livelihood · Soil quality index

19.1 Introduction

Forests provide a wide range of tangible and intangible benefits at local, national, and global levels. Globally, forests cover nearly 4.06 billion hectares land which is around 31% of the total land area. These forests vary in composition and diversity and can contribute substantially to the economic development of any country. Globally, forestry sector has employed 12.5 million people, out of which India accounted for 6.23 million. Between 1700 and 1995, the period of industrialization, 13.9 million km^2 of forest or 9.3% of the world's total area was cleared for industrial uses, cultivation, pastures, and fuelwood. In 1600, approximately one-sixth of India's landmass was under cultivation, and this figure has now gone up to around 60% in 2018. In the colonial period, cultivation expanded rapidly for a variety of reasons such as higher demand of commercial crops like jute, sugar, wheat, and cotton for feeding the increasing urban population and industrial setup. In the early nineteenth century, the forests were considered unproductive, which were converted and utilized for agricultural purposes to generate revenue and enhance the income of the state. So between 1880 and 1920, cultivated area jumped up by 6.7 million hectares. Expansion of railway network in India was also one of the reasons for clearing the forests at a larger scale. As the railway networks spread through India and to fulfill the high demand of wooden sleepers to construct the railway tracks, a larger and larger number of trees were cut. Large areas of natural forests were also cleared to raise tea, coffee, and rubber plantations to meet Europe's growing need for these products.

Post-independence forest policies contributed to an expansion in agricultural production, fulfill the demands of wood-based industries for raw materials, and tightened control of forest lands through restricted access to forests and forest products. In 1952, the government nationalized the forests as well as most of the forest-based industries. Over the years, many rules and regulations were introduced by the government. The Forest Conservation Act was passed in 1980, which specified that the central permission is required to practice sustainable agroforestry in a forest area. Violations or lack of permits was made a criminal offense. These nationalization wave and laws are intended to limit deforestation, conserve biodiversity, and save wildlife. In 1988, India launched its National Forest Policy which led to a program named Joint Forest Management. By 1992, 17 states of India participated in Joint Forest Management programs, bringing about 2 million hectares of forests under protection (Forest Conservation and Utilisation 2001). India's forests are home to hundreds of millions of people, including many Scheduled Tribes, and forests are the main source of subsistence for them. In India, about 250 million people live in and around forests, of which the estimated indigenous *Adivasi* or tribal population stands nearly 100 million. Forests provide livelihood support to them in the form of minor forest produce, water, grazing grounds, and habitat for shifting cultivation.

India is one of the ten most forest-rich countries of the world, and it accounts for 2% of the total global forest area. Globally, forestry sector has employed 12.5 million people, out of which India accounted for 6.23 million. The total forest cover of India in 2001 was 20.55% of the total geographical area. India's forest cover is consistently on the rise. India State of Forest Report (2011) showed an increase of 5188 km^2 of forest and tree cover across the country compared to the ISFR 2017. Presently, nearly 24.5% of India's total land area is under forest and tree cover and still far from the eventual target of 33%, which India has committed to raise to by 2030.

19.2 Status of Forest Resources in India

19.2.1 Pre-independence History of Forests in India

Forests were revered by the people and many religious ceremonies centered on trees and plants. In *Agni Purana*, written about 4000 years ago, it is mentioned that man should protect trees to have material gains and religious blessings. Around 2500 years ago, Gautama Buddha preached that man should plant a tree every 5 years. Sacred groves were marked around the religious places like temples where certain rules and regulations applied. Chandragupta Maurya (around 300 BC) realized the importance of the forests and appointed a high officer to look after the forests. Ashoka orated that wild animals and forests should be preserved and protected. He launched programs to plant trees on a larger scale, and this continued even during the Gupta period.

During the early part of the British rule, large numbers of trees, namely, sal, teak, and sandalwood, were cut for export. But after some time, they realized the importance of trees and began to regulate and conserve. In 1800, a commissioner was appointed to look into the availability of teak trees in the Malabar forests. In 1806, the Madras Government appointed Capt. Watson as the commissioner of forests for organizing the production of teak and other timber tree species suitable for the building of ships. In 1855, Lord Dalhousie framed regulations for conservation of forests in the entire country. Teak plantations were raised in the Malabar hills, as well as acacia and eucalyptus in the Nilgiri Hills. In Bombay, the conservator of forest, Gibson, attempted to introduce rules prohibiting shifting cultivation and plantation of teak forests. During 1865–1894, forest reserves were established to secure material for imperial needs. From the eighteenth century, scientific forest management systems were employed to regenerate and harvest the forest to make it more sustainable. Between 1926 and 1947, afforestation programs were carried out on a larger scale in the Punjab and Uttar Pradesh.

The Indian Forest Act (IFA), 1927 was the first comprehensive act governing the forests, and it serves till date as the basis for forest administration in the country. It describes the forests as village, protected and reserve forests with a difference in their legal status, management, and rights and concessions. IFA describes in detail about forest management, role of forest administration, and forest offences. However, this Act is quite dated now. Many of the provisions of this Act do not address contemporary issues related to forest management in the country, such as people's participation. It does not reflect progressive changes in the forest policy of country.

During World War I, forest resources were severely depleted as large quantities of timber were harvested to build ships and railway sleepers and to pay for Britain's war efforts. World War II made even greater demand on the forests than World War I had done.

19.2.2 Post-Independence Forest Policy in India

The Government of India has framed many national policies and acts for the governance and management of forests. The Forest Policy of 1952 recognized the protective functions of the forest and aimed at maintaining one-third of India's land area under forest. Certain activities were banned and grazing was restricted. In 1952, the government nationalized the forests which were earlier with the *zamindars*. India also nationalized most of the forest wood and non-wood forest product-based industries. Over the years, many rules and regulations were introduced by the Indian Government. In 1980, the Conservation Act was passed to check indiscriminate deforestation and diversion of forest lands for industrial or construction works. The Act was amended in 1988 to further facilitate the prevention of forest destruction.

Forest (Conservation) Act, 1980: This legislation was enacted to control the diversion of forestlands for non-forestry purposes and to slow down deforestation. Under this legislation, the approval of the central government is required for diversion of forestlands for non-forestry purposes. The user agencies have to pay for compensatory afforestation as well as an amount equal to the net present value of the forests diverted. It has substantially brought down diversion of forests for non-forestry purposes.

The National Forest Policy of 1952 was revised in the year 1988 to give a new direction to forest management in the country; for the first time, environmental stability was considered the prime object of forest policy, and direct economic benefits were subordinated to this principal aim. This policy further emphasized the need for biodiversity conservation. One of the most important instruments in forest management in India has been the working plans, prepared at territorial division level, which prescribe the management and silvicultural practices for different forest types and are expected to be in tune with the prevailing forest policy. The policy made no changes to the goal of increasing area under forests to 33% of the total land area in the country as per the earlier policy of 1952.

The main aims of the revised forest policy of 1988 were maintenance of environmental stability; conservation of natural heritage; checking extension of sand dunes, soil erosion, and denudation; substantially increasing forest/tree cover through massive afforestation and social forestry programs; meeting requirements of fuel, wood, fodder, minor forest produce, etc.; and taking steps to create massive people's movement with involvement of women to achieve the objectives and minimize pressure on existing forests. This led to a program named Joint Forest Management, which suggested that the specific villages in association with the forest department will manage specific forest blocks. By 1992, 17 states of India participated in Joint Forest Management programs, bringing about 2 million hectares of forests under protection.

The National Forest Commission was set up in 2003 which submitted its report in March 2006. It highlighted the needs to undertake scientific research to assess the optimum forest/tree cover according to forest type and topography to meet the intended objectives.

Presently, three acts, namely, the Indian Forest Act, 1927, the Wildlife (Protection) Act, 1972 (amended in 2003), and the Forest (Conservation) Act, 1980 (amended in 1988), are the most important legal instruments for forest protection and conservation in India. State-level acts, rules, and regulations to provide for local flexibility, under this umbrella act, are also in place. The Forest (Conservation) Act restricts the de-reservation of forests or use of forest land for non-forest purposes. The Wildlife (Protection) Act provides for the protection of wild fauna and flora, for setting up of PAs, and has categorized wildlife species in six schedules with variable degrees of punishment for possessing and/or transporting them. India as a member nation of the Convention on International Trade in Endangered Species (CITES) regulates the trade in endangered species and cooperates with other members.

The Indian Parliament enacted the Biological Diversity Act in 2002 to promote conservation, sustainable use, and equitable sharing of benefits on India's biodiversity resources. It provides for the establishment of a national biodiversity authority (already set up in 2004), state biodiversity boards, and biodiversity management committees (BMCs) at the level of panchayats (village committees) and municipalities. The BMCs are required also to establish and maintain people's biodiversity registers. The Act will operate side by side with a whole range of other acts, including, in particular, those pertaining to forests, wildlife, Panchayati Raj (village governance) institutions, plant varieties and farmers' rights, and patents. There are a number of potential conflicts in the working of these various acts that need to be resolved carefully to make the Biological Diversity Act 2002 effective (Gadgil 2002). The Indian Parliament has enacted the Scheduled Tribes and Other Traditional Forest Dwellers (Recognition of Forest Rights) Act, 2006, which may have a profound effect on forests and biodiversity (Singh and Kushwaha 2008).

National Agroforestry Policy (2006) recognized that forest laws and formal institutions have undermined traditional community rights and disempowered communities. Such disempowerment has led to the forests becoming open access in nature, leading to their gradual degradation in a classic sense of "tragedy of commons" (MoEF 2006). The policy advocates recognition of traditional rights of communities to "remedy a serious historical injustice."

The Forest Rights Act (FRA) though was enacted in the year 2006, but its implementation started in the year 2008 after elaboration of the implementation guidelines and rules. It recognizes a range of individual and communal rights on forest resources including ownership and management of forest lands, which have been neglected since colonial times. It not only aims to undo the "historical injustice" to the Scheduled Tribes and other traditional forest-dwelling communities but also targets to empower the communities for the "responsibilities and authority for sustainable use, conservation of biodiversity and maintenance of ecological balance" (MoLJ 2007; MoTA 2015) (source: https://www.teriin.org/projects/green/pdf/National-Forestry.pdf).

19.3 Forest Diversity

Biodiversity is the basis for ecosystem services, which constitute the life support system for humans. MEA (2005) recognizes four categories of ecosystem services: (1) provisioning services, goods such as food, fresh water, wood, medicines, fiber, etc.; (2) regulating services, disease control, climate regulation, water purification, flood regulation, erosion control, etc.; (3) cultural services, education, recreation, etc.; and (4) supporting services, nutrient cycling, soil formation, primary productivity, etc.

India is endowed with a great variety of terrain and climate. The range of topography, temperature, and rainfall are responsible for the development of a great variety of macroclimate and microclimate and the resultant-rich biological diversity on the Indian subcontinent. The country has been divided into a number of biogeographic zones based on biota and environmental realms (Rodgers and Panwar 1988). India has nearly all the representative global ecological zones of south Asia. These are (1) tropical rainforest, (2) tropical moist deciduous forest, (3) tropical dry forest, (4) tropical shrubland, (5) tropical desert, (6) tropical mountain, (7) subtropical mountain, and (8) temperate mountain (Forest Conservation and Utilisation 2001). Among them, the largest area is covered by three ecological zones, namely, tropical shrubland, tropical dry forest, and tropical moist deciduous forest.

According to one estimate, India has 16,500 species of flowering plants, 390 species of mammals, 2546 species of fishes, 68,000 species of insects, 17,000 species of fungi and bacteria, 6500 species of algae, 2850 species of bryophytes, 1100 species of pteridophytes, 68,000 species of insects, 5000 species of mollusks, 8000 species of invertebrates, 200 species of amphibians, and 1200 species of birds. Of the reported 16,500 plant species, about 33% are endemic to India. The wet evergreen forests of India which occupy 51,249 km^2 (only 1.5% of India's land surface) hold about 7000 species of flowering plants which is a little less than 50% of the Indian angiospermic flora. The rich diversity of tropical forests can be further appreciated by the fact that the 90 km^2 area of Silent Valley in Western Ghats holds 966 flowering plant species belonging to 559 genera and 134 families (Manilal 1988). An analysis of the endemic flora of India, for example, reveals that the Himalayas including Northeast India (around 4000 endemics), Western Ghats (2000 endemics), and Andaman and Nicobar Islands (250 endemics) are the three biogeographic zones with maximum hyper-diversity areas (NBSAP unpublished). A small pocket of local endemism also occurs in the Eastern Ghats (MacKinnon and

MacKinnon 1986). The floristic diversity of wild relatives of crop plants and related genetic resources is also of a very high order, and these are again scattered in various hot spots in different biogeographic zones (Singh and Kushwaha 2008).

19.4 Forest Types in India

Champion and Seth (1968) have classified Indian forests into five groups based on physiognomy and climate. They are further categorized into 16 type groups and 221 forest types, each with its own biodiversity complement. About 65.6% of the Indian forests are tropical moist to dry deciduous, 8% are tropical wet evergreen, 4% are tropical semievergreen, 9.5% are subtropical, 7% are temperate, and 5.8% are miscellaneous types (Singh and Kushwaha 2008).

19.5 Social Forestry

The term social forestry was coined by Westoby and used in the Ninth Commonwealth Forestry Congress in 1968. It was first recognized as an important component of forestry development and meeting the rural needs in the Interim Report of the National Commission on Agriculture on social forestry during 1972. The Commission emphasized on the socioeconomic importance of social forestry for rural community and in the management of forest resources. It was realized that by taking up the different social forestry activities on the farmers' lands, village community lands, wastelands, and degraded forests close to habitations, it would be possible to meet the requirements of fuelwood, fodder, small timber, etc. of the rural communities. In our country, the concept of social forestry is also found in the preaching of Buddha about 2500 years ago. During early period of British rule, there was a high demand of timber from the forests for industrial expansion and communication requirements. Hence, more attention was given to reserve and demarcate forests for industrial needs. No significance was given to important role of trees to the local population/communities. In the monumental Report on Improvement of Indian Agriculture (1893), Voelcker observed that forests had not been preserved. His observation on keeping aside village forests for the local people was probably the first observation of importance of forests to people's economy.

19.5.1 Objectives of Social Forestry

The objectives of social forestry as defined by the NCA 1976 are as follows:

1. To fulfill the basic needs such as fuel, fodder, small timber, supplementary food, and income from surplus forest products to the rural area and replacement of cow dung

2. To provide employment opportunities in the rural areas and to increase family income considerably
3. To promote cottage industries in the rural areas
4. To organize them in their struggle for socioeconomic development and to integrate economic gains in the distribution of their benefits to the rural society
5. To provide congenial environment and preserve their cultural identity as their life related to forest
6. To indoctrinate the value of village-level self-sufficiency and self-management in the production as well as distribution of forest products with social justice
7. To form the villagers into a well-knit community and an effective functional unit of society which can shape its own destiny
8. To play an important role in the reclamation of degraded lands, conservation of soil and moisture, improvement of agricultural production, as well as prevention of environmental deterioration
9. To increase the natural beauty of the landscape and create recreational forests for the benefit of rural as well as urban population
10. To protect agricultural fields against wind speed and natural calamity;
11. to solve the food problem of the rural area to a great extent
12. To use the available land as per its carrying capacity

19.5.2 Types of Social Forest

The components of social forestry defined by the NCA 1976 include farm forestry, extension forestry, reforestation in degraded forests, and recreation forestry. Social forestry schemes can be categorized into different categories, viz., farm forestry, community forestry, extension forestry, and agroforestry.

19.5.2.1 Farm Forestry

In farm forestry, trees are grown on farmland for commercial and noncommercial purposes. Farmers are encouraged to plant trees on their farmland to meet their domestic needs such as fuelwood, fodder, small timber, etc. A practice of growing trees on farmland already existed in many areas and was the main thrust of most of India's social forestry projects. In addition to providing the basic needs of fuelwood, fodder, small timber, etc., farmers often grow trees on their farmlands as shelterbelts and windbreaks, for soil conservation, or to reclaim wastelands.

19.5.2.2 Community Forestry

The government provides planting materials/seedlings and fertilizer to the community, which is then responsible for nurturing and protecting the plants on community lands. Some communities manage these plantations in a sustainable manner for continual benefit, while others sell the mature timber for a one-time capital divestment.

A forest is governed by local communities called "village forests" or "panchayat forests," reflecting the fact that the administration and resource use of the forest occurs at the village and panchayat levels. Hamlets, villages, and communities of villages may actually administer such a forest (Sinha 2003). Such community forests are usually administered by a locally elected body, usually called the *Forest Protection Committee*, the *Village Forest Committee,* or the *Village Forest Institution.* Such committees are known as *Van Panchayats* in the Kumaon Division of the UK, *Forest cooperative societies* in Himachal Pradesh, and *Van Sanrakshan Samiti* in Andhra Pradesh. Legislation pertaining to communal forests varies from state to state, but typically the state government retains some administrative control over matters like staff appointment and penalization of offenders. Such forests typically conform to the IUCN Category VI Protected Areas, but protection may be enforced by the local communities or the government depending on local legislation.

19.5.2.3 Extension Forestry

The main objective of extension forestry is to increase the area under tree growth. It is the practice of forestry in the areas devoid of tree growth and other vegetation. It includes the following:

1. Mixed forestry: It is the practice of forestry for raising fodder grass with scattered fodder trees, fruit trees, and fuelwood trees on suitable wastelands, panchayat lands, and village community lands.
2. Shelterbelts: It is defined as a belt of trees and/or shrubs maintained for providing shelter from wind, sun, snow drift, etc.
3. Linear strip plantations: These are the plantations of fast-growing tree species on linear strips of land.
4. Rehabilitation of degraded forests: The degraded areas under forests need immediate attention for ecological restoration and for meeting the basic needs of the communities residing in and around such areas.

19.6 Agroforestry

Agroforestry is a land use systems and technology where woody perennial plants (tree, shrubs, herbs, etc.) are deliberately introduced in the same land management practices, along with the agricultural crops and/or livestock, in a spatial temporal sequence. It can be defined by diverse authors as practices which involve "the deliberate integration of trees with agricultural crops and/or livestock either simultaneously or sequentially on the same unit of land" (Nair 1993). The World Agroforestry Centre (WAC) defines the term agroforestry as "a dynamic, ecologically based natural resources management system that, through the integration of trees in farmland and rangeland, diversifies and sustains production for increased social, economic and environmental benefits for land users at all levels."

Agroforestry has several benefits over monoculture systems. Handayani and Prawito (2011) reported that it conserves more than 80% water than monoculture systems. Dahlan Ismail (2009) found that agroforestry systems recover the quality of air by the use of animal and plant waste as organic fertilizers. Nath et al. (2009)

found that agroforestry provides social benefits by working as a defensive system that ensures resource conservation. For a particular area of land, growing of forest crops with agricultural crops is more beneficial to farmers in comparison to agriculture or forest crops alone (Toky 1997). In agroforestry systems, higher productivity and growth of trees could be due to capturing of more growth resources, e.g., water, fertilizers, tillage, light, etc. (Singh et al. 1998; Verma 2008; Chauhan et al. 2012).

Agroforestry research in India is more than 100 years old. However, agroforestry was incorporated into national agricultural and forestry research agendas when the ICAR launched an All India Coordinated Research Project (AICRP) on Agroforestry with 20 centers in 1983 followed by the establishment of the National Research Centre on Agroforestry on 8 May 1988 at Jhansi, U.P. The Centre is now upgraded as the Central Agroforestry Research Institute (CAFRI) from 1 December 2014. At present, there are 37 centers of AICRP on Agroforestry located in 27 state agricultural universities (SAUs) and 9 in the Indian Council of Agricultural Research (ICAR) representing all agro-climates of the country (Handa et al. 2015). In addition to ICAR, the Indian Council of Forestry Research and Education (ICFRE), Dehradun, also conducts research on different aspects of agroforestry (ICFRE 2011). Recognizing agroforestry as a viable venture, many business corporations; limited companies such as ITC, West Coast Paper Mills Ltd., and Hindustan Paper Mills Ltd.; and other institutions initiated agroforestry research with emphasis on production of improved planting material of the fast-growing species (Dhyani et al. 2015) with an objective to meet the demand of wood-based industries. The agroforestry research through the AICRP on Agroforestry was conceptualized with the following six projects:

1. Diagnostic survey and appraisal of existing farming system and agroforestry practices including farmers' preference
2. Collection and evaluation of promising tree species, cultivars of fuel, fodder, and small timber for agroforestry interactions
3. Studies on management practices of agroforestry systems
4. Analysis of economical relation of agroforestry systems
5. Exploring the role of agroforestry in environment protection
6. Studies on postharvest technology, fishery, apiculture, lac, etc. in relation to agroforestry systems

Agroforestry is bound to play a major role in the near future, not only for its importance in food and livelihood security but also for its role in combating the environmental challenges. Agroforestry and trees outside forest will be a key issue in providing a solution to global warming and climate change, enhancing the per unit productivity of the land, and converting degraded and marginal lands into productive areas. The National Agroforestry Policy made several recommendations which will go a long way in stimulating large-scale adoption of the agroforestry by the farmers and will provide the required raw material to wood-based industries on one hand and play its role in energy and environmental security on the other. The major focus of research in the coming years will be on developing agroforestry practices and technologies for critical areas such as arid and semiarid zones and other fragile ecosystems like the Himalayan region and coastal ecosystem to sustain these areas for higher productivity and natural resource management. In agroforestry, silvicultural practices are combined with agricultural crops like legumes, along with orchard farming and livestock ranching on the same piece of land. It is a sustainable land use system that maintains or increases the total yield by combining food crop together with forest tree and livestock ranching on the same unit of land, using management practices that consider the social and cultural characteristics of the local people and the economic and ecological condition of the area.

19.7 National Forest Program (NFP)

The term national forest program (NFP) does not refer to one specific program. It actually incorporates a wide range of approaches that can contribute to the formulation, planning, and implementation of forest policy at national and sub-national levels. As one of the most important outcomes of international forest policy dialogue, this means that NFP is applicable to all countries and to all types of forests. The NFP is a country-specific process that provides a framework and guidance for country-driven implementation of sustainable forest management and forest-related contribution to sustainable development; national implementation of internationally agreed commitments, such as the UN Non-legally Binding Instrument (NLBI) also known as the "Forest Instrument" and UNFCCC decisions on REDD+; and international initiatives such as those related to FLEGT; multilateral and bilateral collaboration, with NFPs being used as a common frame of reference for forest-related international cooperation by the world's major organizations and for most bilateral donors (https://en.wikipedia.org/wiki/National_forest_program).

19.8 Rehabilitation of Forest Dwellers

Scheduled Tribes and Other Traditional Forest Dwellers (Recognition of Forest Rights) Act, 2006

The Forest Rights Act (FRA), 2006 recognizes the rights of the forest-dwelling tribal communities and other traditional forest dwellers to forest resources, on which these communities were dependent on a variety of needs, including livelihood, habitation, and other sociocultural needs. The forest management policies, including acts, rules, and forest policies of participatory forest management policies in both colonial and post-colonial India, did not, till the enactment of this Act, recognize the symbiotic relationship of the STs with the forests, reflected in their dependence on the forest as well as in their traditional wisdom regarding conservation of the forests.

The Act encompasses rights of self-cultivation and habitation which are usually regarded as individual rights; community rights as grazing, fishing, and access to water bodies in forests; habitat rights for PVTGs; traditional seasonal resource access of nomadic and pastoral community; access to biodiversity; community right to intellectual property and traditional knowledge; and recognition of traditional customary rights and right to protect, regenerate, or conserve or manage any community forest resource for sustainable use. It also provides rights to allocation of forest land for developmental purposes to fulfill basic infrastructural needs of the community. In conjunction with the Right to Fair Compensation and Transparency in Land Acquisition, Rehabilitation and Settlement Act, 2013, the FRA protects the tribal population from eviction without rehabilitation and settlement.

The Act further enjoins upon the Gram Sabha and rights holders the responsibility of conservation and protection of biodiversity, wildlife, forests, adjoining catchment areas, water sources, and other ecologically sensitive areas as well as to stop any destructive practices affecting these resources or cultural and natural heritage of the tribes. The Gram Sabha is also a highly empowered body under the Act, enabling the tribal population to have a decisive say in the determination of local policies and schemes impacting them.

Thus, the Act empowers the forest dwellers to access and use the forest resources in the manner that they were traditionally accustomed and to protect, conserve, and manage forests, protects forest dwellers from unlawful evictions, and also provides for basic development facilities for the community of forest dwellers to access facilities of education, health, nutrition, infrastructure, etc.

The Scheduled Tribes and Other Traditional Forest Dwellers (Recognition of Forest Rights) Act, 2006 is a key piece of forest legislation passed in India on 18 December 2006. It has also been called the Forest Rights Act, the Tribal Rights Act, the Tribal Bill, and the Tribal Land Act. The law concerns the rights of forest-dwelling communities to land and other resources, denied to them over decades as a result of the continuance of colonial forest laws in India.

Supporters of the Act claim that it will redress the "historical injustice" committed against forest dwellers while including provisions for making conservation more effective and more transparent. The demand for the law has seen massive national demonstrations involving hundreds of thousands of people (Forest Survey of India 2011).

However, the law has also been the subject of considerable controversy in India. Opponents of the law claim it will lead to massive forest destruction and should be repealed. A little over 1 year after it was passed, the Act was notified into force on 31 December 2007. On 1 January 2008, this was followed by the notification of the rules framed by the Ministry of Tribal Affairs to supplement the procedural aspects of the Act.

19.8.1 Bamboo Resources

Bamboos, known as green gold and poor man's timber, are very important forest resources found in the forests as well as the non-forest areas in the country. They can grow in extreme diverse ranges of soil conditions varying from organically poor to mineral rich and moisture level from drought to flooding which makes them effective for reclaiming degraded lands. Bamboos play an important role in carbon sequestration and biodiversity conversation. Bamboo belongs to the family Poaceae (Gramineae).

In India, the total bamboo-bearing area is estimated to be 15.69 Mha. Indian bamboo endemism is a very high in order. In the deciduous and semievergreen regions of Northeast and the tropical moist deciduous forests of North and South India, the maximum concentration of species is found. India is the second richest country of the world after China in terms of bamboo genetic resources. In India, there are 125 indigenous and 11 exotic species of bamboos belonging to 23 genera (Negi and Naithani 1994). In which nearly 90 species of bamboos are found in the Northeastern hilly states, out of which 41 are endemic to that region. In India, there are three large genera (*Bambusa, Dendrocalamus*, and *Ochlandra*) of bamboos in India with more than ten species each. Among all genera, these three alone represent about 45% of the total bamboo species in India. The most important bamboo species of the semievergreen forests of the Andamans is *Gigantochloa rostrata*. In the eastern region, i.e., West Bengal, Assam, and Northeastern Himalayas, the commercially important bamboos are *B. tulda, D. hamiltonii, and Melocanna baccifera.* There are some genera which are represented by only one species each, e.g., *Ampelocalamus*, *Sarocalamus*, *Chimonobambusa*, *Pseudostachyum*, and *Stapletonia*. In India, bamboos show a great diversity in both their habitat and habit of growth. Their distributions occur in different forest types, ranging from tropical to subalpine zones. However, some of the species are found only in the cultivated state in few gardens. With the growing demand for timber and wood for various purposes, bamboo can be used as a viable substitute for timber in the country. As per productivity and conservation are concerned, there is a need to assemble an adequate information base as a foundation for policy and management decisions. Also, the current market demand and supply gap emphasize the needs for the promotion of bamboo cultivation outside the conventional or natural growth areas (Tewari et al. 2019).

19.9 Forest and Ecosystem Services

19.9.1 Climate Regulation

Forests play a key role within the global carbon cycle, removing carbon dioxide (CO_2) from the atmosphere and converting it to wood as they grow and releasing carbon dioxide back into the atmosphere when trees are burned or decay. Thus, the forests can act as either a source or a sink for carbon, with the potential to sequester carbon and thus reduce net CO_2 emissions. Deforestation contributes 10% of global greenhouse gas emissions and represents the second largest source of annual CO_2 emissions after fossil fuel combustion.

Halting this deforestation and encouraging replanting or sustainable forestry management practices could potentially contribute over one-third of the total emission reductions that scientists say are needed by 2030. Policymakers around the world recognize the potential for forests and natural land area to combat climate change; a total of 97 countries mentioned specific plans to reduce emissions from deforestation or increase forest cover in their Paris Agreement commitments. Forests and the timber they produce sequester and store more carbon than other terrestrial ecosystems and, in line with the emphasis made in Paris, must play an important role in mitigating climate change.

19.10 Forest as Livelihood Option

India has a huge population living close to the forests with their livelihoods critically linked to the forest ecosystems. As per the records of MoEF 2006, there are around 1.73 lakh villages located in and around forests. Though there are no official census figures for the forest-dependent population in the country, different estimates put the figures from 275 million (World Bank 2006) to 350–400 million (MoEF 2010). People living in these forest-fringe villages depend upon forest for a variety of forest products and services such as collection of edible fruits, tubers, roots, and flowers for food and medicines; firewood for cooking; small timber for agricultural implements, house construction, and fencing; fodder for livestock and grazing of livestock in forest; and collection of different marketable non-timber forest products. Therefore, with such a huge population pressure, any over exploitation and unsustainable harvest practice can potentially degrade forest. It has been reported that around 40% of the poor of the country are living in these forest-fringe villages (MoEF 2006). Moreover, a significant percentage of India's tribal population resides in these regions. The forest-fringe communities not just collect these forest products for their own consumption but also for commercial sale, which fetch them some income. It is estimated that the income from sale of the different forest products for households living in and around forest constitutes 40–60% of their total income (Sadashivappa et al. 2006; Mahapatra and Kant 2005; Sills et al. 2003; Bahuguna 2000). A study (Saha and Sundriyal 2012) on the extent of NTFP use in Northeast India suggests that the tribal communities use 343 NTFPs for diverse purposes like medicinal (163 species), edible fruits (75 species), and vegetables (65 species). The dependence for firewood and house construction material is 100, and NTFPs contributed 19–32% of the total household income for the communities under study (Saha and Sundriyal 2012). Forests are not only a source of subsistence income for millions of poor households but also provide employment to poor in these hinterlands. This makes forests an important contributor to the rural economy in the forested landscapes in the country. The widespread poverty and lack of other income-generating opportunities often make these people resort to overexploitation of forest resources. The collection of firewood for sale in the market, though it is illegal, is also extensive in many parts of the forested regions in the country and constitutes the source of livelihood for 11% of the population (IPCC 2007). However, many other forest products have been sustainably harvested by local communities for many years and are a constant source of household income.

The total annual consumption of wood in constructions and furniture—in both commercial and household sectors—as well as for agricultural implements is estimated to be 48.0 million m^3 in round wood equivalent (RWE). However, the total production of timber stands at 45.95 million m^3, showing a gap of 2.05 million m^3 annually (FSI 2011). Firewood constitutes the major source of cooking energy in India, and more than 853 million people use firewood for cooking in India (FSI 2011). As per the 2011 census, 49% of the households in the country use firewood for cooking. The forest-rich states have higher incidence of firewood use for cooking. As the total annual volume of firewood use is concerned, it is estimated to be 216.421 million tons, of which 58.747 million tons (27.14%) are sourced from forests. There have been no estimates for the volume of firewood availability from forests, and the annual availability of firewood from TOF is estimated to be 19.25 million tons.

India's total fodder-consuming livestock population as per the 2007 Livestock Census is estimated to be 518.6 million. Of these, 199.6 millions of livestock depend partially or fully on forest for fodder (Forest Survey of India 2011).

19.11 Indices for Forest Quality

19.11.1 Forest Soils and Site Index

Soil quality is one of the most important factors in forest management decisions. Soils will determine which tree species yield the maximum timber volume, tree rotation age, and ultimately the investment a landowner must make to yield acceptable economic returns from forest management.

Soils vary greatly in their ability to produce merchantable volumes of pulpwood, sawtimber, veneer, poles, piling, or other wood products in a reasonable period of time. In fact, as with any other crop, the better the land, the more productive the forest. Landowners must be aware of soil factors that affect forest production before investing in forest regeneration or management.

19.11.2 Site Index (SI)

The collective influence of soil factors will determine the site index for a particular tree species on a given soil area. Site index is the total height to which dominant trees of a given species will grow on a given site at some index age.

19.11.3 Forest Canopy Density

19.11.3.1 Forest Canopy Cover

It is also known as canopy coverage or crown cover. It is defined as the proportion of the forest floor covered by the vertical projection of the tree crowns. Estimation of forest canopy cover has become an important part of forest inventories. Conventional remote-sensing methods assess the forest status based on qualitative data analysis derived from "training areas." Forest Canopy Density Model is one of the important tools to detect and estimate the canopy density over large area in a time and cost-effective way. This model is based on four indices, i.e., soil, shadow, thermal, and vegetation.

The classification scheme for the purpose of forest cover assessment is described as under:

Class	Description
Very dense forest	All lands with tree canopy density of 70% and above
Moderately dense forest	All lands with tree canopy density of 40% and more but less than 70%
Open forest	All lands with tree canopy density of 10% and more but less than 40%
Scrub	Degraded forest lands with canopy density less than 10%
Non-forest	Lands not included in any of the above classes

19.11.4 Forest Soil Quality Index

Forest soil formation has been influenced by forest vegetation and is generally characterized by deeply rooted trees, significant "litter layers" or O horizons, recycling of organic matter and nutrients including wood, and wide varieties of soil-dwelling organisms. Forest soils harbor an enormous variety of life-forms that principally derive their energy from organic matter produced by photosynthesis and shed from plant structures and from animals. The major biological components of forest soils are plant roots, microbes, and soil animals.

Forest soils are influenced by various factors such as forest vegetation, climate, parent material, and other organisms. Soil provides physical support, supplies nutrients and moisture for vegetation, and stores elements for recycling back to trees. Diverse plant materials comprising forest floors are habitats for animals and microorganisms and facilitate and buffer precipitation inputs. The soil organisms digest organic matter and mix it with mineral soil, contributing to soil structure, porosity, and availability of plant nutrients. The forest floors and soils hold more organic carbon than any other components of any other terrestrial biome. The management practices as well as natural disturbances can affect all of these soil properties (Knoepp et al. 2019).

Assessment of soil quality refers to the measurement of relative changes in soil characteristics over time brought by human management under different land use systems. It may be based either on simple visual observation or involve complex laboratory analyses of soil tests. Evaluation and measurement of soil quality were due to the fact that soil serves multiple functions in maintaining worldwide environmental quality (Doran and Parkin 1994). Soil is the foundation for nearly all land uses. Assessment of soil quality is a holistic approach to view soil in its landscape setting and how it functions within the defined ecosystem. Soil quality indicators are a composite set of measurable physical, chemical, and biological attributes which relate to functional soil processes and can be used to evaluate soil quality status, as affected by management. Soil quality is defined as the capacity of a specific kind of soil to function, within natural or managed ecosystem boundaries, to sustain plant and animal productivity, maintain or enhance water and air quality, and support human health and habitation. Soil organic matter (SOM) is often considered to be the single most important indicator of soil quality and sustainable land management. Soil bulk density varies among soils of different textures, structures, and organic matter content, but within a given soil type, it can be used to monitor degree of soil compaction and puddling. Changes in soil bulk density affect a host of other properties and processes that influence water and oxygen supply. In all forms of land use systems, whether traditional or modern, soil organic matter plays an essential role in sustaining system productivity and preventing land degradation.

19.12 Carbon Sequestration

Carbon sequestration is a process of capturing and long-term storage of atmospheric carbon dioxide to mitigate global warming and to avoid harmful impacts of climate change. It is also explained as the process of removing carbon from the atmosphere and depositing it in reservoirs known as carbon pools. It can be natural or human-induced; examples are forest biomass, wood products, soils, and the atmosphere. Carbon pools in a forest are a complex mix of live and dead organic matter and minerals. Human-induced carbon pools are geological storages of carbon dioxide. The quantity of carbon in a pool is known as carbon stock, and any change may be expelled process of carbon sequestration. The use of forest is also a financially viable technique to reduce emission from atmosphere. It could also bring significant benefits to the local communities involved and consequently helps in reducing poverty at the same time. Forestry projects can bring social, economic, and local environmental benefits to millions of people (Clyde 2012).

The forestry sector cannot only sustain its carbon but also has the potential to absorb carbon from the atmosphere. India has maintained approximately 64 Mha of forest cover for the last decade. The rate of afforestation in India is one of the highest among the tropical countries, currently estimated to be 2 Mha per annum. The annual productivity has increased from 0.7 m^3 ha^{-1} in 1985 to 1.37 m^3 ha^{-1} in 1995. Increase in annual productivity directly indicates an increase in forest biomass and hence higher carbon sequestration potential. The carbon pool for the Indian forests is estimated to be 2026.72 Mt for the year 1995. Estimates of annual carbon uptake increment suggest that our forests and plantations have been able to remove at least 0.125 Gt of CO_2 from the atmosphere in the year 1995. Assuming that the present forest cover in India will sustain itself with a marginal annual increase by 0.5 Mha in area of plantations, it is expected that forests continue to act as a net carbon sink in the future (Lal and Singh 2000).

Carbon dioxide can be injected into depleted oil and gas reservoirs and other geological features or can be injected into the deep ocean; this is known as subterranean injection.

Ever since the Kyoto Protocol, agroforestry has gained increased attention as a strategy to sequester C from both developed and developing nations. The available estimates of C stored in agroforestry range from 0.29 to 15.21 Mg C ha^{-1} per annum above ground and 30–300 Mg C ha^{-1} up to 1 m depth in the soil (Nair et al. 2010). Since the industrial revolution, atmospheric CO_2 has increased by more than 40%, from 280 ppm in 1750 to about 392 ppm in 2012 (Hutchinson et al. 2007). In 3 years from 2010 to 2012, CO_2 emissions increased at an alarming rate of 2 ppm/year or 4.4 Pg C/year. While agroforestry has been recognized as having the greatest potential for C sequestration of all the land uses analyzed in the Land Use, Land-Use Change and Forestry report of the IPCC (2000), our understanding of C sequestration in specific agroforestry practices from around the world is rudimentary at best.

The incorporation of trees or shrubs on farms or pastures can increase the amount of C sequestered compared to a monoculture field of crop plants or pasture (Sharrow and Ismail 2004; Kirby and Potvin 2007). In addition to the significant amount of C stored in aboveground biomass, agroforestry systems can also store C belowground. While most studies report aboveground C sequestration, belowground C and soil C are often not reported from agroforestry systems. The soil C pool, comprising about 2500 Gt, is one of the largest C pools and is larger than the atmospheric pool (760 Gt) (Lal and Singh 2000). The extent of soil C is dependent on a delicate balance between litter and rhizodeposition and the release of C due to decomposition and mineralization. Several other factors such as quality of C input, climate, and soil physical and chemical properties further determine the rate of decomposition and thus stabilization of soil organic C in a particular ecosystem. Since modernization of agriculture in the nineteenth century, soil carbon pool has gradually depleted because of several factors such as deforestation, intensive cropping and biomass removal, soil erosion, and unsustainable agricultural practices. Most of the decline in soil organic matter has been observed in regions under intensive crop production such as continuous row cropping or monocropping. Depletion of soil C has been documented to result in decreased productivity, poor soil physical and chemical properties, and negative secondary environmental impacts. It has been well documented that conversion of degraded agricultural soils into agroforestry systems can rebuild soil productivity (Jose and Bardhan 2012).

19.12.1 Forests as Sink: Afforestation/Reforestation/Plantation/Agroforestry

All the practices work as carbon sink, they take carbon from atmosphere, utilize it in the process of photosynthesis, and store it in the form of biomass or wood. For this process of carbon sequestration to succeed, it is essential that carbon must not return to the atmosphere from burning.

19.12.2 Wetland Restoration

Wetland soil is an important natural carbon pool or sink. Wetlands conserve 14.5% of the soil carbon found in the world. But only 6% of the world's land is composed of wetlands.

19.12.3 Oceans as Sink

Oceans absorb CO_2 from the atmosphere because the concentration of CO_2 in the atmosphere is greater than that in the oceans. This difference in partial pressure of CO_2 results in the gas being absorbed into the world's oceans (Raghuvanshi et al. 2006).

19.13 Strategies for Forest Development in India

India adopted a comprehensive National Action Plan for Climate Change (NAPCC) to address issues related to climate change in 2008 (GOI undated). It has eight missions, which cover a range of sectors and issues important for the country. The National Mission for a Green India, one of the missions, addresses forestry sector issues in this plan. The GIM aims to treat an additional forest and non-forest area of 10 million ha over a period of the next 10 years spread over two national plans starting from the 12th 5-year plan in 2012 (MoEF 2010). It aims to increase the forest or tree cover over 5 Mha of area and improve the quality of cover over another 5 Mha. GIM aims to undertake "holistic view of greening" by focusing on ecosystem restoration and biodiversity conservation rather than merely focusing on plantations (MoEF 2010). It proposes to "shift in mindset from our traditional focus of merely increasing quantity of our forests cover, towards increasing the quality of our forest cover and improving provision of ecosystem services." The mission intends to sequester an additional annual 50–60 million tons of CO_2 by the year 2020. It also aims to improve livelihood and income of three million households living in and around forests (MoEF 2010). The estimated budget of GIM for a period of 10 years is INR 460 billion (MoEF 2010). An autonomous and decentralized governance structure has been proposed to implement and manage the mission. The "revamped" Joint Forest Management Committees (JFMCs) under Gram Sabhas have been conceived as the institutions at grassroots level (MoEF 2010). Similarly, Forest Development Agencies (FDA) have been made as nodal institutions at the district and state level. But GIM has made little progress so far. MoEF has issued implementation guidelines in November 2014 with a proposed budget of INR 13,000 crores for the next 5 years (MoEF 2014). These guidelines detail on the selection of landscapes, institutional structure, monitoring and evaluation, and financial outlay for the program. Much of the proposed budget is drawn through convergence with programs like Compensatory Afforestation Fund Management and Planning Authority (CAMPA) and Mahatma Gandhi National Rural Employment Guarantee Scheme

(MNREGS). It is yet to be seen the effectiveness of this approach (Aggarwal and Chauhan 2015).

About 29 Mha of open forest is considered degraded, of which about 50% has good rootstock, and the rest of the forest either has depleted rootstock or is treeless (MoEF 2006). Areas with good rootstock can be reclaimed through assisted natural regeneration (ANR) and protection under the JFM framework and need to be prioritized as a low-cost strategy. The degraded areas can be brought under multi-purpose tree plantations under a tripartite partnership between the state forest department, JFM committees, and wood-based industries. Similarly, even after accounting for scrubland, encroachments, disputed land titles, and inherent poor productivity, about 11 Mha of wasteland outside traditional forests can be brought under afforestation in collaboration with the private sector. One of the main hurdles in the reclamation of degraded lands is inadequate and delayed release of funds. This hurdle can be overcome by promoting a tripartite partnership between the private sector (wood-based industries), the local community (represented by the FPCs), and the state forest department. In this model, the forest department acts as a facilitator by providing land on joint lease to the FPCs and wood-based industries for a fixed period. The continuation of the lease is subject to adherence to a preapproved technical scheme. The industry invests money to raise plantations, and the harvest is shared between the FPCs and industry according to predefined JFM guidelines. The FPCs are responsible for the day-to-day management and social fencing of the area. However, in order to meet the subsistence needs of the local communities, the industry also invests in a separate forest patch, wherein trees of importance to the local communities are planted. Presently, however, there are no enabling laws or policies to facilitate private sector participation in forestry activities. It is recommended that a policy directive from the central government be issued, enabling private sector involvement of the type suggested above, followed by framing of resolutions and guidelines by individual state governments. The guidelines should also clearly spell out the roles and responsibilities of the three stakeholders and the benefit-sharing mechanisms.

As per land use statistics (2004/05), about 27.29 Mha (13.19 Mha of cultivable wastelands, 10.72 Mha of permanent fallow lands, and 3.38 Mha of land under miscellaneous tree crops) could be potentially brought under plantation/afforestation activities.

19.13.1 Key Elements

India has a well-developed policy and legal framework for the management of forests. The key policies in this regard are:

(a) National Forest Policy 1988
(b) National Conservation Strategy and Policy Statement on Environment and Development 1992
(c) National Environment Policy 2006

The legislative framework is provided by a number of laws, the most important of which are:

(a) Indian Forest Act, 1927
(b) Wildlife (Protection) Act, 1972
(c) Forest (Conservation) Act, 1980
(d) Environment (Protection) Act, 1986
(e) Biological Diversity Act, 2002

Scheduled Tribes and Other Traditional Forest Dwellers (Recognition of Forest Rights) Act, 2006

19.14 Conclusions

Forests and trees outside forests are essential for agricultural production because they protect soil and water, maintain soil fertility, help regulate climate, provide habitat for wild pollinators and the predators of agricultural pests, and constitute a rich store of biodiversity of potential use in agriculture. The forests not only sustain its carbon but also have the potential to absorb carbon from the atmosphere. Agroforestry is bound to play a major role in the near future, not only for its importance in food and livelihood security but also for its role in combating the environmental challenges. Agroforestry and trees outside forest will be a key issue in providing a solution to global warming and climate change and enhancing the per unit productivity of the land and converting degraded and marginal lands into productive areas.

References

Aggarwal A, Chauhan S (2015) Green growth and forestry in India. TERI. The Energy and Resources Institute, New Delhi, p 12

Bahuguna VK (2000) Forests in the economy of the rural poor: an estimation of the dependency level. Ambio 29(3):126–129. https://doi.org/10.1579/0044-7447-29.3.126

Champion HG, Seth SK (1968) A revised survey of forest types of India. Govt of India Press, New Delhi, p 404

Chauhan SK, Sharma R, Sharma SC, Gupta N, Ritu. (2012) Evaluation of poplar (*Populus deltoides* Bartr. Ex Marsh.) boundary plantation based agri-silvicultural system for wheat-paddy yield and carbon storage. Int J Agroforest Silviculture 2(5):239–246

Clyde P (2012) Importance-of-carbon-trading-in-the-global-market, slide share. Available at http://www.slideshare.net/Patricia291Clyde/importance-of-carbon-trading-in-the-global-market

Dhyani SK, Handa AK, Uma. (2015) Agroforestry research in India: present status and future perspective. In: Dhyani SK, Newaj R, Alam B, Dev I (eds) Organized agroforestry research in India: present status and way forward. Boitech Book Publisher, Delhi

Doran JW, Parkin TB (1994) Defining and assessing soil quality. In: Doran JW, Coleman DC, Bezdicek DF, Stewart BA (eds) Defining soil quality for a sustainable environment. Soil Science Society of America Journal, Madison, pp 3–21. https://doi.org/10.2136/sssaspecpub35.c1

Forest Conservation and Utilisation (2001) Food and Agriculture Organisation of the United Nations. http://www.britannica.com/eb/article?eu=121168

Forest Survey of India (2011) India state of forest report. Ministry of Environment and Forests, Government of India, New Delhi, p 286

FSI (2011) Atlas Forest type of India. Forest Survey of India, Dehradun

Gadgil M (2002) India's Biological Diversity Act 2002: an act for the new millennium. J Biosci 28(2):145–147

Handa AK, Dhyani SK, Uma (2015) Three decades of agroforestry research in India: retrospection for way forward. Agric Res J 52(3):1–10

Handayani IP, Prawito P (2011) Agroforestry systems for sustaining rural development and protecting environmental quality. Prosiding Seminar Nasional Budidaya Pertanian, pp 12–23

Hutchinson SA, Cheesman SE, Hale LA, Boone JQ, Eisen JS (2007) Nkx6 proteins specify one zebrafish primary motoneuron subtype by regulating late islet1 expression. Development 134(9):1671–1677

ICFRE (2011) Agroforestry research at ICFRE. Media & Publication Division, Directorate of Extension, ICFRE, New Forest, Dehradun. No. 133 ICFRE Bk - 79/2011

India State of Forest Report (2011) Forest Survey of India. Ministry of Environment & Forests, Government of India, pp 4–5

IPCC (2000) Land use, land-use change and forestry. Cambridge University Press, Cambridge, p 375

IPCC (2007) Climate change 2007: synthesis report. In: Pachauri RK, Reisinger A (eds) Contribution of working groups I, II and III to the fourth assessment report of the Intergovernmental Panel on Climate Change. IPCC, Geneva, p 104

Ismail D (2009) Integrated production systems. Manage Agric Forest Fish Enterp 2:13–19

Jose S, Bardhan S (2012) Agroforestry for biomass production and carbon sequestration: an overview. Agrofor Syst 86:105–111. https://doi.org/10.1007/s10457-012-9573-x

Kirby KR, Potvin C (2007) Variation in carbon storage among tree species: implications for the management of a small-scale carbon sink project. For Ecol Manag 246:208–221

Knoepp JD, Adams MB, Harrison R, West L, Laseter SH, Markewitz D, Callaham MA (2019) History of forest soils knowledge and research. Dev Soil Sci 2019:43–55. https://doi.org/10.1016/b978-0-444-63998-1.00004-5

Lal M, Singh R (2000) Carbon sequestration potential of Indian forests. Environ Monit Assess 60: 315–327. https://doi.org/10.1023/A:1006139418804

MacKinnon J, MacKinnon K (1986) Review of the protected areas system in the Indo-Malayan Realm. Gland, International Union for the Conservation of Nature and Natural Resources, p 284

Mahapatra K, Kant S (2005) Tropical deforestation: a multinomial logistic model and some country-specific policy prescriptions. Forest Policy Econ 7(1):1–24. https://doi.org/10.1016/s1389-9341(03)00064-9

Manilal KS (1988) Flora of silent valley tropical rain forests of India. The Mathrubhumi Press, Calicut, p 398

MEA (2005) Ecosystems and human well-being: synthesis. Island Press, Washington, DC

MoEF (2006) National environment policy 2006. Ministry of Environment & Forests, Government of India

MoEF (2010) National Mission for a Green India (Under the National Action Plan on Climate Change). Draft submitted to Prime Minister's Council on Climate Change [Online]. Available http://moef.nic.in/downloads/public-information/GIM-Report/PMCCC.pdf

MoEF (2014) Implementation guidelines: National Mission for a Green India, Ministry of Environment & Forests, Government of India

MoLJ (2007) The Scheduled Tribes and Other Traditional Forest Dwellers (Recognition of Forest Rights) Act, 2006, Gazette of India, Ministry of Law and Justice [Online]. Available: http://tribal.nic.in/writereaddata/mainlinkFile/File1033.pdf

MoTA (2015) Status report on implementation of the Scheduled Tribes and Other Traditional Forest Dwellers (Recognition of Forest Rights) Act, 2006 [for the period ending 31.01.2015], Ministry of Tribal Affairs, Government of India

Nair PKR (1993) An introduction to agroforestry. Kluwer, Dordrecht, p 491

Nair RM, Whittall A, Hughes SJ, Craig AD, Revell DK, Miller SM, Powell T, Auricht GC (2010) Variation in coumarin content of Melilotus species grown in South Australia. N Z J Agric Res 53(3):201–213

Nath S, Das R, Chandra R, Sinha A (2009) Bamboo based agroforestry for marginal lands with special reference to productivity, market trend and economy. Agroforestry in Jharkhand, Envis Jharkhand News, pp 80–96

National Agroforestry Policy (2006) Ministry of Environment and Forests, Government of India, pp 1–58

Negi SS, Naithani HB (1994) Handbook of Indian bamboos. Oriental Enterprises, Dehradun, p 234

Raghuvanshi SP, Chandra A, Raghav AK (2006) Carbon dioxide emissions from coal based power generation in India. Energy Convers Manag 47(4):427–441. https://doi.org/10.1016/j.enconman.2005.05.007

Rodgers, Panwar (1988) Planning a wildlife protected area network in India. 2 vol. Project FO: IND/82/003, FAO, Dehradun, pp 339, 267

Sadashivappa P, Suryaprakash S, Krishna VV (2006) Participation behaviour of indigenous people in non-timber forest products extraction and marketing in the dry deciduous forests of South India. Conference on International Agricultural Research for Development, University of Bonn, October 11–13, 2006

Saha D, Sundriyal RC (2012) Utilization of non-timber forest products in humid tropics: implications for management and livelihood. Forest Policy Econ 14(1):28–40. https://doi.org/10.1016/j.forpol.2011.07.008

Sharrow SH, Ismail S (2004) Carbon and nitrogen storage in agroforests, tree plantations, and pastures in western Oregon, USA. Agrofor Syst 60:123–130

Sills E, Lele S, Holmes T, Pattanayak SK (2003) In: Sills E, Abt K (eds) Non-timber forest products in the rural household economy. Kluwer Academic Publishers, Dordrecht, pp 259–281Forests in a market economy. Forestry sciences series.

Singh J, Kushwaha SP (2008) Forest biodiversity and its conservation in India. Int For Rev 10(2): 292–304. https://doi.org/10.1505/ifor.10.2.292

Singh K, Ram P, Singh AK, Hussain A (1998) Poplar (*Populus deltoides* Bartram ex Marshall) in forest and agroforestry systems. Indian Forester 114(11):814–818

Sinha H (2003) [https://www.getelectionresult.com Forest and people: understanding the institutional governance, social identity, and people's participation in Indian forest management] Archived 2007-06-10 at the Wayback Machine, Presented at "Politics of the Commons: Articulating Development and Strengthening Local Practices", Chiang Mai, Thailand, July 11–14, 2003

Tewari S, Negi H, Kaushal R (2019) Status of bamboo in India. Int J Econ Plants 6(1):30–39. https://doi.org/10.23910/IJEP/2019.6.1.0288

Toky OP (1997) Poplar an economy booster and eco-friendly agroforestry tree. Agroforest News Lett NRC Agroforest 9:2–3

Verma A (2008) Evaluation of growth behaviour of Populus deltoides Marsh. clones under farm forestry. MSc thesis, PAU, Ludhiana

World Bank (2006) India: unlocking opportunities for forest dependent people in India. Report No. 34481 - IN, World Bank: South Asia Region. 85 pp

Climate

20

Santanu Kumar Bal, Saon Banerjee, Sarathi Saha,
Debasish Chakraborty, and M. A. Sarath Chandran

Abstract

Global food and environmental security are the key challenges before the scientific community in the present era of enhanced climate variability and dwindling resources. Agriculture depends on weather and climate from day-to-day management operations to long-term planning. Uncertainties of weather and climate are posing major threat to food security especially in the developing countries. Therefore, agrometeorology has a vital role to play in increasing agricultural production, decreasing risk/vulnerability, and also taking care of environmental issues. In the present scenario, harvesting the advantage of the benevolent weather and precautions against malevolent weather for maximizing production and minimizing losses, respectively, can be taken only if the farming community is provided with contextual weather-based advisory in time. The need of the hour is also to look at the researchable issues from agrometeorology point of view, so that the region-/district-specific agrometeorological information can be used to prepare development action plan for climate change, identify hotspots, and initiate interdisciplinary collaboration to improve the resource-use efficiency as well as enabling effective mitigation for adverse impacts.

S. K. Bal · M. A. Sarath Chandran
ICAR-Central Research Institute for Dryland Agriculture, Hyderabad, Telangana, India

S. Banerjee (✉) · S. Saha
Department of Agricultural Meteorology and Physics, BCKV, Nadia, West Bengal, India

D. Chakraborty
ICAR Research Complex for North East Region, Umiam, Meghalaya, India

P. K. Ghosh et al. (eds.), *Trajectory of 75 years of Indian Agriculture after Independence*, https://doi.org/10.1007/978-981-19-7997-2_20

Keywords

Agrometeorology · Research · Agromet advisory · Food security · Environmental security

20.1 Introduction

Agricultural meteorology is an interdisciplinary science, where the knowledge of interactions between meteorological and hydrological factors is applied in agriculture. The ultimate goal of agricultural meteorology is to extend and fully deploy knowledge of atmospheric and related processes to optimize agricultural production and hence to increase profitability, decrease risk, and feed an ever-expanding global population. The subject also helps to conserve natural resources and protect our soil, plant, and water resources. Agricultural meteorology uses the basic principles of physical and biological sciences to reduce atmospheric/edaphic stresses (Bal and Minhas 2017) to enhance the production of food, fodder, fiber, and other farm products. Models of atmospheric interactions with plants and soil, made more applicable by expanding historical databases, find increased application in risk management and climate change (Takle 2003).

In India, the study of meteorology and agrometeorology started from time immemorial. In *Upanishads*, one can find the discussion about the processes of cloud formation and rain and the seasonal cycles. Impact of good monsoonal rainfall on better food production was known to our ancestors. The famous novel *Meghdoot*, written by Kalidasa around the seventh century, depicted the date of onset of the monsoon over central India and the path of the monsoon clouds. The scientific study on this subject was started with the establishment of first meteorological observatory in Kolkata by the British East India Company in 1785.

With the development of agricultural sciences, the importance of weather and climate, as basic input or resources in agricultural planning, became well established. Every plant process related with growth, development, and yield of a crop is affected by weather. Similarly, every farm operation, such as plowing, irrigation, and manuring, is affected by weather. In present days, although lots of improvement in agronomy, genetics, field machinery, plant protection, and allied fields are observed, still the crop productivity is not sustainable. Soil health is deteriorating and climatic variability is a threat to agriculture. Hence, it is high time to relook the research activities in the field of agrometeorology. A highly trained workforce in agricultural meteorology is also needed to address future needs for global food security in a changing global climate.

20.2 Climatic Trends and Variability

The word "trend" generally refers to change or rate of change of any parameter over a period of time. Researches in the field of climatology often require the detection, estimation, and prediction of trends along with the statistical analysis. The trend may show a liner or nonlinear pattern. Simple linear regression and Mann-Kendall (M-K) test along with the Theil-Sen robust estimate are some of the common approaches to delineate the trend (NCAR 2014). On the other hand, climate variability indicates the variations in the mean state of the climate and its basic features. The time frame of these changes is usually longer than individual weather events but smaller than climate change condition. Random variability or noise does not follow any systematic pattern, whereas periodic variability maintains regular and distinct modes or patterns (Rohli and Vega 2018). Short-term fluctuations occurring for a time period, ranging from few months to 30 years, can be denoted as climate variability. In most of the cases, climate variability is caused by some natural processes. Several studies have been conducted to find out the trend and variability of Indian climate through the past years. The time frame under different studies varied, but the findings helped to understand the long-term trends.

Temperature, rainfall, and tropical cyclone intensities are some of the prime parameters. Temperature is one of the most important weather parameters influencing the behavior of other factors including general circulation over a place. A rise of 0.60, 1.0, and 0.18 °C per 100 years in the annual mean, maximum, and minimum temperatures, respectively, has been reported for the entire India during 1901–2010. North, central east, and northeastern India have experienced greater warming than Peninsular India during 1981–2010 period. Highest increasing trend has been observed for post-monsoon and winter season (Srivastava et al. 2017). India's economy is heavily dependent on monsoon rainfall. Monsoon rainfall is almost 80% of annual rainfall over most parts of India and contributes to the groundwater buildup. Guhathakurta and Revadekar (2017) mentioned that the analysis of rainfall over India is generally based on the quality-controlled district rainfall data for the period. Any long-term trend in the southwest monsoon rainfall was not observed for 1901–2010 in the country as a whole. Jharkhand, Chhattisgarh, and Kerala were noted with significant decreasing trends, whereas Gangetic West Bengal, West Uttar Pradesh, Jammu and Kashmir, Goa, Madhya Maharashtra, Rayalaseema, Coastal Andhra Pradesh, and North Interior Karnataka had increasing trends of southwest monsoon rainfall. Mallya et al. (2016) conducted an experiment to find out the epochal and decadal variation in drought characteristics over monsoon rainfall in India. They observed that monsoon had taken up the nature of turning into extreme and regionally varied. The droughts were also to be seen to become more regional. Both intensity and frequency of drought during 1971–2004 period showed an increasing trend. The tendency of drought occurrence shifted from west to east and the Indo-Gangetic Plain. Ali et al. (2019) mentioned in their study that extreme precipitation events have been showing an increasing trend for most of the river basins in the Indian subcontinent during 1966–2005 and there is strong probability of a rapid boost in the multiday flood events in the future than the single-day events. Tropical cyclone is a common phenomenon in Indian subcontinent during the transition period of two seasons. Previously, areas adjacent to the Bay of Bengal used to be frequently affected by cyclones. Western part of the country residing beside the Arabian Sea used to seldom experience cyclones. Amphan, Yaas, and Nisarga are some of the severe super cyclones generated over the Bay of Bengal and

Arabian Sea, respectively, in recent times. Investigations of trend of tropical cyclone provide guidance to manage its effects. Long-term trend analysis by Singh (2001) revealed that a decrease of six to seven disturbances and one to two cyclonic storms per hundred years was occurring over the Bay of Bengal and the Arabian Sea during southwest monsoon season. Niyas et al. (2009) observed the linear trend of tropical cyclone frequency over the North Indian Ocean as a whole, the Bay of Bengal, and the Arabian Sea during various seasons for 1891–2008 timescale. They observed a significant decreasing trend on an average but an increasing trend over the Bay of Bengal for May and November months. Greatest rate of reduction of frequency was experienced during the monsoon months.

20.3 Crop Production in Relation to Weather

Raising crops is probably one of the most holistic and oldest professions found on Earth. A good harvest ensures the stability in economy. Success in farm production relies on some direct factors like breed, soil health, nutritional status, moisture condition, and weather. Among these, weather is the most uncertain one. It is always variable and cannot be controlled by the farmers. Natural abnormalities can lead to drastic loss of farm produce. Mavi (1994) estimated that weather variability causes almost three quarters of annual losses in farm production. Solar radiation, temperature, soil moisture, relative humidity, and bright sunshine hours are few important weather elements that influence the crop life cycle during growing season (Makone et al. 2015). Studies on crop weather relationships have been considered as a lucrative field of research to the Indian scientists for decades. Establishment of agricultural meteorology division at the IMD in 1932 had widened the path of further research works in this subject area. Earlier research works oriented toward macroscale level analysis of rainfall distribution and variability, evaporation estimation, as well as microclimatic requirements of crops. Jacob first attempted the incorporation of statistical methods in the study of crop weather relationship with special reference to wheat crop grown in Punjab. He observed the beneficial effects of rains for autumn and spring crops (Ramdas and Kalamkar 1938). Crop micro meteorological studies became popular in Indian science in the 1980s, once eminent scientist J.L. Monteith joined ICRISAT. Long before that, crop weather relationship on rainfed crops in India was initiated by Prof. L.A. Ramdas in 1926. Yield response to different weather conditions was portrayed for several rainfed crops like rice, sorghum, finger millet, ground nut, castor, and sunflower under different parts of the country (Rao and Vijayalakshmi 1986; Victor et al. 1991). There is a close relationship between yield and rainfall (Kumar et al. 2019). The dry spells and rainfall deficit within crop season play vital roles in determining productivity of rainfed crops (Bal et al. 2022d). There are several studies which simulated a reduction in total crop duration (Chandran et al. 2021; Chandran et al. 2022) and yield (Rao et al. 2022) due to the increase in temperature under various emission scenarios. The impact of severe heat wave 2022 on crops and other sectors of Indian agriculture has been characterized and reported by Bal et al. (2022c). Apart from temperature and rainfall,

Table 20.1 Cool weather damage to rice

Growth stage	Types of cool weather damage to rice plant
Before nursery	Retarded cultivation
Nursery stage	Inferior germination and growth, withering and seedling rot, delay in transplanting due to freezing immediately after removing a cover on the protected nursery
Early stage: Transplanting, tillering, panicle formation	Delay in transplanting, poor rooting, discoloration of leaves, decrease in tiller number, delay in growth and formation of young panicles, reduction in size of panicles
Panicle initiation to booting	Degeneration of rachis branches, decrease in spikelets, cessation of spikelet growth, delay in heading, non-heading, browning of leaf sheath
Heading stage	Delay of heading
Flowering stage	Delay of flowering, non-fertilization, and nonflowering of lower spikelets
Ripening stage	Incomplete ripening, discoloration of unhulled rice grains, cessation of ripening due to early frost

(Adapted from Nishiyama 1984)

solar radiation influences the yield mostly. Solar energy specially ranging between 400- and 700-nm wavelengths is the source of photons initiating the photosynthetic process in plants. UV, visible, and NIR radiations impose photosynthetic, genetic, thermal, and morphological effect on crops. Table 20.1 shows the harmful effect of aberrant weather.

20.4 Livestock Management and Weather

Climate is the main factor for suitability of any form of life to a particular place on Earth. The management of weather and climate risks in agriculture has become an important issue nowadays due to climate change. Animal agriculture is a major contributor to climate change, responsible for 18% of the greenhouse gas emissions. Livestock systems based on grazing and mixed farming (in most prevalent India) is expected to be affected more by climate change. Greenhouse gas emissions by the livestock sector could be cut as much as 30% through the wider use of existing best practices and technologies. In the event of climate change, the biggest challenge will be how to balance a huge number of livestock or the productivity per head, at the same time improving the sustainability of livestock sector. Another challenge would be to improve the efficiency of proper water utilization and optimization of nutrient footprint for the livestock sector. The direct effects from air temperature (heat/cold stress), humidity, wind speed, and other climatic factors influence animal performance such as growth, milk production, wool production and reproduction, etc. Pawar et al. (2016) have reported that with the rising global temperature and lower heat tolerance of modern poultry genotypes, heat stress has emerged as a major concern causing significant economic losses to the poultry industry.

The impact of climate on livestock production has been categorized by Rotter Rotter and Van de Geijn (1999) as (a) availability of feed grain; (b) pasture and forage crop production and quality; (c) health, growth, and reproduction; and (d) disease and spread. Animal health is also affected by climate change in four ways: heat-related diseases and stress, extreme weather events, adaptation of animal production systems to new environments, and emergence or re-emergence of infectious diseases which are critically dependent on environmental and climatic conditions. Furthermore, while vulnerability to climate has hardly been documented in the context of India, experimental studies have been conducted on effects of season and climate on production performance and other physiological parameters of dairy animals. Thermal stress on Indian livestock particularly cattle and buffaloes has been reported to decrease estrus expression and conception rate (Upadhyay et al. 2007). Further, the length of service period and dry period of all dairy animals increased during drought (Maurya 2010). In addition, the hot-humid weather conditions were found to aggravate the infestation of cattle ticks like *Boophilus microplus*, *Haemaphysalis bispinosa*, and *Hyalomma anatolicum* (Basu and Bandhyopadhyay 2004).

Livestock's efficiencies depend on the quality and quantity of feed and water that animals need to survive, produce, and reproduce. About 10% of the croplands is used for producing animal feed, and other agricultural land provide crop residues used for feeding livestock. The influence of the climate on the distribution of plant variety and type is complex. The effects of climatic interaction with soil characteristics and direct effect on plants influence the distribution of various other biological components of the agroecosystem—pest, diseases, herbivorous animals, pollinators, soil microorganisms, etc. – all of which in turn influence plant communities. In production systems where animals are fed on concentrates, rising grain prices (may be driven by climate change) increase the pressure to use animals that efficiently convert grains into meat, eggs, or milk. Thus, within such systems, climate change may lead to greater use of poultry and pigs at the expense of ruminants and greater focus on the breeds that are the best converters of concentrate feed under high external input conditions.

The management strategies for livestock related to weather include:

Water supply: Animals must have access to large quantities of water during periods of high environmental temperatures. Much of the water is needed for evaporative heat loss to help them cool off. Hence, provision must be made for supply of continuous clean, fresh, and cool water to the animals.

Feeding time: Providing feed to the animals during cool period, i.e., evening or night, will improve the feed intake by the animals. Further, regularly cleaning the feeding trough and providing fresh feed will encourage the animals to take more feed.

Shade: The use of shades is an effective method to help cool animals. Shades can cut the radiant heat load from the sun by as much as 40%. The best shades have white or reflective upper surfaces. Shifting the animals to cool shaded area during the hot climatic conditions should always be practiced for the betterment of their health.

Ventilation: Increasing the ventilation or air circulation in the animal sheds will aid the animals in effective dissipation of heat. The air circulation inside the shed can be increased by keeping half side wall, i.e., open housing system, use of fan, increasing the height of the building, etc.

20.5 Water Management Toward Better Agriculture

A major part of the developed global water resources is used for food production. The estimated minimum water requirement per capita is estimated at 1200 m^3 annually of which 55 m^3 for domestic use and 1150 m^3 for food production (FAO 1994). In most countries, 60–80% of the total volume of developed water resources is used for agriculture and may reach well over 80% for countries in the arid and semiarid regions. Rainfall in most parts of the world for at least some part of the year remains insufficient to grow crops, and rainfed food production is heavily affected by the annual variations in precipitation. Irrigation is an obvious option to increase and stabilize crop production. Major investments have been made in irrigation over the past 30 years by diverting surface water and extracting groundwater. The irrigated areas in the world have, over a period of 30 years, increased by 25% with, in particular, a period of accelerated growth during the 1970s and early 1980s (FAO 1994). The expansion rate has slowed down substantially because a major part of the surface waters has already been developed, while groundwater resources have become overexploited at an alarming rate. Sustainable food production will depend on the judicious use of water resources as fresh water for human consumption and agriculture is becoming increasingly scarce. To meet future food demands and growing competition for clean water, a more effective use of water in both irrigated and rain-fed agriculture will be essential. Options to increase water use efficiency include the conservation of rainfall, the reduction of irrigation water losses, and the adoption of cultural practices that increase production per unit of water. Water use for crop production depends on the interaction of climatic parameters that determine crop evapotranspiration. Bal et al. (2022a) has reported that the water demand of maize will be altered under future climate in India. The compilation, processing, and analysis of meteorological information for crop water use and crop production will therefore constitute a key element in developing strategies to optimize the use of water for crop production and to introduce effective water management practices.

Serious water shortages are developing in many countries, and water for agriculture is becoming increasingly scarce in the light of growing water demands from different sectors. To meet basic food requirements, effective strategies need to be developed to optimize crop production per unit of water both in rainfed and in irrigated agriculture. The analysis of climatic information for crop water use is therefore a key element in developing appropriate strategies to face the global water crisis and looming food shortages. This is particularly applicable for rainfed areas, where successful crop production is highly dependent on timely sowing, which in turn is dependent on the availability of rainfall (Bal et al. 2022b). A better understanding of the intricate interactions between climate, water, and crop growth needs to be a priority area in agrometeorological studies as well as timely availability of agrometeorological information and tactical decisions in increasing and sustaining agricultural productions. The main strategic decisions for which the information is needed include assessment of crop production potential, identification of appropriate regions for a specific crop, choice of crop, and their effective management practices so that the water use can be maximized.

20.6 Energy Management with Special Reference to Solar Energy

Energy sector is one of the key sectors contributing significantly to the growth of country's economy, and it's also very much important from agricultural point of view. Energy sectors need to play a more useful role in defining, formulating, and implementing the research projects with close involvement of all utilities such that the benefit reaches the ultimate consumer. The increase in energy consumption, particularly in the past several decades, has not only benefitted the agricultural sector for its management, but huge consumption of fossil fuels has caused visible damage to the environment in various forms.

Solar energy is the energy derived from the sun through the form of radiation. India is endowed with rich solar energy resource. India's need to increase energy provision for its population and fast-growing economy poses a formidable challenge which is perceived as both a great opportunity and a necessity for the country to increase the share of renewable in the overall energy mix. At present, India is the sixth largest country in the world in electricity generation, having aggregate capacity of 177 GWs, out of which 65% is from thermal, 21% from hydro, 3% from nuclear, and the rest about 11% from renewable energy sources. The average intensity of solar radiation received over India is 200 MW/square km (megawatt per kilometer square) with 250–300 sunny days in a year. Solar is an important, although currently underutilized, energy resource in India with the potential to offer an improved power supply (especially in remote areas) and increase the security of India's energy supply. Solar energy intensity varies geographically with Western Rajasthan receiving the highest annual radiation energy and the northeastern regions receiving the least. India has a good level of solar radiation, receiving the solar energy equivalent of more than 5000 trillion kWh/year. The annual global radiation varies from 1600 to 2200 kWh/m^2, which is comparable with radiation received in the tropical and subtropical regions. The equivalent energy potential is about 6000 million GWh of energy per year.

The Government of India has set an ambitious target of achieving 100,000 MW of solar photovoltaic (PV)-based power generation capacity in the country and doubling the farmer's income by the year 2022. Considering the plentiful availability of solar insolation in terms of both duration and intensity in India (5.3–7.0 kWh m^{-2} day^{-1}) and particularly in arid parts of India, i.e., in Rajasthan, solar irradiations are available in abundance for almost 300 days clear sky (Poonia et al. 2021). Agrivoltaic system, which is an integration of PV generation and crop production, has the potential to achieve the above-said two targets by 2022. Agri-voltaic system produces food and generates renewable energy from a single land unit. The concept of integrating both food production and energy generation on a single land unit has evolved in recent times due to ever-increasing demands for the land resources. Production of food occurs by conversion of solar energy to food through photosynthetic process, whereas PV-based energy generation occurs through conversion of solar energy to electric energy through photovoltaic process. Both these processes require land as a basic natural resource (Santra et al. 2021). Therefore, competition for land may arise in the future for agricultural use and PV-based electricity generation. There is possibility that solar PV-based electricity production will be preferred over agriculture because of its higher efficiency of converting solar energy. Agri-voltaic system provides opportunity to generate electricity from farmers' field and thus can increase farmers' income.

20.7 Weather Forecasting and Agromet Advisory Service System

The subject "Meteorology and Weather Forecasting" got importance worldwide as early as the seventeenth century after the invention of the thermometer and the barometer and the formulation of laws governing the behavior of atmospheric gases. In India also, scientific works on this area were started with publication of treatise on the Indian summer monsoon. British scientist Halley published this treatise in 1636, where the seasonal reversal of winds due to differential heating of the Asian land mass and the Indian Ocean was discussed. The British East India Company established several meteorological observatories. Calcutta and Madras weather stations were established in 1785 and 1796, respectively. Pioneer works on tropical cyclone were carried out by Captain Henry Piddington and published in series in the *Journal of the Asiatic Society*. The tropical cyclone affecting Southern West Bengal in 1864 and consecutive monsoon failure in 1866 and 1871 were the major climatic events of that period. Thus, the history of different weather phenomenon over Indian subcontinent was well documented, and considering the importance of weather in our day-to-day life, the Government of India established the India Meteorological Department (IMD) in the year 1875. Afterward, IMD has progressively expanded its infrastructure for meteorological observations, communications, forecasting, and weather services. In its history of almost 147 years, IMD expanded its application to newer areas and adopted updated technology. India was the first developing country in the world to have its own geostationary satellite, INSAT, for continuous weather monitoring of this part of the globe, and particularly for cyclone warning. With the help of advanced satellite technology and numerical weather prediction models, the weather forecasting in India has also advanced a lot.

It is a well-known fact that the weather prediction involves two important components, namely, data collection and modeling. Data about weather is collected over land surfaces (rain gauges, weather stations, etc.), in oceans (weather buoys), in the lower atmosphere (weather balloons and sensors on airplanes), and from space (satellites). The IMD operates approximately 1400 automatic rain gauges (ARGs) and 727 automatic weather stations (AWSs) in the country (IMD 2021). Along with the network of weather data collection systems, computer-based numerical weather prediction (NWPs) models are also needed for more accurate forecasts. To fulfill the need, NCMRWF was established in 1988.

The agrometeorological information can be applied to improve and protect the livelihood of farming community. Such improvements include yield quantity, quality, and income while safeguarding the agricultural resource base from degradation. The agromet advisory services provide a very special kind of inputs to the farmers as advisories through which farmers can take the advantage of favorable weather and minimize the negative impact of adverse weather. The provision of timely and accurate agromet advisories assumes great importance in this era of climate variability and change. The advisory service can play a big role in food security and poverty alleviation. Forecast on extreme weather events also plays an important role in farm management (Table 20.2).

Based on the medium-range weather forecast, AAS bulletins have been prepared for 636 districts and issued on every Tuesday and Friday. State composite bulletins (23) and national AAS bulletins have also been issued simultaneously. Value addition to medium-range weather forecast by regional meteorological centers (RMCs) and meteorological centers (MCs) makes the forecast very accurate. To help the farmers to cope with climate risks and also to effectively use seasonal to inter-annual climate forecasts, IMD in collaboration with Central Research Institute for Dryland Agriculture, Hyderabad, continued issuing AAS bulletins based on extended-range weather forecast and monthly weather forecast to fulfill the needs of different users including planners at state and national levels and farmers (Chattopadhyay and Chandras 2018).

20.8 History of Coordinated Research on Agrometeorology in India

In the year 1983, the Indian Council of Agricultural Research, Government of India started the All India Coordinated Research Project on Agrometeorology (AICRPAM), with Hyderabad as the head quarter, to initiate coordinated researches on various agrometeorological aspects to understand the relationships between weather and crop production systems. The project was started with 10 cooperating centers and later expanded to 25 centers by the end of eighth 5-year plan covering all state agricultural universities at that time (Fig. 20.1). The AICRPAM undertakes research and extension work on five themes, viz., agroclimatic characterization, crop-weather relationships, crop growth modeling, weather effects on pests and diseases, and agromet advisory services. In 1998, an agrometeorological data bank was established with the support of the Department of Science and Technology (DST). This is a repository of weather data from agromet observatories maintained

Table 20.2 Forecast on extreme weather events

Extreme weather	IMD forecast available	Key data and data source
Cyclone	Included in countrywide daily weather warning bulletin, with track and intensity forecast (6 h to 5-day lead time)	Data: Wind, rainfall intensity, ocean-atmosphere interaction source: Weather buoys, ships and airplanes, satellites
Thunderstorm	Included in countrywide daily weather warning bulletin (for coming week); nowcast based on location to be initiated in 2019 (with few hours lead time)	Data: Precipitation, cloud formation Source: Radar, lightning sensors, satellites
Heat wave	Countrywide heat wave warning bulletin issued, as needed (5-day lead time)	Data: Temperature, particularly maximums and minimums source: Weather stations
Drought	Countrywide weekly drought outlook; weekly, biweekly, and monthly aridity maps through the monsoon season	Data: Rainfall, evapotranspiration, soil moisture source: Agromet observatories
Hailstorm	Included in countrywide weather warning bulletin (5-day lead time); to be included in all-India thunderstorm nowcast (few hours lead time)	Data: Precipitation, temperature, wind speed, dew point Source: Radar, upper air observatories
Dust storm	Included in countrywide weather warning bulletin (5-day lead time)	Data: Temperature, wind speed, humidity, evaporation, soil moisture Source: Agromet observatories

(Adopted from Gopalakrishnan and Subramanian 2020)

by state agricultural universities, ICAR institutes and their regional stations, field experiment data generated by AICRPAM centers, etc.

The AICRPAM has conducted extensive agroclimatic characterization of the states where its cooperating centers are working and have come out with agroclimatic atlases. Crop-weather relationships were quantified by all the cooperating centers and are being published as agrometeorology of respective crops. The AICRPAM has also improved IMD's crop weather calendar by adding more components to it, viz., standard meteorological week-wise optimum range of weather parameters for obtaining higher yield and conducive range of weather parameters of incidence of pests and diseases. Further, the project has developed "Dynamic Crop Weather Calendar," a software which guide for favorable sowing and irrigation decisions based on soil moisture dynamics based on historical, real-time, and forecast weather. The DCWC is proposed to be linked to the decision support system (DSS) of India Meteorological Department (IMD) for automation of agromet advisory services (Vijaya Kumar et al. 2021). Under crop growth modeling, location- and crop-specific genetic coefficients are being calibrated and validated for various crops, and impact of projected climate was assessed using crop simulation models like DSSAT and InfoCrop. Location-specific thumb rules are being developed for forewarning of pest/disease incidence. In collaboration with IMD, the

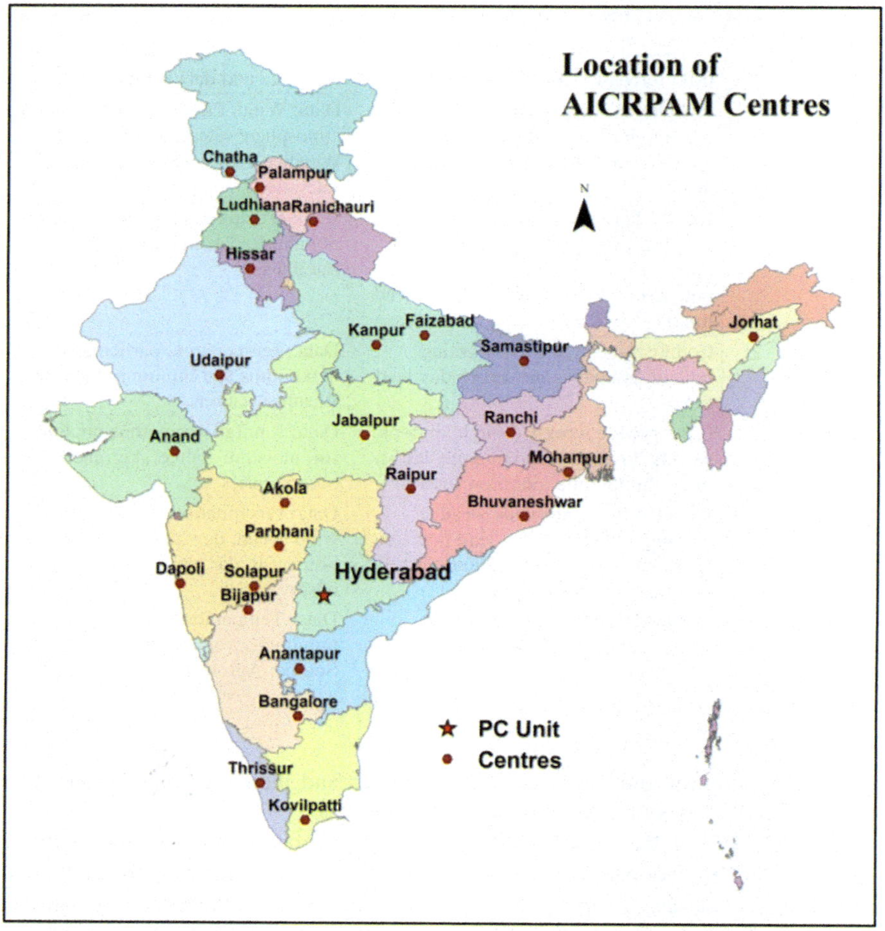

Fig. 20.1 AICRPAM network in India

AICRPAM is also issuing National Agromet Advisory Services' bulletin (NAAS) based on extended-range weather forecast (ERFS) on every Friday.

20.9 Contingent Crop Planning under Uncertain Weather

As the extreme weather events have enhanced manifold under enhanced global warming situation, the role of contingent crop planning has become very much important. Unless major reductions in carbon dioxide (CO_2) and other greenhouse gas emissions occur in the coming decades, global warming of 1.5 °C and 2 °C will be exceeded during the twenty-first century (IPCC 2021). Agriculture plays a pivotal role in India's economy with about 19.9% share in GDP during 2020–2021. Several forecasts for the coming decades projected an increase in atmospheric CO_2 concentration and temperature and changes in precipitation pattern, resulting in more frequent droughts and floods, cold and heat waves, and other extreme hazardous events in India (Srinivasa Rao et al. 2016). Those weather anomalies have an impact on the stability of food production and the required growth rate, not only in rainfed conditions but also in other agricultural production areas. Devising mitigation and adaptation strategies should be the approach to make agricultural production shock-proof against climate variability. Moreover, a timely agromet advisory can save money on fertilizers, seeds, plant protection agents, and other inputs, as well as labor and the output itself (especially at the harvest time after the crop reaches physiological maturity) (Bal and Chandran 2021). The India Meteorological Department (IMD) is issuing short-range, medium-range, and long-range weather forecasts, with improved accuracy with the advancement in computing skills. The National Council of Applied Economic Research (NCAER) has assessed the impact of AAS of Gramin Krishi Mausam Seva (GKMS) of IMD and reported that the average annual income of farming households worked out to be Rs. 2.43 lakh for those who modified one to four crop management practices, Rs. 2.45 lakhs for those who modified five to eight practices, and Rs. 3.02 lakhs for those who adopted all the nine changes, against Rs. 1.98 lakh for farmers who did not adopt AAS (NCAER 2020). The All India Coordinated Research Project on Agrometeorology (AICRPAM) under the Indian Council of Agricultural Research (ICAR) is issuing microlevel agromet advisories in selected locations using block-level weather forecast issued by IMD (Subba Rao et al. 2017).

Agriculture contingency planning is one of the most important tactics for dealing with unpredictable weather disasters that cannot be avoided. Real-time crop contingency measures are necessary to minimize the crop losses, to safeguard allied sectors, and to improve the production efficiencies (Bal et al. 2021). The ICAR-CRIDA has developed district-level agricultural contingency plans (DACP) for the entire country. In the face of increasing climate variability, adoption and implementation of the district-level contingency plans are priorities for many state governments. District-level contingency plans are technical documents containing integrated information on agriculture and allied sectors (i.e., field crops, horticulture, livestock, poultry, fisheries) and technological solutions for all major weather-related aberrations including extreme events, viz., droughts, floods, heat wave, cold wave, untimely and high-intensity rainfall, frost, hailstorms, pest, and disease outbreaks, and are aimed to be utilized by district authorities. A template of DACP prepared by the ICAR-CRIDA consists of two parts:

- District agricultural profile of a district with information on resource endowments such as rainfall, soil types, land use, irrigation sources, more dominant crops, and cropping systems along with their sowing windows; livestock, poultry, and fishery information; production and productivity statistics; major contingencies faced by the district and digital soil and rainfall maps.
- The detailed strategies for weather-related contingencies anticipated in crops/ cropping systems such as delay in onset of monsoon of different duration;

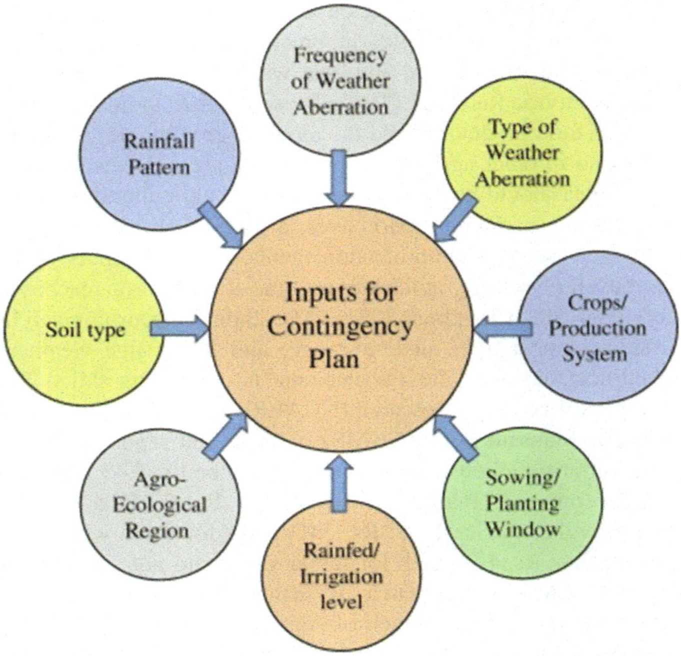

Fig. 20.2 Inputs required for DACP (Srinivasarao et al. 2020)

mid-season monsoon breaks resulting in drought both in rainfed and irrigated situations and adaptation strategies for weather-related extreme events.

These contingency plans include information on alternate crop varieties/crops to choose in the event of a monsoon delay or early-season drought, as well as agronomic measures for mid- season and late-season droughts. In addition, strategies for cattle, poultry, and fisheries have been incorporated (Rajendra Prasad et al. 2013). The inputs required for DACP is depicted in Fig. 20.2.

20.10 Operational District-Level Agricultural Contingency Plans (DACP)

The preparation for operationalization of crop contingency plans begins with the long-range weather forecast (LRF) issued by IMD, which provides an overall idea whether southwest monsoon will be above normal/normal/below normal. If the forecast is below normal, steps to implement crop contingency plans are taken up. The events of mid-season drought can be identified using extended-range weather forecast issued by IMD, which is valid for 2 weeks. Regular agricultural operations can be planned according to the medium-range weather forecast. Nowcast is the latest forecast issued by IMD in the event of hailstorms, cloud bursts, lightning, etc., which can be used by farmers to protect the livestock and high-value crops.

The implementation of DACP requires planning at district and state levels, and the overall coordination and facilitation are provided by the Government of India. The seasonal forecast for southwest monsoon made by IMD and South Asian Climate Outlook Forum and other international agencies was utilized to make an analysis of the prospects of monsoon and the measures to be taken up to face the scenarios. In order to make plans relevant on near real-time basis, interface meetings were organized with concerned line departments of the state government before the commencement of *kharif* season. The meetings are conducted jointly by DAC and ICAR in various states, and participants include senior officials of departments of agriculture, KVKs, SAUs, seed agencies, and other stakeholders. The deliberations cover drought management, suggesting alternative crops, seed availability, and preparedness strategies of line departments. The DACPs, due to its wide coverage in terms of sectors and weather aberrations, could be implemented through different developmental initiatives such as NMSA, IWMP, PMKSY, etc.

20.11 Future Strategies on Agrometeorological Research in India

Agricultural meteorology studies the meteorological, hydrological, pedological, and biological elements that influence agricultural production, as well as the relationship of agriculture and the environment. Its goals are to understand these effects and to help farmers by incorporating supportive knowledge and information into agrometeorological practices and through agrometeorological services. Despite the advances in agricultural technology in the past, agricultural production is still largely depending on weather. Because of the challenges to agricultural production posed by increasing climate variability, associated extreme events, and climate change, agrometeorological research and services are more important than ever.

20.11.1 Characterizing Agroclimatic Environment

For efficient land use planning, determining suitable crops for a region, risk analysis of climatic hazards, profit calculations in farming, production or harvest forecasts, adoption of farming methods, choice of farm machinery, etc., thorough characterization of the agroclimate is the need of the hour as climate change has become a new normal. How climate variability and change will affect the length of crop-growing season can also be explored using it. Thus, potential utilization of a cultivable region can be ensured by characterizing the agroclimate, considering the aspects of climate variability and change.

20.11.2 Establishing Crop-Weather Relationships

The rapid changes in climatic normal and variability in weather have created enormous scope to explore the influence of weather parameters on growth, phenology, yield, and yield attributes of crops. There is a renewed interest in this area as increase/decrease in temperature, altered precipitation pattern, and increase in frequency of extreme weather events are posing greater threat to crop production. Devising adaptation and mitigation strategies to combat the negative effects of weather/climate variability/change on crop growth and development needs to be explored. How the microclimate within a crop field changes under climate variability is also a point of interest. Changes in fluxes of water vapor, CO_2/O_2, and other GHGs between crop fields and atmosphere need to be explored using long-term data.

20.11.3 Crop-Weather-Pest/Disease Interaction Studies

The natural habitats, hosts, and life cycle of pests and pathogens are also affected by climate variability/change. The world has witnessed the emergence of new pest like fall army worm in maize. At the same time, some existing pests/diseases are spreading their horizon and started affecting new crops. How the changing climate is altering the pest-disease-host interaction is an interesting area to be explored.

20.11.4 Opportunities in Agromet Advisory Services

Provision of timely and accurate agromet advisory services (AAS) can augment crop production to a great extent by helping the farmers in crop selection, timely crop management operations, saving inputs, and the produce itself. As the world is witnessing a quantum leap in the computational facilities, it is expected that weather forecast will become more accurate. It is also envisaged that dissemination methods of AAS will also be improved with the advancements in information communication technologies. This is an area of applied social research in the time to come. Researches on improvement in real-time and contextual agromet advisories need to be strengthened.

20.11.5 Preparing for Anticipated Climate

To assess the impact on crops in projected climatic scenarios, the available Global Circulation Models (GCMs) need to be evaluated first on a local scale using conventional statistical measures for selecting best performing GCMs/RCMs. Then a multi-model ensemble (MME) using the selected GCMs should be made. Available downscaling techniques may be employed to reduce model bias or uncertainty at a local scale for using as an input in crop simulation models for estimating the impacts in forecast climate.

20.11.6 Shift in Mesoclimate/Microclimate

Microclimate of the crops is changing due to changes in greenhouse gases (GHGs). Events like mass scale residue burning is affecting the gaseous and particulate composition of the atmosphere, impacting the photosynthetic pathways in case of crops/trees. Similarly, large-scale changes in land-use pattern alter the microclimatic environment and impact the circulation pattern at microscales. Though not in a short time, it will have impact on the agriculture as a whole. Therefore, it needs to be assessed thoroughly.

20.11.7 Concern for Environmental Security

Since the 1990s, the scope of environmental security has broadened to include multiple subsets, including food security, energy security, and water security, as well as emerging notions of adaptation and resilience to hazards, e.g., climate security. Environmental security considers the abilities of individuals, communities, or nations to cope with environmental risks, changes or conflicts, or limited natural resources. For example, climate change can be viewed a threat to environmental security. Human activity impacts CO_2 emissions, impacting regional and global climatic and environmental changes and thus changes in agricultural output. Environmental security such as food security, water security, and energy security-related aspects can be attempted using agrometeorological interventions for achieving Sustainable Development Goals. Therefore, the principal aim is to consider securitization of the environment and natural resources and assess new conceptions of environmental security in the context of global change using the science "agrometeorology."

20.11.8 Weather Aspects of Land Degradation

Land degradation, the loss of the land's biological productivity, caused by human as well as climatic change is another environmental issue concerning agriculture. The main reasons are erosions caused by water and wind. Water erosion is mainly influenced by intensity of rainfall and edaphic characteristics like slope, soil texture, soil structure, etc. Wind erosion is influenced by surface cover, wind speed and direction pattern, and soil structure/type. Sustainable development of these degraded lands can only come about through concerted efforts based on sound understanding of climate resources and the risk of climate-related or climate-induced natural disasters for a given region for prescription of effective land management practices.

20.11.9 Use of Geospatial Technologies

There is enormous scope for application of advanced technologies, viz., remote sensing and geographic information systems (GIS), in crop production. The increased availability of high-resolution satellite imageries in the public domain enhances the applicability of remote sensing data in different areas of crop production, viz., crop acreage estimation, biotic and abiotic stress management, crop loss estimation, etc. Remote sensing together with GIS platform can be utilized for planning regional-level research studies pertaining to agricultural meteorology. Open-source platforms like Google Earth Engine have thrown out a plethora of opportunities to undertake research in agricultural meteorology using remote sensing data.

20.11.10 Crop Simulation Models

Crop models have been extensively used for evaluating the impact of projected climate on individual crops. Much work has not been done in modeling cropping systems for assessing the impact of climate change. Another area of interest is the integrated modeling where crop model is combined with economic models which offer a comprehensive analysis. Using remote sensing data as input for the crop models is another area to look at regional scale.

20.11.11 Mining of Temporal-Spatial Agromet Data

Application of machine learning, big data analytics, etc. in reliable weather forecasting, crop yield prediction, and crop-weather relationships may also be explored. A large number of algorithms are available for classification and prediction. Programming languages like Python and R provide excellent platforms for exploring these possibilities. This list is not exhaustive, and there are many other areas in agricultural meteorology that can further be explored.

References

Ali H, Modi P, Mishra V (2019) Increased flood risk in Indian sub-continent under the warming climate. Weather Clim Extremes 25:100212

Bal SK, Chandran MAS (2021) Minimizing weather risk in agricultural management through Agromet advisory services (AAS). In: Mallappa et al (eds) Climate change and resilient food systems; issues, challenges, and way forward. Springer Nature, Singapore, pp 245–260. https://doi.org/10.1007/978-981-33-4538-6_9

Bal SK, Chandran MAS, Madhavan SV, Rao AV, Manikandan N, Praveen Kumar R, Parameswaran PV, Attri SD, Singh P, Mohanty A, Singh VK (2022a) Water demand in maize is projected to decrease under changing climate in India. Sustainability 14(3):1419. https://doi.org/10.3390/su14031419

Bal SK, Manikandan N, Sandeep VM, Kumar PV, Lunagaria MM, Rao AS, Pramod VP, Singh VK (2022b) Criteria based decisions for determining agroclimatic onset of the crop growing season. Agric For Meteorol 317:108903. https://doi.org/10.1016/j.agrformet.2022.108903

Bal SK, Minhas PS (2017) Atmospheric stressors: challenges and coping strategies. In: Minhas PS et al (eds) Abiotic stress management for resilient agriculture. Springer Nature Singapore Pte. Ltd., pp 9–50. https://doi.org/10.1007/978-981-10-5744-1_2

Bal SK, JVNS P, Singh VK (2022c) Heat wave 2022 – causes, impacts and way forward for Indian agriculture. ICAR-CRIDA Technical Bull 1:p50

Bal SK, Rao KV, Chandran MAS, Sasmal S, Singh VK (2021) Weather forecast, agriculture contingency plan and agromet–advisory services for climate resilient agriculture. Indian. J Agron 66(5th IAC Special Issue):S1–S14

Bal SK, Sandeep VM, Kumar PV, Rao AS, Pramod VP, Manikandan N, Rao CS, Singh NP, Bhaskar S (2022d) Assessing impact of dry spells on the principal rainfed crops in major dryland regions of India. Agric For Meteorol 313:108768. https://doi.org/10.1016/j.agrformet.2021.108768

Basu AK, Bandhyopadhyay PK (2004) The effect of season on the incidence of ticks. Bull Anim Health Prod Afr 52(1):39–42

Chandran MAS, Banerjee S, Mukherjee A, Nanda MK, Mondal S, Kumari VV (2021) Evaluating the impact of projected climate on rice–wheat-groundnut cropping sequence in lower Gangetic plains of India: a study using multiple GCMs, DSSAT model, and long-term sequence analysis. Theor Appl Climatol 145(3):1243–1258

Chandran MAS, Banerjee S, Mukherjee A, Nanda MK, Kumari VV (2022) Evaluating the long-term impact of projected climate on rice-lentil-groundnut cropping system in lower Gangetic plain of India using crop simulation modelling. Int J Biometeorol 66(1):55–69

Chattopadhyay N, Chandras S (2018) Agrometeorological advisory services for sustainable development in Indian agriculture. Biodiversity Int J 2(1):13–18. https://doi.org/10.15406/bij.2018.02.00036

Takle ES (2003) Agricultural meteorology and climatology. In: Holton JD (ed) Encyclopedia of atmospheric sciences, pp 54–60

FAO (1994) Water for Life. World Food Day 1994, Rome

Gopalakrishnan T, Subramanian K (2020) Agrometeorological advisory Services in India: an assessment. Centre for Science and Environment, New Delhi

Guhathakurta P, Revadekar J (2017) Observed variability and long-term trends of rainfall over India. In: Rajeevan M, Nayak S (eds) Observed climate variability and change over the Indian region. Springer Geology, Springer, Singapore

IMD (2021) ARG Network. http://aws.imd.gov.in/

IPCC (2021) Summary for policymakers. In: Masson-Delmotte V, Zhai P, Pirani A, Connors SL, Péan C, Berger S, Caud N, Chen Y, Goldfarb L, Gomis MI, Huang M, Leitzell K, Lonnoy E, Matthews JBR, Maycock TK, Waterfield T, Yelekçi O, Yu R, Zhou B (eds) Climate change 2021: the physical science basis. Contribution of working group I to the sixth assessment report of the intergovernmental panel on climate change. Cambridge University Press, Cambridge

Kumar Y, Singh R, Kumar A, Dagar CS (2019) Quantification of crop weather relationship and the effect of different planting dates on growth and yield of potato cultivars in a sub-tropical environment at Hisar. J Appl Nat Sci 11(1):17–22. https://doi.org/10.31018/jans.v11i1.1965

Makone P, Patel JG, Desai CK, Das S, Pal V, Paramar JK (2015). Influence of weather parameters on summer green gram (*Vigna radiata* (L.) Wilczek) at Sardar krushinagar) J Agrometeorol 17(1):142–144

Mallya G, Mishra V, Niyogi D, Tripathi S, Govindaraju RS (2016) Trends and variability of droughts over the Indian monsoon region. Weather Clim Extremes 12:43–68

Maurya RK (2010) Alternate dairy management practices in drought prone areas of Bundelkhand region of UP (Doctoral dissertation, IVRI, Izatnagar)

Mavi HS (1994) Introduction to agrometeorology. Oxford and IBH, New Delhi

NCAER (2020) Estimating the economic benefits of investment in monsoon mission and high-performance computing facilities. National Council of Applied Economic Research, New Delhi, India, Report No. 20200701: 76

NCAR (2014) The climate data guide: Trend Analysis: National Center for Atmospheric Research. https://climatedataguide.ucar.edu/climate-data-tools-and-analysis/trend-analysis
Nishiyama I (1984) Climatic influence of pollen formation and fertilization. In: Tsunoda NT (ed) Biology of Rice. Japan Scientific Society, Tokyo
Niyas NT, Srivastava A, Hatwar HR (2009) Variability and trend in the cyclonic storms over the North Indian Ocean, Meteorological Monograph No. Cyclone Warning 3:1–34
Pawar SS, Sajjanar B, Lonkar VD, Nitin KP, Kadam AS, Nirmale AV, Brahmane MP, Bal SK (2016) Assessing and mitigating the impact of heat stress in poultry. Adv Animal Veterinary Sci 4(6):332–341. https://doi.org/10.14737/journal.aavs/2016/4.6.332.341
Poonia S, Santra P, Jat NK, Jain D, Meena HM (2021) Agri-voltaic system: a novel technology for doubling the farmer's income in hot arid ecosystem of western Rajasthan. India Indian Farmer 8(5):351–360
Prasad R, Ravindra Chary G, Rao KV, Prasad YG, Srinivasa Rao C, Ramana DBV, Sharma NK, Rao VUM, Venkateswarlu B (2013) District level contingency plans for weather aberrations in Himachal Pradesh. CSK HPKV, Palampur, HP and central research Institute for Dryland Agriculture. Hyderabad-500 059, India. P.222
Ramdas LA, Kalamkar RJ (1938) Statistical investigations on crop-weather relationships in India. Sankhyā: Indian. J Stat 3(3):285–290
Rao UMB, Vijayalakshmi K (1986) Rainfall-yield relationships in rainfed sorghum in India. Mausam 37:529–532
Rao AVMS, Chandran MAS, Bal SK, Pramod VP, Sandeep VM, Manikandan N, Raju BMK, Prabhakar M, Islam A, Kumar SN, Singh VK (2022) Evaluating area-specific adaptation strategies for rainfed maize under future climates of India. Sci Total Environ 836:155511
Rohli RV, Vega AJ (2018) Climatology, 4th edn. Jones & Bartlett Learning, Burlington. ISBN 9781284126563
Rotter R, Van de Geijn SC (1999) Climate change effects on plant growth, crop yield and livestock. Clim Chang 43(4):651–681
Santra P, Meena HM, Yadav OP (2021) Spatial and temporal variation of photosynthetic photon flux density within agrivoltaic system in hot arid region of India. Biosyst Eng 209:74–93
Singh OP (2001) Long term trends in the frequency of monsoonal cyclonic disturbances over the North Indian Ocean. Mausam 52(4):655–658
Srivastava AK, Kothawale DR, Rajeevan MN (2017) Variability and long-term changes in surface air temperatures over the Indian subcontinent. In: Rajeevan M, Nayak S (eds) Observed climate variability and change over the Indian region. Springer Geology. Springer, Singapore, pp 17–35. https://doi.org/10.1007/978-981-10-2531-0_2
Srinivasa Rao CH, Gopinath KA, JVNS P, Singh AK (2016) Climate resilient villages for sustainable food security in tropical India: concept, process, technologies, institutions, and impacts. Adv Agron 140:101–214
Srinivasarao CH, Rao KV, Gopinath KA, Prasad YG, Arunachalam A, Ramana DBV, Mohapatra T (2020) Agriculture contingency plans for managing weather aberrations and extreme climatic events: development, implementation and impacts in India. Adv Agron 159:35–91
Upadhyay RC, Singh SV, Kumar A, Gupta SK, Ashutosh A (2007) Impact of climate change on milk production of Murrah buffaloes. Ital J Anim Sci 6(sup2):1329–1332
Victor US, Srivastava NN, Ramana Rao BV (1991) Application of crop water use models under rainfed conditions in seasonally arid tropics. Int J Ecol Environ Sci 17:129–1237
Vijaya Kumar P, Bal SK, Dhakar R, Sarath Chandran MA, AVM SR, Sandeep VM, Pramod VP, Malleswari SN, Sudhakar G, Solanki NS, Shivaramu HS (2021) Algorithms for weather-based management decisions in major rainfed crops of India: validation using data from multilocation field experiments. Agron J 113(2):1816–1830
Vijaya Kumar P, Subba Rao AVM, Sarath Chandran MA, Rao VH, Srinivasarao Ch VUM (2017) Micro-level Agromet advisory services using block level weather forecast – a new concept-based approach. Curr Sci 112:227–228

Sustainable Production Systems

21

T. K. Das, Sunil Kumar, Anup Das, M. A. Ansari, Rishi Raj, and Sourav Ghosh

Abstract

A sustainable agriculture system is one that maintains its productive capacity consistently on a temporal and spatial scale. The Green Revolution during the late 1960s could be accredited for solving India's food crisis and transforming the country from 'begging bowl status' to a food surplus nation. However, unabated and injudicious resource (fertilizer, irrigation, pesticides) usage over the years has jeopardized the agriculture system and created multiple problems specially in the Green Revolution belts. Loss of landraces, unproductive soils, groundwater depletion, pest resistance and pesticide residues in foods and environment have been among the most prominent issues. Sustainable agriculture endorses 'going back to basics' by inculcating the virtues of ancient agriculture and principles of indigenous knowledge. Organic farming and natural farming in areas of inherent advantages are being popularized worldwide. Conservation agricultural practices including zero tillage, residue retention and crop diversification have been

T. K. Das (✉) · R. Raj
ICAR-Indian Agricultural Research Institute, New Delhi, Delhi, India
e-mail: tkdas@iari.res.in

S. Kumar
ICAR-Indian Grassland and Fodder Research Institute, Jhansi, Uttar Pradesh, India

ICAR-Indian Institute of Farming System Research, Modipuram, Uttar Pradesh, India

A. Das
ICAR Research Complex for Eastern Region, Patna, Bihar, India

M. A. Ansari
ICAR-Indian Institute of Farming System Research, Modipuram, Uttar Pradesh, India

S. Ghosh
ICAR-Central Research Institute for Jute and Allied Fibres, Barrackpore, West Bengal, India

P. K. Ghosh et al. (eds.), *Trajectory of 75 years of Indian Agriculture after Independence*, https://doi.org/10.1007/978-981-19-7997-2_21

recommended in the rice-wheat growing belts to jointly combat soil degradation and climate change and enhance input use efficiency. Farming system approach through integration of allied enterprises like crops, dairy and fishery, along with multipurpose trees and component for water/nutrient recycling, could be effectively implemented for better resource recycling and doubling farmer's income. Under rainfed agro-ecosystems, agri-horticulture, silvi-pasture and horti-pastoral systems can reap better dividends than sole agri-enterprise. Sustainable agriculture systems have several ecosystem services including lowering of carbon footprint and global warming potential. Thus, sustainable production system must be focused for location-specific zones and upscaled to sustain productivity and natural resource base and enhance income and employment in changing climate.

Keywords

Agri-horti-silvi-pastoral system · Conservation agriculture · Integrated farming system · Indigenous technical knowledge · Precision nutrient management

21.1 Introduction

The human population of India has increased to 1210.2 million at a growth rate of 1.76% in 2011 and is estimated to increase further to 1530 million by 2030. There are projections that demand for food grains would increase from 250 million tonnes to 345 million tonnes in 2030. The Indian economy is predominantly rural and agricultural, and the declining trend in size of landholding poses a serious challenge to the sustainability and profitability of farming. In view of the decline in per capita availability of land from 0.5 ha in 1950–1951 to 0.15 ha by the turn of the century and a projected further decline to less than 0.1 ha by 2020, it is imperative to develop strategies and agricultural technologies that enable adequate employment and income generation, especially for small and marginal farmers who constitute more than 80% of the farming community. Sustainability refers to the perpetual ability and is a system issue. Sustainability as a goal seeks permanence for an activity or a system where a system is a conglomeration of different elements, components, subsystems or constituents knit together into an integrated whole that help in the system's organization and maintenance of its integrity (von Bertalanffy 1968). A sustainable system is one that survives or persists through emerging stresses and shocks owing to the interactive nature of its components (Costanza and Palten 1995). Raman (2006) delineated four groups of determinants – physical, biological, socio-economic and cultural – as important in the structuring and management of agricultural systems for sustainability. Physical determinants include climate (temperature, rainfall and radiation), land availability and soil quality parameters such as physical, chemical and biological properties, erosion and water-holding capacity, depth, and slope. Biological determinants include energy and biotic factors such as insect pests and predators, weeds, plant diseases, functional diversity of soil biotic community, photosynthetic efficiency of the crops to be grown, crop rotations and cropping patterns. Socio-economic and cultural determinants are factors such as population density, economic and market opportunities (capital, credit, prices of inputs and outputs), policy issues, political relationships, availability, accessibility and affordability of technology, inputs, labour, diets, work culture, traditional knowledge systems, sociocultural beliefs and ethics. Sustainable agriculture seeks permanence to agricultural production systems to enable food security for all in an ecologically sound, economically viable and socially responsible manner. Sustainable agriculture requires the simultaneous maintenance of several complex dimensions, productivity, ecological integrity, economic viability and social relevance. Sustainable systems start with the adoption of best management practices (BMPs). Once in place, BMPs lead to maximum economic yields (MEY), and together, they lead to sustainability both economically and environmentally (Hegde et al. 2012).

The greatest challenges for sustaining the present status and to meet the necessities of future generations are the diminishing carrying capacity of the ecosystem services of our planet in all its ecologies. Indications of dangers are loud and clear through the effects of climate change manifestations. Humans were appropriating more than 40% of all the net primary product of terrestrial photosynthesis (Vitousek et al. 1986). Almost 40% of crop land had been degraded in less than four decades, and almost 25 billion tonnes of top soil was being lost to erosion every year. Severe depletion of fossil fuel and water shortages in many countries became a cause for worry. Clear indications of anthropogenic climate change led to the constitution of the Intergovernmental Panel on Climate Change (IPCC) in 1988 to study the problem. All these factors, combined with excessive loss of biodiversity, signalled an obvious threat to agricultural production and the life support systems provided by nature. Uncontrolled population growth and asymmetric economic growth and environmental degradation also subverted the socio-economic system and imposed various kinds of socio-economic stress arising from glaring inequalities in empowerment, standards of living, health, livelihood opportunities, food security and so on. The danger of these developing into threat to regional and global security and peace became a cause for concern. In this chapter, the evolution of sustainable production systems has been discussed under different periods, viz. (i) pre-independence era, (ii) Green Revolution era and (iii) post-Green Revolution era.

21.2 Crop Production Status in Pre-Independence Period

Indian agriculture dated back to 9000 BC in northwest part of India with subsistence type of farming and domestication of crops and animals. Plants and animals were considered essential to their survival by the Indians. Introduction of tools, implements and small machineries along with development of advanced techniques led to higher productivity and surplus production. By pre-mughal period, Indian products had reached the world via existing trading networks, and foreign crops were introduced to India. During the medieval period, irrigation channels were invented which reflected in the economies of other regions of the world. Indian rulers also developed land and water management systems for assured production and profitable farming. Before 1947, under British rule, more than 90% of the national income relied on the Indian agriculture sector and majority of the country's population resided in rural areas where agriculture was the primary source of livelihood. Records of rice, wheat, millets and pulses cultivation as food crops in ancient as well as medieval India indicated that ancient India was self-sufficient in food production and exported a large range of produces from crops like spices, rice and cotton and silk textiles (Randhawa 1980). Higher food grain production was mainly due to the favourable agro-climatic conditions, dependable monsoon rains and high soil fertility which enabled harvesting of two crops in a year. The medieval period (particularly the post-Akbar), however, started experiencing a decline in agricultural production, which caused reduction in the availability of food, ultimately resulting in several famines, viz. the great famine of Madras in 1647, and that of Gujarat in 1666 quickly followed by two more famines (in Gujarat) in the years 1718 and 1744 (Randhawa 1982). The problem of food shortage was accelerated with the beginning of British rule during which the emphasis shifted towards production of commercial crops like tea, coffee, indigo, opium, jute, cotton and sugarcane (Bayly 1985). For example, from 1901 to 1947, the food production declined—whereas the population rose by 38% and the increase in cultivated area was to the extent of 18% only. The crisis was most acute in Bengal, where food output declined at an annual rate of 0.7% from 1921 to 1946, whereas population grew at an annual rate of 1% (Kumar 2019). Between 1893–1894 and 1945–1946, production of commercial crops increased by 85% and that of food crops fell by 7%, which resulted in regular famines. Cash crop increase was 0.57% per annum, and there was a decline of food grain production by (−) 1.14% per annum.

The British invasion resulted in total commercialization of India's agriculture. On the eve of independence, the once most prominent sector of this country suffered from stagnation and constant degradation. Several contributable factors included zamindari system, forced commercialization (production of crops only for sale against Indian system of self-consumption), lack of supervision and negligence in reforms, fragmented land ownership (scattered land owned by different individuals which made it even harder for cultivation), outdated technology (lack of machines and old-fashioned methods of cultivation), low productivity (due to the lack of innovativeness and poor field and crop management), feud among landowners and cultivators, higher dependence on rain and subsistence cultivation. Blyn (1966) revealed that between 1891 and 1947, aggregate grain output in British India increased at an average rate of 0.11% per year. In fact, in the latter half of the period, the growth rate was a negligible 0.03%. Rice output, constituting half of the total output, actually declined over the 56-year period at an average annual rate of 0.09%. Despite low population growth rate (0.67%) between 1911 and 1941, per capita availability of food grain – taking into account international trade flows – declined by as much as 26%. Food import dependency was up to 50% of all capital expenditures. Decline in food grains was around 1.14% per capita per annum, while cash crops increased though at 0.57% per person per annum.

Some positive effects of British rule were also witnessed in the form of increased gross cropped area in most regions between 1870 and 1920, usually led by marketable crops such as wheat, cotton, oilseeds, sugarcane and tobacco. There were some efforts made by the British rulers to expand irrigation in the form of wells or canals (Randhawa 1983); even then, the total annual production of food grains remained far below the requirement to feed the nation. Regional specialization of crop production was based on climatic conditions, soil, etc. Deccan districts of Bombay presidency grew cotton, Bengal grew jute and Indigo, Bihar grew opium, Assam grew tea, and Punjab grew wheat. However, it can be seen that negative effects of commercialization of agriculture were outweighing positive effects of commercialization. Various reports mention loss of approximately 60 million lives in several famines in pre-independent India, and the country was called as a 'Hungry Nation' (Siegel 2018). The recurrence of famines opened the eyes of British rulers, and the then British Government decided to establish six agricultural colleges/research institutions during the fag end of the nineteenth century and early twentieth century. This was followed by a new impulse to develop agriculture (in India) by the appointment of 'The Royal Commission on Agriculture' in the year 1926 which recommended the creation of the Imperial Council of Agricultural Research to promote, guide and coordinate agricultural research throughout India, and the same was set up on 23 May 1929.

21.3 Indigenous Technical Knowledge

The advent of the concept of sustainable agriculture in the late 1980s in Indian agricultural scenario has evoked the interest on indigenous technical knowledge (ITK) that has the element of using natural products to solve problems pertaining to agriculture and allied activities. The indigenous technical knowledge (ITK) is socially desirable, economically affordable, and sustainable, involves minimum risk and focuses on efficient utilization of eco-friendly resources. Indigenous technical knowledge plays an essential role in sustainable grassroots innovations (Sow and Ranjan 2021). The traditional practices have been classical examples of nonexploitative and non-polluting methods of natural resources leading to sustainable agriculture. These traditional knowledge and technologies can play a significant role in the overall socio-economic development of the communities. Many farmers in India traditionally adopted water harvesting measures such as farm ponds, check dams, shallow wells dug in depressions to collect rainwater, diversion channels (khuls), brick lined tank, channels constructed through hilly rivers (pynes), tank cascades, Bandhara, Khadin, Nalla check, Pat system, Phad system, Chauka system and Haveli system, etc. (Sow and Ranjan 2021). In northeastern region of India, ITKs like Zabo farming, terrace farming or panikheti, bamboo drip irrigation, pond-based farming systems and rice + fish farming are excellent example of resource conservation and sustainable farming (Das et al. 2013). Traditional farmers have found ways of improving soil structure, water-holding capacity and nutrient and water availability without the use of artificial inputs. Farmers in Eastern Gangetic Plains follow different ITKs for nutrient management in crops and cropping systems. To mention a few are preparation of FYM from cow dung and litters, in situ decomposition of greases and other weeds in soil, water hyacinth compost, peening sheep in fields, use of jute leaves as manure and many others (Ghosh et al. 2001).

21.4 The Green Revolution: Solving India's Food Crisis

The Green Revolution was started in the 1960s in order to increase food production, to alleviate extreme poverty and malnourishment in the country and to feed millions. We distinguish the first GR period as 1966–1985 and the post-GR period as the next two decades. Large public investment in crop genetic improvement was built on the scientific advances already made in the developed world for the major staple crops – wheat, rice and maize – and adapted those advances to the conditions of developing countries. High-yielding monohybrid crops were introduced as a part of Green Revolution, since the major problem with indigenous seeds was their inherent inability to respond to chemical fertilizers. On the contrary, new varieties were created to produce higher yields in conjunction with the use of chemical fertilizers and very intense irrigation. Much of the success was caused by the combination of high rates of investment in crop research, infrastructure and market development and appropriate policy support that took place during the first Green Revolution. There has been a remarkable shift in India in the cropping patterns for both wet season and winter crops since the Green Revolution. Rice and wheat replaced pulses, bajra (pearl millet) and jowar/sorghum as dominant food crops while cotton rose as the key cash crop. The main wet season crops in 1965–1966 comprised bajra (46%), rice (13%) and sorghum (12%); however, in 1995–1996, rice (34%) was the major crop followed by bajra (27%) and cotton (24%). For winter crops, wheat has increased in production as major crop from 43% in 1965–1966 to 64% in 1995–1996. In Haryana, the yields of rice and wheat have increased considerably (Singh 2000). The per capita net availability of food grains increased over the years. The food grains in India increased from pre-Green Revolution (50.82 million tonnes) in 1950–1951 to 303.34 million tonnes in 2020–2021 (Fig. 21.1). Productivity of food grains also increased as well, from 522 kg/ha in 1950–1951 to 2377 kg/ha in

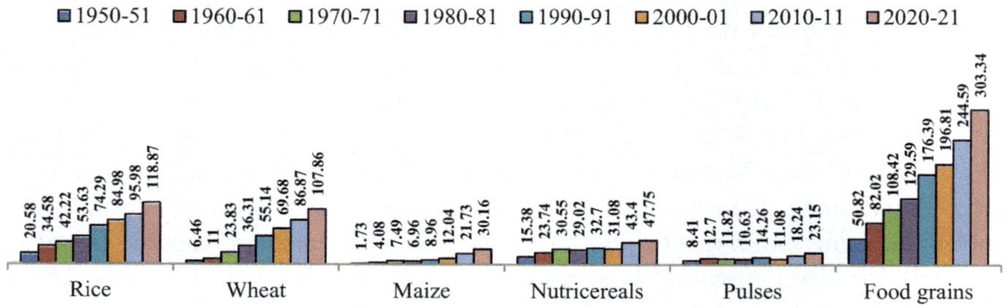

Fig. 21.1 Trends of food grain production (million tonnes)

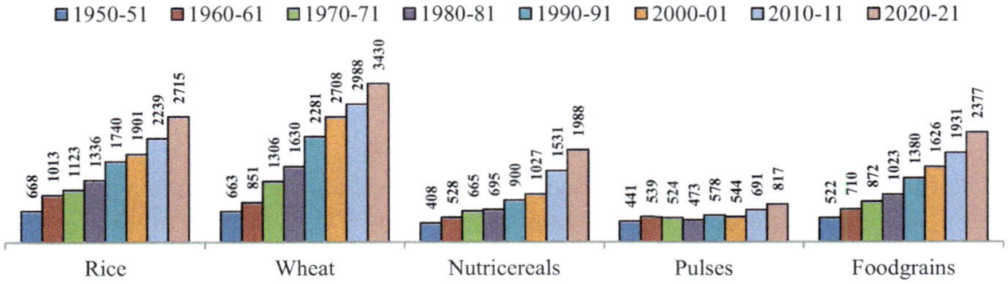

Fig. 21.2 Trends of food grain productivity (kg/ha)

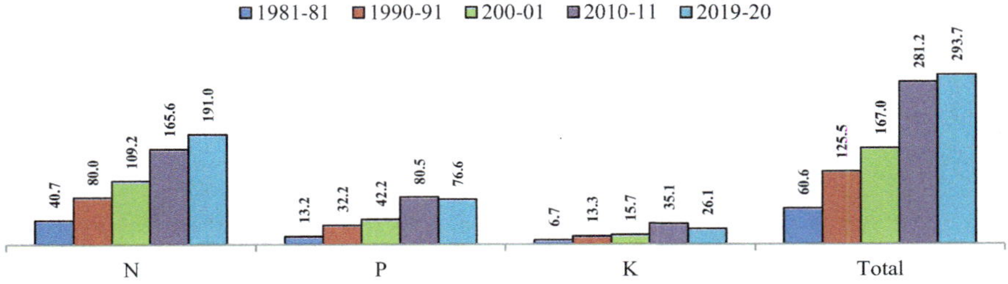

Fig. 21.3 Consumption of fertilizers (in Lakh tonnes)

2015–2016 (Fig. 21.2). The food grain production during 2021–2022 have been recorded as all-time highest of 314.51 million tonnes, which is about sixfold increase since independence.

The amount of chemical fertilizers used post-advent of the Green Revolution was quite high, as well as the increase in the consumption of chemical fertilizers for the cultivation of crop (Fig. 21.3). Fertilizer consumptions presented in Fig. 21.3 elucidate the steep increase in the use of fertilizers since 1981–1982. The overuse of chemical fertilizers to get high yield causes physical and chemical degradation of the soil by altering the natural microflora and increasing the alkalinity and salinity of the soil (Singh 2000). The excessive use of groundwater for irrigation depleted the water table in many parts of the country. The increase in consumption of major cereals such as rice and wheat along with pulses and the decrease in the addition of coarse cereals, foods of animal origin and fruits and vegetables in the diet lead to deficiency of micronutrients such as iron, zinc, calcium, vitamin A, folate and riboflavin among the population causing anaemia, keratomalacia, blindness and infertility in severe cases. Surveys conducted by the National Nutrition Monitoring Bureau and others also conclude the same that the Indian diets based on cereal pulse are qualitatively deficient in micronutrients (NNMB 2003).

Increasing pressure of the population on the land dictates the need for potential utilization of all available land. However, large parts of the land are degraded by desertification, soil salinity, waterlogging, floods and droughts, due to inefficient agricultural practices, and deforestation has caused excessive soil erosion (Gill

1992). There is a marked change in soil fertility caused by changes in agricultural practices during the Green Revolution. To exemplify, 3% of soil in 1980 had a low P content, and by 1995, a low P content was found in 73% of soil, while the area of soil with a low N content only increased from 89 to 91%. Soils with a high K content have decreased from 91% in 1980 to 61% in 1995. The wheat-rice rotation is disturbing the balance of available nutrients in the soil and also causing a deficiency of micronutrients, particularly zinc and copper (Gill 1992). Improving production-specific soil-water-fertilizer-crop management practices need to be evolved based upon comprehensive soil resource information. Research of various aspects of pollution of soil, water and environment, fertilizers, salinity, insecticides, pesticides and industrial effluents should be intensified. Some critical environmental issues including soil erosion, generation of dust and soil compaction due to heavy machinery have also emerged. The major ecological and societal impacts of GR were loss of landraces that were indigenous to our country and the loss of soil nutrients making it unproductive, and the excessive use of pesticides increases the presence of its residues in foods and environment.

21.5 The Post-Green Revolution Era: Sustainable Production

21.5.1 Integrated Farming Systems

Declining size of landholdings without any alternative income-augmenting opportunity is resulting in fall of farm income and causing agrarian distress. To meet the multiple objectives of poverty reduction, food security, competitiveness and sustainability, several researchers have recommended the farming system approach to research and development. In post-Green Revolution (GR) era, individual or crop-centric systems changed into integrated farming system to integrate the farming system modules or to enhance system productivity, resource-use efficiency and sustainable livelihood security. Farming system approach requires involvement of agriculture, horticulture, soil conservation, forestry, fisheries, animal husbandry (piggery and poultry) and allied enterprises (apiculture, etc.). Integrated farming system is a reliable way of obtaining high productivity with substantial nutrient economy in combination with maximum compatibility and replenishment of organic matter by way of effective recycling of organic residues/wastes, etc. obtained through integration of various land-based enterprises (Fig. 21.4). Intensive integrated farming system (IIFS) refers to sustainable production on the one hand and livelihood security on the other, wherein all the components of agriculture, horticulture, forestry, livestock, poultry and fishery can be integrated in a complimentary way besides soil conservation measures, vermicompost, mushroom production, apiculture, liquid manure preparation, etc. For example, under integrated farming systems, straw and stover collected from farm is used for compost preparation which in turn is used for the production of crops, vegetable and fruits and fish production. In the same way, poultry and piggery manure can be prepared and used. The excess produce or waste materials from horticulture can be used as feed

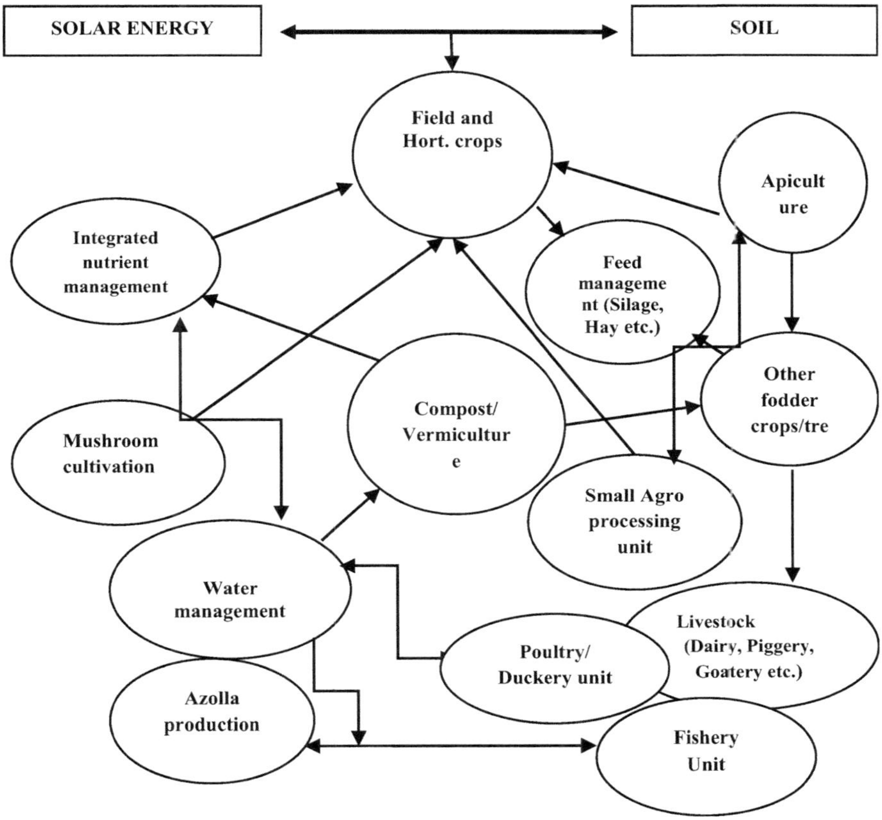

Fig. 21.4 Model of integrated farming system developed by ICAR RC for NEH Region, Manipur Centre, Imphal (Ansari et al. 2013, 2014; Prakash et al. 2017)

materials for piggery (Ansari et al. 2013, 2014, 2021). Rainwater harvesting is playing a pivotal role for sustainable production especially in rainfed farming system (Table 21.1). On station farming data of different agro climatic zones are presented in Table 21.1.

21.5.2 Conservation Agriculture: A New Paradigm in Agricultural Research

Producing food more and more to feed the ever-burgeoning population of India still remains a challenge before the agricultural scientists and farmers even after almost 50 years of the Green Revolution and 75 years of India's independence. Declining factor productivity and inefficient resource utilization are limitations with the conventional agriculture. Besides, increasing fossil fuel usage, crop residue burning and

Table 21.1 On-station farming systems developed in various regions of India, mean data of 6 years (from 2011–2012 to 2017–2018) regarding (gross return, net returns (Rs), man days and resource-saving) (Panwar et al. 2021a, b)

Sl. no.	NARP zone (location of centre)	Farming system	Area (ha)	Gross return (Rs)	Net returns (Rs)	Man days	Resource-saving
1. Western Himalayas							
2	Palampur (Himachal Pradesh)	Cropping systems (0.75 ha) + dairy (2 cows) + horticulture (0.17 ha) + vermicompost unit (0.01 ha)	1.00	3,66,098	1,75,459	468	25
3	Pantnagar (Uttarakhand)	Cropping systems (0.52 ha) + dairy (2 cows) + horticulture (0.19 ha) + agroforestry (0.23 ha) + vermicompost (0.005 ha)		5,27,409	2,34,730	450	23
2. Eastern Himalayas							
4	Umiam (Meghalaya)	Cropping systems (0.70 ha) + horticulture (0.20 ha) + poultry (600 broiler +50 layer) + piggery (3 nos.) + fishery (0.05 ha)	1.00	4,12,368	1,54,050	310	6
5	Jhorhat (Assam)	Cropping systems (0.65 ha) + horticulture (0.095 ha) + dairy (0.11) + fisheries (0.092)	1.00	4,50,027	2,30,324	438	25
3. Lower Gangetic Plains							
	Kalyani (West Bengal)	Cropping systems (0.42 ha) + dairy (2 cows) + horticulture (0.11 ha) + fishery (0.09 ha) + vermicompost (0.018 ha)	0.66	1,90,599	73,697	295	28
4. Middle Gangetic Plains							
7	Patna (Bihar)	Cropping systems (0.40 ha) + dairy (2 cows) + horticulture (0.10 ha) + fishery (0.12 ha) + duck (35 nos.) + vermicompost (0.01 ha) + biogas (2 m^3)	0.80	2,92,800	94,320	239	–
5. Upper Gangetic Plains							
8	Kanpur (Uttar Pradesh)	Cropping systems (0.38 ha) + dairy (2 buffalo +1 cow) + horticulture (0.30 ha) + vermicompost (0.01 ha) + biogas (1.5 m^3) + mushroom (0.002 ha)	0.70	2,78,302	1,08,117	343	36
6. Trans-Gangetic Plains							
9	Ludhiana (Punjab)	Cropping systems (0.64 ha) + dairy (2 cows) + horticulture (0.20 ha) + fishery (0.20 ha) + agroforestry (0.03 ha) + vermicompost (0.005 ha) + apiary (5 boxes)	1.00	6,34,160	3,09,157	293	18

7. Eastern plateau and hills							
11	Raipur (Chhattisgarh)	Cropping systems (0.60 ha) + dairy (2 cows) + horticulture (0.22 ha) + fishery (0.072 ha) + poultry (0.003 ha) + vermicompost (0.003 ha) + biogas (1.5 m^3) + mushroom (0.003)	1.00	2,86,652	1,40,444	308	16
8. Central plateau and hills							
13	Jabalpur (Madhya Pradesh)	Cropping systems (0.90 ha) + dairy (3 cows) + horticulture (0.0024 ha) + fishery (0.06 ha) + poultry (300 nos.) + vermicompost (0.0039 ha) + mushroom (0.0024 ha)	1.00	2,46,670	1,57,715	443	17
9. Western plateau and hills							
14	Akola (Maharashtra)	Cropping systems (0.70 ha) + horticulture (0.25 ha) + goat (12 nos.) + vermicompost (0.004 ha) + mushroom (0.0024 ha)	1.00	1,95,624	55,247		14
10. Southern plateau and hills							
17	Coimbatore (Tamil Nadu)	Cropping systems (1.02 ha) + dairy (2 buffaloes +1 cow) + horticulture (0.16 ha) + vermicompost (0.005 ha)	1.20	7,01,664	2,81,726		25
19	Rajendra Nagar (Andhra Pradesh)	Cropping systems (0.70 ha) + dairy (3 buffaloes) + horticulture (0.20 ha) + poultry (20 nos.) + goat (35 nos.) + vermicompost (0.015 ha)	1.00	4,99,176	1,35,650		25
11. East Coast plains and hills							
22	Bhubaneswar (Odisha)	Cropping systems (0.32 ha) + horticulture (0.19 ha) + dairy (2 cows) + fishery (0.33 ha) + poultry (380 nos.) + duck (20 nos.) + agroforestry (0.094 ha) + vermicompost (0.0033 ha) + mushroom (0.010 ha) + biogas (0.0048 ha) + apiary (2 boxes)	1.25	2,98,938	1,03,139		15
12 West coast plains and hills							
23	Karjat (Maharashtra)	Cropping systems (0.50 ha) + horticulture (0.40 ha) + dairy (3 cows) + poultry (90 nos.) + goat (6 nos.) + vermicompost (0.0018 ha)	1.00	5,88,144	1,60,382		20
13. Gujarat plains and hills							

(continued)

Table 21.1 (continued)

Sl. no.	NARP zone (location of centre)	Farming system	Area (ha)	Gross return (Rs)	Net returns (Rs)	Man days	Resource-saving
24	S.K. Nagar (Gujarat)	Cropping systems (0.70 ha) + dairy (2 buffaloes) + horticulture (0.25 ha) + vermicompost (0.01 ha)	1.00	4,35,551	2,39,444		44
14. Western dry region							
25	Durgapura (Rajasthan)	Cropping systems (1.00 ha) + dairy (2 cows) + horticulture (0.25 ha) + poultry (50 nos.) + goat (6 nos.)	1.45	4,18,665	2,62,544		23

changing land use lead to a considerable global warming and climate change. This implicates that sustainable food production system should inculcate efficient use of natural and man-made resources. Climate smart agronomic management tools and approaches are identified to support this stride. Conservation agriculture (CA) aims at higher productivity and profitability through rational and sustainable use of available resources on a long-term basis. CA-based crop management practices are being increasingly advised for adoption by the farmers in the rice-wheat belt of the Indo-Gangetic Plains (IGPs) to address many problems of the rice-wheat system. CA strives to achieve acceptable profits together with high and sustained production levels. It has potential to emerge as an effective strategy to mitigate the climate change through reduction of the emissions of GHGs and increase the amount of carbon sequestered in soil or above-ground. There are three interdisciplinary and interrelated/interlinked principles in a perfect CA system, which include minimum soil disturbance (no tillage and minimum traffic for agricultural operations), permanent soil cover (leave and manage crop residues on soil surface as cover/mulch) and diversified crop rotations including a legume (for cover crops, green manures, brown manures). Long-term benefits of adopting CA practices include yield increase and reduction in yield variability, enhancing resource-use efficiency and improvements in soil health and environmental quality.

21.5.2.1 Conservation Agricultural Practices

Direct Seeded Rice

Direct seeded rice (DSR) is a resource-efficient technology in which seeds of rice either in simple or in pre-germinated are sown directly in dry or wet conditions. Depending on water and labour availability, this method is divided into (i) wet DSR, in which sprouted rice seeds are broadcast or sown in lines on wet/puddled soil or in standing water, and (ii) dry DSR, in which dry rice seeds are drilled or broadcast on unpuddled soil either after dry tillage or zero tillage or on a raised bed (Kumar and Ladha 2011). This technology does not affect rice quality and can be practised in different ecologies, including upland, medium and lowland, deep water and irrigated areas by large as well as small farmers. DSR has great potential for minimizing the cost of production, soil health hazards and negative impacts on the succeeding crops. Further, reduction in CH_4 emission from rice paddies is observed due to the adoption of resource conservation technologies (DSR), particularly in rice-wheat production system (Singh et al. 2021). Therefore, DSR is technically and economically a feasible alternative to conventional puddled transplanted rice. The important agronomic practices for successful cultivation of DSR are mentioned below.

Direct seeded rice crop should be sown 10–12 days before the historical trends for the onset of monsoon. This would help in the timely establishment of rice crop before rains to reduce seeding mortality due to submergence and make efficient use of rainwater and timely planting of succeeding crop of wheat after rice harvest. For cultivars with medium fine grain, seed rate of 25–30 kg/ha is optimum for DSR crop. Generally, high-tillering rice cultivars are most suitable for DSR. Scented rice basmati type and hybrids rice generally perform better in direct seeded conditions.

After sowing, irrigations can be given at 7–10 days interval so that sufficient moisture remains in soil. Subsequent irrigation can be given when fine hair-line cracks appear on the surface. Water stress must be avoided at tillering, panicle initiation and grain-filling stages, which are very critical for good yield. Weed is a major limiting factor in dry DSR due to the absence of standing water. Before sowing of DSR, the existing and newly germinated weeds can be knocked down with timely and judicious use of glyphosate or paraquat at 1.0 kg/ha using 400–500 L water/ha or mechanically by 1–2 shallow ploughing (zero harrowing). Pre-emergence herbicides such as pendimethalin (1.5 kg/ha), pretilachlor (S) (0.5–0.75 kg/ha) and butachlor (1.5 kg/ha) may be used in direct seeded rice for controlling annual grasses, sedges and broadleaved weeds. Post-emergence herbicides for controlling mixed populations of weeds (grass and broadleaved ones) include bispyribac-Na (20–25 g/ha) at 20–25 DAS, pyrazosulfuron-ethyl (20–25 g/ha) at 15–20 DAS or Almix (4 g/ha) at 20–25 DAS.

Bed Planting (Furrow Irrigated Raised Bed System (FIRBS))
Bed planting is a system where crops are grown on the raised beds alternated by furrows for conserving inputs like seed, fertilizer, water, etc. Beds are usually made at 0.6–1.2 m apart, a minimum of two to three rows of crops are sown on the beds, and irrigation water is applied in the furrows. The height of the beds is maintained at about 15 to 20 cm. During the last decade, practice of raised bed planting has emerged with a greater pace in Indo-Gangetic Plains (IGPs). The major concern of this system is to enhance the productivity and save the irrigation water. The potential agronomic advantages of beds include improving soil structure due to reduced compaction through controlled trafficking and reduced waterlogging and timely machinery operations due to better surface drainage. The beds also create the opportunity for mechanical weed control and improved fertilizer placement. In rice-wheat systems in Asia and Australia, permanent beds also provide the opportunity for diversification to waterlogging-sensitive crops not suited to conventional flat sowing. Typical irrigation savings range from 18% to 30–50% in bed-planted crops (Hobbs and Gupta 2003; Jat et al. 2005). This is a water-saving technology, which saves 30–40% water for growing wheat depending upon the soil type. In addition to water saving, this technology also saves nitrogen by about 25% due to higher nitrogen-use efficiency. In this technology after preparation of land, all three activities, namely, bed formation, placement of fertilizer and sowing of wheat, are done in a single operation. Crop cultivars are known to vary significantly in their performance on raised beds. It is suitable for seed production also because of bolder grain and easier rouging. It reduces the herbicide dependence due to mechanical weed control with the same bed planter fitted with interculture tines with simultaneous placement of fertilizer. In situations where sowing can be delayed due to pre-sowing irrigation, dry seeding can be done on raised beds followed by irrigation immediately after seeding. Irrigation can also be given at grain-filling stage, which is generally avoided by the farmers for fear of crop lodging.

21.5.2.2 Impact of CA on Resource-Use Efficiency (Water, Energy, Solar Radiation)

Conservation agriculture can increase radiation use efficiency, which helps in accumulating higher crop biomass. Zero tillage helps in timely sowing of wheat and thus reduces terminal heat effects (~65 kg ha^{-1} day^{-1}) compared to conventional till wheat (~77 kg ha^{-1} day^{-1}) even under late planting condition (after 21 November to 20 December). Zero tillage with residues keeps canopy temperatures lower by 1–1.5 °C during grain-filling stage (cooling due to transpiration) owing to sustained soil moisture availability to the plants (Jat et al. 2009). The lowered canopy temperature leads to temperature moderation in CA-based system under residue cover on soil surface. Higher canopy temperature in CA during morning and lower canopy temperature during evening were due to insulating property of residue. Malik et al. (2004) reported that ZT in 0.25 m ha in IGP could save 75 million m^3 water, leading to saving of 1029 million m^3 of water every year from almost 3.43 million ha of wheat under ZT. Similarly, Rodell et al. (2010) viewed that 13–17 billion m^3 of groundwater is lost from the northwestern plains of Punjab, Haryana and western Uttar Pradesh. It was also found that a triple ZT practice in rice-wheat-mung bean system resulted in 25–30% savings in irrigation water and 91% higher system water productivity (SWP) than conventional rice-wheat system (Mohammad et al. 2018). The soil water potential (SWP) was highest in zero-till broad bed with residue under wheat-based systems and cotton-wheat had higher SWP than pigeon pea-wheat and maize-wheat systems (Aggarwal et al. 2017). The CA-based maize-wheat-mung bean system could be beneficial in terms of energy utilization and efficiency in the irrigated northwestern IGPs (Saad et al. 2016). Also, triple ZT system with three crop (rice, wheat, mung bean) residues led to a saving of almost 60 kg N/ha in rice and wheat crops in a year. Cotton-wheat system under ZT permanent broad and flat bed with residue performed better under 75% N than 100% N and could save 67.5 kg N ha^{-1} in cotton and wheat in a year. Advantages of CA-based practices like ZT, residue retention, brown manuring (Das et al. 2019a, 2019b), etc. over CT with regard to crop productivity and profitability (Das et al. 2020) and pest dynamics particularly weeds and nematodes (Baghel et al. 2020) have also been reported.

21.5.2.3 Impact of CA on Soil Properties (Physical, Chemical and Biological)

Decomposing crop residue forms soil organic matter (SOM), which can hold water up to 20 times in its weight and improve soil aggregation and soil structure. Bhattacharyya et al. (2013) and Das et al. (2013) reported that ZT with residue resulted in higher aggregate stability, higher aggregate sizes and total organic carbon in soil aggregates than CT. There was improvement in water distribution, NH_4^+-N and NO_3^--N in the root zone of wheat under CA (Shafeeq et al. 2020). Higher available N (20%) in soil was associated with CA over CT. CA had higher TN, organic fractions of TN and NH_4-N. ZT and residue retention could increase total C by 45% and soil microbial biomass by 83% at 0–50 cm depth in a 20-year-old experiment in Brazil over CT (Balota et al. 2004). Similarly, there was a fourfold increase in earthworm population in ZT over CT (Radford et al. 1995). Singh et al. (2018) also reported similar improvement in soil enzyme activities, glomalin content and aggregate stability under ZT in the IGP. Higher dehydrogenase and fluorescein diacetate activities resulting from CA practices compared to CT were owing to OM accumulation through crop residue that could increase microbial activity in soil. Most of the cultivated Indian soils have <5 g/kg SOC compared with 15 to 20 g/kg in uncultivated virgin soils (Lal 2004). The CA-based rice-wheat-mung bean system with residue resulted in almost 13% higher total SOC concentration than conventional rice-wheat in the 0–5-cm soil layer (Bhattacharyya et al. 2015). There was an increase of almost 396 kg/ha/yr. in total SOC stock (Mg ha^{-1}) achieved in this CA system over TPR-CTW system.

21.5.2.4 GHG Emission and Climate Change Adaptation and Mitigation

The CA system can reduce overall GHG emission and play a role in adaptation and mitigation to climate change. Preventing residue burning and improving nitrogen-use efficiency would help to reduce CO_2 and other GHG emissions. One-tonne rice straw burning can emit CO_2-C, 280 kg; CH_4-C, 3 kg; N_2O-N, 0.07 kg; and GWP, 1118 kg CO_2 equiv. Direct seeded rice and wheat under raised beds or ZT could reduce global warming potential (GWP) compared to the TPR-CTW system (Jat et al. 2009; Gupta et al. 2016). There was 34% reduction in GWP upon shifting to DSR from TPR. The temporal variation in N_2O emission implied that crop residue retention had higher emissions of N_2O than without residue. However, the PBB + R treatment had significantly lower GHG intensity (0.20–0.23-kg CO_2 kg^{-1} grain) than the others. The higher GHG intensity values in CT plots indicated that higher GHGs were emitted per kg of grain produced. Increased N_2O emission and reduced CO_2 emission in the ZT system with reverse trends observed in CT system led to GWP values comparable in both treatments. Ghosh et al. (2022) obtained 64% lower-yield scale C footprint (304 kg CO_2-eq. t^{-1}) under triple ZT rice-wheat-mung bean system with residue retention compared to conventional till rice-wheat system.

21.6 Precision Nutrient Management

Plant nutrients play an important role in crop production and the realization of food security. But, often, the farmers' decision to apply fertilizer is resource-driven rather than science-driven as they have no tools to measure the available nutrients present in the soil. They often have no idea of the soil's inherent capacity to supply nutrients. These practices not only result in low crop productivity but also adversely affect the environment. In spite of the high cost and unwanted environmental impact of fertilizer, we cannot overcome the challenges of increasing global population and decreasing arable land without its use. Hence, the importance of knowledge on the efficient use of fertilizer is essential for economic reasons as well as for limiting the effect of its excess usage on the environment.

Precision or site-specific nutrient management (SSNM) is a science-based approach by which crops receive nutrients as and when needed, according to specific field conditions in a given cropping season. Precision nutrient management (PNM) works on principles of 4R, i.e. right fertilizer source (integration of plant sources with inorganic fertilizers) at the right rate (decisions based on soil nutrient supply and plant demand), at the right time (decisions based on the dynamics of crop nutrient uptake) and in the right place (manage spatial variability within the field to meet site-specific crop needs and limit potential losses).

Precision nutrient management is made possible by new technologies like Global Positioning system (GPS), sensors, geographic information systems (GIS) and advanced software and precision application equipment. It aims to modify fertilizers spatially and temporally for cost efficiencies and productivity and environmental gains. Matching fertilizer inputs to site-specific field conditions requires measurement and understanding of soil spatial variability and crop nutrient status and its relation to crop response. Apart from these, the following tools are very useful for precise fertilizer recommendation:

Leaf colour chart: The Leaf Colour Chart (LCC) is used to determine the N fertilizer needs of rice, wheat and maize crops. LCC has four green strips, with colour ranging from yellow green to dark green. It determines the greenness of the rice leaf, which indicates its N content.

SPAD meter: The SPAD (soil plant analysis development) meter is a nondestructive tool and was developed as an alternative to nitrogen nutrition index. Under various climatic conditions and diverse field situations, the N recommendation based on SPAD meter increased the efficiency on N in rice and wheat and also ensure high yields and with economic benefits.

Nutrient expert: Nutrient Expert® (NE) is a nutrient decision support software that uses the principles of site-specific nutrient management (SSNM) and enables farmers to develop fertilizer recommendations tailored to a specific field or growing environment. NE takes into account the most important factors affecting nutrient management recommendations and uses a systematic approach of capturing information, which is important for developing a location-specific recommendation. Nutrient Expert® allows users to draw required information from their own experience, farmers' knowledge of the local region and farmers' practices. NE can use experimental data, but it can also estimate the required. With NE, parameters can be estimated using proxy information, which allows farm advisors to develop fertilizer guidelines for a location without data from field trials.

GreenSeeker: GreenSeeker (GS) is an optical sensor which measures visible and near-infrared (NIR) spectral response from plant canopies to detect N stress that is rapidly increasing. The GreenSeeker™ canopy sensor is a commercially available and widely used active optical sensor that employs red (650 ± 10 nm) and NIR (770 ± 15 nm) wavebands. While chlorophyll contained in the palisade layer of the leaf controls reflectance of the visible light, reflectance of the NIR electromagnetic spectrum depends upon the structure of mesophyll tissues, which reflect as much as 60% of all incident NIR radiation. Measured spectral reflectance is expressed as spectral vegetation indices such as the normalized difference vegetation index (NDVI) calculated as:

$$NDVI = (NIR - VIS)/(NIR + VIS)$$

21.7 Organic Farming

Organic farming is a holistic production management system which promotes and enhances agro-ecosystem health, including biodiversity, biological cycles and soil biological activity. It emphasizes the use of management practices in preference to the use of off-farm inputs, taking into account that regional conditions require locally adapted systems. IFOAM defines 'organic agriculture as a production system that sustains the health of soils, ecosystems and people'. It relies on ecological processes, biodiversity and cycles adapted to local conditions, rather than the use of inputs with adverse effects. The USDA defines it as follows: 'organic farming is a system which avoids or largely excludes the use of synthetic inputs (such as fertilizers, pesticides, hormones, feed additives etc.) and to the maximum extent feasible rely upon crop rotations, crop residues, animal manures, off-farm organic waste, mineral grade rock additives and biological system of nutrient mobilization and plant protection'. Organic crop production in the organic farming system is gradually gaining momentum worldwide. Global links have been forged in all continents as organic agriculture has been seen to be an effective rural development option.

According to the latest FiBL survey on certified organic agriculture worldwide, as of the end of 2015, data on organic agriculture was available from 179 countries (172 in 2014) (Willer and Lernoud). There were 50.9 million hectares of organic agricultural land in 2015, including inconversion areas. The regions with the largest areas of organic agricultural land are Oceania (22.8 million hectares, which is almost 45% of the world's organic agricultural land) and Europe (12.7 million hectares, 25%). Latin America has 6.7 million hectares (13%) followed by Asia (four million hectares, 8%), North America (three million hectares, 6%) and Africa (1.7 million hectares, 3%). The countries with the most organic agricultural land are Australia (22.7 million hectares), Argentina (3.1 million hectares) and the United States currently; 1% of the world's agricultural land is organic. The highest organic shares of the total agricultural land, by region, are in Oceania (5.4%) and in Europe (2.5%). In the European Union, 6.2% of the farmland is organic. However, some countries reach far higher shares: Liechtenstein (30.2%) and Austria (21.3%). In 11 countries, 10% of the agricultural land or more is organic.

Organic farming in India differs from conceptions around the world in the smaller size of farms managed by each household which, therefore, includes greater levels of own labour, the farmer's more intimate knowledge of field conditions and fewer animals managed per household which bring it closer to environmentally sustainable agriculture. As one of the world's largest producers of rice, tea, fruits and vegetables, various spices, pulses, medicinal plants and cashew nuts, India's first internationally certified organic products emerged in the mid-1970s, supported by UK's Soil Association. India has evolved into a rich history of agricultural practices and continues to adapt technologies like biodynamic and other systems into its organic practices. India's organic farmers have been at the forefront of developing field-based technologies ranging from vermicomposting to integrated livestock practices that facilitate their ability to improve soil fertility even in semi-arid or barren areas. Different parts of India have developed their own local or regional systems for ecological agriculture such as agnihotra and panchakavya that are now gathered in one umbrella term: 'Jaivic Krishi' or 'Jaivik Kheti'. Overcoming early doubts on the potential of organic systems to meet growing needs including in experiments conducted in ICAR'S Network Project on Organic Farming, it has been established that in appropriate technologies in hill areas and rainfed tracts with poor and marginal soils and relatively virgin soils, organic farming system is superior to conventional agriculture in productivity, resource-use efficiency and profitability.

In January 1994, 'Sevagram Declaration' for promotion of organic agriculture in India, organic farming has grown many folds, and a number of initiatives at the government and nongovernment level have given it a firm direction. While the National Programme for Organic Production (NPOP) defined its regulatory framework, the National Project on Organic Farming (NPOF) has defined the promotion strategy and provided necessary support for area expansion under certified organic farming. Before the implementation of NPOP during 2001 and introduction of accreditation process for certification agencies, there was no institutional arrangement for assessment of organically certified area. Initial estimates during 2003–2004 suggested that approximately 42,000 ha of cultivated land were certified organic. By 2009, India had brought more than 9.2 million ha of land under certification, and it has been increased further (Ravisankar et al. 2021). By March 2021, the cultivated area under organic farming had increased to 2.66 million ha (APEDA 2021), while ~0.73 million hectares had been brought under the Participatory Guarantee System (PGS). As a result, approximately 2.4% of the net farmed land is currently under certification or in the process of conversion to organic farming. India produces a diverse range of crops under organic management, with oilseeds, sugar crops, fibre crops, cereals, millets and pulses accounting for the majority of the agricultural basket. The Indian government has set a goal of converting at least 4% of net cultivated land to organic farming by March 2026. In terms of total arable area under organic farming, India is presently the ninth largest in the world and the largest in terms of the total number of organic producers. India produces roughly 2.75 million tonnes of certified organic products in 2019–2020, with an export volume and value of 0.6389 million tonnes and INR 46,860 million (689 million USD). However, organic farming promotion is hampered by a lack of yield, as well as the availability of inputs for nutrition, weed, pest and disease management (Table 21.2). Organic nutrient management practices for various cropping systems are presented in Table 21.2.

Table 21.2 Organic nutrient management practices identified for various cropping systems in different agro-ecosystem

Location (state)	Cropping system (s)	Sources to meet nutrients
Coimbatore (Tamil Nadu)	Cotton-maize-green manure (GM) Chillies-sunflower-green manure	Farmyard manure (FYM) + non-edible oil cakes (NEOC) + Panchagavya (PG)
Raipur (Chhattisgarh)	Rice-chickpea	Enriched compost (EC) + FYM + NEOC + biodynamic (BD) + PG
Dharwad (Karnataka)	Groundnut-sorghum Maize-chickpea	EC + VC + green leaf manure (GLM) + biodynamic and PG spray
Ludhiana (Punjab)	Maize-wheat-summer green gram	FYM + PG + BD in maize, FYM + PG in wheat, and FYM alone in moong
Bhopal (Madhya Pradesh)	Soybean-wheat Soybean-chickpea Soybean-maize	FYM + PG + BD
Pantnagar (Uttarakhand)	Basmati rice-wheat-green manure Basmati rice-chickpea Basmati rice-vegetable pea	FYM + VC + NC + EC + BD + PG
Ranchi (Jharkhand)	Rice-wheat-green manure	VC+ karanja cake + BD+ PG

Ravisankar et al. 2021

21.8 Natural Farming

'Natural farming' or 'eco-agriculture' or 'eco-friendly agriculture' is suggested as a neoteric approach to improve both traditional and modern agricultural practices, which aims to safeguard the environment, public health and communities (Mishra 2013). Sustainable agricultural practices enable food production without compromising the needs of future generation. The meaning of zero-budget natural farming (ZBNF) can be understood with the word 'budget' referring to credit and expenses; thus, the phrase 'zero budget' means without using any credit and without spending any money on purchased inputs, and natural farming means farming with nature and without chemicals (NAAS 2019). It encourages the natural symbiosis of soil microflora and crop plants. It is assumed that it has no or least side effects on the health of soil and environment. ZBNF is based on four pillars, viz. (i) *Jeevamrit*, nectar of life (consisting of microbes) that is prepared from dung and urine of indigenous Kapila cow (not other animals like exotic or cross-bred cows, bulls or buffaloes); (ii) *Beejamrit*, the seed treatment; (iii) *Acchadana* (mulching); and (iv) *Waaphasa* (soil aeration/moisture). Plant protection measures include a mixture of butter milk, cow milk, pepper powder, neem seed and green chilli (Palekar 2016). The inputs used for seed treatments and other inoculations are locally available in the form of cow dung and cow urine. ZBNF farmers thus have lower cost of inputs and thus have better capacity to increase incomes. At the same time, ZBNF crops helps in retaining soil fertility and is climate change-resilient. Thus, ZBNF is low-input, climate-resilient type of farming that encourages farmers to use low-cost locally sourced inputs, eliminating the use of chemical fertilizers and industrial pesticides. ZBNF recommends the use of only traditional varieties/land races with inherently low-yield potential which respond to low level of soil-applied nutrients. A policy paper on ZBNF (NAAS 2019) reviewed that Beejamrit used in ZBNF will not increase the yield potential of high-yielding modern varieties and what it can only do is to protect the Beejamrit-treated varieties from some diseases and the microbes can add a few plant growth-promoting substances which can have incremental effect and will not have significant impact on quantum jump in their yield. ZBNF is said to be 'zero-cost or zero-input natural farming', and therefore, whatever quantity is harvested is treated as net profit to the farmer.

The Government of Andhra Pradesh has launched a scale-out plan (on 2 June 2018) to transition six million farms/farmers cultivating eight million hectares of land from conventional synthetic chemical agriculture to zero-budget natural farming (ZBNF) by 2024, making Andhra Pradesh India's first 100% chemical-free agriculture farming state. However, studies initiated by the ICAR-Indian Institute of Farming Systems Research (IIFSR), Modipuram, at several locations in the country have clearly indicated that yield levels were drastically reduced in rice-wheat cropping system (the backbone of national food security) by 59% in wheat and 32% in basmati rice. Results of a 3-year natural farming experiment by the University of Agricultural Sciences (UAS), Dharwad, indicated a yield decline of at least 30% in soybean-wheat, groundnut-sorghum and maize-chickpea cropping systems, while it was 17% in cotton-groundnut intercropping system. These trials have clearly established that food security will be seriously challenged along with the farmers' income, if ZBNF is adopted (NAAS 2019).

21.9 Food-Fodder-Based System

The profitability of livestock largely will depend upon the availability of quality feed and forages as it accounts to 65–70% of cost in their rearing. It is estimated that the demand will reach to 1012 million tonnes of green fodder and 631 million tonnes of dry forage by the year 2050. At the current level of growth in forage resources, there will be 18.4% deficit in green fodder and 13.2% deficit in dry fodder in the year 2050 (IGFRI Vision 2050). The area under forages is stagnating around 4 of net cultivated area since last decades, and the option for further expansion or horizontal spread is meagre. The availability of green fodder will be more difficult under changing climatic conditions and preference of food and commercial crops for human consumption. This necessitates focus towards inclusion of forages in existing food-based cropping systems in temporal or space dimension. Fodder-based crop sequences have the potential for realization of higher monetary returns over food-based crop rotation in majority of the agro-ecological regions. Inclusion of forages in any of the season was equal or superior than food-based systems (Kumar et al. 2012).

Fig. 21.5 Food-fodder-based system for rainfed areas: trispecific hybrid + *Sesbania*-sorghum (fodder) + cowpea-barley (Kumar et al. 2021)

In Bundelkhand region, maize (grain)-berseem cropping sequence recorded maximum wheat equivalent yield as compared to groundnut-wheat + mustard and black gram-wheat + mustard. Another crop sequence of cowpea-wheat sequence proved more remunerative (38.8 t green fodder, 4.63 t wheat grain and 5.02 t wheat straw) than grain sorghum-wheat cropping system which enriched the soil fertility and economized fertilizer N to the extent of 40-kg nitrogen per hectare. In food-fodder production systems, fodder cowpea (13.6 q/ha dry fodder) and annual *Sesbania* (18.9 q/ha dry fodder) with grain sorghum were marked as better legumes. Inter-row spacing of pigeon pea could be utilized effectively by intercropping of sorghum, teosinte, maize, bajra, cowpea, clusterbean and deenanath grass yielding green forage of 5 to 12 t/ha without significant reduction in pigeon pea grain yield. In *rabi* season, barley in association with senji proved to be quite promising and mixtures of safflower with oat and barley resulted in higher fodder yields. Inclusion of berseem in rice rotation saved 100 kg N/ha at the expense of only 20 kg P_2O_5/ha in the system (140 kg N + 140 kg P_2O_5/ha) as compared to wheat-rice (240 kg N + 120 kg P_2O_5/ha). Intercropping of berseem and senji improved the physical properties of the soil by increasing the infiltration rate (4.82 mm/h) and reducing the bulk density (1.26 g/cm^3). In intercropping with forage cereals like sorghum, the green forage, dry matter and crude protein yields of cowpea were highest in paired row spacing followed by additive and replacement series. Sowing of berseem and lucerne in border method at every fourth skipped row produced 7.8 and 4.6 t/ha green fodder without any adverse effect on the yield of wheat.

Under rainfed semi-arid region (up to 500-mm rainfall), *Pennisetum* trispecific hybrid (TSH) + *Sesbania* (planted on beds) + (sorghum + cowpea (fodder)-barley/chickpea (grain)-planted in furrows) cropping system produced highest forage equivalent yield (101–105 t/ha/year) with in situ moisture conservation (CA-based practices) and lifesaving irrigation through field pond as compared to conventional rainfed system (87 t/ha/year). This technology is capable of prolonging the fodder availability up to May–June (dry period). This model can sustain three to our adult cattle inits (ACU) if fed 100% DM based on forage diet and eight to nine ACU if fed 1/3 DM-based fodder diet with quality fodder (Kumar et al. 2021) (Fig. 21.5).

Forages (perennial grasses, viz. Bajra Napier hybrid, TSH, *Setaria* and Guinea grass) on bunds/farm boundaries/irrigation channels/terrace risers can produce 7.0–11.0 q green fodder per 100 meter bund length. In spite of additional farm productivity, it also works as a guard crop for main crop, reduces run-off loss of water and controls soil erosion. Technology can sustain 1 milch animal and reduces milk production cost. If only 10% bunds are kept under fodder production, we can produce approximately 17.9 tonnes green fodder annually. The technology has shown significant impact under NIFTD (National Initiative on Fodder Technology Demonstration) programme operated in 100 KVKs of the country (Ghosh et al. 2016). Furthermore, there exists a scope for improving the basket of feed resources through the use of nonconventional/underutilized feed resources like spineless cactus, *Lathyrus*, sugar beet and *Moringa*. These can be incorporated in unutilized lands/marginal soils/degraded areas along with other existing options of forages/grasses. Spineless cactus is well suited for arid and semi-arid region in degraded lands, pasture/Nalla bunds and field boundaries and thrives well in drought-prone/scanty rainy areas. *Moringa* and sugar beet can be incorporated well for energy−/protein-rich feed supplements in milk-shed areas. Spineless cactus can be successfully fed to livestock as supplementary feed (30% replacement) along with green fodder, hay and straw, but not as sole feed (Kumar et al. 2021).

21.10 Agri-Horti- and Silvi-Pasture-Based Systems

A combination of livestock and crop production in a unit land-use management system has been an integral part of the farming pattern in India. The practice is more prolific in hilly tract and in rainfed lands. Based upon the last four to five decades of research on rainfed farming system, agroforestry systems (alley cropping, agri-horticultural systems, horti-pastoral systems) in arable lands and tree farming/wood lot, range/pasture management, silvi-pastoral systems, timber and fibre system) in cultivable wasteland/marginal lands are suggested. Suitable tree/shrubs, grasses and legumes have been identified for use in different agroforestry systems in varied habitats of the country. The developed models of silvi-pasture and horti-pasture are region-specific and have the ability to sustain the system on an ecological and monetary basis (Ghosh et al. 2012; Kumar et al. 2017a, 2017b).

21.10.1 Silvi-Pasture

Rainfed/arid regions of the country have degraded land and constraints in arable farming. Animal husbandry is the main occupation of farmers after rainfed crop cultivation in these areas. Livestock is dependent on forage produced in rangeland areas. In such areas, large-scale grazing of animals results in quite low pasture production restricting availability during monsoon only. To solve the above-said problem, silvi-pasture models (MPTS + pasture) have been developed that produce higher forage per unit area per unit time as well as round the year than open pasture.

In degraded lands under semi-arid tropical environment at IGFRI, Jhansi reported that on average of 7 years, *C. ciliaris* and *P. maximum* produced 4.20 and 6.84 t/ha/year dry matter yield under tree canopies of *A. tortilis* and *L. leucocephala*, respectively. Some of the promising silvi-pasture for different regions has been presented below (Table 21.3) and (Fig. 21.6):

21.10.2 Horti-Pasture System

Under horti-pastoral system forages are grown in wide inter-row spaces of fruit trees for economic utilization of orchard lands. Attempts have been made to identify suitable fruit trees for different agro-ecological zones of the country for their use in horti-pastoral and agri-horticultural plantations. There are ample scope and many opportunities for introducing fodder crops in existing orchards. Horti-pasture system integrates pasture (grass and/or legumes) and fruit trees to fulfil the gap between demand and supply of fruit, fodder and fuel wood through utilizing moderately degraded land. Ber-, aonla-, guava-, bael-, *Moringa*-, mulberry-, etc.-based horti-pasture systems have been experienced for enhancing forage productivity at IGFRI. During a period of 10 years, it gives B:C ratio of 4:6 and supports 2–3 ACU/ha in a year. Perennial grasses, viz. NB hybrid, TSH, *Setaria*, Guinea grass, *Cenchrus ciliaris* and range legumes *Stylosanthes hamata*, are potential fodder types that can be integrated in existing orchards in tropical and subtropical regions. In hilly regions, apple/peach/apricot/plum/almond/mango/guava/litchi orchards can be intercropped with potential grasses like tall fescue/rye grass/timothy/orchard grass with white and red clovers. According to an estimate, approximately 19.0 and 17.9 MT of green fodder can be obtained from mango and coconut orchards of the country (30% area of the total orchard area of these orchards). For calculation of this estimate, a conservative figure of 40 t/ha/year productivity was taken. From other orchards (guava, sapota, litchi and apple) contribution could be 4.0 MT of green fodder. Such model can be implemented through convergence of government schemes like the National Horticulture Mission, Coconut Development Board and National Mission on Oilseeds and Oil Palm (Fig. 21.7).

21.11 Hill and Mountain Farming

Mountains play a crucial role in sustaining about 10% of the world's population directly. In addition, mountains sustain life of people living in the plains as they are the major source of water supply as majority of rivers originate from these ecosystems. The major mountain ranges in India are the Himalayas and the Western Ghats. They traverse an arc of about 2500 km between the Indus and the Brahmaputra rivers. In India, the mountain ecosystem is spread over 13 states/UTs of India: Jammu and Kashmir, Ladakh, Himachal Pradesh, Uttaranchal, Sikkim, Assam, Arunachal Pradesh, Manipur, Meghalaya, Mizoram, Nagaland, Tripura and West Bengal. They are inhabited by >51 million people, covering 18% of the

Table 21.3 Promising silvi-pastures for different regions of India

S. no.	Ecological region	Suitable trees, grasses and range legumes	Fodder potential (DM t/ha)	Fodder availability period and sustaining ACU/ha	Suitable rainfed region of India
1.	Semi-arid	*Ficus*/*Hardwickia binata*/*acacia*/*Morus*/neem-based silvi-pasture with grass (*Chrysopogon fulvus*, *Cenchrus ciliaris* and *Panicum maximum*) + legume (*Clitoria ternatea* and *Stylosanthes seabrana*)	12.3	Grasses and legume (July to December (65–70%) and tree leaves (march to June (30–35%) and 2–3 ACU/ha	Forest area, degraded lands in semi-arid region (rainfall 400–700 mm)
2.	Arid	*Prosopis cineraria*/*acacia*/*Ziziphus nummularia-based* system along with *Cenchrus ciliaris*, *Lasiurus sindicus*	1.5–2.5	Grasses and legume (July to October (50–55%) and tree leaves (November to June (45–50%) and 1 ACU/ha	CPRs, grazing lands of arid region (rainfall 200–250 mm)
3.	Coastal	*Acacia auriculiformis*/*Casuarina equisetifolia*/*Acacia nilotica* and *Prosopis chilensis*/*Prosopis juliflora* along with *C. ciliaris*, *Chrysopogon*, *Heteropogon*, *Panicum maximum*, *Stylosanthes hamata*	8–10	Grasses and legume (July to October (60–65%) and tree leaves (November to June (35–40%) and 2–3 ACU/ha	CPRs, grazing lands, forest fringes of arid region (rainfall 1500–3500 mm)
4.	Western Himalayan region	*Arundinella nepalensis*/*Morus alba*/*Grewia optiva*/*Agrostis stolonifera*/*bauhinia variegate* along with *Bothriochloa pertusa*, *Dactylis glomerata*, *red and white clovers*	2–5	1–2 ACU/ha	Pastures and grasslands in low, mid and upper hills

(continued)

Table 21.3 (continued)

S. no.	Ecological region	Suitable trees, grasses and range legumes	Fodder potential (DM t/ha)	Fodder availability period and sustaining ACU/ha	Suitable rainfed region of India
5.	Northeastern hill region	MPTs (*Alnus nepalensis*, *Michelia champaca*, *Schima wallichii*, *Parkia roxburghii*, *Artocarpus heterophyllus*) along with broom grass (*Thysanolaena maxima*), Congo signal grass (*Brachiaria rosenesis*), hybrid Napier (*Pennisetum typhoides* x *P. purpureum*) and Guinea grass (*Panicum maximum*)	18–28	–	Degraded lands, sloping land field bunds, farm fences, etc.

Hybrid napier on terrace risers in NEH region

Hardwickia binnata + *sylosanthes hamate Cenchrus ciliaris*

Fig. 21.6 Silvi-pastoral systems

geographical area and 6% of India's population. The rich environmental heritage of the Himalayan region is under pressure from natural and human-induced stresses.

The hill agriculture has its own unique characteristics, and the growth potential of hill agriculture has remained under-exploited due to various mountain specifies like undulating topography, lack of system-specific technologies, poor marketing and

Fig. 21.7 Horti-pastoral systems

processing infrastructure and underdeveloped supporting institutions (Wani 2011). Resource degradation in terms of soil and water loss through erosion, nutrient mining, low cropping intensity, low-yielding local varieties, etc. is some of the major constraints for sustainable production in hills (Mandal and Sharda, 2013). However, hilly states have a lot of potential to accelerate agricultural growth through diversification from low- to high-value crops. There is a very good potential for adoption of location-specific IFS models, conservation of agriculture and organic farming for sustainable food and nutritional security and conservation of natural resource base for the posterity. Soil loss to the tune of more than 40 t/ha has been reported as against the national average of 16 t/ha from NER. Excessive tillage operations, residue removal/burning and monocropping are the major factors contributing to land degradation in hill farming systems. Thus, conservation agricultural practices compromising no-till (NT) or minimum tillage (MT), residue retention/cover crops and rational crop rotations have been found very promising for optimum productivity of rice- and maize-based cropping systems and improving soil properties in NER (Das et al. 2020). There are ample opportunities to harness the production potential of surface water and agro-climatic diversities that favours cultivation of fruits, vegetables and crops of industrial importance like ginger, turmeric, king chilli, strawberry, passion fruits, etc.

Ghosh et al. (2010) reported that double NT practice in rice-based system was cost-effective, restored SOC stock, favoured biological activity, conserved water and produced better yield, which were 70.8, 46.7 and 49.0% higher compared to CT. Adoption of NT, along with retention of maize stalks and application of *Ambrosia* spp. mulch at 5 Mg ha^{-1}, resulted in the maximum improvement in soil quality parameters and enhanced rapeseed (*Brassica napus* subsp. napus) yield in maize-rapeseed cropping system in the Eastern Himalayas. Mulching treatment, comprising of the retention of maize stock cover along with application of *Ambrosia* spp. mulch at 5 Mg ha^{-1}and poultry manure at 5 Mg ha^{-1}, enhanced mean SOC concentration by 30.4% and mean weight diameter of aggregates by 100% as compared with those under control. Thus, NT and mulching are recommended measures for protecting surface soil and improving soil quality in the EHR (Das et al. 2014).

Table 21.4 Ecosystem-specific sustainable soil management practices in NER

Valley lands/ lowlands	Uplands/ terraced lands	Sloping lands	Shifting cultivation
Conservation tillage (MT/NT)	Conservation tillage (MT/NT)	Agroforestry systems	Strip/intercropping
Crop residue recycling	Mulching/ residue retention	Hedge rows on contours	Fast-growing multipurpose tree species (MPTs) like *Alnus nepalensis*
Organic manure	Cover crops	Cover crops	Improved varieties
Weed biomass	Green manuring	Perennial fodder crops	Foliar nutrition
Green leaf manure (GLM)/ green manure	Integrated nutrient management	Medicinal and aromatic grasses	Biofertilizer
Integrated nutrient management	Perennial fodder crops	Vegetative barrier through aromatic and medicinal grasses	Seed treatments
Crop diversification with pulses/ oilseeds	Crop diversification	Permanent soil cover	Hedge row barriers on contours
Integrated farming system (IFS)	Rotation with legumes	Protected cultivation of high-value organic crops	Residue retention/mulching
Intensification of rice fallow areas	Biochar application	Across the slope cultivation	MT/NT
	Hedge rows on terrace risers/ boundaries		
	MPTs on fences		
	Watershed-based IFS		

Saha et al. 2021

The strategy of sustainable agriculture comprising soil and water management, nutrient cycling and soil health management varies from location to location as per soil conditions, climate, land-use change, fertility status, existing land management practices, crops and cropping systems and resources available with the farmers. Based on these factors, different conservation effective best management practices (BMPs) have been developed/standardized and recommended to the stakeholders (Table 21.4). Crop rotation, minimum tillage, no-till (NT), intercropping, mulching, organic farming, cover cropping and management of crop residues contribute to SOC build-up. Lal (2003) reported that approximately 1.14 Pg (10^{15} g) of C is released into atmosphere through global water erosion processes.

Minimum soil disturbances through the adoption of MT and NT along with covering the land surface through crop residues, organic manures, leaf litter and other vegetative materials and its proper management are the most effective measure to prevent soil from blowing, beating action of rainfall, splash erosion and loss of sediment in run-off. In a field experiment conducted at northwestern Himalayas, the performance of no-till (NT) was assessed in terms of SOC build-up. It was reported that surface soil of NT plots had 11% higher total SOC than conventional tilled plots after 4 years in rice-wheat cropping system (Bhattacharyya et al. 2013). Studies at ICAR-IISWC (formerly CSWCRTI), Dehradun, showed that green manure mulching was beneficial which produced 18 to 26% higher grain yield of maize (2.52 to 2.69 t/ha) over control (2.14 t/ha). Furthermore, the results also revealed that sunn hemp + *Leucaena* mulch was more beneficial which yielded 2.69 t/ha grain as compared to sunn hemp (2.59 t/ha) and *Leucaena* mulch (2.52 t/ha) alone. Mulching also reduces evaporation, keeps down weeds, improve soils structure and infiltration rate and ultimately increases crop yields (Mandal et al. 2017). Live mulch applied at 4–5 t/ha in the inter-row space found to be beneficial in reducing run-off from 42 to 22% and soil loss from 12 t/ha to 3 t/ha in central Himalayan region. Similarly, at 4% slope, the minimum tillage + crop residue treatment recorded about 11% higher yield over conventional tillage at northwestern Himalayan region. The highest SOC buildup rate was observed under *Gliricidia* + GFS, ranging from 1.35 t/ha/yr. at 1 m to 0.35 t/ha/y in Odisha hill region (Mand al 2018). Averaged over 3 years of observation, the efficacy of *Indigofera* + grass filter strip (GFS) was found better with lowest run-off (8.9%) and soil loss (5.0 t/ha), followed by *Gliricidia* + GFS (10.7% run-off, 6.3 t/ha soil loss). It might be possible that short and compact growth habit and better soil binding efficiency of *Indigofera* roots resulted in lower run-off and soil loss, when integrated with GFS. Provision of only a row of GFS can reduce run-off by 14% and soil loss by about 23%. *Gliricidia* alone reduced run-off by 26% and soil loss by 34%, whereas GFS supplement further improved the conservation efficacy (run-off reduced by 33% and soil loss by 35%) (Dass et al. 2011). Even 1-year-old hedgerows could reduce run-off by 63–71% and soil loss by 82–86% (Lin et al. 2009). Ghosh et al. (2012) palmarosa + vegetative barrier of (i.e. VS + MT + live mulch) has been observed as the best treatment with respect to resource conservation and produced 28% higher yield than *Panicum* (Ghosh et al. 2012).

21.12 Managing GHG Emission in Farming

During 1970–2010, GHG emissions from Indian agriculture have increased by about 75%. The increasing use of fertilizers and other agri-inputs and the rising population of livestock are the major drivers for this increase in GHG emission (Pathak et al. 2014). Increased use of production inputs, such as mineral fertilizer, has made Indian agriculture more greenhouse gas (GHG)-intensive. Agricultural production is a major emitter of GHGs, currently accounting for 18% of total GHG emissions in India (INCCA 2010). With a population of ~1.3 billion, it is evident that the food system in India will be central to the global challenge of providing sufficient nutritious food while minimizing GHG emissions. Livestock and rice production were found to be the main sources of GHG emissions in Indian agriculture with a country average of 5.65 kg CO_2 eq. kg^{-1} rice, 45.54 kg CO_2 eq. kg^{-1} mutton meat and 2.4 kg CO_2 eq. kg^{-1} milk. Production of cereals (except rice), fruits and vegetables in India emits comparatively less GHGs with <1 kg CO_2 eq. kg^{-1} product (Sylvia et al. 2017). On a *per* ha basis, GHG emissions for major food crops in India are generally lower than those in Europe and North America, with GHG emissions for cereals twofold to threefold greater in Europe (2000–3000 kg CO_2 eq. ha^{-1} $year^{-1}$) (Carlton et al. 2012). Continuous flooding generates the highest CH_4 emissions, while longer and more frequent periods of water drainage reduce emissions. For example, changing the water regime from continuously flooded to multiple drainage periods reduces CH_4 emissions by ninefold (data not shown). High emissions on a *per* ha basis correspond with high GHG emissions *per* kg rice. With the exception of flooded rice, the major source of variation in GHG emissions for crops is due to variation in fertilizer application (Sylvia et al. 2017). The GWP has been found to range from 240 kg CO_2 eq. ha^{-1} in pulses to 3700 kg CO_2 eq. ha^{-1} in continuously flooded rice. Agriculture has the potential to mitigate GHGs cost-effectively through the adoption of low carbon in agricultural technologies and management practices (Pathak et al. 2014). The fluxes of CO_2, CH_4 and N_2O can be reduced by efficient management of carbon and nitrogen flows in the agricultural ecosystems. Any practice that increases the photosynthetic input of carbon and/or slows the return of stored carbon to CO_2 via respiration, fire or erosion will increase carbon reserves, thereby 'sequestering' carbon or building carbon 'sinks'.

The mitigation of GHG emission from agriculture can be achieved by sequestering C in soil and reducing emissions of methane and nitrous oxide from soil through changes in land-use management. Changing of crop mixes to include more plants that are perennial or have deep root systems increases the amount of carbon stored in the soil. Cultivation systems that leave residues and reduce tillage, especially deep tillage, encourage the build-up of soil carbon. The changes in crop genetics and proper management of irrigation, fertilizer use and soils can reduce emission of both nitrous oxide and methane. Such options are important not only for global warming mitigation but also for improving soil fertility. In addition, GHG emissions can be reduced by the substitution of fossil fuels for energy production by the agricultural feedstocks (e.g. crop residues, dung, and dedicated energy crops). Alternate wetting and drying/mid-season drainage, drip and sprinkler irrigation, nitrification inhibitor/urea super-granules, site-specific N management/use of LCC, direct seeding of rice, crop diversification (e.g. rice to maize) and adoption of CA in rice have been recommended for GHG emission mitigation in rice and wheat crops in IGPs (Pathak et al. 2014).

21.13 Conclusions

Sustainability is a perpetual issue in time scale with associated dynamism in resource base and outputs in terms of variety and quantity. Sacrificing sustainability in agriculture will pose grave threat to the basic food security of the countries, primarily dependent on agriculture. For decades, input and energy-intensive agriculture being practised in Northern and Northwestern Indo-Gangetic Plains has exhausted the soil and water resources, and the agriculture system has faced numerous challenges that may lead to virtual breakdown and desertification in agriculture. Scientists advocated several sustainable agriculture principles to farming for better soil, environment and food quality. Integrated nutrient management with 5R nutrient stewardship could replace conventional approach of excessive fertilizer application. Several long-term 'on-station' and 'on-farm' research trials have recommended conservation of agriculture-based diversified cropping practices/rotations for improving resource-use efficiency, soil health and environmental quality. Selected organic and natural farming belts can be developed based on soil, climate and market suitability. Small and marginal farmers especially under rainfed agro-ecosystems should adopt recommended farming system approach for better farm production and risk alleviation. Under marginal and less productive lands, alternate land-use systems such as agri-horti-pasture, horti-pasture and silvi-pasture have been recommended for better productivity and climate resilience. Cropping systems and land-use practices must be aligned based on principles of sustainability for sustainable productivity and better insurance against biotic and abiotic hazards.

References

Aggarwal P, Bhattacharyya R, Mishra AK, Das TK, Šimůnek J, Pramanik P, Sudhishri S, Vashisth A, Krishnan P, Chakraborty D, Kamble KH (2017) Modelling soil water balance and root water uptake in cotton grown under different soil conservation practices in the Indo-Gangetic plain. Agric Ecosyst Environ 240:287–299

Ansari MA, Prakash N, Baishya LK, Punitha P, Sharma PK, Yadav J, Kabui GP, Ki Levis C (2014) Integrated farming system: an ideal approach for developing more economically and environmentally sustainable farming systems for the eastern Himalayan region. Indian J Agric Sci 84(3):356–362

Ansari MA, Prakash N, Baishya LK, Punitha P, Yadav JS, Sharma PK, Blessa S, Ansari MH (2013) Comparative study on conventional and improved integrated farming systems for sustainable production, income generation and employment opportunity among the tribal farmers in hilly regions of Manipur. Indian J Agric Sci 83(7):765–772

Ansari MA, Sharma SK, Roy SS, Ramakrishna Y, Datt S, Ningombam A, Singh NA, Luiram S, Prakash N (2021) Documenting the agriculture based indigenous traditional knowledge in Manipur state of north eastern India. Indian J Tradit Knowl 20(4):1065–1074

APEDA (2021). Agricultural and Processed Food Product Export Development Authority. https://apeda.gov.in/apedawebsite/organic/organicproducts.htm

Baghel JK, Das TK, Mukherjee PI, Nath CP, Bhattacharyya R, Ghosh S, Raj R (2020) Impacts of conservation agriculture and herbicides on weeds, nematodes, herbicide residue and productivity in direct-seeded rice. Soil Tillage Res 201:104634

Balota EL, Filho AC, Andrade DS, Dick RP (2004) Long-term tillage and crop rotation effects on microbial biomass and C and N mineralization in a Brazilian Oxisol. Soil Tillage Res 77:137–145

Bayly CA (1985) State and economy in India over seven hundred years. Economics History Rev 38: 583–596

Bhattacharyya R, Das TK, Pramanik P, Ganeshan V, Saad AA, Sharma AR (2013) Impacts of conservation agriculture on soil aggregation and aggregate-associated N under an irrigated agro-ecosystem of the Indo-Gangetic Plains. Nutr Cycl Agroecosyst 96:185–202

Bhattacharyya R, Das TK, Sudhishri S, Dudwal B, Sharma AR, Bhatia A, Singh G (2015) Conservation agriculture effects on soil organic carbon accumulation and crop productivity under a rice–wheat cropping system in the western Indo-Gangetic Plains. Eur J Agron 70:11–21

Blyn G (1966) Agricultural trends in India, 1891-1947; output, availability and productivity. University of Pennsylvania Press, Philadelphia, U.S.A

Carlton R, West J, Smith P, Fitt BL (2012) A comparison of GHG emissions from UK field crop production under selected arable systems with reference to disease control. Eur J Plant Pathol 133:333–351

Costanza R, Palten BC (1995) Defining and predicting sustainability. Ecol Econ 15:193–196

Das A, Lal R, Patel DP, Ramkrushna GI, Layek J, Ngachan SV, Ghosh PK, Bordoloi J, Kumar M (2014) Effects of tillage and biomass on soil quality and productivity of lowland rice cultivation by small scale farmers in north eastern India. Soil Tillage Res 143:50–58

Das A, Layek J, Ramkrushna GI, Babu S, Thoithoi Devi M, Dey U, Suting D, Gulab GS, Lyngdoh DB, Prakash N (2019b) Integrated organic farming system (IOFS): an innovative approach for enhancing productivity and income of farmers in north eastern hill region of India. Indian J Agric Sci 89(8):1267–1272

Das TK, Bhattacharyya R, Sharma AR, Das S, Saad AA, Pathak H (2013) Impacts of conservation agriculture on total soil organic carbon retention potential under an irrigated agro-ecosystem of the western Indo-Gangetic Plains. Eur J Agron 51:34–42

Das TK, Ghosh S, Nath CP (2019a) Brown manuring optimization in maize: impacts on weeds, crop productivity and profitability. J Agric Sci 157:599–610

Das TK, Nath CP, Das S, Biswas S, Bhattacharyya R, Sudhishri S, Raj R, Singh B, Kakralia SK, Rathi N, Sharma AR, Dwivedi BS, Biswas AK, Chaudhari SK (2020) Conservation agriculture in rice-mustard cropping system for five years: impacts on crop productivity, profitability, water-use efficiency, and soil properties. Field Crops Res 250:107781

Dass A, Sudhishri S, Lenka NK Patnaik US (2011) Runoff capture through vegetative barriers and planting methodologies to reduce erosion and improve soil moisture, fertility and crop productivity in southern Orissa. India Nutr Cycl Agric Ecosyst 89: 45–57

Ghosh BN, Dogra P, Bhattacharyya R, Sharma NK, Dadhwal KS (2012) Effects of grass vegetation strips on soil conservation and crop yield under rainfed conditions in the Indian sub-Himalayas. Soil Use Manag 28:635–646

Ghosh DC, Mandal KG, Hati KM, Ghosh PK (2001) Indigenous nutrient management practices for some ecoregions of West Bengal Plains. In: Achraya CL, Ghosh PK, Subba Rao A (eds) Indigenous nutrient management practices - wisdom alive in India. Indian Institute of Soil Science, Nabibagh, Bhopal, India, pp 285–300

Ghosh PK, Das A, Saha R, Enboklang K, Tripathi AK, Munda GC, Ngachan SV (2010) Conservation agriculture towards achieving food security in north East India. Curr Sci 99:915–921

Ghosh PK, Sunil K, Pathak PK, Singh AK, Kumar V (2016) National initiative on fodder technology demonstration: a Giant stride towards self-sufficiency. ICAR-IGFRI, Jhansi, pp 1–139

Ghosh S, Das TK, Rana KS, Biswas DR, Das DK, Singh G, Bhattacharyya R, Datta D, Rathi N, Bhatia A (2022) Energy budgeting and carbon footprint of contrasting tillage and residue management scenarios in rice-wheat cropping system. Soil Tillage Res 223:105445

Gill JS (1992) Land use, conservation, management and development of land resources of Punjab. Report of Department of Soil Conservation and Engineering, Chandigarh, Punjab

Gupta DK, Bhatia A, Kumar A, Das TK, Jain A, Tomer R, Malyan SK, Fagodiya RK, Dubey R, Pathak H (2016) Mitigation of greenhouse gas emission from rice–wheat system of the Indo-Gangetic plains: through tillage, irrigation and fertilizer management. Agric Ecosyst Environ 230:1–9

Hegde DM, Babu SSN, Rana DS, Vyas KS (2012) Agronomic research on oilseeds in India: an overview. Indian J Agronomy 57:58–63

Hobbs PR, Gupta RK (2003) Resource conserving Technologies for Wheat in Rice-wheat systems. In: Ladha JK, Hill J, Gupta RK, Duxbury J, Buresh RJ (eds) Improving the productivity and sustainability of rice-wheat systems: issues and impact. ASA, Spec. Publ. 65, chapter 7, SA Madison, WI USA, pp 149–171

IGFRI Vision (2050) Indian Grassland and Fodder Research Institute, Jhansi, Uttar Pradesh, India

INCCA (2010) Indian network for climate change assessment, Ministry of Environment and Forests, government of India. Greenhouse Gas Emissions, India

Jat ML, Gathala MK, Ladha JK, Saharawat YS, Jat AS, Kumar V, Sharma SK, Kumar V, Gupta RK (2009) Evaluation of precision land leveling and double zerotill systems in Rice-wheat rotation: water use, productivity, profitability and soil physical properties. Soil Tillage Res 105:112–121

Jat ML, Sharma SK, Gupta Raj K, Sirohi K, Chandana P (2005) Laser land leveling: the precursor technology for resource conservation in irrigated eco-system of India. In: Abrol IP, Gupta RK, Malik RK (eds) Conservation agriculture-status and prospects. CASA, New Delhi, pp 145–154

Kumar R (2019) India's green revolution and beyond. Economics and Political Weekly. August 24, 2019 pp. 41–48

Kumar S, Dev I, Agarwal RK, Dixit AK, Ram SN (2012) Agronomic research on forages in India: an overview. Indian J Agronomy 57:92–104

Kumar S, Kumar KT, Singh JB, Prasad M, Ghosh PK (2021) Effect of food-fodder farming on sustainable resources use and carbon dynamics in rainfed semi-arid tropics of India. Commun Soil Sci Plant Anal:1–14. https://doi.org/10.1080/00103624.2021.1921188

Kumar S, Palsaniya DR, Kiran Kumar T, Dixit AK, Das MM, Ghosh PK (2017a) Farming systems: issues and strategies. SSPH Publisher, New Delhi, India, pp 1–571

Kumar S, Palsaniya DR, Kiran Kumar T, Misra AK, Ahmed S, Kantwa SR, Ghosh PK (2017b) Non-conventional fodder resources. In: Ghosh PK, Mahanta SK, Singh JB, Vijay D, Kumar RV, Yadav VK, Kumar S (eds) Approaches towards fodder security in India. Studera Press, Delhi, pp 35–56

Kumar V, Ladha JK (2011) Direct seeding of rice: recent developments and future research needs. Adv Agron 111:296–413

Lal R (2003) Soil erosion and the global carbon budget. Environ Int 29(4):437–450

Lal R (2004) Soil carbon sequestration impacts on global climate change and food security. Science 304:1623–1627

Lin C, Tu S, Huang J, Chen Y (2009) The effect of plant hedgerows on the spatial distribution of soil erosion and soil fertility on sloping farmland in the purple-soil area of China. Soil Till Res 105:307–312

Mandal D, Sharda VN (2013) Appraisal of soil erosion risk in the eastern Himalayan region of India for soil conservation planning. Land Degrad Dev 24:430–437

Mandal D, Srivastava P, Giri N, Kaushal R, Cerda A, Alam NM (2017) Reversing land degradation through grasses: a systematic meta-analysis in the Indian tropics. Solid Earth 8(217–233):2017. https://doi.org/10.5194/se-8-217-2017

Malik RK, Yadav A, Gill GS, Sardana P, Gupta RK, Piggin C. (2004). Evolution and acceleration of no-till farming in rice-wheat system of the Indo-Gangetic Plains. 4th international crop science congress held in Brisbane, Australia from September 26 to October 1.73 p

Mishra M (2013) Role of eco-friendly agricultural practices in Indian agriculture development. Int J Agric Food Sci Technol 4(2):11–15

Mohammad A, Sudhishri S, Das TK, Singh M, Bhattacharyya R, Dass A, Khanna M, Sharma VK, Dwivedi N, Kumar M (2018) Water balance in direct-seeded rice grown under a conservation agriculture-based rice-mustard cropping system in north-western Indo-Gangetic Plains of India. Irrig Sci 36:381–393

NAAS (2019) Zero budget natural farming - a myth or reality? Policy paper no. 90. National Academy of Agricultural Sciences, New Delhi, p 20

NNMB (2003) National nutrition monitoring bureau. National Institute of Nutrition, Indian Council of Medical Research, Hyderabad

Palekar S (2016) Zero budget natural farming. http://palekarzerobudgetspiritualfarming.org

Panwar AS, Prusty AK, Shamim M, Ravisankar N, Ansari MA, Singh R (2021b) Nutrient recycling in integrated farming system for climate resilience and sustainable income. Indian J Fertilizers 17(11):1126–1137

Panwar AS, Ravisankar N, Singh R, Prusty AK, Shamim M, Ansari MA, Noopur K (2021a) Potential integrated farming system modules for diverse ecosystems of India. Indian J Agronomy 66:S15–S32

Pathak H, Bhatia A, Jain N (2014) Greenhouse gas emission from Indian agriculture: trends, mitigation and policy needs. Indian Agricultural Research Institute, New Delhi

Prakash N, Ansari MA, Roy SS, Punitha P, Baishya LK, Sailo B, Sharma SK, Nagchan SV (2017) Farming system approaches for Sustainable Hill agriculture in Manipur, 167 pages. ICAR Research Complex for NEH Region, Manipur Centre, Lamphelpat, Imphal. (ISBN: 978-81-7622--396-6); Publisher: Biotech Books, 4762-63/23, Ansari Road, Darya Ganj, New Delhi - 110002)

Radford BJ, Key AJ, Robertson LN, Thomas GA (1995) Conservation tillage increases soil water storage, soil animal populations, grain yield and response to fertiliser in the semi-arid sub-tropics. Aust J Exp Agric 35:223–232

Raman S (2006) Agricultural sustainability: principles, Processes and Prospects. Food Products Press, New York, USA, p 474

Randhawa MS (1980) A history of agriculture in India, vol I. Indian Council of Agricultural Research, New Delhi, pp 483–490

Randhawa MS (1982) A history of agriculture in India, vol II. Indian Council of Agricultural Research, New Delhi, pp 342–343

Randhawa MS (1983) A history of agriculture in India. Vol. III. 96, 197-283. Indian Council of Agricultural Research, New Delhi

Ravisankar N, Ansari MA, Panwar AS, Aulakh CS, Sharma SK, Suganthy M, Suja G, Jaganathan D (2021) Organic farming research in India: potential technologies and way forward. Indian J Agronomy 66:S142–S162

Rodell M, Isabella V, James SF (2010) Satellite-based estimates of groundwater depletion in India. Nature 460:999–1002. https://doi.org/10.1038/nature08238.www.nature.com/nature

Saad AA, Das TK, Rana DS, Sharma AR, Bhattacharyya R, Lal K (2016) Energy auditing of maize-wheat-greengram cropping system under conventional and conservation agriculture in irrigated north-western Indo-Gangetic Plains. Energy 116:293–305

Saha S, Yadav D, Layek A, Babu S, Ansari MA, Das B, Chowdhury S, Singh R, Rajkhowa DJ, Ghosh PK, Das A (2021) Potential of conservation agriculture in hill ecosystems: impact on resource sustainability. Indian J Agronomy 66:S128–S141

Shafeeq PM, Aggarwal P, Krishnan P, Rai V, Pramanik P, Das TK (2020) Modeling the temporal distribution of water, ammonium-N, and nitrate-N in the root zone of wheat using HYDRUS-2D under conservation agriculture. Environ Sci Pollut Res 27(2):2197–2216

Siegel BE (2018) Hungry nation: food, famine, and the making of modern India. Cambridge University Press, Cambridge, pp 21–49

Singh RB (2000) Environmental consequences of agricultural development: a case study from the green revolution state of Haryana, India. Agric Ecosyst Environ 82(1–3):97–103

Singh G, Bhattacharyya R, Das TK, Sharma AR, Ghosh A, Das S, Jha P (2018) Crop rotation and residue management effects on soil enzyme activities, glomalin and aggregate stability under zero tillage in the Indo-Gangetic Plains. Soil Tillage Res 184:291–300

Singh R, Babu S, Avasthe RK, Das A, Praharaj CS, Layek J, Kumar A, Rathore SS, Kancheti M, Kumar S, Yadav SK, Pashte V (2021) Organic farming in North–East India: Status and strategies. Indian J Agronomy 66:S163–S179

Sow S, Ranjan S (2021) Indigenous technical knowledge (ITK) for sustainable agriculture in India. Agric Food e-News Letter 3(1):31–35. Available from: https://www.researchgate.net/publication/349278516_Indigenous_Technical_Knowledge_ITK_for_Sustainable_Agriculture_in_India [accessed Jul 07 2021]

Sylvia HV, Sapkota TB, Hillier J, Stirling CM, Macdiarmid JI, Aleksandrowicz L, Green R, Joy EJ, Dangour AD, Smith P (2017) Greenhouse gas emissions from agricultural food production to supply Indian diets: Implications for climate change mitigation. Agric Ecosyst Environ 237: 234–241

Vitousek PM, Ehrlich PR, Ehrlich AH, Matson PA (1986) Human appropriation of the products of photosynthesis. Bioscience 36:368–373

von Bertalanffy L (1968) General system theory: foundations, development, applications. George Braziller, New York

Wani MH (2011) Hill agriculture in India: problems and prospects of mountain agriculture rapporteur: Ind. J Agric Econ 66(1):1–3

Dryland Agriculture 22

Gajjala Ravindra Chary, Kodigal A. Gopinath,
Bellapukonda Murali Krishna Raju, Chitiprolu Anantha Rama Rao,
Venugopalan Visha Kumari, Priyabrata Santra, Vinod Kumar Singh,
and Suryanarayana Bhaskar

Abstract

Dryland agriculture is as old as India's agriculture. Before and at the time of independence, agriculture was the primary source of national income and occupation. Droughts had been recurrent making dryland agriculture more risk-prone and limiting production in dryland agriculture. Farmers practiced dryland farming that was built on decades of experience, mainly as subsistence farming. Prior to independence, the Indian agriculture sector exhibited primarily three traits, low production, high risk, and instability, and the production was alarmingly low. Agricultural policy underwent massive agrarian reforms between the 1950s and the mid-1960s. Droughts in the mid-1960s spurred the Indian government to invest significantly in dryland agriculture research. In 1970, the Government of India through the Indian Council of Agricultural Research initiated All India Coordinated Research Project for Dryland Agriculture in the country to address the problems in dryland agriculture. During the 1970s and 1980s, many soil-, water-, and crop-based dryland technologies were developed which were integrated into soil and water conservation programs in the states. This enabled the development and implementation of watershed programs at national level leading to 47 model watershed projects that resulted in productivity, income, and resource conservation gains in the dryland areas of the country. During and after the 1990s, many dryland agroecology-specific natural resource management and crop-based technologies were developed and integrated into various district, state,

G. R. Chary (✉) · K. A. Gopinath · B. M. K. Raju · C. A. R. Rao · V. V. Kumari · V. K. Singh
ICAR - Central Research Institute for Dryland Agriculture, Hyderabad, Telangana, India

P. Santra
ICAR - Central Arid Zone Research Institute, Jodhpur, Rajasthan, India

S. Bhaskar
Natural Resource Management Division, ICAR, New Delhi, India

P. K. Ghosh et al. (eds.), *Trajectory of 75 years of Indian Agriculture after Independence*, https://doi.org/10.1007/978-981-19-7997-2_22

and national programs resulting in a sustainable production system. Contingency crop planning practices and strategies were developed and implemented to cope with delayed onset of monsoon and agricultural drought. The sustainability of dryland agriculture is impacted by climate risks and challenged by many biophysical and socioeconomic constraints. The research, development, and policy focus should be to achieve sustainable and climate-resilient dryland agriculture. Some future strategies in this direction are suggested in this chapter.

Keywords

Dryland agriculture · Watersheds · Real-time contingency planning

22.1 Introduction

Dryland agriculture is as old as agriculture in India. Expanding the domain of dryland farming, today rainfed agriculture accounts to 52% of the net sown area, is home to two-thirds of livestock, and supports 40% of the human population. It is crucial to country's economy and food security since it contributes to about 40% of the total food grain production (85%, 83%, 70%, and 65% of coarse cereals, pulses, oilseeds, and cotton, respectively), supports two-thirds of livestock and 40% of the human population, further also influences livelihoods of 80% of small and marginal farmers, and is most vulnerable to monsoon failures. Even if full irrigation potential is created, still 40% of net cultivated area will remain as rainfed agriculture which would continue to be a major food grain production domain.

22.1.1 Pre-independence and Independence Period

In 1880, the Government of India appointed the First Famine Commission to suggest measures to tackle drought and also to suggest preventive measures to avoid famine in the future. The Royal Commission, India, in 1920 noted that "the problems of cultivation of crops in tracts entirely dependent upon rainfall deserve far closer attention." The first systematic and scientific approach to the problem of dryland farming started in 1923 by establishing the Dryland Research Station at Manjri near Pune. The first systematic scientific approach to tackle the problems of dry farming areas was initiated by Tamhane in 1923 on small plots of Manjari farm near Pune, and the work passes on to Kanitkar in 1926. A comprehensive research scheme was drawn up by Kanitkar with financial support from the ICAR. Subsequently, the dryland agriculture research was initiated in 1933 at five ICAR-sponsored dry farming research centers, viz., Sholapur and Bijapur in 1933, Hagari and Raichur in 1934, and Rohtak in 1935. A decade of work (1933–1943) was mainly on rainfall analysis, physicochemical properties of soils, physiological studies in millets, and agronomic aspects which resulted in a series of dry farming practices commonly known as the Bombay dry farming practices, Hyderabad dry farming practices, and Madras dry farming practices. These practices stressed the need for contour bunding to conserve soil and water, deep plowing for improved water intake and storage, interculture operation of crops for efficient use of the limited moisture in the soil, application of FYM, low seed rate with wide spacing, mixed cropping, and crop rotation. However, the marginal return of 15–20% over the base yield of 200–400 kg/ha did not enthuse the farmers very much to adopt these research results.

Agriculture was the main source of national income and occupation at the time of independence. Agriculture and allied activities contributed nearly 50% to India's national income. Around 72% of the total working population was engaged in agriculture. During the pre-independence period, the concept of cropping system received little consideration except for beneficial effects of legumes/oilseeds as mixed/intercropping green manuring and sequential cropping, while little attention was paid for efficient use of natural resources. The true concept of cropping system gained momentum during the 1960s for bringing self-sufficiency in food through enhanced cropping intensity besides breaking technology and policy barriers. The advent of high-yielding, photo-insensitive, input-responsive varieties of rice and wheat totally revolutionized the Indian agriculture in less than a decade and brought a paradigm shift in farming with respect to tillage, planting time and method, fertilizer use, irrigation, pest control, and harvesting. This resulted in a major shift toward cereal-based cropping systems by relegating less productive, risk-prone legumes and oilseeds to marginal dryland areas.

The Indian agricultural sector before independence displayed mainly three characteristics which gave the lay down to the backwardness and stagnation of the economy. Low productivity, high level of risk, and instability were key three characteristics. Due to a lack of a proper and efficient canal network, farmers were primarily reliant on rainfed agriculture. Crop failure was a major issue. The resource-poor farmers in dryland regions practiced mixed cropping, and perennial crops were mixed with small and large ruminants. Farmers used local, traditional, or landrace varieties of minor and major crops at this period. Agricultural performance was poor during the interwar years (1918–1939). The annual growth rate of all crop output was 0.4% from 1891 to 1946, and food grain output remained virtually unchanged (Roy 2006).

Traditional system of land, water, crop, and energy management Most of the land management practices included organic compost, inclusion of livestock, agroforestry, etc. Some of the available literatures for traditional system of land, water, and crop management in dryland are given in Table 22.1. Apart from crop production, crop protection aspects were also given importance. Summer plowing, use of ash, traditional method of grain storage, cover crop, etc. were followed to protect the crops from various pests and diseases.

22.1.2 After Independence

In 1953, a Central Soil Conservation Board was established to give impetus to soil and water conservation. In 1954, toward increasing the productivity of drylands, the Indian Council of Agricultural Research (ICAR) established the Soil Conservation

Table 22.1 Traditional systems of land, water, and crop management

Traditional agricultural practices	Features	State	References
Kanabandi	Farmers build barrier by using pieces of small dead wood or local vegetation to check wind velocity within safer limits	Rajasthan	Mathur (1995)
Terraces or bun cultivation	It is a slope- and valley-type cultivation which is useful for improving crop production and retaining moisture for soil conservation	Meghalaya	Jeeva et al. (2006)
Live bunding/ vegetative bunding	Bushes of *Subabul* and *shevri* and grasses like *vetiveria* are planted between the bunds of field across the slope for soil conservation against water erosion	Uttar Pradesh	Srivastava and Pandey (2006)
Livestock panning and fallowing	Farmer used panning of cow, sheep, and goat and fallowing the field at the end of winter to improve fertility of the field for the next crop	Madhya Pradesh and Uttar Pradesh	Singh and Sureja (2008)
Utera cropping system	Under rainfed agroecosystem, the next crop is sowing before harvesting to utilize the soil moisture of the previous crop	Madhya Pradesh	Singh and Sureja (2008)
Pannendu Pantalu	It is known as 12-crop system, in which millets, pulses, oil crop, and various vegetables are grown on the single piece of land	Andhra Pradesh	Bhushan et al. (2001)

Training and Demonstration centers at eight locations. The studies at the Indian Agricultural Research Institute (IARI) and other ICAR institutes laid emphasis on biological component of dryland farming technology, viz., development of high-yielding varieties, and also provided information on new soil management and moisture conservation practices and recommended specific crops for different rainfed areas in the country and also new dry farming technology for the Telangana area of Andhra Pradesh (IARI 1970).

During the 1950s to the mid-1960s, agricultural policy witnessed tremendous agrarian reforms, institutional changes, development of major irrigation project, and strengthening of cooperative credit institution (Tripathi and Prasad 2009). Land reforms were important in increasing agricultural production during this phase. The Community Development Program, decentralized planning, and the Intensive Area Development Programs were also initiated for regenerating Indian agriculture that had stagnated during the British period. In order to encourage the farmers to adopt better technology, incentive price policy was adopted in 1964, and the Agricultural Price Commission was set up to advice the Government on the fixation of support prices of agricultural crops. Despite the institutional changes and development programs introduced by the Government during this phase, India remained dependent upon foreign countries for food to feed the rising population.

There was a need to replace the traditional crops due to changed agroclimatic situations. During 1971–1972, a new crop, i.e., sunflower, was newly introduced almost at all locations of the AICRPDA which performed exceedingly better even when sown late by 1 or 2 months in the Deccan *rabi* tract. Efforts were mounted through the integrated dryland development (IDLAD) projects to popularize the crop. This crop proved an alternative to *kharif* groundnut which was risky in the Deccan tract. At Bellary, Bijapur, and Sholapur, rainfed wheat was replaced with safflower, and the area under sole safflower was increased ultimately augmenting the oilseed production (Randhawa and Venkateswarlu 1980).

Agricultural growth is one of the main facets of India's economic development and national food sufficiency policies. The aggregate agricultural output increased annually at 2.6% during period from 1950–1951 to 2006–2007. Further disaggregating of agriculture into subsectors, crop, livestock, forestry, and fishing, shows that fisheries and livestock were the main sources of the acceleration in growth rate of agricultural output in the 1980s. The growth rate of aggregate agricultural output turned up to 3.29% during the initial years of reforms, which was 0.43% point higher than the previous period. However, the situation of agriculture turned adverse during post-WTO period, and this covered all the subsectors of agriculture. The growth rates in output of all crops decelerated from 2.93% to 1.57%. The livestock declined from 4.21% to 3.40%. The fisheries declined from 7.48% to 3.25%. Only forestry witnessed a sharp increase from 0.09% to 1.82% (Tripathi and Prasad 2009).

22.2 Decadal Growth Shift/Trend of Rainfed Area (1960–2016)

Gross rainfed area (million ha) and share (%) of different crops from TE 1960–1961 to TE 2015–2016 (four to five decades) in India are furnished in Table 22.2. It was observed that share of rice and total pulses in rainfed area remained more or less constant but share of total cereals and millets (excluding rice and wheat) has shown declining trend over the period. Share of total oilseeds in rainfed area doubled from 10% to 21% in the last four to five decades. Among the oilseeds, share of soybean doubled (6–12%) in 15 years from TE 2000–2001. Fodder crops share though increased over the decades but not at the pace of growth in livestock population in rainfed areas. Cotton's share in rainfed area has increased by about 3%. Share of spices and condiments in rainfed area increased up to 2000–2001 and then started declining. Share of sugarcane in rainfed area decreased gradually over the period. In rainfed regions, shift is happening toward soybean, cotton, and maize from sorghum, pearl millet, groundnut, sunflower, and other cereals and millets.

Irrigation potential created in the country over the years vide various schemes, and programs such as Drought-Prone Area Program (DPAP), Dryland Development Program (DDP), Prime Minister *Krishi Sinchayee Yojana* (PMKSY), etc. led to increase in area sown under irrigation from 18% in 1960–1961 to 50% in 2018–2019 leaving the dryland area as half of the net sown area (Table 22.3). Though growth in irrigated area is more or less steady, the jump of about 10 million ha during the

Table 22.2 Gross rainfed area (in million ha) and share (%) of different crops over the decades in India

Year/crops	TE 1960–1961	TE 1970–1971	TE 1980–1981	TE 1990–1991	TE 2000–2001	TE 2010–2011	TE 2015–2016
Jowar/sorghum	14.27	13.36	12.92	11.43	8.44	6.69	5.48
Bajra/pearl millet	8.97	9.97	8.87	9.00	8.19	8.14	7.01
Maize	3.17	3.81	3.88	3.96	4.76	6.02	6.47
Ragi/finger millet	1.68	1.72	1.89	1.70	1.55	1.22	1.10
Rice	17.63	18.41	19.39	19.27	18.88	17.72	17.81
Wheat	7.23	6.60	5.90	3.97	3.17	2.27	1.89
Barley	1.57	1.07	0.73	0.39	0.27	0.18	0.15
Other cereals and millets	4.12	4.01	3.42	2.12	1.36	0.82	0.68
Bengal gram/chickpea	7.13	4.97	4.79	4.66	4.63	5.44	5.16
Red gram/pigeon pea	1.99	2.13	2.28	2.86	3.12	3.32	3.36
Other pulses	9.24	9.59	10.39	10.54	10.64	10.74	9.96
Groundnut	5.10	5.57	5.24	5.85	5.34	4.38	3.68
Sesamum	1.39	1.44	1.33	1.78	1.47	1.82	1.63
Rapeseed and mustard	0.85	0.78	0.92	1.52	1.73	1.52	1.28
Linseed	1.08	0.89	1.05	0.83	0.54	0.27	0.19
Other oilseeds	1.87	2.05	2.70	5.46	9.50	12.70	14.34
Cotton	5.46	5.23	4.82	4.24	5.52	6.36	8.27
Tobacco	0.24	0.28	0.25	0.21	0.20	0.23	0.19
Sugarcane	0.57	0.54	0.54	0.41	0.33	0.32	0.24
Total condiments and spices	0.88	0.94	1.06	1.07	1.21	1.03	0.89
Total fruits and vegetables	1.73	2.21	2.49	3.36	3.33	3.52	3.53
Fodder crops	3.81	4.41	5.14	5.38	5.82	5.30	6.68
Soybean	–	–	–	–	5.77	9.31	11.86
Sunflower	–	–	–	–	0.98	0.93	0.46
Total rainfed area (million ha)	121.81	123.22	120.54	118.76	107.71	103.12	99.09

Source: C.A. Rama Rao, CRIDA, 2021 (Information submitted CRIDA-RAC)

decade of TE 1990–1991 to TE 2000–2001 was very high. Substantial area (about 7 million ha) was brought under irrigation during a span of 8 years from TE 2010–2011 to TE 2018–2019.

Table 22.3 Changes in net sown, net irrigated, and percentage net irrigated area of the country over years

Year	Net area sown (million ha)	Net irrigated area (million ha)	% Net irrigated area
TE 1960–1961	132.66	24.03	18.12
TE 1970–1971	138.96	30.10	21.66
TE 1980–1981	140.72	38.43	27.31
TE 1990–1991	142.37	46.96	32.98
TE 2000–2001	141.72	56.72	40.03
TE 2010–2011	140.88	63.08	44.78
TE 2018–2019	139.32	69.89	50.17

Source: DES, DA&FW, GoI

Table 22.4 Drought before and after independence in the country

Drought severity	Before independence (1876–1947)	After independence (1948–2010)	Total
Slight	11	5	16
Moderate	8	7	15
Severe	0	4	4
Calamitous	3	0	3
Total	22	16	38

Source: Fertilizer Statistics 2016–2017, The Fertilizer Association of India, New Delhi

22.3 Climate Variability, Climate Shift, Droughts, and Extreme Events (1891–2021)

India experienced 26 large-scale droughts with increasing frequencies during the periods 1891–1920, 1965–1990, and 1999–2012. India experienced 26 severe droughts: 1891, 1896, 1899, 1905, 1911, 1915, 1918, 1920, 1941, 1951, 1965, 1966, 1972, 1974, 1979, 1982, 1986, 1987, 1988, 1999, 2000, 2002, 2009, 2012, 2014, and 2015. Droughts have become a common feature in southern, western India, and some pockets in other parts of the country. Important concern has been that traditionally high rainfall regions are affected with droughts particularly inter-droughts and drought-prone regions experiencing high-intensity rains or cyclones. Over the past two centuries, India experienced several droughts (Table 22.4), and some are very severe causing human mortality. Negative impacts of drought are more pronounced in dryland areas on account of fragile resource base and poor investment capacities of dryland farmers. The country faced 22 droughts (of 72 years before independence) during 1876–1947 and 16 droughts (of 63 years after independence) during 1948–2010 with drought incidence probability of 0.31 and 0.25 per year, respectively. The country witnessed three calamitous droughts before independence, but there were four severe droughts after independence.

The occurrence of drought is very frequent in the meteorological subdivisions like West Rajasthan, Tamil Nadu, Jammu and Kashmir, and Telangana (NRAA 2013). The all-India mean annual temperature has increased by 0.5 °C in the period 1901–2003. The high inter- and intra-seasonal variability in rainfall distribution, rainfall events, and extreme temperatures are causing crop damages and losses to farmers (Sikka et al. 2016). Studies pertaining to India show enough evidence of rising mean temperature during post-1970 period. Greater warming of 0.21 °C/10 years during post-1970 period as compared to 0.51 °C/100 years during the past century has been reported. In many parts of India, the frequency of occurrence of cold nights declined, while the frequency of occurrence of warm nights and warm days significantly increased. Besides, the country experienced 15 deficit and 6 excess monsoon years in post-1960 period in comparison to only 27 deficit and 20 excess monsoon years during 1871–2014. Extreme positive departures from the normal maximum temperature result in heat wave in different parts of the country. The maximum number of heat waves occurred over East Uttar Pradesh followed by Punjab, East Madhya Pradesh, and Saurashtra and Kutch in Gujarat. During the decade 1991–2000, a significant increase in the frequency, persistency, and spatial coverage of heat wave/severe heat wave was observed in comparison to that during the earlier decades 1971–1980 and 1981–1990 (Pai et al. 2004). The Intergovernmental Panel on Climate Change (IPCC) in its Fifth Assessment Report observed that "warming of climate system is unequivocal", and since the 1950s, many of the observed changes are unprecedented over decades to millennia. India's average temperature has risen by around 0.7 °C during 1901–2018. By the end of the twenty-first century, the average temperature over India is projected to rise by approximately 4.4 °C relative to the recent past (1976–2005 average), under the RCP 8.5 scenario (Krishnan et al. 2020). Krishna Kumar et al. (2011) observed greater warming (mean annual surface air temperature) of 0.21 °C/10 years during post-1970 period as compared to 0.51 °C/100 years during the past century.

Raju et al. (2013) revisited climatic classification at district level (Fig. 22.1) with the help of moisture index (MI) computed using precipitation and potential evapotranspiration data for period 1971–2005 and highlighted the climatic shifts occurred in the districts post-1970 period compared to the climatic classification of Krishnan (1988) which was based on pre-1970 climatic data. The study revealed climatic shifts in about 27% of the geographical area in the country. Significant reflections were a substantial increase of arid region in Gujarat and a decrease of arid region in the state of Haryana increase in semi-arid region in Madhya Pradesh, Tamil Nadu and Uttar Pradesh were observed due to shift of climate from dry sub-humid to semi-arid. Notable observation was that the moist subhumid pockets in Chhattisgarh, Orissa, Jharkhand, Madhya Pradesh, and Maharashtra states have turned dry subhumid to a larger extent.

District-level analysis of vulnerability and risk to climate change was done following the framework given in the Fifth Assessment Report of the Intergovernmental

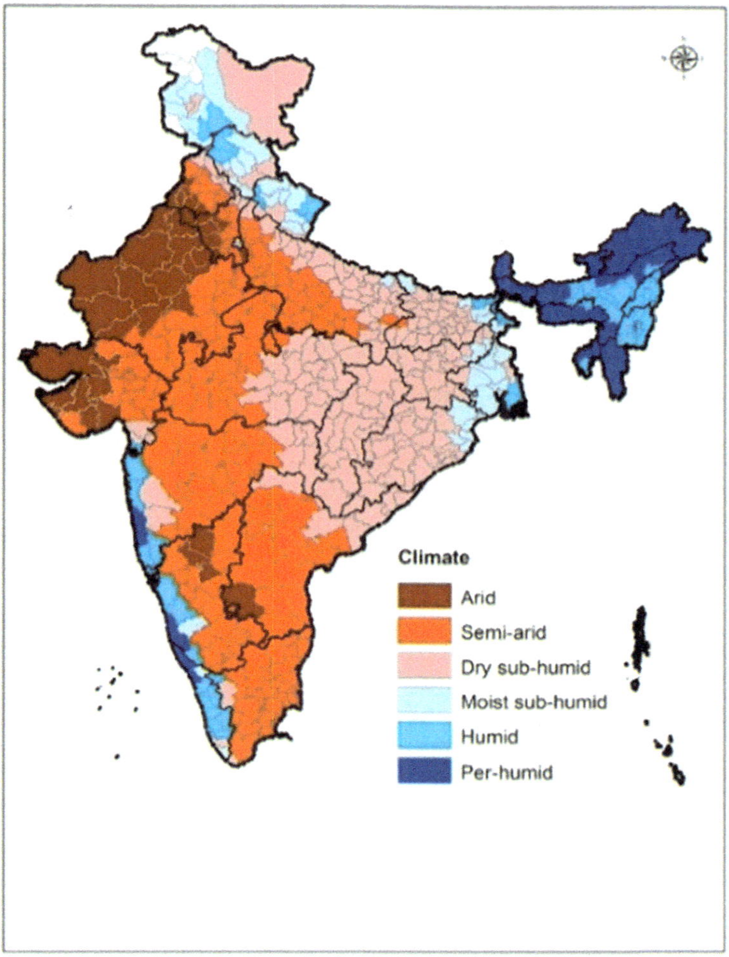

Fig. 22.1 Climatic classification at district level (1971–2005). (Source: Raju et al. 2013)

Panel on Climate Change. According to this conceptualization, risk is a result of interaction between exposure, hazard, and vulnerability, and vulnerability is a function of sensitivity and adaptive capacity of the system of interest. In this analysis, hazard was constructed as a combination of historical hazard (based on historical incidence of cyclones, drought, and floods) and future hazard which was represented as a combination of indicators, viz., change in annual rainfall, July rainfall, maximum temperature, minimum temperature, drought incidence, incidence of dry spells, extreme rainfall events, etc. The changes in these indicators as obtained by an ensemble of CMIP5 climate projections for RCP 4.5 for the period 2020–2049 with respect to the baseline 1976–2005 were used. Similarly, a number of indicators were

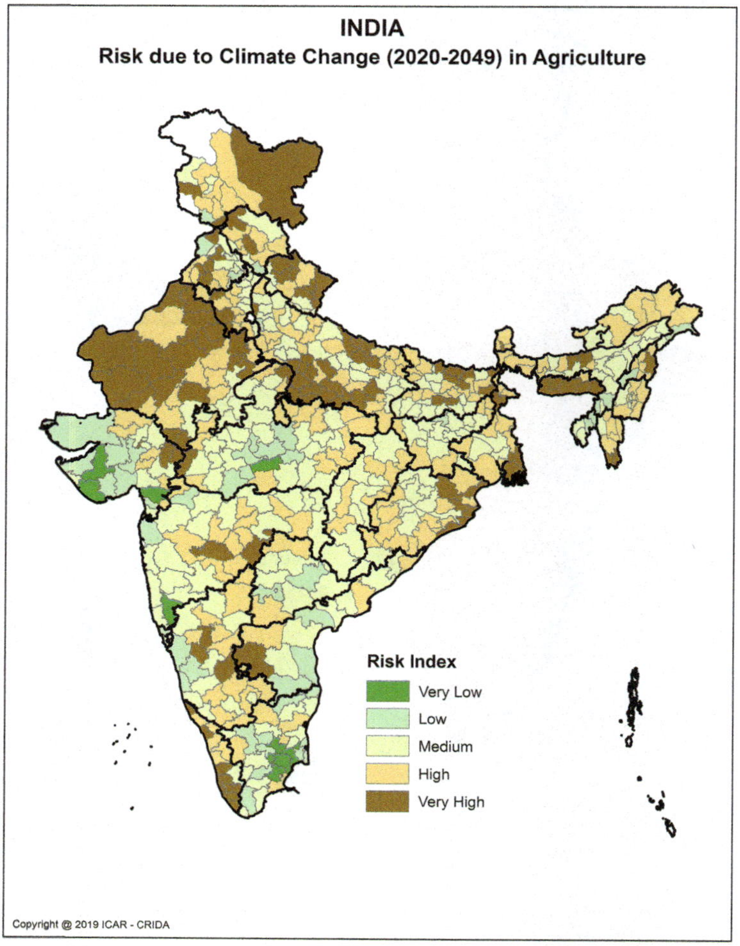

Fig. 22.2 Risk due to climate change (2020–2049) in agriculture. (Source: Rama Rao et al. 2019)

chosen to represent exposure and vulnerability components of risk. Using the data on indicators, indices for exposure, vulnerability, and hazard were computed, and finally an index of risk was computed. The districts were categorized into different degrees of risk (Fig. 22.2). The analysis indicated "very high" risk for 109 districts in Uttar Pradesh (22), Rajasthan (17), Bihar (10), Kerala (8), Uttarakhand (7), Odisha (6), Punjab (5), and the remaining states of West Bengal, Karnataka, Haryana, Gujarat, Mizoram, Assam, Himachal Pradesh, etc. Most of the 201 districts with "high" risk are in Uttar Pradesh, Madhya Pradesh, Chhattisgarh, Karnataka, Rajasthan, Bihar, Odisha, Maharashtra, etc. Districts with high risk are to be given higher priority for investment and intervention planning for enhancing resilience to climate change (Rama Rao et al. 2019).

22.4 Green Revolution (GR) and Post-GR Era

The Green Revolution in the mid-1960s, though a boon to Indian agriculture, ushered in era of wide disparity between productivities of irrigated and dryland agriculture, and thus, the "green revolution" era had largely bypassed the dryland agriculture including dryland areas in eastern region of the country. Several development programs were initiated for improving dryland farming. The "everything, everywhere" approach of taking up all major interventions uniformly across all regions of the country has not paid much dividend. The developmental approach in dryland areas did not fully capture aspects like livelihood, soil resources, reliability of irrigation, socioeconomic profile, infrastructure, etc. neglecting region-specific interventions befitting to the natural resource endowment, social capital, infrastructure, and economic condition (NRAA 2012).

The droughts of the mid-1960s catalyzed the Government of India to invest on dryland research significantly, and in 1970, the ICAR launched the All India Coordinated Research Project for Dryland Agriculture (AICRPDA) at Hyderabad, in collaboration with the Canadian International Development Agency (CIDA) with 23 centers across the country. In 1976, the Dryland Operational Research Projects were started for action research. The ICAR launched the National Agricultural Research Project (NARP) in 1979 which strengthened the regional research capabilities of agricultural universities with location-specific research including on dryland agriculture. In 1983, 47 model watersheds were started under the ICAR, and in 1984, watershed development programs were initiated in four states with assistance of the World Bank. The milestone in dryland farming research in India was the upgradation of AICRPDA to the status of an institute, i.e., the Central Research Institute for Dryland Agriculture (CRIDA) in 1985 to carry out basic and strategic research, while network research under AICRPDA continued in applied and adaptive research. The ICAR launched the National Agricultural Technology Project (NATP) in 1999, and under this project, CRIDA undertook projects on Production Systems Research, Mission Mode, and Technology Assessment and Refinement-Institute Village Linkage Programme (TAR-TAR). In 2007, ICAR launched the National Agricultural Innovation Project (NAIP) in which CRIDA undertook a program on enhancing rural livelihoods through natural resource management and sustainable farming systems. Another significant milestone in dryland research was launching of the National Initiative on Climate Resilient Agriculture (NICRA) at CRIDA in 2010.

During 1970–1987, sound foundations of systematic and location-specific research were laid out across 23 AICRPDA centers (Singh 1987; Ravindra Chary et al. 2016). The research during this period resulted in developing technologies related to "low monetary input," basic crop production practices like time of seeding and plant population geometry in relation to rainfall, weed management, identifying regions where intercropping was feasible and worthwhile to increase cropping intensity, best cultivars for the component crops, determining the optimum N and P doses for intercropping systems (Chetty 1983), efficient methods for crop lifesaving irrigation, in situ moisture conservation by tillage, crop husbandry for weather aberrations and alternate or multiple land use, double cropping, manipulation of sowing and harvesting dates and plant populations, and fertilizer use (AICRPDA 2003). These technologies amply demonstrated that yield of dryland crops could be increased by at least 100% with improved varieties and sowing methods and higher yields with advancement of sowing dates, particularly post-rainy period, minimizing the risk with split application of N and alternate crops for aberrant weather situations. Dry seeding was recommended for the locations/soil types where the conditions (soil) do not permit sowing operations with the onset of monsoon. Low monetary inputs like improved seed, timely sowing, and timely weeding helped bridge the yield gaps in dryland crops. In black soil regions, with 500–1000 mm rainfall, the productivity of upland rainy season crops could be substantially improved by providing furrows graded to 0.2–0.3% slope, to transmit excess rainwater (Verma 1982). In case post-rainy season crops grown on conserved soil moisture, the available soil moisture in the profile at the sowing time decided the choice of crops. With the advent of high-yielding and input-responsive varieties to suit different situations, the agriculture became more *production-oriented*.

In chronic drought-prone areas, deep tillage (20–30 cm) was found specifically to be applicable to soils having textural profiles or hard pans. Under unimodal (<500 mm) rainfall situation in semiarid regions with shallow Alfisols, sowing across the slope and ridging later were useful. In black soils, deep tillage combined with compartmental bunding was found to be the most effective soil management practice. With rainy season-cropped Vertisols (unimodal rainfall regions), water surplussing is an integral part of in situ moisture conservation. In bimodal medium (500–750 mm) rainfall representing semiarid Alfisol, graded border strips were found advantageous. Surface mulching with crop residues prevented moisture loss and prolong the moisture retention period and enhanced yield of crops (Patil et al. 1981). The *year-round tillage* as a means of controlling weeds and conserving soil moisture was crystallized into a concept of great significance to drylands. Deep tillage increased the yields of crops across the climates and soil types. The ridges and furrows always increased the yield, however with more effectiveness during moderate drought (AICRPDA 2003).

The intercropping systems with additive series were most successful with base crops as sorghum, maize, pearl millet, pigeon pea, safflower, and wheat with higher land equivalent ratios (with average of 23% more) and output and returns, spread of labor peaks, maintenance of soil fertility (with inclusion of legume), and stability in production. Intercropping of fast-growing legumes like cowpea and green gram as cover crops benefitted the base crop in better resource efficiency (Vishnumurthy and Vijayalakshmi 1993). Stable and economically viable double and intercropping systems were evolved with predominant crops of the region as base crops and pulses and oilseeds as intercrops for various rainfall and soil-type regions with potentials of higher productivity and income that increased cropping intensity to 150–200% (AICRPDA 2003; Rafey and Verma 1988). *Crop substitution* concept was evolved in which the performance of various new crops was evaluated vis-a-vis traditional crops, e.g., in Vertisols of Bellary, sorghum was efficient than cotton (Singh 1987). The cropping intensity could be increased considerably depending on the soil types and moisture availability period. However, the duration of the crop cultivars influenced the selection of a cropping system. In the high rainfall (>1000 mm) regions of Orissa, Eastern Uttar Pradesh, and Madhya Pradesh, a second crop could be grown in the residual moisture after a 90-day duration variety of upland rice than 120-day duration; similarly, in the Vertisols of Malwa (Madhya Pradesh) and Vidarbha (Maharashtra), a change of 140- or 150-day sorghums to about 90- or 100-day cultivars provided an opportunity to grow chickpea or safflower in sequence. Double cropping was possible only in areas receiving more than 750-mm rainfall with a soil moisture storage capacity of more than 200 mm. A new concept of fertilizer use was evolved to tailor to the available soil moisture status at least in post-rainy season areas.

The alternate land use systems developed were tree farming, ley farming (*Stylosanthes hamata* with sorghum rotation), silvopasture (*Leucaena leucocephala* + *Stylosanthes hamata* + *Cenchrus ciliaris*) and agro-horticulture (guava/custard apple/pomegranate/ber-based) (Korwar et al. 1997). *Leucaena leucocephala* is the most popular tree species to serve as hedgerow in the alley cropping system (Hegde et al. 1988). Studies at AICRPDA during 1975–1983 revealed that *Dichanthium*, *Sehima*, and *Lasiurus* are suitable for severe drought-prone areas, while *Cenchrus ciliaris*, *Panicum maximum*, and *Urochloa* were for moderate drought-prone areas. *Stylosanthes hamata*, a pasture legume, was identified for improvement of soil fertility and as quality fodder for Alfisols of Hyderabad (Reddy and Hampaiah 1982).

The rainwater harvesting and utilization to annual crops and fruit crops proved to be beneficial (Radder et al. 1995). Harvesting runoff water and storage in farm ponds could be a distinct possibility in red soils of Karnataka. Nearly 50% of the stored water can be used for protective irrigation (Havanagi 1982). The yield increased due to protective/supplemental irrigation to dryland crops under severe, and moderate drought was up to 25%. Response was mostly noted in crops to supplemental irrigation after withdrawal of monsoon (Vittal et al. 2005). The benefits of proven intercropping/double cropping systems provided ample scope for crop diversification tagging with rainwater management practices, particularly in dryland rice production system areas. On-farm demonstration of in situ moisture conservation practices on more than 2000 farmer's fields on cotton, oilseeds, *rabi* sorghum, and pearl millet gave additional returns of 20–25% over farmers' practice depending on the crop/cropping system. Chhattisgarh and Jharkhand demonstrated opportunities for additional income through introduction of pulse- and oilseed-based intercropping systems in place of dryland rice on uplands and to increase cropping intensity during *rabi* on medium lands through a second crop of gram or vegetables by adoption of moisture conservation practices. This technology, if adopted on community basis, can bring large areas of *rabi* fallows in eastern India under productive use. During 2001–2005, an entirely new approach of *crop planning as per soil-site suitability* was conceptualized under NATP-Mission Mode Project on Land Use Planning for Management of Agricultural Resources in Dryland Agroecosystem wherein 400 interventions were demonstrated on 132 soil subgroups on varying topo-sequences in 16 micro-watersheds by 13 AICRPDA centers. This provided much needed land use diversification from the traditional dryland utilization and indicated microlevel variations of soils and management practices on a topo-sequence (Ravindra Chary et al. 2008) are the prime factors influencing land productivity which increased from 30% to 50% and in few cases more than double. The soil-site suitability criteria were developed for 41 field, horticulture, and high-value crops.

22.5 Land and Water Management and Their Productivity

Earlier in dryland areas, most of the land and water management have been at the community level, relying upon harvesting rainwater in tanks and small underground storage structures. The major landmark in the evolution of the watershed approach in India includes the launching of a centrally sponsored scheme of soil conservation in the catchments of river valley projects in 1974. The watershed development programs that were initiated by the Government of India in Seventh Five-Year Plan (1985–1990) and subsequent watershed development programs, about 20,000 micro-watershed projects, aimed at resource conservation, productivity, stability, and livelihood improvement of rural poor and environmental protection in fragile dryland areas. An amount of Rs. 286 billion was invested on various watershed development projects since inception. The Indian Council of Agricultural Research (ICAR) developed 47 model watershed projects in the dryland area of the country in the year 1982. Of these, 30 model watersheds were monitored by AICRPDA centers and CRIDA. During the Seventh Five-Year Plan of India, high priority was given to watershed approach to develop dryland agriculture. The first to adopt a "watershed approach" for dryland development were the drought-prone area program (DPAP) and the desert development program (DDP) in 1987. In 1989, the Integrated Wasteland Development Project (IWDP) scheme followed suit. Subsequently, a fourth major program based on the watershed concept was initiated through the Ministry of Agriculture's National Watershed Development Project for Rainfed Areas (NWDPRA). The National Watershed Development Project (rainfed areas) was launched in 1991 in 25 states and 2 union territories for restoration of ecological balance and to sustain biomass production. All of these programs shared a common objective of land and water resource management for sustainable production. In recognition of their shared objectives but disparate approaches, in 1994, the GOI established a common set of operational guidelines, objectives, strategies, and expenditure norms for these "watershed development" projects. The Guidelines for Watershed Development Projects became operational in 1995, and there has been a massive countrywide increase in the number and financing for community-based projects for micro-watershed development since then. During the year 1998–1999, the food grain of the country crossed 200 million tons. The 63% of the rainfed area of the country contributed to 45% of the output. For the 11th Five-Year Plan (2007–2012), a total of some USD 2.4 billion of financing was provided to the Ministries of Agriculture's and Rural Development's programs in support of watershed development. In 2010, the Ministry of Rural Development (MoRD) consolidated its various watershed development programs into one comprehensive Integrated Watershed Management Program (IWMP). Since then, the average annual financing for the IWDP's watershed development projects has increased by almost 43%.

A meta- analysis of 636 micro-watersheds in the country revealed that watershed programmes provided multiple benefits of augmenting income, employment, productivity, cropping intensity, soil conservation, building social capital, reducing poverty. The analysis also reported a mean internal rate of returns (IRR) was 27.4 % indicating a vast potential to upgrade watershed programme in the country. The study also indicated that the benefits were more in the poor income regions and in high rainfall zones (700–1000 mm) with available technologies and with more people participation. The study also suggested that the watershed program is the most important strategies to bring socioeconomic change in the rainfed areas with proper institutional arrangement and with a holistic, participatory, and business model approach and, further, to develop and identify suitable technologies for low (<700 mm) and high (>1000 mm) rainfall regions (Joshi et al. 2008).

Ex situ rainwater management Ex situ rainwater management involves collection of runoff water from treated or untreated land surface/catchments and stored in farm pond, percolation tanks, or closed water tanks/reservoirs and other water harvesting structures and recycled using energy-efficient lifting pumps and micro-irrigation systems for protective/lifesaving/supplemental/presowing irrigation to crops. The traditional rainwater harvesting structures commonly found in different regions of India are tank, *Talab*, *Tanka*, *Ahar*, *Zabo*, *Khadin*, *Nadi*, *Ad-bandh*, etc. The ponds and tanks prevailing in southern India give ample evidence about the knowledge of man regarding harvesting and utilization of runoff in low rainfall regions. Hydrological studies conducted at some of the AICRPDA centers suggest that even in arid and semiarid environments, rains are sometimes received as heavy downpours resulting in runoff. This runoff of about 10–30% could be capitalized for water harvesting and runoff recycling. For selecting an ideal site of a farm pond, selection of a catchment which can yield fairly good amount of runoff during the rainy season becomes imperative. The amount of runoff is governed by factors such as rainfall amount and intensity, land topography mainly land slope and length of slope, soil type, and land use pattern. Location of farm pond in the middle of the cultivated field will be ideal so that a sizable amount of runoff could be collected and irrigation given through gravity flow. Soils with low permeability minimize seepage losses and are, therefore, preferred for having farm pond.

Nala bunds and percolation tanks *Nala* (a natural water course) bunds and percolation tanks are structures constructed across *nalas*. In fact, both terms are synonymous although used alternately at different places. The advantages of *nala* bunding are the following:

1. Checks erosion.
2. Silted up *nala* bed can be brought under cultivation.
3. Acts as a percolation tank increasing the water supply of wells on downstream side, even in years of inadequate rainfall when the minor irrigation works may fail.
4. Part of the *nala* basin gets flooded and soaked, ensuring success of a crop or two.
5. Since no water distribution system is involved, the cost is low, and land acquisition is minimum.

Efficient use of harvested rainwater Strict water budgeting for sustainable use of the harvested rainwater needs to be followed allowing only low duty crops. Sugarcane, rice, and wheat should be avoided at the same time encouraging pulses and oilseeds. Water can be harvested or transported into them in the arid ecosystem. Such water should be treated as of immense value for sustaining tree vegetation during the post-rainy periods. Irrigation cover for the first 2 years is essential for trees and fruit trees providing proper basins.

Artificial recharge of groundwater The surface of artificially increasing the rate and amount of water entry into the groundwater reserves is known as artificial recharge. The increasing demand for groundwater as a source of fresh water has brought into focus the need for artificial recharge. Different surface methods used for artificial recharge are flooding method, basin method, ditch and furrow method, natural channel method, irrigation method, percolation tanks, and channel binding methods. The groundwater recharging technology developed by AICRPDA Center Parbhani comprises of a model filtration unit for artificial recharging of open wells. This model unit mainly consists of an efficient filtration unit constructed near the open well. Filtration needs construction of three blocks, viz., primary filter with dimension of 0.6 × 0.6 m which is combined with layer of stones, sand, and metals. The filter collects major sediments from runoff water. Primary filter is joined by 4″ diameter pipe to silt trapping unit I energy dissipation unit (1.1 m × 1.5 m) having a rectangular notch opening in the main filter unit. The third block acts as a main filtration tank with dimensions as 2 m × 2 m × 2 m having three layers of different filter material such as 30-cm sand, 30-cm metal, and 30-cm stones which act as filter material. The main filter unit is joined to open well by the 4″ diameter. PVC pipe. The filter unit is constructed with walls and cement concrete. The filtration efficiency of the main filter is 92–95%. The availability of 25–35% surface runoff for well recharging is useful to enhance groundwater level to the tune of 0.3–3.4 m.

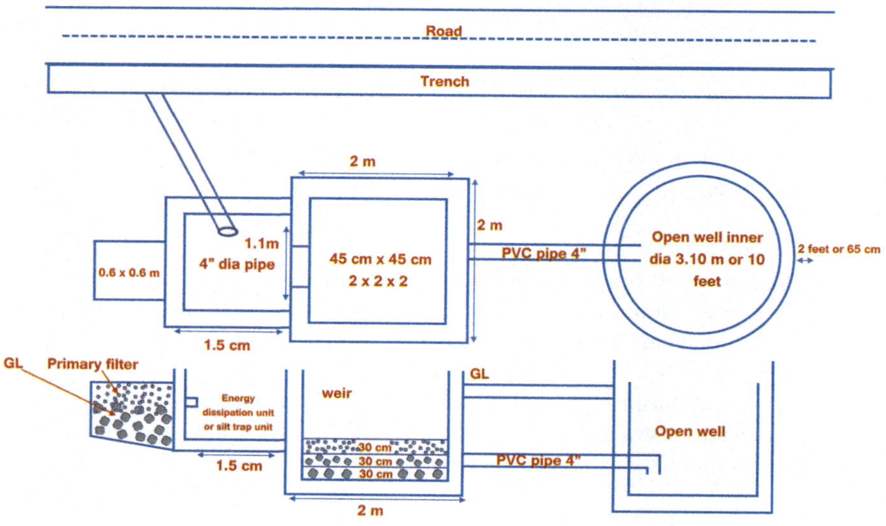

Schematic diagram of model infiltration unit for artificial recharging of open wells

22.6 Crop Production Strategies: Climate-Resilient Variety Cropping Systems

Under changing climate conditions, introduction of high-yielding, drought-resistant/ drought-tolerant varieties holds the promise for getting higher yields. As a general rule, dryland crops are sown early with the onset of monsoon to realize higher yields. And any delay in monsoon beyond normal period affects sowing of many crops of longer duration or narrow sowing window. The crops with wider sowing windows can still be taken up till the cutoff date without major yield loss, and only the change warranted could be the choice of short-duration cultivars. Beyond the sowing window, the choice of alternate crops or cultivars depends on the farming situation, soil, rainfall, and cropping pattern in the location and extent of delay in the onset of monsoon. For example, pulses and oilseeds are preferred over cereals with respect to water requirement and for delayed *kharif* sowing. Cluster bean, moth bean, and horse gram are better choice for low rainfall areas as compared to other *kharif* season pulses. For cultivation on conserved soil moisture during *rabi* season, chickpea and lentil are preferred over peas and French bean. Similarly, among oilseeds, groundnut, castor, sesame, and niger perform well under dryland conditions during *kharif* season. In the rapeseed-mustard group, taramira is the best choice for light-textured soil with low-moisture storage capacity, followed by Indian mustard. Among the *kharif* cereals, coarse cereals (millets, ragi, and sorghum) are better choice over maize and rice. Similarly, in *rabi* season, barley does well under conserved soil moisture than wheat. Among the millets, *Setaria* is the most suited for late-sown condition without any serious effect on productivity.

Since dryland agriculture is risk-prone, adoption of agroecology-specific cropping systems is an essential prerequisite for drought mitigation and enhanced productivity of dryland crops. In arid regions where annual/crop seasonal rainfall is <500 mm and length of growing period is <90 days, short-duration drought-tolerant pulses such as moth bean and cereals of 10–12 weeks' duration such as pearl millet and minor millets are suggested. Under changing climate scenario, many conventional practices and cropping systems are becoming redundant and ineffective and thus need revalidation and modification in accordance to changing climate and soil-site conditions. This also calls for critical examination of important modifiers of cropping systems, viz., soil-type, rainfall pattern, length of growing season, temperature regimes, etc. so that available farm resources are effectively used. Broad guidelines for potential dryland cropping systems for varying range of rainfalls, soils, and agricultural drought situations are suggested in Table 22.5.

Different intercropping and double systems were demonstrated in AICRPDA-NICRA villages, and agroecology-specific best performing/risk-resilient cropping systems were identified (Table 22.6).

Traditionally, double cropping including relay cropping is practiced in dryland regions with sufficient rains (usually >750 mm) and good soil moisture holding capacity (>150 mm). However, some more areas could bring under double cropping through the use of available dryland technologies, viz., rainwater management,

Table 22.5 Potential cropping systems and agricultural drought vulnerability based on rainfall and soil types

Mean annual rainfall (mm)	Major soil order	Growing season (weeks)	Suitable cropping system	Agricultural drought (frequency)
350–650	Alfisols, shallow Vertisols, Aridisols, and Entisols	15	Single rainy season	Severe drought (once in <5 seasons)
350–650	Deep Aridisols and Inceptisols	20	Either rainy or post-rainy season crop	Moderate drought (once in 5–10 seasons)
350–650	Deep Vertisols	20	Post-rainy season crop	Moderate drought (once in 5–10 seasons)
650–800	Alfisols, Vertisols, Inceptisols	20–30	Intercropping	Less prone to drought (once in 10–20 seasons)
800–1100	Deep Vertisols, Alfisols, and Entisols	30	Double cropping	Less prone to drought (once in 10–20 seasons)
>1100	Deep Alfisols, Oxisols	30+	Double cropping	Nil to less prone to drought (once in >20 seasons)

Source: Modified from CRIDA (1997)

choices of crops, short-duration varieties, and agronomic practices. Out of the two crops, one could be short durations (usually legumes) and another medium duration (usually cereals) for optimum use of available growing season. For example, a second crop could be successfully grown in high rainfall regions of Odisha, Eastern Uttar Pradesh, and Madhya Pradesh by replacing medium- to long-duration (>120 days) rice variety with short-duration (<100 days) rice variety, while another crop of chickpea or safflower could be taken in Malwa (MP) and Vidarbha (Maharashtra) regions by substituting sorghum variety of 140–150 days with 90–100-day cultivars (Table 22.7). Similarly, relaying a short-duration and fast-growing crop in standing principle crop provides good opportunity for the efficient use of growing period.

Nutrient management Soils in drought-prone regions are universally deficient in nitrogen. Contrary to the past belief on excessive risks associated with the use of chemical fertilizers, research findings conclusively established that there is vast potential for increasing crop yields through fertilization, typically N fertilizers, across contrasting dryland environments. The cost-benefit ratio is highly favorable when crops are fertilized. A fertilized crop is able to withstand drought better than a non-fertilized one. A stressed crop can also recover faster if it is fertilized following relief from stress. Pest attack was low. Split application ensures against loss in nutrients and deep root system for reaching subsurface stored moisture in the

Table 22.6 Best performing cropping systems

AICRPDA center/climate/soil type/state	Intercropping system
Arjia/semiarid (hot dry)/Vertisols/ Rajasthan	Maize + black gram (2:2); groundnut + sesame (6:2); sorghum + green gram (2:1)
Anantapuram/arid (hot)/Alfisols/ Andhra Pradesh	Groundnut + pigeon pea (11:1); pigeon pea + pearl millet (1:1)
Hisar/arid (hyper)/Inceptisols, Haryana	Pearl millet + green gram (6:3); pearl millet + cluster bean (4:4)
Kovilpatti/semiarid (hot dry)/ Vertisols, Tamil Nadu	Cotton + radish/onion (1:2); cotton + sesame (2:2)
Rajkot/semiarid (hot dry)/ Vertisols, Gujarat	Cotton + sesame (1:1); cotton + fodder maize (1:1); cotton + soybean (1:1); groundnut + castor (3:1)
SK Nagar/semiarid (hot dry)/ Entisols, Gujarat	Castor + green gram (1:1); castor + sesame (1:1)
Solapur/semiarid (hot dry)/ Vertisols, Maharashtra	Pigeon pea + sunflower (1:2); pigeon pea – soybean (1:3); pearl millet + pigeon pea (2:1)
Vijayapura/semiarid (hot dry)/ Vertisols, Karnataka	Pearl millet + pigeon pea (2:1); pearl millet + castor (2:1); pigeon pea + groundnut (4:2)
Akola/semiarid (hot moist)/ Vertisols, Maharashtra	Cotton + green gram (1:1); soybean + pigeon pea (4:2) Cotton + sorghum + pigeon pea + sorghum (3:1:1:1)
Bengaluru/semiarid (hot moist)/ Alfisols, Karnataka	Pigeon pea + field bean (1:1); finger millet + pigeon pea (8: 2); castor + finger millet (1:2)
Indore/semiarid (hot moist)/ Vertisols, Madhya Pradesh	Soybean + pigeon pea (4:2); maize + soybean (2:4)
Parbhani/semiarid (hot moist)/ Vertisols, Maharashtra	Soybean + pigeon pea (4:2); cotton + green gram (1:1); cotton + pigeon pea (6:1)
Rakh Dhiansar/semiarid (moist dry)/Inceptisols, Jammu	Maize + black gram (2:1)
Ballowal Saunkhri/subhumid (hot dry)/Inceptisols, Punjab	Maize + green gram (2:1)
Chianki/subhumid (hot moist)/ Inceptisols, Jharkhand	Pigeon pea + okra/sorghum (1:1); pigeon pea + groundnut (1:2)
Faizabad/subhumid (hot dry)/ Inceptisols, Uttar Pradesh	Pigeon pea + maize/black gram (1:1); black gram + sesame (1:1)
Rewa/subhumid (hot dry)/ Vertisols, Madhya Pradesh	Soybean + pigeon pea (4:2); sesame + pigeon pea (4:2)
Varanasi/subhumid (hot dry)/ Inceptisols, Uttar Pradesh	Pigeon pea + rice/okra (1:1); maize + black gram (1:2)
Jagdalpur/subhumid (hot moist)/ Inceptisols, Chhattisgarh	Sorghum + okra (1:2); sorghum + sesame (1:1); maize + cowpea (1:1); rice + pigeon pea (4:1)
Phulbani/subhumid (hot moist)/ Alfisols, Odisha	Maize + cowpea/pigeon pea (2:2)
Biswanath Chariali/per humid (hot) Inceptisols, Assam	Sesame + black gram (2:2)

Table 22.7 Efficient double cropping systems

Agroclimatic zone/ state	Moisture availability period (days)	Double cropping system
Malwa plateau, Madhya Pradesh	210–230	Soybean-wheat/wheat, maize-chickpea/ safflower
	190–210	Sorghum-safflower/chickpea, soybean-safflower
Baghelkhand, Madhya Pradesh	210–230	Rice-chickpea/lentil
	190–210	Sorghum-chickpea, black gram/green gram-wheat
		Groundnut-chickpea
Bundelkhand, Uttar Pradesh	190–220	Sorghum-chickpea, black gram-mustard/ safflower, fodder cowpea-mustard
Vidarbha, Maharashtra	190–210	Groundnut-safflower, sorghum-safflower
	170–190	Green gram-safflower
Southern Maharashtra	160–180	Green gram-sorghum/safflower
Southern Rajasthan	160–180	Green gram-safflower
Central Karnataka	130–150	Cowpea-sorghum, green gram-safflower

Source: AICRPDA annual reports

aberrant weather more so against loss in nutrients with occurrence of above normal rainfall during crop growth period. Studies have established the value of a number of naturally occurring nutrients containing (organic manures) and generating (biofertilizers) sources to augment overall nutrient turnovers for soil fertility management. Green manure was found to be a dependable source of several plant nutrients. Typically, it could meet half the N requirements of a crop. Inclusion of legumes in a rotation benefitted the succeeding crop equivalent to 10–30 kg N ha^{-1}. Integrated nutrient management (INM) in combination with legume crop is recommended for higher productivity. INM besides nutrient supplementation, also has the ability to hold additional water and result in favourable soil biological interactions. Fertilizer cost can be reduced by substitution of fertilizer with organics. Application of crop residues in combination with chemical fertilizer resulted in higher sustainable yield and maintained higher levels of nitrogen, phosphorus, and organic carbon. Green leaf manure proved promising in increasing the sustainability in yield and improving the organic carbon, infiltration rate, and hydraulic conductivity of the soil. The crops should be supplied essential nutrients considering fertilizer use efficiency, off-season losses in nutrients, and nutrient requirement. Limiting nutrients also need to be supplemented so that crop completes its life cycle without serious limitation. This can be done by a balanced mixture of inorganic, organics, neem, etc. as nitrification inhibitors, VAM, biofertilizers, FYM, and residues for soil acidity adjustment.

Soil carbon management Carbon sequestration studies in rainfed production system and soil types at AICRPDA centers indicated that conjunctive use of chemical fertilizers and organic manure resulted in higher sustainable yield index (SYI) over unfertilized control and sole application of either chemical fertilizers or organic manures. The mean annual C input were recorded maximum in soybean system followed that in rice and groundnut systems. The carbon footprints (Tg CE/ha/year) were higher in cereal cropping systems followed by oilseed and pulse systems. Implementation of appropriate land use and management practices which can maintain and enhance both carbon storage and other ecosystem services is an important strategy for climate moderation and also for enhancing the provisioning services from the rainfed systems.

Crop residue recycling Crop residues are a principal source of C, which constitutes about 40% of the total biomass on dry weight basis. Generally, farmers burn crop residues like stalks of pigeon pea and cotton without recycling them. Therefore, shredding of crop residues is to be mechanized. This will help in shredding stalks of pigeon pea, cotton, and green biomass into small pieces to facilitate easy mixing of the residue in the soil or compost/vermicompost pits. Crop residues have several positive impacts on physical, chemical, and biological properties of soil through either incorporation or retention on the surface. There are various methods of crop residue application to soil both in situ and ex situ. The in situ system is an efficient method of recycling organic residues, since crop residues and animal waste are properly utilized directly into the soils with no nutrient loss. Permanent crop cover with recycling of crop residues is a prerequisite and integral part of conservation agriculture. However, sowing of a crop in the presence of residues of preceding crop is a problem. But new variants of zero-till seed cum fertilizer drill/planters such as rotary disc drill have been developed for direct drilling of seeds even in the presence of surface residues (loose and anchored up to 10 tons ha^{-1}). These machines are very useful for managing crop residues for conserving moisture and nutrients as well as controlling weeds in addition to moderating soil temperature.

Foliar nutrition of crops In dryland agriculture, occurrence of frequent dry spells leads to low soil moisture; hence, application of fertilizers in right quantity and at the right time may not be possible. Under these situations, application of fertilizers through foliar spray results in efficient absorption. Foliar nutrition does not address any specific nutrient deficiency but supplies a small amount of all nutrients to keep leaf growth lush. Though foliar spray is not a substitute to soil application, it certainly is considered as a supplement to soil application. The plant nutrients are not only required for better plant growth and development but also helpful to alleviate different kinds of abiotic stresses like drought stress. Foliar nutrition enhances overall nutrient level in the plant and increases sugar production during stress period. In foliar sprays, products with higher specific analysis is designed to address specific deficiencies that have been identified. It is not practical to attempt to supply nutrients that plants require in large amounts like nitrogen, phosphorus, and potassium or those that are relatively immobile such as calcium or boron, solely by foliar nutrition. Foliar applications can supplement soil nutrients and provide a small boost in growth and yield particularly under dry conditions. It is advantageous over soil application in specific occasions in certain crops.

Farm mechanization Farm mechanization in dryland agriculture plays a vital role in reducing cost of crop production and drudgery and increasing cropping intensity. In the past five decades, CRIDA-AICRPDA mechanization research helped in evolving new technologies with location-specific recommendations that increased the cropping intensity apart from meeting the timeliness and precision in dryland agriculture. Since many years, farm machinery has become one of the important inputs that improve the efficiency in field operations at lower cost with high precision, ensuring comfort with reduced drudgery. It has been widely proved that the mechanization of various field operations increases the overall productivity by 12–34%, facilitates enhancement of cropping intensity by 5–22%, and increases gross income to farmers by 29–40%. The use of improved farm equipment has shown yield advantage of 15–20%. Deficit farm power is one of the impediments in timely completion of operations in drylands. Diminishing number of draught animal and diversion of agricultural labor to nonagricultural activities during peak operation season has created a farm power crisis in dryland farming. Hence, meeting farm power requirements with mechanical power sources like tractors and power tillers has become essential for sustenance of drylands. Although tractors are adapted in many dryland regions including Malwa (Madhya Pradesh), parts of Punjab, Uttar Pradesh, and Haryana, other dryland regions are still suffering from farm power deficit.

Alternate land use/agroforestry systems Agroforestry is a combination of land use management and technologies where woody perennial are combined in some temporal and spatial arrangement. However, with food shortages and increased threats of climate change, agroforestry has been identified being having the potential to address various on-farm adaptation needs and mitigation pathways. There are many success stories where agroforestry has proved its potential in degraded and unutilized wastelands for providing food, fodder, and fuel. The National Forest Policy aims to develop new strategy of forest conservation to meet out the huge pressure by human beings and livestock. Further policy option should also give more emphasis for identifying and adapting underutilized trees species like *Prosopis* having mutlistress-tolerant capability. The role of private sector cannot be excluded and hence could be strengthened further to explore more areas including biofuels.

In agri-silviculture system, nitrogen-fixing tree species offer immense possibilities of supplementing the nitrogen requirement of crops grown in association, besides providing rich organic matter and atmospheric nitrogen, improving soil structure, and preventing land degradation. In agri-horticulture/agri-silvi-horticulture system, short-duration arable crops are raised in the interspace of fruit trees. The agricultural crops provide seasonal revenue, while fruit trees are managed for 30–35 years giving regular returns of fruit and in some cases fuelwood and fodder from pruned biomass. The system works best in medium to deep soil with good water-holding capacity. Individual farm ponds and supplemental watering in the off-season will certainly improve the scope of fruit farming in drylands. Silvipastoral land use systems are significant where agroforestry crop farming systems are not feasible owing to low rainfall and lack of water. In horti-pastoral system, there is an integration of fruit trees with pasture/grass. To improve wasteland/degraded land through horti-pastoral system, the first step is to protect from biotic interference followed by selection of suitable top feed species which have fast growth and good coppicing ability and are highly palatable. In addition, the tree species which have the ability to withstand browsing, trampling and intensive lopping and are resistant/ tolerant to drought and extreme climatic conditions should be selected. Alley cropping, also called hedge row intercropping, integrates the benefits of fallow period directly into the cropping period. Crops are sown in alleys between rows of trees, usually leguminous. The trees are pruned at regular intervals, at about 0.6 m above ground level, and the prunings are used for mulch or animal meal. It is a quite flexible technology that benefits crop and livestock activities and can, through a modification of tree management techniques, provide fuel for the household. Farm boundaries, bank, river banks, etc. may also be suitably utilized for lean period fodder. Live hedges are lines of trees or shrubs planted on farm boundaries or on border of home compounds, pastures, fields, or animal enclosures. The purpose is different from other agroforestry technologies based on trees planted in lines such as boundary planting, contour strips, or hedge row intercropping. Besides their main function to control human and animal movement, they provide fodder, fuelwood, and food, act as wind break, and enrich the soil depending on the species used.

Land capability-based productive farming systems are identified (Fig. 22.3) for drought-prone regions based on land capability, rainfall, and soil orders and the outcome of research information generated at AICRPDA centers (Vittal et al. 2003). This would help for planning and implementation decisions at regional level on agroforestry or land use systems. In these systems, livestock will be an integrated component. Microlevel highly diversified farming systems are practicable only in drylands due to resilience in adoption of diversification from crop through tree to animal.

22.7 Strategies for Climate-Resilient Farming

Risk and vulnerability assessment at sub-district level Climate resilience in dryland agriculture can be better addressed through risk and vulnerability assessment at sub-district level.

Developing climate-resilient varieties and cropping systems A well-planned complementary cropping systems and intercropping have great potential to reduce the risk and uncertainties in production through effectively exploiting differential performance of crops to moisture and temperature stresses and intermittent and terminal

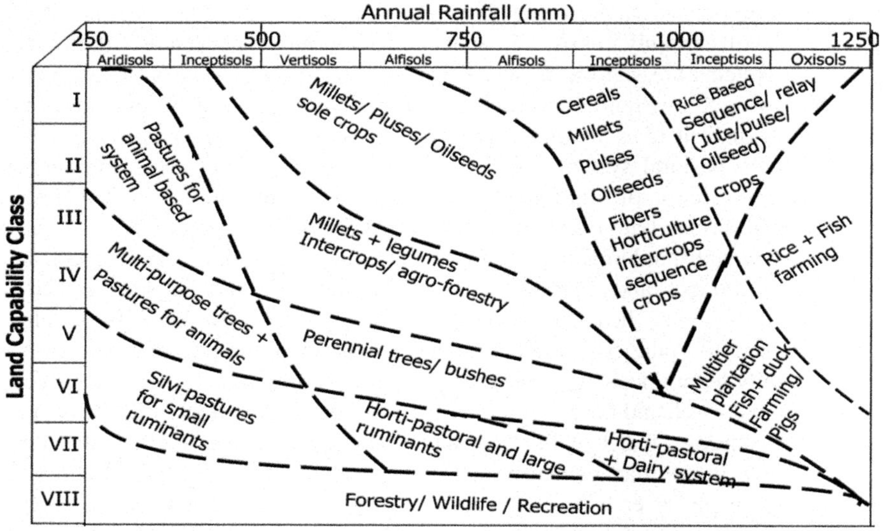

Fig. 22.3 Productive farming system matrix in rainfed agriculture (Vittal et al. 2003)

droughts in different agroecological settings. Similarly, double and relay cropping systems with suitable component crops/varieties with resource conservation technologies have greater scope for climate-resilient agriculture. Further, breeding-resilient crop varieties that fit in emerging cropping systems are very important for enhancing adaptation capabilities. It is essential to characterize agroecology/ resource domain (climate and soil) of these varieties for wider adoption. However, all these efforts will have little significance if it is not supported with real-time strategies to handle farming emergencies emerging due to enhanced weather vagaries. The important issues include developing zone-wise climate-resilient cropping system packages along with associated management practices, agroforestry-based cropping systems, evaluation of crop varieties in cropping system mode, evaluating cropping system modules in farming system perspective, and demonstration of proven systems in cluster approach for large-scale adoption.

Real-time contingency planning During 1972–1973, large-scale scarcity of rainfall was experienced all over the country, particularly in the scarcity region of Maharashtra, Karnataka, and Andhra Pradesh. Roving seminars were organized by the ICAR at different locations, at the end of which *new phrases* were coined, viz., *contingent crop planning and midseason correction*. As a follow-up, the AICRPDA (All India Coordinated Research Project for Dryland Agriculture) centers at Solapur and Bijapur collected data on these two aspects and after analysis of weather data for the past 100 years listed the weather aberrations: *(1) delayed onset of monsoon,*

(2) early withdrawal of monsoon, (3) intermittent dry spells of various durations, (4) prolonged dry spells causing changes in the strategy, and (5) prolonged monsoon (AICRPDA 1983). *Contingency plans*, for each region, was a conceptual approach unique from AICRPDA project in developing location-specific contingent crop strategies which were first published in 1977 (Ravindra Chary et al. 2012), and with further refinements and updating in crops and varieties, the first document was brought out by AICRPDA in 1983 on *contingent crop production strategy in rainfed areas under different weather condition* (CRIDA 1983). The AICRPDA network centers developed crop contingency plans for each center's domain (Subba Reddy et al. 2008; Ravindra Chary et al. 2012). Further, during 2009–2010, AICRPDA centers prepared contingency measures considering weather aberrations, seasons, and the predominant *kharif* and *rabi* crops with appropriate crop management strategies. The Central Research Institute for Dryland Agriculture (CRIDA) with information available at AICRPDA centers and SAUs prepared district-level agriculture contingency plans for 650 districts in collaboration with the Ministry of Agriculture, GoI, ICAR institutes, agricultural universities, Krishi Vigyan Kendras (KVKs), and the state line departments. These plans essentially suggest coping strategies/measures in agriculture, horticulture, livestock, fisheries, and poultry sectors in the event of delayed onset of monsoon, seasonal drought, unseasonal rainfall events, floods, cyclones, hailstorm, and heat/cold wave (Venkateswarlu et al. 2011).

The real-time contingency planning (RTCP) is conceptualized in All India Coordinated Research Project for Dryland Agriculture (AICRPDA) as "any contingency measure, either technology related (land, soil, water, crop) or institutional and policy based, which is implemented based on real time weather pattern (including extreme events) in any crop growing season" as two-pronged approach: (1) preparedness and (2) implementing contingency measures on real-time basis. The RTCP aims first to establish a crop with optimum plant population during the delayed onset of monsoon, ensure better performance of crops during seasonal drought and extreme events, enhance performance, improve productivity and income, and enhance the adaptive capacity of the small and marginal farmers. The preparedness emphasizes on a combination of tolerant variety/crop system and rainwater/soil/crop/nutrient management practices along with timely availability of inputs, while real-time basis implementation focuses on the crop/soil/moisture/nutrient management measures to cope with delayed onset of monsoon, seasonal drought, floods, and other extreme events (AICRPDA-NICRA Annual Report 2013). The must-to-do practices for initial preparedness for RTCP implementation are shown below:

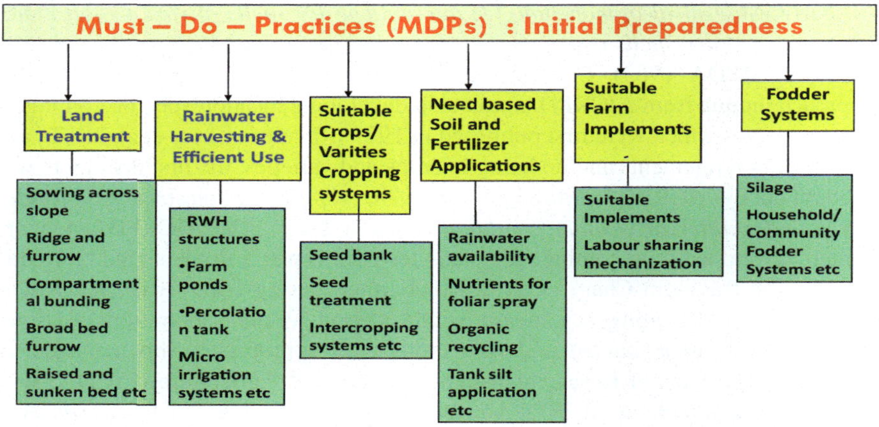

In rainfed areas, as a general rule, early sowing of crops with the onset of monsoon is the best-bet practice that gives higher realizable yield. Major crops affected due to monsoon delays are those crops that have a narrow sowing window and therefore cannot be taken up if the delay is beyond this cutoff date. Crops with wider sowing windows can still be taken up till the cutoff date without major yield loss and only the change warranted could be the choice of short-duration cultivars. Beyond the sowing window, the choice of alternate crops or cultivars depends on the farming situation, soil, rainfall, and cropping pattern in the location and extent of delay in the onset of monsoon. During 2011 to 2021, under delayed onset of monsoon conditions, improved varieties of major rainfed crops were evaluated by AICRPDA centers (Table 22.8) under NICRA, and best performing varieties were identified (Ravindra Chary et al. 2020). Across the rainfed crops, these varieties gave about 15–30% higher yields compared to local farmers' varieties.

The key real-time interventions that helped to cope with delayed onset of monsoon were introduction and identification of most suitable crops and short-duration drought-tolerant varieties and demonstration of cropping systems (Ravindra Chary et al. 2020), resowing within a week to 10 days with subsequent rains when germination is less than 30%, thinning, etc.; to cope with early season drought were repeated interculture to break soil crust and remove weeds, avoiding top dressing of fertilizers till favorable soil moisture, in situ moisture conservation through opening conservation furrows; to cope with midseason drought were avoiding top dressing of fertilizers till favorable soil moisture, in situ moisture conservation through opening conservation furrows, surface mulching with crop residues, foliar spray of 1% KNO_3 or 0.5% water-soluble fertilizers, and providing protective irrigation, if available; and to cope with terminal drought were harvesting crop at physiological maturity with some realizable yield or harvest for fodder, providing protective irrigation if available, or preparing for *rabi* sowing in double-cropped areas. Introduction of drought-tolerant varieties gave about 15–35% higher yields compared to local farmers' varieties. The interventions to mitigate early season drought helped in adaptation of crops and realizing improved yields by 16–30% compared to no

Table 22.8 Improved varieties of major rainfed crops

AICRPDA center/climate/soil type/state	Crop/variety/hybrid
Arjia/semiarid (hot dry)/Vertisols/ Rajasthan	Maize PM-3; black gram T-9; horse gram AK-41, AK-42; groundnut TG 37-A
Anantapuram/arid (hot)/Alfisols/ Andhra Pradesh	Pigeon pea, LRG-30; pearl millet, ICTP-8203, ICMV-221
Hisar/arid (hyper)/Inceptisols/ Haryana	Pearl millet HHB-226, HHB-67; cluster bean HG-365, HG 2-20; green gram, Muskan, Basanti, MH-421
Kovilpatti/semiarid (hot dry)/ Vertisols/Tamil Nadu	Maize, COH(M)7, NK 30; green gram, VBN-4, CO-6; cotton, RCH-530, Tulasi-117
Rajkot/semiarid (hot dry)/ Vertisols/Gujarat	Groundnut, GG-20, GJG-9; black gram, Urad-1; cotton, G. cotton Hy-8, GTHH-49; pigeon pea, BDN-2; castor, GCH-7
SK Nagar/semiarid (hot dry)/ Entisols/Gujarat	Pearl millet, GHB-558; maize, HQPM-1; green gram, GM-4; black gram, GU-1; castor, GCH-7
Solapur/semiarid (hot dry)/ Vertisols	Pigeon pea, Vipula; sunflower, Bhanu, Phule Bhaskar; black gram, TPU-1
Vijayapura/semiarid (hot dry)/ Vertisols/Karnataka	Pearl millet, ICTP-8203; pigeon pea, TS-3R; horse gram, GPM-6
Akola/semiarid (hot moist)/ Vertisols/Maharashtra	Soybean, JS-95-60; cotton, PKV-Rajat; pigeon pea, PKV-Tara, Vipula
Bengaluru/semiarid (hot moist)/ Alfisols/Karnataka	Finger millet, GPU-28; pigeon pea, BRG-2, TTB-7; field bean, HA-3, HA-4
Indore/semiarid (hot moist)/ Vertisols/Madhya Pradesh	Soybean, JS-95-60, JS-93-05, RVS-2001-4; pigeon pea, Pusa-992; horse gram, AK-42
Parbhani/semiarid (hot moist)/ Vertisols/Maharashtra	Soybean, MAUS-81; pigeon pea, BDN-711, BDN-708; green gram, BM-4
Jhansi/semiarid (hot moist)/ Inceptisols/Uttar Pradesh	Sesame, JTS-08; black gram, Azad-2, Sekhar-1
Rakh Dhiansar/semiarid (moist dry)/Inceptisols/Jammu	Maize, Double Dekalb; black gram, Pany U-19, Uttara
Ballowal Saunkhri/subhumid (hot dry)/Inceptisols/Punjab	Maize, PMH-1, PMH-2; black gram, Mash-114; sesame, RT-346, Pb Til-1
Faizabad/subhumid (hot dry)/ Inceptisols/Uttar Pradesh	Rice, NDR-97, Baranideep; pigeon pea, NDA-1, NDA-2
Rewa/subhumid (hot dry)/ Vertisols/Madhya Pradesh	Soybean, JS-95-60, JS-93-05; pigeon pea, Asha
Varanasi/subhumid (hot dry)/ Inceptisols/Uttar Pradesh	Rice, NDR-97, Vandana; sesame, GT-1; pearl millet, Pusa-322
Jagdalpur/subhumid (hot moist)/ Inceptisols/Chhattisgarh	Upland rice, Sahabhagi Dhan; finger millet, GPU-28
Phulbani/subhumid (hot moist)/ Oxisols/Odisha	Rice, Vandana, Sahabhagi Dhan; maize, Nirmal-51, Navjot; black gram, PU-30, Sekhar
Biswanath Chariali/per humid (hot)/Inceptisols/Assam	Upland rice, Gitesh, Ranjit, TTB-404, Dishang

Source: Ravindra Chary et al. (2020)

contingency measures. RTCP measures of foliar sprays of water-soluble NPK (19:19:19) and KNO_3 in mitigating midseason dry spells gave 15–25% higher yield in different crops compared to no spray. The effect of midseason and terminal drought on different crops was mitigated mostly by providing supplemental irrigation from harvested rainwater in farm ponds and foliar sprays. Supplemental irrigation improved yields by 20–25% in cotton, 40% in groundnut, and 40–55% in soybean at different locations. Similarly, foliar spray of 1% KCl in rice during dry spell at flowering-milking stage increased yield by 25% compared to no spray (AICRPDA-NICRA Annual Reports 2013, 2014, 2015, 2016, 2017, 2018, 2019, 2020). The experiences of RTCP measure implementation were included in the policy documents such as Revised Manual for Drought Management, Ministry of Agriculture and Farmers' Welfare, GoI; National Agriculture Disaster Management Plan, Ministry of Home Affairs, GoI; and Revised Common Technical Guidelines for New Generation Watershed Management developed by the National Rainfed Area Authority (NRAA) and Department of Land Resources, Ministry of Rural Development and Panchayati Raj, GoI.

The experiences of RTCP implementation indicated many opportunities for developing adaptation strategies both in terms of preparedness and real-time response such as production of seed of alternate crops/varieties; seed corporations, agricultural universities, KVKs, etc. have greater roles to provide suitable drought-tolerant and short-duration seed material during the event of delayed onset of monsoon; and establishment of community/village seed banks for production and distribution of quality seeds, rainwater harvesting, and storage structures are capital and labor-intensive, thus can be converged with national/state/district programs, and promote farm mechanization through custom hiring of implements.

Building resilience at agriculture landscape level On-farm participatory action research was undertaken to build resilience at landscape level in Kavalagi micro-watershed, Karnataka, in Southern India by the AICPDA center, Vijayapura. The watershed area is predominantly rainfed and cultivated with post-monsoon *rabi* crops, viz., chickpea and sorghum. The watershed was characterized for land and soil resources and was mapped at 1:4000 scale. The soil units were delineated into soil conservation units (SCUs) considering soil physical parameters such as slope, depth, texture, and erosion, and soil quality units (SQUs) were delineated considering parameters such as EC, OC, pH, and calcareousness. These two layers of SCUs and SQUs together were further delineated into homogeneous land management units (LMUs). The on-farm trials on chickpea (cv. JG-11) and sorghum (cv.BJV-44 and cv.M-35-1) were conducted on LMUs I, III, V, and VII for 3 years wherein the seasonal rainfall varied and was deficit up to 30% compared to normal seasonal rainfall. Though the rainfall was the same for the entire watershed in three seasons, the yield of crops varied due to spatial variability in characteristics of LMUs. The yield of both chickpea and sorghum was higher on LMU I > LMU III > LMU V > LMU VII due to limitations in soil physical and chemical characteristics with LMU I having more favorable land characteristics for crop growth (AICRPDA-NICRA Annual Reports 2018, 2019, 2020). The results revealed that the performance of crops varies due to spatial variability in land and soil characteristics in a landscape though the rainfall is uniform. Therefore, it is emphasized that building resilience of agriculture is to be achieved at landscape level with suitable cropping in a favorable land and soil environment.

22.8 Participatory and Community Approaches

Participation of communities is the prerequisite for attaining inclusive development of dryland agriculture at watershed or village level. Establishing village-level institutions (VLIs) is an integral part of any participatory approach. A village-level institution (VLI) is a formal body intended to ensure sustainable agriculture and rural development in India. The very purpose of forming VLI is to provide people ownership of any development project by making them an integral part of decision-making, giving them control over their resources, autonomy to implement the project, and carry on the process even after the completion of such projects.

The real-time contingency measures have been implemented in the adopted villages by AICRPDA centers by establishing the VLIs like Village Climate Risk Management Committee (VCRMC), custom hiring center (CHC), seed bank, fodder bank, etc. which have a greater role to play in the initial preparedness for the implementation of real-time contingency planning. A VCRMC, a unique institution at the village representing all categories of farmers, is formed with the approval of the *gram sabha* (village-level decision-making body). The VCRMCs have a greater role in identifying and implementing need based climate risk-resilient interventions such as renovation and/or establishing new farm ponds/percolation tanks/other water harvesting structures for the creation of water assets for drought proofing, crop-based interventions, establishing and efficient functioning of custom hiring centers, etc. Timeliness of agricultural operations is crucial to cope with climatic variability, especially for sowing and intercultural operations as soil moisture status provides a limited sowing window, particularly in rainfed agroecosystems. With increasing climatic variability, timely sowing or resowing is important, which is required for ensuring optimum plant population and better crop growth. However, small holding farmers often are not able to take up sowing in time for want of labor and implements. Though farm machines are used for completing farm operations, their access is limited. A Custom Hiring Management Committee (CHMC) needs to be constituted in villages for timely decisions on hiring and maintenance of implements. Village seed bank has the role to provide seeds of improved and stress tolerant varieties to farmers on time. This is one of the important interventions for imparting resilience against climate variability. The purpose of a village seed bank is to serve as an emergency seed supply when farmers experience shortage of seeds, where there is a need for resowing of crop. The farmer groups need to be trained and given seed and guidance to multiply the seed. Fodder bank consists of planting high-quality fodder species which can provide high biomass in a short time and bridge the forage scarcity during the annual dry seasons and also during the long dry spells within the growing season. Planting of high biomass yielding and fast-growing grasses and shrubs suitable for fodder not only increases fodder availability but also reduces erosion and landslides that originate in these areas. These fodder banks also help in the preservation and storage of surplus fodder and availability of nutritious fodder during the period of fodder scarcity and enhance nutritive value of crop residue and other cellulosic wastes for animal feeding. The concept of "nutrient bank" evolved wherein the essential manures and fertilizers, soil amendments, foliar spray chemicals, biofertilizers, etc. are maintained locally and made available in time by the local community. The nutrient banks help farmers in restoring the productive capacity of soils and local environments. These nutrient banks are run by SHGs in conjunction with *gram panchayats*. During intermittent droughts, farmers can approach nutrient bank to avail foliar sprays like KNO_3 spray, which enhances drought tolerance. Similarly, when there is a shortage of fertilizers, nutrient bank plays a vital role in providing the farmers with the required amount of fertilizers.

22.9 Solar Farming: Green Energy Use in Dryland Agriculture

Agriculture is one industry that accounts for roughly 7–8% of India's total energy use. Pumping irrigation water, using heavy machinery for various farm tasks, processing and value addition of farm produce, and so on are all key energy-consuming activities in the agriculture industry. There has been a tremendous increase in energy use in agriculture as the food production system has progressed from rural to a future technology-driven system. India has an estimated renewable energy potential of about 900 GW from commercially exploitable sources. Among the total renewable potential, wind power potential is about 102 GW at 80-m mast height, solar power potential of about 750 MW assuming 3% wasteland is made available, and bioenergy potential of 25 GW, and the rest is by other renewables. Some of the technologies that can be helpful in rainfed areas (NAAS 2018) of the country are given below.

Agrivoltaic system From a single land unit, an agrivoltaic system produces food and generates sustainable electricity. Due to ever-increasing demands for land resources, the concept of integrating both food production and energy generation on a single land unit has grown in recent years. Energy from the sun is required for both food production and photovoltaic-based renewable energy generation. Food is produced by converting solar energy to food through the photosynthetic process, which has a conversion efficiency of 3%, whereas PV-based energy is produced by converting solar energy to electric energy through the photovoltaic process, which has a conversion efficiency of 15%. Both of these processes necessitate the use of land as a primary natural resource. The interspaces of an agrivoltaic system are used to grow suitable crops that are preferably short in height, low in water demand, and shade-tolerant. Mung bean (*Vigna radiata*), moth bean (*Vigna aconitifolia*), and cluster bean (*Cyamopsis tetragonoloba*) are suitable crops for interspace area in arid western Rajasthan and Gujarat during kharif season and cumin (*Cuminum cyminum*), isabgol (*Plantago ovata*), and chick pea (*Cicer arietinum*) during rabi

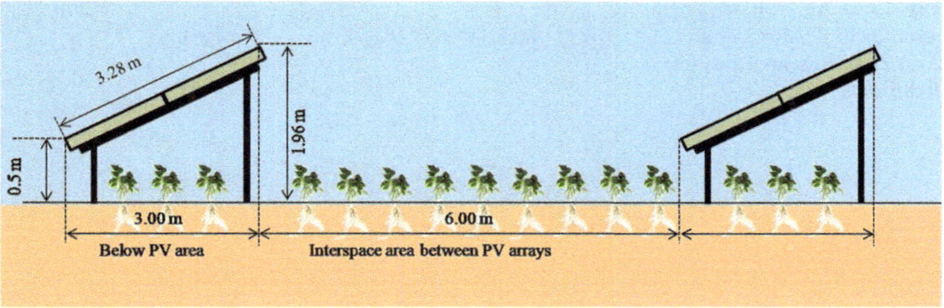

Fig. 22.4 Design of agrivoltaic system. (Source: ICAR-CAZRI)

season. In addition to these arable crops, medicinal plants such as gwarpatha (*Aloe vera*), sonamukhi (*Cassia angustifolia*), and shankhpushpi (*Convolvulus pluricaulis*) can be planted in the interspace.

AVS integrating both food production and PV generation from a single land use system was designed and developed at ICAR-Central Arid Zone Research Institute, Jodhpur, with a system capacity of 105 kW_p and installed in about 1 acre land (26° 15′27.82″ N and 72°59′34.57″ E). In the AVS, crops are cultivated at interspace areas between PV arrays as well as at below PV areas. Another important feature of the AVS is the rainwater harvesting from top of PV module to a water storage tank, which is recycled for cleaning of PV module and for providing supplemental irrigation to crops.

The agrivoltaic system was installed with three designs each at a separate block of 35 kW_p in an area of 32 × 32 m. PV array designs in three blocks of the AVS was different, and these are (1) single-row PV array, (2) double-row PV array, and (3) triple-row PV array. The single-row model consisted of one row of PV module in the array, whereas double- and triple-row model consisted of two rows and three rows of PV module in the array, respectively. Performance evaluation of the systems showed that double-row AVS is the most feasible system. Design parameter of the double-row PV module installations is shown in Fig. 22.4. Inclination of PV module was kept as 26°, which is almost equal to the latitude of the experimental site.

The double-row model has the bottom row with full density, whereas top row has a PV density of 60%, and following this design 400 kW_p AVS can be installed in 1 ha area. A field view of the established AVS at ICAR-CAZRI, Jodhpur, is given in Fig. 22.5

The interspace area and the below PV area are used for cultivation of suitable crops. Height of crops is a key parameter for selection of crops for agrivoltaic system because tall-growing crops may create shade on PV modules and thus reduce the PV-based electricity generation. Therefore, crops with low height (preferably shorter than 50 cm) and which tolerates certain degree of shade and require less amount of water are most suitable for solar farming system. Electricity generated by the PV modules of the system is sold to local grid through net metering system. Electricity

Fig. 22.5 Agrivoltaic system installed at ICAR-Central Arid Zone Research Institute, Jodhpur

generation capacity of the system at Jodhpur is 4-5 kWh per day. Top surface of PV modules of the system is utilized for harvesting rainwater to a water reservoir, which is recycled for cleaning PV module and providing supplemental irrigation to crops.

The AVS may bring huge opportunity to dryland farmers since rainfed-based crop production is risky because of uncertainty and scarcity of rainfall. Ecological and social benefits of agrivoltaic system are increased income from farmland and land productivity (land equivalent ratio may be increased up to 1.41, reduction in soil erosion by wind and decreases dust load on PV module, improvement in microclimate for crop cultivation and optimum PV generation, recycling of harvested rainwater for cleaning PV modules and irrigating crops (1.5 lakh L/acre and can provide 40 mm irrigation in 1 acre land), soil moisture conservation by reducing the wind speed on ground surface, reduction in GHG emission (598.6 tons of CO_2 savings/year/ha). Few perceived challenges of agrivoltaic system are: high aeolian activity in arid zone during hot summer months posesing challenges in PV generation, power grid network at farmer's field is limited for evacuation of PV generated electricity, fesability of large capacity off-grid agrivoltaic system, safety of field workers engaged in agricultural activity, ownership issue: Farmer and solar power plant functionary and high capital investment during initial establishment.

The concept of AVS system may be implemented in farmers' field in arid and semiarid region of the country where land productivity is inherently low and solar energy is available in plenty. To make the AVS as a potential option to increase farmers' income in fragile arid and semiarid ecosystem, it is necessary to provide grid network nearby farmers' field to create selling opportunity of PV-generated electricity from the system.

22.10 Future Projection

Climate change and climate variability impact Indian agriculture in general and more pronounced on rainfed agriculture. Climatic risks like droughts and floods, poor water and nutrient retention capacity of soil, and low soil organic matter (SOM) impact rainfed agriculture highly vulnerable, requiring a different outlook and strategy. Risk is also to be addressed in terms of building resilience of crops, soils, and farmers. Resilience to climate change will depend on increasing agricultural productivity with available water resources, refining technologies, and timely deployment of affordable strategies to accomplish potential levels of arable land and water productivity. The projected area, production, and yield of cereals under various production systems in India for 2030 and 2050 indicated that while irrigated production systems can contribute an additional yield of 15%, the rainfed production systems could remain the same. Thus, there would be a need for a strategic mix of better technology adoption, institutional innovations, and incentive system to enhance productivity of rainfed cereals (Table 22.9).

The large yield gaps remain in several rained crops and regions between yields obtained at research stations and on farmers' fields (Ravindra Chary et al. 2012). Although the average per hectare productivity levels have increased from 0.6 tons in the 1980s to 1.2 tons at present in rainfed areas, large gaps still remain in several rained crops and regions between yields obtained at research stations and on farmers' fields. The nation's demand is average productivity of 2 tons/ha from rained agriculture. With current gains in the existing rainfed production systems, there is a need for more profitable land use systems. In several disadvantaged areas, the yield gaps will continue to remain large even in 2050, both due to non-adoption of technologies and nonavailability of tailor-made agroecology-specific package of

Table 22.9 Projected area, yield, and production of cereals under different production systems of India

Cereals	2030	2050
Rainfed area (million ha)	40	36
Rainfed yield (tons/ha)	1.8	2.0
Rainfed production (million tons)	73	72
Irrigated area (million ha)	57	62
Irrigated yield (tons/ha)	4.3	4.6
Irrigated production (million tons)	248	285
Total area (million ha)	97	98
Total yield (tons/ha)	3.3	3.7
Total production (million tons)	321	357

Source: Observations, projections, and impacts—India 2011 UK Met office report. Eroding demand for some of the rainfed commodities is another challenge while other crops are enjoying favorable market demand. Therefore, technologies are to be developed for those crops which have expanding markets but are difficult to be grown in rainfed conditions. Wherever appropriate, steps must be taken to contribute to the formulation of policies for promoting crops that are not so resource-demanding but at the same juncture have better food, nutrition, and fodder value practices. While evolving strategies for bridging yield gaps, due attention must be given to regional imbalances in terms of natural resources and technology intake capacity of farmers. For yield maximization, selecting genotypes with wide adaptability and resilience to climate variability remains a challenge.

As the demand for water from non-farm sectors increases and availability to agriculture declines, the conflicts between upstream and downstream users may increase over time. A fallout of such process is the possible conversion of existing productive irrigated lands to rainfed lands (Sikka et al. 2016). A 4 °C rise in temperature and a 10% decrease in rainfall are expected to reduce the streamflow by 33% in mean annual, 15% in pre-monsoon, 35% in monsoon, 32% in post-monsoon, and 21% in winter seasons. The National Water Mission, institutionalized under the National Action Plan for Climate Change, has set the target to improve the efficiency of water use by at least 20%. In a given land use setting, climatic variables especially temperature and rainfall regulate the irrigation water demand. At present, in India, blue and green water availability is above the 1300 m^3/capita/year threshold. However, with climate change, blue-green water availability is estimated to decrease to less than 1300 m^3/capita/year, implying that by 2050, all of India could be exposed to water stress.

22.11 Future Strategies

1. *Rainfed agroecosystem zoning* considering the natural resources and socioeconomic parameters for identifying the domain areas for developed rainfed agroecology-specific technologies.
2. *Climate risk assessment at sub-district level* for risk prioritization, developing, and implementation of adaptation plans for drought proofing
3. *Enhancing soil quality and productivity*: Availability of high-resolution maps will help to develop suitable land use planning strategies for sustainable use of natural resources. Soil carbon management through enhancing carbon sequestration in rainfed soils is very essential.
4. *Enhancing water productivity*: Refining in situ conservation and rainwater harvesting technologies to suit different environments/situations. Watersheds would be units for the study. Based on the experiences of AICRPDA and NICRA, adaptation plans for climate variability based on water resource availability would be developed in a holistic manner for implementation of district authorities for rainfed regions.
5. *Agroecology-specific potential crop zoning* for crop alignment as per agroecology specificity, commodity crop-specific strategies, and crop diversification.
6. *Diversifying within farm for sustainable intensification*: The crop-tree-animal-based systems to be evolved by strengthening predominant traditional rainfed farming systems in prioritized rainfed districts that enhance resource use efficiency and livelihoods by providing risk resilience, food and nutritional security, staggered employment, and income.

Rainfall zone (mean annual rainfall)	Traditional rainfed farming systems	Agroecology-specific components along with efficient in situ and ex situ rainwater management practices
<500 mm	Livestock-crop-based	Small ruminants, nutritious cereals/millets
500–750 mm	Crop-horticulture-livestock-based	Small/large ruminants, predominant rained crops, and dryland horticulture
750–1000 mm	Crop-horticulture-livestock-poultry-based	Predominant rained crops, dryland horticulture, agri-hortisystems, rainfed vegetable crops, small/large ruminants, improved breeds of poultry
>1000 mm	Multiple enterprise based on multiple water use	Predominant rained crops, lowland rice with water-saving technologies, dryland horticulture, vegetable crops, other high-value crops, agri-hortisystems, small/large ruminants, improved breeds of poultry, fish and other income-generating enterprises like seed production, apiary, mushroom cultivation, etc.

Diversifying to non-farm linked to agriculture The non-farm income can be generated through seed production and processing units, agriclinics and custom hiring services, seedling nursery supply units, vermicomposting, bee keeping, mushroom production, artisans, farm equipment repair and maintenance, food processing, etc.

Sustainable fodder production Each farmer should at least allocate 10% of their land for fodder production, integrating ruminants with trees in the form of silvipastoral, agri-silvipastoral, and horti-pastoral systems. Fodder production strategies include:

1. *Tank beds—common pool resources for fodder production*: growing of two or more annual fodder crops as sole crops in mixed stands of legume (*Stylo* or cow pea or hedge lucerne, etc.) and cereal fodder crops like sorghum and ragi in rainy season followed by berseem or lucerne, etc. in *rabi* season in order to increase nutritious forage production round the year; perennial deep-rooted top feed fodder trees and bushes such as *Prosopis cineraria*, *Hardwickia binata*, *Leucaena leucocephala*, and *Acacia nilotica* trees; and modified plants of cactus which are highly drought-tolerant and produce top fodder
2. *Fodder production systems in homesteads: Azolla, fodder production as contingency plan*—days and can be sown immediately after rains under rainfed conditions in arable lands during *kharif* season

Smart farm mechanization Development of farm implements/machinery equipment for small farm holdings for land management practices, resource conservation systems, various intercropping systems, etc.

Drought early warning systems, monitoring, and development of real-time advisory mechanisms for rainfed regions Combining real-time data and forecast of weather information, advisory to be developed at sub-district level by utilizing the dynamic crop weather calendar information. This would lead to the development of district adaptation plans. The experiences of real-time contingency plans of AICRPDA are also to be considered. The district adaptation plans are to be linked with drought monitoring systems for autogeneration of agroadvisories and dissemination through ICT tools. State-of-the-art tools such as machine learning, AI, and public domain datasets would be used to map the vulnerable areas of drought. There is a need for value-added weather management services which include delineation of climate vulnerable zones at microlevel, real-time agromet-advisories, operationalization of contingency plans at microlevel, climate predictions, and pest and disease forewarning systems.

Capacity building By involving communities in disaster management planning at the local-level preparedness planning, vulnerability mapping while preparing the community-level drought management plans, local governments, and disaster management can be enabled to gain better understanding about the vulnerability of the communities.

Policy and institutional support Support for promotion of community/village seed banks, fodder banks, and nutrient banks, custom hiring centers for farm implements, creation of water assets and safety nets for risk cover, for mainstreaming/convergence of the government programs and for scaling out climate-resilient dryland technologies such as PMKSY, IWMP, MGNREGA, RKVY, NFSM, etc.

22.12 Way Forward

Dryland agriculture extends in a wide array of agroecologies and socioeconomic settings; thus, concerted efforts are needed (1) to place rained agriculture research and development in the agroecological context and (2) to set priorities on the basis of carrying capacity of natural resources through participatory methods. The sustainability of dryland agriculture would be the hallmark of future research and development. A system approach in program mode is necessary to ensure sustainability of agriculture and replicability of site-specific technologies to other locations as area action plans. Therefore, a paradigm shift is needed in research from commodity-*centric to rained agroecosystem-centric*. The research agenda of rainfed agriculture also needs to address the national priorities and international commitments such as (1) enhanced production and profitability to meet the growing needs of the population and diversified requirements of the country by leveraging science and by sustainable utilization of natural resources by minimizing the impact of climate and other related constraints/risks; (2) neutralizing the land degradation through suitable interventions and quantification of accrued benefits of land and water conservation programs; (3) promotion of rainfed agroecosystems for generation of environmental benefits to honor national commitments; (4) reducing the water and energy footprints of rainfed systems through improving productivity enhancement; (5) meeting the envisaged goals of the government such as doubling farm incomes, promoting organic farming, and guiding investments in natural resources in such a way to realize maximum/optimal benefit from the investments; and (6) focusing on the commitments made by India as part of international agreements such as the Paris Agreement, UNCCD, SDGs, etc. In view of the pervasive influence of the climate change, there is a need to move from individual farmer to landscape perspective, with either village or watershed as a unit. Strategies that enhance climate-resilient agriculture are the most appropriate starting point for sustainable agriculture and sustainable intensification. A broad, inter-institutional, and multidisciplinary approach is needed to address the challenges of enhanced food security, improved natural resource management, biodiversity protection, climate change mitigation, and energy security and meets the demand for environmental goods and services. A number of ongoing national/state development programs have provisions that can facilitate the adoption of labor-intensive technologies within rainfed agroecosystem. However, it has to be ensured that adequate technological backstopping flows to design and implementation of such interventions. Mainstreaming resilient technologies through strong convergence with government schemes and appropriate policy interventions; strong preparedness for weather aberration (based on long term experiences or trends) along with actually responding to the situation and capacity building of primary and secondary stakeholders are very important.

References

AICRPDA (1983) Improved agronomic practices for dry land crops in India. All India Co-ordinated Research Project for Dry land Agriculture (AICRPDA), Hyderabad, pp 1–63

AICRPDA (2003) Annual reports 1971-2001. Eldoscope Electronic document. All India Co-ordinated Research Project for Dry land Agriculture (AICRPDA), CRIDA, Hyderabad, pp 63–57

AICRPDA-NICRA (2013) Managing weather aberrations through real time contingency planning. In: Srinivasa Rao C, Ravindra Chary G, Nagarjuna Kumar R, Maruthi Sankar GR, Venkateswarlu B (eds) Annual report (2012–13). AICRPDA, CRIDA, Hyderabad. 190 p

AICRPDA-NICRA (2014) Managing weather aberrations through real time contingency planning. In: Ravindra Chary G, Gopinath KA (eds) Annual report (2013–14). AICRPDA, ICAR-CRIDA, Hyderabad, p 241

AICRPDA-NICRA (2015) Managing weather aberrations through real time contingency planning. In: Srinivasa Rao C, Ravindra Chary G, Gopinath KA (eds) Annual report (2014–15). AICRPDA, CRIDA, Hyderabad, p 185

AICRPDA-NICRA (2016) Managing weather aberrations through real time contingency planning. In: Ravindra Chary G, Gopinath KA, Narsimlu B, Srinivasa Rao C (eds) Annual report (2015–16). AICRPDA, CRIDA, Hyderabad, p 226

AICRPDA-NICRA (2017) Managing weather aberrations through real time contingency planning. In: Ravindra Chary G, Gopinath KA, Narsimlu B, Srinivasa Rao C (eds) Annual report (2016–17). AICRPDA, CRIDA, Hyderabad, p 190

AICRPDA-NICRA (2018) Managing weather aberrations through real time contingency planning. In: Ravindra Chary G, Gopinath KA, Narsimlu B, Sammi Reddy K (eds) Annual report (2017–18). AICRPDA, CRIDA, Hyderabad, p 176

AICRPDA-NICRA (2019) Managing weather aberrations through real time contingency planning. In: Ravindra Chary G, Gopinath KA, Narsimlu B (eds) Annual report (2018–19). AICRPDA, CRIDA, Hyderabad, p 184

AICRPDA-NICRA (2020) Managing weather aberrations through real time contingency planning. In: Ravindra Chary G, Gopinath KA, Narsimlu B (eds) Annual report (2019–20). AICRPDA, CRIDA, Hyderabad, p 189

Bhushan C, Srinidhi A, Kumar V, Singh G (2001) Rising to the call: good practices of climate change adaptation in India. Centre for Science and Environment, New Delhi

Chetty CKR (1983) Research on intercropping systems in drylands - a review of decade's work (1971–72 to 1980–81). AICRPDA, CRIDA, Hyderabad. 124 p

CRIDA (1983) Contingent crop production strategy in rainfed areas under different weather conditions. In: Venkateshwarlu J, Vishnumurthy TV, Padmanabhan MV (eds) Project bulletin no. 5. CRIDA, Hyderabad. 76 p

CRIDA (1997) CRIDA perspective plan – vision 2020. CRIDA, Hyderabad, p 80

Government of India (1994) Guidelines for watershed development projects

Havanagi GV (1982) Water harvesting and lifesaving irrigation for increasing crop production in red soil areas of Karnataka. In Proceedings of the Symposium on Rainwater and Dryland Agriculture, 3 October, 1980. Indian National Science Academy, New Delhi, pp 159–170

Hegde BR, Krishnagowda KT, Parvathappa HC (1988) Improved dry land practices. In: Somani LL (ed) Recent advances in dryland agriculture. Part 2. Scientific Publishers, Jodhpur, pp 277–285

IARI (1970) A new technology for dry land farming. Indian Agricultural Research Institute, New Delhi

Jeeva SRDN, Laloo RC, Mishra BP (2006) Traditional agricultural practices in Meghalaya, North East India. Indian J Tradit Knowl 5(1):7–18

Joshi PK, Jha AK, Wani Suhas P, Sreedevi TK, Shaheen FA (2008) Impact of watershed program and conditions for success: a meta-analysis approach. In: Global theme on agroecosystems report no. 46. International Crops Research Institute for the Semi-Arid Tropics, Andhra Pradesh. 24 p

Korwar GR, Rao JV, Osman M, Pratibha G (1997) Agro forestry work at CRIDA - a brief note. In: Workshop of AICRP on Agro forestry, 28–30 January 1997. ANGRAU, Hyderabad

Krishna Kumar K, Patwardhan SK, Kulkarni A, Kamala K, Koteswara Rao K, Jones R (2011) Simulated projections for summer monsoon climate over India high by a resolution regional climate model (PRECIS). Curr Sci 101:312–326

Krishnan A (1988) Delineation of soil climatic zones of India and its application in agriculture. Fertil News 33:11–19

Krishnan R, Gnanaseelan C, Sanjay J, Swapna P, Dhara C, Sabin TP, Jadhav J, Sandeep N, Choudhury AD, Singh M, Mujumdar M, Parekh A, Tewari A, Mehajan R (2020) Introduction to climate change over the Indian region. In: Krishnan et al (eds) Assessment of climate change over the Indian Region Ministry of Earth Sciences (MoES), Government of India. Springer Nature, Singapore. https://doi.org/10.1007/978-981-15-4327-2. ISBN 978-981-15-4326-5, ISBN 978-981-15-4327-2 (eBook)

Mathur BL (1995) Kana bundi indigenous method of controlling wind erosion. Honey Bee 12:7

NAAS (2018) Renewable energy: a new paradigm for growth in agriculture. Strategy paper no. 10. National Academy of Agricultural Sciences, New Delhi. 20 p

NRAA (2012) Prioritization of rainfed areas in India. Study report 4. National Rainfed Area Authority, New Delhi. 100 p

NRAA (2013) Contingency and compensatory agriculture plans for droughts and floods in India - 2012. Position paper no. 6. National Rainfed Area Authority, New Delhi. 87 p

Pai DS, Thapliyal V, Kokate PD (2004) Decadal variation in the heat and cold waves over India during 1971-2000. Mausam 55(2):281–292

Patil ND, Umrani NK, Shende SA, Manake BS, Kale SP, Shingte AK (1981) Improved crop production technology for drought prone areas of Maharashtra. All India Coordinated Research Project for Dryland Agriculture, Maharashtra. 96 p

Radder GD, Belgaumi MI, Itnal CJ (1995) Water harvesting procedures for dry land areas. In: Singh RP (ed) Sustainable development of agriculture in India. Scientific Publishers, Jodhpur, pp 195–205

Rafey A, Verma UK (1988) Production potential of legumes, oilseeds and cereals in intercropping system with pigeonpea. Ind J Soil Conserv 58(6):433–436

Raju BMK, Rao KV, Venkateswarlu B, Rao VMS, Rama Rao CA, Rao VUM, Rao B, Ravi Kumar N, Dhakar R, Swapna N, Latha P (2013) Revisiting climatic classification in India: a district-level analysis. CRIDA, Hyderabad 500 059, India. Curr Sci 105(4):25

Rama Rao CA, Raju BMK, Islam A, Subba Rao AVM, Rao KV, Ravindra Chary G, Nagarjuna Kumar R, Prabhakar M, Sammi Reddy K, Bhaskar S, Chaudhari SK (2019) Risk and vulnerability assessment of Indian agriculture to climate change. ICAR-CRIDA, Hyderabad, p 124

Randhawa NS, Venkateswarlu J (1980) Indian experience in semi-arid tropics prospects and retrospect. In: Proceedings of the International Symposium on development and transfer of technology for rainfed agriculture and the SAT Farmer, 28 August–1 September 1979, ICRISAT Center, India. ICRISAT, Andhra Pradesh

Ravindra Chary G, Vittal KPRV, Maruthi Sankar GR, Ramamurthy V, Sharma RA, Dubey DP, Patil MN, Sharma KL, Pratibha G (2008) Assessment of influence of Soil-site characteristics n soybean productivity in swell-shrink and associated soils of semiarid tropics. Ind J Agric Res 23(2):14–23

Ravindra Chary G, Venkateswarlu B, Sharma SK, Mishra JS, Rana DS, Kute G (2012) Agronomic research in dryland farming in India: an overview. Ind J Agron 57:157–167

Ravindra Chary G, Srinivasa Rao C, Gopinath KA, Sikka AK, Kandpal B, Bhaskar S (2016) Improved agronomic practices for rainfed crops in India. In: All India Coordinated Research Project for Dryland Agriculture (AICRPDA). ICAR-CRIDA, Hyderabad, p 292

Ravindra Chary G, Gopinath KA, Bhaskar S, Prabhakar M, Chaudhari SK, Narsimlu B (2020) Resilient crops and cropping systems to cope with weather aberrations in rainfed agriculture. In: All India Coordinated Research Project for Dryland Agriculture. ICAR-CRIDA, Hyderabad, p 56

Reddy NV, Hampaiah R (1982) Alternate land use in Dryland Agriculture. In: A decade of dryland agriculture research in India, 1971–1980. All India Coordinated Research Project for Dryland Agriculture, Hyderabad, pp 1–124

Roy T (2006) Agricultural prices and production, 1757–1947. In: Wolpert S (ed) Encyclopedia of India, vol 1. Thomson, Gale, Detroit, MI, pp 20–22. ISBN 0-684-31350-2

Sikka AK, Kandpal BK, Islam A, Dhyani SK (2016) Climate change and wheat. Geogr You 15(94): 53–58

Singh SP (1987) History of Indo-Canadian dryland research. Central Research Institute for Dryland Agriculture and Agriculture Canada, Hyderabad. 28 p

Singh RK, Sureja AK (2008) Indigenous knowledge and sustainable agricultural resources management under rainfed agro-ecosystem. Indian J Tradit Knowl 7(4):642–654

Srivastava SK, Pandey H (2006) Traditional knowledge for Agroecosystem management. Indian J Tradit Knowl 5(1):122–131

Subba Reddy G, Ramakrishna YS, Ravindra Chary G, Maruthi Sankar GR (2008) Crop and contingency planning for rainfed regions of India - a compendium by AICRPDA. AICRPDA, ICAR-CRIDA, Hyderabad, p 174

Tripathi A, Prasad AR (2009) Agricultural development in India since Independence: a study on progress, performance and determinants. J Emerg Knowl Emerg Markets 1(1):63–92

Venkateswarlu B, Singh AK, Prasad YG, Ravindra Chary G, Srinivasa Rao C, Rao KV, Ramana DBV, Rao VUM (2011) District level contingency plans for weather aberrations in India. Central Research Institute for Dryland Agriculture, Natural Resource Management Division, ICAR, Hyderabad, p 136

Verma GP (1982) Review of Soil Research in India, Part-II. In: 12th International Congress on Soil Science, 8–16 Feb, 1982, New Delhi, India. pp 666–675

Vishnumurthy T, Vijayalakshmi K (1993) Intercropping sorghum with clusterbean or greengram in Alfisols of Andhra Pradesh. Indian J Dryland Agric Res Dev 8(1):54–56

Vittal KPR, Singh HP, Rao KV, Sharma KL, Victor US, Ravindra Chary G, Sankar GRM, Samra JS, Singh G (2003) Guidelines on drought coping plans for rainfed production systems. AICRPDA, CRIDA, ICAR, Hyderabad. 39 p

Vittal KPR, Rao KV, Sharma KL, Victor US, Ravindra Chary G, Maruthi Sankar GR, Samra JS, Singh G (2005) Agricultural drought zonation, cropping practices and amelioration paths for rainfed production systems. Indian J Dryland Agric Res Dev 20(2):155–202

Part IV

Biotic Stress Management for Sustainable Agriculture

Biotic Stress Management 23

Bhimeshwari Sahu, Vijay K. Choudhary, M. P. Sahu, K. Kiran Kumar, G. K. Sujayanand, R. Gopi, V. Prakasam, J. Sridhar, J. Mallikarjuna, H. K. Singh, K. C. Sharma, P. N. Sivalingam, and P. K. Ghosh

Abstract

In India yield loss to crops due to various biotic stresses is estimated to be Rs. 225,000 crores/year. Globally, it is estimated that plant diseases and invasive insect pests cost around $290 billion. During the pre-independence era, several crop pests and diseases and their causal agents were recorded. Some of the diseases also reached to the level of causing severe yield loss in an epidemic proportion. During independence and the green revolution era, introduction of high-yielding varieties led to the emergence of various new pests and diseases, and they were mostly managed through pesticide spray. During this period, establishment of institutional setup, identification of many viral and virus-like diseases, identified unknown aetiologies, etc happened. Later use of biopesticides, eco-friendly management of pests, Integrated Pest Management

B. Sahu · J. Sridhar · J. Mallikarjuna · K. C. Sharma · P. N. Sivalingam (✉) · P. K. Ghosh
ICAR-National Institute of Biotic Stress Management, Raipur, Chhattisgarh, India

V. K. Choudhary · M. P. Sahu
ICAR-Directorate of Weed Research, Jabalpur, Madhya Pradesh, India

K. K. Kumar
ICAR-Central Citrus Research Institute, Nagpur, Maharashtra, India

G. K. Sujayanand
ICAR-Indian Institute of Pulses Research, Kanpur, Uttar Pradesh, India

R. Gopi
ICAR-Sugarcane Breeding Institute, Research Centre, Kannur, Kerala, India

V. Prakasam
ICAR-Indian Institute Rice Research, Hyderabad, Telangana, India

H. K. Singh
Department of Plant Pathology, Indira Gandhi Krishi Vishwavidyalaya, Raipur, Chhattisgarh, India

P. K. Ghosh et al. (eds.), *Trajectory of 75 years of Indian Agriculture after Independence*, https://doi.org/10.1007/978-981-19-7997-2_23

(IPM)/Integrated Weed Management (IWM) and biological control of pests, developing resistant varieties, usage of nanotechnology, identifying molecular mechanism of host-pathogen interaction, etc. gained focus. Progress has been made in weed management from hand weeding to herbicide usage. After the green revolution, the usage of newer pesticides for the management of pests and diseases has increased tremendously along with institutional setup for research and development including education. The main challenges for the future are emerging pest resistance to pesticide, molecular understanding of host-pest reaction, biosecurity for transboundary pests, usage of biopesticides and biocontrol agents, etc. The achievements and development in the field of plant protection in India, future challenges, and issues to be addressed have been compiled and presented.

Keywords

Biotic stress · Insect · Disease · Nematode · Weed · Management · Pesticide · Herbicide

23.1 Introduction

Insects, pathogens, nematodes, weeds, and others are the pests of agriculture crops. These pests cause serious threat to crop production and the loss is estimated to be 20–40% of crop production. The Food and Agriculture Organization (FAO) estimates that a total annual loss of 20–40% of global crop production is due to pests. Each year, plant diseases cost the global economy around $220 billion, and invasive insects around US$70 billion. In India yield loss to crops is estimated at 26, 20, 6, and 8% by insects, plant pathogens, rodents, and others, respectively, with an annual loss of Rs. 225,000 crores annually (Kumar et al. 2021). Weeds usually require external efforts to be excluded from the cropping system and eventually reduce the economy, deteriorate the environment, human health, and amenity. Weeds alone account for 37% of total annual crop yield loss in India. During pre- and post-independence and before the green revolution period, establishment of scientific institution, studying the life cycle of the pests and their perpetuation, and managing pests through cultural practices with minimal use of pesticides were given major emphasis. After the green revolution, the usage of pesticides for the management of pests and diseases has increased tremendously along with institutional setup for research and development, research and education on different fields of entomology, mycology and plant pathology, nematology, weed science, use of biocontrol agents, integrated pest management, etc. Off late many new challenges are emerging like pesticide resistance to insects and pathogens, molecular understanding of pests and their reactions with host plants, biosecurity for transboundary pests, and usage of biopesticides and biocontrol agents. The chronological development of achievements in the field of plant protection has been presented in this chapter.

23.2 Plant Pathogens and Disease Management

23.2.1 Pre-independence Era

Plant diseases are known even before the development of modern agriculture in India as they were recorded in Vedas (1500–500 BC), Buddhist literature (Jataka of Buddhism), Kautilya's Artha-sastra (32–180 BC), Krishi-Parashara sangam literature, Agni-purana in Tamil (500–700 AD), Brhatsamhita of Varaha Mihira, etc. The *Vrikshayurveda* is the book in which the first mention of plant disease is seen, and the cause was believed to be abiotic factors (Nene 2003). The foundation for Plant Pathology in India started with the appointment of Sir E. J. Butler, the 'father of plant pathology in India' as the first imperial mycologist at the Indian Agricultural Research Institute (IARI) in 1905. During the pre-independence period, mostly the diseases of different crops and outbreak and their preventive methods have been reported for the diseases such as wilt of pigeon pea, rust of wheat, red rot and mosaic disease of sugarcane, late blight of potato, wilt of red gram, rice diseases, life cycle of smut diseases, root wilt of coconut, spike disease of sandal, small cardamom, mosaic disease, yellow leaf disease of arecanut palm, cotton stenosis, tristeza disease of citrus, and yellow vein mosaic disease of *bhendi* (Rishi 2009). The discipline of plant pathology in India has got a speedy growth due to the emergence of three disease epidemics. These epidemics are red rot of sugarcane in Bihar and Uttar Pradesh caused by *Colletotrichum falcatum* during 1938–1942, Bengal famine in 1942 by *Helminthosporium oryzae* (causing blight of rice), and wheat rust by *Puccinia graminis tritici* which created shortage of food grains in Madhya Pradesh and undivided Bengal during 1946–1947. The other notable disease outbreaks during this period were coffee rust, coconut wilt, varietal breakdown due to wheat rust, outbreak of sugarcane red rot, and the problem of phytophthora in potato. Considering the importance of coffee, United Planters Association of South India, Coonoor, has developed several rust-resistant clones of coffee and management procedures to contain the disease in Tamil Nadu and Karnataka (Nagarajan 2010). During this period, certain new diseases were identified or introduced in India which in certain cases had created epidemic and severe economic losses in Indian agriculture. These include phytophthora disease in rubber tree and blast disease of rice. In India, the first disease that was suspected to be caused by bacteria was 'bangle blight' of potato in Bombay (now Mumbai). It was reported in 1892 to be prevalent in Pune and other places in Mumbai and considered the causal organism to be a fungus (Mollison 1901). The other bacterial diseases included brown rot or wilt of potato (Rangaswami 1962), canker disease of citrus (Fawcett and Jenkins 1933), black arm of cotton (Patel and Kulkarni 1948), red stripe of sugarcane, and black rot of cabbage. These diseases became serious since the recent replacement of resistant, diploid, rainfed, indigenous cottons (*Gossypium arboreum* and *G. herbaceum*) with high-yielding but susceptible, irrigated, tetraploid cottons (*G. hirsutum* and *G. barbadense*) (Verma and Singh 1971a, b). Certain varieties of *G. hirsutum*

(101–102B, Reba. B-50, BJA-592, HG-9, and P.14T.128) have been demonstrated to be resistant to all the races present in India (Singh and Verma 1971). It has been demonstrated for the first time that certain systemic fungicides (plantvax and vitavax) checked the secondary spread of the disease and to some extent aid in curing the plants (Uppal 1948). During the nineteenth century, plant viruses did not receive much attention in India. One of the earliest records is the spike disease of sandalwood in 1899, followed by mosaic disease of small cardamom. The systematic research work in plant virus in India was started at IARI, New Delhi, under the leadership of R. S. Vasudeva (1905–1987) and S. P. Raychaudhury (1916–2005). Major breakthrough in the field of plant virology was started with the investigation of tobacco leaf curl virus transmitted by whitefly in the 1930s (Pal and Tandon 1937). The next two decades mostly focused research on dissemination and management of a number of virus disease of economic importance. During this period, many new virus diseases were observed in different cereals, pulses, plantation, and vegetable crops. A plant virus research laboratory was first established in Pune in 1938. In early days, mostly conventional approach was followed to select the superior varieties by farmers to protect the crop against diseases. Plant breeding also played a significant role in developing varieties for improving yield and tolerance to existing and emerging biotic stresses. The major objective of resistant breeding was to improve disease resistance in economically important crops through induced mutation/mutation breeding and development of screening techniques. A systematic work on wheat improvement in India started only during the first decade of the twentieth century when the Agricultural Research Institute was established by Albert Howard and GLC Howard in 1905 at Pusa Bihar. Howard and Howard (1910) had developed some improved varieties of wheat. A systematic work on breeding for rust resistance started at IARI in 1935. After 18 years of research at IARI, the variety NP 809, resistant to all the three rusts, was evolved under a planned programme. The uses of a cultural method started with early civilization and their development. Since the British colonies were spread around the world, several plant species were introduced into India and the plant materials also moved out. Soon it was realized that insect and plant pathogens were getting introduced unintentionally along with the planting materials. To combat such issues of introduction of pathogens, the Government of India enforced the Plant Quarantine Act in 1914. Several pests and diseases in India got introduced during this period such as late blight of potato, banana bunchy top, bacterial blight, streak diseases of paddy, etc. and they were widespread. Some others like golden nematode and wart disease of potato and downy mildew of onion are still localized in certain parts of the country. In India, copper sulphate was the first fungicide used against sorghum smut. Later in the 1950s, mancozeb, cupravit, and other formulated fungicides came in the market and greatly benefitted the farmers. The Indian Type Culture Collection (ITCC) was established in 1936 with a view to furnish the knowledge on living fungi. It is the largest fungal genetic resource centre established in India.

23.2.2 Independence Era

On the basis of earlier lessons, this era was devoted to identifying the factors favouring disease development and their effective management. The climatic factors that were correlated with the brown spot epidemic in Bengal in 1942 were above-normal rainfall, high humidity, prolonged periods of cloudy weather, and a favourable temperature warmer than average during the flowering and grain filling stages (Webster and Gunnell 1992). Sanitation of the nursery site and healthy planting materials were recommended against the diseases of transplanted crops. Mundkur (1948) worked on the management of cotton wilt through resistance breeding. Later in the 1950s, mancozeb, copper oxychloride, and other formulated fungicides came on the market and greatly benefitted the farmers. Antibiotics use for the control of plant diseases was introduced during this period. The increasing importance of plant pathology led to the foundation of the Indian Phytopathological Society in 1947 by B.B Mundkur, with the focus on the field of Mycology, Plant Pathology, Bacteriology, Virology, Phytoplasmology, and Nematology. It provided a unique platform to the scientists working in the field of plant pathological research to share their work. The society published its official journal *Indian Phytopathology*, a leading plant pathology research journal. Patel (1948), the father of plant bacteriology in India, established the School of Plant Bacteriology at the College of Agriculture, Pune, and reported 40 new species of plant pathogenic bacteria. During this period, the causal agent of potato leaf rot and tuber rot was identified (Dastur 1948) and also life cycle of cereal rusts and linseed rusts in India, biochemistry of host-parasite relation using cotton wilt pathogen, developed the concept of vivotoxins, identified large number of smut fungi, etc were established. Several bacterial diseases have long been known in India, such as citrus canker, leaf spot of mango, and black arm of cotton and they might have originated there. Although bacterial diseases were important and widespread, very little work was done until 1950. Severe black-vein disease of cabbage (caused by *Xanthomonas campestris* pv. *campestris*) was later reported in Bombay (Patel 1948) and West Bengal (Chattopadhyay and Mukherjee 1955). Two virus research institutes were established at Shimla and Kalimpong. Prutbi and Samuel (1937) observed tobacco leaf curl virus on tomato plants in India. Later in Northern India the severe nature of leaf curl disease in tomato was first reported by Vasudeva and Samraj (1948). This was followed by yellow mosaic disease in lima bean and Dolichos yellow mosaic disease of mungbean (Nariani 1960). Next decades mostly concentrated on the dissemination, control as well as virus-vector relationship and sources of resistance. A notable feature of these findings is the elucidation of the cause of some serious diseases of complex aetiology such as citrus die-back, coconut root wilt, and sandal spike (Raychaudhuri et al. 1969).

23.2.3 Green Revolution Era

The Green Revolution was an endeavour initiated by Norman Borlaug in 1970, who won Nobel Peace Prize for his work in developing HYVs of wheat. This is often credited with having transformed India from 'a begging bowl to a breadbasket'. The independent India rated food and agriculture as its priority sectors and therefore emphasized on irrigation, chemical fertilizers, and pesticides to maximize food production. During this period, there was corresponding increase in pests and diseases, improved method of management measures, understanding of pest and pathogen life cycle and unknown aetiology, institutional setup for higher education and training, dissemination of technologies, etc. The major achievements are National Wheat Disease Survey and Surveillance Programme in 1967 for monitoring and epidemiological studies of wheat rusts in India. This led to the establishment of 'Puccinia Path' in the Indian sub-continent and from where the wheat breeding for resistance to rust disease started. Gummosis of sugarcane caused by *X. vasculorum*, a very serious disease of the crop, was only reported from Madras in 1960 (Rangaswami 1960). With the introduction of the rice variety TN-1, the bacterial leaf blight became a serious bacterial disease in rice crop (Srivastava et al. 1967). At present the bacterial blight is a major hurdle in stepping up rice yields through intensive cultivation of high-yielding varieties. Major epidemic of this disease occurred during 1979–1980 in the most important rice growing states of India such as Andhra Pradesh, Punjab, Haryana, and western Uttar Pradesh. The pioneering work was done on *Xanthomonas campestris* pv. *malvacearum, a* causative agent of bacterial blight of cotton by Varma at ICAR_IARI, New Delhi (1955) and laid a solid foundation of Indian Plant Bacteriology. Gummosis of sugarcane caused by *X. vasculorum*, a very serious disease of the crop, was only reported from Madras in 1960. Several viral diseases also recorded during this period particularly whitefly transmitted viruses (Varma 1955).

Antisera and particle morphology have been used in the identification and detection of plant viruses in recent years, and fundamental studies on purification and serology have received attention with the introduction of the ultracentrifuge in most laboratories. A number of viruses, such as mosaic streak of wheat and *tungro* of rice (Raychaudhuri et al. 1969), various cucurbit mosaics, cowpea mosaic (Chenulu et al. 1968), coconut root wilt (Summanwar et al. 1969), barley mosaic (Dhanraj and Raychaudhuri 1969), and sunn hemp mosaic (Nariani et al. 1970; Raychaudhuri 1947), have been purified, their particle morphology studied, and antisera prepared. Tissue culture techniques opened the way for investigations concerning host-parasite relationships and inhibition studies, as well as the production of virus-free plants by meristem-tip culture. A number of viruses such as chilli mosaic, sunn hemp mosaic, tobacco mosaic, and potato virus X have been successfully cultivated in tissue culture (Raychaudhuri and Mishra 1965), and attempts are in progress to get virus-free plants of citrus by tissue culture technique.

Increased monoculture, reduced crop diversity and rotation, and use of herbicide have increased vulnerability to diseases. Pathogens tend to develop resistance to pesticides, requiring a higher use of pesticides to sustain productivity. Inappropriate and excessive pesticide use led to increased and unnecessary disease outbreaks and additional loss because of the inadvertent destruction of natural enemies of diseases, disease resistance, and resurgence of secondary diseases. A number of ways have been suggested by different authors to achieve durability of resistance of which gene deployment, guided distribution of genes in space and time, and multilines were used for gene deployment in India (Adugna 2004). The developed wheat multiline varieties were resistant to yellow rust and brown rust. These were: Kalyansona lines KSML3, MLKS11, KML7406/Bithoor and *Sonalika* lines, MLSKA-9, MLSKA-12 (Harahap and Silitonga 1988). Ultimately, overuse of pesticides in crop production was substantially reduced (Kumar 2014). Site-specific systemic fungicides which included several major groups such as benzimidazoles, pyrimidines, phenylamides, sterol biosynthesis inhibitors, and dicarboximides were introduced in the mid-1960s onwards. During the past decade, more novel compounds with different modes of action notably phenylpyrroles, anilinopyrimidines, strobilurins, spiroxamines, phenylpyridylanines, quinolines, etc. have been developed with bio-efficacy against diverse plant diseases (Thind 2016). In the contemporary period, to mitigate the negative effect of pesticides, the Insecticide Act (1968) of the Government of India came into force to cover various issues related to import, manufacture, biosafety, withdrawal, and quality control aspects of the pesticides marketed or tested in India. Under this act, a Registration Committee was constituted to register pesticides only after satisfying itself regarding their efficacy and safety to human beings and animals. There is a network of 49 State Pesticide Testing Laboratories and 2 Regional Pesticide Testing Laboratories of the Central Government, and the Central Insecticides Laboratory, established to test and analyse the quality of pesticides (https://icar.org.in/files/Agril-Legislation.pdf). The All India Co-ordinated Research Project (AICRP) on Biological Control of Crop Pests and Weeds was established in 1977 with 10 centres under the aegis of the ICAR for carrying out biological control research in different parts of the country. Another important research was on the forecasting of diseases for apple scab, wheat rust, brown leaf spot of rice, coconut root wilt, etc. (http://ecoursesonline.iasri.res.in). Another landmark achievement was the identification of the causal agent of *Khaira* disease of rice as Zn deficiency by DrYL Nene and awarded the International Rice Research Prize in 1967 by the FAO.

23.2.4 Post-Green Revolution Era

The main objective in recent times is to gain food security through implementation of scientific principles in cultivation. Numerous diseases like papaya ringspot and *phytophthora* diseases have taken serious proportions in recent years. In 2008, there was a flare up of neckblast and brown plant hopper of rice in Haryana causing crop losses up to 40% due to persistent and unprecedented rains at the fag end of the monsoon season. Likewise, diseases with unfamiliar or complex aetiology such as mango malformation and parawilt and grey mould of cotton and other diseases such as sheath blight of rice/maize, bract mosaic of banana, downy mildews in maize and several other crops have become devastating by virtue of evolution of virulent races or resurgence capabilities (Kumar and Gupta 2012).

There are a number of fungal pathogens that have been found to develop resistance against fungicides due to the introduction of systemic fungicides and their indiscriminate use without following the recommended application procedures (Thind 2016). From the mid-1970s, researchers have started concentrating on the biocontrol of plant pathogens using *Trichoderma*, *Pseudomonas*, *Bacillus*, etc. which brought a paradigm shift in the management of diseases (Ram et al. 2021). The Indian government is promoting research, production, registration, and adoption of biopesticides with open hands, through various rules, regulations, policies, and schemes. The National Farmer Policy (2007) has strongly recommended the promotion of biopesticides for increasing agricultural production and sustaining the health of farmers and the environment (Kumar 2014). In India, an overview of literature on biocontrol of plant diseases clearly indicated a spurt in their exploration and exploitation for the management of diseases of crop plants as indicated by a significant increase in the number of publications (Sharma et al. 2014). Current efforts are on to characterize the new resistance genes present in various crops and applying molecular tools to characterize the host resistance gene. These efforts have permitted going in for marker-aided selection of disease-resistant material in segregating population. During the last three decades, plant pathology has rapidly developed with the advent of advanced molecular techniques that have led to the identification of a large number of plant pathogens, most importantly viruses. Accurate and quick diagnosis of viral pathogens is imperative for the success of mapping of epidemics, breeding for resistance, and developing quarantine and other control measures (Rishi 2009). Eco-friendly measures such as marker-aided selection of disease resistance is being developed for transgenic pigeon peas resistant to sterility mosaic virus at ICRISAT, Hyderabad. Pathogen-derived resistance, antiviral agents, and biocontrol of insect vectors should be tried vigorously at much larger scales for long-term disease management (Rishi 2009). At Punjab Agricultural University (PAU), three BLB resistant genes *xa5*, *xa13*, and *Xa21* were pyramided in PR106 and Pusa 44 background and two of the PR1106 have been included in All India Coordinated trials during 2002. A similar work has also been successfully carried out in Central Rice Research Institute (Now NRRI) to pyramid three genes *xa5*, *xa13*, and *Xa21* into the elite rice cultivars Lalat and Tapaswini (Joshi and Nayak 2010). Narayanan et al. (2002) improved an elite indica rice line IR50 by pyramiding blast resistance gene *Piz5* and bacterial blight resistance gene *Xa21* through marker-assisted selection and genetic transformation. Pyramiding with *Xa2*1 and Wx (waxy) gene showed durable resistance to bacterial leaf blight and high amylose content. All combinations of the three resistant genes were pyramided using STS markers. The first report of marker-assisted transfer of genes conferred multiple resistances to three different diseases in rice wherein genes *xa13* and *Xa21* are for bacterial blight resistance and *Pi54* for blast resistance (Singh et al. 2012). PCR-based molecular markers were used in a backcross breeding programme to introgress three major bacterial blight resistance genes (*Xa21*, *xa13*, and *xa5*) into Samba Mashuri from a donor line (SS1113).

India has experienced the Green Revolution in wheat and rice during the 1970s which made us self-sufficient in food grains production. The situation has changed significantly in the twenty-first century due to climate change and increasing population pressure. Post-green revolution advances have made research progress in biotechnology, which paved the way for the development of high-yielding, stress- and disease-resistant genetically modified (GM) varieties in various crops. More than 20 crops are under various stages of research and field trials for genetic modification in India, namely cotton, rice, wheat, maize, brinjal, potato, sorghum, mustard, groundnut, cauliflower, okra, chickpea, pigeon pea, castor, sugarcane etc. for the traits viz. insect resistance, herbicide tolerance, drought tolerance, salinity tolerance, virus resistance, quantitative traits (yield increase), nutrition improvement, etc. (Shukla et al. 2018). The process of fungicides discovery has undergone a noteworthy change over the years. After the era of broad-spectrum multisite and site-specific systemic fungicides, several novel action fungicides of different chemical classes have been developed in the past two decades. These are more eco-friendly products and used at a much lower dose rates as compared to the earlier compounds. Most noted among these are the strobilurins (QoIs), derived from *Strobiluru stenacellus*, a wild mushroom. These are analogues to strobilurins-A and have a broad range of disease control. Azoxystrobin was the first strobilurin introduced in 1996, and currently nine strobilurins compounds are available. Other important fungicides introduced for the control of diverse diseases in the last decade are Oxazolidinediones (famoxadone), Phenoxyquinolines (quinoxyfen), Anilinopyrimidines (cyprodinil, pyrimethanil), Valinamides (iprovalicarb, benthiocarb), Mandelamides (mandipropamid), phenylpyrroles (fenpicloil, fludioxonil), MBIs (carpropamid), Spiroketalamines (spiroxamine), benzamides (mandipropamid), Cyanoimidazoles (cyazofamid), Thicarbamates (ethaboxam), Amdoximes (cyflufenamid), Phenoxyquinolines (quinoxyfen), Imidazoles (fenanmidone), and Benzamides (fluopicolide, zoxamide) representing different chemistries and mode of action. The majority of these have been developed for use against oomycete pathogens. The main advantages of new generation compounds are ecologically safer profiles and are required to be used at much lower rates than their earlier counterparts. A few of the lately developed fungicides have been registered for use in India and a good number of novel action fungicides are currently under evaluation. Azoxystrobin and fenamidone have been registered for use against grape downy mildew and potato blight. Prominent among those being tested against different diseases are mandipropamid. iprovalicarb, benthiavalicarb, fluopicolide, famoxadone, cyazofamid, pyraclostrobin, and kresoxim methyl. These recently introduced fungicides represent major advances in technology, potency against target diseases, selectivity, safety, and rate reduction but these fungicides are site-specific and cause target site resistance among the pathogens (Leadbeater 2012). Thus, it is very important to proactively design and implement resistance management strategies through recommendations of new fungicide classes, as well as maintaining existing products. Swift development of resistance to strobilurins is now well documented and these are now categorized under high-risk fungicides.

Their use has to be regulated and FRAC guidelines adopted so as to sustain their efficacy levels (Kumar and Gupta 2012).

In recent days, nanotechnology plays a major role in plant disease management due to its eco-friendly nature and potentiality. Nanoparticles can be used such as nanoparticles of silver, copper, silicon, and zinc and have been used in disease management and also serve as a carrier to transport chemical pesticides and biocontrol agents to the target site. It is estimated that 90% of chemicals are lost in the environment during or after their application. Therefore, nanotechnology would provide green and efficient alternatives for the management of plant disease without harming the nature. Rapid detection of diverse plant pathogens such as bacteria, virus, fungus, and nematodes can be done by the use of nanoparticles as a diagnostic tool (Yao et al. 2009). In India, initiatives have been taken by ICAR and SAUs, and NIPHM, Hyderabad, has initiated a Plant Health Newsletter to focus and strengthen sustainable agriculture as a new trend for plant health and disease management to reduce the pesticide load by 40% by 2018 compared with the load level of 2011.

Future progress in mycology and plant pathology will be dictated by new developments in allied sciences as in agronomy, genetics, molecular biology, geography, bio-informatics, data science, robotics and related branches of science. Trade-related issues focusing on quarantine, risks, international pathology, and post-harvest diseases research would draw greater attention. Some of the important aspects initiated by R&D are biotechnological approaches for host plant resistance, modern diagnosis, biopesticides, IDM, RNAi, nanotechnology, image analysing, artificial intelligence, drone application of chemicals, etc.

Plant Quarantine regulatory measures are operative through the 'Destructive Insects & Pests Act, 1914' (Act 2 of 1914) in the country. The purpose and intent of this Act is to prevent the introduction of any insect, fungus, or other pest, which is or may be destructive to crops. The import of agricultural commodities is presently regulated through the Plant Quarantine (Regulation of Import into India) Order, 2003, the provisions of New Policy on Seed Development, 1988. Further, the significance of Plant Quarantine has increased in view of globalization and liberalization in international trade of plants and plant material in the wake of Sanitary and Phytosanitary (SPS) Agreement under WTO. The phytosanitary certification of agricultural commodities being exported is also undertaken as per International Plant Protection Convention (IPPC), 1951. The pest risk analysis is mandatory for all the plants/plant materials prior to its import into India as per PQ Order, 2003. As per Plant Quarantine (Regulation of Import into India) Order, 2003, a total of 94 entry points including 46 seaports, 24 airports, and 24 land custom stations are notified as the points of entry for import of plants and plant materials to India. Besides, 77 Inland Container Depot/Container Freight Station and 11 Foreign Post Offices have also been notified for the entry of plants/plant material under the PQ Order, 2003. National Bureau of Plant Genetic Resources (NBPGR) in New Delhi and its regional station at Hyderabad are involved in processing of germplasm, seed, plant material of agricultural, horticultural, and silvicultural crops of all the institutions of ICAR functioning in the country.

The Indian Type Culture Collections (ITCC) marks an important achievement. India is an affiliate member of the World Federation for Culture Collections (WFCC), and ITCC is registered with the World Data Centre for Microorganisms (WDCM, registration number 430). The Ministry of Environment and Forest, Government of India, has designated ITCC to act as the biological repository during 2007 under the Biological Diversity Act, 2002. The main objectives of ITCC are to act as repositories for fungal and bacterial cultures, to provide services viz., supply of authentic fungal and bacterial cultures, and identification of fungi and bacteria to technocrats and scientists working in research Institutions, universities, and industries. There are more than 3800 cultures of fungi of all groups, viz., Oomycetes, Zygomycetes, Ascomycetes, Basidiomycetes, and Deuteromycetes, which include plant pathogens, biocontrol agents, fungi for medical and industrial use including mushroom and yeast, and 145 cultures of plant pathogenic bacteria. These are being conserved and maintained at ITCC. ICAR-National Bureau of Agriculturally Important Microorganism (NBAIM) was established for preservation and storing microbial diversity in India. The Bureau is engaged in multifarious activities in the area of microbial diversity, biological control, microbial genomics, preservation and maintenance of microbial cultures. NAIMCC is a designated microbial repository for agriculturally important microorganisms (AIMs) under the National Biodiversity Act, 2002, and is a member of WFCC. Currently, NAIMCC holds accessions of 6907 AIMs including 2595 bacteria, 3981 fungi, and 331 cyanobacteria. The Indian system has developed a procedure to register the novel genetic stocks with NBPGR, New Delhi, for any resistant stocks against pests and diseases.

Biological control is the way of controlling plant diseases by the application of fungi, bacteria, actinomycetes, and viruses (bacteriophages). It has been estimated that rupees 60,000 crores worth of crop are lost each year due to plant diseases. Indiscriminate and non-judicious use of synthetic pesticide for preventing and controlling crop diseases adversely affects the environment, microbiome, development of resistance, and hormoligosis in several plant pathogens. This also acts as a serious non-tariff barrier to the trade of farm commodities. Among agriculturally important microbes *Trichoderma viride*, *T. harzianum*, *Pseudomonas fluorescens*, and *Bacillus subtilis* are most efficient antagonistic bioagents of plant diseases, producers of biologically active metabolites, elicitors, and inducers of systemic resistance. A better understanding of the application of genomics and genetic modification techniques opened new doors for multifaceted traits enhancement in strains of bioagents. The knowledge of the diversity spectrum and genetic structure of microbial population has direct implication in the formulations of potential bioagents. The AICRP on Biological Control of Crop Pests and Weeds was established in 1977 with 10 centres under the aegis of ICAR. This AICRP was elevated to Project Directorate of Biological Control in 1993. Pesticides are regulated in India through the Insecticides Act, 1968, and Insecticides Rules, 1971. The experiences in administering this Act over the past five decades have exposed certain gaps. In this context, the union cabinet has recently approved the Pesticides Management Bill, 2020. Later, some report described the field cases of fungicide resistance in India: *Venturia inaequalis*, *Gloeosporium ampelophagum*, and *Aspergillus flavus* to benzimidazoles; *Drechslera oryzae* and *Pyricularia oryzae*, to edifenphos; *Plasmopara viticola*, *Phytophthora parasitica*, *P. infestans*, and *Pseudoperonospora cubensis* to metalaxyl; *Phytophthora infestans* to oxadixyl, and *Uncinula necator* to triadimefon. AICRP on Pesticide Residues, later re-designated as the All India Network Project on Pesticide Residues (AINP-PR), was established by the Government of India under the aegis of the Indian Council of Agricultural Research during the year 1984–1985 to overcome the problem of widespread contamination of pesticides in various food commodities and environmental factors such as water and soil in India. Society of Pesticide Science, India (SPS India), was formally launched and inaugurated on 16 November 1989.

Integrated disease management (IDM) to manage the ill effects of pesticides and using all other management measures such as host plant resistance, cultural practices, and biological control was evolved during this period. National Research Centre for Integrated Pest management (NCIPM) has the mandate to develop and promote Integrated Pest Management (IPM) technologies for major crops with minimum ecological implications; generate national database on all aspects of pest management, advice national priorities and pest management policies; establish linkages with national and international institutes in the area of IPM, and to extend technical consultancies. Survey and surveillances were conducted for major diseases of agricultural and horticultural crops. An Android App, viz. 'PESTPREDICT', was developed for insect pest and disease management in rice, pigeon pea, groundnut, and tomato. Mobile app-based insecticide and fungicide calculator was also developed for tomato, groundnut, rice, and pigeon pea. IDM/IPM strategies were validated for different cops in the farmers' fields of different ecosystem. India managed well on the Ug99 race of wheat stem rust pathogen as our pathotype analyses have revealed that pathotypes within the Ug99 lineage do not occur in India or in neighbouring countries, and that Sr31 is still effective against the stem rust pathogen in India (Prasad et al. 2018). Moreover, India has a proactive system to combat these threats. Despite the absence of this group of stem rust pathotypes in India, we remain vigilant and ongoing rust surveillance, pathotype monitoring, and varietal deployment are in place and there is effective preparedness to any rust threat to wheat.

The National Centre for Mushroom Research and Training (NCMRT) came into existence at Solan (later renamed as National Research Centre for Mushroom) in 1983 during the VIth Five Year Plan under the auspices of ICAR. By the concerted efforts of the scientists of ICAR-DMR, Solan, mushroom productivity in the country has almost doubled while production has registered a sixfold increase. National Certification System for Tissue Culture Raised Plants (NCS-TCP) is being implemented by the Department of Biotechnology (DBT), Govt. of India, since 2006, as per the Gazette of India Notification under the Seeds Act, 1966. NCS-TCP has been instrumental in building capacities of the tissue culture companies for producing quality planting materials and also enhancing their market reach through a certification process. The purpose of NCS-TCP provides guidance on requirements and procedures for Accreditation of Test Laboratory for virus diagnosis and genetic fidelity/uniformity testing of tissue culture raised plants and accreditation of Tissue

Culture production facility. A Referral Center for Virus Indexing has been established at the Advanced Centre for Plant Virology, Division of Plant Pathology, Indian Agricultural Research Institute (IARI), New Delhi (https://dbtncstcp.nic.in). ICAR-National Bureau of Agriculturally Important Microorganisms (ICAR-NBAIM) is a constituent organization under the aegis of ICAR. The Bureau started functioning from July 2001. During the XIth plan, NBAIM was designated as a Recognized Repository under the National Biodiversity Act (2002) for storing microbial wealth in India and in 2014, the Bureau acquired ISO 9001:2008 certification. NBAIM has established and strengthened NAIMCC which is an affiliated member of WFCC. NAIMCC has state-of-the-art facilities for short-term and long-term conservation of microorganisms with more than 6500 preserved accessions of fungi, bacteria, actinomycetes, cyanobacteria, and archaea. The National Institute of Plant Health Management (NIPHM) in Hyderabad is an ongoing scheme from the XIth Plan to develop human resources in various aspects of Plant Health Management and Biosecurity issues. NIPHM is engaged in creation of Master Trainers in the field of Plant Health Management, Pesticide Management, and Biosecurity besides extending policy support to Central and State Governments (https://niphm.gov.in). The institute known as ICAR-National Institute of Biotic Stress Management, Raipur, undertakes basic and strategic studies on plant protection in India. Plant Quarantine (Regulation of Import into India) Order, 2003, regulates import and prohibition of import of plants and plant products into India. The specified planting material for propagation (viz., cuttings, saplings, bud woods, etc.) requires growing under Post Entry Quarantine (PEQ) for a specified period. The import permit for such planting material is granted based on a certificate from Designated Inspection Authorities for conducting further PEQ inspections and the final clearance is granted based on the PEQ Inspection Report. The pest risk analysis is mandatory for all the plants/plant material prior to its import into India as per PQ Order, 2003 (https://plantquarantineindia.nic.in). The risk of exotic pests and diseases is minimized by identifying the potential pests which can get into the country with the specified commodity and seeking export certification for their freedom/pest free area status, etc. from the exporting country.

In India the improved variety Pusa Basmati-1 was developed by using conventional breeding method along with MAS to incorporate two bacterial resistant gene (*xa13* and *Xa21*). Another hand MAS is used to improve other non-Basmati rice in India. PCR-based molecular marker used a backcross breeding method to introgress three major gene (*Xa21*, *xa13*, and *xa5*) against bacterial blight of rice into samba Mashuri at DRR (presently IIRR) from a donor line (SS1113) in which all three genes present homozygous condition, Swarna and IR 64 with *Xa4*, *xa5*, *xa13*, *Xa21* at CRRI India (Kumar et al. 2017). Apart from that, some of the bacterial blight resistant lines which have been developed in India are IR 24, Lalat, and Tapaswani. Major problems associated with gene stacking are for finding resistance gene. However, if anyhow it is found then there is a problem of linkage drag and many times it is also seen that there is an adverse effect on yield of the crops. Tomato leaf curl New Delhi virus (ToLCNDV) used marker-assisted selection (MAS); six genes Ty-1, Ty-2, Ty-3, Ty-4, ty5 and ty-6 are being utilized to address the tomato leaf curl

New Delhi Virus (ToLCNDV). The Ty gene(s), alone and in pyramided combinations from donors, were introgressed to IARI varieties, viz. Pusa Rohini, Pusa Ruby, Pusa Sadabahar, and Pusa Sheetal through hybridization (https://icar.org.in/sites/default/files). The gene-pyramiding strategy is deployed for various rice varieties against different pests and diseases to create broad-spectrum horizontal resistance. For rice blast, three major genes Pi1+Pi2+Pi33 were pyramided and proved their durable resistance against blast disease. In India, the rice variety ADT43 pyramided with a combination of four blast resistance genes *Pi1+Pi2+Pi33+Pi54* for blast resistance. Pyramiding for blast (*Pi5* and *Pi54*) + bacterial leaf blight (*xa13+xa21*) and brown plant hopper (*Bph3*, *Bph17* and *Bph18*) were performed in Pusa basmati 1121 and Pusa basmati 6 (Singh et al. 2011). Pyramiding of rust resistance, different rust resistance genes are available in various backgrounds. In many cases, useful resistance is associated with genetic drag. Because rust resistance genes such as *Lr9*, *Lr19*, *Lr26*, and *Yr9* have been rendered ineffective, other genes are needed in desirable backgrounds to combat wheat rusts (Bhardwaj 2011). To overcome these challenges, an extensive programme was initiated in Shimla to pyramid useful resistance genes in acceptable agronomic backgrounds (Bhardwaj 2011). Wherever required, marker-assisted confirmation of resistance is used. Consequently, 35 rust-resistant genetic stocks were developed and registered with NBPGR, New Delhi.

During the course of our research, it was found that certain varieties out of 22 varieties developed resistant to leaf rust causing fungus strain Ug99. Out of these 22 wheat varieties, some such as DBW 17, PBW 550, and Lok 1 are being cultivated in wheat-growing states in India. DWR is a member of the US-based Borlaug Global Rust Initiative to combat Ug99 disease globally a fungal disease originated in Uganda in 1999, belongs to a race of black stem rust, which brings 100 per cent crop loss unlike other rusts that partially affect the yield.

23.3 Insect Pests and Their Management

23.3.1 Pre- and Post-independence Era

The mention of insects in vedic period is evident from terms like *pathanga*, *satpada*, *pipilika*, *makshika*, *bharamara* etc., in Sanskrit writings. The term *satpada* (an equivalent of hexapoda) refers to honeybee in the book, *Amarakosha* (first century AD). Umaswati (0–100 AD) was the first person who performed detailed classification of insects. Further, Charaka (1200–1000 BC), an Indian physician, classified the bees based on morphology. Susrutha (100–200 AD), a physician by profession, had classified ants (*pipilika*) into six groups and he described the structure and habits of these ants. In ancient period, honeybees (*madhumakshika*) and lac (*laksha*) were mentioned in *chanakya sutra*, *arthasastra*, *sakuntala*, and *mahabarata* (David and Ramamurthy 2012). This indicates people of ancient India were aware of honeybees and were consuming honey and utilizing the bees wax and lac for dye purpose. The art of silkworm propagation and silk cloth manufacture were known 6000 years ago (Husain 1938). Records show that the Persian rulers were gifted with silken stuff by Indian rulers during 3870 BC, and there is a mention of silk secretion and cocoon formation by silkworm in Yajnavalkya. In *Atharvanaveda*, there is a mention of harmful insects for man and food crops.

In 1758, Carl Linnaeus published *Systema Naturae* which contained 28 species of Indian insects. During 1788 the Asiatic society of Bengal was started and many papers were published. In 1769, J.G. Koenig collected many insects from *coromandel* area and southern peninsular India, which were studied by Linnaeus. He had published the special account of termites of Thanjavur district in south India. Linnaeus recognized his work and named red cotton bug as *Dysde rcuskoenigi*. During 1883, Bombay natural history society was started. Numerous insect related works were published in the journal of the Bombay natural history society. 'Fauna of British India' series was commenced under the editorship of W.T. Blandford during 1883. The Entomological part of the 'Fauna of British India' (now Fauna of India) series started with Sir George Hampson's contribution. He authored the first four volumes on the moths of India during 1892. During 1897, Bingham issued volumes on 'Hymenoptera' (ants, bees, and wasps). Since then volumes on other groups of insects like Coleoptera (beetles), Hemiptera (bugs), Odonata (dragonfly and damselfly), etc., were published. Later in 1889 Indian Museum, Calcutta, published the Indian Museum Notes in five volumes. Rothney published on Indian Ants in 1893. This is the earliest record of biological pest control in India, which identified that the stationary items where white ants were attached were kept free by red ants. On 25 August 1897, Sir Ronald Ross discovered malarial parasite in a dissected Anopheles mosquito in Begumpet at Secunderabad. In 1909, the central malaria bureau was established at Kasauli which was later on developed into the National Institute of Communicable Diseases. Daniel Austin and Ronald Ross studied extensively on houseflies, sand flies, fleas, lice, and bed bugs (David and Ramamurthy 2012).

During 1901, Lionel de Nicevelle was appointed as the first entomologist to the Government of India. This marked the birth of professional entomology in India. Further, Maxwell Lefroy in 1903 was appointed as the first Imperial entomologist at IARI, Pusa (Bihar), in 1905. Subsequently this institute was shifted to New Delhi and the post was re-designated as Head of the Division of Entomology. Subsequently, the states also started entomological work. Madras, Punjab, and Uttar Pradesh appointed their first State/Provincial Government Entomologists in 1912, 1919, and 1922, respectively. In 1906, Forest Research Institute (FRI) was established and E.P. Stebbing became the first forest zoologist which was later re-designated as forest entomologist in 1922. In 1926, S.K. Sen was appointed at IVRI, Mukteswar, to study the vectors of rinderpest. He was succeeded by JT Edwards. In 1942, CFC Beeson published his monumental work 'The ecology and control of the forest insects of Indian and the neighbouring countries'. The nineteenth century marks the major progress and expansions in the field of applied entomology. In 1916, the Natural History section of Indian museum was formed as Zoological Survey of India. In 1921, Indian central cotton committee was set up to study the cotton insect pests. In 1925, Indian Lac Research Institute was established.

The systematic study of entomology was started with the formation of Entomological Society of India (ESI) on 1 January 1937 in Calcutta and started its first publication of *Indian Journal of Entomology* from 1939 onwards. Similarly, All India Beekeepers Association, Pune, started publishing the *Indian Bee Journal* since 1937. In 1966, Cecidological Society of India was formed at Allahabad. T.B. Fletcher was the first government entomologist of the Madras state, and he published a book on *some south Indian insects* in 1914 (David and Ramamurthy 2012).

There was a lack of systematic biological control efforts to manage insect pests until the seventeenth century. The apple woolly aphids entered into India through apple rootstock importation from England (Gupta 2015). Apple San Jose scale also got introduced in the 1960s. In 1957, the Commonwealth Institute of Biological Control (CIBC) was established with 23 need-based substations from Srinagar in North to Palghat in South. Further, in 1977, AICRP on Biological Control of crop pest and weeds was established with 10 centres. Due to the success of biological control programme, it was elevated as Project Directorate of Biological Control (PDBC) at Bangalore in 1993. In India, up to 2003, 166 exotic biological control agents were introduced, of which 33 could not be released in field, 71 recovered after release, 6 providing excellent control, 7 substantial control, and 4 partial control (Singh 2004). During XIth plan 2007–2012 the PDBC was upgraded to National Bureau of Agriculturally Important Insects (NBAII). To manage insect pest introduction to India, plant quarantine act was enforced in 1912. Similarly, in 1914 destructive insect pest act was enforced. Historically there was a lack of any research evidence for resistant variety screening in vedic period or pre-independence era. The rice gall midge (*Orseolia oryzae*) (Wood-Mason reported in 1880) which was a minor pest became a major pest after introduction of high-yielding varieties like IR5 and IR8. The varieties like PTB 18, PTB 21, CR 56-17, and Leuang 152 were identified as resistant donors and were crossed with the high-yielding varieties (Banerjee 1971). The insecticides, DDT and HCH, were introduced during 1947 and 1949 respectively for controlling several public health and agricultural insect pests. Since then, the first report of insecticide resistance was reported in mosquitoes, *Culex fatigans*, vectoring filariasis in U.P. and Bombay during 1952. Since then, reports of DDT and HCH resistance in mosquitoes were reported throughout the country. The appearance of lindane resistance (30-fold) in veterinary insect *Boophilus microplus* was first reported in 1963 from IVRI, Muktheswar. In the same year, agricultural pest resistance was also reported from singhara beetle, *Galerucella birmanica*, from Delhi to HCH and DDT. This insect was followed by tobacco caterpillar, *Spodoptera litura*, from Rajasthan in 1965 and diamond back moth, *Plutella xylostella*, in 1968 from Punjab. Pyrethroid resistance to gram pod borer *Helicoverpa armigera* Hubner in various parts of country was reported in 1986. The first report of resistance in stored product insect was from flour beetle *Tribolium castaenum* against DDT and malathion from Delhi in 1971 (Mehrotra 1989). This was followed by rice weevil *S. oryzae* and lesser grain borer *R. dominica* in 1973 and 1976.

During this period, Indian farmers possessed large wealth of Indigenous Technical Knowledge (ITKs) to manage insect pests such as (1) digging one feet trench around groundnut field and placing calotropis leaves in them to ward off red hairy caterpillar (*Amsacta albistriga*), (2) keeping fire torch at night to attract and destroy red hairy caterpillar adult moths, (3) spraying notchi (*Vitex negundo*) leaf extract during milky stage of rice to control earhead bug (*Leptocorisa acuta*), (4) Calotropis leaves were broadcasted in rice field to reduce hopper incidence, (5) applying neem seed kernel extract to check nymphal emergence and tungro virus incidence in rice, (6) fixing a circular tin of 1 inch at 8 feet aboveground level in coconut tree to avoid rat damage, (7) placing a scarecrow in field to ward of birds, (8) farmers use 'y' shaped stick with leather string known as '*kavan*' to ward off birds, and (9) neem leaves (*Azadiracta indica*), *Vitex negundo*, and *Pongamia pinnata* were dried and were mixed with grain bulk to ward off storage pest such as rice weevil, pulse beetle, etc. similarly smearing of oil with pulse to contain pulse beetle (Baskaran and Narayanasamy 1995).

23.3.2 Green Revolution and Post-Green Revolution Era

Biotic stresses have emerged and re-emerged as highly virulent strains and caused huge economic losses due to luxurious plant growth and susceptibility. Even the plant resistance to a particular biotic stress has been knocked down due to evolutionary changes in genetic makeup or mutations in the insects and pathogens. Over-dependency on a very few high-yielding varieties combined with prophylactic application of chemical pesticides rendered the country's rice farmers vulnerable to plagues of insecticide-induced resurgent pests with special reference to brown plant hopper (Thorburn 2015). Similar experiences in wheat and other crops in which biotic stresses have emerged and caused huge economic losses. To combat losses caused by the emerging and re-emerging biotic stresses in crops, novel insecticide molecules, novel molecular tools, good agricultural practices, environmentally safe biological and microbial control techniques, nanotechnology, development tolerant/resistant varieties etc. have been widely explored after the green revolution (Kumar et al. 2021).

23.3.2.1 Emerging and Invasive Biotic Stresses Recorded in India

The insect pests like *Helicoverpa armigera* (Hubner), *Plutella xylostella* (L.), *Pectinophora gossypiella* (Saunders), *Bemisia tabaci* (Gennadius), *Heliothist heivora* Waterhouse, *Nilaparvata lugens* (Stal), *Earias vitella* (Fabricius), *Tribolium castaneum* (Herbst), *Sitophilus oryzae* (L.), *Callosobruchus chinensis* (L.), *Paracoccus marginatus*, *Macrosiphum miscanthi* Takahashi, *Orselia oryzae* (Wood-Mason), *Liriomyza trifolii* Burgess, *Nilaparvata lugens* Stal, *Sogatella furcifera*, *Pyrilla perpusilla* Walker, *Sesamia inferens* Walker, *Ceratova cunalanigera*, *Phenacoccus solenopsis*, *Tuta absoluta*, *Spodoptera frugiperda*, *Paraleyrodes bondari* (Josephrajkumar et al. 2019), and *Aleurodicus rugioperculatus* (Srinivasan et al. 2016) have emerged as economically important pests in agricultural and horticultural ecosystems after the green revolution due change in the cropping systems, intensive agriculture, cropping sequences, climate change, evolution, etc. (Rathee and Dalal 2018; Murali Baskaran et al. 2019). These invasive pests have been posing serious threat and causing economic losses to crops in India. Continuous efforts of the scientific and farming community to suppress these pest populations had paved way for devising new avenues of pest management.

Monitoring and surveillance of biotic stresses have been shifted from traditional to modern techniques. Recent advances in robotics and space research have enabled to monitor pest populations in remote and inaccessible areas for desert locust. Machine learning algorithm was also effective to identify desert locust breeding sites in East Africa using key climatic factors, temperature and rainfall as well as edaphic factors like sand and moisture contents along with greenness. The prediction model developed based on these factors could predict oviposition and breeding sites accurately in East African countries, Kenya, Uganda, Sudan, and Saudi Arabia which would aid in preventing pandemics by destroying eggs and immature stages in small breeding sites. Such novel technological interventions can save crops from devastating losses caused by desert locust pandemics across the continents. Recently, innovative harmonic radar has been developed and utilized for tracking the movement of *Vespa velutina,* a pest of honeybees, and to identify its nests by fitting on insects (Maggiora et al. 2019). This technology can be very well adopted to other insect pests such as locusts, bees, butterflies, moths, bumble bees, dragonflies, etc.

Resistant varieties to overcome the economic losses caused by the insect pests were also developed. Breeding varieties for resistance/tolerance to insect pests have been found to be the best strategy. One of the classical developments with respect to the development of resistant varieties is Bt cotton governing resistance to bollworms. The cotton production was challenged by pests like bollworms, *H. armigera, E. vitella,* and *P. gossypiella,* and caused huge economic losses. Bt cotton was created by genetically altering the cotton genome to express a microbial protein from the bacterium *Bacillus thuringiensis*. In short, the transgene inserted into the plant's genome produces toxin crystals that the plant would not normally produce which, when ingested by a certain population of organisms, dissolves the gut lining, leading to the organism's death (Kranthi and Stone 2020). Then Monsanto and Mahico came out with wonderful genetically modified Bt cotton in the year 2002 in India which has become boon to cotton stake holders. Initially, Bollgard-I was released for commercial cultivation to suppress *H. armigera* and *S. litura* and subsequently Bollgard-II governing resistance to *P. gossypiella* also. In 2003–2004, India was the world's third largest cotton producer and seventh biggest exporter of cotton in the world. Overall, India was a net importer of cotton. With the introduction of Bt cotton, India in a decade rose to become not only the biggest producer of cotton bales but also the second biggest exporter. Bt cotton cultivation is the strong indicator of initial reductions in pesticide use. Today more than 92% area under cotton cultivation is Bt cotton. However, bollworms have slowly broken down the resistance and caused outbreak of pink bollworm especially in central India during 2015 and also in 2018–2019 especially in Maharashtra and Gujarat

(Buradikatti 2015). To combat this issue, Bt-resistant management strategies were developed by popularizing refuge strategy which delayed the onset of resistance. Very interestingly, a series of resistant rice varieties were bred to control the brown plant hopper *Nilaparvata lugens* and also in many other crops. Sincere efforts are made to develop Bt brinjal resistant to brinjal soot and fruit borer, *Leucinodes arbonalis*. However, due to strong resistance from NGOs and public, this Bt crop has not yet surfaced under condition in India.

Subsequently, environmental issues have started mounting up because of the indiscriminate use of hazardous pesticides to control insect pests in India. Indiscriminate use of insecticides at higher dose leads to resistance, resurgence, and residue problems in Indian agriculture. Historically, cotton and vegetables have accounted for more than 50% of insecticide usage in India. Organophosphates (OPs) and organochlorine insecticides had been gradually replaced by pyrethroids during the late 70s and 80s and then subsequently, the OPs and pyrethroids have been replaced by neonicotinoids and other compounds of novel chemistry during the late 90s, worldwide. Several field problems such as poor selection of chemicals and sub-standard application practices exacerbated the control failures of insecticides against *B. tabaci* in India (Naveen et al. 2017). Slowly resistance to key insecticide molecules has been reported in sucking pests like whitefly, brown plant hopper, thrips, etc. As of 2015, 42 pesticides or pesticide formulations were banned or withdrawn from the Indian market. Chlorpyriphos, one of the wonderful broad-spectrum molecules, has also been found to have adverse effects on human health specially children. Hence, the USA had banned the usage of chlorpyriphos as chlorpyrifos and chlorpyrifos-methyl have been identified as possible causes of neurological damage in children. India had imposed ban on usage chlorpyriphos in vegetables from 2007 onwards. Recently, the Ministry of Agriculture and Farmers Welfare also proposed to ban 27 insecticide/pesticide molecules in India considering their potentially hazardous nature to the environment, human beings, and non-target organisms in 2020.

In India, the Central Insecticides Board and Registration Committee (CIBRC) is the governing body that regulates biopesticides under the Insecticides Act of 1968. After the stage was fully set, all the provisions of the Insecticides Act were brought into force with effect from 1 August 1971. This board advises the central and state governments on technical matters related to the manufacturing, sale, transportation, and distribution of all insecticides to ensure safety to humans and animals. The registration committee of the CIBRC grants licenses after scrutinizing their formulations and verifying claims made by the importer or the manufacturer. Manufacturers can register newly developed products under section 9-(3B) (provisional registration) or 9(3) (regular registration) section of the Insecticides Act. The Ministry of Agriculture and Farmers Welfare and the Department of Biotechnology provide funds to promote biopesticide research and commercialization in India. The 35 Central Integrated Pest Management Centers located in 29 states, National Research Centre for Integrated Pest Management, and National Bureau of Agriculture Insect Resources (NBAIR) are also involved in monitoring the quality, production, and application of biopesticides. Currently, there are 361 bio-control laboratories in India associated with the production of biopesticides (DPPQS 2017). Other agencies like the National Agricultural Research System (NARS) and National Board of Accreditation (NBA) are responsible for evaluating quality standards and research in biopesticides. Additionally, these organizations provide training to officials of state agriculture departments concerning quality control of biopesticides (Singh et al. 2016). The National Policy for Farmers (2007) implemented by the federal Department of Agriculture and Cooperation currently recommends the use of biopesticides to farmers for increasing farm productivity and maintaining environmental and human health.

Hence, tremendous research efforts were made to develop eco-friendly pest management technologies such as biological control including parasitoids, predators, entomopathogens such as fungi, bacteria, virus, and nematodes. A total of nearly 15 biopesticide formulations are registered by CIBRC for use in India. The biopesticide market in India has seen tremendous improvement by increasing its share in the total pesticide consumption and usage (projected to be 7.3%). Although microbial insecticides provide several advantages over traditional pest control products, they have not achieved the desired level of commercial development and application in India. Several factors limit the Indian microbial pesticide market. Product quality control issues, including low microbial count that results in poor performance in the field, is cited as a primary limitation (Gupta and Dikshit 2010). Studies conducted by NBAIR revealed that between 50 and 70% of entomopathogenic fungal-based products in India had either fewer colony propagules than listed on the label, excessive moisture content in solid formulations, or contaminants and, hence, did not meet published CIBRC standards. Limited shelf life may constrain the adoption of some microbial biopesticides, especially in rural areas where access to fresh products and refrigerated storage is limited. Some talc-based formulations of entomopathogenic fungi have a shelf life of 3–4 months, but many other formulations become ineffective due the lack of proper storage conditions (Ramanujam et al. 2014; Kumar et al. 2019).

The Pesticide Management Bill, 2020, was introduced in Rajya Sabha to regulate the manufacture, import, sale, storage, distribution, use, and disposal of pesticides, in order to ensure the availability of safe and good quality pesticides and minimize the risk to humans, animals, and environment. The Bill seeks to replace the 53-year-old Insecticides Act, 1968. The current state of regulation of pesticides in India, using the extant law called Insecticides Act, 1968, has not caught up with post-modern pest management science nor has taken cognizance of a huge body of scientific evidence on the ill effects of synthetic pesticides. The death due to acute pesticide poisoning and hospitalizations that Indian farm workers and farmers. It is not just human beings but wildlife and livestock that are poisoned routinely by indiscriminate use of toxic pesticides across the country. The key features of the bill are complete information, compensation provision, registration requirement, and strict regulatory norms towards achieving sustainable developmental goals.

23.3.2.2 IPM Modules Across the Country

Soon after the green revolution, pest problems have mounted up in Indian agriculture. Research was oriented towards the development of viable and eco-friendly integrated pest management modules to farmers in the country. Validation and promotion of numerous location-specific IPM module in rice, wheat, maize, cotton, pulses, oil seeds, and horticultural crops have been done and the same has been disseminated to the farmers across the country. Pest forecasting models have developed as part of IPM to forewarn and alert the farmers to safeguard crops from pest attack (Malik et al. 2021). Smart agricultural technologies have been developed to address some of the challenges in plant protection.

23.3.2.3 Biosafety and Biosecurity

Research has been focussed on invasive, quarantine and transboundary pests after the green revolution by strengthening quarantine regulations. Rapid globalization and advancements in transport, travel, tourism coupled with liberalization of trade pose increased risk of introduction of exotic and invasive pests into the country. Plant pests which gained entry into India are causing significant economic damage to agricultural production. Pest risk analysis (PRA) is a science-based tool to tackle the alien pests of concern to any nation while facilitating international trade. PRA is a process which helps to assess the risks of entry, establishment, and spread potential of exotic pests. PRA is to identify the options to prevent the entry and management options in the event of pest establishment. The international standards brought out by IPPC serve as guidance for carrying out PRA. Plant biosecurity is of paramount importance to any country especially India to safeguard food security, sustainability of agricultural/horticultural production, and also in protecting the livelihood of people. The recent concern in the national agriculture scenario is to manage many types of invasive alien pest (insects/mites/nematodes/diseases, etc.) species that cross over from across the globe.

23.3.2.4 Institutional Developments

The Directorate of Plant Protection Quarantine & Storage (DPPQS) was established in the year 1946 on the recommendation of Woodhead Commission as an apex organization for advising the Government of India and state governments on all the matter related to Plant Protection. Plant protection activities encompass activities aimed at minimizing crop losses due to pests through integrated pest management, plant quarantine, regulation of pesticides, locust warning and control, and training in desert areas besides training and capacity building in plant protection. It is attached with the Office of Ministry of Agriculture and Farmers Welfare. It has various Sub-Offices throughout India and the Plant Protection Network of Directorate of PPQ&S in India.

23.3.2.5 NIPHM, Hyderabad

In order to bestow greater functional flexibility, and broader reach in delivery against the emerging challenges in the field of Plant Health Management, a very crucial area for enhancing our country's agricultural production, the Department of Agriculture

& Cooperation of the Ministry of Agriculture, Govt. of India, had taken a decision for transforming this Institute into an autonomous body during 2008. It is a national level institute under the administrative control of the Department of Agriculture & Cooperation, Ministry of Agriculture & Farmer's welfare, Government of India, established in the year 1966 as Central Plant Protection Institute (CPPTI) at Hyderabad. The Institute was rechristened subsequently as National Plant Protection Training Institute (NPPTI). It became an autonomous body in the year 2008 with the expanded scope of promoting environmentally sustainable Plant Health Management practices in diverse and changing agro-climatic conditions and Plant Biosecurity Management and Pesticide Management through capacity building programmes, besides providing inputs for policy formulation on plant health management, plant biosecurity, invasive alien species, market access, pesticide management etc. at state and national level. In the year 2009, the National Bureau of Agriculturally Important Insects (NBAII) (formerly known as Project Directorate on Biological Control) Bangalore was established.

23.3.2.6 NCIPM New Delhi

The National Research Centre for Integrated Pest management (NCIPM) is a premier research institute of Indian Council of Agriculture Research (ICAR). The institute was established in February 12, 1988, to develop and promote Integrated Pest Management (IPM) technologies for major crops with minimum ecological implications; generate national database on all aspects of pest management; advice national priorities and pest management policies; establish linkages with national and international institutes in the area of IPM; and to extend technical consultancies.

23.3.2.7 NIBSM, Raipur

The institute known as ICAR-NIBSM was created after recommendations of the Veerappa Moily Oversight Committee on the implementation of the reservation in higher educational institutions for expansion, inclusion, and excellence. The foundation stone of the institute was laid on 7 October 2012 as the 99th research institute under ICAR, New Delhi. NIBSM is entrusted with a unique research mandate of devising novel mitigation measures of biotic stresses in the farming sector through revamping the inadequacy of the ongoing national programmes for assuring national food and nutrition security menaced by vibrant pests/pathogens in the pursuit of climate change, pathogenic mutations, intervention to keep a clean environment, and global regulations under WTO/IPR regimes. Buildings were inaugurated by honourable Prime Minister of India on 28 September 2021.

23.4 Nematode and Their Management

23.4.1 Pre- and Post-independence Era

The history of Indian Nematology dates back to 1901 when Dr. C.A. Barber first reported the occurrence of root knot nematode on tea in the then Devala State of South India. Subsequently, several nematode diseases *viz.*, the 'Ufra' disease of rice in East India caused by *Ditylenchus angustrus* in 1913 and 'White tip' disease of rice by *Aphelenchoides besseyi* in 1936 were reported which caused concerns. The discoveries of 'Molya disease' of wheat and barley in Rajasthan caused by cereal cyst nematode *Heterodera avenae* in 1958 and the potato cyst nematode *Globodera rostochiensis* in the Nilgiri Hills of Tamil Nadu in 1961 prompted the Indian Government to strengthen the science of plant nematology in the country (Seshadri 1986; Dhawan et al. 2001). Therefore, comprehensive research on plant parasitic nematodes (PPN) was initiated by IARI at its two divisions, viz. Entomology and Mycology & Plant Pathology. In India, Ayyar (1933) was the first one to attempt the control of root knot nematodes by (1) burning of sorghum stalks followed by spading, (2) growing a trap crop knol-khol with egg plant, and removing and destroying it before the emergence of new nematodes, (3) applying chemicals including kerosene, carbolic acid, potassium cyanide, formalin, carbon disulphide, lime, sulphuric acid, and ranicide, and (4) crop rotation with ragi, maize, sorghum, or red gram. Fumigants such as DD and EDB were used in early trials for the control of root knot nematodes (Sen 1960; Nirula 1961), the golden nematode of potato in the Nilgiris (Seshadri and Kumarswami 1964), and against nematodes infesting sugarcane, potato, and wheat. During this period, there had been very little progress in the field of nematology in the country.

23.4.2 Green Revolution Era

An organized study on PPN was started towards the early 1960s in the country. By realizing the importance of PPN as limiting factors in agricultural production, the ICAR has sanctioned the establishment of the Division of Nematology at IARI in 1966. In the year 1967, the FAO assigned B. Weischer at CPCRI Regional Station, Kayangulam, Kerala, for survey of nematodes associated with the root wilt of coconut caused by burrowing nematode, *Radopholus similis*. This led to the establishment of the Division of Nematology at the Station in 1972 (Seshadri 1986). Exposure of soil to solar heat by ploughing of field provides good control of nematodes (Krishnamurthy and Elias 1969; Seshadri and Sethi 1978). Rotation with graminaceous hosts (wheat, barley, ragi, maize, sorghum) and dicots (mustard, sesame, asparagus, and African marigold) for 1–2 years suppressed root knot nematode population (Gaur 1975; Sundaresh and Setty 1977; Roy 1978). Amendments of soil with non-edible oil cakes such as neem, karanj, and mahua have been found effective against root knot nematodes (Singh and Sitaramaiah 1973; Khan et al. 1973). The use of DD (a mixture of 1,2-dichloropropane, 1,3-dichloropropene, and other chlorinated hydrocarbons)/DBCP (1,2-Dibromo-3-chloropropane) has reported spectacular increase of yield in various crops. In India, the nematophagous fungi (*Dactylariae udermata* and *Dactylellae inopaya*) were first recorded. Subsequently, a large number of fungal species and other microbial agents have been recorded in the country.

23.4.3 Post-Green Revolution Era

From 1980, several chemicals such as organophosphates (fensulfothion, phosphamidon, dichlorofenthion, fenamiphos, chlorpyriphos, ethoprophos) and carbamates (aldicarb, carbofuran, dimethoate, methomyl, thiodemeton, oxamyl) have been reviewed and preferred for the management of nematodes (Sen and Dasgupta 1982). However, at present, majority of these chemicals have been either banned or withdrawn due to their toxicity to human health and hazard to the environment (Kumar et al. 2019). From 1980, the country initiated new programmes to breed a broader range of crops with nematode resistance coupled with other desirable characters and reported resistant cultivars/lines against important PPN in vegetables, pulses, and cereals. However, few cultivars are in use on a large scale by growers for one or other reason (Dhawan et al. 2001). Several biocontrol agents such as fungi (*Pochonia chlamydosporia*, *Paecilomyces lilacinus*, *Trichoderma harzianum*, etc.) and bacteria (*Pseudomonas fluorescens*, *Pasteuria*) have been successfully isolated and some of them have been formulated as commercial agents for the management of different life stages of PPN (Kumar 2020; Mohan et al. 2020). With the increasing domain knowledge, several new innovations and approaches were made in reducing the damage caused by various plant pathogens in agricultural and horticultural crops. RNAi-mediated suppression of genes essential for nematode parasitism, survival, and development offers novel targets for nematode management (Banerjee et al. 2017). Host-induced gene silencing (HIGS) has revolutionized the possibilities to knockdown the activity of many structural, developmental, or housekeeping genes of a nematode by introducing dsRNA or siRNA molecules into the host plant. This strategy ensures the continuous delivery of dsRNA/siRNA to the feeding nematodes (Dutta et al. 2015a; Shivakumara et al. 2017; Chaudhary et al. 2019). Yadav et al. (2006) first proved host-generated RNAi in tobacco to knockdown both integrase and splicing factor genes in *M. incognita*. A number of studies have reported the efficacy of HIGs in cyst nematodes (Dutta et al. 2020) and root knot nematodes (Dutta et al. 2015b; Kumar et al. 2017). Nanotechnology is an emerging and promising technique due to controlled delivery of functional molecules and strong inhibitory activity against PPN (Kalaiselvi et al. 2019). Thakur et al. (2018) noted that incubation of *M. incognita* juveniles in water with *Bacillus licheniformis*-derived GoldNPs for 3 h resulted in 100% mortality of juveniles. Recently, Baronia et al. (2020) noted that the application of experimental AgNP formulation (1 μg/ml) resulted in negligible to complete suppression of gall formation caused by *M. graminicola* compared to the control on rice (Pusa basmati 1121) under glasshouse conditions. Similar results were obtained in field soil assays at the rate of 3 μg/ml, lower than the commercial formulation, Silvox® (150 μg/ml). The changes in agricultural situations and climate have resulted in the emergence of new nematode problems in India which include *M. graminicola* on rice in Mandya district of Karnataka, West Bengal, Odisha, and Assam; *M. graminicola* on onion and garlic in Karnataka, Haryana, West Bengal, and Gujarat; floral malady in tuberose caused by *Aphelenchoides besseyi* in West Bengal and Odisha; *M. incognita* on gerbera and carnation in polyhoses in southern parts of the country;

M. indica on Kagzi lime and cotton in Gujarat; *M. incognita* on pomegranate in Maharashtra, Karnataka, Gujarat, and Andhra Pradesh; *M. enterolobii* on Guava in Tamil Nadu and Andhra Pradesh, which are few examples of current nematode threats in the country (Walia and Khan 2018).

Nematode management in future are (1) to carry out extensive surveys of diseases of national and international importance in order to have endemic nematodes data, (2) identification and characterization of resistant sources against PPN of different crops, (3) identification, evaluation, and development of formulations of new fungal and bacterial antagonists of plant parasitic nematodes, (4) strategies on utilization of multiple biocontrol agents with complementary and synergistic modes of action and integration of biocontrol agents with other management methods such as chemicals, crop rotation, and use of resistant cultivars could be highly effective against PPN. (5) Recently, a number of CRISPR/Cas9 genome editing protocols have been established in *C. elegans*, which could open doors to perform these studies in PPN during their interaction with host plants.

23.5 Weed Management

Weeds are major biotic constraints in agriculture. It causes an economic yield loss of about 11 billion US$ in ten major field crops (Gharde et al. 2018). In general, weed management is location-specific and integrated weed management (IWM) is considered to be effective to minimize crop-weed competition to enable crops to optimally use resources such as space, light, carbon-di-oxide (aboveground) nutrient, and water (belowground) for attaining the optimal harvestable crop yield (Choudhary and Dixit 2018). In the past 50 years, several weed management technologies have been developed and disseminated; still, weeds continue to be a major problem. Continuous monitoring of weeds in an ecosystem and appropriate and dynamic weed management practices are required to tackle weeds of various ecosystems (Rao 2018).

23.5.1 Pre-independence Era

Before the independence of India, cultural practices, hand weeding, and mechanical weeding were the major weed management practices. During the period there was not much information available on weed management in general and herbicides in particular. Among various available options, physical methods contributed to the larger area followed by smaller weeding tools. During the period, agriculture was mainly practiced for subsistence purposes. In the pre-independence era, more availability of manpower and lesser wages helped in adopting the physical method of weed control. With the advancement in time, in 1920 AD, mechanical tools were invented and were used for weed management. These included cultivators, blades, harrows, finger weeders, rotary hoes, rod weeders, etc. along with hand hoeing, digging, cutting, dredging, and chaining operations through hoes, axes sickle etc.

A majority of the farmers used to burn the field after the crop harvest; this killed the weed seeds and is the cheapest method of weed control. In general, physical methods and cultural practices (like tillage, planting, fertilizer application, irrigation, crop rotation, summer fallowing, etc.) were done for the reduction in weed density and severity, which favoured more towards crops. However, this method was not sufficient to completely control the field weeds; the use of herbicides started just before independence. Several cultural practices like tillage, planting, fertilizer application, irrigation crop rotation summer fallowing etc., are employed for creating favourable conditions for the crop. These practices if used properly help in controlling weeds. Cultural methods alone cannot control weeds, but help in reducing the weed population.

23.5.2 Post-independence Era

During this period, there was a rapid increase in weed population, although the cultivation area was limited. These created a huge pressure to import food grains. Along with the imported food grains, several weed species were introduced and established in the country. Subsequently, it proliferated and distributed in a major chunk of the area. Ending the pre-independence era, a new approach to chemical weed control came into the existence in India and the world. The earliest attempt to control weeds in India with herbicides was made in 1937 in Punjab for controlling *Carthamus oxyacantha* by using sodium arsenite. 2,4-D was registered as the first selective herbicide for the control of broad-leaved weeds. It was first tested in India in 1946. Since then, several chemicals have been imported and tested. Some of them were quite effective in controlling certain weeds. This compelled to initiate herbicide-based research in India from 1948 onwards with 2,4-D. After independence, in 1951, phenyl substituted urea group of herbicides, e.g., monuron and diuron were introduced. Triazine group of herbicides were developed in 1955. Uses of dinitroanilines were started in agriculture in 1960. Glyphosate came into existence for weed control as a systemic and non-selective herbicide in 1970. A PL 480 research project with USAID was undertaken (1964–1969) on the ecology of ten common noxious weeds including *Chenopodium album*, *Cyperus rotundus*, *Eichhornia crassipes*, *Anagallis arvensis*, *Spirodela polyrrhiza*, *Portulaca oleracea*, *Cassia tora*, *Eleusine indica*, *Amaranthus spinosus*, and *Eleocharis palustris*. Although quarantine law had been imposed, but still due to improper imposition, some of the weeds could trash pass and got distributed across various parts of the country with imported materials. Later, the quarantine law was further improvised and various kinds of filters were placed to avoid the entry and distribution of weed seeds. Since then several types of research have been conducted in sole herbicides between 1980 and 1989. The major emphasis is given to herbicide-based weed management (such as alachlor, atrazine, bifenox, butachlor, 2,4-D, dicamba, diquat, fluchloralin, fluroxypyr, glyphosate, methabenzthiazuron, metoxuron, nitrofen, paraquat, propanil, simazine, terbutryne, and sethoxydim) for weed management. The efficacy of herbicides in managing weeds in different crops, herbicide efficacy interaction with irrigation, fertilizers, effect of herbicides sprayed in one crop on the succeeding crops, and tolerance of crop cultivars to herbicides were certain aspects of herbicide-based studies. Mechanical weeders like hand hoe, blade hoe, and paddy weeder were found equally effective in managing weeds and were found more economical than hand weeding (Singh et al. 1976). During the period, major research was done on a critical period of crop-weed competition, weed flora survey, the physiological study of perennial weeds, ecology of *Parthenium hysterophorus* (Tiwari and Bisen 1984), effect of herbicide on soil microflora (Mukhopadhyay 1980), integrated Striga control in sorghum (Choudhari et al. 1980); biology and control of *Oxalis latifolia* were reported (Muniyappa et al. 1983), competitive cultivars and allelopathy. This provided excellent weed control, but continuous use of the same herbicides for a prolonged period has created another problem like herbicide-resistant weeds (*Phalaris minor* against Isoproturon) during the 1990s (Malik and Singh 1993). These compelled to re-synthesize the weed management practices. The research has been started with the combined use of two herbicides (tank mix and/or pre-mix) for broad-spectrum weed control. Still, these practices were not efficient in controlling weeds. Over the period, research on the sole use of herbicides decreased and emphasis was given to research on integrated weed management (IWM). Biotechnological interventions were used to understand molecular diversity of *Phalaris minor* populations in wheat (Dhawan et al. 2008) and mechanism of resistance of *Phalaris* to isoproturon (Singh et al. 2004). Weed utilization such as Lantana and Eupatorium as green manure (Mankotia et al. 2006) and weed biomass for nitrogen substitution in rice-rice system (Rajkhowa 2008) was assessed.

Research and adoption on IWM gradually increased 9% (1980–1989) to 36% (2010–2018). Based on the research article published about 39% are from herbicides alone and comparing herbicides performance with hand weeding and other methods. Emphasis has been given to control weed adopting soil solarization along with crop husbandry practices like tillage with and without irrigation; crop residue/straw incorporation (Das and Yaduraju 2008); and irrigation and nitrogen management to enhance weed management efficacy. Likewise, usage of mechanical weeders (rotary weeder) increased; it saved 57% of total labour use for weeding (Subudhi 2004). The total saving of female labour is by 4.85 times and male labour by 6.6 times over hand weeding (Remesan et al. 2007). Publications on integrated weed management included a combination of herbicides with manual weeding (Singh and Singh 2004), trash burning, inter-cultivation (Subramanian and James 2006), tillage (Sarma and Gautam 2006), and rotation (Singh 2006). The importance of decision-making tools was brought to light (Babu and Yaduraju 2000).

23.5.3 Post-Green Revolution Era and Present

In the last decade the major emphasis is given to IWM (Rao and Nagamani 2010); conservation agriculture and weed management (Bhullar et al. 2016); aquatic weed management (Sushilkumar 2011); climate change and weed management (Singh et al. 2011); control of Orobanche (Punia 2014); weedy rice management (Abraham and Nimmy 2015); crop-weed-fertilizer-water (Kaur et al. 2018); mulching (Choudhary and Kumar 2014); intercropping (Choudhary et al. 2016); non-chemical methods (rotary weeder use) for managing weeds in rice (Deshmukh 2012); and use of black polythene mulch (25 μm thickness UV resistant) for managing weeds (Ram et al. 2017). Herbicide compatibility has been evaluated (Choudhary and Dixit 2021), weed management in Bt cotton (Ramachandra et al. 2016), and solarization for reducing weed seed bank in soil. Harnessing the benefit of conservation agriculture using happy seeder and herbicides in rice-wheat-green gram cropping system in vertisol (Singh et al. 2017) to facilitate timely sowing in standing stubbles, minimize weed infestation, lower cost of production, improve fertilizer/water-use efficiency, and improve soil health (Choudhary and Choudhury 2018). However, weeds are very clever and have adapted to varied climatic conditions and accumulated more biomass resulting in less susceptibility at recommended herbicide doses. Therefore, there is an urgent need to develop effective, economical, and ecologically safe integrated weed management strategies through interdisciplinary research. A total of 60 herbicides of different modes of action are registered in our country. More than 700 formulations of herbicides are currently available on the market. These include combination formulations of two different herbicides, which are becoming popular amongst farmers due to broad-spectrum weed control.

23.6 Future Issues and Challenges

23.6.1 Challenges

(1) Need for establishment of national referral laboratories for quick identification/diagnosis and containment in order to prevent the wide spread of pests. A well-defined and coordinated mechanism should be developed for safeguarding our country; (2) Artificial Intelligence for monitoring, tracking breeding sites, migration of movement of locusts, and aerial application of pesticides for site-specific management of transboundary quarantine pests; (3) exploitation of advanced breeding techniques, including agricultural biotechnology, to introduce insect and disease resistance into major crop plants. Much success has been achieved for enhancing yield traits using such technologies but research is limited to breeding biotic stress resistant varieties against key pests; (4) establishing Public-Private Partnerships to contribute to control biotic stresses; (5) modelling of crop pest epidemics and disease outbreaks for major pests across the crops should be developed; (6) development and implementation of a national crop protection strategy founded based on the principles of integrated management; (7) investment in training provision among researchers and users to improve use of new technologies and innovation; (8) a nationwide programme on robust monitoring and advisory system for quarantine, invasive, transboundary biotic stresses need to be developed; (9) robust forewarning systems for timely management of invasive, quarantine pests are to be developed in

tune with changing climate, Extrapolation model, INDO-BLIGHTCAST being employed for controlling late blight disease for other biotic stresses in India; (10) exploration of semiochemicals/pheromones or allelochemicals for the management of sucking pests is a big challenge ahead of scientific community. Utilization of such chemicals in trapping technology would be of great help to farming community in overcoming losses caused by vector and vector-borne viral diseases; (11) there is a need to put in lot of efforts on development of suitable consortia of microbes for the management of biotic/abiotic stresses and reduced cost of cultivation; (12) the biocontrol agents need to be assessed for their compatibility with commonly used agrochemicals periodically; (13) induction of higher tolerance in strain/isolates of the bio-agents against need-based pesticides for control of soil-borne diseases; (14) compatibility studies of biopesticides with biofertilizers and fast-tracking of biopesticide registration and (15) early warning system and pest advisory need to be developed and strengthened in view of changing climatic scenario; (16) climate risk management through developing DSS (Decision Support System) and improving agro-advisory network services; (17) develop simulation models and forecasting system to predict the outbreak of diseases; (18) extensive field studies under various conditions should be needed to understand the interaction between nanoparticles, microorganisms, soil and plants against PPNs; (19) strategies on utilization of multiple biocontrol agents with complementary and synergistic modes of action and integration of biocontrol agents with other management methods such as chemicals, crop rotation, and use of resistant cultivars could be highly effective against PPNs; (20) there is a need to develop strict regulatory mechanisms to maintain the quality and availability of biocontrol agents at affordable cost to farmers; (21) weed management policy needs to be formulated on an urgent basis to control the severity, spread, and threat to the environment. Empower the community to manage weeds under the policy.

23.6.2 Issues

There is an urgent need for National Policy on Agrochemicals with an emphasis on safe use of pesticides, fast-track transparent time-bound online registration system by revamping the current registration process based on the recommendations of duly constituted independent expert committees, guidelines for registration of newer molecules shall also be harmonized, complete check on sale and use of spurious pesticides, high-quality well-equipped quality control accredited national pesticides testing laboratories across the country should be established with international standards, biosafety data for new molecules to be generated preferably through notified National Accreditation Board for Testing and Calibration Laboratories (NABL), pesticides of biological origin, growth regulators whose toxicity levels are very low can be relaxed from stringent regulatory ambit, strengthen and encourage manufacturing of new molecules under Make in India initiative, capacity building and skill development in assessing unregistered pesticides in imported commodities need to be strengthened, and an urgent need to move forward to innovations in biopesticides of biological origin, green chemistry pesticides, etc. A centre of excellence (CoE) on agrochemicals with multifaceted wide-spectrum and modern bioscreening facilities needs to be established urgently, and greater thrust needs to be given to develop low-cost technologies for mass production and bulk availability of biocontrol agents and biopesticides.

References

Abraham CT, Nimmy J (2015) Weedy rice invasion and its management. Indian J Weed Sci 47: 216–223

Adugna A (2004) Alternate approaches in deploying genes for disease resistance in crop plants. Asian J Plant Sci 3:618–623

Ayyar PNK (1933) Madras Agric J 21:97–107

Babu MBBP, Yaduraju NT (2000) Decision support systems: a tool for sustainable weed management. Indian J Weed Sci 32:119–128

Banerjee SN (1971) Recent progress in rice insect research in India. In: In: Symposium on rice insects. Proceedings of a symposium on tropical agricultural researches. Japan International Research Center for Agricultural Sciences, Tsukuba, pp 19–24

Banerjee SA, Banerjee SS, Gill OP, Gupta A, Dahuja PK, Jain, Sirohi A (2017) RNA interference: a novel source of resistance to combat plant parasitic nematodes. Front Plant Sci 8:834

Baronia RP, Kumar SP, Singh, Walia RK (2020) Silver nanoparticles as a potential nematicide against Meloidogyne graminicola. J Nematol 52:e2020

Baskaran V, Narayanasamy P (1995) Traditional pest control. Caterpillar Publications, Annamalainagar, p 90

Bhardwaj SC (2011) Wheat rust pathotypes in Indian subcontinent then and now. In: Wheat-productivity enhancement under changing climate. Univ. of Agri. Sciences, Dharwad

Bhullar MS, Pandey M, Kumar S, Gill G (2016) Weed management in conservation agriculture in India. Indian J Weed Sci 48:1–12

Buradikatti K (2015) Pink bollworm a nightmare for Bt cotton growers. The Hindu (5 December 2015)

Chattopadhyay SB, Mukherjee K (1955) Black vein disease of cabbage in West Bengal. Sci Cult 21:107–108

Chaudhary S, Dutta K, Shivakumara TN, Rao U (2019) RNAi of esophageal gland-specific gene Mi-msp-1 alters early stage infection behaviour of root-knot nematode, Meloidogyne incognita. J Gen Plant Pathol 85:232–242

Chenulu VV, Sachchidananda J, Mehta SC (1968) Studies on mosaic disease of cowpea from India. Phytopathol Z 63:381–387

Choudhari SD, Muzaffar S, Upadhyay UC (1980) Integrated Striga control in sorghum. Indian J Weed Sci 12:15–20

Choudhary VK, Choudhury BU (2018) A staggered maize-legume intercrop arrangement influences yield, weed smothering and nutrient balance in the Eastern Himalayan Region of India. Exp Agric 54(2):181–200

Choudhary VK, Dixit A (2018) Herbicidal weed management on weed dynamics, crop growth and yield in direct seeded rice. Indian J Weed Sci 50(1):6–12

Choudhary VK, Dixit A (2021) Bio-efficacy of sequential herbicide application for weed management in dry direct seeded rice. Indian J Agric Sci 91(1):79–83

Choudhary VK, Kumar PS (2014) Influence of mulching on productivity, root growth and weed dynamics of maize (Zea mays)-based cropping systems. Indian J Agron 59(3):364–370

Choudhary VK, Dixit A, Chauhan BS (2016) Resource use maximization through legume intercropping with maize in eastern Himalayan region of India. Crop Pasture Sci 67:508–519

Das TK, Yaduraju NT (2008) Effect of irrigation and nitrogen levels on grassy weed competition in wheat and comparative eco-physiology of Phalaris minor Retz. and Avenasterilis ssp. ludoviciana Dur. in wheat. Indian J Weed Sci 39:178–184
Dastur JF (1948) Phytophthora spp. of potatoes (Solanum tuberosum L.) in the Simla hills. Indian Phytopathol 1:19–12
David BV, Ramamurthy VV (2012) Elements of economic entomology. Namrutha Publication, Chennai, p 147
Deshmukh G (2012) Rotary weeder for drudgery reduction of women during weeding in rice field. Indian J Weed Sci 44:132–134
Dhanraj KS, Raychaudhuri SP (1969) A note on barley mosaic in India. Plant Dis Rep 53:967–968
Dhawan SC, Gaur Pankaj HS, Kaushal KK, Ganguly S, Chawla G, Singh RV, Sirohi A (2001) Indian nematology: progress and perspectives division of nematology. New Delhi, IARI, p 174
Dhawan RS, Singh R, Punia SS, Dhawan AK, Dudeja SS (2008) Molecular diversity of little seed canary grass (Phalaris minor Retz.) populations from wheat growing belts of India. Indian J Weed Sci 40:101–108
DPPQS (2017) Directorate of plant protection quarantine and storage. Ministry of Agriculture and Farmers Welfare, Government of India. http://ppqs.gov.in/divisions/integrated-pest-managment/bio-control-labs. Accessed 30 June 2017
Dutta TK, Bankar P, Rao U (2015a) The status of RNAi based transgenic research in plant nematology. Front Microbiol 5:760
Dutta TK, Papolu PK, Banakar P, Choudhary D, Sirohi A, Rao U (2015b) Tomato transgenic plants expressing hairpin construct of a nematode protease gene conferred enhanced resistance to root-knot nematodes. Front Microbiol 6:260
Dutta TK, Papolu PK, Singh D, Sreevathsa R, Rao U (2020) Expression interference of a number of Heterodera avenae conserved genes perturbs nematode parasitic success in Triticum aestivum. Plant Sci 301:110670
Fawcett H, Jenkins AE (1933) Records of citrus canker from herbarium specimens of the genus Citrus in England and the United States. Phytopathology 23:820–2451
Gaur HS (1975) PhD thesis. IARI, New Delhi
Gharde Y, Singh P, Dubey RP, Gupta PK (2018) Assessment of yield and economic losses in agriculture due to weeds in India. Crop Prot 107:12–18
Gupta R (2015) Management and comparative efficacy of various treatments against *Eriosomala nigerum* Hausmann on apple trees (*Malus domestica* BORKH.) in Jammu province of Jammu & Kashmir state. Int J Adv Biotechnol Res 5(4):319–321
Gupta S, Dikshit AK (2010) Biopesticides: an ecofriendly approach for pest control. J Biopest 3: 186–188
Harahap Z, Silitonga TS (1988) Breeding for resistance against major pests and diseases of rice. In: Proceedings of international symposium workshop (ISW'88), Kuala Lampur, Malaysia, pp 110–112
Howard A, Howard GLC (1910) Wheat in India: its production, varieties and improvement. Calcutta, Thacker, Spink and Co., p 288
Husain MA (1938) Entomology in India, past, present and future. Curr Sci 6(8):422–424
Josephrajkumar A, Mohan C, Babu M, Krishna A, Krishnakumar V, Hegde V, Chowdappa P (2019) First record of the invasive Bondar's nesting whitefly, Paraleyrodes bondari Peracchi on coconut from India. Phytoparasitica 47(3):333–339
Joshi RK, Nayak S (2010) Gene pyramiding-A broad spectrum technique for developing durable stress resistance in crops. Biotechnol Mol Biol Rev 5(3):51–60
Kalaiselvi D, Mohankumar A, Shanmugam G, Nivitha S, Sundararaj P (2019) Green synthesis of silver nanoparticles using latex extract of Euphorbia tirucalli: a novel approach for the management of root knot nematode, Meloidogyne incognita. Crop Prot 117:108–114
Kaur S, Kaur R, Chauhan BS (2018) Understanding crop-weed-fertilizer-water interactions and their implications for weed management in agricultural systems. Crop Prot 103:65–72
Khan MW, Khan AM, Saxena SK (1973) Acta Bot Ind 1:49–54

Kranthi KR, Stone GD (2020) Long-term impacts of Bt cotton in India. Nat Plants 6(3):188–196
Krishnamurthy GVG, Elias NA (1969) Tobacco News 2(7):17–20
Kumar S (2014) Trichoderma: mass production, formulation, quality control, delivery and its scope in commercialization in India for the management of plant diseases. Afr J Agric Res 9(53): 3838–3852
Kumar KK (2020) Fungi: a bio-resource for the control of plant parasitic nematodes. In: Yadav AN, Mishra S, Kour D, Yadav N, Kumar A (eds) Agriculturally important fungi for sustainable agriculture, volume 2: functional annotation for crop protection. Springer, Cham, pp 285–311
Kumar S, Gupta O (2012) Expanding dimensions of plant pathology. JNKVV Res J 46(3):286–293
Kumar A, Kakrana A, Sirohi A, Subramaniam K, Srinivasan R, Abdin MZ, Jain PK (2017) Host-delivered RNAi-mediated root-knot nematode resistance in Arabidopsis by targeting splicing factor and integrase genes. J Gen Plant Pathol 83:91–97
Kumar KK, Sridhar J, Murali-Baskaran RK, Senthil-Nathan S, Kaushal P, Dara SK, Arthurs S (2019) Microbial biopesticides for insect pest management in India: current status and future prospects. J Invertebr Pathol 165:74–81
Kumar J, Murali-Baskaran RK, Jain SK, Sivalingam PN, Mallikarjuna J, Kumar V, Sharma KC, Sridhar J, Mooventhan P, Dixit A, Ghosh PK (2021) Emerging and re-emerging biotic stresses of agricultural crops in India and novel tools for their better management. Curr Sci 121(1):26–36
Leadbeater A (2012) The role of FRAC in resistant management. J Mycol Plant Pathol 42(1):25
Maggiora R, Saccani M, Milanesio D, Porporato M (2019) An innovative harmonic radar to track flying insects: the case of vespa velutina. Sci Rep 9:1–10
Malik RK, Singh S (1993) Evolving strategies for herbicides use in wheat: resistance and integrated weed management. In: Proceedings international symposium. Indian Society of Weed Science, Jabalpur, pp 225–238
Malik M, Sehgal M, Chakraborty A (2021) Use and dissemination of information communication technology to farming women is Boon in Indian Agriculture. Biot Res Today 3(5):332–334
Mankotia BS, Thakur RC, Kumar J, Kumar S (2006) Use of lantana and eupatorium as green manure in rainfed maize-wheat system in north-west Himalayas. Indian J Weed Sci 38:263–266
Mehrotra KN (1989) Pesticide resistance in insect-pests Indian scenario. Pestic Res J 1(2):95–103
Mohan S, Kumar KK, Sutar V, Saha S, Rowe J, Davies KG (2020) Plant root-exudates recruit hyperparasitic bacteria of phytonematodes by altered cuticle aging: implications for biological control strategies. Front Plant Sci 11:763
Mollison I (1901) A text book of Indian Agriculture. Advocate of India Steam Press, Bombay, p 283
Mukhopadhyay SK (1980) Effects of herbicides and insecticides alone and their combinations on soil microflora. Indian J Weed Sci 12:53–60
Mundkur BB (1948) Fungi and plant diseases. Macmillan & Co. Ltd., London, p 246
Muniyappa TV, Prasad TVR, Krishnamurthy K (1983) Biology and chemical for control of Oxalis latifolia Linn. Indian J Weed Sci 15:182–187
Murali Baskaran RK, Jain SK, Sivalingam PN, Mooventhan P, Sridhar J, Sharma KC, Kaushal P, Kumar J (2019) National status of biotic stresses of crops. ICAR-National Institute of Biotic Stress Management, Raipur, p 174
Nagarajan S (2010) Plant pathology and Indian agriculture past, present and future. Indian Phytopathol 53(2):121–128
Narayanan NN, Baisakh N, Vera Cruz CM, Gnanamanickam SS, Datta K, Datta SK (2002) Molecular breeding for the development of blast and bacterial blight resistance in rice cv. IR50. Crop Sci 42:2072–2079
Nariani TK (1960) Yellow mosaic of Mung (Phaseolus aureus L.). Indian Phytopatol 13:24–29
Nariani TK, Kartha KK, Prakash N (1970) Purification of southern sannhemp mosaic virus using butanol and differential centrifugation. Curr Sci 39:539–541
Naveen NC, Chaubey R, Kumar D, Rebijith KB, Rajagopal R, Subrahmanyam B, Subramanian S (2017) Insecticide resistance status in the whitefly, Bemisia tabaci genetic groups Asia-I, Asia-II-1 and Asia-II-7 on the Indian subcontinent. Sci Rep 7(1):1–15

Nene YL (2003) Crop disease management practices in ancient medieval, and pre-modern India. Asian Agri-History 7:185–201
Nirula KK (1961) Indian Potato J 3:72–75
Pal BP, Tandon RK (1937) Types of leaf-curl in Northern India. Indian J Agric Sci 7:363–393
Patel MK (1948) Xanthomonas uppalii sp. nov. pathogenic on Ipomoea muricata. Indian Phytopathol 1:67–69
Patel MK, Kulkarni YS (1948) Xanthomonas malvacearum (E. F. Smith) Dowson on exotic cottons in India. Curr Sci 17:243–244
Prasad P, Bhardwaj SC, Savadi S, Kashyap PL, Gangwar OP, Khan H, Singh SB, Kumar S (2018) Population distribution and differentiation of Puccinia graminis tritici detected in the Indian subcontinent during 2009–2015. Crop Prot 108:128–136
Prutbi HS, Samuel CK (1937) Entomological investigations on the leaf curl disease of tobacco in North Bihar. I. Transmission experiments with some suspected insect vectors. II. An alternate host of the virus and the insect transmitter. Indian J Agric Sci 7:659–670
Punia SS (2014) Biology and control measures of Orobanche. Indian J Weed Sci 46:36–51
Rajkhowa DJ (2008) Utilization of weed Biomass for nitrogen substitution in rice (Oryza sativa) - rice system. Indian J Weed Sci 40:27–32
Ram P, Sreenivas G, Rani PL, Madhavi A (2017) Weed management to improve productivity and nutrient uptake of Rabi maize. Indian J Weed Sci 49:90–92
Ram RM, Pandey AK, Singh HB (2021) Biocontrol research in India. In: Satyanarayana T, Deshmukh SK, Deshpande MV (eds) Progress in mycology. Springer, Singapore. https://doi.org/10.1007/978-981-16-2350-9_13
Ramachandra D, Ramamohan G, Bhan A, Suresh PJ (2016) Weed management in cotton: the potential of GM crops. Indian J Weed Sci 48:136–143
Ramanujam B, Rangeshwaran R, Sivakumar G, Mohan M, Yandigeri MS (2014) Management of insect pests by microorganisms. Proc Indian Natl Sci Acad 80(2):455–471
Rangaswami G (1960) Studies on two bacterial diseases of sugarcane. Curr Sci 29:318–319
Rangaswami G (1962) Bacterial plant diseases in India. Asia Publishing House, Bombay, p 163
Rao AN (2018) The historical and future perspective of weed Science research in India. In: Sushilkumar A, Mishra JS (eds) Fifty years of weed research in India. Indian Society of Weed Science, Jabalpur, pp 1–23
Rao AN, Nagamani A (2010) Integrated weed management in India. Indian J Weed Sci 42:1–10
Rathee M, Dalal P (2018) Emerging insect pests in Indian agriculture. Indian J Entomol 80(2): 267–281
Raychaudhuri SP (1947) A note on the mosaic virus in sannhemp (Crotalaria juncea Linn.) and its crystalization. Curr Sci 16:26–28
Raychaudhuri SP, Mishra MD (1965) Cultivation of some plant viruses in callus tissue cultures. Indian Phytopathol 28:50–53
Raychaudhuri SP, Nariani TK, Lele VC (1969) Citrus dieback problem in India. Proc First Int Citrus Symp Riverside 3:1433–1437
Remesan R, Roopesh MS, Remya N, Preman PS (2007) Wet land paddy weeding- a comprehensive comparative study from South India. Agric Eng Int 9:7–11
Rishi N (2009) Significant plant virus diseases in India and a glimpse of modern disease management. J Gen Plant Pathol 75:1–18
Roy AK (1978) Indian J Nematol 8:156–158
Sarma CK, Gautam RC (2006) Effect of tillage; seed rate and weed control methods on weeds and maize (Zea mays L.). Indian J Weed Sci 38:58–61
Sen AK (1960) Indian Agric 4:113–116
Sen K, Dasgupta MK (1982) Pesticides 16(1):14–17
Seshadri AR (1986) Nematology in India: achievements and prospects. In: Swarup G, Dasgupta DR (eds) Plant parasitic nematodes of India: problems and prospects. IARI, New Delhi, pp 1–9
Seshadri AR, Kumarswami T (1964) Madras Agric J 51:92–93
Seshadri AR, Sethi CL (1978) Wheat research in India. ICAR, New Delhi, pp 158–187

Sharma P, Sharma M, Raja M, Shanmugam V (2014) Status of Trichoderma research in India: a review. Indian Phytopathol 67(1):1–19

Shivakumara TN, Chaudhary S, Kamaraju D, Dutta TK, Papolu PK, Banakar P, Sreevathsa R, Singh B, Manjaiah KM, Rao U (2017) Host-induced silencing of two pharyngeal gland genes conferred transcriptional alteration of cell wall-modifying enzymes of Meloidogyne incognita vis-à-vis perturbed nematode infectivity in eggplant. Front Plant Sci 8:473

Shukla M, Al-Busaidi KT, Trivedi M, Tiwari RK (2018) Status of research, regulations and challenges for genetically modified crops in India. Biotechnol Agric Food Chain 9(4):173–188

Singh SP (2004) Some success stories in classical biological control in India. In: Asia-Pacific Association of Agricultural Research Institutions (APAARI). FAO-RAP, Bangkok, p 73

Singh R (2006) Effect of cropping sequence, seed rate and weed management on weed growth and yield of Indian mustard in Western Rajasthan. Indian J Weed Sci 38:69–72

Singh R, Singh B (2004) Effect of irrigation time and weed management practices on weeds and wheat yield. Indian J Weed Sci 36:25–27

Singh RS, Sitaramaiah K (1973) Pant Agric Tech India 6:289

Singh RP, Verma JP (1971) Control of bacterial blight of cotton. Pesticides 48:60–62

Singh SR, Singh OP, Singh Y, Sharma HC, Singh M (1976) Mechanical weeding in direct-sown rice. Indian J Agric Sci 46:507–509

Singh DV, Gaur AK, Mishra DP (2004) Biochemical and molecular mechanisms of resistance against isoproturon in Phalaris minor: variations in protein and RAPD profiles of isoproturon resistant and sensitive Phalaris minor biotypes. Indian J Weed Sci 36:256–259

Singh RP, Ramesh KS, Singh MK (2011) Impact of climate and carbon dioxide change on weeds and their management–a review. Indian J Weed Sci 43:1–11

Singh A, Singh VK, Singh SP et al (2012) Molecular breeding for the development of multiple disease resistance in Basmati rice. AoB Plants 2012:29. https://doi.org/10.1093/aobpla/pls029. PMID: 23125910; PMCID: PMC3487461

Singh HB, Keswani C, Bisen K, Sarma BK, Kumar P (2016) Development and application of agriculturally important microorganisms in India. In: Singh HB, Sarma BK, Keswani C (eds) Agriculturally important microorganisms: commercialization and regulatory requirements in Asia. Springer, New York, pp 167–182. https://doi.org/10.1007/978-981-10-2576-1-10

Singh PK, Sondhia S, Dubey RP, Sushilkumar Kumar B, Gharde Y, Choudhary VK (2017) Adoption and impact assessment of weed management technologies in wheat and greengram under conservation agriculture system in central India. Indian J Weed Sci 49:23–28

Srinivasan T, Saravanan PA, Josephrajkumar A, Rajamanickam K, Sridharan S, David PMM, Natarajan N, and Shoba N (2016) Invasion of the rugose spiralling whitefly, Aleurodicus rugioperculatus Martin (Hemiptera: Aleyrodidae) in Pollachi tract of Tamil Nadu, India

Srivastava DN, Rao YP, Durgapal JC, Jindal JK, Singh W (1967) Screening rice varieties for resistance to bacterial blight. Indian Farm 17:25–28

Subramanian E, James MG (2006) Effect of chemical, cultural and mechanical methods of weed control on wet seeded rice. Indian J Weed Sci 38:218–220

Subudhi ECR (2004) Evaluation of weeding devices for upland rice in the eastern Ghat of Orissa, India. Int Rice Res Notes 29(1):78–80

Summanwar AS, Raychaudhuri SP, Jagadish Chandra K, Prakash N, Lal SB (1969) Virus associated with coconut root (wilt) disease. Curr Sci 3:208–210

Sundaresh HN, Setty KGH (1977) Curr Sci 6:157–158

Sushilkumar (2011) Aquatic weeds problems and management in India. Indian J Weed Sci 43:118–138

Thakur RK, Dhirta B, Shirkot P (2018) Studies on effect of gold nanoparticles on Meloidogyne incognita and tomato plants growth and development. BioRxiv 428144

Thind TS (2016) Development of fungicide resistance in plant pathogens with reference to Indian scenario. Kavaka 47:54–62

Thorburn C (2015) The rise and demise of integrated pest management in rice in Indonesia. Insects 6(2):381–408

Tiwari JP, Bisen CR (1984) Ecology of Parthenium hysterophorus. Indian J Weed Sci 16:203–206
Uppal BN (1948) Diseases of cotton in India. Indian Central Cotton Comm Bombay 32:170
Varma PM (1955) Ability of the whitefly to carry more than one virus simultaneously. Curr Sci 24: 317–318
Vasudeva RS, Samraj J (1948) A leaf curl disease of tomato. Phytopathology 38:364–369
Verma IP, Singh RP (1971a) Epidemiology and control of bacterial blight of cotton. Proc Indian Natl Sci Acad 1971:191
Verma IP Singh RP (1971b) Problems in controlling bacterial blight of cotton. In: 2nd Int Symp Plant Pathol New Delhi, pp 109–10
Walia RK, Khan MR (2018) Compendium of nematode diseases of crop plants. In: ICAR-All India coordinated research project on nematodes in cropping systems, New Delhi, India, p 110
Webster RK, Gunnell PS (eds) (1992) Compendium of rice diseases. APS Press, St. Paul
Yadav BC, Veluthambi K, Subramaniam K (2006) Host generated double stranded RNA induces RNAi in plant parasitic nematodes and protects the host from infection. Mol Biochem Parasitol 148:219–222
Yao KS, Li SJ, Tzeng KC, Cheng TC, Chang CY, Chiu CY et al (2009) Fluorescence silica nanoprobe as a biomarker for rapid detection of plant pathogens. Adv Mater Res 79:513–516

Invasive and Transboundary Pests

24

R. K. Murali-Baskaran, Palanisamy Mooventhan, Pankaj Kaushal, and P. K. Ghosh

Abstract

A total of 33 invasive pests including insect and noninsect pests have been reported to enter and establish in various crop habitats in India during pre- and post-independence era and Green and Post-Green Revolution era. It continued till date with the second decade of the twenty-first century and has caused heavy yield losses, leading to a threat for food security. Out of the invasive pests encountered so far, 18 insect species belong to Hemiptera (six whiteflies, five mealybugs, three aphids, three scale insects, one psyllid), six Lepidoptera, three Coleoptera, two Hymenoptera, one Diptera, and three noninsect pests (one Acarina, one Mollusca, one root-knot nematode). Trade policies, free trade, and globalization invited maximum number of invasive pests into India during the Post-Green Revolution era and the second decade of the twenty-first century. The Covid-19-induced lockdown is responsible for few insects to attain dreadful status in causing significant damage and yield losses in crops. Two insects, namely desert locust and carambola fruit fly, are considered as transboundary pests in India. The ICAR institutes/SAUs have developed efficient Integrated Pest Management (IPM) for almost all invasive pests of India, which are available for practice in many public domains. The Plant Quarantine Organization is the apex body in India to monitor the spread, distribution, and outbreak of invasive and transboundary pests, besides providing plant protection advisories through obtaining information from early warning systems (EWSs) on subscription/ request.

R. K. Murali-Baskaran (✉) · P. Mooventhan · P. Kaushal · P. K. Ghosh
ICAR-National Institute of Biotic Stress Management, Raipur, Chhattisgarh, India
e-mail: rk.baskaran@icar.gov.in

P. K. Ghosh et al. (eds.), *Trajectory of 75 years of Indian Agriculture after Independence*, https://doi.org/10.1007/978-981-19-7997-2_24

Keywords

Invasive pests · Transboundary pests · Early warning systems · Plant quarantine

24.1 Introduction

When invasive pests occupy new ranges/new habitats, they have the potential to manipulate ecological and economic systems on a global basis by altering ecosystems, tritrophic interactions, and threating biodiversity (Noar et al. 2021). Additionally, substantial reduction in host plant fitness, lack of coevolution between natural enemies and pest species, and wide spread mortality are recorded, which capitalize the ecological, cultural, and economic losses (Rigling and Prospero 2018). The invasive insects-induced global crop losses is estimated for up to 1.4 trillion USD annually (Bradshaw et al. 2016). Increasing international trade has created diverse pathways for the entry/outbreak of the invasive species, which accelerate the rate at which introductions occur. As globalization intensifies, the worldwide economic and ecological burden of invasive species is expected to increase. In India, 8 invasive pests during pre- and post-independence era, 14 during Green and Post-Green Revolution era, and 11 during second decade of the twenty-first century were recorded in various crop habitats.

Transboundary plant pests (TPPs) are those migratory insects, plant diseases, and weeds that can spread to several countries and reach epidemic proportions, cause significant losses to crops, threaten food security, damage the local biodiversity and environment, and create serious socioeconomic and public health consequences. TPPs-induced decline in agricultural productivity contributes to poverty and hunger, particularly of small farmers, besides acting as the barriers to trade. Three major reasons for the increased risk of TPPs that spread to challenge the world's food security include (1) global movement of agricultural goods, (2) global movement of tourists and migration, and (3) global climate change.

TPPs contribute substantially in pest and disease-induced global food loss, which is approximately one-third of annual food production. Across the world, more specifically in the Asia-Pacific region, a number of TPPs that occur as epidemics or are endemic within national boundaries face severe threats. Climate change is creating new ecological platform for the entry and establishment of pests and diseases from one geographical region to another. This expansion will continue to result in huge financial losses, therefore require large eradication programs and effective control measures. In India, desert locust and carambola fruit fly are reported to be the major TPPs, which are under surveillance in coordination with FAO. This chapter retrieves on various invasive insect and noninsect pests and transboundary insect pests, which had/have been posing problem to India during pre- and post-independence era, Green and Post-Green Revolution era, and second decade of the twenty-first century.

24.2 Invasive Pests of Pre- and Post-Independence Era

24.2.1 Giant African Snail (GAS), *Achatina fulica* (Ferussac) (Mollusca: Achatinidae)

GAS is a native to East Africa and was introduced from East Africa to India through living specimens from Mauritius to Calcutta in 1847 and in 1907. It multiplied in large numbers in Calcutta and spread to Bihar, Assam, Madhya Pradesh, Maharashtra, Kerala, Tamil Nadu, and Karnataka in gardens and ornamental nurseries. In 1946–1948, it appeared as epidemic in Balasore, Odisha in vegetable fields, especially cabbage, cauliflower, and pumpkin (Behura 1955).

24.2.2 San Jose Scale (SJS), *Quadraspidiotus perniciosus* (Comstock) (Hemiptera: Diaspididae)

SJS is a native to Northern China but it attained threat status at San Jose in California in 1873. It entered into India through Kashmir state in 1879 along with the flowering plants of *Cydonia japonica* and became a serious pest of temperate fruits, especially apples of North-Western India, namely Jammu and Kashmir, Himachal Pradesh, and Uttar Pradesh in 1921. It was reported to attack 51 hosts including apple, plum, peach, and peas in Jammu and Kashmir, Himachal Pradesh, Uttar Pradesh, Tamil Nadu, Karnataka, West Bengal, Sikkim, Assam, and Meghalaya, besides causing 100% fruit infestation in Khunmoh, Srinagar district (Tuhan et al. 1979).

24.2.3 Coffee Green Scale (CGS), *Coccus viridis* (Green) (Hemiptera: Coccidae)

CGS is a major pest of Arabian coffee and was introduced in India in 1889 via Ceylone (Singh and Rao 1977). It is very common in coffee-growing areas of Karnataka, Tamil Nadu, and Kerala, besides attacking guava, mango, orange, and so on.

24.2.4 Potato Tuber Moth (PTM), *Phthorimaea operculella* (Zeller) (Lepidoptera: Gelechiidae)

PTM is a native to the Western South America and was introduced in India in 1900, along with seed potatoes from Italy. The actual damage was first reported in 1906 in East Bengal. PTM became destructive on potato both in field (51.5–56.4%) and storage (70%) conditions in Karnataka and Uttar Pradesh (Lefroy 1968).

24.2.5 Apple Woolly Aphid (AWA), *Eriosoma lanigerum* (Hausmann) (Hemiptera: Aphididae)

AWA is a native to the America and was introduced to Coonoor, Tamil Nadu, India in 1889 and in 1909 on the nursery stock imported from England to Shimla district, Himachal Pradesh and other apple-growing areas of India (Thakur and Dogra 1980).

24.2.6 Diamond Back Moth (DBM), *Plutella xylostella* L. (Lepidoptera: Yponomeutidae)

DBM is a native to Southern Europe. It appeared destructive on cruciferous crops (Gandhale et al. 1982). In India, it was reported in 1914 and distributed in all crucifer-growing areas to cause average crop damage of 52% (David and Kumaraswami 1978).

24.2.7 Cottony Cushion Scale (CUS), *Icerya purchasi* Maskell (Hemiptera: Margarodidae)

CUS is a native to Australia and was a threat to citrus industry of Southern California (Caltagirone and Doutt 1989). It was first recorded in 1928 at Nilgiris and Madurai districts of Tamil Nadu and believed to be introduced through imported orchard stocks or flowering plants, besides through apple cuttings imported from Ceylone (Rao and Cherian 1944). A further spread was reported in many parts of Tamil Nadu, Karnataka, Kerala, and Maharashtra and reported to feed on 38 hosts including *Acacia*, *Casuarina*, *Citrus* sp., and *Rosa* sp. (Rao 1944).

24.2.8 Sugarcane Woolly Aphid (SWA), *Ceratovacuna lanigera* Zehntner (Hemiptera: Aphididae)

SWA was reported for the first time in West Bengal in 1958 on sugarcane (Basu and Banerjee 1958). Both nymph and adults suck the phloem sap and affect photosynthesis by excreting copious honeydew, which in turn results in the development of sooty mold fungus.

24.3 Invasive Pests in Green and Post-Green Revolution Era

24.3.1 Pine Woolly Aphid (PWA), *Pineus pini* (Macquart) (Hemiptera: Adelgidae)

PWA was first recorded in Nilgiris, Tamil Nadu on pine trees in 1970 and entered from Western and Central Europe (McAvory et al. 2007).

24.3.2 Litchi Longhorn Beetle (LLB), *Aristobia reticulator* (Viet) (Coleoptera: Cerambycidae)

LLB is a native to China and was first reported in Tamil Nadu in 1997 (Shylesha et al. 2000). It attacks guava, China rose, redgram, citrus, and mango. It was found attacking litchi in Arunachal Pradesh, causing 76% loss to the crop.

24.3.3 Subabul Psyllid (SP), *Heteropsylla cubana* Crasford (Homoptera: Psyllidae)

SP is a native to temperate South America. It is reported to enter India through wind currents from Sri Lanka in 1988 (Singh and Jalali 1990). Its infestation was first reported in Kattupakkam, Chengalpattu, Tamil Nadu in February 1988. Thereafter, it moved to Bengaluru, Pune, and Nagpur cities. It was further spread to Andhra Pradesh and Nepanagar in Madhya Pradesh (Singh and Bhandari 1989). It is reported to cause heavy damage to subabul.

24.3.4 Codling Moth (CM), *Cydia pomonella* (L.) (Lepidoptera: Tortricidae)

CM is a serious pest of temperate fruits including apple, walnut, peach, pear, apricot, etc. and was first introduced in the Ladakh region of Jammu and Kashmir from Pakistan in 1989 (Pawar and Kapil 1982), causing 95.6% loss to apple.

24.3.5 Coffee Berry Borer (CBB), *Hypothenemus hampei* (Ferrai) (Coleoptera: Scolytidae)

CBB is a native to Central Africa and occupied several coffee-cultivating countries including Central and South America, South East Asia, and Pacific countries (Waterhouse and Norris 1987). In India, it attained the pest status in 1990 in the estates of Gudalur of Nilgiris district, Tamil Nadu. It is believed that the route of spread of this pest through infested coffee berry brought from Sri Lanka by Tamil refugees (Kumar et al. 1990). It entered into Wayanad district of Kerala and was also detected in Kutta estate of Kodagu district in Karnataka in February 1991 and Siddapur areas in August 1993. The CBB infestation affected coffee berry export of India worth US$ 300 million (Sreedharan 1995).

24.3.6 Serpentine Leaf Miner (SLM), *Liriomyza trifolii* (Burgress) (Diptera: Agromyzidae)

SLM is a native to Florida and Caribbean Islands. It entered into India (Hyderabad) through infested Chrysanthemum cut flowers during 1991 and occupied in Karnataka, Andhra Pradesh, Maharashtra, Gujarat, and Delhi. It was recorded on cotton in Karnataka, then spread into tomato and cucurbits, causing damage to the extent of 30–40%. Severe infestation of SLM was reported to attack ridge gourd and tomato in Andhra Pradesh; cotton, tomato, and castor in Karnataka; and okra, smooth gourd, field bean, hybrid cotton, and castor in Tamil Nadu (Pawar and Kapil 1982).

24.3.7 Spiralling Whitefly (SWF), *Aleurodicus disperses* Russell (Hemiptera: Aleyrodidae)

SWF is a native to the Caribbean region and Central America. It was first recorded in cassava in November 1993 in Trivandrum district of Kerala. It spread to tapioca and rubber, and it occupied 72 hosts in Kerala (Prathapan 1996) and 96 hosts in Karnataka (Murali Krishna 1999) in different years, besides attacking guava, pomegranate, banana, and ornamental crops.

24.3.8 Coconut Eriophyid Mite (CEM), *Aceria guerreronis* (K.) (Acarina: Eriophyidae)

In India, CEM was first reported in 1997 in the coconut plantations of Ernakulam district of Kerala (Sathiamma et al. 1998). It was reported in Alleppey, Kottayam, and Trichur of Kerala, Udumalapet and Pollachi of Tamil Nadu, and coconut-growing areas of Karnataka and Andhra Pradesh (Shivarama Reddy and Naik 2000). It is reported to spread through the winds and transport of infested nuts.

24.3.9 Silver Leaf Whitefly (SLWF), *Bemisia tabaci* B-biotype (Homoptera: Aleyrodidae)

SLWF is a native to Israel and its occurrence was first reported in Kolar district of Karnataka in vegetable crops during 1999; thereafter, the spread was noticed in cauliflower, tomato, and chillies of Hosur area, Tamil Nadu. It is a highly dangerous sucking pest as it spreads virus diseases to crops indirectly (Mishra et al. 2020).

24.3.10 Eucalyptus Gall Wasp (EGW) or Blue Gum Chalcid, *Leptocybe invasa* Fisher and La Salle (Hymenoptera: Eulophidae)

EGW was first introduced into Karnataka and Tamil Nadu in 2001 from Australia. It causes extensive damage to leaves of eucalyptus trees by producing galls (Jacob et al. 2007).

24.3.11 Lotus Lily Midge (LLM), *Stenochironomus nelumbus* Tok et Kur (Diptera: Chironomidae)

LLM was reported in Thirunavaya, Kerala in 2005. It was introduced from China. The newly hatched grub causes leaf rotting in lotus by mining of subcuticular parenchyma (Deepu and Habeeburrahman 2008).

24.3.12 Erythrina Gall Wasp (EGW), *Quadrastichus erythrinae* Kim (Hymenoptera: Eulophidae)

EGW was introduced from Tanzania, East Africa into Kerala, India in 2006 (Faizal et al. 2006).

24.3.13 Cotton Mealy Bug (CMB), *Phenococcus solenopsis* Tinsley (Hemiptera: Pseudococcidae)

CMB was first recorded in cotton crop of Gujarat in 2006. It was introduced from Central America (Tanwar et al. 2007).

24.3.14 Papaya Mealybug (PMB), *Paracoccus marginatus* Williams and Granara de Willink (Hemiptera: Pseudococcidae)

PMB is a native to Mexico and was first recorded in Coimbatore, Tamil Nadu in 2007 in papaya gardens (Tanwar et al. 2010). A classical biological control for this pest was launched in 2010 by importing an encyrtid parasitoid, *Acerophagus papayae*, from Puerto Rica, Latin America with the help of United States Department of Agriculture—Animal and Plant Health Inspection Services (USDA-APHIS) (Shylesha et al. 2011).

24.4 Invasive Pests in the Second Decade of the Twenty-First Century

24.4.1 Jack Beardsley Mealybug (JBMB), *Pseudococcus jackbeardsleyi* Gimple and Miller (Hemiptera: Pseudococcidae)

JBMB is a native to the United States and was first recorded in Karnataka in 2012 in banana crop (Shylesh 2013).

24.4.2 Madeira Mealybug (MMB), *Phenococcus madeirensis* Green (Hemiptera: Pseudococcindae)

MMB is a Neotropical native and was recorded in hibiscus plants in Karnataka in 2012 (Shylesha and Joshi 2012).

24.4.3 South American Tomato Moth (SATM), *Tuta absoluta* (Meyrick) (Lepidoptera: Gelechiidae)

SATM had been first reported in 2014 in Karnataka (Sridhar et al. 2014) and turned a regular pest of India due to its fast spread and fecundity. During the subsequent years, threat of *T. absoluta* was noticed at several parts of India including Bengaluru (Karnataka), Rajendranagar (Telangana) (Kumari et al. 2015), Ludhiana (Punjab), and Varanasi and Mirzapur (Uttar Pradesh) and attained the key pest status.

24.4.4 Coconut Spindle Leaf Beetle (CSLB), *Wallacea* sp. (Coleoptera: Chrysomelidae)

CSLB was introduced from the Oriental region, Australia into the coconut plantations of Andaman Islands in 2014 (Prathapan and Shameem 2015).

24.4.5 Rugose Spiralling Whitefly (RSW), *Aleurodicus rugioperculatus* Martin (Hemiptera: Aleyrodidae)

RSW, a native to America, was first noticed on coconut from Pollachi, Tamil Nadu and Palakkad, Kerala in 2016, and subsequently occupied almost all the southern parts of India including Tamil Nadu, Karnataka, Kerala, and Andhra Pradesh, besides extending its habitats in banana, mango, sapota, guava, cashew, maize, ramphal, oil palm, Indian almond, water apple, jackfruit, and many ornamental plants (Anonymous 2019). Aphelinid parasitoid, *Encarsia guadeloupae* Viggiani (Hymenoptera: Aphelinidae), has been reported to cause 56–82% parasitization on spiralling whitefly under natural condition. Growing banana and *Canna indica* along with coconut can conserve the parasitoid. Entomopathogenic fungus, *Isaria fumosorosea*, under various field testing in Andhra Pradesh, Karnataka, Kerala, Tamil Nadu, West Bengal, and Maharashtra is evident to cause 91% control on RSW (Anonymous 2020c).

24.4.6 Fall Armyworm (FAW), *Spodoptera frugiperda* (J. E. Smith) (Lepidoptera: Noctuidae)

FAW was first reported in Shivamoga, Karnataka in 2018 (Kalleshwaraswamy et al. 2019) in maize and established as a regular insect pest of maize with its attack also reported on sorghum and bajra. It spreads rapidly in various parts of India because of its peculiar behaviors of fast flying capacity, voracious, and polyphagus in nature. During May–July 2018, it occupied the entire Karnataka, Tamil Nadu, Telangana, and Andhra Pradesh. In Maharashtra, Madhya Pradesh, Chhattisgarh, and Odisha, the spread took place in August 2018; Gujarat, West Bengal in September 2018; Jharkhand in October 2018; Tripura and Mizoram in November 2018; Nagaland in March 2019; Sikkim, Meghalaya, and Manipur in May 2019; Uttar Pradesh, Bihar, and Assam in June 2019; Uttarakhand in July 2019; and Haryana and Punjab in August 2019, while the FAW incidence was not reported in Kerala, Himachal Pradesh, and Union Territories (Andaman and Nicobar Islands, Dadra and Naga Haveli, Daman and Diu, Chandigarh, Lakshadweep, Puducherry, Delhi, Ladakh, and Jammu & Kashmir). The neighboring countries including Bhutan, Pakistan, and Afghanistan are still free from FAW occurrence (Fig. 24.1).

Hymenopteran parasitoids, namely *Telenomus* sp. (Platygastridae), *Trichogramma* sp. (Trichogrammatidae), *Chelonus* sp., *Glyptapanteles creatonoti* (Viereck), and *Cotesia* sp. of Braconidae, *Phanerotoma* sp. and *Campoletis chlorideae* Uchida of Ichneumonidae (Shylesha et al. 2018), and Nuclear polyhedrosis virus (NPV of Sf) (Sivakumar et al. 2020) and *Nomuraea rileyi* of *S. frugiperda* larvae, have been documented (Shylesha et al. 2018) to have good biocontrol potential on FAW. Web-based FAW portals of FAO (FAO 2020); CABI-based news, research, practical extension materials, videos and other resources; mobile app on FAW Monitoring and Early Warning System (FAMEWS); Global Action for Fall Armyworm Control in 2020 (CABI 2020b); and ASEAN Action Plan on FAW Control (Anonymous 2020e) are the recently established systems for early detection and management of FAW.

24.4.7 Nesting Whitefly (NWF), *Paraleyrodes minei* laccarino (Hemiptera: Aleyrodidae)

NWF is a native to Syria and was first reported in Kerala in 2018 on coconut.

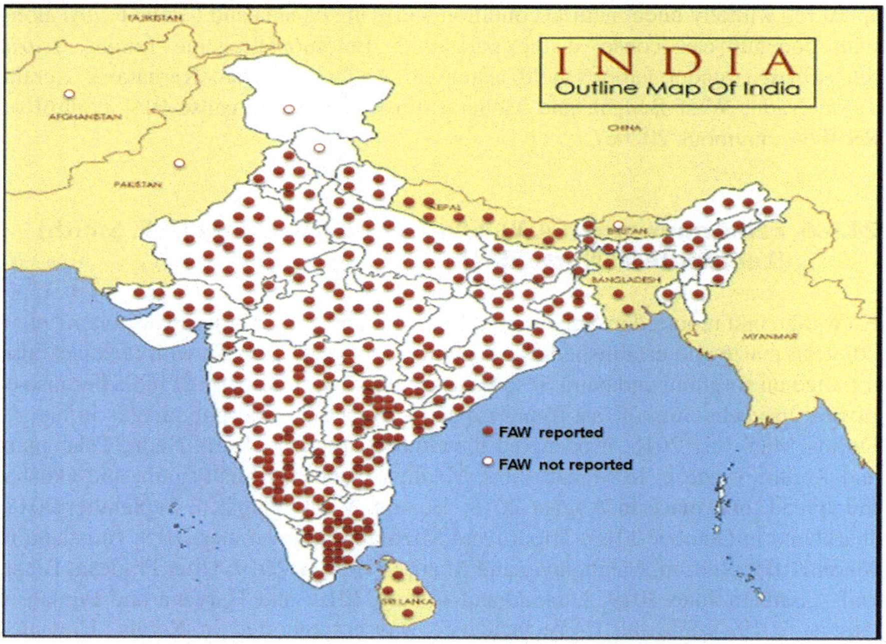

Fig. 24.1 Distribution of FAW in South East Asia

24.4.8 Bondar's Nesting Whitefly (BNSF), *Paraleyrodes bondari* Peracchi (Hemiptera: Aleyrodidae)

BNSF is a native to Central America and was reported on coconut at Kerala in 2018.

24.4.9 Neotropical Whitefly (NTWF), *Aleurotrachelus atratus* Hempel (Hemiptera: Aleyrodidae)

NTWF was recorded on coconut for the first time in Mandya, Bengaluru in 2019. It is a native to Brazil.

24.4.10 Cassava Mealybug (CMB) *Phenacoccus manihoti* Matile-Ferrero (Hemiptera: Pseudococcidae)

Invasions of CMB was first recorded in Thrissur, Kerala followed by its widespread infestations in the neighboring state of Tamil Nadu in Southern India (Anonymous 2020d). CMB-specific encyrtid parasitoids, *Anagyrus lopezi* De-Santis (Hymenoptera: Encyrtidae) Karyani et al. 2016) and *Anagyrus papayae* (Shylesha et al. 2018), have been reported to be in association with CMB under natural field condition for parasitization.

24.4.11 Root-Knot Nematode (RKN), *Meloidogyne enterolobii* (Tylenchida: Heteroderidae)

RKN is posing a serious threat to guava (*Psidium guajava*) cultivation throughout the world. More recently, it was reported in the guava orchards of Tamil Nadu (Suresh et al. 2019), Uttarakhand (Kumar and Rawat 2018), and Luni Basin of Rajasthan (Bhati and Parashar 2020) where trees declined, showing yellowing, and stunted growth. It was also identified in the guava orchards of Andhra Pradesh and Telangana. This nematode causes severe root galling, thereby reduces the nutrient uptake ability of plants.

24.5 Transboundary Pests of India

24.5.1 Desert Locust (DL), *Schistocerca gregaria* Forskål (Orthoptera: Acrididae)

DL is considered as one of the pandemics across the world due to its intercontinental flight rate, high fecundity, and voraciousness, which can cause a huge damage to several economically important crops (Cressman 2016). Out of several desert locust plagues occurred in India between 1964 and 2010, the most severe one was those occurred during 2020, resulting huge crop losses. Till now since 1993, no such severity was faced by India. In 2020, India experienced widespread and frequent cyclone-related unusual rainfall-induced DL attacks, which had a profound negative impact on all sectors of the Indian economy, including agriculture, resulting in dearth of Indian economy substantially (Anand 2020).

The excess rainfalls during March–April 2020 in India favored rapid multiplication of desert locust, with 20% excess population, which could facilitate desert locust to form three to four plaques and reached up to Central India including Maharashtra, Madhya Pradesh, and the bordering districts of Chhattisgarh, which had never been experienced for past 27 years in Central India. The excess rainfall-induced locust attacks were compounded with inadequate control of locust due to lockdown-related restricted movement of officials, extension officers, farmers, and spray-men, as well as limited access of plant protection inputs, resulting in uncontained locust populations (Anonymous 2020a, b).

Aerial application of ultra-low volume (ULV) formulations of pesticides (Malathion 96%) is in practice in several countries; however, due to huge side effects on the environment and nontarget organism, an Integrated Pest Management (IPM) program using biopesticides from plants and microorganisms is greatly reliable for successful control of desert locust (Kooyman 2003; Githae and Kuria 2021).

Early research on the use of fungi as a biocontrol agent against desert locust was focused on *Metarhizium flavoviride* (Youssef 2014; Reda et al. 2018). The oil formulation of *M. flavoviride* was reported to reduce the hopper bands tremendously (Lomer 1997). However, the negative effect of this formulation on bees and other nontarget organisms discouraged its widespread use against desert locust control (Ball et al. 1994).

The fungus, *Metarhizium acridum* (Ascomcota: Hypocreales), isolated from Africa and Australia during 1990s was considered as the best alternative to *M. flavoviride*. *Metarhizium anisopliae* var. *acridum*, commonly known as *Metarhizium acridum* (Bischoff et al. 2009), is an environmentally safer commercial biopesticide that has been developed for ultra-low volume spraying (Van Huis 2007). The biopesticide kills about 70–90% of treated locusts within 14–20 days, with no measurable impact on nontarget organisms (Lomer et al. 2001).

In 2020, FAO recommended the commercial product of *M. acridum* (Green Muscle™) as a biological solution for DL in Somalia, Kenya, and Ethiopia (CABI 2020a). It is imperative that India imports and uses the right formulation of *M. acridum* for effective management of DL. Although biological control of desert locust is a safe method, it will take several days to bring under control. Rational use of pesticides along with biological control of desert locust is suggested, considering natural-risk management plans for locust outbreaks as well as the benefit and cost of proposed control measures and their environmental and health impact.

24.5.2 Carambola Fruit Fly (CF), *Bactrocera carambolae* Drew and Hancock (Diptera: Tephritidae)

CF is a native to Asia and specifically to Malaysia, southern Thailand, and western Indonesia (Muller 2008). In past years, this species has invaded South America via the trade of fruits from Indonesia. It is a polyphagous species that attacks 100 different host plants including avocado, guava, mango, papaya, and orange, among several others (Pasinato et al. 2019).

The application of green muscardine fungi, *Metarhizium* spp., in the soil, the wrapping of fruits in newspaper, a brown paper bag, or a sleeve, and the use of bait spray are some of the eco-friendly methods used to control this pest (Muller 2008). The occurrence of CF has been reported in severe form in Bangladesh, Vietnam, Malaysia, and the Andaman and Nicobar Islands; localized in India, Thailand, and Indonesia; and absent in Afghanistan, Pakistan, Nepal, Bhutan, Burma, and few north eastern states of India (Fig. 24.2). The Indian Council of Agricultural Research (ICAR), New Delhi is in vigil in monitoring the movement of CF in boundaries of India especially in the geo-boundaries between India and Bangladesh.

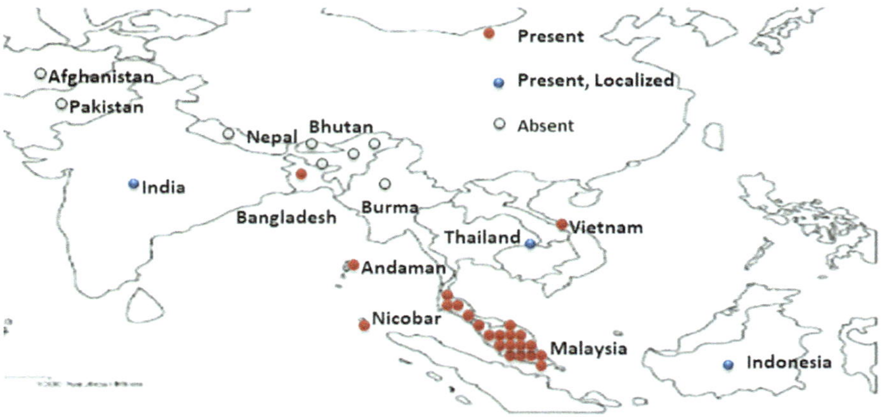

Fig. 24.2 Distribution of Carambola fruit fly, *Bactrocera carambola*

24.6 Future Projection

Though pest biology- and ecology-based efficient IPMs are available in public domains for mitigation of invasive and transboundary pests, early warning on their spread, distribution, and outbreak is more important. Several early warning systems (EWSs) are currently functioning globally, which include (1) European and Mediterranean Plant Protection Organization (EPPO), (2) the North American Plant Protection Organization (NAPPO), (3) PestLens, (4) ProMED, (5) International Plant Protection Convention (IPPC), and (6) International Plant Sentinel Network (IPSN). The networking of various EWSs is summarized in a flow chart (Fig. 24.3).

In India, the Plant Quarantine Organization, Ministry of Agriculture and Farmers Welfare, Government of India, is the apex body for implementation of plant quarantine regulations and it provides plant protection advisories concerning the country through obtaining information from EWSs. The Department of Plant Protection, Quarantine and Storage (DPPQS), India has a national network of 29 plant quarantine stations at different sites, which include airports (10), seaports (10), and land frontiers (9). Two categories of materials are imported in total under the PQ Order, 2003: (a) bulk consignments for consumption and sowing/planting and (b) samples of germplasm in small quantities for research purposes. The plant quarantine stations under the DPPQS undertake quarantine processing and clearance of consignments of the first category (Fig. 24.4).

The National Bureau of Plant Genetic Resources (NBPGR), ICAR, New Delhi undertakes the quarantine processing of all plant germplasm and transgenic planting material under exchange for which it has well-equipped laboratories and green house complexes, and recently, a containment facility has also been established for processing transgenic. The NBPGR also has a well-equipped quarantine station in

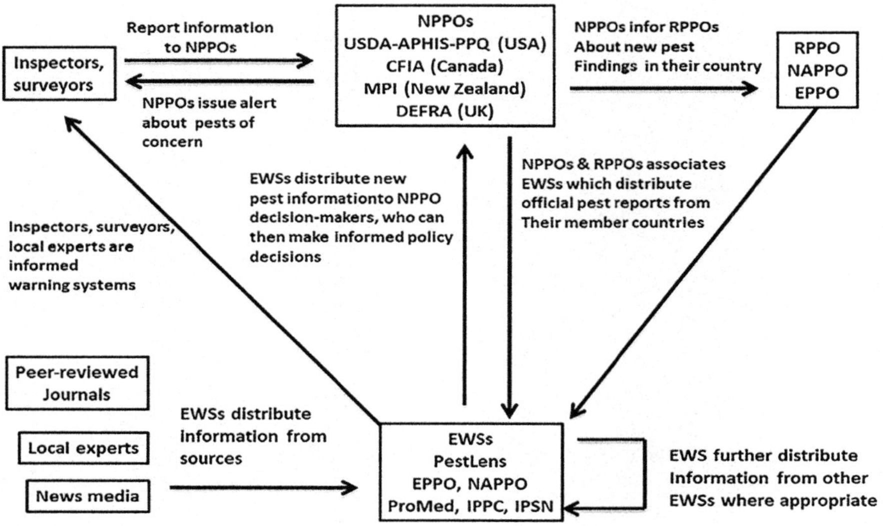

Fig. 24.3. Information flow between early warning systems and plant protection organizations. Source: Inspired from Noar et al. (2021). *NPPO* National Plant Protection Organization, *USDA-APHIS-PPQ* United States Department of Agriculture Animal Plant Health Inspection Service Plant Protection and Quarantine, *CFIA* Canadian Food Inspection Agency, *MPI* Ministry of Primary Industries, *DEFRA* Department of Environment, Food and Rural Affairs, *EWS* Early warning system, *EPPO* European Mediterranean Plant Protection Organization, *NAPPO* North American Plant Protection Organization, *IPPC* International Plant Protection Convention, *IPSN* International Plant Sentinel Network, *RPPO* Regional Plant Protection Organization

Hyderabad, which mainly deals with export samples of the International Crops Research Institute for the Semi-Arid Tropics (ICRISAT), Telangana.

FAO established an Emergency Prevention System (EMPRES) for Transboundary Animal and Plant Pests and Diseases in 1994, with a goal of addressing world food security and fighting transboundary animal and plant pests and diseases. The core principles of the EMPRES program are (1) early warning, detection, and reaction, (2) contingency planning, (3) promotion of environmentally sound control technologies, and (4) close collaboration and partnership with affected countries, national and international agricultural research centers, and other international institutions.

During the annual Meeting of G20 Agricultural Chief Scientists (MACS-G20) meet in April 2019, three major decisions were taken to monitor and management of TPPs across the world, which include (1) designation of National Reference Laboratories (NRLs) for early detection and correct diagnosis of TPPs and the establishment of networks among them, (2) improvements in bio-vigilance or global surveillance systems for identification and mitigation of potential threats before they impact the agricultural sector, and (3) the challenges and future directions of research collaborations and contributions to the International Year of Plant Health 2020.

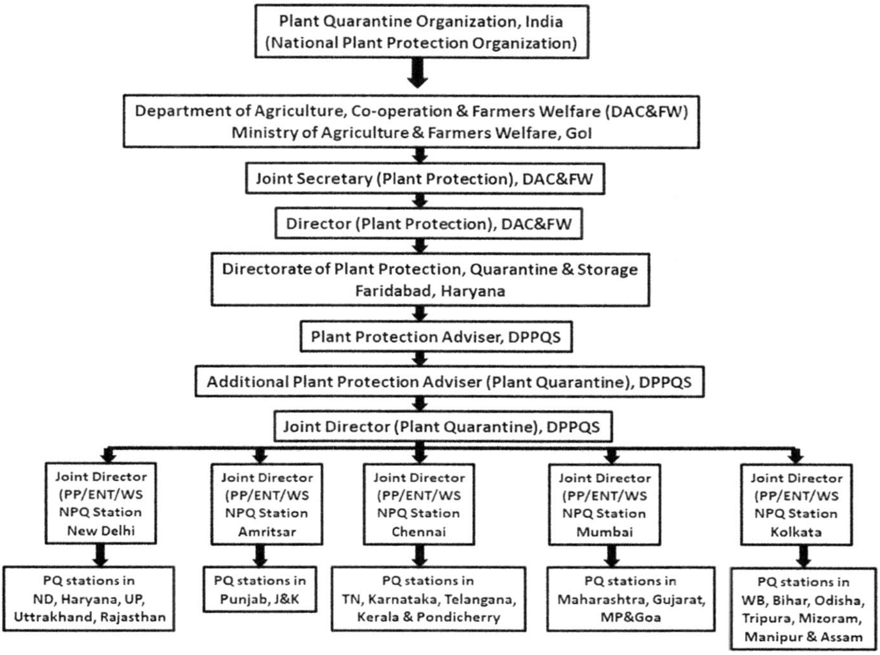

Fig. 24.4 Indian setup for monitoring and early detection of invasive and transboundary pests

24.7 Conclusions

Invasive pests and TPPs are threat to plant health, which has a global impact on food security. Thirty-three invasive insect and noninsect pests have already entered, occupied new crop habitats, and established in India so far since 1847. The Covid-19-induced lockdowns aggravated the spread and outbreaks of certain invasive pests including fall armyworm, cassava mealybug, and root-knot nematode due to restricted movement of field level plant protection workers, nonavailability of insect mitigation inputs such as pesticides, biopesticides, and botanicals. The ICAR institutes/State Agricultural Universities of India have developed well-equipped Integrated Pest Management (IPM), pest-wise based on pest-biology and -ecology of the invaded pests in the new habitats, which are readily available in public domains for the stakeholders to adapt. However, rigorous/rampant monitoring of the entry of new invasive pests through sea- and airports should be a continuous exercise of the officers and extension and field level workers, as the free trade policies and globalization are favorable for frequent entry of invasive pests.

Biocontrol agents including parasitoids, predators, entomopathogenic fungi, bacteria, and virus are reported to have good control potential on the invasive pests. Many such natural enemies, which had/have been moving along with host insects into new ranges/habitats, are being identified and utilized for biological control. However, the natural enemies of many invasive pests are not reported to move along with host insects, wherein the importation of potential natural enemies from the native/origin of invasive pests is recommended.

The desert locust and Carambola fruit fly are considered as potential transboundary insect pests of India. The Locust Warning Organization, Faridabad is the premier agency to monitor and forewarn the movement and occurrence of desert locust in India, using the geospatial technologies including GPS, GIS, and remote sensing. The Carambola fruit fly reported to prevail in Bangladesh may be likely to invade India, which requires rigorous monitoring of geo-boundaries. In addition, early monitoring, early detection, and early/fore warning on the movement of already existing and/or likely to invade in future in India are prerequisites. Few early warning systems (EWSs) are functioning globally, which disseminate the information on emerging, invasive, and transboundary pests to the member countries in the form of e-mail, bulletins, websites, newsletters, etc. on subscription/request. In India, Plant Quarantine Organization has been gathering information from International Plant Protection Convention (IPPC) and alerting stakeholders, government authorities, and policymakers about contemporaneous high-priority plant pest news. The successful early warning in biosecurity requires close collaboration between developers and end users to ensure that generated warnings are duly considered by decision-makers, reflect best practice, scientific understanding, and the working environment facing frontline actors.

The plant stress phenotyping using Artificial Intelligence (AI), with cascading steps of Deep Learning (DL), Active Learning (AL), Machine Learning (ML), and Transfer Learning (TL) (Singh et al. 2021), and development of electronic-nose with biosensors (Al-Dayyeni et al. 2021), handheld smartphone-based sensors, field-portable sensors, etc. (Das and Mohar 2020) for in situ detection of damages caused by emerging/invasive pests, can be the recent developments that need to be explored in developing countries like India.

References

Al-Dayyeni WS, Al-Yousif S, Taher MM, Al-Faouri AW, Tahi NM, Jaber MM, Ghabban F, Najm IA, Alfadli IM, Ameerbakhsh OZ, Mnati MJ (2021) A review on electronic nose: coherent taxonomy, classification, motivations, challenges, recommendations and datasets. IEEE Access 9:88535. https://doi.org/10.1109/ACCESS.2021.3090165

Anand S (2020). https://www.deccanherald.com/opinion/covid-19-impact-on-agriculture-varied-and-devastating-828390.html

Anonymous (2019) Rugose spiralling whitefly, *Aleurodicus rugioperculantus* (Hemiptera: Aleyrodidae). ICAR-National Bureau of Agricultural Insect Resources. https://www.nbair.res.in/node/833

Anonymous (2020a). https://www.thehindu.com/sci-tech/agriculture/the-hindu-explains-why-is-the-locust-surge-posing-a-threat-to-agriculture-in-india/article31712573.ece

Anonymous (2020b). https://www.thehindubusinessline.com/economy/agri-business/twin-pest-attacks-pose-a-swarm-of-trouble-for-farmers-already-hit-by-covid/article32105679.ece

Anonymous (2020c) ICAR-NBAIR curtails dangerous invasive rugose spiralling whitefly through innovative biocontrol strategies, ICAR-National Bureau of Agricultural Insect Resources. https://www.nbair.res.in/node/833

Anonymous (2020d) NBAIR pest alert, occurrence of cassava mealybug, *Phenococcus manihoti* Matile-Ferrero in India. ICAR-NBAIR, Bengaluru. https://www.nbair.res.in/index/php/node/1260

Anonymous (2020e) Response to pests and diseases. https://www.growasia.org/fall-armyworm-control

Ball BV, Pye BJ, Carreck NL, Moore D, Bateman RP (1994) Laboratory testing of a mycopesticide on non-target organisms: the effects of an oil formulation of *Metarhizium flavoviride* applied to *Apis mellifera*. Biocontrol Sci Tech 4:289–296

Basu AN, Banerjee SN (1958) Aphids of economic plants of West Bengal. Indian Agric 2:89–112

Behura BK (1955) Depredations of the giant African land snail, *Achatina fulica* (Ferussac) in Balasore (Orissa). J Bombay Nat Hist Soc 54(2):287–288

Bhati SS, Parashar A (2020) First report and morphological description of *Meloidoggyne enterolobii* infecting guava (*Psidium guajava* L.) in transitional plain of Luni basin of Rajasthan. Int J Curr Microbiol App Sci 9(9):2267–2274

Bischoff JF, Rehner SA, Humber RA (2009) A multilocus phylogeny of the *Metarhizium anisopliae* lineage. Mycologia 101:512–530

Bradshaw CJ, Lerory B, Bellard C, Roiz D, Albert C, Fournier A, Barbet-Massin M, Salles JM, Simard F, Courchamp F (2016) Massive yet grossly underestimated global costs of invasive insects. Nat Commun 7:12986

CABI (2020a) Green muscle providing strength against devastating locusts in the horn of Africa. Centre for Agriculture and Bio-science International. https://www.cabi.org/news-article/green-muscle-roviding-strength-againsdevastating-locusts-in-the-horn-of-africa/

CABI (2020b) Invasive species compendium-fall armyworm. Centre for Agriculture and Bioscience International. https://www.cabi.org/isc/fallarmyworm

Caltagirone LE, Doutt RL (1989) The history of the vedalia beetle importation to California and its impact on the development of biological control. Annu Rev Entomol 34:1–14

Cressman K (2016) Desert locust. In: Shroder JF, Sivanpillai R (eds) Biological and environmental hazards, risks, and disasters. Elsevier, London, pp 87–105

Das T, Mohar M (2020) Development of a smartphone-based real time cost-effective VOC sensor. Heliyon 6(10):e05167. https://doi.org/10.1016/j.heliyon.2020.e05167

David BV, Kumaraswami T (1978) Elements of economic entomology. Popular Book Depot, Madras

Deepu M, Habeeburrahman PV (2008) Invasion and severe outbreak of the lotus lily midge, *Stenochironomus nelumbus* Tok et Kur (Diptera: Chironomidae) in Indian freshwater ecosystems. Curr Sci 94(12):1569–1570

Faizal MH, Prathapan KD, Anith KN, Mary CA, Lekha M, Rini CR (2006) Erythina gall wasp, *Quadrastichus erythinae*, yet another invasive pest new to India. Curr Sci 90(8):1061–1062

FAO (2020) Fall armyworm: global action for fall armyworm control. Food and Agricultural Organization of the United Nation, Rome. https://doi.org/10.7717/peerj.5796

Gandhale DN, Patil AS, Awate BG, Naik IM (1982) Effectiveness of newer insecticides for diamond back moth, *Plutellla xylostella* L. Pesticides 16(9):25–27

Githae EW, Kuria EK (2021) Biological control of desert locust (*Schistocerca gregaria* Forskal). CAB Rev 16:13. https://doi.org/10.1079/PAVSNNR202116013

Jacob JP, Devaraj RN, Natarajan R (2007) Outbreak of the invasive gall inducing wasp, *Leptocybe invasa* on eucalyptus in India. Invasives 8:4

Kalleshwaraswamy CM, Asokan R, Mahadevaswamy H (2019) First record of invasive fall armyworm, *Spodoptera frugiperda* (J. E. Smith) (Lepidoptera: Noctuidae) on rice (*Oryza sativa*) from India. J Entomol Zool Stud 7:332–337

Karyani RD, Maryana N, Rauf A (2016) Host specificity test of parasitoid, *Anagyrus lopezi* (De Santis) (Hymenopter: Encyrtidae) on four mealybug species associated with cassava. Indonesia J Entomol 13:30–39

Kooyman C (2003) Report of the workshop on the use of Green Muscle® (*Metarhizium anisopliae* var. *acridum*) and desert locust adult pheromone (Phenylacetonitrile: PAN) to control desert locust hopper bands. Food and Agriculture Organization of the United Nations, Roma

Kumar S, Rawat S (2018) First report on the root-knot nematode, *Meloidogyne enterolobii* (Yang and Eisenback, 1988) infecting guava (*Psidium guajava*) in Udham Singh Nagar of Uttarakhand, India. Int J Curr Microbiol App Sci 7(4):1720–1724

Kumar VPK, Prakasam CB, Vijayalakshmi CD (1990) Coffee berry borer, *Hypothenemus hampei* (Coleooptera: Scolytidae): first record from India. J Coffee Res 20(2):161–164

Kumari DA, Anitha G, Anitha V, Lakshmi BKM, Vennila S, Rao NHP (2015) New record of leaf miner, *Tuta absoluta* (Meyrick) in tomato. Insect Environ 20:136–138

Lefroy HM (1968) The potato moth. Agric J India 2(3):294–295

Lomer CJ (1997) *Metarhizium flavoviride*: recent results in the control of locusts and grasshoppers. In: New strategies in locust control. Basel, Birkhäuser, pp 159–169

Lomer CJ, Bateman RP, Johnson DL, Langewald J, Thomas M (2001) Biological control of locusts and grasshoppers. Annu Rev Entomol 46:667–702

McAvory TJ, Zilahi-Balogh GMG, Salom SM, Kok LT, Guoliang Z (2007) Development and feeding of *Tetraphleps galchanoides*, a predator of the hemlock woolly adelgid. Biol Control 52:339–350

Mishra GP, Dikshit HK, Ramesh SV, Tripathi K, Kumar RR, Aski M, Dasgupta U, Kumar A, Praveen S, Nair RM (2020) Yellow mosaic disease (YMD) of Mungbean (*Vigna radiate* (L.) Wilczek): current status and management opportunities. Front Plant Sci 24:918. https://doi.org/10.3389/fpls.2020.00918

Muller AV (2008) Carambola fruit fly situation in Latin America and the Caribbean. Proc Caribbean Food Crops Soc 44(1):135–144

Murali Krishna M (1999) Bio-ecology, host range and management of spiralling whitefly, *Aleurodicus disperses* Russell, M.Sc. (Ag.) thesis, University of Agricultural Sciences, Bengaluru

Noar RD, Chelsea J, Jahant-Miller, Emerine S, Hallberg R (2021) Early warning systems as a component of integrated pest management to prevent the introduction of exotic pests. J Integr Pest Manage 12(1):1–7

Pasinato J, Redaelli LR, Botton M, de Jesus-Barros CR (2019) Biology and fertility life table of *Bactrocera carambolae* on grape and acerola. Entomologia 63(3):217–223. https://doi.org/10.1016/j.rbe.2019.06.001

Pawar AD, Kapil N (1982) Monitoring of codling moth, *Cydia pomonella* (L.) in Kinnaur district (Himachal Pradesh) by sticky traps baited with codling moth sex pheromone. Plant Protect Bull 34(3):13–14

Prathapan KD (1996) Outbreak of the spiralling whitefly, *Aleurodicus disperses* Russel (Homoptera: Aleyrodidae) in Kerala. Insect Environ 22:36–38

Prathapan KD, Shameem KM (2015) *Wallacea* sp. (Coleoptera: Chrysomelidae)- a new spindle infesting leaf beetle on coconut palm in the Andaman and Nicobar Islands. J Plan Crops 43(2): 162–278

Rao VP (1944) *Acerya purchase* Maskell in South India. Indian J Entomol 43:246–247

Rao YR, Cherian MC (1944) The fluted scale, *Icerya purchase* Mask., as a pest of wattle in South India, and its control by the biological methods. Madras Agric J 22(3):92–102

Reda M, Mashtoly TA, El-Zemaity MS, Abolmaaty A, Abdelatef GM, Marzouk AA (2018) Susceptibility of desert locust, *Schistocerca gregaria* (Orthoptera: Acrididae) to *Bacillus cereus* isolated from Egypt. Arab Univ J Agric Sci 26:725–734

Rigling D, Prospero (2018) *Cryphonectaria parasitica*, the causal agent of chestnut blight: invasion history, population biology and disease control. Mol Plant Pathol 9:7–20

Sathiamma B, Nair CP, Koshy PK (1998) Outbreak of a nut infesting Eriophyid mite, *Eriophyes guerreronis* (K.) in coconut plantations in India. Indian Coconut J 1998:1–3

Shivarama Reddy L, Naik SL (2000) Spread of Eriophid mites into the coconut gardens of Chittoor district, Andhra Pradesh. Indian Coconut J 30(12):8–9

Shylesh AN (2013) Host range of invasive Jack Beardsley mealybug, *Pseudococcus jackbeardsleyi* Gimpel and Miller in Karnataka. Pest Manage Hort Ecosyst 19(1):106–107

Shylesha AN, Joshi S (2012) Occurrence of Madeira mealybug, *Phenococcus madeirensis* Green (Hemiptera: Pseudococcidae) on cotton in India and record of associated parasitoids. J Biol Control 26(3):272–273

Shylesha AN, Thakur NS, Ramchandra A (2000) Incidence of litchi trunk borer, *Aristobia testudo* Voet (Coleoptera: Lamiidae) on guava in Meghalaya. Pest Manage Hort Ecosyst 6(2):156–157

Shylesha AN, Joshi S, Rabindra RJ, Bhumannavar BS (2011) Classical biological control of the Papaya mealybug. ICAR-NBAIR, Bengaluru

Shylesha AN et al (2018) Studies on new invasive pest, *Spodoptera frugiperda* (J. E. Smith) (Lepidoptera: Noctuidae) and its natural enemies. J Biol Control 32:145–151

Singh P, Bhandari RS (1989) Further spread of Leucaena Psyllid, *Heteropsylla cubaba* in India. Indian Forest 115(5):303–309

Singh SP, Jalali SK (1990) Spread of *Heteropsylla cubaba* Crawford, Subabul psyllid to Karnataka. Myforest 26(2):153–156

Singh SP, Rao NS (1977) Field evaluation of fish oil insecticidal rosin against soft green scale, *Coccus viridis* on citrus. J Plant Protect 7(2):208–209

Singh A, Jones S, Ganapathysubramanian B, Sarkar S, Mueller D, Sandhu K, Nagasubramanian K (2021) Challenges and opportunities in Machine-augmented plant stress phenotyping. Trends Plant Sci 26(1):10. https://doi.org/10.1016/j.tplants.2020.07.010

Sivakumar G et al (2020) Isolation and characterization of indigenous nucleopolyhedrovirus infecting fall armyworm, *Spodoptera frugiperda* (J. E. Smith) (Lepidoptera: Noctuidae) in India. Curr Sci 119:860–864

Sreedharan K (1995) Integrated management of coffee berry borer. Indian Coffee 59(10):9–10

Sridhar V, Chakravarthy AK, Asokan R, Vinesh LS, Rebijith KB, Vennila S (2014) New record of the invasive South American tomato leaf miner, *Tuta absoluta* (Meyrick) (Lepidoptera: Gelechiidae) in India. Pest Manage Hort Ecosyst 20:148–154

Suresh P, Poornima K, Kalaiarasan P, Nakkeeran S, Vijayakumar RM (2019) Characterization of guava root knot nematode, *Meloidogyne enterolobii* occurring in Tamil Nadu, India. Int J Curr Microbiol App Sci 8(9):1987–1998

Tanwar RK, Jeyakumar P, Monga D (2007) Mealybugs and their management. TEC Bull 19:1–16

Tanwar RK, Jeyakumar P, Vennila P (2010) Papaya mealybug and its management strategies. Tech Bull 22 National Centre for Integrated Pest Management, New Delhi

Thakur JR, Dogra GS (1980) Woolly apple aphid, *Eriosoma lanigerum*, research in India. Trop Pest Manage 26(1):8–12

Tuhan NC, Pawar AD, Parry M (1979) Biological control of San-Jose scale, *Quadraspidiotus perniciosus* (Comstock) (Hemiptera: Diaspididae). Plant Protect Bull 31(4):73–79

Van Huis A (2007) Strategies to control the Desert Locust *Schistocerca gregaria*. In: Area-wide control of insect pests. Springer, Dordrecht, pp 285–296

Waterhouse DF, Norris KR (1987) Biological control: pacific Prospects. Inkata Press, Melbourne

Youssef NA (2014) Effect of certain entomopathogenic fungi and nematode on the desert locust *Schistocerca gregaria* (Forskal). Ann Agric Sci 59:125–131

Biopesticides 25

Sujoy Saha, Ratna Thosar, Sumant Kabade, Komal Pawar,
and Kaushik Banerjee

Abstract

Food production in a sustainable mode is input-intensive. The Green Revolution resulted in a large-scale use of chemical pesticides and fertilizers to escalate production but anthropogenic activities had negative impacts on the environment and ecosystems. In order to overcome these challenges and meet the requirements for food and supplies, the productivity and sustainability of agricultural practices should be improved and novel strategies must be found. Enhanced agricultural productivity can be achieved in many ways, such as through increasing crop yield by providing manure and organic-based treatments, including biopesticides, or by limiting yield loss due to extreme environmental conditions. Incidence of pest and diseases are the most limiting factors in achieving higher yield. In case of crop protection, development of pesticide resistance is one of the major issues caused due to the heavy and frequent application of fungicides. The alternatives of chemical pesticides became a mandate so that they can be used individually or in combination with standard pesticides for the management of pests. Biopesticides, which are pest management agents based on living microorganisms or natural products, offer a great promise in controlling yield loss without compromising the quality of the product. They are formulations made from naturally occurring substances that control pests by nontoxic mechanisms and in an eco-friendly manner. Biopesticides are cheaper and do not leave a massive mark on natural resources. Eco-friendliness, target specificity,

S. Saha (✉) · R. Thosar · S. Kabade · K. Pawar
ICAR-National Research Centre for Grapes, Pune, Maharashtra, India

K. Banerjee
National Research Centre on Grapes, Pune, Maharashtra, India

P. K. Ghosh et al. (eds.), *Trajectory of 75 years of Indian Agriculture after Independence*, https://doi.org/10.1007/978-981-19-7997-2_25

biodegradability, and requirement of minimal quantity are the striking characteristics of biopesticides. In India, due to rising popularity for organic farming and awareness of the harmful effect of chemical pesticides, the demand for biopesticides had been increasing with time. This chapter encompasses the information regarding biopesticides with respect to its scope, registration process, formulation mechanisms, government initiatives, and policy perspectives in India.

Keywords

Microbes · Bioefficacy · Formulation · Safety · Registration

25.1 Introduction

Agriculture has been the backbone of the Indian economy and it will continue to remain so for a long time. Indian agriculture sector accounts for 18% of India's gross domestic product (GDP) and provides employment to 50% of the countries work-force. India is the world's largest producer of pulses, rice, wheat, spices, and spice products (Madhusudhan 2015).

India is an agricultural country and so the agriculture was the main source of national income at the time of Independence. Hence, the agricultural development was the essential part of overall economic development. The chronology of events needs an understanding for essentiality of crop protection vis-à-vis biopesticides.

25.2 Pre-independence and Independence Era

The agricultural sector in pre-independence India was very disposed to disintegrating and highly unstable. There was no steady production rate for various crops. The main reason for this was improper infrastructure for perennial irrigation (https://www.padmad.org/2017/03/state-of-indian-agriculture-farmers-in.htm).

In the mid-1911–41, population growth was higher than agricultural growth, indicating declining per-capital food availability (0.72% per annum). A decline in foodgrains was around 1.14% per capita per annum, while the sector of cash crops increased at 0.57% per person per annum. Therefore, it shows that condition of agriculture was very poor in the pre-independence India (https://abhishekdudhal.wordpress.com/indian-economy-pre-independence-era-agriculture).

At the time of independence, around 72% of the total working population was engaged in agriculture and this indicated that Indian economy was undeveloped. After 61 year of independence, the share of agriculture in total national income declined from 50% in 1950 to 18% in 2007–08 (Tripathi and Prasad 2010). Around 50 years before the independence, the growth rate of Indian agriculture was about 1% per annum, which was then increased about 2.6% per annum in the post-independence era. In the period after independence, India has made enormous progress with respect to food security. India was depending on food aid to meet national requirements before the mid-1960s. After that, India suffered from severe drought in 1965 and 1966, which influenced the government to modify its agricultural policies and that it could not depend on overseas aid and imports for food security. Previously, India implemented significant policy developments focused on the goal of food grain self-subsisting. This facilitated to India's "Green Revolution." It began with the decision to implement superior, high yielding, and disease-resistant wheat varieties in combination with better farming knowledge to improve productivity. There was a need of site to try such a new crop. For this purpose, the Indian government selected "Punjab" state, as it has a reliable water supply and a history of agricultural success. The cascade of events that followed ushered a new era of agricultural triumph.

25.3 Pesticide Consumption Trend Since 1950

The cultivated crops would suffer from pests and diseases affecting yield loss. There was thus a great incentive to find ways of overcoming the problems caused by pests and diseases. The first recorded use of insecticides is about 4500 years ago by Sumerians who used sulphur compounds to control insects and mites, while about 3200 years ago the Chinese were using mercury and arsenical compounds for controlling body lice (https://agrochemicals.iupac.org/index.php?option=com_sobi2&sobi2Task=sobi2Details&catid=3&sobi2Id=31).

The Green Revolution resulted in a large-scale use of pesticides and synthetic nitrogen fertilizers, giving rise to improved irrigation tactics and crop varieties. The main objective was to gain food security through scientific methods (Behal 2020). Since the past four decades, the demand of agrochemical for pest control (i.e., pesticides) has been increased rapidly and these pesticides dominated the traditional methods used to reduce crop damages due to insect pests, diseases, and weeds. Though the pesticides contributed significantly to food security by reducing the postharvest losses, it has potential ill-effects on human and environmental health, water pollution, and food contamination.

For the purpose of increasing agricultural production, there are instances of usage of pesticides in an uncontrolled manner, which directly affected agriculture, human health, animals, and environment. This also affected the fertility of agricultural lands. Moreover, there is reduction of nutrient content in agriculture products. There is a direct as well as indirect effect of pesticides on human health. The cases of irritation on eye and skin, weakness, paralysis, cancer, and many diseases are increasing continuously. Hence, the use of pesticides should be according to scientific guidelines. In international trade, none of the pesticides that are genotoxic in nature (which damages DNA and cause mutations or cancer) are approved for use on food production and processing. Adverse effects from these pesticides occur only above a certain safe level of exposure.

As soon as people come into contact with pesticides from diverse sources, it may cause acute poisoning or long-term health effects, including cancer and adverse effects on reproduction (WHO 2018). There are some examples of acute health effects, e.g., stinging eyes, rashes, blisters, blindness, nausea, dizziness, diarrhea, and death. Chronic effects are likely to cause cancers, birth defects, reproductive harm, immunotoxicity, neurological and developmental toxicity, and disruption of the endocrine system (https://www.pesticidereform.org/pesticides-human-health). There are some groups of pesticides that are extremely hazardous to human body and cause chronic effects. Such group consists of organophosphates and carbamates, which are like nerve gas and they attack and damage the nervous system.

25.4 Pesticide Residue and Its Impact

The impact of pesticides is manifold, which can be briefly summarized in the following sections.

25.4.1 Water Pollution

Water is one of the most treasurable natural resources in the world. All living organisms depend on water throughout their life. Water resources are under risk from water pollution, water scarcity, and water conflict. In recent years, there have been concerns about pesticides entering both surface and ground water in various countries. Both surface water and ground water need to be protected from the introduction of pesticides. Contamination of surface water is less serious than ground water. Most of the surface water bodies except deep lakes have a rapid turnover rate as it dilutes the concentration of the contaminants rapidly (Aydinalp and Porca 2004). Groundwater is commonly used for the purpose of irrigation and drinking. Therefore, groundwater contamination by pesticides is becoming much more serious problem as pesticides do not degrade there as rapidly as in other environmental components. There are many studies that examined the effects of pesticide residues on water bodies. For example, Agarwal et al. (2015) revealed chlorinated hydrocarbons, organophosphates, and their derivatives as the major contaminants since they persist long in the environment.

25.4.2 Food Contamination

Pesticide residues include the fraction, which may remain on or in food after they are applied to food crops. The extreme permitted levels of these residues called as Maximum Residue Limits (MRLs) of foods are regularly stipulated by regulatory bodies. In September 2008, the European Union (EU) issued new and revised MRLs for roughly 1100 pesticides ever used in the world.

In fruits and vegetables, most of the pesticide residues are retained on peel surface. Pesticides may be applied on fruits and vegetables during different agricultural practices such as before blooming, growing, and after harvesting (Bajwa and

Sandhu 2014). The humans and animals get affected by the presence of pesticide residues in their food and feed, respectively. Therefore, the reason of pesticide residue found in dairy products is their feed, which is contaminated with pesticide. In addition, the pesticides from the environment can be transferred to milk. Contamination of milk and dairy products with pesticide residues is a matter of serious public health concern, since they are widely consumed by infants, children, and adults (Bedi et al. 2018).

The exposure to synthetic pesticides might have relations with a higher risk of Parkinson's disease and could alter specific genes involved in its development (https://www.healthline.com/nutrition/pesticides-and-health). Pesticides are generally used in modern agriculture and food production at pre- and postharvest stages. A considerable fraction of pesticide residues is expected to degrade during processing, although in certain processes, e.g., dehydration, the residues might get enhanced. In addition, there are some tactics to reduce the exposure of pesticide residue loads in food, such as cooking, washing, peeling, and so on, that either degrade/dissipate or physically remove the residues.

25.4.3 Human and Environmental Health

Globally, one of the major concerns of consumers today is the presence of pesticide residues in food. These residual chemicals might be persistent in nature and cause serious lethal effects on human health when exposed beyond certain levels. Cereal crops are frequently attacked by insect-pests during storage, and the insecticides are sprayed to preserve cereal grains for more than 1 year without any attack by pests and insects (Inobeme et al. 2020). Several studies have reported that pesticide residues are mainly found in the outer covering of cereal grains. Hence, processing methods such as milling and related processes are practiced, which can bring down the load of pesticide residues in cereals. Some specific pesticides such as cypermethrin is found in seed coats of pulses. Occasionally, insecticides and other pesticides are noticed to have contaminated various milk and related products originated from fodder and feed that remained linked to the fat portion. Sometimes, mycotoxins such as aflatoxins (AF) also contaminate the cereals, spices, and milk products. It was found that when animals consume the AF-contaminated feed, the milk might detect for aflatoxin M1, which is the hydroxylated metabolite of aflatoxin B1 (Murshed 2020).

The detrimental effects of chemical pesticides on different domains of life instigated thoughts of alternative uses in the period following the Green Revolution.

25.5 Post Green Revolution Era

The Green Revolution was initiated in the 1960s to address the issue of malnutrition in the developing world. The technology of the Green Revolution involved bioengineered seeds that worked in conjunction with chemical fertilizers and heavy irrigation to increase crop yields. The technology was readily adopted in many states in India, and for some people it was a great success; however, it has unintended, harmful consequences on agriculture and human health. Development of pesticide resistance is one of the major issues caused due to the heavy and frequent application of fungicides.

25.5.1 Pesticide Resistance

Resistance is the genetic ability of some individuals to survive an application or applications of pesticides. The pesticide(s) no longer effectively kills a "high" number or percent (>90%) of individuals in the pest population. Resistance develops at the population level and is an inherited trait. The surviving pests can genetically pass such traits to their next generation, thus enriching the gene pool with resistant genes. In addition, resistance indicates a change in the genetic composition of pest population in response to selection by a pesticide over time.

India was one of the first countries to initiate a large-scale use of synthetic pesticides for the control of insect pests of public health and agricultural importance. The modern era of vector control and plant protection in India started with the introduction of DDT in 1947 (Mehrotra 1985a), followed by HCH in 1949 (Mehrotra 1985b), organophosphates in 1953, and carbamates a little later (Banerjee 1979). Though these pesticides have brought immense benefits to the country, they also had environmental consequences (Mehrotra 1983, 1989a, b).

The pesticide (DDT) resistance in India was first noticed in Uttar Pradesh and Bombay (now Mumbai) in insect pests of public health importance, namely mosquitoes transmitting malaria and Culexfatigans, a transmitter of filaria, respectively, in 1952 and the concern about it led to an International Conference organized jointly by the World Health Organization and the Government of India in 1958 at New Delhi (Pal et al. 1952; Anonymous 1958). This was because of the large scale use of DDT in the National Malaria Control Programme/National Malaria Eradication Programme. In case of agricultural pest, Singhara beetle, *Galerucella birmanica*, resistant to DDT and HCH was reported from Delhi region in 1963 (Mehrotra 1985a). Pesticide resistance in stored grain pests was first reported in flour beetle, *Tribolium castaneum*, in 1971 against DDT and malathion from Delhi (Bhalia 1971; Rajak et al. 1973).

25.5.2 Reasons for Developing Pesticide Resistance

1. Frequent application of pesticides with the same mode of action increases the "Selection pressure" for resistance.
2. Recombination of genes at the time of sexual reproduction in fungus.
3. Plant pathogens with very broad host range, having more chances of developing resistance.

4. Use of lower doses of pesticides than the lethal dose builds the proportion of the population with a higher level of resistance.

Use of narrow mode of action product, application of tank-mixing, or alternating products with the same mode of action and using a product with multiple modes of action, which can come from a single active ingredient or a premix of two or more active ingredients, are the viable options for avoiding the resistance development.

The alternatives of chemical pesticides became a mandate so that they can be used individually or in combination with standard pesticides for the management of pests.

25.6 Biopesticides

Biopesticide is a formulation made from naturally occurring substances that controls pests by nontoxic mechanisms and in an eco-friendly manner (Kumar 2012). Biopesticides may be derived from animals (e.g., nematodes), plants (e.g., Chrysanthemum, Azadirachta), and microorganisms (e.g., *Bacillus thuringiensis, Trichoderma*, nucleopolyhedrosis virus), and include living organisms (natural enemies), their products (phytochemicals, microbial products), or by-products (semiochemicals), which can be used for the management of pests injurious (Mazid et al. 2011).

At present, India ranks 12th in the world for the use of pesticides, with an annual production of 90,000 tons. A vast majority of the population in India (56.7%) is engaged in agriculture and is therefore exposed to the pesticides used in agriculture (Government of India 2001; Gupta 2004). Exposure to pesticides either occupationally or environmentally causes a range of human health problems (Indira Devi 2007). Approximately, 10,000 deaths reported annually due to use of chemical pesticide worldwide, with about three-fourths of these occurring in developing countries (Horrigan et al. 2002). Harmful health hazards and tremendous harm to the environment occurs due to inappropriate and excessive pesticide use and application. Mostly, the chemical pesticides are not easily degradable and hence linger in soils, leach to the ground water, and consequently enters the food chain and affect the human health indirectly. Scientific reports has confirmed the direct and indirect relationship between exposure of pesticide use and several illness signs (Kishi et al. 1995). The health effects of pesticide use is a serious public health concern during the past decades. In 1962, Rachel Carson first mentioned the sudden death of birds caused by indiscriminate spraying of pesticides (DDT) in her book *Silent Spring* (Roy 2016). Indiscriminate use of pesticides leads to suffering from different types of health problems such as eye irritation, fatigue, skin irritation, back pain, headache, vomit, and dizziness in farmers of Kullu and Shimla districts of Himachal Pradesh, and also death of insects and bees (Kumari and Sharma 2012). Most of the chemicals are mutagenic, linked to the development of cancer, birth defects, chromosomal aberration, and so on (Galloway et al. 1987; Leiss and Savitz 1995; Colborn 1996; Arbuekal and Server 1998). Unbridled use of pesticides develops resistance in pests and plant diseases to many chemical pesticides, so alternatives are necessary. At the same time, the public is increasingly demanding chemical-free crops as awareness about the health implications of ingesting chemicals grows.

Biopesticides are cheaper and do not leave a massive mark on natural resources. Eco-friendliness, target specificity, biodegradability, and requirement of minimal quantity are the striking characteristics of biopesticides (Singh and Kumar 2021). Efficient use of biopesticides by the end users lowers the residues of chemical pesticides and helps in eliminating contamination problems associated with it. Biopesticides may either directly kill harmful organisms or interfere with reproduction or simply repels pests with substances they do not like and hence suppressed pest population so that they can be managed over time. Its multisite mode of action behavior makes biopesticides less vulnerable to pests-building resistance. The use of biopesticides improves root and plant growth by enhancing the soil microflora and results in increased agriculture production. They have a zero or low reentry and handling interval. Some microbial biopesticides can reproduce on or near to the target pest/disease, giving some self-perpetuating control.

25.6.1 Biopesticide Consumption and Market

In India, due to rising popularity for organic farming and awareness of the harmful effect of chemical pesticides, the demand for biopesticide has been increasing day by day. For example, the demand for biopesticide increases from 10,447 MT (metric tons) in 2016–17 to 11,054 MT in 2020–21(http://ppqs.gov.in/statistical-database). According to the FAO, the country had 1.9 million hectares of land under organic cultivation in 2018, which was 0.2 million hectares more than the previous year. Such a vast organic farming industry is fueling the biopesticide market in India, and it is anticipated to continue in the future. As per the Indian Ministry of Agriculture, in the last 10 years, consumption of biopesticides increased by 23% (https://www.mordorintelligence.com/industry-reports/indian-biopesticides-market). However, in 2012–13 during the 15th Lok Sabha, the standing committee on chemicals and fertilizers mentioned in its report that the use of biopesticides had increased beyond the expectations from 123 MT in 1994–95 to 8110 MT in 2011–12 (Mishra et al. 2020).

Data obtained from the Directorate of Plant Protection, Quarantine & Storage (DPPQS), Government of India indicated that consumption of biopesticides has increased by 20% from 2016–17 to 2020–21 in India (Fig. 25.1). Rajasthan, West Bengal, and Maharashtra are the leading states with 1021, 1017, and 934 MT biopesticides consumption, respectively. This data also highlight a low reach of biocontrol programs in north-eastern regions of the country in comparison to southern (http://ppqs.gov.in/statistical-database).

A total of 970 biopesticides companies are currently registered with India's Central Insecticides Board and Registration Committee (CIBRC), which screens potential biopesticides for biosafety (Padma 2019). However, only 15 biopesticides have been registered so far, which include six bacterial, two viral, six fungal, and two plant products (Anonymous 2021). *Bacillus thuringiensis*, *Trichoderma viride*,

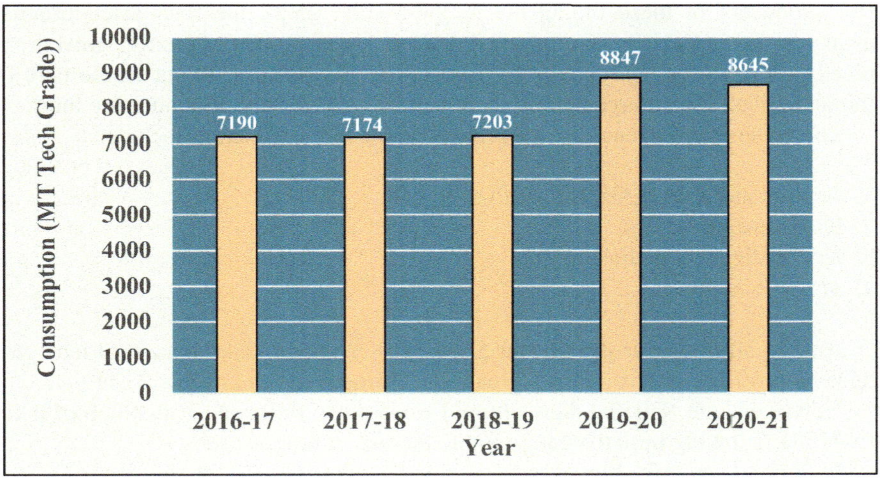

Fig. 25.1 Bio-pesticide consumption (MT) in India during 2016–17 to 2020–21

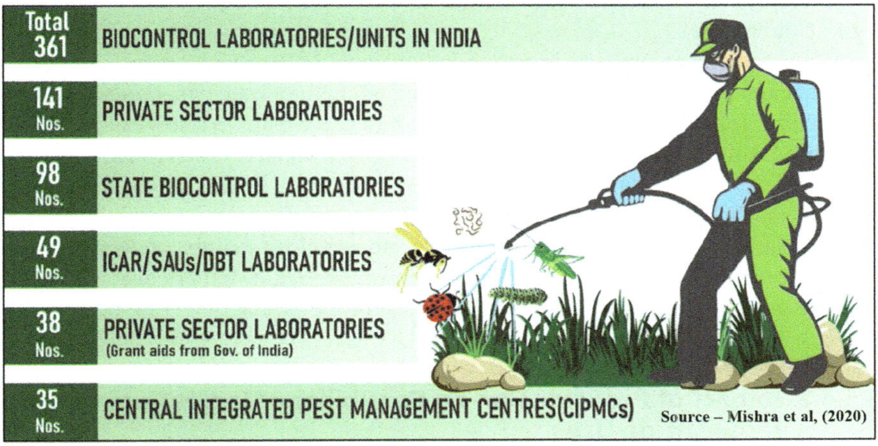

Fig. 25.2 An overview of the current structure of biocontrol laboratories and units working in India. (Source: Mishra et al. (2020))

Metarrhizium, and *Beauveria bassiana* nuclear polyhedrosis virus and neem are currently used to protect plants (Rao et al. 2007).

The current market for biopesticides in India is approximately INR 10.0 billion and estimated to grow at a CAGR of 18% per annum (Anonymous 2021). The share of biopesticides segments in Indian biopesticide industry comprise 66% fungal, 29% bacterial, 4% viral, and 1% other (plant-based pheromones) biopesticides. Mishra et al. (2020) mentioned some agencies involved in biopesticide production and marketing in India, which are given below. Although there are more than

350 producers of these bio-agents in India (Fig. 25.2), the formulated product available for farmers for various treatments is about 30,000 MT only. Total combined production is just enough to meet the requirement of 4% of arable land of Indian agriculture. A huge gap exists in demand and supply of quality products.

The government agencies involved in commercial production include:

1. Central and state agricultural universities.
2. ICAR institutes.
3. Krishi Vigyan Kendras (KVK).
4. State biocontrol lab.

Among all these, Krishi Vigyan Kendras (KVK) and state biocontrol labs produce biopesticide on local demand with government support (Table 25.1)

Nowadays, the National Agricultural Cooperative Marketing Federation of India (NAFED) is taking initiatives to promote the use of biopesticides.

In India, most of the biopesticide production takes place in public sector units. It is estimated that the public sector contributes to almost 70% of the biopesticides production. Major companies involved in production are Biotech International Ltd. (New Delhi), International Panaacea Ltd. (New Delhi), and T. Stanes & Company Ltd. (Coimbatore).

Some foreign companies in collaboration with Indian companies initiated the biopesticide production and have also entered in biopesticides production. For example, Lupin Agrochemicals, a Bombay-based company, is now working with US-based Abbott Laboratories. Sugar and distillery companies such as KCP Sugar & Industries Corporation Ltd. (Andhra Pradesh), Rajshree Sugars & Chemicals Ltd. (Tamil Nadu), New Swadeshi Sugar Mills (Bihar), and Bannari Amman Sugars Ltd. (Tamil Nadu) have also started to produce biopesticides mainly of *Pseudomonas fluorescens* and *Trichoderma harzianum*. However, lack of experience and quality control are the major issues still plaguing most of the products.

25.6.2 Impact of Pesticide Ban on Possibility of Increasing Biopesticide Use

While the country continues to battle Covid-19, plant diseases and pests continue to pose threat to food production and farmers' livelihood. New pests and diseases continue to surface on plants affecting production. Twenty-seven pesticides proposed for ban by government in May 2020 constitutes almost 25% of the total generic pesticide market in India. The pesticides along with their constituents from 134 formulations are registered to control diseases and other pests. Timing of the notification, however, has shaken the industry, farmers, and plant protection scientists equally, given the uncertainty due to Covid-19 crisis and also due to onset of the Kharif crop season when protection of crop from diseases and pest is considered crucial to ensure crop security. Series of discussions, deliberations, and suggestions are engaged into by industry, academia, and scientific associations for a

Table 25.1 Description of some common biopesticides

Sr. No.	Source	Type	Organism	Pest type	Target crop	Reference(s)
1	Bacteria	Insecticide	*Bacillus thuringiensis* var. *kurstaki*, *B. thuringiensis* var. *tenebrionis*	Caterpillars, fungi (*Botrytis*) elm leaf beetle, alfalfa weevil	Vegetables, fruits, ornamentals, cereals, potato	Koul (2011), Bravo et al. (2007) and Saberi et al. (2020)
		Fungicide	*Bacillus subtilis*	*Botrytis* spp.	Vegetables, fruits, and ornamentals	Koul (2011) and Bravo et al. (2007)
2	Fungi	Insecticide	*Beauveria bassiana*	Whitefly	Protected edible and ornamental plant production	McGuire and Northfield (2020)
		Fungicide	*Coniothyrium minitans*, *Trichoderma harzianum*	*Sclerotinia* spp., *S. sclerotiorum*	Outdoor edible and nonedible crops and protected crops Starwberry crops	Gams et al. (2004) and Dolatabadi et al. (2011)
		Herbicide	*Chondrostereum purpureum*	Cut stumps of hardwood trees and shrubs	Forestry	Bailey (2014)
		Nematicide	*Paecilomyces lilacinus*	Plant–parasitic nematodes in soil	Vegetables, soft fruit, citrus, ornamentals, tobacco, and turf	Moreno-Gavíra et al. (2020)
3	Virus	Insecticide	*Cydia pomonella* granulovirus	Codling moth	Apples and pears	Kadoi'c et al. (2020)
4	Oomycetes	Herbicide	*Phytophthora palmivora*	*Morenia orderata*	Citrus crops	Lala et al. (2019)
5	Neem (*Azadirachta indica*)	Insecticide	Azadirachtin	Aphids, scale, thrips, whitefly, leafhoppers, weevils	Vegetables, fruits, herbs, and ornamental crops	Chaudhary et al. (2017)

(continued)

Table 25.1 (continued)

Sr. No.	Source	Type	Organism	Pest type	Target crop	Reference(s)
6	Plant extracts	Fungicide	*Reynoutria sachalinensis* (giant knotweed) extract	Powdery mildew, downy mildew, botrytis, late blight, citrus canker	Protected ornamental and edible crops	Marrone (2002)
		Herbicide	Plant essential oils	Ragwort, many arthropods	Grassland	Isman (2020)
		Nematicide	*Quillaja saponaria*	Plant parasitic nematodes	Vineyards, orchards, field crops, ornamentals, and turf	Guerra and Sepúlveda (2021)
7	*Talaromyces flavus*; *Clitoria ternatea* (butterfly pea); *Trichoderma harzianum*; *bacillus thuringiensis* var. *tenebrionis*; *lactobacillus casei* fermentation products	Biopesticides	–	*Glomerella cingulata* and *Colletotrichum acutatum*, *Helicoverpa* spp., *fusarium oxysporum*, *Agelastica alni*, *Spodoptera litura*, *Helicoverpa armigera*, *Aphis gossypii*, *Xanthomonas fragariae*, *Spodoptera littoralis*, and others	Strawberry, cotton, gladiolus hybrids, alder leaf, hazelnut, and other economically important plants and trees	Ishikawa (2013), Mensah et al. (2014), Kirk and Schafer (2015), Eski et al. (2017), Dubois et al. (2017), Pavela et al. (2017) and El-Abbassi et al. (2017)
8	Semiochemical	Attractant	Citronellol	Tetranychid mites	Apples, cucurbits, grapes, hops, nuts, pears, stone fruit, nursery, and ornamental crops	Mauchline et al. (2018) and Mossa et al. (2017)
		Attractant	Multicomponent sex pheromone, such as (E,E)-8,10- dodecadien-1-ol	Codling moth	Fruits such as apples and pears	El-Sayed et al. (1999)

9	Arbuscular mycorrhizal fungi	Mutual inhabitant in the roots	Fungi	*Fusarium verticillioides*; pathogens affecting below ground plant organs	*Zea mays*	Olowe et al. (2018), Bharadwaj and Sharma (2006) and Mukerji and Ciancio (2007)
10	Microalgae	Filamentous cyanobacterium; single-celled green algae	*Nostoc piscinale*, *Chlamydopodium fusiforme*, *Chlorella vulgaris*	–	–	Ranglová et al. (2021)
		–	*Anabaena laxa*, *Calothrix elenkinii*	Increase in fungicidal activity	Coriander, cumin, and fennel	Kumar et al. (2013)
	Nanobiopesticide	Silver nanobiopesticide	None	*Alternaria alternata*, *A. solani*	*Alternaria* leaf blight and leaf spot diseases in tomato, pepper, and potato	Narware et al. (2019)
		Sargassum muticum-derived NPs	None	*Ariadne merione*, a lepidopteran pest	–	Narware et al. (2019) and Rodrigues et al. (2019)
		Caulerpa scalpelliformis and *Mesocyclops longisetus*-derived NPs	None	*Culex quinquefasciatus*	–	Narware et al. (2019)

Source: Kumar et al. (2021)

science-led decision instead of imposing irrational blanket ban endangering food security and raising cost of cultivation. During last more than five decades, the area under bio-intensive IPM remained below 3% with biopesticides constituting only around 3% of pesticide market in the country. The present volume of around 27,000 tons of licensed production of registered biopesticides is a miniscule quantity for a large arable area of 142 million hectares in India. Nevertheless, biopesticides cannot qualify as alternative to chemical pesticides but can serve as an important component of disease management. Recently, a gazette notification was released about the use of biocontrol agents where some of the genera of biocontrol agents were recently notified as biostimulants by the Government of India (Gazette Notification 2021). Therefore, chronic toxicology tests may not be required for these formulations to enable easy and speedy registration of these bio-products.

25.6.3 Registration

In India, the registration of biopesticides is regulated by "The Insecticide Act, 1968" (amended in 2000). This Act constitutes Central Insecticides Board (CIB) and Registration Committee (RC) under Sections 4 and 5, respectively. This Act came into force to regulate the import, manufacture, sale, transport, distribution, and use of insecticides including biopesticides, with a view to prevent risk to human beings or animals and for matters connected therewith (https://legislative.gov.in/sites/default/files/A1968-46.pdf). For import or manufacturing of any pesticides, one should apply to the Registration Committee. Two types of biopesticides included in schedule to the Insecticides Act, 1968.

1. Microbial pesticides:
   This includes antagonistic fungi and bacteria (e.g., *Bacillus species*, *Pseudomonas species*, *Trichoderma species*), entamopathogenic fungi (e.g., *Metarrhizium anisopliae*, *Beuveria bassiana*, *Verticillium lecanii*), and viruses (e.g., Nuclear Polyhedrosis virus (NPV), Granulosis viruses). A total of 38 microbial pesticides are included in the Insecticide Act, 1968.
2. Plant origin pesticides:
   This includes neem-based pesticides, pyrethrum, karanjin, eucalyptus leaf extract, and so on.

   In 2015, the government also passed a bill known as the Insecticides (Amendment) Bill, 2015. The Bill added a modification in Section 9 of the Insecticide Act (1968), inserting sub-sections of nanotechnology-based pesticides after sub-Section (3C).

In exercise of the powers conferred by Section 36 of the Insecticides Act, 1968 (46 of 1968), the Central Government, after consultation with the CIB made insecticides Rules, 1971, which governs the manufacture, grant of a license, expiry of the license, product labeling, packaging and sale, and use of insecticides. The RC grants registrations, only after the data is provided on the efficacy and safety of products to human beings and animals. The rule also assures that the samples of pesticides should be regularly checked for quality purposes. In the case of biopesticides, the following criteria are taken into consideration.

25.6.3.1 Chemistry and biological characteristics

It includes systemic name, strain, detail composition, and specifications of the product, test procedure, contaminants, manufacturing process, shelf life, and coding (in case of formulation)

25.6.3.2 Bioefficacy

In this aspect, data obtained by performing laboratory tests, field trials with proposed formulation on ICAR, SAU, CSIR, ICMR institute, phytotoxicity data, and two-year or two-season data on effect of nontarget organisms, i.e., natural parasites and predators, is required.

25.6.3.3 Toxicity

Toxicity data on single exposure studies, skin and mucous membrane irritation, toxicity to birds, fish, honey bees, silk worm, and earthworm with formulation is required.

25.6.3.4 Packaging

Type and manner of packaging, container contents compatibility, labels and leaflets related to the product and instruction for storage, and use and disposal of empty containers should also be mentioned.

25.6.3.5 Registration

There are three types of registrations present in India:

- *Regular (original) Registration u/s 9(3)*—after submission of complete data to the satisfaction of the Registration Committee.
- *Provisional Registration u/s 9(3b)*—if an insecticide is being introduced for the first time in India, Registration Committee may register it provisionally for a period of 2 years.
- *Repeat or 'Me Too' Registration u/s 9(4)*—on the basis and same conditions as those of original registration. This is for product and source, which are already registered in India. Section 9(4) is the easiest method of registration and if all the documents submitted to CIB are in proper order, then registration is issued within 6 months of submitting the application.

In today's scenario, online registration of pesticides is also possible. Computerized Registration of Pesticides (CROP) is a web-based Application for Registration of Pesticides developed for the Central Insecticides Board and Registration Committee. The application is supposed to automate the entire registration procedure.

The highlights of the application are summarized below:

- Online Application for Registration of Pesticides under Sections 9(3) and 9(3b).
- Online status checking.
- Frequent email notification to the applicant at certain milestone of the registration process till Deficiency Reporting and Approval.
- Online application for the grievances.

Efforts have been made by Indian government to root out spurious and unregulated biopesticides in market and to promote the use of biopesticides in crop protection. To ensure the availability of quality product from genuine source and proper registration of the product, the Government of India through its regulatory body, CIBRC, introduced new registration guidelines in 2017. According to this guidelines, the following registration certificates (RCs) are required for sale of agrochemicals in India.

1. Technical Indigenous Manufacturing Certificate (TIM).
2. Technical Import (TI).
3. Formulation Indigenous Manufacturing (FIM).
4. Formulation Import.

Steps of Registration Process

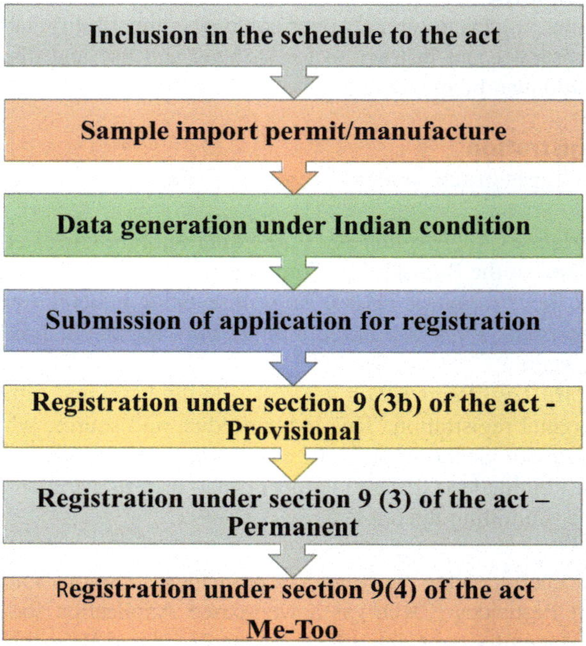

Source: https://www.slideshare.net/subhashchaudhary2

Undoubtedly, the new registration guidelines make the TI registration process more expensive and time-consuming. However, the new guidelines are well defined and reduce confusion about Indian registration process. However, there are still areas of improvement and CIB is already working in making registration process more transparent and smoother. For companies planning to enter India and invest money in registrations, this is the right time to do so. With transparency and clear requirements, it has become easier for genuine sources to get TI registration, and with stringent guidelines it has become more difficult for illegitimate sources to register new products or renew old registrations.

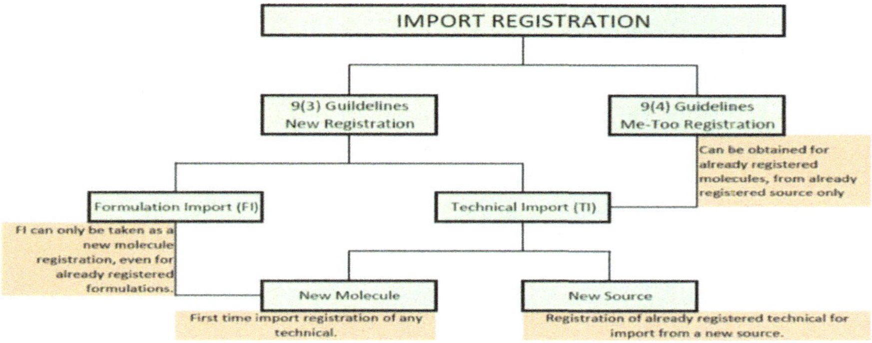

Source: http://www.dextrainternational.com

Based on the guidelines of the Organization for Economic Co-operation and Development (OECD), the CIB has streamlined the guidelines and data requirements for registration and minimum infrastructural facilities required for the production of biopesticides. Guidelines and data requirements for minimum infrastructure facilities and the same for the registration of biocontrol products under Sections 9 and 9(B) are being governed by RC of CIB.

25.6.4 Formulation Mechanism

Formulation refers to the preparation of a product from an active ingredient by the addition of certain active (functional) and non-active (inert) substances (Grewal 2005). Formulation of biopesticides comprises the blending of a biologically active metabolite or microorganism with a carrier or solvent, as well as adjuvants to develop a product, which can be delivered to the target weed, pathogen, or pest using existing equipment. Formulated organisms are suspended in a suitable carrier, which is supplement by additives to maximize survival in store, optimize application to the target, and protect the organisms after application. Biopesticides are live and proteinous, hence sensitive to environment and storage conditions.

Function of Formulations

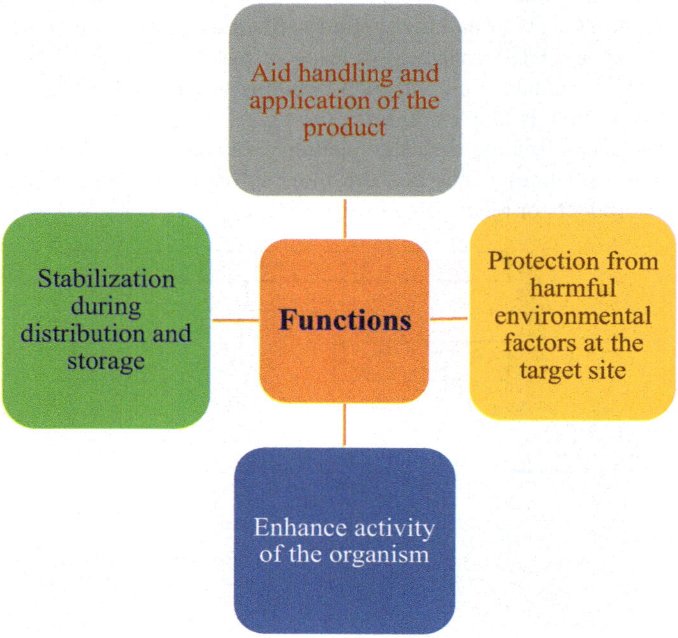

Regarding their physical state, biopesticide formulations can be divided into two categories:

Liquid formulations
Dry formulations

Liquid formulations can be water-based, oil-based, polymer-based, or combinations. Water-based formulations (suspension concentrate, suspo-emulsions, capsule suspension, etc.) require adding of inert ingredients, such as stabilizers, stickers, surfactants, coloring agents, antifreeze compounds, and additional nutrients.

Dry formulations can be produced using different technologies, such as spray-drying, freeze-drying, or air-drying either with or without the use of fluidized bed. They are produced by adding binder, dispersant, wetting agents, etc. (Tadros 2005; Brar 2006; Knowles 2008).

Types of Formulations Available

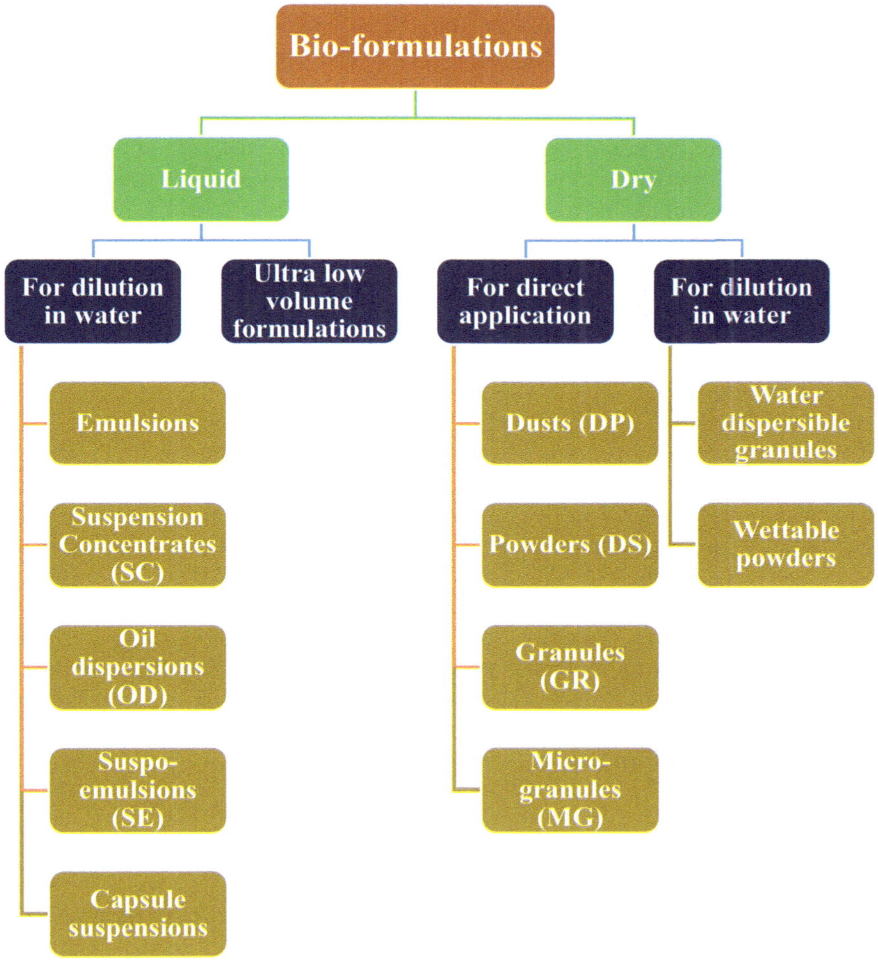

25.6.5 Requirements and Challenges for Biopesticide Formulation

The biological activity or mode of action of a pesticide is determined by its biologically active microorganism or metabolite. The active ingredient in its raw form requires formulation to facilitate its handling, storage, and application. A challenge for the formulation of microbial biopesticides is the presence of additional fermentation materials besides the microbes themselves. These materials may adversely react with formulation ingredients. Another challenge facing microbial biopesticide formulators is the tendency for gram-positive bacteria such as *Bacillus* spp. to produce a slime layer in connection with endospore formation. Suspension of these bioformulations for spraying can lead to excessive sliminess and the blockage of sprayer filters and nozzles. While the slime layer is removable, e.g., by centrifugation, it may contain valuable components such as lipopeptides, which may contribute to the biopesticidal effect. Shorter shelf life of biopesticides is the challenge in biopesticides use. Significant formulation efforts are being targeted at the long-term stabilization of microbial and other biopesticides. Shelf life of 1–2 years is desirable for biopesticides. The correct storage of these products is vital especially for non-spore-forming, gram-negative bacteria. Shelf life and viability may be extended by reducing storage temperature through freezing or refrigeration. While refrigeration is not an option in most cases, exceptions do exist for high-value crops, and there appears to be a growing tendency for growers to treat biopesticides as perishable high-value products. Microorganisms may be frozen or refrigerated in culture broth or an appropriate buffer. However, depending on the microorganism, storage in culture broth may impact long-term cell viability, and extensive testing and development in this area is currently a priority. Besides storage, longer shelf life may be obtained by increasing the number of microorganisms in the product to ensure viability despite a subsequent decline in their population. Fermentation parameters such as the rate of dehydration and the physiological state of the bacteria on harvest from the medium will also affect the shelf life of the final formulation.

Semiochemicals are difficult to formulate due to their volatility. Their encapsulation in traditional polymer matrixes or natural biological microcapsules such as yeast cells may reduce their volatility and degradation by facilitating their controlled release to the environment. However, as microcapsules are about the same size as pollen grains, microencapsulated biopesticides may pose a hazard to foraging bees. These may carry the microcapsules back to their hives where the biopesticide is released, with the risk of harming the bees.

25.6.6 Government Initiatives and Policy Perspectives

For the first time in India, the National Agriculture Policy (2000) was announced on July 28, 2000. The policy emphasized on adequate and timely supply of certified seeds, fertilizers, plant protection chemicals, and biopesticides to farmers. For the systematic development of organic agriculture in the country, the National Programme for Organic Production (NPOP) was launched in the year 2001 by the Agricultural and Processed Food Products Export Development Authority (APEDA) under the Ministry of Commerce and Industry, Government of India. The institutional framework for accreditation and certification of organic agriculture was the major highlight of NPOP, which gained recognition agreements with the European Union and the United States Department of Agriculture (USDA) (NPOP 2018). Further, for strengthening the wave of organic mission and minimizing the risk of chemicals, the Department of Agriculture & Cooperation, Ministry of Agriculture introduced Organic Farming Policy in 2005. The policy included organic sources of nutrients such as biofertilizers, organic manures, compost, and biocontrol agents (biopesticides) as certified inputs for organic farming. National Bank for Agriculture and Rural Development (NABARD) also started its capital investment subsidy scheme "Strengthening and Modernization of Pest Management Approach in India" (SMPMA), which provided financial assistance for setting up of biofertilizers/biopesticides units as back-ended subsidy of 25% restricted to 4 million rupees. The National Mission for Sustainable Agriculture (NMSA) under the National Action Plan on Climate Change (NAPCC), which was related to the issues of "Sustainable Agriculture," was launched in the year 2010. The third mission intervention of NMSA was related to pest management, and it emphasized incentivizing research, commercial production, and marketing of biopesticides. The major focus was to develop new biopesticides and technologies by incorporating sterile insect techniques, transgenic insects, applications of novel botanicals, semiochemicals, and endophytic microbial metabolites for pest control and disease forecasting (NMSA 2010). Apart from these, Soil Health Management (SHM) and "Paramparagat Krishi Vikas Yojana" (PKVY) have also been initiated for promoting organic farming through the adoption of organic villages by cluster approach and Participatory Guarantee System (PGS) certification (Reddy 2018). In the PGS, farmers pledge that their produce is free from all types of synthetic chemicals such as fertilizers, pesticides, and hormones, which is done with a self-regulatory support system that includes a local group of five or more organic farmers. The quality assurance standards of produce are harmonized by the PGS Organic Council, which permits the use of its PGS label on a product as a mark of quality (https://www.pgsorganic.in). In the last 5 years, the government has also taken appropriate steps for promoting nationwide use of biopesticides.

The "Zero-Budget Farming," which has got huge success in southern India, is already being practiced in some other states of the country. The farming practice is also stated as Zero Budget Natural Farming (ZBNF) by the Food and Agriculture Organization of the United Nations (UNFAO) and stresses on bypassing the unnecessary burden of farming inputs such as procurement of high-cost seed, chemical fertilizers, and pesticides (https://www.fao.org/agroecology/detail/en/c/443712/). Instead of using such expensive tools, it encourages the use of farmer's own seeds and locally available natural fertilizers and biopesticides for organic farming.

ICAR institutes in the country have established Institute Technology Management Units (ITMU) since 2008, which are supported by five Zonal Agro-Technology Management Centers (ZTMCs) at the zonal level and an Intellectual Property and Management Unit at the ICAR headquarters in New Delhi, governing overall IP, technology commercialization, and policy issues [40]. Public sector organizations in India, including ICAR, license technologies to stakeholders on "nonexclusive basis" only. These ITMUs at all levels have been given the responsibility of facilitation of commercialization of technologies.

25.6.7 Role of Concerned Authorities/Retailers and Growers

Due to the subsidy element/incentive on conventional pesticides, the existing agro-industry is reluctant to undertake research and production of biopesticides. However, in the last few decades, restrictions on heavy usage of chemical pesticides and phasing out and banning of a few dangerous chemicals aggravated pressure on manufacturers and big industry houses to produce biopesticides for commercial applications. However, the percentage share of biocontrol products is still far below in comparison to chemicals. Measures such as training to the potential entrepreneurs, provision of institutional credit, subsidies, insurance, and exemption from taxes and duties can stimulate the production of biopesticides. Promotion by the government for the use of biopesticides and declaration of no pesticide zones may also improve the situation in favor of the bio-products. For example, guidelines laid down in the NPOP relating to the organic mission were more critically implemented in "Sikkim Organic Mission" (SOM), and by converting around 75,000 hectares of agricultural land, it is now India's first organic state (http://www.sikkimorganicmission.gov.in/about-us/organic-movement/). By studying the SOM model, it was observed that authorities and growers were advised to use organic inputs instead of synthetics. Now, states of Mizoram, Arunachal Pradesh, and Kerala are trying to follow the same model for becoming organic.

25.6.8 Future Directions

The gap in knowledge, skill, and attitude of farmers toward biopesticides has to be comprehensively addressed. The state agriculture and horticulture departments have to include these biopesticides and biofertilizers in their package of practices. The frontline extension staff of state agriculture and line departments has to be oriented toward these eco-friendly products and suitably address farmers' queries and apprehensions. Extension folders and ready-to-use study materials and reference materials have to be prepared and made available in all Indian languages. Orientation of academia and researchers also needs to be changed from chemical intensive agriculture to use of alternative sustainable methods and products. The students and agriculture graduates have to be trained on use of these green products through inclusion in course curriculum at different levels. A diploma course on these bio-products at SAUs may create a batch of certified para-extension professionals. A fundamental shift in focus at the universities and R & D organizations governance level from an education-, research-, and publication-based reward system to a commercial activity-based one, which entails patenting and licensing of technologies, is required.

Government may also promote start-ups on production and sale of already registered biopesticides. Several ICAR institutes and a few SAUs have recently established Agri-Business Incubator Cells (ABIs) to promote start-up agribusiness ventures and commercialization with technological backstopping. However, these ABIs are also managed by in-house scientific/technical experts. Management experts need to be roped in to run/assist these units on business model. ABIs can act as bridge between R & D institutions and entrepreneurs/start-ups/industry to promote these technologies.

Governments have to identify regions/areas niche for promotion of organic agriculture and these biopesticides may be promoted as supplementary and complimentary inputs in a big way in these areas. For instance, arid and semi-arid regions, where the use of chemical inputs is relatively low, may be targeted for promotion of these biopesticides. Arid western Rajasthan is a suitable example in this case.

The production of the already registered biopesticides has to be decentralized using licensing/sublicensing and other mechanisms. Small scale industries, farmer producer organizations, and innovative farmers have to be involved for enhancing large scale production and diffusion. Decentralized production and sale points will address the issue of short life span of these biopesticides and enhance the confidence and trust of the farming community. Social certification by the farming community also acts as a kind of quality control measure.

The R & D institutions have to collaborate with other stakeholders/organizations from early stages of R & D of product development to develop the data required by the regulatory agencies related to bioefficacy, toxicology, biosafety, and environmental concerns. This will help the potential industries who acquire new technologies to register the products and in production and marketing.

ITMU cells are managed by in-house scientific/technical experts. The ITMU staff has to be trained or oriented on the regulatory issues or business professionals have to be hired to run/assist these ITMUs.

The benefits of biopesticides have to be looked not just at farmers' level but also at the broader levels of sustainable eco-friendly agriculture/agroecosystem and ecosystem services.

In India, research on biocontrol is still having some fundamental issues, hence lacking in the development of quality products. Field testing, registration, and licensing itself take a lot of time before a new product is commercialized and available for field application. Hence, the current system of registration needs to be streamlined in a way that it should favor the registrants. Similarly, the requirement of the budget for fulfilling all these formalities is also too high, which restrains an entrepreneur from investing. The provision of tax holidays, easy loans, and small incubation center/start-ups at regional levels should be introduced for strengthening financial support and marketing of the product.

For making the R & D system more robust, schemes and budgetary support are needed from the government or big industrial houses. Being low in pollution index and carbon footprint, these bio-products should be encouraged in comparison to harmful chemical pesticides.

It has also been noticed that in India most of the company's manufacturing biopesticide products contain only single microbial strain. These products have limited applications and can only be used against phytopathogens within a narrow range. Building products in a broad portfolio (such as using a consortium of efficient strains), which are broad-spectrum covering several phytopathogens, is going to be a better strategy and such products will be economical and have more trust among the end users.

Shorter shelf life of active ingredients in biopesticides has remained a major concern, which also affects commercial production. Shorter shelf life not only reduces the efficiency of biopesticides against the target pest but also impacts their competitiveness with chemicals. Technological breakthroughs in this segment are urgently needed. There are some reports where the coating of active ingredient with suitable biopolymer enhanced the shelf life of formulation (Aziz Qureshi et al. 2015; Sharma et al. 2019). Arora and Mishra (2016) also suggested that adding some additives such as secondary metabolites, precursor molecules along with suitable carriers may extend shelf life and activity of biocontrol products. However, unfortunately, the commercialization of these techniques has not been done so far (Mishra and Arora 2018).

While talking in terms of patent landscaping, India is far behind other countries where biologicals are mainly used for crop protection (Mittal and Singh 2006). In India, research related to nanotechnology-based biopesticide formulations is only at its early stage, and commercially none of the nanotechnology-based biopesticides have been developed (Chhipa and Joshi 2016; Hashem et al. 2018).

India still lacks a clear perspective for using genetically engineered microorganisms (GEMs) in the form of biocontrol products. A series of guidelines related to the safety assessment of genetically modified organisms (GMOs), i.e., research, confined field trials, food safety assessment, and environmental risk assessment, are now jointly adopted under Rules, 1989 notified under the Environment (Protection) Act, 1986 (Ahuja 2018). Limited efforts have been made regarding research on GMO applications in the field of biocontrol, and only Bt cotton is commercially approved (Shukla et al. 2018). Hence, it can be concluded that as of now the status of biopesticides in India still requires a lot of support and improvement so as to compete and phase out chemical pesticides.

Research of biopesticide production, formulation, and delivery could greatly assist in commercialization of biopesticides. It seems likely that biopesticides will have a wider use in the future as their application methods improve and better and cheaper choice of different inerts has been identified for various formulations. The use of biopesticides with adjuvants has been shown to enhance their activity and that fact has opened new opportunities for further development in that area. Selection of an appropriate formulation can improve product stability, enhance and extend activity, and may reduce inconsistency of field performance of many potential bio-agents. Biopesticides are seen as a tool for developing a more rational pesticide use strategy and future products should have improved balance between efficiency and cost (El-Sayed 2005; Rao et al. 2007; Glare 2012; Khater 2012). Trends relating to the type of biopesticide formulations would probably go the way from wettable powders and suspension concentrates to water dispersible granules, and from dusts to granules, for safety reasons, and from single component to multicomponent formulations. In addition, an increased number of controlled release formulations can be expected to optimize their biological effects, while new types of formulations, such as nanoemulsion, nanosuspension, and nanocapsule suspension, will result from newly developed nanotechnology (Rao et al. 2007; Glare 2012). Significant progress has been made in developing formulations and application methods but much work still remains to be done regarding the use of biopesticides for plant protection. Further improvement of techniques and multidisciplinary research of plant pathologists, formulation chemists, and agricultural engineers are likely to provide good, safe, effective, and inexpensive products for plant protection.

25.7 Conclusions

Since decades, biopesticides are being used for preventing crops from pests and pathogens. However, their market and place among agrochemicals are still way behind in comparison to conventional chemicals. Scientific and technological interventions related to the development of biopesticides are also deficient as proven by lack of understanding of mechanisms at ground level, a dearth of reliability on bio-products (by the end users), and absence of penetration in the market of pest control (in India). The challenges in broad usage of biopesticides in India are related to the efficacy, shelf life, production methods, narrow range of host or target pathogens/pests, poor performance in the field, problems in the delivery system, economics, and regulations. Apart from these technical issues, there is a need of push from the authorities for promoting commercialization of biopesticides.

References

Agarwal A, Prajapati R, Singh O, Raza PSK, Thakur LK (2015) Pesticide residue in water—a challenging task in India. Environ Monit Assess 187(2):1–21

Ahuja V (2018) Regulation of emerging gene technologies in India. BMC Proc 12(8):14. https://doi.org/10.1186/s12919-018-0106-0

Anonymous (1958) Indian J Malariol 12(4):261

Anonymous (2021) Insecticides/pesticides registered under section 9(3) of the insecticides act, 1968 for use in the Country. http://ppqs.gov.in/sites/default/files/insecticides_registered_under_section_-93.pdf

Arbuekal TE, Server LE (1998) Pesticides exposures and fetal death: a review of the epidemiologic literature. Crit Rev Toxicol 28:229–270

Arora NK, Mishra J (2016) Prospecting the roles of metabolites and additives in future bioformulations for sustainable agriculture. Appl Soil Ecol 107:405–407. https://doi.org/10.1016/j.apsoil.2016.05.020

Aydinalp C, Porca MM (2004) The effects of pesticides in water resources. J Cent Eur Agric 5(1):5–12

Aziz Qureshi A, Vineela V, Vimala Devi PS (2015) Sodium humate as a promising coating material for microencapsulation of Beauveria bassiana conidia through spray drying. Dry Technol 33:162–168. https://doi.org/10.1080/07373937.2014.938814

Bailey KL (2014) The bioherbicide approach to weed control using plant pathogens. In: Abrol DP (ed) Integrated pest management. Academic, Cambridge, MA, pp 245–266

Bajwa U, Sandhu KS (2014) Effect of handling and processing on pesticide residues in food-a review. J Food Sci Technol 51(2):201–220

Banerjee SN (1979) Agricultural production through increased pesticide usage. Pestic Inf 4(4):17–19

Bedi JS, Gill JPS, Kaur P, Aulakh RS (2018) Pesticide residues in milk and their relationship with pesticide contamination of feedstuffs supplied to dairy cattle in Punjab (India). J Anim Feed Sci 27(1):18–25

Behal A (2020) The Green Revolution and a dark Punjab. https://www.downtoearth.org.in/blog/agriculture/thegreen-revolution-and-a-dark-punjab-72318

Bhalia SK (1971) Focus on the spread of insecticide resistance in stored product pests. Entomologists' Newsl 1:41–42

Bharadwaj A, Sharma S (2006) Role of arbuscular mycorrhizae in pest management. Int J Ecol Environ Sci 32:193–205

Brar SK, Verma M, Tyagi RD, Valero JR (2006) Recent advances in downstream processing and formulations of Bacillus thuringiensis based biopesticides. Process Biochem 41(2):323–342

Bravo A, Gill SS, Soberón M (2007) Mode of action of *bacillus thuringiensis* cry and Cyt toxins and their potential for insect control. Toxicon 49:423–435

Chaudhary S, Kanwar RK, Sehgal A, Cahill DM, Barrow CJ, Sehgal R, Kanwar JR (2017) Progress on *Azadirachta indica* based biopesticides in replacing synthetic toxic pesticides. Front Plant Sci 8:610

Chhipa H, Joshi P (2016) Nanofertilisers, nanopesticides and nanosensors in agriculture. In: Ranjan S, Dasgupta N, Lichtfouse E (eds) Nanoscience in food and agriculture. Springer International Publishing, Cambridge, pp 247–282

Colborn T (1996) Hormonal sabotage. (synthetic chemicals and the endocrine systems)(includes related article on p-nonylphenol). Nat Hist 105(3):42–47

Dolatabadi KH, Goltapeh EM, Varma A, Rohani N (2011) In vitro evaluation of arbuscular mycorrhizal-like fungi and Trichoderma species against soil borne pathogens. J Agric Technol 7:73–84

Dubois C, Arsenault-Labrecque G, Pickford J (2017) Evaluation of a new biopesticide against angular leaf spot in a commercial operation system. Acta Hortic 1156:757–764

El-Abbassi A, Saadaoui N, Kiai H, Raiti J, Hafidi A (2017) Potential applications of olive mill wastewater as biopesticide for crops protection. Sci Total Environ 576:10–21

El-Sayed W (2005) Biological control of weeds with pathogens: current status and future trends. J Plant Dis Prot 112(3):209–221

El-Sayed A, Bengtsson M, Rauscher S, Lofqvist J, Witzgall P (1999) Multicomponent sex pheromone in codling moth (Lepidoptera: Tortricidae). Environ Entomol 28:775–779

Eski A, Demir D, Sezen K, Demirbag ZA (2017) New biopesticide from a local *bacillus thuringiensis* var. *tenebrionis* (Xd3) against alder leaf beetle (Coleoptera: Chrysomelidae). World J Microbiol Biotechnol 33:95

Galloway SM, Armstrong C, Reuben S, Colman B, Brown C, Cannon AD, Bloom F, Nakamura M, Ahmed S, Rimpo DJ, Margolin BH, Resnick MA, Anderson MJ, Zeiger BE (1987) Chromosome aberrations and sister chromatid exchanges in Chinese hamster ovary cells, evaluation of 108 chemicals. Environ Mol Mutagen 10:1–175

Gams W, Diederich P, Poldmaa K (2004) In: Mueller GM, Bills GF, Foster MS (eds) Fungicolous fungi. In biodiversity of fungi. Academic, Cambridge, MA, pp 343–392

Glare T, Caradus J, Gelernter W, Jackson T, Keyhani N, Köhl J, Stewart A (2012) Have biopesticides come of age. Trends Biotechnol 30(5):250–258. pmid:22336383

Government of India (2001) Tenth five-year plan: 2002–2007. Planning Commission of India, New Delhi, pp 513–566

Grewal PS (2005) Formulations and quality control of entomopathogenic nematodes. Department of Entomology, Ohio-State-University Wooster

Guerra F, Sepúlveda S (2021) Saponin production from Quillaja genus species. An insight into Its applications and biology. Sci Agric 78:305

Gupta PK (2004) Pesticide exposure—Indian scene. Toxicology 198(1–3):83–90

Hashem AS, Awadalla SS, Zayed GM, Maggi F, Benelli G (2018) Pimpinella anisum essential oil nanoemulsions against Tribolium castaneum—insecticidal activity and mode of action. Environ Sci Pollut Res 25(19):18802–18812. https://doi.org/10.1007/s11356-018-2068-1

Horrigan L, Lawrence RS, Walker P (2002) How sustainable agriculture can address the environmental and human health harms of industrial agriculture? Environ Health Perspect 110(5): 445–456

Indira Devi P (2007) South Asian network for development and environmental economics. SANDEE working paper no. 20-07, Kathmandu, Nepal

Inobeme A, Mathew JT, Okonkwo S, Ajai AI, Jacob JO, Olori E (2020) Pesticide residues in food: distribution, route of exposure and toxicity: in review. MOJ Food Process Technol 8(3): 121–124

Ishikawa S (2013) Integrated disease management of strawberry anthracnose and development of a new biopesticide. J Gen Plant Pathol 79:441–443

Isman MB (2020) Bioinsecticides based on plant essential oils: a short overview. Z Naturforsch 75: 179–182

Kadoi'c BM, Bažok R, Mikac KM, Lemic D, Paja'c ŽI (2020) Pest management challenges and control practices in codling moth: a review. Insects 11:38

Khater HF (2012) Prospects of botanical biopesticides in insect pest management. J Appl Pharm Sci 02(05):244–259

Kirk WW, Schafer RS (2015) Efficacy of new active ingredient formulations and new biopesticides for managing fusarium root rot disease of gladiolus hybrids. Acta Hortic 1105:55–60

Kishi M, Hirschhorn N, Djajadisastra M, Satterlee LN, Strowman S, Dilts R (1995) Relationship of pesticide spraying to signs and symptoms in Indonesian farmers. Scand J Work Environ Health 21:124–133

Knowles A (2008) Recent developments of safer formulations of agrochemicals. Environmentalist 28(1):35–44. https://doi.org/10.1007/s10669-007-9045-4

Koul O (2011) Microbial biopesticides: opportunities and challenges. Perspect Agri Vet Sci Nutr Nat Resour 6:1–26

Kumar J (2012) Biopesticides: a need for food and environmental safety. J Fertil Pestic 3:4. https://doi.org/10.4172/2155-6202.1000e107

Kumar M, Prasanna R, Bidyarani N, Babu S, Mishra BK, Kumar A, Adak A, Jauhari S, Yadav K, Singh R et al (2013) Evaluating the plant growth promoting ability of thermotolerant bacteria and cyanobacteria and their interactions with seed spice crops. Sci Hortic 164:94–101. https://doi.org/10.1016/j.scienta.2013.09.014

Kumar J, Ramlal A, Mallick D, Mishra V (2021) An overview of some biopesticides and their importance in plant protection for commercial acceptance. Plan Theory 10:1185. https://doi.org/10.3390/plants10061185

Kumari S, Sharma H (2012) The impact of pesticides on Farmer's health: a case study of fruit bowl of Himachal Pradesh. Int J Sci Res, 2319-7064

Lala F, Cahyaningrum H, Hadiarto A, Brahmantiyo B (2019) Implementation of Biopesticides to Control *Phytophthora palmivora* Butl. and *Conopomorpha cramerella* Snell. on Cacao in South Halmahera. In: Proceedings of the 5th International Conference on Food, Agriculture and Natural Resources (FANRes, 2019), Ternate, Indonesia. Atlantis Press, Amsterdam, pp 17–19

Leiss JK, Savitz DA (1995) Home pesticide use and childhood cancer: a case-control study. Am J Public Health 85:249–252

Madhusudhan L (2015) Agriculture role on Indian economy. Bus Econ J 6(4):1

Marrone P (2002) An effective biofungicide with novel modes of action. Pestic Outlook 13:193–194

Mauchline AL, Hervé MR, Cook SM (2018) Semiochemical-based alternatives to synthetic toxicant insecticides for pollen beetle management. Arthropod Plant Interact 12:835–847

Mazid S, Kalida JC, Rajkhowa RC (2011) A review on the use of biopesticides in insect pest management. Int J Sci Advan Technol 1:169–178

McGuire AV, Northfield TD (2020) Tropical occurrence and agricultural importance of *Beauveria bassiana* and *Metarhizium anisopliae*. Front Sustain Food Syst 4:6

Mehrotra KN (1983) Benefits and hazards of pesticides to society. Pestic Inf 9(3):38–42

Mehrotra KN (1985a) Use of DDT and its environmental effects in India. Proc Indian Natl Sci Acad B51:169–184
Mehrotra KN (1985b) Use of HCH (BHC) and its environmental effects in India. Proc Indian Natl Sci Acad K51:581–595
Mehrotra KN (1989a) Pesticide resistance in insect-pests Indian scenario. Pestic Res J 1(2):95–103
Mehrotra KN (1989b) Pesticide residues in food. Consumer Education and Research Centre, Ahmedabad, pp 42–58
Mensah R, Moore C, Watts N, Deseo MA, Glennie P, Pitt A (2014) Discovery and development of a new semiochemical biopesticide for cotton pest management: assessment of extract effects on the cotton pest *Helicoverpa spp*. Entomol Exp Appl 152:1–15
Mishra J, Arora NK (2018) Secondary metabolites of fluorescent pseudomonads in biocontrol of phytopathogens for sustainable agriculture. Appl Soil Ecol 125:35–45. https://doi.org/10.1016/j.apsoil.2017.12.004
Mishra J, Dutta V, Arora NK (2020) Biopesticides in India: technology and sustainability linkages. 3 Biotech 10:210. https://doi.org/10.1007/s13205-020-02192-7
Mittal R, Singh G (2006) Patenting scenario in agriculture: Indian perspective. Curr Sci 90(3):296–300. https://doi.org/10.1504/IJIPM.2014.062788
Moreno-Gavíra A, Huertas V, Diánez F, Sánchez-Montesinos B, Santos M (2020) *Paecilomyces* and its importance in the biological control of agricultural pests and diseases. Plan Theory 9: 1746
Mossa ATH, Abdelfattah NAH, Mohafrash SMM (2017) Nanoemulsion of camphor (*eucalyptus globules*) essential oil, formulation, characterization and insecticidal activity against wheat weevil, *Sitophilus granaries*. Asian J Crop Sci 9:50–62
Mukerji KG, Ciancio A (2007) Mycorrhizae in the integrated Pest and disease management. In: Ciancio A, Mukerji KG (eds) General concepts in integrated pest and disease management. integrated management of plants pests and diseases, vol 1. Springer, Dordrecht
Murshed S (2020) Evaluation and assessment of aflatoxin M1 in Milk and Milk products in Yemen using high-performance liquid chromatography. J Food Qual 2020:8839060
Narware J, Yadav RN, Keswani C, Singh SP, Singh HB (2019) Silver nanoparticle-based biopesticides for phytopathogens: scope and potential in agriculture. In: Koul O (ed) Nano-biopesticides today and future perspectives. Academic, Cambridge, MA, pp 303–314
Olowe OM, Olawuyi OJ, Sobowale AA, Odebode AC (2018) Role of arbuscular mycorrhizal fungi as biocontrol agents against *fusarium verticillioides* causing ear rot of *Zea mays* L. (maize). Curr Plant Biol 15:30–37
Padma TV (2019) What is preventing the widespread adoption of biopesticides in India? Mongabay series: conserving agro-biodiversity. https://india.mongabay.com/2019/07
Pal R, Sharma MID, Krishnamurthy BS (1952) Studies on development of resistant strains of houseflies and mosquitoes. Indian J Malariol 6:303–316
Pavela R, Waffo-Teguo P, Biais B, Richard T, Mérillon JM (2017) *Vitis vinifera* canes, a source of stilbenoids against *Spodoptera littoralis* larvae. J Pest Sci 90:961–970
Rajak RL, Ghagte M, Krishnamurthy K (1973) Bioassay technique for resistance to malathion of stored product insects. Pest Contr 15:11–13
Ranglová K, Lakatos GE, Câmara Manoel JA, Grivalský T, Suárez Estrella F, Acién Fernández FG, Molnár Z, Ördög V, Masojídek J (2021) Growth, biostimulant and biopesticide activity of the MACC-1 chlorella strain cultivated outdoors in inorganic medium and wastewater. Algal Res 53:102136
Rao GVR, Rupela OP, Rao VR, Reddy YVR (2007) Role of biopesticides in crop protection: present status and future prospects. Indian J Plant Prot 35(1):1–9
Reddy AA (2018) Impact Study of Paramparagath Krishi Vikas Yoyana National Institute of Agricultural Extension Management (MANAGE) Hyderabad Organic Agriculture Area (lakh ha)
Rodrigues D, Costa-Pinto AR, Sousa S, Vasconcelos MW, Pintado MM, Pereira L, Rocha-Santos TAP, Costa JPD, Silva AMS, Duarte AC et al (2019) *Sargassum muticum* and *Osmundea*

pinnatifida enzymatic extracts: chemical, structural, and cytotoxic characterization. Mar Drugs 17:209

Roy M (2016) Impact of pesticide use on the health of farmers: a case study in Bankura, W.B., India. Int J Innov Res Sci Eng Technol 5(7):12370–12374

Saberi F, Marzban R, Ardjmand M, Pajoum SF, Tavakoli O (2020) Optimization of culture media to enhance the ability of local bacillus thuringiensis var. tenebrionis. J Saudi Soc Agric Sci 19: 468–475

Sharma A, Jaronski S, Reddy GVP (2019) Impact of granular carriers to improve the efficacy of entomopathogenic fungi against wireworms in spring wheat. J Pest Sci. https://doi.org/10.1007/s10340-019-01161-1

Shukla M, Al-Busaidi KT, Trivedi M, Tiwari RK (2018) Status of research, regulations and challenges for genetically modified crops in India. GM Crops Food 9(4):173–188. https://doi.org/10.1080/21645698.2018.1529518

Singh L, Kumar V (2021) Need of biopesticides triumph over chemical pesticides. Int J Curr Microbiol App Sci 10(01):93–101

Tadros T (2005) Surfactants in nano-emulsions. In: Applied surfactants: principles and applications, pp 285–308. https://doi.org/10.1002/3527604812

Tripathi A, Prasad AR (2010) Agricultural development in India since independence: a study on progress, performance, and determinants. J Emerg Knowl Emerg Mark 1(1):8

Part V

Agricultural Policies for Sustainable and Secured Nation

Trajectory of Market Reforms

26

Raka Saxena, Shivendra K. Srivastava, Neha Joshi, Rohit Kumar,
and Muruga Subramanian Raman

Abstract

The transformation of the agricultural economy triggered new market architecture in India. Postindependence policies in relation to agriculture marketing aimed at making it profitable for farmers, available and affordable for consumers. This demanded the development of various sectors of agriculture marketing such as market regulation, market infrastructure, market functions, market information, value addition, price forecasting, and agribusiness. Reforms made since independence can be classified as institutional reforms, structural reforms, and legislative reforms. Given the surplus food production amid the huge inadequacy of storage and other logistics infrastructure, improved storage infrastructure would reduce distress sale by the farmers, price crash during harvest season due to glut and postharvest losses. Agricultural export is emerging as an important avenue for disposition of surplus output. Effective agri-logistics and efficient supply chains must result in the desired quality as per the international standards.

Keywords

Agricultural commodities · Reforms · Producers' share · Price realization · Exports · Untapped potential

R. Saxena (✉) · S. K. Srivastava · M. S. Raman
ICAR—National Institute of Agricultural Economics and Policy Research, New Delhi, India
e-mail: raka.saxena@icar.gov.in

N. Joshi
G.B. Pant University of Agriculture and Technology, Pantnagar, Uttarakhand, India

R. Kumar
Department of Dairy Business Management, Sanjay Gandhi Institute of Dairy Technology, Bihar Animal Sciences University, Patna, India

P. K. Ghosh et al. (eds.), *Trajectory of 75 years of Indian Agriculture after Independence*, https://doi.org/10.1007/978-981-19-7997-2_26

26.1 Introduction

The agricultural ecosystem has been stagnant for too long, and developmental efforts were predominantly focused on production growth. Emphasis on postproduction activities is coming to the fore in recent years. The transformation of the agricultural economy triggered a new market architecture. The mission of doubling farmers' income (DFI) is an integrated and dedicated program to fulfill this goal. DFI strategies accord due emphasis on the realization of remunerative prices by the farmers through appropriate marketing interventions and reforms besides the emphasis on productivity enhancement, diversification, intensification, and nonfarm diversification. Thus, it is a prerequisite that agriculture should emerge as a profitable and remunerative enterprise. This requires an acceleration in the pace of reforms and structural transformation of the agricultural sector to unleash the inherent potential. Many high-powered bodies and committees highlighted the need and stressed the effective implementation of agricultural marketing reforms to benefit the farming community directly (Srivastava and Saxena 2021). Postindependence policies in relation to agriculture marketing aimed at making it profitable for farmers, available, and affordable for consumers as per their customized need of the particular product. This demanded the development of various sectors of agriculture marketing such as market regulation, market infrastructure, market functions, market information, value addition, price forecasting, and agribusiness.

26.2 Chronology of Agriculture Market Reforms in India

The conventional system of agricultural marketing was not organized, leading to high marketing costs due to unauthorized charges, involvement of big wholesalers and traders in between, low level of accessibility for the farmers, and many such issues. The need was felt to safeguard the interest of the producers by promising them basic marketing facilities by providing them an organized and regulated platform for marketing. Many committees were set up to recommend different types of reforms for the improvement of the sector and many studies reported the effectiveness of these reforms over the period. Reforms made since independence can be classified under the following heads (Box 26.1):

1. Structural reforms.
2. Institutional reforms.
3. Legislative reforms.
4. Innovations.

Box 26.1 Classification of Agriculture Market Reforms in India

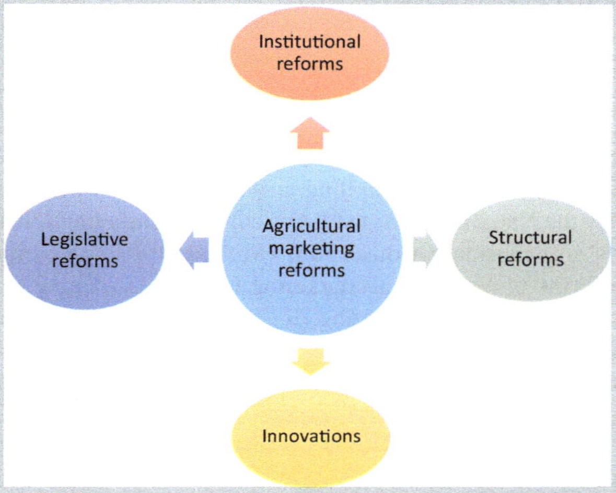

26.2.1 Structural Reforms

The inception of the regulated market dates back to the nineteenth century aiming at the availability of raw materials at affordable prices. The Karanjia cotton market, 1886, was the first regulated market directed to secure the supply of pure cotton to textile mills in Manchester. The first legislation was the Berar Cotton and Grain Market Act of 1897, which played the role of the model Act for other legislations in the country. This turned over a new leaf in the development of the marketing sector of the country. Afterward, to further strengthen the sector, the Royal Commission on Agriculture, 1928, under the chairmanship of Lord Linlithgow, submitted the report on agricultural and rural economy progress. The Central Banking Enquiry Committee, 1931, along with the Royal Commission on Agriculture, suggested a system for the regulation of the agricultural marketing system. Consequently, the Central Marketing Department was established in 1935 in India. Subsequently, its merger with the inspection directorate under the Ministry of Food led to the endowment of the Directorate of Marketing and Inspection (DMI) in 1935. A network of central and regional laboratories was also founded by DMI under AGMARK for the certification of 157 agricultural commodities. The number has risen to a spanning 224 different commodities at present.

Later in 1938, DMI encouraged the state governments to play an active role in regulating marketing practices by harboring the rights and needs of the producers and, owing to liberalization and globalization, to explore new ventures for strengthening the marketing system, Government of India appointed an Inter-Ministerial Task Force. It was aimed to review the then marketing system and provide suggestions for the development of the same. Committee recommended the infrastructural development of market yards, promotion of competition by providing alternating market structures, and prevention of farmers' exploitation. This pointed toward the formulation of model legislation on agricultural marketing. The draft model legislation was titled the State Agricultural Produce Marketing (Development and Regulation) Act, 2003. Since agriculture is a state subject, the agricultural markets are regulated under the State APMCs Act. The Act has been responsible for ensuring complete transparency in the pricing system, promotion of value addition, processing-related activities and promotion of public–private partnership, publication of the data regarding arrivals, and many more functions. DMI, accompanied by the National Institute of Agricultural Marketing (NIAM), 1988, and Small Farmers' Agribusiness Consortium (SFAC), 1994, was established under institutional reforms of agricultural marketing to provide training to promote agriculture marketing. The state marketing departments of states were set up as the component of the central marketing department. The establishment of state marketing boards was directed toward properly coordinating the regulated markets set up in different states. However, the requirement was felt for the supervision and guidance of these state marketing boards. Therefore, an apex body was established in 1988 under the name of the Council of State Agricultural Marketing Boards (COSAMB).

Another phase of the development of agricultural marketing takes into account the requirement of a sound market intelligence system. Therefore, DMI embarked on an IT project under the name of AGMARKNET in alliance with the National Information Center (NIC). This project has linked all the state agricultural marketing boards, APMCs, DMI, and its regional offices to disseminate information regarding agricultural marketing. It keeps the general public aware of price, arrivals, storage facilities, safety standards, laws, and regulations about the market under different heads mentioned on the official site.

26.2.2 Institutional Reforms

Apart from the structural reforms for the improvement of agriculture marketing, many institutional reforms have also taken place. These reforms were in collaboration with the structural reforms leading to the establishment of several institutions guiding the marketing process in India. Several institutions were set up with specific objectives to perform and specific goals to achieve. The establishment of some of the institutions is discussed above in relation to structural reforms.

Foundation of Food Corporation of India (FCI) Legislation was passed to establish the corporation in 1965 to procure, distribute, and timely release food grains to minimize their price fluctuations. To cope with the global trade and maintenance of foreign exchange, agricultural trade in the form of export and imports are crucial components. The Ministry of Commerce laid the foundation of the Agricultural and Processed Food Products Export Development Authority (APEDA) in the APEDA Act, 1985. APEDA was set up in 1986 to cover all the export-related issues of agricultural commodities. The Marine Products Export Development Authority (MPEDA) was laid down in 1972 to promote the export of marine products. In the same way, to encourage the development of the horticultural sector, an autonomous body was established in 1984 under the name of the National Horticulture Board (NHB). NHB is held responsible for technical and financial assistance to different market organizations, dissemination of market information by building a database, and promotion activities for commercial production of horticultural crops. Along with crops, livestock also shares a significant part in agriculture trading, so for the development of production, marketing, and promotion of dairy products in 1965.

Several specific commodity boards were also set up under these reforms, such as Tea Board, under Tea Act 1953; Coffee Board, under Coffee Market Expansion Act, 1940; Rubber Board, under Rubber Board Act, 1947; Tobacco Board, under Tobacco Board Act, 1947; Spices Board, under Spices Board Act, 1987; Coir Board, 1953; Silk Board, under Central Silk Board Act, 1949; National Meat and Poultry Processing Board, 2009; and many more.

To control the quality of the agricultural commodities, DMI imposed the grading standards under the head of AGMARK, 1937. Agriculture Produce (Grading and Marking) Act was one of the legal instruments legislated in acts to improve agricultural marketing status. For the standardization of manufactured products, the Indian Standards Institution (ISI) was set up in 1947 under ISI Certification Marks Act to prepare, promote, and certify the products. Later, ISI was renamed the Bureau of Indian Standards (BIS) in 1987. It was established to perform all the functions performed by ISI along with more emphasis on consumer protection and quality improvement. Likewise, Food Safety and Standards Act (FSSA), 2006, has been established as a statutory body for affirming the scientific standards for manufactured, processed, and consumable products regarding their distribution, sale, and import of food. In addition, there are other marks indicating the quality and constituency of the product such as the Eco mark, FPO (Fruit Products Order) mark, and mark to identify vegetarian and nonvegetarian food products.

Price administration is one of the major components of agriculture marketing. The Commission for Agricultural Cost and Prices (CACP) was formulated to administer pricing policy recommendations for agricultural crops to the government. It was initially set up under the name of Agricultural Prices Commission in 1965 as per the recommendations of Shri L.K. Jha committee. It is an advisory body to the government regarding different prices such as procurement prices, minimum support prices, statutory minimum prices, fair and remunerative prices, and issue prices. To incorporate the fruits and vegetables under the price support scheme, Market Intervention Scheme (MIS) was implemented to protect the benefit of growers and maintain their price stability during peak and lean seasons. Loses incurred by the growers are also covered under the scheme on a 50:50 basis by central and state government.

26.2.3 Legislative Reforms

The Government of India has passed many legislatures over the decades to regulate, control, promote, and develop the agricultural market in India. Some of the important laws were regarding regulating and functioning of APMCs in state and union territories; regulating quality, grading, and standardization; regulating weights, measures, and packaging; regulating the storage and warehousing of agricultural commodities, Consumer Protection Act; regulating supply, demand, and prices; and setting up of commodity corporations and commodity boards, Seeds Act, Insecticides Act, Fertilizer Control Order, Protection of Plant Varieties and Farmer's Rights (PPV&FR) Act.

26.2.4 Innovations

An innovation in marketing is the implementation of new marketing to perform the functions of the marketing system. Several alternative systems of marketing have emerged over the past few years. e-Business is overtaking in India in the form of e-auction, e-tailing, and electronic spot markets. In the era of technology, transformation has been encountered in every sector of the economy. The application of ICT brought a revolution among the farmers. These applications provide facilities to link farmers with the markets, digital platforms for the consumers, and dissemination of information such as digital mandi, e-NAM, Kisan Call Centres, etc.

26.2.5 Miscellaneous Reforms

The history of cooperative marketing stemmed from the Co-operative Marketing Societies Act, 1912, led to the establishment of cooperative societies for marketing. The first five-year plan laid the foundation of the apex cooperative marketing society, National Agricultural Cooperative Marketing Federation (NAFED) in 1958 and for promoting the development of cooperative societies, National Cooperative Development Corporation (NCDC) in 1963.

Various promotion activities in the form of incentives were provided to promote external trade until economic liberalization, which took place in 1991. The introduction of these policy reforms in India has relaxed many restrictions in quantitative terms also, over agricultural trade leading to the simplification of global trade. The export–import policy, implemented in 1992–97 and 2002–07, further propelled the liberalization. As the Indian economy is small holding based, the marketable surplus is not always sufficient to market, leading to heavy distress sales. This led to the foundation of a legal entity, the Producer Organization. It can be registered under any of the provisions such as the Co-operative Societies Act, Multi-State Cooperative Society Act, and public trusts.

26.3 Need for Further Reforms in Agricultural Marketing

After attaining self-sufficiency in the number of cereals and other crops, managing surplus food has emerged as a major challenge in the country. This drives the attention to channelizing the agricultural output to more lucrative destinations and places where greater demand exists. The same is evident through the aggregate demand and supply estimates by various scholars. The projections for the future also indicate that these gaps will continue to widen, especially in some commodities such as rice, wheat, and cotton. The country is witnessing the accumulation of a surplus stock of commodities such as rice and sugar. This entails transforming the agricultural output, which fits into desired demand frame. The country also lacks the infrastructure to handle the increasing quantum of different commodities.

Continuous emphasis on diversification and high-value agriculture also necessitates a specific kind of infrastructure for reducing postharvest losses and handling extreme price situations. The changing consumption pattern away from staple to high-value food commodities warrants the production basket to be responsive to the shifting food demand. However, biasedness toward the traditional cropping systems, relatively higher marketing inefficiency of high-value crops along with infrastructural constraints resist the adoption of high-value crops. Thus, incentivizing farmers for diversification through various policy interventions such as price incentives, support infrastructure, trade facilitation, and vertical market integration is essential. The interventions are required to attract modern capital and investments into logistics and food value chains, thereby assisting in laying the foundation for the promotion of better marketing infrastructure in the country. This will also ensure the participation of the various stakeholders at different value chains and thus promote agri-entrepreneurship.

26.3.1 Stylized Fact 1: Producer's Share in Consumer Rupee Is Low

It has been established that growth in real price would ensure higher income growth than the mere reliance on technology-led reforms. The output growth will drive sufficient growth for farmers only if the marketing arrangements are efficient enough to support the income growth. Consequently, the price is the resultant outcome of any commodity system (Chand et al. 2015). The commodity and marketing systems have always been intended for the realization of remunerative prices by the farming community. Unfortunately, the farmers' share in the consumer rupee for most of the agricultural commodities in the country is lower, especially for horticultural commodities, which is indicative of the various inefficiencies of the agricultural marketing structure. The meta-analysis of several studies conducted at the regional/state level provides evidence of the temporal and spatial variation in price realization across various commodities and regions (Fig. 26.1). The price realization by farmers (% share in consumer price) indicates that it is the lowest for fruits and vegetables after pulses. It further varies according to the marketing channel adopted within a given geography for the selected commodity. Though onion is a vital staple vegetable in pan-India, farmers' share remains only around 44%. Banana trading seems

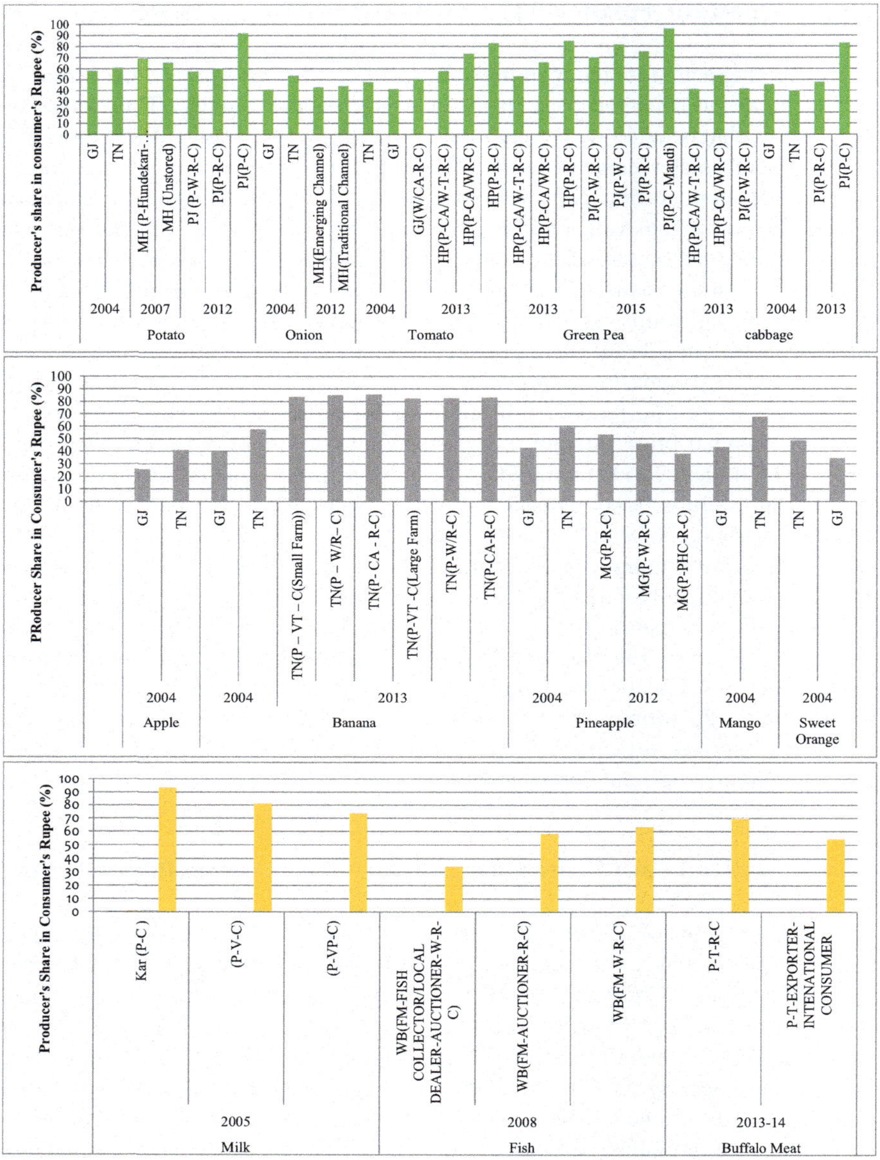

Fig. 26.1 Producer's share for major horticultural and livestock products

more efficient as only 15–18% of the consumer's price is eroded in the marketing process.

Raising the farmers' share in consumer prices is a pressing need, and the policy interventions facilitating the efficient marketing of agricultural commodities are of paramount importance. Many innovative marketing arrangements have shown that farmers' shares can be magnified, and marketing costs and margins of the chain can be managed efficiently through appropriate interventions.

26.3.2 Stylized Fact 2: Much of the Trade Happens Outside APMC in Most of the Commodities with Greater Reliance on Local Private Traders

The APMCs were created to keep the farmers away from the exploitative and dominant local systems. APMCs provided a legal and facilitating environment for the domestic trade of agricultural commodities. As the country moves through a more global, integrated, and competitive trade environment, it is the opportune time for farmers to be provided with better and remunerative choices to take advantage of emerging opportunities. Numerous studies and surveys, including the extensive surveys by the National Sample Survey Office (NSSO), reveal that a large volume of trade takes place outside APMCs, which are largely unregulated and noncompetitive. More than half of the marketable surplus is sold outside the *mandis* (Fig. 26.2). Most farmers are forced to sell through local traders who act as aggregators and further channel the produce to APMCs. The traders (both local and in APMCs) provide a lot of informal finance to the farmers for meeting their cultivation and postproduction requirements. Thus, the farmers are bound to sell through these agencies due to a lack of scale or indebtedness.

It is desirable that these transactions are legalized to incentivize both sellers (farmers) and buyers to participate through these organized marketing systems. This would improve marketing efficiency through increased competition among the buyers, improve the bargaining power of the farmers, and facilitate the realization of remunerative prices. The APMCs will be required to improve their efficiencies and services to attract trade.

26.3.3 Stylized Fact 3: Agricultural Policies Are Effective in the Selective Domain and Crops

Examination of the price policy implementation reveals limited penetration across geographies and crops. Over the last four and a half decades, the price policy implementation has boosted mainly wheat and rice crop among food grains and sugarcane and cotton, among other crops (Chand 2003). The procurement has also been very limited and confined to two to three states; approximately 40–50% of rice and wheat procurement during 2013–20 were from Punjab and Haryana only. The recent years have witnessed diversification in procurement efforts, and states like Madhya Pradesh and Chhattisgarh are also added to the procurement basket. The situation created an imbalance in the demand and supply of important agricultural commodities such as pulses, oilseeds, and coarse cereals. The Shanta Kumar Committee Report on the restructuring of FCI recommended that pulses and oilseeds

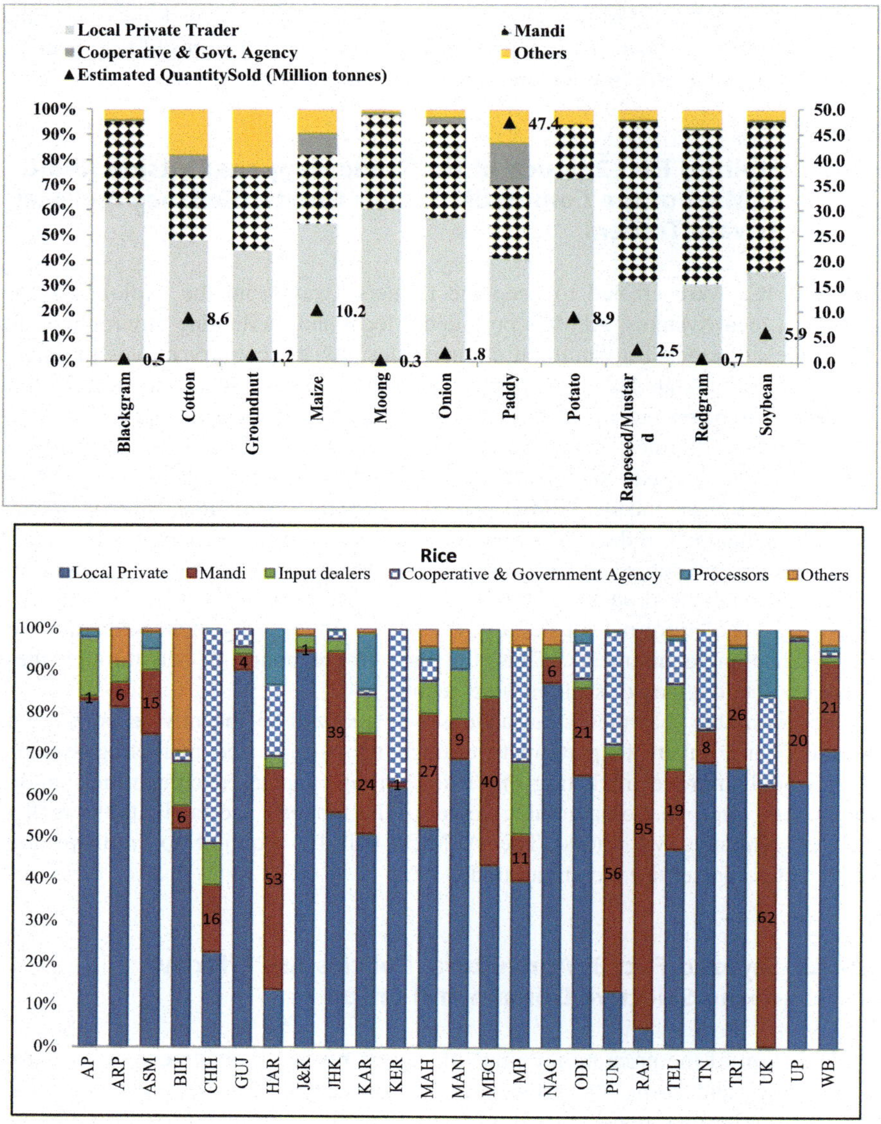

Fig. 26.2 Agency-wise marketing shares for major kharif crops

deserve priority, and Government of India (GoI) (2018) must provide better price support operations for them and dovetail their MSP (minimum support price) policy with trade policy so that the landed costs of imports are not below the domestic MSP rate. The report further states that in 2012–13, only 6% of total farmers in the country gained from selling wheat and paddy directly to any procurement agency. The prices

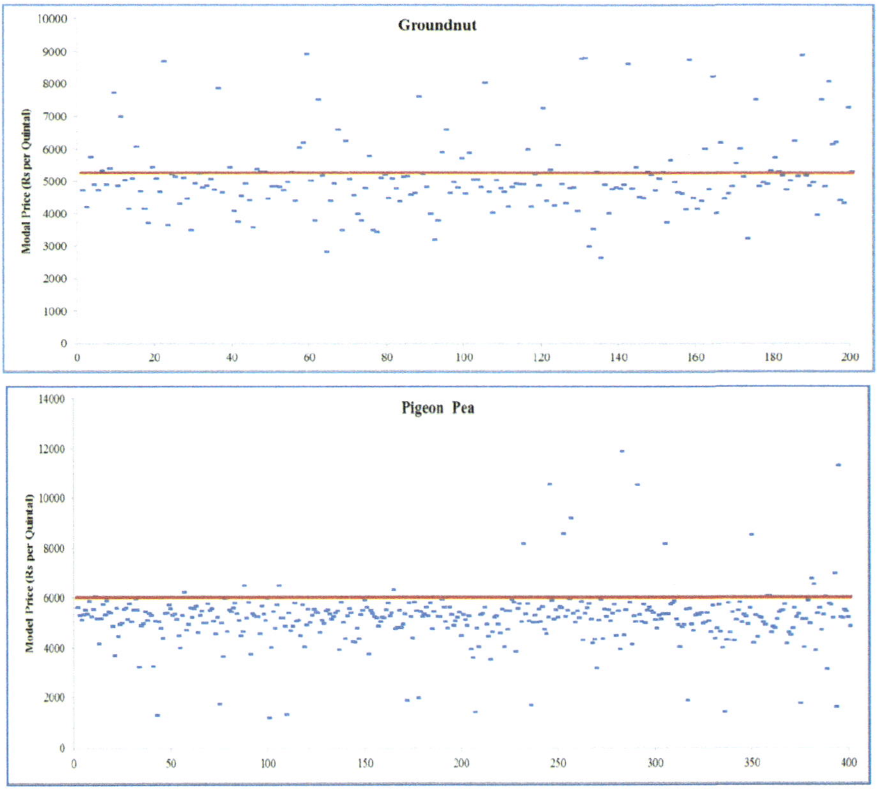

Fig. 26.3 Average wholesale prices across major markets

received by farmers are often below the MSP in many crops and in many markets where it is not supported by effective procurement (Planning Commission, 2007). Effective price discovery options need to be considered to buttress markets and assure market price to farmers, along with the development of infrastructural and market intelligence facilities for agriculture, to allow the farmers to transcend into a powerful market force in their own right.

In general, most of the trade in other crops occurs below MSP, and the market prices remain below MSP due to a lack of an effective price support mechanism. Figure 26.3 illustrates the recent situation in the case of pigeon pea and groundnut in December 2020 and January 2021, where the market prices remained lower than MSP in the majority of the markets.

The last 3-year estimates of MSP for rice and wheat indicate that majority of the trade, except for a few in all the states, has occurred at a price above MSP. The market prices for rice and wheat were about 71–92% and 3–15% higher than the MSP. Nonetheless, in the major wheat-producing state of Uttar Pradesh, the market price was lower than MSP in a few months (Fig. 26.4).

Fig. 26.4 Price realization in major rice- and wheat-producing states

26.3.4 Stylized Fact 4: Ensuring Small-Scale Processing and Investment in Infrastructure Would Be Critical for the Sustenance of Agriculture

DFI Committee places a strong emphasis on promoting small-scale processing and value systems that integrate multiple supply chain activities for greater efficiency and quality outcomes. The existing infrastructure, however, is inadequate to effectively absorb and efficiently monetize the farm output leading to seasonal and regional variations in demand–supply balance (GoI 2018). Overall, value addition for the total food produced in India has been estimated at just about 10% (varying from 4% in fresh agro-produce to 50% in milk) as compared to about 65% in developed countries like the United States and 23% in developing countries like China. Moreover, inadequate postharvest agri-logistics and processing facilities lead to huge losses (about 92,000 crore in 2013 as per the estimates of ICAR-Central Institute of Post-Harvest Engineering and Technology) and high-price variability.

The area and value pyramid for various agricultural categories establishes the importance of horticulture as one of the high-value sectors. Further, the rising share of non-foodgrains in the gross-cropped area and crop output signals increasing diversification toward high-value crops. On the other hand, the evidence indicates that there has been negative growth in the GVA Food Manufacturing by Household Sector in the case of fruits, vegetables, meat, which signifies that the farmers are not able to take direct advantage of government programs and schemes for promoting the small-scale processing. The recently launched package of Rs 1 lakh crore will help establish the desired infrastructure through lending to FPOs, SHGs, JLGs, and other organizations.

The challenges emanating from the current agribusiness environment heighten the need for strong backward and forward linkages to facilitate optimal input use and effective monetization of agricultural output. This requires the creation of a conducive business environment and support through developmental initiatives, technological dissemination, and favorable policies. Furthermore, providing farm inputs and services to the farmers will increase the adoption of modern technologies and good agricultural practices for producing quality products demanded by the consumers.

Given the surplus food production amid the inadequacy of storage and other logistics infrastructure, improved storage infrastructure would reduce distress sale by the farmers, price crash during harvest season due to glut, and postharvest losses. Better inventory management will also help accelerate farm produce processing and export. Improving agri-logistics and processing infrastructure and linking farmers to processors hold the key to accelerating diversification toward high-value food commodities. Improving postharvest infrastructure requires fostering public and private investment.

26.3.5 Stylized Fact 5: Reforms Much Needed to Fillip Tapping of Untapped Potential in Agricultural Trade

The country has attained self-sufficiency in the case of cereals. In the case of rice and wheat, we have been suffering from the problem of plenty. Agricultural export is emerging as an important avenue for the disposition of surplus output. Agricultural exports account for approximately 10–13% of the country's merchandise exports, and the exports of various commodities have increased significantly in recent years. Nonetheless, the agricultural trade suffers from numerous issues pertaining to agricultural supply chains, starting from the initial stage of farm production until reaching the final destinations. These issues relate to traceability, sanitary and phytosanitary considerations, effective handing, and proper agri-logistics. Though the share of agricultural exports in agricultural output is rising, the value stands only at around 8%. It is emerging that the country will be required to sell 20–25% of the incremental agri-food production in the overseas markets in the coming years, considering the emerging demand and supply scenario (Chand 2020). This accentuates enhancing the competitiveness of agricultural commodities in the international markets, which will build on the premise of more efficient agricultural supply chains. Effective agri-logistics and efficient supply chains must result in the desired quality as per the international standards.

Box 26.2 Current Status of Agricultural Exports (Value in Million Dollars) and Untapped Potential (%) for Major Commodities

Major exportable items	TE 2019–20	Untapped potential
Rice (HS 1006)	499.0	17–60
Cotton (HS 52)	470.5	6–85
Tea (HS 0902)	53.2	51–64
Coffee (HS 0901)	37.2	32–87
Fish and crustaceans (HS 03)	438.8	47–85
Meat and offal (HS 02)	254.4	49–51
Onion (HS 0703)	31.5	56–58
Potato (HS 0701)	4.9	58–68
Grapes (HS 0806)	24.2	64–71

Source: INTRACEN

Further, Indian agricultural commodities have huge export potential, much of which is still untapped (Box 26.2). In the case of marine products, the country's single largest export category, as large as 47–85% potential, can be tapped through effective supply chain management and policy interventions. Similarly, other commodities indicate substantial untapped potential. Export promotion and enhancement remain a primary agenda under the new agricultural export policy, which aims to double the agricultural exports from US$ 30+ billion to US$ 60+ billion in the next 5 years. The policy also aims to diversify the export basket, providing an institutional mechanism for pursuing market access, tackling barriers and dealing with sanitary and phytosanitary issues, and enabling the farmers to benefit from export opportunities in the overseas market.

26.3.6 Stylized Fact 6: Price Volatility Remains a Major Concern for Agricultural Commodities

The prices of agricultural commodities have remained high for the past few years, with food inflation being higher than non-food items. Perishable commodities are the

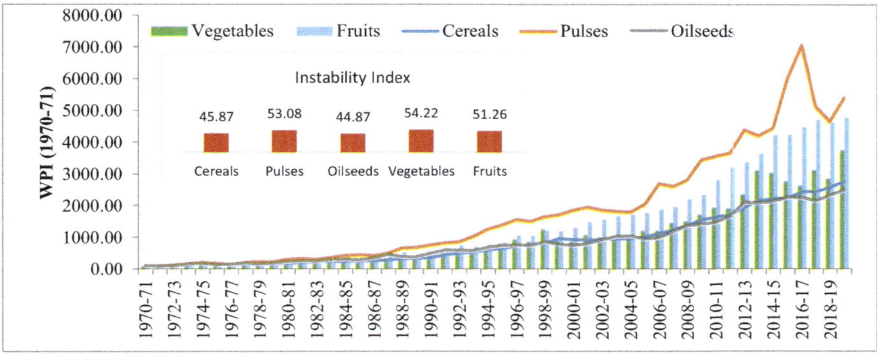

Fig. 26.5 WPI and Cuddy-Della Valle instability index for commodity group

prime component contributing to food inflation. With demand outstripping the supply and the lack of an efficient supply system and logistics, this inflationary pressure is expected to continue. The low price and income elasticity at times with unstable supply also contributes to higher price volatility. The price volatility for horticultural commodities is relatively higher, as evidenced by the instability index for various crop categories estimated for 1970–71 to 2019–20. The instability index for vegetables and fruits is estimated at 54.22% and 51.22%, respectively (Fig. 26.5). Besides, the instability index for pulses is also very high. A highly volatile price brings uncertainty and risk at both producer and consumer end, affecting social welfare.

26.3.7 Stylized Fact 7: Constraints in the Marketing of Livestock Products

The significantly rising contribution of the livestock sector to the Indian economy is apparent; the sector contributes around 28.63% to the total GVA of the agriculture and allied sectors (2018–19). Among the allied activities, the contribution of the dairy and fisheries sectors is immense in sustaining the income of people, especially for the marginalized and vulnerable communities. India is the largest milk producer in the world, producing around 198.4 million tons of milk during 2019–20. Animal husbandry infrastructure development fund of Rs 15,000 crore aims to support the private investment in dairy processing, enabling value addition and improved cattle feed infrastructure.

Consequent to a budget announcement on the inclusion of the livestock sector in the Kisan Credit Card in February 2020, 1.5 crores of dairy farmers of milk cooperatives and milk producer companies were targeted to provide Kisan Credit Cards (KCC) as part of the Prime Minister's Atma Nirbhar Bharat Package. A persistent high volume of postharvest loss eradicates a significant amount of the product from the market. For fisheries, the postharvest losses in developing countries like India are up to 25% (FAO 2015). Researchers have raised the importance of private investment in improving the domestic supply chains through value addition and transportation sectors (Sathiadhas et al. 2011). India, accounting for 7.58% of the global production, is the second largest fish-producing country in the world. The sector contributed 7.28% to the country's agricultural GVA (2019–20). There are immense opportunities in the sector to enhance the country's export and livelihood generation and sustain the income of around 28 million people.

India, especially in the marginalized and vulnerable communities, has promoted meaningful socioeconomic development. Realizing the potential, scope, and importance of the fisheries sector, the new flagship scheme, Pradhan Mantri Matsya Sampada Yojana (PMMSY), was launched in May 2020 as a part of the Atma Nirbhar Bharat Package by the Government of India with an estimated investment of Rs 20,050 crores comprising of central share of Rs 9407 crores, state share of Rs 4880 crores, and beneficiaries contribution of Rs 5763 crores for 5 years from FY 2020–21 to FY 2024–25. PMMSY aims to enhance fish production to 220 lakh metric tons by 2024–25 at an average annual growth rate of about 9%. The ambitious scheme will result in doubling export earnings to Rs 100,000 crores and generate direct and indirect employment opportunities of about 55 lakhs in the fisheries sector over the next 5 years. PMMSY further intends to increase aquaculture production.

26.4 Implementation Pathways

Enhancing farmers' incomes warrants the market-led agricultural transformation. The government is trying to promote the FPOs, SHGs, cooperatives, etc. for their effective linkages with upstream buyers and financial institutions. It also requires facilitating arrangements to improve infrastructure and services in the APMCs. The evidence indicates that the considerable export potential remains untapped. Government can facilitate the promotion of trade by effectively integrating the local markets into the global market and will help increase the farmers' income by exploring the trade opportunities in the international markets. We may also document other developing nations' success stories and learn lessons. For example, the market reforms contributed significantly to China's agricultural development and helped smooth the transformation from the planned to a market economy; the country moved from the system of state purchases and sales to relying primarily on the private market. Indian agriculture is dominated by informal value chain finance, traders being the primary source of lending to the farmers. Dependence on traders for financial requirements at various stages of crop cultivation restricts the farmers from benefiting from new marketing avenues. Therefore, interventions to increase access to the cumbersome free formal financial assistance will break the traditional informal financial assistance and thus promotes the farmers' participation in the new marketing channels. The backbone of all agricultural trade is agri-logistics. The transformation will come through the concomitant development of village-level logistics units to service agricultural trade, which should be undertaken in the name of the farmer–producer–owner. Many interventions and steps in the Union Budget 2021 are already aligned with the reform agenda. The efforts include improved allocations on market infrastructure and price support schemes (Srivastava et al. 2020). Mapping of the infrastructure facilities is a prerequisite to framing the policy interventions. The role of area and market-specific infrastructure facilities is significant in the smooth functioning of a market. Therefore, infrastructure mapping and prioritization for states and commodities are required.

References

Chand R (2003) Government intervention in food grain markets in the new context. In: Policy paper no. 19, ICAR-National Institute of Agricultural Economics and Policy Research, New Delhi

Chand R (2020) New farm acts: understanding the implications. In: NITI working paper series 1/2020. NITI Aayog, New Delhi. https://niti.gov.in/sites/default/files/2020-11/New_Farm_Acts_2020.pdf

Chand R, Saxena R, Rana S (2015) Estimates and analysis of farm income in India, 1983–84 to 2011–12. Econ Polit Wkly:139–145

FAO (2015) Global initiative on food loss and waste reduction. In: Save food, vol 25. Food and Agriculture Organization of the United Nations, New Delhi, p 2018. http://www.fao.org/3/ai 40 68e.pdf

Government of India (GoI) (2018) Reports on doubling farmers income by 2022. Ministry of Agriculture and Farmers Welfare, New Delhi. https://agricoop.nic.in/hi/doubling-farmers-income-0

Sathiadhas R, Narayanakumar R, Aswathy N (2011) Efficiency of domestic marine fish marketing in India—a macro analysis. Indian J Fish 58(4):125–131

Srivastava SK, Saxena R (2021) New farm acts, 2020: rationale and challenges. Curr Sci 120(6): 981

Srivastava SK, Singh J, Kumar NR, Singh NP, Ahmad N (2020) Changing agricultural labour market and its effects on farm economy in India. Indian J Agric Econ 75(4):469–480

Agribusiness and Entrepreneurship 27

Ritambhara Singh, Vishita Khanna, K. B. Ramappa, and Tulika Kumari

Abstract

The business baton in a nonfarm family is transferred from one generation to another, while in a farm family, this is not often the case. Most of the farmers do not wish their next generation taking up this baton from them. In addition, the next generation is not keen to hold this baton and move forward. One of the major reasons behind this is poor income from farming and allied activities. Long back in 1928, the report of Royal Commission on Agriculture mentioned, "*The Indian peasant is born in debt, lives in debt and bequeaths the indebtedness to his successors*." And the change is very slow. Realizing the miseries of Indian farmers, the Indian Government called for doubling farmers' income and set up an inter-ministerial committee that submitted its report in 2018. The committee devised several strategies and recommendations to double farmers' income eventually insisting on the need of moving agriculture from being taken up as a subsistence activity to a profitable enterprise. From profitable, it means that the receipt from farming and allied activities should be good enough to cover the cost, meet family expenses, and still there should be a surplus that can be invested in adoption of good technologies by the farmer. Therefore, to increase farm incomes and to bring in farm income revolution, the country is committed to turn farming and allied activities into "enterprise" and farmer as an "entrepreneur." This gets support from a supportive policy environment that favors the growth of agribusiness systems. Several programs are also in place to support

R. Singh (✉) · K. B. Ramappa · T. Kumari
School of Agribusiness and Rural Management, Dr. Rajendra Prasad Central Agricultural University, Samastipur, Bihar, India

V. Khanna
International Agribusiness Management Institute, Anand Agricultural University, Anand, Gujarat, India

P. K. Ghosh et al. (eds.), *Trajectory of 75 years of Indian Agriculture after Independence*, https://doi.org/10.1007/978-981-19-7997-2_27

rural and agricultural entrepreneurship. Needless to say, agribusiness and entrepreneurship strategies could bring in much needed stability in rural India in terms of growth and development. Hence, "agribusiness" approaches and "agripreneurship" must be promoted aggressively.

Keywords

Agribusiness · Secondary agriculture · Entrepreneurship · Postproduction activities · Agripreneurship

27.1 Agriculture: History and Evolution

The history of agriculture dates back to as good as the history of humankind. As the humankind evolved, so did agriculture. It is to the knowledge of everyone that prior to realizing the systematic cultivation of plants and animals for consumption, humans were hunters and gatherers. It is true that wild grains were gathered and consumed in Middle Stone Age (280,000 years ago to 50–25,000 years ago), archaeological evidence indicates that the domestication of plants for food purpose is only 10,000 years old. Wheat, barley, peas, lentil, chickpea, flax, rice, and potato are few of the earliest cultivated crops by humans for consumption purpose in the Eastern Mediterranean region. Slowly, agriculture spread to the rest of the world, and in India along the Indus River, the intensification began in Bronze Age. After the discovery of sea routes, the crops too moved from one continent to another, and what we see today is more the result of a globalized world. Potato from the Americas travelled to the Europe and wheat, barley, and rice travelled from the Mediterranean to the Americas. Agriculture further intensified with the British Revolution between seventeenth and nineteenth centuries. The aim of the British Agricultural Revolution was to enhance production via extensive use of inputs such as irrigation and fertilizer. The practice of crop rotation was made popular by the British Revolution only. This was the period in Britain when agricultural production rose faster than the population. Since the beginning of twentieth century, the agriculture started shaping up in a different way. Following the success of the British Agricultural Revolution, other developed nations also started getting engaged in pushing up agricultural productivity and mechanization. And along with it, the production of synthetic fertilizers and pesticides also progressed. The surpluses produced by these countries were traded with those less in resource base.

India being agriculturally and culturally rich, faced challenges as those who explored the trade routes to India exploited the resource base and left miseries behind. The British, the last of the rulers, were no exception when they left in 1947, the "Golden Bird" was not "Golden" anymore. There was poverty, there was hunger, and there was a broken socioeconomic fabric. The country had to rely on imports to fulfill stomachs. There are only the consistent efforts of the farmers of this nation and several policy reforms that today India is the net exporter of food and is among the largest producers of almost 17 products worldwide. Today the focus of the country and the government is to increase total factor productivity and to raise farmers' income. Agribusiness and entrepreneurship are one of the key solutions to achieve these goals. Agribusiness approaches have been hailed worldwide for their uniqueness in establishing sustainable food systems. Another important contribution of agribusiness cannot be ignored as it encourages many across the world to setting up new enterprises, promoting innovative thinking, and opening up the doors to entrepreneurship, especially in developing countries. Such an approach can uplift the economic status of any country by contributing significantly to all the three sectors of any economy. The Indian Government acknowledges the role of "agribusiness" and "entrepreneurship" to enhance farm incomes and create employment opportunities. In addition, the Committee on Doubling Farmers' Income (DFI) called for promoting secondary agriculture and strengthening postproduction activities to develop agriculture on the lines of agribusiness. The Government of India's holistic policy approach and favorable investment environment has attracted youth to take up agribusiness enterprises and become job providers rather than job seekers. This chapter focusses on prospects of "agribusiness" and "entrepreneurship," as well as the policy environment in greater detail. The chapter also highlights a few success stories, especially from the Covid-19 pandemic period to highlight how the pandemic has thrown immense opportunities for the development of sector.

27.2 The Agribusiness Approach to Farming

The term "agribusiness" was coined by J.H. Davis and R.A. Goldberg, two Harvard Business School professors, in 1957, when in their book "*A Concept of Agribusiness*," they defined it as

> *The sum total of all operations involved in the manufacture and distribution of farm supplies; production operations on the farm; and the storage, processing, and distribution of farm commodities and items made from them.*

These two professors favored large-scale farming to bring in revolution in American agriculture sector, through more involvement of corporate in the agricultural value chains, thereby reducing the dependency of agriculture on state. The recommendations they suggested were after their thorough research and deeper observations in field when they found strong linkages up and down the agricultural value chains. They noticed that a trader is dependent on manufacturer for supply and the manufacturer is dependent on the trader to connect that supply to the market. In between them are several institutional arrangements such as finance, storage, and government policies to facilitate transactions. Hence, the economists duo supported that private sector must be responsible for bringing in stability in agriculture sector rather than the government. The professors believed that the flaws in the American agriculture sector at that time were the results of government intervention via "New Deal" policy. The "New Deal" policy reforms were a series of reforms in the United States to uplift the economy from the Great Depression. In 1920s, the United States produced huge surplus, thanks to mechanization and the aggressive use of other inputs. However, in the same decade, "The Great Depression" hits America and the problem of plenty emerged. Due to depression, almost all businesses failed, and there was no demand for the commodities as there was no money with the people to pay for the same. As a result, the grains and meat had to rot in fields and in slaughter houses, respectively, despite millions were hungry at homes and on streets. The US Govt felt that in order to revive economy, agriculture must be revived and so farmers must realize better prices. Therefore, the government enacted "Agricultural Adjustment Act" to artificially raise farm prices by creating artificial scarcity, among several other agendas to bring profit to farmers. The American public did not find the Act pleasing and the Supreme Court called it "Unconstitutional," stating that the government shall not regulate agricultural production and it has no powers to control so. Davis and Goldberg's keen observation and research led to the conclusion that government shall not interfere in the agricultural sector, rather facilitate by introducing proper arrangements in the value chain. As the commodity is not just produced, it moves along the actors in the value chain and there are strong linkages in this chain, and all activities are interlinked. They argued that end-to-end approach is a must and everything from input supply to consumption of any commodity must be addressed as a system, as a business model. Hence, the amalgamation of agriculture and business formed the much needed "agribusiness."

Of late, India also realized that agriculture needs to be promoted on agribusiness lines to bring income revolution in farming. The small and marginal holdings should be promoted for collective production and marketing of produce, input and the output should be linked through appropriate linkages, and private investment should be increased in the agriculture and allied sectors to induce capital formation, which in turn will support production systems. Transforming agriculture into agribusiness will generate large-scale employment, promote uniform regional development and equitable distribution of wealth and power, catalyze forward and backward linkage for the development of the economy, and will boost exports. Several models of entrepreneurship rising from secondary agriculture and that of Farmer Producer Organization could be an effective way to engender stable employment and income opportunities in rural India. In addition, the government has several schemes to facilitate entrepreneurship and develop agriculture on agribusiness lines to make farming a viable and lucrative enterprise.

The upcoming decade is expected to be highly crucial for agribusiness as the Indian Council of Agricultural Research (ICAR) estimates that by 2030, the demand for food grains will increase to 345 million tons. In addition, an increase in population, income, and the effect of globalization tend to increase the demand for quantity and quality of food, emphasizing the nutritive value of the produce which urges for innovation in the sector.

27.3 Evolution of Agricultural Education and Research

The earliest records suggest that one of the first agricultural schools was established in 1868 at Saidapet, Chennai. The school was later relocated to Coimbatore and today we see an expansion of the same in the form of the very reputed Tamil Nadu Agricultural University. Some more agricultural colleges were established between 1869 and 1906 at Kanpur, Pune, and Sabour. In this period, the country also witnessed some of the worst famines in the Indian history, and so Lord Curzon, then Viceroy of India, felt a strong need to focus on Indian agriculture. As a result of his thinking and persuasion, a seed was sown for Agricultural Research Institute in Pusa, Bihar. Later, in 1928, the Royal Commission on Agriculture (RCA) highlighted that India was far behind in terms of research in agriculture. The commission thus recommended that a central organization must be in place to play an important role to lead agricultural research and rural development. The commission also called for an agricultural advisor to the Government of India and insisted that a new organization must be set up to connect Pusa and the provincial research institutes. It was in this context that the RCA proposed the setup of the Imperial Council of Agricultural Research (now Indian Council of Agricultural Research), whose major objectives were to promote, guide, and coordinate agricultural research in the country.

Not just research, the network of agricultural schools and colleges was also very poor in the pre-independence era. It was only in 1950, when upon the recommendations of Radhakrishnan Commission (University Education Commission) that the country should focus on setting up of educational institutions/ universities in rural area, a team went to the United States to study the Land Grant pattern. Upon return, the team suggested that India must go ahead with set up of rural universities on the similar pattern. The first such university was established in the state of erstwhile Uttar Pradesh, famously known as G.B. Pant University of Agriculture and Technology, and the rest is history. Today, through its National Agricultural Research System, India has one of the largest network of agricultural universities and agricultural research centers worldwide. The regulation of agricultural education, research, as well as extension is the mandate of Indian Council of Agricultural Research, Veterinary Council of India (Veterinary subdiscipline), and Indian Council of Forestry Research and Education (Forestry subdiscipline). Courtesy this network, the agricultural research and education evolved in India via several reforms. A strong team of scientific professionals have made this country witness Green, White, Blue, Pink, and Golden Revolutions. Not to exaggerate, today we are in surplus and at exportable levels. Miles we have arrived at from where we were at the time of independence, and miles to go. Reforms must go on as per the changing environment and needs. Agricultural universities constitute around 9% of all the universities in the country, but the enrollment in these is less than 1% of the total enrollment in higher education in India. This itself leaves a bigger question for the education system, to pause and think and take appropriate actions to attract the best minds in the country to train in agricultural science. The National Education Policy (NEP) 2020 called for revival of agricultural education with allied disciplines. The policy suggested the strengthening of the capacity and quality of agricultural and allied disciplines to increase agricultural productivity and uplift the economic status of the Indian farmers. It also suggested to develop professionals in agriculture and veterinary science through integrated education programs. These professionals must have thorough understanding of ground level realities and must be able to bring a change through their innovative minds to use local and traditional knowledge of the farmers and blend the same with appropriate technologies to produce a market centric product. At the same time, they must be conscious enough of the critical issues of declining land productivity, climate change and sustainability, food and nutritional security, and poverty and hunger. To benefit the local community, the NEP also suggested to establish Agricultural Technology Parks to endorse technology incubation and dissemination and encourage sustainable methodologies. In light of this policy, the fact cannot be ignored that we need well-trained agribusiness professionals.

27.4 Need for Agribusiness and Entrepreneurship

Agribusiness seems to be the only answer to all the questions relating to food insecurity and low revenue farming. Agribusiness combines agriculture and entrepreneurship to help resolve some of the pressing issues of the world. Agribusiness helps to pave the way to sustainable development by creating jobs, empowering young people, and protecting the environment. According to Sherrie Silver, advocate for the rural youth at International Fund for Agricultural Development, "Farming is seen as unglamorous but it is not just about working on the field rather it's about entrepreneurship as well." Initially, we used to cultivate fruits and vegetables according to the seasons, but today due to the advancement in technology, we can get all-year-round crop using the technologies of greenhouse, modified atmosphere, and many more. This has made India's agriculture sector very sturdy to show resilience to crisis. As they say that every crisis leads to opportunity, we saw many start-ups coming up in pandemic times to facilitate and establish stable supply chains, several innovations in terms of product, and services lined up in this phase only. Hence, agribusiness has greater scope than ever. It is a growing vibrant field nowadays and is an excellent example of private and academia working together. Following four pillars shall be considered while evaluating prospects of agribusiness in India.

27.4.1 Robust Demand

As we talk about food, we tend to think of agriculture. So every living creature needs food and thus the dependency on agriculture will never reduce for humans as well as other beings. Almost 50% of an average Indian's total spending is on food. Agricultural production is thus going to drive industrial growth and service sector. This should reduce the inflation in urban areas and increase the employment in both farm and nonfarm activities (Modi 2013).

27.4.2 Attractive Opportunities

Due to urbanization, India is creating pockets of concentrated demand for high-value products. The country needs cost-effective technologies that are safe for the environment and sustainable for life. A plethora of new technologies such as Remote Sensing, Cloud-based Data Mining, and Artificial Intelligence are a ray of hope for the upcoming era. These technologies can minimize the unpredictability and risk of agribusiness, which will motivate the young generation to participate actively in the sector and bring in more innovative and sustainable models.

27.4.3 Competitive Advantage

India has a great benefit of abundant number of livestock and is one of the biggest market for international brands. Alongside, low wage rate and high availability of labor makes it very cost-effective to survive in the competitive market. Similarly, our resource base for production of commodities is very large. All we need is to combine it effectively and raise our "total factor productivity."

27.4.4 Policy Support

The focus of the government toward agriculture sector has always been very specific and goal centric. Indian policies have evolved from production centric to market centric now. A holistic policy environment is good enough to attract long-term investment and increase "gross capital formation" in agriculture. For example, in 2020, the Indian Government announced an economic package worth Rs 20 lakh crore to help various sectors recover from the adverse impact of Covid-19. Out of this, the government solely dedicated Rs 1 lakh crore to agriculture and allied activities. The fund is now utilized in infrastructure creation such cold chain, warehouse creation, and postharvest management function. Along with this, the government has also declared sector-wise funding, i.e., Rs 10,000 crore for micro food enterprise to invest on marketing and branding, Rs 15,000 crore for dairy infrastructure, Rs 4000 crore for herbs-based agriculture, and Rs 500 crore for bee-keeping ventures.

Census 2011 revealed 116 million people seeking jobs and highest employment level prevails among graduates/diploma holders/postgraduates. Additionally, around 60% of the labor force in the local labor markets do not have the requisite job skills. To create new jobs and enterprises, the Indian Government has taken several measures through schemes such as MNREGA, Make in India, Skill India, and Startup India initiatives, Standup India, Atal Innovation Mission, and so on. To enhance employment and to bring about structural changes, the government also aims to promote entrepreneurship and liberalize the business environment in the country. In the changing economic scenario, several market opportunities have emerged/are emerging. At the same time, such changing scenario poses new challenges before the agricultural value chains. The growth of agriculture in the near future has to come from improvements in productivity, region-specific diversified farming systems, established linkages between production and marketing, farm-level agroprocessing, secondary agriculture, and aggregating small and marginal farmers. In addition, agricultural extension systems need reorientation. Extension must be demand-driven, farm business-oriented, and should involve multiple players from public and private sectors. All this calls for a new era of professionalism such as team work, creating alliances, supportive leadership, change in personal behavior, and attitude. To inculcate the value of professionalism in students through knowledge sharing, skill development, and to make them able and experienced, there are various other components that the agricultural education covers such as Student Ready Program and Experiential Learning. Strengthening of experiential learning and entrepreneurship/skill development through hands-on-training in agriculture and allied sciences is the need of the hour. ICAR's Fifth Deans' committee highlighted the need for promotion of entrepreneurship and recommended curricula reforms for undergraduate/diploma programs across various faculties in State Agricultural Universities. The objective was to increase the employability and employment potential, and enabling graduates to turn into job providers rather than job seekers. Hence, the committee introduced *Entrepreneurship Development and Business Management* as one of the common courses across the degree programs in agricultural universities. Five other courses are:

- Environmental Studies and Disaster Management
- Communication Skills and Personality Development
- Information and Communication Technology
- Agricultural Informatics, and
- Economics and Marketing.

All these courses help the professional development in students as per the changing needs, and train them to be job providers rather than job seekers.

27.5 Acknowledging Agripreneurship

There is no denial to the fact that the government is committed toward the welfare of the farmers, and has called for income revolution in the farming sector. Bringing in the income revolution, the aim is to increase the purchasing power of farmers, thereby increasing their consumption and the level of investment in the farming sector. A farmer must have timely access to the effective resource base and at the same time he must be able to sale the output at the desired price. Not just this, income revolution also means that a farmer must be able to generate enough receipts from the surpluses so that he can adopt effective and innovative technologies, adopt new business models, and is able to make any other kind of investment required by an enterprise.

Today, the changing environment due to various reasons such as trade liberalization, market fluctuations, varying consumer lifestyle and consumption pattern, enhanced ecological regulations, new demand for quality products, food and nutrition security, and advance in technologies has necessitated agripreneurship in the country. These entrepreneurship developments are leading to more innovations, portfolios, and creativity in all areas of farming business. In order to enhance entrepreneurship activities in agriculture, the Government of India has initiated several programs and schemes to develop entrepreneurial skills in people, infusing them with desirable knowledge and expertise to run businesses. Many private organizations, NGOs, financial institutions, and governments (both state and center) have implemented several programs to promote entrepreneurship across the country. In addition to training, many schemes also made provision of seed funding, grants, startup loans, and subsides to the highly innovative, creative, sustainable, and viable projects.

A farmer has to create his own uniqueness through marketing strategy or innovation on some other front in order to make the produce as special and marketable as possible. The government is constantly making efforts to step forward and minimize the risks of such enterprise by organizing trainings, sponsoring mentoring programs, awarding innovative ideas, and to make startups more lucrative in many other ways. Many farmers face issues of investment as they are from low income group, have lack of resources, and are limited access to land. Such farmers can now opt for a collateral-free loan at very low interest rates if they enter into agri-entrepreneurship model. Promoting an enterprise approach to farming, developing it into agribusiness, and establishing rural industries will also arrest migration from rural areas. Agripreneurship thus plays a vital role in strengthening the national economy in various ways.

27.6 Evolution of Policy Environment

Immediately after the independence, the major policy focus was to bring in Agrarian land reforms which was characterized by abolishing zamindari and freeing up land markets, giving the titles to the cultivators. This period also witnessed the emergence of agricultural credit cooperative societies. Following this, a major thrust was made on production front. India announced "Grow More Food Campaign," which was assisted by High Yielding Varieties and is popularly known as "Green Revolution" and was a production-centric approach as it was the need of the hour to produce and reduce dependency on food imports. This was the peak when agricultural research, extension, fertilizer, and irrigation network and supply all witnessed huge support by the government. As a result, India was able to achieve self-sufficiency in food grains. In the next stage, focus shifted to diversification and the success of Green Revolution was replicated to other sectors such as meat and poultry (Pink Revolution), fisheries

(Blue Revolution), milk (White Revolution), oilseeds (Yellow Revolution), fruits and vegetables (Golden Revolution), and the very recent in honey (Sweet Revolution). The series of revolutionary measures were meant to get the country secure in carbohydrates, proteins, fibers, fats, vitamins, and minerals. Thus, the country aimed at achieving not only food but nutrition security too. The policy focus shifted to develop markets, improve infrastructure, facilitate credit facilities and develop banking network, promote sustainable consumption and production systems, and improve rural livelihoods. The beginning of twenty-first century was marked by series of market-centric reforms such as Agricultural Produce Marketing Committee (APMC) Act 2003, Agricultural Produce and Livestock Marketing (Promotion and Facilitation) Act 2017, setting up of futures market to minimize risk and discover fair prices, Foreign Direct Investment (FDI), and opening up Contract Farming in agriculture in order to attract foreign investors in agriculture sector. To attract much needed capital formation and boost ground level credit in agriculture, government announced schemes such as Animal Husbandry Infrastructure Development Fund (AHIDF), Fisheries Infrastructure Development Fund (FIDF), Agri Marketing Infrastructure Development Fund (AMIDF), Dairy Processing and Infrastructure Development Fund (DIDF), Micro Irrigation Fund, Pradhan Mantri Krishi Sinchayi Yojana (PMKSY), Soil Health Card Scheme (SHC), and Paramparagat Krishi Vikas Yojana (PKVY). Besides this, several other schemes of the Government of India such as Pradhan Mantri Matsya Sampada Yojana, promoting 10,000 FPOs and 500 Fish FPOs, e-NAM, Kisan Credit Card, Pradhan Mantri Formalization of Micro food processing Enterprises (PMFME), Operations Total, and Electronic Warehouse Receipt—all are aimed at strengthening postproduction systems and open vast entrepreneurial opportunities in agriculture. The Green Revolution Krishonnati Yojana, which is an Umbrella scheme comprising 12 subschemes, reflects government's holistic approach toward agricultural food systems. These 12 subschemes are as follows:

1. Mission for Integrated Development of Horticulture (MIDH)
2. National Mission on Oilseeds and Oil Palm (NMOOP)
3. National Food Security Mission (NFSM)
4. National Mission for Sustainable Agriculture (NMSA)
5. Sub-Mission on Agriculture Extension (SMAE)
6. Sub-Mission on Seeds and Planting Material (SMSP)
7. Sub-Mission on Agricultural Mechanization (SMAM)
8. Sub-Mission on Plant Protection and Plant Quarantine (SMPPQ)
9. Integrated Scheme on Agricultural Census, Economics and Statistics
10. Integrated Scheme on Agricultural Cooperation
11. Integrated Scheme on Agricultural Marketing (ISAM)
12. National e-Governance Plan in Agriculture (NeGPA).

Three mega programs such as National Food Security Mission (NFSM), Rashtriya Krishi Vikas Yojana (RKVY), and National Horticulture Mission (NHM) have really changed the fate of agriculture scenario in the country by diverting their focus other than production aspects. Moreover, these programs are also made so flexible that the state government can make changes in their allocation as per the requirement. One of the important objectives of the scheme is to create employment opportunities for skilled, unskilled unemployed youths of the country. Additionally, the economic liberalization in the 1990s shifted the focus of the policy toward trade. India developed its very first Agricultural Export Policy in 2018 to aim at doubling exports to the tune of USD 60+ billion by 2022. The government aims at achieving so via boosting high value addition and promoting exports of perishables, promoting novel, traditional, and nontraditional agriproducts, providing institutional support to pursue market access, and connecting farmers to overseas markets.

The Rashtriya Krishi Vikas Yojana (RKVY), initiated in the year 2007, was one of the flagship schemes of Indian Government. The scheme aimed at promoting agripreneurship and agribusiness. The aim of scheme was to strengthen agri and allied infrastructure to encourage agripreneurship and agribusiness through much needed financial support. The scheme also aimed to nurture business incubation ecosystem. In 2017–2018, RKVY scheme was approved for continuation as Rashtriya Krishi Vikas Yojana—Remunerative Approaches for Agriculture and Allied Sector Rejuvenation (RKVY-RAFTAAR). The major objective of RKVY-RAFTAAR is reforming farming as a lucrative economic phenomena through promotion of agripreneurship. In this process, the scheme started strengthening of existing agribusiness incubators and set up new R-ABI (RKVY-RAFTAAR Agri-business Incubators). The seed stage funding of R-ABI Incubatees, agripreneurship orientation, and idea/pre-seed stage funding of agripreneurs are the other components in the process of promoting agribusiness. Besides this, the R-ABIs supports the innovative ideas of agripreneurs that could enhance the productivity in agriculture and ultimately enhance the farmers' income and rural economy. The R-ABIs also facilitate aggregation and networking of various stakeholders through innovators, investors, agripreneurs, and other institutions.

Startup India is another program launched by the Government of India in 2016, with an objective to build an ecosystem that fosters the innovation and startup for sustainable economic growth and employment generation. The program aimed at transforming India into a job-creator country instead of job seekers. Under the Startup India initiative, it provides support for simplification and handling of startups, funding and incentives to eligible startups, and incubation and industry–academia partnership. With 14,740 recognized startups and 1.7 lakh jobs in the year 2020, the program acted as a catalyst for the growth of entrepreneurs. Atal Innovation Mission (AIM) of NITI Aayog, is another flagship initiative by Indian Government to promote entrepreneurship and innovation. It was initiated in 2016 with the objective of holistic development through problem-solving approach and encouraging innovative mindset in schools and also to create an entrepreneurship ecosystem in universities and research institutions. AIM launched the Atal Tinkering Lab (ATL) program at school level to culture curiosity and innovation in young students through latest technology such as Internet of Things (IoT), Robotics, and 3D paintings. AIM also launched Atal Incubation Centres (AICs) at university level to support innovative and emerging entrepreneurs who wants to build sustainable enterprises.

It is evident that the continuous evolving schemes/programs/policies provide support to the emerging agribusiness and improve the socioeconomic status of the society.

27.7 Role of Agri Incubators

As the agriculture is moving forward toward commercialization to reduce the poverty and boost the prosperity, agri incubators play crucial role to encourage the entrepreneurship. Agribusiness incubators are organizations that help small and emerging agripreneurs to grow and expand. They act as catalyst in developing technology and value chains that flourish small agribusinesses. The basic role of any incubator is to support the nascent companies by providing the organizations with all necessary support that includes technical, financial, and marketing access. In general, incubators provide following services:

1. Capacity building, training, and mentoring services
2. Technology transfer and intellectual property policy advisory services
3. Provides technological and strategic advice
4. Collaboration at national and international level
5. Seed funding
6. Links to investors and other financial institutions
7. Infrastructure and other facilities
8. Consultancy services
9. Product development

Along with these services, the agribusiness incubators combine entrepreneurial drive of startups with the available resource for the new ventures. The team of researchers and financers working for incubators analyzes the business model carefully and understands the type and level of risk involved prior to funding or supporting these startups. They help in planning and minimizing risk and increase the ratio of surviving startups by supporting them during seeding issues. Next, these startups increase employment opportunity for the local economy by commercializing technologies. Incubators facilitate startups by providing them all required resources and additional services such as business planning and financing. Hence, the incubators are key contributors for boosting the prosperity and reducing poverty in rural as well as urban areas. The financial support to the agribusiness incubators is offered by the Government of India under RKVY-RAFTAAR, Startup India, and Atal Innovation Mission (AIM), as discussed previously. These schemes/programs provide funds and nurture the incubation ecosystem. Till date, approximately 56,633 startups are recognized by the Department for Promotion of Industry and Internal Trade (DPIIT) of Ministry of Commerce and Industry under Startup India scheme. Along with this, more than 1800 startups are supported by Atal

Incubation Centre under AIM. Under RKVY-RAFTAAR, 10% of total annual outlay is devoted to support the incubation centers.

27.8 The Risks are Easy to Manage: We have Many Stories to Tell

Any agripreneur will have to mitigate five major risks to survive in the competitive market and earn profit. With each risk comes an opportunity to convert it into an Unique Selling Proposition for the enterprise. There are thousands of success stories in India and abroad to lead the way for budding agripreneur in India. The Covid-19 pandemic has opened up several opportunities for innovations. Here onward, we highlight the success mantra of overcoming these barriers by enterprises who found bigger opportunities in different types of risks, and throw lot of positivity for budding entrepreneurs.

27.8.1 The Story of Yinka Adesola in Nigeria Who Managed Production Risk Successfully

This production risk involves the possibility of getting yields lower than the projected yield. There could be many reasons leading to production loss such as extreme weather conditions, attack from disease or pest, and natural calamities (Sciabarrasi 2021). Yinka Adesola, a progressive woman farmer of Oyo state of Nigeria, sets a perfect example of risk mitigation by diversification. She owns 20 ha land out of which she cultivates on 12 ha (1 ha each month) to get an all-year-round crop. She grows 8–10 varieties of crops on this land, which consists of sweet corn, pepper, cucumber, and eggplant. The rest of the land is engaged in animal production and other agribusiness activities. She uses an integrated farming technique, crop rotation technique, and all the waste generated from animals are utilized on the farm itself. Due to her practices and management techniques, she is renowned in Nigeria, and a small agripreneur approaches her for comprehensive 3-month training on farming and agribusiness. She can attain five times the yield that her fellow farmers in the region get from the same area of land. This journey of Yinka was not so smooth as it seems. Like other farmers, she also suffered. She invested her complete asset in farming and started with Cassava where she incurred a huge loss due to very high rainfall that year. She kept up her hopes and cropped maize next year and again lost it. She was curious to know why it was happening. She saw various videos of successful farmers and learnt the methods of integrated farming and crop rotation. This became a game changer for her and she was able to successfully overcome production risks by implementing management tools (Udegbunam 2021).

27.8.2 The Story of Maharashtra's "Farm to Kitchen" Who Managed Marketing Risk Successfully

Any product undergoes this type of risk especially when it is seasonal and perishable like Agricultural Produce. The price could be lower than expected, the quality could be poor, and so the market may not have its demand or even competitors may capture the market share by quoting a lower price. Marketing risk depends on various factors such as price, quality, demand, access to markets, and competitors. It is very difficult for a farmer to control these factors and gain profits from the market. To get the expected profits, he needs to develop some strategies that may be true to his market and his produce (Sciabarrasi 2021). "Farm to Kitchen" managed this risk successfully in the Covid-19 pandemic. "Farm to Kitchen" is an e-Commerce initiative facilitated by KVK-Jalna, Department of Agriculture and Umed ZP Jalna. This service was launched in April 2020. Mr Pravin Ghaghav, the founder of Sarvadnya Foods, saw the opportunity in the problem that farmers were facing due to Covid-19. During the lockdown, farmers were not able to reach the mandis and consumers were also reluctant to come out of their house. This initiative assisted the citizens to stay in their house and provided them home delivery of fruits and vegetables direct from the farms, enabling farmers to market their goods (Indian Council of Agricultural Research 2020).

27.8.3 The Story of Grape Farmers in Ahmednagar Managing Financial Risk Successfully

Financial risk involves cash crunch, generation of lower profit than expected loss of assets, etc. Usually, this type of risk is caused due to the above-mentioned risks. Along with this, the financial risk may occur due to some unavoidable constraints such as government policies, inflation rates, medical emergencies, and fluctuation rates. Ahmednagar is predominantly known for its export quality of grapes. In 2020, due to Covid-19, farmers were unable to export their goods, and marketing in the domestic market was also very difficult due to restrictions on travel. In 2019, grapes were sold at an average price of Rs 50–80 per kg, whereas during 2020 pandemic-induced lockdown, prices plummeted to Rs. 7–8 per kg. With the help and guidance of KVK Ahmednagar, these farmers started processing grapes into raisins to avoid distress selling. They successfully processed 220 tons of harvested grapes and prevented a net loss of Rs 88 lakhs. The story suggests that to mitigate financial risks, the farmers must have a backup plan (Indian Council of Agricultural Research 2020).

27.8.4 The Stories of Waste to Wealth Models for Managing Environmental Risk Successfully

Stubble burning issues in Punjab, Ludhiana, and western UP are on a rise year by year. Stubble burning is harmful as it produces gases such as carbon dioxide, carbon monoxide, and oxides of sulphur. The number of farm residue fires were 60,536 in 2018 and 46,649 in 2019. Between September and November 2020, the incidences crossed 77,700 in Punjab and Haryana. The number of farm fires were highest in the year 2020 since 2012. Several startups have come forward to provide solution to this grave problem. The Confederation of Indian Industry is already working in the fields of Punjab and Haryana. The machines such as rotavator and mulcher were provided to the farmers in time. These two machines help the farmers clean fields in time. Farm to energy, a Punjab-based startup, has offered solutions to convert residue into bioenergy. It has set up bioenergy plants and at the same time it trains farmers to manage crop residues. It purchases residue from the farmers and uses it for the production of biocoal. The government also directed NTPC to purchase biocoal due to which the whole supply chain has strengthened. This has brought several behavioral changes in the farming society (The Time of India 2020).

Another example comes from Dr Rajendra Prasad Central Agricultural University, Pusa, Samastipur, Bihar, whose "Sukhet Model" of using agri crop residues along with kitchen and animal waste has been applauded by none other than Hon'ble Prime Minister Shri Narendra Modi in 80th episode of the *Mann ki Baat* program. The model is unique in the sense that it provides incentives to the farm households by connecting them with Pradhan Mantri Ujjwala Yojana. The university has also developed several other technologies for converting waste to wealth, promoting "type C" of Secondary Agriculture as recommended by the committee on DFI. These technologies, if adopted, have the capacity to increase farm income by 10–15%.

27.8.5 The Story of Amul, Managing Risk Arising out of Human Resources, Successfully

This risk is associated with humans and their family members, farm labors, and employees. Human Resource Risk can arise due to lack of management skills and poor communication. AMUL's largest plant in Mumbai, which is situated in Taloja, was facing manpower crunch during lockdown due to Covid-19, as 50 contractual labors did not turnup for the shift after PM Modi announced 3 weeks lockdown. Milk is an essential commodity for millions of household in India, and the plant had to be kept working in order to manage the supply. The government had already suggested them to work with 50% of the total staff, and on the top of it laborers were reluctant to come to plant as they were scared about their safety. The plant manager with the help of other staff convinced the laborers by ensuring their safety. They installed wash basins with sanitizers on the premises. All the workers were provided with masks, gloves, and other safety wears. There were clear demarcations of distance to be maintained while loading the truck, and the drivers were only allowed inside with identification card for contact tracing. Thus, human resource management is also an art and requires spontaneous actions in order to ensure their safety and gain trust (The Economic Times 2020).

27.8.6 The Story of Akash Chaurasia, a Marginal Farmer from Bundelkhand Who Adopted Innovative Farming Technique

Akash Chaurasia, a marginal farmer from Bundelkhand in Madhya Pradesh, earns nearly 15 lakhs per annum from his 2.5 acres (1 ha) farm by practicing multilayer cropping and other allied agriculture practices such as vermicomposting, biopesticide, and milk. With low cost and innovative farming technique, one crop below the soil like ginger, one crop above the soil which is usually any green leafy vegetable, on the top of it lies a shrub/creeper like ivy gourd, and then papaya is grown at the top. Due to the green leafy vegetable, there are no weeds and so insects are less. The crop like ginger benefits the soil automatically as it gets ploughed by harvesting. This kind of farming can be done only in shade, so they use grass to cover the farm and create a shade to give moderate sunlight retaining the moisture needed for the crop. On shade, he puts creeper with the support of bamboo. This helps him get an all-year-round income by this model. In addition, the shade protects the crop from sudden heat or cold weather due to climate change. This method helps saving water, as in same quantity requirement of water, he gets four to five crops and is able to balance the soil nutrition (Anonymous 2019).

27.8.7 The Story of Farmizen Building up Sustainable Food Systems

The basic concept of this startup was to create a viable and sustainable kitchen garden for the residents of metropolitan cities and get the expertise of farmers on growing methods, inputs, and with flexible time. Along with this, it ensures that you eat only what you grow yourself and end-to-end traceability of the goods is ensured. On the customer's end the need was adequate produce of assured quality, and on the farmer's end the need was assured price and minimizing the risk. Now by this model, a customer can rent a piece of land and grow vegetables of his choice there. He is even free to visit the farm whenever he wants and if he wants to help on the farm; that is also possible according to the mutual understanding with the farmers. The rest of the work at farm will be done by the farmer and the technical support, and input will be supplied by Farmizen. Farmers are associated with Farmizen as partners and they get a predictable fixed income so as to enhance decision-making about the crops to be grown and inputs to be used. Farmers are responsible for land, labor, water, electricity, and so on, and Farmizen is responsible for organic inputs, organic pest control mechanisms, customer support, marketing, and deliveries. With a subscription of Rs 2500 per month including rent, seed, sapling cost, and delivery charges that are all pretty affordable, Farmizen is doing wonders. Thus, Farmizen is a very innovative initiative to build a food ecosystem that is 100% transparent and traceable for consumers, farmers, and the environment (The Times of India 2018).

27.9 The Way Forward

As discussed, agribusiness approaches are the right way to make farming profitable. Understanding entrepreneurship is a key driver for economic development, and India's policies have been rightly inclined toward creating more agripreneur. They may bring more innovations in agriculture and establish a thriving economy through a systematic approach. Hence, the future focus areas of agribusiness entrepreneurship may be summarized as follows. In addition, it must be taken care that the solutions are not limited to this only.

27.9.1 Promotion of Entrepreneurship Development Programs and Schemes

Educating interested entrepreneurs through various entrepreneurship developmental programs, Startups and Incubation Centers need to be continued to promote agripreneurship, so that they are able to set up various agri-clinics and agri-business centers, to cater the needs of the farming community effectively, at nominal and competitive prices. In these programs, both behavioral and technical skill-based trainings should be imparted to the candidates, as well as to the officers of the implementing agencies, for successful implementation of the programs/schemes. In addition to training, the viable projects should be supported through seed funding, grants, subsidies, and handholdings.

27.9.2 Promotion of Collectivization Through Institutions

To help the small and marginal farmers to purchase inputs at wholesale price and sell their goods at retail prices, there is a need for promotion of collective groups, associations, Farmer Producer Organizations, Farmer Producer Companies, Cooperative Institutions, and so on. These institutions are a viable tool for effective transfer of government benefits to the farming community, especially in the case of capital intensive equipment and infrastructure facilitation. The collective enterprise will also help in reducing the input costs and promote scientific temperament.

27.9.3 Adoption of Climate-Smart Agriculture (CSA)

In the changing climate scenario, the agripreneur should try to adopt the CSA approach to transform and reorient agricultural systems in such a way that they are suitable to their local conditions. This will directly help to meet the sustainable developmental goals (SDGs)/global goals. As an entrepreneur, they may provide climate risk information and tools to minimize crop losses on account of disasters. In addition, supply chains innovations may be provided where the technology be used to reduce emissions effectively in agribusiness systems.

27.9.4 Managing the Problem of Plenty

The country has been supporting and framed several policies toward enhancing agricultural production. Now, the need is to shift toward protecting the farmers from the price volatilities, especially during bumper supply in the market. Warehousing is a solution to guard against these price fluctuations for a short period but there is a lack of other warehousing services such as lack of information on number of warehouses available, location, available capacity for storage, storage rents, type of storage, when to lift the stock, how to avoid distress sale, collaterals for raising loans, warehouse management, and reducing spoilage, which can really help the farming community to make the informed decisions. It is an opportunity for the new agripreneur to get into such kind of services.

27.9.5 Promotion of Organic Farming

Due to the changing consumption pattern and increased awareness on health and nutritional aspects of the foods, there is a greater demand for organic farming across the country as well as worldwide. These practices are also in line with the sustainable agriculture. Hence, the agripreneur can stress on organic production, establishing a market linkage, creating a brand, and follow the grading and standardization by making use of the available incentives under different schemes of the government, as discussed earlier.

27.9.6 Innovative Marketing Strategies to Increase Farm Income

As it is learnt during the Covid-19 pandemic, there is a need for innovative marketing strategies. This is possible by deep understanding of the agricultural value chains. Skill upgrading in product marketing and value addition processes needs a continuous improvement; even e-commerce is an emerging opportunity with huge potential in the present days.

27.9.7 Reduced Input Costs Through Adoption of Advance Techniques

There is no doubt that the modern technologies have substantial improvement over the agricultural production and sustainability. Their adoption in any such sensors, weather tracking, satellite imaging, automation, RFID, GIS, GPS, nanotechnology, robotics and analytics, blockchain technology, vertical farming, which may be in the form either technology or practices, will bring both environmental and economic benefits to farmers, community, and the society. Either the adoption or promotion of these technologies may become an opportunity for our agripreneur.

27.9.8 Agricultural Innovation and Inclusive Value Chain Development

It is highly important to understand the evolving consumer. The innovations are needed not only on farm but also along the value chain. Changing tastes and preferences of consumer, globalization, and increasing incomes provide huge opportunities for budding entrepreneurs.

References

Anonymous (2019) Akash Chaurasia, a millionaire farmer from Bundelkhand. Down to Earth. https://www.downtoearth.org.in/video/agriculture/akash-chaurasia-a-millionaire-farmer-from-bundelkhand-64365. Accessed 20 Jun 2021

Indian Council of Agricultural Research (2020) Innovative agri-solutions during COVID-19. In: Singh SK (ed) Farm to kitchen – an e-commerce initiative by young entrepreneur. ICAR, New Delhi. https://krishi.icar.gov.in/jspui/handle/123456789/37070. Accessed 20 Jun 2021

Modi A (2013) Robust agricultural growth is key to India's economic growth prospects. India Today. https://www.indiatoday.in/business/india/story/agricultural-growth-economic-growth-gdp-inflation-kharif-crop-rabi-crop-211249-2013-09-17. Accessed 1 Aug 2021

Sciabarrasi M (2021) The big five risks faced by farmers. UMass Amherst New England Vegetable Guide. Nevegetable.org. https://nevegetable.org/big-five-risks-faced-farmers. Accessed 20 Jun 2021

The Economic Times (2020) How an Amul plant is kept going through the lockdown. https://economictimes.indiatimes.com/news/politics-and-nation/how-an-amul-plant-is-kept-going-through-the-lockdown/articleshow/74985716. Accessed 21 Jun 2021

The Time of India (2020) Stubble burning: how farmers are using innovative solutions to save the environment – YouTube. Accessed 26 July 2021

The Times of India (2018) MY STARTUP IDEA: Shameek Chakravarty's Farmizen helps you rent a farm and grow your own food. https://timesofindia.indiatimes.com/people/my-startup-idea-shameek-chakravartys-farmizen-helps-you-rent-a-farm-to-grow-your-own-food/articleshow/67000466.cms. Accessed 30 Sep 2021

Udegbunam O (2021) Women in agriculture: how i overcame huge losses - farmer. Premium Times Nigeria. https://www.premiumtimesng.com/agriculture/agric-news/455793-women-in-agriculture-how-i-overcame-huge-losses-farmer.html. Accessed 19 Jun 2021

Food Processing Industry 28

M. L. Nithyashree and Raka Saxena

Abstract

The food processing industry (FPI) development role in India is one of the prioritized activities for the economy's growth. In addition, the food processing sector's various direct and indirect effects have attracted more investors and other policymakers in recent years. As a result, there is an incremental capital flow from domestic and international sources. Nevertheless, in general, this sector's progress still needs acceleration in terms of the level of processing. Indian food processing firms are primarily tiny and scattered across the states. The recent growth pattern of the food industry and its subsector composition indicated the structural changes in the sector. Indicating changes in the subsector contribution, particularly toward high-value commodities in the food industry, requires changes in terms of technology upgradation and nutritional safety concerns. FPI over the past decades depicted a growth rate of 6.64%, but the hike in production cost is also significant, accounting for the material cost. The industry also shows comparatively lower profit rates and wages than the average figures of the manufacturing sector. In order to raise the level of processing significantly, focusing on efficiency gain by improving the material cost conversion ratio, the financial status of the firms (liquidity ratio) is vital. Thus, appropriate instruments that facilitate working capital flow, particularly to the small- and medium-sized firms, as well as efficiency enhancement in the agricultural sector, are required for the development of the food processing industry.

M. L. Nithyashree (✉)
Division of Agricultural Economics, ICAR-Indian Agricultural Research Institute, New Delhi, India

R. Saxena
ICAR-National Institute of Agricultural Economics and Policy Research, New Delhi, India

P. K. Ghosh et al. (eds.), *Trajectory of 75 years of Indian Agriculture after Independence*, https://doi.org/10.1007/978-981-19-7997-2_28

Keywords

Food processing industry · Value-added growth · Liquidity ratios · Productivity and profitability

28.1 Introduction

India's food processing sector is receiving more attention because of its various effects. Some of the significant and direct implications of the food processing industry (FPI) development lie in creating employment opportunities, particularly in the rural nonfarm sector, reducing harvest and postharvest losses, stabilizing the farm price, and rising farmer's income in the long run. On the other hand, increasing demand for quality and nutritional food driven by urbanization and changing food consumption patterns necessitates the technology-driven growth of FPI (Chengappa 2004). Though processing and preservation of food items have been typical in the Indian kitchen from time immemorial, positioning FPI as a critical manufacturing commercial activity is taking place only along with the modern practices of agriculture and diversification. India is one of the major producers and suppliers of food grains, milk, fruits and vegetables, nuts, and spices. Significant horticultural production in the past decade has also witnessed the changing path of commercially oriented agriculture sector development. This diversification and available surplus are yet to harness the benefit of value addition using processing.

Recent progress in the growth of FPI is still needed acceleration and the level of processing needs to be raised. Presently, the level of processing in the country is around 10%, which is less compared with other countries such as the United States (80%), Malaysia (80%), France (70%), Thailand (30%), and Australia (25%). This low level of processing has also been reflected in the global share of India's processing sector, i.e., India held the 17th position in the exports of processed food in the world and had a share of just about 2% in the aggregate global exports (EXIM 2017). In addition, harvest and postharvest losses in the country account for Rs 92,651 crore in 2012–13 (Jha et al. 2015). In the long run, high-value commodities such as processed food export are perceived as most beneficial for the country's development (Gopinath and Carver 2002). FPI needs its strategy not only to deal with perishable commodities but also to attract investors as compared with the other manufacturing sectors while meeting the safety standards. With this background, the present chapter deliberates the role and growth pattern of the FPI and its developmental path, particularly in the postindependence, and discusses the opportunities and critical challenges that this industry faces. The study relies on secondary data, mainly published by the Ministry of Statistics and Programme Implementation (MoSPI), Government of India (GoI), and other secondary data sources and publications, focusing more on the organized sector. Section 28.2 presents some of the salient features of the FPI structure in terms of composition, geographical distribution, and its vital role in the economy. Performance of the FPI in terms of two border indicators—productivity, profitability—and financial health of the firms in the industry are described in Sect. 28.3. In Sect. 28.4, the developmental path, growth drivers, key opportunities, and challenges of the FPI are discussed, and finally, Section 28.5 presents the conclusions.

28.2 Indian Food Processing Industry: Structure, Composition, and Regional Spread

The food processing industry in India spreads across the organized and unorganized sectors. It can be seen from Table 28.1 that the processing firms are predominantly located in the unorganized sector, and they constitute around 98.43% in 2015–16. The sector also contributes significantly to employment, i.e., 63.60%. The importance of the organized food processing sector is complemented by its relative share in the industry through investment and value addition. FPI market size in 2015–16 was around Rs 1226.52 billion, of which 70% of the contribution has come from the organized sector. The significance of this sector can also look at from the point of investment, wherein one-third of the investment share has translated as two-thirds of the output generation. Irrespective of the sector, around 85% of the food processing firms spread across 12 states. States such as Andhra Pradesh (14.75%), Tamil Nadu (12.78%), and Telangana (9.99%) have the highest share of processing firms in the organized sector. Uttar Pradesh (14.26%), West Bengal (13.11%), and Maharashtra (9.32%) have the maximum share of the firms in the unorganized sector (Fig. 28.1). Underlining the importance of the organized sector to scale up the value addition, the study further focused more on the features of productivity and profitability that we discuss in detail in the next section.

The food processing industry has a wide range of products. Based on a four-digit classification by National Industrial Classification (NIC)—2008, FPI is grouped into 11 subsectors. The subsectors' contribution and sectoral change in the number of firms, workers, investment, and output in the industry are presented in Table 28.2. It shows that the industry value-added shares in constant terms indicated that the traditional sectors, namely grain mill and starch products, dairy products, oils and

Table 28.1 Structure and relative share of organized and unorganized sectors in the food processing industry in India, 2015–16

Particulars	Unorganized sector	Organized sector	Total
Firms (No.)	2,274,396	37,098	2,311,494
	(98.43)	(1.67)	
Employment (No.)	2,795,534	1,599,818	4,395,352
	(63.60)	(36.40)	
GVA (Rs billion)	367.99	858.53	1226.52
	(30.00)	(70.00)	
Fixed capital (Rs billion)	2959.75	1780.95	4740.70
	(62.43)	(37.57)	

Source: Authors' computation using ASI and NSSO database
Note: Figures in parentheses are percentages of the total

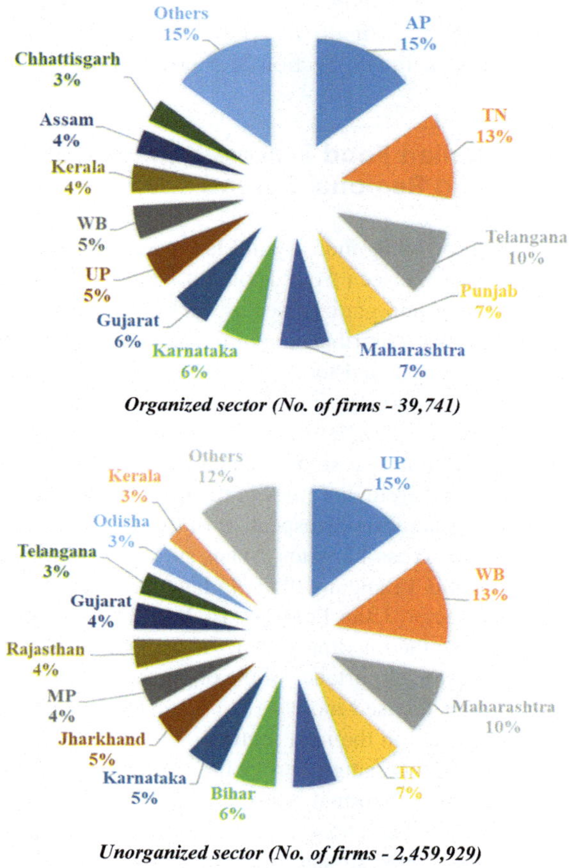

Fig. 28.1 Regional spread of food processing firms across the sector. (*Source*: ASI, NSSO, 2015–16)

fats, and the sugar industry contributed around 51.79% to the total output of the industry. Additionally, the share of an emerging product group—prepared meals and dishes (canned/cooked/ready to eat products)—added 19.60%. Employment generation in the food processing sector is always a topic of interest to the various stakeholders in the country. Within the FPI, almost half of the persons employed in the subsector prepared meals and dishes (25.39%) and grain mill and starch products (20.35%), followed by sugar, dairy, and bakery products with 12.53, 9.70, and 8.45% share, respectively.

With the rising demand for value-added products, there is a structural change in the industry that can be seen as subsectors composition within FPI (Table 28.2). It reveals that from 1991 to 2018, the high-value commodities Gros Value Added (GVA) share is almost doubled in total GVA in the FPI. Meat, fish, processed fruits and vegetables, milk, and bakery products share of GVA in FPI GVA increased from 0.49, 2.84, 0.68, 6.88, and 3.59% in 1991–92 to 1.6, 4.4, 3.82, 9.11, 12.64, and 8.77% in 2017–18, respectively. The increasing share of GVA in these sectors is in

Table 28.2 Changes in the sectorial composition of organized food processing industry, 1991–2018 (in %)

Key indicators	No. of factories			No. of workers			Fixed capital			Gross value added		
Year/product groups	1991–92	2003–04	2017–18	1991–92	2003–04	2017–18	1991–92	2003–04	2017–18	1991–92	2003–04	2017–18
Meat products	0.16	0.24	0.48	0.33	0.91	1.52	0.54	2.08	1.24	0.49	1.09	1.6
Fish products	0.96	1.38	1.42	1.16	2.93	5.36	0.75	2.49	2.55	2.84	2.67	4.44
Fruits and vegetables	1.14	2.2	3.32	1.48	2.54	4.51	1.68	3.1	4.03	0.68	2.19	3.82
Oils and fats	16.75	10.91	8.05	8.88	6.23	5.74	16.77	12.31	7.97	15.45	13.36	9.11
Dairy products	2.44	4.01	5.46	4.26	5.84	9.70	10.44	7.67	12.72	6.88	15.21	12.64
Grain and starch	50.15	53.92	51.77	21.72	22.56	20.35	10.22	11.8	13.38	13.54	14.46	16.09
Bakery products	3.8	3.95	4.65	2.9	3.22	8.45	2.36	2.32	4.07	3.59	4.44	8.77
Sugar	6.96	3.85	1.93	30.47	20.1	12.53	41.85	42.38	33.82	30.12	24.13	13.95
Sugar confectionery	0.71	1.11	1.86	0.69	0.93	2.65	0.93	1.81	3.59	1.75	3.73	3.41
Miscellaneous food	15.35	16.37	18.45	27.33	33.24	25.39	13.59	11.93	12.54	22.75	15.25	19.6
Animal feed	1.58	2.06	2.61	0.77	1.5	3.81	0.86	2.11	4.09	1.91	3.47	6.57
FPI	100	100	100	100	100	100	100	100	100	100	100	100

Source: Author's calculations based on ASI database

line with rising capital investment and the number of firms, which is also reflected in their employment share increment. There is a significant reduction in the GVA of the sugar industry share to 13.95% in 2017–18 from 30.12% in 1991–92. Though there is a reduction in the employment share in the sugar industry over the period, it still accounts for 12.53% of the industry's total workers. It ranks third place after the miscellaneous food and grain industry. Thus, these changes in the subsector contribution to the food industry highlight the changes needed in terms of technology upgradation and nutritional safety concerns.

28.3 Performance of the Indian Food Processing Industry

The performance of any industry can be approached by analyzing the industry performance over a period of time or comparing its performance with other industry progress. Here, to assess the performance of FPI, we relied on both approaches. Market size and growth of FPI was analyzed and presented in Table 28.3. It indicates that the average value of GVA for the industry grew by 6.64% from 1980 to 2018. The highest growth of more than 10% was registered in the case of high-value products such as meat, processed fruits and vegetables, and feed industry. Except for the sugar industry, more or less, all the subsector growth was in line with the FPI growth. However, a temporal trend in the ratio of GVA to the output of the FPI remains the same with the fluctuating trend from 1980 to 2018 (Fig. 28.2). As an exception, the ratio peaked at 14.45% in 1982–83 and 14.80% in 1994–95. It reflects that the material cost or total input consumed relative to the value of total output generated over a period of time remains the same. In other words, producing output

Table 28.3 Subsectors output growth in the food processing industry

Particulars	TE 1991–92	TE 2004–05	TE 2017–18	1980–2018 Trend growth (%)
	Rs billion (at 2011–12 prices)			
	Gross value added			
Meat products	0.7	3	15.2	12.99
Fish products	2.7	6.7	31.2	8.92
Fruits and vegetables	1.2	7.1	37.1	11.65
Oils and fats	24	37.2	73.6	6.34
Dairy products	14.2	40.5	101.8	8.92
Grain mill and starch	18.7	42.8	139.2	8.18
Bakery products	5.9	11.7	54.4	8.71
Sugar	51.3	73.2	120.2	3.57
Sugar confectionery	2	9.3	31.4	13.8
Miscellaneous food	38.1	44.3	172.3	5.53
Animal feed	2.4	8.8	40.9	11.53
FPI	161.2	284.6	817.2	6.64

Source: Author's calculations based on ASI database

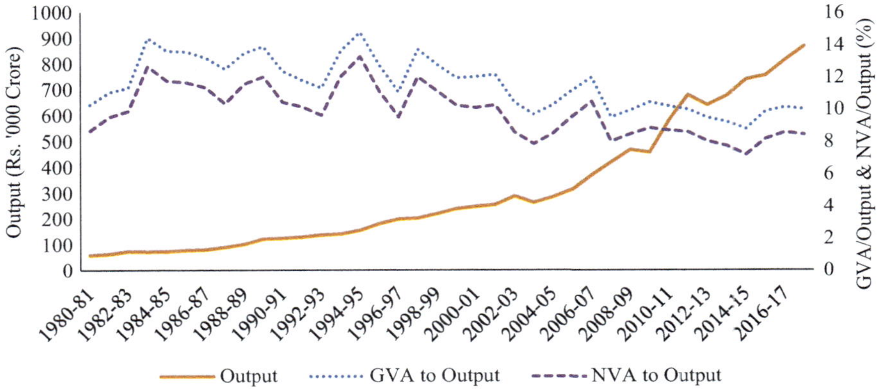

Fig. 28.2 Trends in total output, ratio of GVA and NVA to output in the FPI, 1980–2018

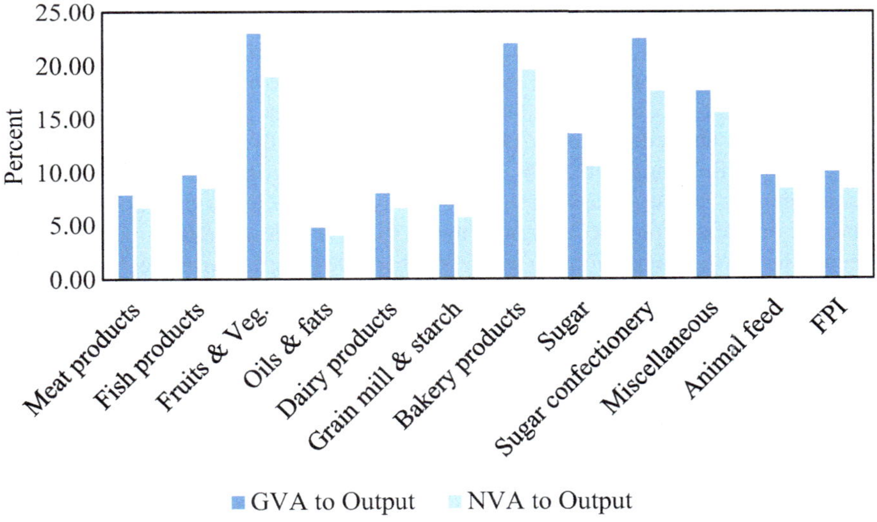

Fig. 28.3 Ratio of gross and net value added to the output of FPI and subsectors, TE 2017–18

worth Rs 100, almost 90 rupees has to be consumed as material. From Fig. 28.3, it is visible that, across the products, meat, fish, oil and fats, dairy, grain, and feed industry showed a high material cost. On the other hand, processed fruits and vegetables, bakery, sugar confectionery, and miscellaneous food items reached a higher ratio of 22.98, 22.05, 22.50, and 17.67%, respectively. It indicates their value-intensive production nature and relatively low-cost inputs.

Further, to understand the composition of various cost components, the cost of production of FPI and subsectors was obtained by adding the cost of material, users' cost of capital, workers wage and salaries, and fuel, which enters into the production process during the year. The proportion of each component in the total cost of production is calculated for the periods TE 1991–92, TE 2004–05, and TE 2017–18 and presented in Table 28.4. It reveals that the cost of production of FPI in TE 1991–92 was Rs 3444.9 billion. For this share, material cost constituted 87.28%, followed by the cost of capital (5.05%), wages and salaries (4.57%), and fuel (3.59%). Over the period, the production cost has increased to Rs 81,466.4 billion in TE 2017–18. More importantly, the share of production components over the period reveals that material cost continues to dominate. Its share has marginally increased from 87.28 to 88.90% in TE 1991–92 and TE 2017–18, respectively. On the other side, the share of the user cost of capital, wages and salaries, and fuel have marginally reduced to 4.58, 3.59, and 2.93% in TE 2017–18 from 5.05, 4.57, and 3.10%, respectively in TE 1991–92. Among the products, most of the subsectors have shown a similar pattern and spent more on purchasing materials. The meat industry cost of materials has significantly increased to 92.55% in TE 2017–18 from 77.56% in TE 1991–92. Contrarily, some of the products, namely grain and starch, bakery, miscellaneous food, and feed industry indicated a reduction in the share of material cost and increasing share of users cost of capital during 1991–2018.

The study has compared outstanding loans, interest paid, and profitability of FPI with the manufacturing sector to analyze the performance of financial indicators. As shown in Fig. 28.4, the trend in the profit share of FPI in the manufacturing sector raised initially and reduced after 2001–02, though it slightly improved after 2003–04 but continues fluctuating. In 2015–16 the share of profit from the food industry to the manufacturing industry was 4.58%, with a comparatively lower profit rate (12.14%) than the manufacturing sector profit rate (19.01%). The lower rate of profit has also been reflected in remuneration to workers, where the manufacturing sector wages for workers in real terms were Rs 1.30 and Rs 1.03 lakh per annum in the case of FPI. Similarly, the corresponding nominal wage was Rs 1.49 and Rs 1.17 lakh per annum in 2017–18 (Nithyashree and Pal 2020).

The financial performance of the FPI was analyzed using some of the critical liquidity parameters and presented in Table 28.5. It revealed that the computed current asset ratio for the FPI and its subsectors indicated a value of more than 1. This indicates, on average, a firm in the industry can pay its current liabilities within a year. Exceptionally, in the recent period, the sugar industry's financial performance was not promising, with the value of the current ratio being less than 1 (0.95). An alternative and reliable measure of short-term liquidity, i.e., the quick-acid ratio, indicated that FPI in general and subsectors (except processed fruits and vegetables) turn to be weak (with the ratio value less than 1) in liquidating its assets, which is a cause of concern.

The receivable days and payable days measure the average number of days the company takes to collect revenue after a sale has been made and how long it takes a company to pay its invoices to suppliers, respectively. Lesser the days, the better will be the financial health; generally, less than 90 days is an acceptable benchmark. In the case of FPI generally, it takes 39 and 19 receivable and payable days, respectively. However, as an exception, the sugar industry indicated a delay in payment to the supplier, which is also in line with a low-current and quick-acid value. How best

Table 28.4 Production cost and its components in the food processing industry, 1991–2018 (TE: Triennium Ending)

Key indicators	Material			Cost of capital			Wages and salaries			Fuel			Cost of production		
	Percent												(Rs billion)		
Year/industry group	TE 1991–92	TE 2004–05	TE 2017–18	TE 1991–92	TE 2004–05	TE 2017–18	TE 1991–92	TE 2004–05	TE 2017–18	TE 1991–92	TE 2004–05	TE 2017–18	TE 1991–92	TE 2004–05	TE 2017–18
Meat products	77.56	82.84	92.55	7.40	6.21	2.22	9.51	4.62	2.74	5.52	6.33	2.50	9.9	138.4	2111
Fish products	90.65	90.14	90.95	4.22	4.09	3.16	2.70	2.53	3.54	2.43	3.24	2.35	63.0	480.6	3498.4
Fruits and vegetables	71.98	70.40	76.57	13.86	12.22	10.12	9.89	9.19	8.75	4.27	8.18	4.56	17.7	138.3	1420.4
Oils and fats	92.33	94.04	94.60	3.12	2.57	2.14	1.56	0.89	1.23	2.98	2.51	2.02	999.4	3930.2	15,333.6
Dairy	88.08	89.11	91.47	3.58	3.13	2.93	4.96	4.03	3.35	3.38	3.73	2.25	351.8	1839.5	13,052.2
Grain and starch	91.29	90.56	90.35	3.81	3.60	4.43	2.49	2.29	2.18	2.41	3.55	3.04	716.0	3262.8	19,368.6
Bakery	80.83	81.27	77.54	4.77	4.52	5.65	9.34	6.82	10.16	5.06	7.39	6.65	75.5	344.9	2139.7
Sugar	78.18	72.77	81.18	9.68	17.71	10.50	9.68	7.07	5.21	2.45	2.45	3.11	738.6	2407.2	9859.9
Sugar confectionery	76.45	78.24	79.00	10.50	8.34	8.35	8.26	7.04	8.72	4.79	6.38	3.93	17.3	147.4	1328.8
Miscellaneous food	84.70	81.59	83.59	4.76	5.41	5.14	5.22	6.45	6.70	5.32	6.54	4.56	391.9	1325.4	8933.8
Animal feed	91.80	91.90	91.64	2.89	2.63	3.05	3.16	2.66	3.15	2.15	2.81	2.16	63.7	525.2	4420.1
FPI	87.28	86.98	88.90	5.05	5.93	4.58	4.57	3.56	3.59	3.10	3.53	2.93	3444.9	14,540.1	81,466.4

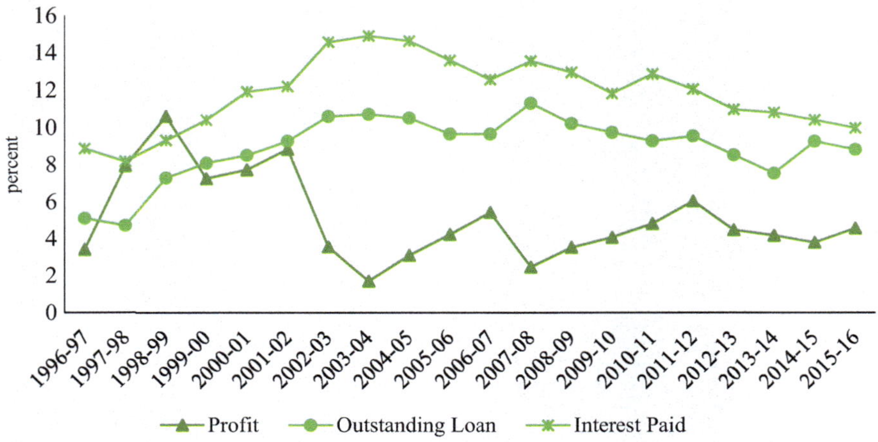

Fig. 28.4 Trend in profit, interest, and outstanding loan share of FPI in the manufacturing sector, 1996–2016

the firm will manage its inventory and how fast it converts inputs into cash flow are captured through inventory days and cash conversion cycle, respectively. Over the period, the length of these two indicators has extended from 2000 to 2016, particularly for the grain, starch, and sugar industries.

Technology-focused productivity being the primary driver of output growth, various scholars have analyzed the pattern of resource use and their productivity growth in the FPI in the Indian context. Studies on Total Factor Productivity Growth (TFPG) from the 1980s to till mid-2000s indicated a low rate of technological growth. Kumar and Basu (2008) and Ali et al. (2009) indicated 1.03 and 1.04% growth during TFP 1988–89 to 2004–05 and 1980–81 to 2001–02, respectively. The pattern of technological gain varied within the FPI. Among the subsectors, the feed, meat, fruits and vegetables, grain, starch, and confectionery industry indicated positive growth in the postliberalization period and reduced TFP in fish, vegetable oils, sugar, bakery, and miscellaneous foods (Ali et al. 2009). Negative growth of sugar companies with 0.99% TFP during 2005–06 to 2013–14 was reported by Singh (2016). Ali (2007) indicated that the rise in capital investment led to significant growth in the meat industry, while Singh (2016) mentioned that fluctuating sugar prices and financial crisis pushed 65% of the firms into a loss. These studies show that the food industry growth, particularly in the organized sector, is mainly driven by incremental input use and scale efficiency. Significant reduction in the resource use efficiency and technological regress indicates the untapped potential of the FPI in India. In addition, variation in the growth patterns across the subsectors in the FPI is primarily attributed to capital investment. Capital investment is the most crucial factor in accelerating the processed food industry growth. Researchers indicated that encouragement of import and R&D investment (Kumar and Basu 2008) and trained human resources and management of quality standards (Kumar 2008) are necessary to speed up technological growth.

Table 28.5 Performance of FPI and subsectors in terms of liquidity ratio

	TE 2000–01	TE 2005–06	TE 2010–11	TE 201516	TE 2000–01	TE 2005–06	TE 2010–11	TE 2015–16
	Current ratio				Quick ratio			
Meat products	3.15	1.89	1.70	2.09	2.58	1.27	1.15	1.65
Fish products	1.59	1.43	1.43	1.40	0.96	0.81	0.81	0.90
Fruits and vegetables	1.47	1.57	1.52	2.11	0.66	0.77	0.71	1.55
Oils and fats	1.38	1.23	1.25	1.37	0.74	0.64	0.62	0.88
Dairy products	1.35	1.26	1.30	1.45	0.83	0.69	0.71	0.95
Grain and starch	1.32	1.31	1.52	1.52	0.65	0.76	0.55	0.59
Bakery products	1.26	1.37	1.65	1.35	0.70	0.92	1.13	0.93
Sugar	1.11	1.16	1.17	0.95	0.19	0.42	0.33	0.25
Sugar confectionery	1.23	1.25	1.35	0.89	0.48	0.68	0.71	0.39
Miscellaneous foods	1.56	1.41	1.50	1.57	0.68	0.74	0.84	0.98
Animal feed	1.54	1.50	1.63	1.40	0.89	0.84	0.81	0.84
FPI	1.30	1.30	1.29	1.26	0.60	0.75	0.53	0.62
	Ratio of working capital (WC) to annual sales				Receivable days			
Meat products	9.45	12.57	14.95	14.51	7.30	45.20	48.77	47.96
Fish products	4.88	10.73	10.75	8.99	20.30	50.13	44.29	37.86
Fruits and vegetables	5.67	19.99	30.65	49.31	3.24	55.47	60.19	153.54
Oils and fats	3.15	7.93	9.71	11.07	15.44	46.51	38.40	51.04
Dairy products	1.89	6.36	6.09	8.80	20.48	33.84	15.40	27.95
Grain and starch	0.42	9.81	26.47	19.16	9.00	48.79	51.02	48.72
Bakery products	2.94	10.29	9.54	5.77	9.24	31.60	16.17	19.32
Sugar	0.87	9.43	12.11	−4.66	4.17	34.54	19.53	25.17
Sugar confectionery	3.85	5.99	1.45	−3.52	7.13	29.14	15.74	17.97
Miscellaneous foods	3.63	9.69	16.57	15.63	8.17	37.21	30.63	40.42
Animal feed	5.23	10.12	10.40	8.15	16.09	44.77	21.27	31.51
FPI	0.96	9.74	13.14	9.80	8.44	48.05	30.67	39.40
	Creditor days				Inventory days			
Meat products	26.70	29.72	33.99	20.39	31.85	47.86	31.37	20.38
Fish products	7.12	34.88	31.63	25.98	17.79	63.84	54.26	38.64
Fruits and vegetables	0.05	47.86	67.52	53.19	32.06	88.16	141.36	94.34
Oils and fats	17.10	46.02	58.15	56.62	19.68	57.82	59.25	54.21
Dairy products	17.17	34.36	20.22	30.77	15.58	46.90	35.24	34.40
Grain and starch	14.03	47.15	38.38	36.82	34.74	61.43	144.19	124.48
Bakery products	13.95	38.22	24.26	23.61	23.91	46.68	24.39	24.78
Sugar	31.22	59.73	88.82	119.95	107.37	141.95	228.79	249.16

(continued)

Table 28.5 (continued)

	Creditor days				Inventory days			
Sugar confectionery	8.53	32.68	31.12	35.62	22.75	46.98	34.02	53.14
Miscellaneous foods	12.96	41.61	29.81	36.84	30.44	62.04	54.70	56.89
Animal feed	8.98	38.18	21.72	20.65	20.57	52.88	37.60	32.96
FPI	4.81	10.39	18.74	19.41	45.15	61.56	101.51	87.94

	Cash conversion cycle (CCC)			
Meat products	12.45	63.35	46.15	47.95
Fish products	30.97	79.09	66.92	50.52
Fruits and vegetables	35.25	95.77	134.03	194.70
Oils and fats	18.02	58.31	39.50	48.63
Dairy products	18.89	46.37	30.42	31.58
Grain and starch	29.72	63.07	156.83	136.38
Bakery products	19.20	40.06	16.30	20.49
Sugar	80.32	116.77	159.50	154.38
Sugar confectionery	21.34	43.44	18.64	35.49
Miscellaneous foods	25.64	57.65	55.52	60.47
Animal feed	27.68	59.47	37.15	34.67
FPI	48.78	99.21	113.44	107.92

28.4 Developmental Path, Opportunities, and Challenges of the Food Processing Industry

Various policy initiatives by the Indian Government during the 1970s and 1980s reveal the nation's effort to bring the economy to the forefront. However, it was only after the liberalization of 1991 that significant economic changes, particularly in the industrial sector, were reported. Rodrik and Subramanian (2005) describe the process of effort in terms of policy changes of the 1980s and 1990s to boost the manufacturing productivity in India as pro-business and pro-market reforms, respectively. A pro-business focused on raising the profitability of the existing industrial and commercial establishments, whereas a pro-market orientation aimed to remove market bottlenecks and facilitate new entrants. By and large, all these initiatives continued to encourage more capital-intensive industries.

Pertinent to agro-industry development, the importance of villages, agriculture, and agro-industry was much emphasized by Mahatma Gandhi many years ago; as a result, the country followed different strategies for the food industry growth. They are broadly described under their phases, namely Mahatma Gandhi-Swadeshi Phase (1947–50), Nehru-Mahanalobis Phase (1950–84), and Modernization Phase (1984 onward). Though the country has set up a separate ministry dedicated to food processing in 1988, in the modern phrase, the sector could not benefit much from these policies until the introduction of the Foreign Exchange Management Act (FEMA) in 1999 (Reserve Bank of India 2021).

At this juncture, the country's transitional phase in production basket expansion into commercial cultivation and diversification toward high-value crops and structural transformation of employment patterns witnessed the need for expansion of the food processing industry to have a multiplier effect on the Indian economy. On the other hand, rising per capita income and changing consumption patterns, particularly in the middle and high-income groups (Bhalla and Hazell 1998; Chengappa 2004), also gear up the demand for high-value products in terms of nutritional value. Thus, processed food exports are perceived as most beneficial, and developed countries tend to export more processed food, further increasing their trend (Gopinath et al. 1996; Gopinath and Carver 2002). This has led to a significant change in the trade structure and composition of agri commodities. It is reported that the share of processed food in total food export increased to 63% from 41% and the contribution to this from the developing countries tripled during 1980–2006 (Jongwanich 2009).

Realizing the untapped potential, investment flow into FPI was given priority, and in the country, many policy measures have been taken to attract investment, both domestic and foreign capital. Some of the major ones are the introduction of Pradhan Mantri Kisan Sampada Yojana (PMKSY) and the approval of 100% FDI through an automatic route. In addition, several fiscal incentives, the establishment of food parks and agri export zones, the establishment of food processing and training centers, and the establishment of laboratories to ensure quality standards are the critical policy and regulatory measures undertaken in the FPI. All these efforts result in capital accumulation at the rate of 9.31% growth per annum from 1980 to 2018, with faster growth after the 2000s (Nithyashree and Pal 2020). Recent significant policy developments by the Indian Government, namely doubling farmers' income by 2022 and the agriculture export policy of 2018 for doubling agricultural exports by 2024, emphasize enormous significance for the promotion of the food processing sector. Therefore, the FPI sector has been identified as one of the major strategies to achieve the above two policy objectives.

The approach of institutional support in terms of Farmer Producer Company (FPC) and upward revision of Micro Small Medium Enterprises (MSMEs) opened a new avenue to the progress of the firms in the FPI. In mobilizing small firms in the unorganized sector, the role of FPC alliance with food processing firms is noteworthy (Satyasai and Singh 2021). Besides this, the FPI progress is insufficient for the country, which contributes to a double loss—resources use inefficiency of the sector itself and underutilization of the raw material from the primary production and human capital. The agriculture sector in the country itself can improve the regulatory framework to usher in more investment and increase the competitiveness for the betterment of value addition by channeling the resources better. Growth and innovation in the food manufacturing sector are also primarily influenced by the reforms and efficiency gains in the agriculture sector. While efficiency gains in agriculture will make the material cheaper, industry efficiency will increase the demand and make the price competitive. Though better efficiency is of equal importance to bring competitiveness in both sectors, it is reported that efficiency gains in agriculture are relatively more important and have a more substantial effect than food processing (Gopinath et al. 1996). In addition, to captivate the technology coming from the source of FDI, it is necessary to expand the absorptive capacity of the domestic industry.

To integrate the unorganized tiny firms, importance should be given to enabling the competitive environment to operate efficiently, mainly access to finance. This indicates formal institutional credit accounts for 16% of total credit demand by MSMEs (Pradhan and Agarwal 2021). Along with some of the other constraints as indicated in the survey by FICCI (2010), addressing the actual and desired level of employment of skilled labor, infrastructure development and safety measures are equally essential to accelerate the developmental path of the food processing industry.

28.5 Conclusions

The various effects of the food processing industry development in India are very much relevant during the phase of structural change. Food processing firms are primarily tiny and spread across the organized and unorganized sectors, with the presence across the states. However, the significance of organized sector and scale expansion can be seen from the point of investment, wherein one-third of the investment share has translated as two-thirds of the output generation in the FPI. This refers to the need to bring firms from the unorganized to the organized sector to scale up the level of processing. The changes in the subsector contribution to the FPI reveal the need for technology upgradation in terms of nutritional value and competitive price for processed produce. Industry performance reveals the higher growth of high-value products such as meat, processed fruits and vegetables, and the feed industry. At the same time, the material cost conversion ratio is found to be low. Thus, value-intensive production nature and relatively low-cost inputs are critical. The effort to improve the financial health of FPI is equally essential, which is reflected in liquidity indicators. Growing capital intensity reflects that the food industry is attracting investment at the national and international levels. The need for appropriate facilitating factors such as working capital accessibility, particularly to the small- and medium-sized firms, and efficiency enhancement in the primary agriculture sector will speed up the level of processing that will allow this sector to develop in a much bigger way.

References

Ali J (2007) Productivity and efficiency in Indian meat processing industry: a DEA approach. Indian J Agric Econ 62(4):637–648. https://www.researchgate.net/publication/292627450_Productivity_and_efficiency_in_Indian_meat_processing_industry_A_DEA_approach

Ali J, Singh SP, Ekanem E (2009) Efficiency and productivity changes in the Indian food processing industry: determinants and policy implications. Int Food Agribus Manag Rev 12(1):43–66. https://doi.org/10.22004/ag.econ.53745

Bhalla GS, Hazell P (1998) Foodgrains demand in India to 2020: a preliminary exercise. Econ Polit Wkly 32(52):150–154. https://www.epw.in/journal/1998/10/discussion/demand-foodgrains-2020.html

Chengappa PG (2004) Emerging trends in agro–processing in India. Indian J Agric Econ 59(1): 55–74. https://www.semanticscholar.org/paper/Emerging-Trends-in-Agro-Processing-in-IndiaChengappa/d6eb7c69744479258543947ea4d01f45b5f85e70#:~:text=Emerging%20Trends%20in%20AgroProcessing%20in%20India.%20Agricultural%20transformation,Green%20revolution%20technologies%20are%20strengthening%20our%20food%20

EXIM (2017) International trade in processed foods: an Indian perspective. In: Working paper no.61, Export-Import Bank of India. https://www.eximbankindia.in/Assets/Dynamic/PDF/PublicationResources/Research Papers/79file.pdf

FICCI (2010) Bottlenecks in Indian food industry–survey 2010. Federation of Indian Chambers of Commerce and Industry, New Delhi, pp 1–32. http://ficci.in/Sedocument/20073/Food–Processing–Bottlenecks–study.pdf

Gopinath M, Carver J (2002) Total factor productivity and processed food trade: a cross–country analysis. J Agric Resour Econ 27(2):539–553. https://doi.org/10.2307/4098785

Gopinath M, Roe TL, Shane MD (1996) Competitiveness of U.S. food processing: benefits from primary agriculture. Am J Agric Econ 78(4):1044–1055. https://doi.org/10.2307/1243860

Jha SN, Vishwakarma RK, Ahmad T, Rai A, Dixit AK (2015) Report on assessment of quantitate harvest and post-harvest losses of major crops and commodities in India. ICAR-All India Coordinated Research Project on Post-Harvest Technology, ICAR-CIPHET, PAU, Ludhiana. https://doi.org/10.13140/RG.2.1.3024.3924

Jongwanich J (2009) The impact of food safety standards on processed food exports from developing countries. Food Policy 34(5):447–457. https://doi.org/10.1016/j.foodpol.2009.05.004

Kumar P (2008) Structure and performance of food processing industry in India. J Indian School Polit Econ 22(1–4):127–116. https://www.researchgate.net/publication/308294866_Structure_and_performance_of_food_processing_industry_in_India

Kumar M, Basu P (2008) Perspectives of productivity growth in Indian food industry: a data envelopment analysis. Int J Product Perform Manag 57(7):503–522. https://doi.org/10.1108/17410400810904001

Nithyashree ML, Pal S (2020) Rising capital intensity and employment potential of Indian food processing industry. Indian J Agric Econ 75(4):518–533. ISSN: 2582–7510. http://isaeindia.org/wp-content/uploads/2021/02/Vol.-75-No.-4-October-December-2020.pdf

Pradhan HK, Agarwal PK (2021) MSME financing in India: key to the 5 trillion economy. Inclusive Finance India Report 2020. pp. 137–157. Tamal Bandyopadhyay, Access Assist. ISBN 978-81-949303-1-0

Reserve Bank of India (2021) Foreign direct investment flows to India. Monthly bulletin. India's Central Bank, Government of India. pp. 997–1017. https://rbidocs.rbi.org.in/rdocs/Bulletin/PDFs/02BAT090512.pdf

Rodrik D, Subramanian A (2005) From "Hindu growth" to productivity surge: the mystery of the Indian growth transition. IMF Econ Rev 52:193–228. https://doi.org/10.2307/30035894

Satyasai KJS, Singh A (2021) The food processing industry in India: regional spread, linkages and space for farmer producer organisations. In: Bathla S, Kannan E (eds) Agro and food processing industry in India. India studies in business and economics. Springer, Singapore. https://doi.org/10.1007/978-981-15-9468-7_4

Singh SP (2016) Technical change and productivity growth in the Indian sugar industry. Proc Econom Financ 39:131–139. https://doi.org/10.1016/s2212-5671(16)30257-x

Transformation of Agricultural Extension System 29

P. Mooventhan, Rajarshi Roy Burman, Rabindra Nath Padaria, and Souvik Ghosh

Abstract

In order to realize agricultural potential and to increase agricultural yields, India's extension system has experienced major conceptual, structural, and institutional changes over a period of time beginning with Green Revolution. In India, major changes in the provision and financing of agricultural extension services emerged in the late 1990s. The reform programs of the late 1990s were motivated by the failure of the World Bank funded Training and Visit (T&V) model to promote agricultural development by providing effective and efficient extension services in a timely and sustainable manner. Further, two of the most prominent research and extension reform initiatives were the World Bank funded 1998–2004 Diversified Agricultural Support Project (DASP) and the 1999–2005 National Agricultural Technology Project (NATP). In addition to these programs, another prominent reform initiative is the ongoing e-Choupal initiative of the Indian Tobacco Company (ITC). The private-sector program employs information and communication technologies (ICTs) as instruments for improving agricultural extension service provision in terms of outreach (cost), effectiveness, efficiency, transparency, and accountability. Policies need to provide incentives for the private and third sectors to actively participate in agricultural research and extension, either individually or in partnership with public service providers. The list of the factors that could strengthen the agricultural extension system is determined by the lack of evidence-based performance assessments of the existing reform initiatives and is therefore not exhaustive. Hence, reporting standards need to be defined that help to identify the efficiency and effectiveness

P. Mooventhan (✉) · R. R. Burman · R. N. Padaria · S. Ghosh
ICAR-National Institute of Biotic Stress Management, Raipur, Chhattisgarh, India
e-mail: p.mooventhan@icar.gov.in

P. K. Ghosh et al. (eds.), *Trajectory of 75 years of Indian Agriculture after Independence*, https://doi.org/10.1007/978-981-19-7997-2_29

of programs, highlight problem areas in the implementation and budgeting of programs, and thus help to define corrective actions (Raabe 2008).

Keywords

Extension system · ATMA · T&V · FFP · NAIp · KVK · NICRA · ICT initiative

29.1 Introduction

Approaches to agricultural extension in India and worldwide are still evolving continuously. Since the Green Revolution in the 1970s and 1980s and the acknowledged unsustainability of the Training and Visit (T&V) program (Moore 1984), agricultural extension, with its focus on increasing production via technology transfer, has adopted decentralized, participatory, and demand-driven approaches. While the call for demand-driven agricultural extension has existed for several decades now, new modes of reaching out to farmers could have significant impact in India, as they might better reflect the local information needs of farmers. The diverse nature of the Indian subcontinent, with its wide variety of agroclimatic regions and broad range of socioeconomic conditions in the rural population, calls for agricultural extension approaches is situation-specific. As more than 81% of Indian farmers cultivating an area of 2 ha or less (India, Directorate of Economics and Statistics 2009; NSSO 2006), there is an increasing need for stronger extension advisory system for small-holder farmers. Further, progress in poverty and hunger reduction crucially depends on the increased productivity and profitability of these farmers, which in turn depends on the successful delivery of agricultural extension.

India has made significant achievement in agriculture by increasing food production by four folds during last six decades. Among many drivers to accomplish this task, the policy, research, and extension support have played crucial role (Reddy and Swanson 2006). Public extension played a major role in ushering Green Revolution in Indian agriculture. Extension education has now developed as a full-fledged discipline having its own philosophy, objectives, principles, methods, and techniques, which must be understood by every extension worker and others connected with the rural development.

29.2 Extension Systems in India

Agriculture development in India is basically a state subject. However, the crucial role the agricultural sector plays from the perspective of ensuring food security of its large population, the Union Government plays a major role in formulating policies that has direct bearing on the growth of agricultural sector. The Union Government mainly provides roadmap through its policies, programs, and budgetary support to the sector. The programs conceived at national level are mainly implemented by the states through its development departments. Besides, states also formulate region-specific development programs.

Similarly, Indian Council of Agricultural Research (ICAR) is an apex body at the national level that supports research and extension activities to evolve effective Transfer of Technology (TOT) models. The State Agricultural Universities also contemplate to develop extension models suitable to take up transfer of technology besides implementing the models evolved by ICAR system.

29.2.1 Pre-Independence Era: 1866–1900

Iinitiatives	Purpose
Establishment of Agriculture Department in 1871 as a recommendation of Famine Commission (1866)	To collect information on cotton program for the mills in Manchester
Famine Commission (1880) recommended revival of Department of Agriculture (DOA) under control of secretary (1881) and all provinces to have DOA (1882)	To facilitate establishment of experiments headed by inspector general of agriculture
Famine commission of 1901 recommended: Setting up of Imperial agricultural research institute, Pusa (Bihar)	Beginning of organized agricultural research
Government of Indian act of 1919 empowered transfer of all departments of Agriculture and Rural Development to states.	Agriculture became a state subject
Royal Commission Report recommended to cease IAS to function	IAS closed in 1924

Source: Dimension of Agricultural Extension, Singh et al. (n.d.)

29.2.2 Lessons of Pre-Independence Efforts

- Individual attempts.
- Leader driven.
- Not replicable.
- Situation specific.
- Provided insight for future organization.

29.3 Post-Independence Period

The first planned attempt started with the launching of Community Development Programme in 1952, followed by the National Extension Service in 1953. These programs were able to educate responsive farmers to take up improved methods of farming across the country. The other important Area-Based Special Programmes were Intensive Agricultural District Programme (IADP 1960), Intensive Agriculture Area Programme (IAAP 1964), and High Yielding Varieties Programme (HYVP 1966), besides Farmers Training Centers (1967) to train farmers on high-yielding varieties and improved methods of farming to back up the above programs. The cumulative effect of these programs resulted in increased productivity, which made way to usher in *Green Revolution* in Indian agriculture during late 1970s.

However, these programs widened gap between resource-rich and resource-poor farmers. In order to enable resource-poor farmers to take benefit of improved farm technology, many client-based programs were introduced. The most important ones being Small Farmers Development Agency (SFDA 1969), Marginal Farmers and Agricultural Labourers Programme (MFAL 1969), District Rural Development Agency/Society (DRDA 1976), Integrated Rural Development Programme (IRDP 1978), and Lab to Land Programme sponsored by ICAR (LLP 1979). Although these programs were able to improve the socioeconomic conditions of beneficiaries, they were isolated in a given time and implemented in a phased manner.

By the middle of 1980s, it was observed that extension services in the developing countries were suffering from a number of weaknesses, including the dissipation of extension workers' energies on low-priority tasks, the lack of single as well as clear line of command, and low level of agricultural knowledge and skill among field level functionaries. As a means of reforming and strengthening the extension service, a reorganized agricultural extension system, known as "Training and Visit" (T&V) system, was introduced in the country.

29.3.1 Training and Visit System

This system was introduced in India in 1974 with the World Bank assistance. Training and Visit system became the dominant method of restructuring the extension services in over 60 countries in Asia, Africa, and Latin America. The system aimed to achieve change in production technologies of farmers through professional assistance for the contact farmers from well-trained extension personnel on agricultural research and supported by supply, service, and marketing facilities, which were earlier lacking in National Extension Service. The important features of T&V system are: (1) professionalism, (2) single line of command, (3) concentration of efforts, (4) time-bound work, (5) field and farmer orientation, (6) regular and continuous training, and (7) linkage with research.

29.3.2 Broad-Based Extension System (BBES)

This system was aimed to rectify the defects of T&V system in some of the states. In the Broad Based Extension System (BBES): (a) the role of subject matter specialists was amplified and they were invited to formulate messages suitable to their land-based activities (agriculture, sericulture, animal husbandry, horticulture, human resource development, creating agriculture infrastructure, etc.), (b) village extension workers had full-time job by offering messages during lean season as well, and (c) the concept of broad-based education laid emphasis on formulating and delivering composite messages to the farmers to meet the needs of their full agricultural environment.

29.3.3 District Level Agriculture Technology Management Agency (ATMA) Model

In a country like India, where agroclimatic zones widely differ besides significant variation in socioeconomic status of farmers, uniform extension service is not a panacea for all regions. It was realized that public extension system will have to be placed in new decentralized institutional arrangements, which are demand-driven, farmer-accountable, bottom-up, and have farming system approach. To address these issues, the ATMA model was envisaged as an alternate public extension institution.

The ATMA extension mechanism/model was firstly implemented in Andhra Pradesh, Bihar, Himachal Pradesh, Jharkhand, Maharashtra, Odisha, and Punjab, covering four districts in each state in 1998 under the guidance of National Institute of Agricultural Extension Management (MANAGE), Hyderabad.

The Government of India is funding the ATMA program in all districts across the country. ATMA is a registered society of key stakeholders (farmers, line/development departments, nongovernment organizations, input dealers, mass media, agribusiness companies, farmers organizations, etc.) involved in agriculture activities for sustainable agricultural development in the district. Though the State Department of Agriculture serves as a nodal agency for implementing ATMA, the program aims to increase coordination and integration among developmental departments. Emphasis has been laid on providing flexible working environment and establishing effective integration of all the stakeholders at the district level, thereby increasing input into program planning and resource allocation, especially at the block level, and thereby increasing accountability of stakeholders.

Every district has to prepare the Strategic Research and Extension Plan (SREP) for implementing ATMA in respective districts. The SREP is prepared through participatory methodologies such as Participatory Appraisal Techniques involving all stakeholders and farmers. In June 2010, revised guidelines for ATMA were released to address some of the constraints identified since the implementation began in 2005, especially the lack of qualified personnel at all levels, the absence of formal mechanisms to support extension delivery below the block level, inadequate infrastructure support to State Agricultural Management and Extension Training Institutes (SAMETIs), and the lack of convergence with other central and state schemes. Additional personnel will therefore be added at the district and block levels exclusively for ATMA, and a farmer friend (FF), a progressive farmer for two villages, will work to support extension at the village level. Other schemes, such as the agri-clinics and RKVY (National Agricultural Development Project), will work with and through the ATMA structure.

29.3.4 Initiatives of ICAR

The Indian Council of Agricultural Research (ICAR) took up number of extension programs over the years. The first program was National Demonstration Scheme (1964) initiated during 1964–65 to demonstrate the production potentiality of major crops in the farmers' field. The Operational Research Projects (ORPs) was started in 1975 to identify technological as well as socioeconomic constraints and to formulate and implement the problem-solving technology modules on an area/watershed/target group basis in the operational area. Lab-to-Land program was launched in 1979 to transfer low-cost technologies in agriculture and allied enterprises. As part of technology mission on oilseeds and pulses, the council started frontline demonstration in 1990–91.

29.4 Extension Activities of Commodity Boards, Financial Institutions, Input Agencies, Nongovernment Organizations, and Media Organizations

Commodity Boards (Coffee Board, Spice Board, Tobacco Board, Coconut Development Board, etc.) extend crop/commodity-specific technical know-how to the farmers to a limited extent, as many of these boards do not have grass-root level functionaries throughout the country. Financial institutions normally provide assistance in the preparation of agriculture project proposals by their technical staff to the farmers and others.

Besides providing critical inputs such as seeds, planting materials, fertilizes, and plant protection chemicals, agricultural input agencies also sponsor/organize training program to educate farming community. The media organizations (print and electronic media) are disseminating timely information on weather, technical information, and marketing information. Various committed nongovernmental organizations and philanthropists are also rendering rural extension services to the rural community in the field of agriculture and allied sectors, health, sanitation, education, and water supply across the country.

29.5 Post Green Revolution Era

29.5.1 NAIP

Agriculture is and will continue to be the main driver of country's economic growth with social justice. Our agriculture did extremely well and it was on the ascendancy till the mid-1990s but after that the growth slowed down. Since 1996–97 the growth rate of agricultural GDP has been, on an average, 1.75% per year in contrast with the rate of 4% that is required. On the other hand, the farmer has been facing rising input costs, declining returns from the inputs, uncertain market, increasing role of market in agriculture, and blurring of distinction between the domestic market and the international market. To assist the farmer in these changing contexts, new strategies and innovative solutions are urgently required, which in turn will require technological support. Hence, the agricultural research system that generates technologies must conduct the business of agricultural research in an innovative way. The World Bank-aided National Agricultural Innovation Project (NAIP) has been conceived to pilot this innovation in conducting agricultural research.

29.5.1.1 The Basic Principles

1. To give the agricultural research and technology development system an explicit development and business perspective through innovative models. In other words, the agricultural research system should be able to support agriculture as a business venture and also as a means of security of livelihood of the rural Indian while maintaining excellence in science.
2. To make the National Agricultural Research System a "pluralistic" system where every organization having stake in agricultural research, public, private, or civil society, has to play a role.
3. Working in well-defined partnership groups with clear common goals and understanding on sharing responsibilities and benefits.
4. Funding through competition so that a wide choice of excellent innovative ideas come in from the stakeholders themselves.
5. Work with focus, plan, and time frames.
6. Develop well-tested models for application of agricultural research and technology for profitability of farming, income generation, and poverty alleviation.

29.5.2 NATP

The National Agricultural Technology Project (NATP), a World Bank-aided project, is being implemented by the ICAR and the Department of Agriculture and Co-operation (DAC) since November 1998. It has three major components, namely Organization and Management (O&M) System, Research, and Innovations in Technology Dissemination (ITD). The ICAR executes the O&M, Research, and a part of ITD. The overall goal of the NATP is to increase income and reduce extreme poverty and hunger by improving agricultural technology development.

29.5.3 ARYA

Realizing the importance of rural youth in agricultural development of the country, ICAR has initiated a program on "Attracting and Retaining Youth in Agriculture" (ARYA).

29.5.3.1 Objectives

1. To attract and empower the youth in rural areas to take up various agriculture, allied, and service sector enterprises for sustainable income and gainful employment in selected districts.
2. To enable the farm youth to establish network groups to take up resource and capital intensive activities such as processing, value addition, and marketing.
3. To demonstrate functional linkage with different institutions and stakeholders for convergence of opportunities available under various schemes/programs for sustainable development of youth.

ARYA project will be implemented in 25 states through Krishi Vigyan Kendras (KVKs), one district from each state. In one district, 200–300 rural youths will be identified for their skill development in entrepreneurial activities and establishment of related microenterprise units. KVKs will involve the Agricultural Universities and ICAR Institutes as technology partners. At KVKs, one or two enterprise units will be established as well so that they serve as entrepreneurial training units for farmers. The purpose is to establish economic models for youth in the villages so that youths get attracted in agriculture and overall rural situation is improved. The concurrent monitoring, evaluation, and midterm correction will be an integral part of project implementation.

29.5.4 Krishi Vigyan Kendra

Krishi Vigyan Kendra (Farm Science Centre) is an innovative institution of ICAR established at district level. The first KVK was established in 1974 and has grown as a largest network in the country with 611 KVKs in 2011. KVKs are funded by ICAR and administered by ICAR Institutes/SAUs/Deemed Universities/Nongovernment Organizations or State Department of Agriculture.

KVKs play a vital role in conducting on-farm testing to identify location-specific agricultural technologies and demonstrate the production potential of crops at farmers' fields through frontline demonstrations. They also conduct need-based training programs for the benefit of farmers and farm women, rural youths, and extension personnel to update their knowledge and skills and to orient them in the frontier areas of technology development. KVKs are creating awareness about improved agricultural technologies through large number of extension programs (Adhiguru et al. 2009).

29.5.4.1 Impact of KVK

In the present context of achieving food and nutritional security, farm sector is a serious concern for the policymakers. Therefore, it was pertinent to assess a vast network of KVKs so that this science-based district level institution could be strengthened to play a more proactive role in the development of this vital sector (Fig. 29.1).

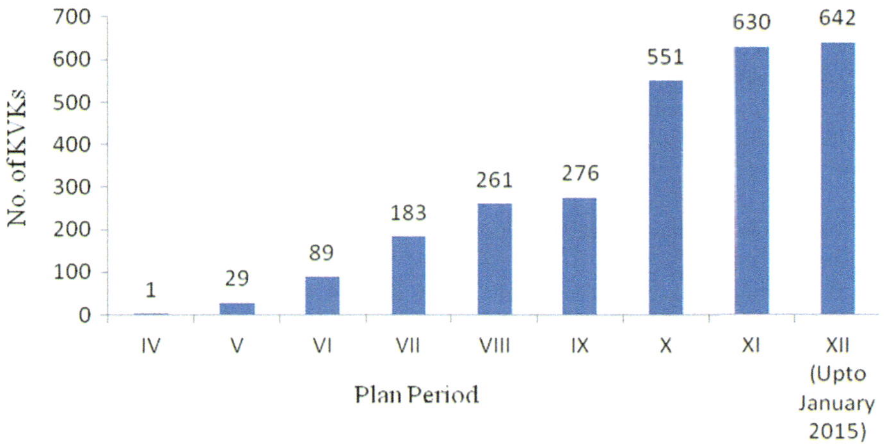

Fig. 29.1 Cumulative growth of KVKs in India. (Source: NILERD)

Currently, the impact of KVK activities is measured by the number of participants attending each learning workshop, the percentage of participants that adopt the technology/skill after the workshop, and changes in the participants' income. Like most of the public extension system, the focus of the centers is assessment and refinement of technologies through learning programs, on-farm testing, and field demonstrations. Information on market access and consumer demands is rarely considered. KVKs have been criticized for reaching limited numbers of farmers, and largely those within close range of the center.

29.5.5 Farmers' FIRST Program (FFP)

The Farmer FIRST as a concept of ICAR is developed as farmer in a centric role for research problem identification, prioritization, and conduct of experiments and its management in farmers' conditions. The focus is on farmer's Farm, Innovations, Resources, Science and Technology (FIRST). Two terms "enriching knowledge" and "integrating technology" qualify the meaning of Farmer FIRST in the Indian context. Enriching knowledge signifies the need for the research system as well as farmers to learn from each other in context to existing farm environment, perception of each other, and interactions with the subsystems established around. Technology integration is looked from the perspective that the scientific outputs coming out from the research institutions many times do not fit as such in the farmers' conditions and thus, certain alterations and adaptations are required at field level for their acceptance, adoption, and success.

29.5.5.1 Aims and Objectives of FFP

"Farmer FIRST" program aims at enhancing farmer–scientist interface for technology development and application. It will be achieved with the focus on innovations, technology, feedback, multiple stakeholder's participation, multiple realities, multi-method approaches, vulnerability, and livelihood interventions. The specific objectives are:

- To enhance farmer–scientist interface, enrich knowledge, and facilitate continued feedback.
- To identify and integrate economically viable and socially compatible technological options as adoptable models for different agroecological situations.
- To develop modules for farm women to address drudgery reduction, income enhancement, and livelihood security.
- To study performance of technologies and perception of the farmers about agriculture as a profession in the rural settings.
- To build network of linkages of organizations around the farm households for improving access to information, technology, input, and market.
- To institutionalize Farmer FIRST process.

29.5.5.2 Components of FFP

1. Enhancing farmer–scientist interface.
2. Technology assemblage, application, and feedback.
3. Partnership and institution building.
4. Content mobilization.

29.5.6 NICRA-Model Village

National Innovations in Climate Resilient Agriculture (NICRA) – a National Network Project of the Indian Council of Agricultural Research (ICAR) with the objectives to enhance the resilience of Indian agriculture to climate change and climatic vulnerability through various components such as strategic research on adaptation and mitigation, technology demonstration on farmers' fields to cope with current climate variability, sponsored and competitive research grants to fill critical research gaps, and capacity building of different stakeholders. The rationale for the Technology Demonstration Component (TDC) is based on the premise that an array of technologies are available to cope with different types of climate-related vulnerabilities in the National Agricultural Research System. The National Innovations in Climate Resilient Agriculture (NICRA) was launched in 2011 to address the challenges of climate variability and climate change along with farmers' need to adopt quickly increasing frequency of drought, flood, and other extreme events by application of science and technology.

29.5.7 ICT and Extension Service

The importance of agricultural extension in providing the relevant information, technology, and knowledge to the farmers and creating the enabling environment to increase production and productivity is quite clear, as mandated under the National Agricultural Extension Policy (NAEP). The General Directorate of Extension and Agriculture Development (GDEAD) has been developing many ICT innovations in linking research-extension-education for agriculture. These include Farmers Call Centers, Digital Library, Development of ICT applications and website, Video technology, Post Graduate Diploma program and Extension Service Messages.

Information and communications technology (ICT) extension services involve the transfer of practical knowledge and exchange of market information through ICT platforms. These solutions are relevant to agricultural and rural transformation processes, especially for smallholders. While traditional media such as radio and television continue to play a major role in extension and development communication, growth in the use of internet and mobile technology for communication is perceived to be a game changer in the extension services space. ICT extension service providers offer a range of information services to the smallholder farmers, from preharvest stage to postharvest stage. They help the farmers understand and adopt agricultural best practices on crop selection, input management, land selection and preparation, finance, transportation, packaging processing, and marketing of the agricultural produce. The enterprises provide these services via radio and television shows, mobile applications, digital video disks (DVDs), and interactive voice response (IVR) technology. Enterprises that provide information services can help improve agricultural yields and guide farmers in procuring and using the right inputs and participating in commercial value chains (Saravanan 2010).

29.5.8 ICT Model

Providing tools for preharvest efficiency SEs have adopted ICT to support farmers in achieving efficiency through information systems. *Indiagriline* is a web-based portal that enables farmers to forecast demand, access records of their previous transactions with the company, register their sugarcane area, submit payment information, and monitor demand, among other services. ICT enables remote agriculture extension where farmers can use email and digital cameras to reach experts and seek crop diagnostics support. SEs such as Cojengo provide a smartphone-based diagnostic tool for animal health to improve disease diagnosis, surveillance, and treatment of cattle in sub-Saharan Africa.

Virtual City, a private Kenyan technology start-up, has developed AgriManagr software, which is used by collection centers to manage the process of buying agricultural produce from farmers. AgriManagr has several benefits for both the procurer and the farmer. It eliminates the manual transcription that inevitably results in record-keeping errors or fraud. It speeds procurement and ensures clarity and

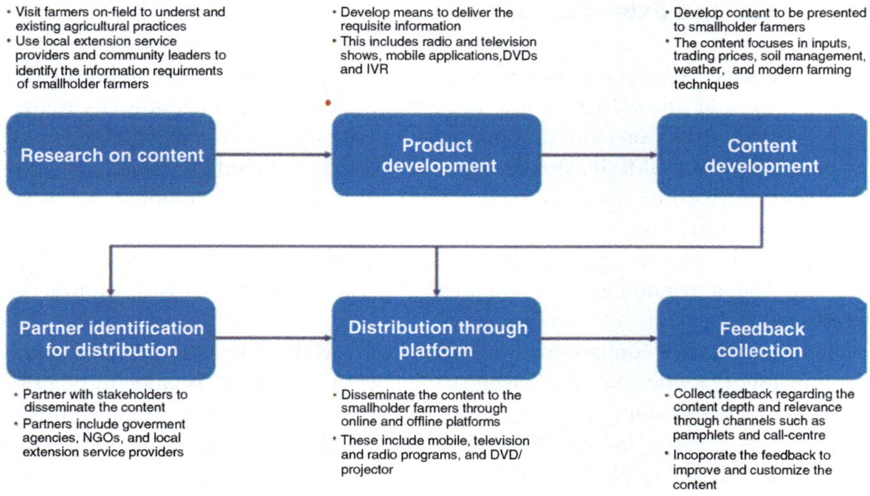

Fig. 29.2 Process of model. (Source: USAID, MANAGE report)

accuracy of information, which increases the buyers' ability to respond rapidly to bottlenecks or opportunities. *eKutir* has developed several innovative productivity enhancement solutions to improve the livelihoods of smallholder farmers (Srivastava 2018). For example, it has developed a tool called *mrittikka*, which provides soil nutrient analysis and recommendations. Another tool called *ankur* assists farmers with better seed selection to achieve increased productivity and farm yield. On the basis of the agro-climatic analysis and the type and condition of farmland, the tool recommends the best seed type/variety for the crop localized for each region (Fig. 29.2).

29.6 Strategic Plan for Popularization of Technology

- Diversification and intensification of existing farming systems.
- Improvement of productivity/income from different enterprises/commodities in exiting farming systems. (Sustainability of natural resources and enabling the farming community (male and female) to command the extension system is to be built into these components.)
- Sustainability of the production system.
- Capacity building of extensionists, researchers, farmers, market players, and other partners like NGOs, etc.
- Dovetailing and redesigning of various ongoing schemes of agriculture and other line departments and research institutions in the public, private, and NGO sector.
- Market-led extension for enhancement of profits with the focus on postharvest technologies and value addition.
- Promotion and use of ICT in extension.

- Promotion of public–private partnership.
- Mainstreaming gender concern (empowerment).
- Any other program component considered necessary for the project/area.

29.7 Strategic Planning for Agriculture Development

During the last decade, a number of management tools have been developed, which are helpful in facilitating stakeholder's involvement in an effective manner. Based on these tools, a participatory methodology has been worked out for preparing Strategic Plans (Research and Extension) for the district. The ultimate objective of both research and extension system is to increase agricultural production. Formulating extension and research agenda based on producers' requirement results in technology that will be more acceptable to users. This also helps in allocation of resources to both extension and research activities to be taken up in the district. Therefore, the Strategic Plan for each district is the need of the hour to address specific problems of the farming community, especially resource-poor and other disadvantaged groups. The Strategic Plan is the process of identifying research and extension gaps in agriculture and allied sectors. It suggests an appropriate Strategic Plan for agricultural development of the district.

29.8 The Contents of Strategic Plan

The extension and research interventions would differ across the Agro-Ecological Zone (AEZ) between crops, livestock, and farming systems as affected by roads, markets, input supply outlets, service facilities, and between farm households as a reflection of their resource endowment and socioeconomic status. Therefore, in formulating a Strategic Plan, the following guiding principles should be kept in view. The Strategic Plan will have two sections.

- Identify and build on important farming system innovations or success stories that may intensify or diversify existing farming systems and, thereby, increase farm household income.
- Increase farmers' access to markets, technologies, and resources through farmers' groups and organizations.
- On-farm collaborative technology development, testing, and refinement to address serious technological gaps in the existing farming systems.
- Promote appropriate natural resource management (NRM) plan for building and maintaining sustainable production systems within each Agro-Ecological Situations (AESs).

References

Adhiguru P, Birthal PS, Kumar BG (2009) Strengthening pluralistic agricultural information delivery systems in India. Agric Econ Res Rev 22:71–79

Moore M (1984) Institutional development, the World Bank, and India's new agricultural extension programme. J Dev Stud 20(4):303–317

Raabe K (2008) Reforming the agricultural extension system in India: what do we know about what works where and why? IFPRI discussion paper 00775. International Food Policy Research Institute, Washington, DC

Reddy MN, Swanson B (2006) Strategy for up-scaling the ATMA model in India. In: Vreyens JR (ed) Proceedings of the Association for International Agricultural and Extension Education. AIAEE 22nd Annual Conference, Clearwater Beach, FL

Saravanan R (2010) India. In: Saravanan R (ed) ICTs for agricultural extension: global experiments, innovations and experiences. New India Publishing Agency, New Delhi

Singh AK, Singh L, Burman RR (n.d.) Dimensions of agricultural extension. Rama Publishing, London

Srivastava A (2018) Using mass media and ICT for agricultural extension: a case study. Int J Sci Eng Res 9(2):1–9

Agricultural Reforms in India

30

Rohit Kumar, Muruga Subramanian Raman, and Sonia Chauhan

Abstract

Indian government has always diligently worked to enhance farmers' livelihoods by announcing and implementing agricultural reforms as agriculture is the backbone of the economy of the country. The announcement of the Kisan Credit Card to facilitate credit facility, Soil Health Card to improve soil health, PMKSY to cover more area under irrigation, PMFBY to cover insurance of the crops and neem-coated urea are some of the reforms faced by Indian agriculture. For the purpose of enhancing backward and forward linkages and ensuring that farmers receive fair prices, a number of market changes have been implemented. The Farmers' Produce Trade and Commerce (Promotion & Facilitation) Act of 2020, the Farmers (Empowerment & Protection) Agreement on Price Assurance and Farm Services Act of 2020, and modifications to the Essential Commodities Act of 1955 are designed to increase farmers' income by boosting market realization, lowering price risk, and bolstering agricultural supply systems. The Pradhan Mantri Annadata Aay Sanrakshan Abhiyan (PM-AASHA) is a new umbrella program that the government has established to broaden its procurement across crops and regions. Atma Nirbhar Bharat launched as framework to start-ups, farmer producer groups, and agricultural cooperative societies. The Pradhan Mantri Kisan Sampada Yojana (PMKSY) seeks to build agro-processing clusters and improve agro-processing. Along with the Pradhan Mantri Kisan Maan Dhan Yojana (PMKMY), which offers a social safety net for small and marginal farmers, the Pradhan Mantri Kisan Samman Nidhi (PM-KISAN) program was

R. Kumar (✉)
Department of Dairy Business Management, Sanjay Gandhi Institute of Dairy Technology, Bihar Animal Sciences University, Patna, India

M. S. Raman · S. Chauhan
ICAR-National Institute of Agricultural Economics and Policy Research, New Delhi, India

P. K. Ghosh et al. (eds.), *Trajectory of 75 years of Indian Agriculture after Independence*, https://doi.org/10.1007/978-981-19-7997-2_30

also created with the goal of providing income assistance to all farmer families throughout the nation. The use of artificial intelligence and Smart Krishi in resolving farmers' problems and claims definitely enhances their income and makes agriculture a more profitable and tension-free occupation.

Keywords

Doubling farmers' income · Market reforms · Micro-irrigation · Risk management

The government has diligently worked to increase farmers' incomes by double during the past 3 years and used the fiscal year 2015–2016 as the base year for recording the income of the farmers. Improvements in crop productivity, enhanced livestock productivity, resource use efficiency, increased cropping intensity, diversification toward high-value crops, increase in the real prices received by farmers, and a shift from farming to non-farming occupations have been identified as the seven sources of growth. It is observed that due to government interference, the capital is increasingly being directed toward the agriculture sector by higher budgetary allotment, mobilization of non-budgetary resources, improvement in institutional credit, and encouragement of corporate investments. Specific credit requirements are being prioritized for even smallholders. In recent years, farm credit disbursement has surpassed goals. Fishing and animal husbandry-related activities are now covered under the Kisan Credit Card (KCC) program. The government has announced a Rs 1 lakh crore fund to support agricultural infrastructure projects for an inexpensive and commercially viable postharvest management infrastructure at the farm gate and aggregation. Additionally, Operation Greens, which includes the three commodities (Tomatoes, Onions, and Potatoes) with the most price volatility, has been given a 500 crore rupees extension (GoI, 2021).

A lot of effort is being placed on productivity-based advances, with a focus on some important crops including pulses, oilseeds, and nutri-cereals (millets). The technological advancements and supportive governmental environment have significantly improved the production of pulses in the country. A plan for expanding the area and production of nutrition grains has also been developed. By 2022–2023, output set under these strategies is 21 million tons. A sub-mission on nutri-cereals has been initiated by the government. A plan to overcome the shortfall of oilseeds is also being developed. The production of oilseeds reached to 32.3 million tons in 2018–2019 from 27.5 million tons in 2014–2015 to (GoI, 2018a). Likewise, productivity also increased from 1075 kg/ha in 2014–2015 to 1265 kg/ha in 2018–2019. The amount of edible oil production increased from 9.8 million tons in 2014–2015 to 12.9 million tons in 2018–2019.

30.1 Reforms for Doubling Farmers' Income

The Mission on Integrated Development of Horticulture has expanded the area under horticultural crops. Positive outcomes and improvement in cropping intensity are being seen from diversification toward high-value crops. In order to manage risks effectively, the Pradhan Mantri Fasal Bima Yojana (PMFBY) was introduced in 2016. It offers insurance coverage for all phases of the agricultural cycle, including postharvest hazards in certain circumstances. According to studies, micro-irrigation had a favorable effect on crop output, water conservation, resource use efficiency, and cost reduction, while Soil Health Cards (SHC) and neem-coated urea helped producers to achieve greater productivity and reduction in costs. The area covered by micro-irrigation under the PMKSY has risen from 5.5 lakh ha in 2015–2016 to 11.7 lakh ha in 2019–2020. An 8–10% decrease in the usage of chemical fertilizers is reported by the impact research of soil health cards (Reddy, 2017). As a result of applying fertilizers and micronutrients in accordance with SHC guidelines, 5–6% overall improvement in crop production was also noted (Reddy, 2017).

With the use of micro-irrigation, irrigation costs were reduced by 20–50%, power usage decreased by about 31%, fertilizer use was reduced by 7–42%, and average fruit and vegetable yield increased by approximately 42.3 and 52.8%, respectively (GoI, 2014). Farmers' average income growth ranged from 20 to 68%. Crop production and productivity improvements are closely correlated with an increase in farm mechanization. The Sub-Mission on Agriculture Mechanization distributed agri-equipment and established custom hiring centers, hi-tech hubs, and farm machinery banks, increasing farm mechanization from 1.94 kW/ha (2012–2013) to 2.02 kW/ha energy utilization in 2018 (GoI, 2018b). The goal for 2022 is 4 kW/ha.

The Agricultural Infrastructure Fund (AIF) was established to develop agri-logistics, and funds have been granted to boost National Agriculture Market (e-NAM) and Grameen Agricultural Markets (GrAMs). For the purpose of enhancing backward and forward linkages and ensuring that farmers receive fair prices, a number of market changes have been implemented. A few of the most recent changes include the Farmers' Produce Trade and Commerce (Promotion & Facilitation) Act of 2020, the Farmers (Empowerment & Protection) Agreement on Price Assurance and Farm Services Act of 2020, and modifications to the Essential Commodities Act of 1955. These Acts are designed to increase farmers' income by boosting market realization, lowering price risk, and bolstering agricultural supply systems. The government has implemented new criteria for the announcement of Minimum Sport Price (MSP) for 23 commodities. Consequently, the terms of trade have improved. The government is also targeting promotion of 10,000 FPOs (Farmer Produce Organizations) by 2024.

The Pradhan Mantri Annadata Aay Sanrakshan Abhiyan (PM-AASHA) is a new umbrella program that the government has established to broaden its procurement across crops and regions. For effective price discovery in the market, the e-NAM was designed to be a market-based system. The e-NAM program has covered almost 1000 markets in 18 States and three Union Territories. Farmers now have access to registered warehouses and FPO storefronts for direct sales and shipping. Enhancing agricultural export continues to be a priority, and the Agri-Export Policy aims to quadruple agricultural exports by 2024. In terms of agricultural exports, the main contributing commodities are seafood, rice, cotton, beef, and horticulture goods. The Agri-Export Policy projects US$ 60 billion in exports by 2022. Despite the constraints and effects of the epidemic, exports increased significantly in 2020.

To support agriculture infrastructure projects at farm gates and aggregation points (Primary Agricultural Cooperative Societies, Farmers Producer Organizations, Agriculture entrepreneurs, Start-ups, etc.) a financing facility worth Rs. 1,000,000 crore will be provided. An amount of Rs 10,000 crore has been set aside for the formalization of micro-food businesses and the implementation of cluster-based agricultural strategy to be used. Fishermen will receive Rs 20,000 crore under the PM Matasya Sampadana Yojana. An amount of Rs 13,000 crore has been set aside to ensure that all cattle, buffaloes, sheep, goats, and pigs are fully vaccinated. The infrastructure for animal husbandry has received a boost of Rs 5000 crore. To promote the growing of herbs, Rs 4000 crore has been set aside. The initiative seeks to cover 10 lakh hectares with herbal farming in 2 years. An amount of Rs 500 crores has been set aside for beekeeping activities.

The Pradhan Mantri Kisan Sampada Yojana (PMKSY) seeks to build agro-processing clusters and improve agro-processing. Farmers' incomes will increase as associated businesses are promoted and secondary agriculture is expanded. With the implementation of the National Bamboo Mission in 2018–2019, specific funding has been designated for the promotion of aromatic and medicinal plants. Along with the Pradhan Mantri Kisan Maan Dhan Yojana (PMKMY), which offers a social safety net for small and marginal farmers, the Pradhan Mantri Kisan Samman Nidhi (PM-KISAN) program was also created with the goal of providing income assistance to all farmer families throughout the nation.

30.2 Market Reforms Through e-NAM

States must make changes in order to facilitate both market unification and online trading, including (i) a single license that is valid throughout the state, (ii) a single point levy of market fee, and (iii) the inclusion of electronic auction as a method of price discovery. Additionally, the e-auction platform must be promoted by the State Marketing Boards/APMCs. The states will need to ensure that the mandis that are integrated with NAM makes provision for requisite online connectivity, hardware, and assaying equipment. Research on e-markets in Karnataka, where e-markets have been operational since 2012, demonstrates unequivocally that farmers realize higher prices and that market arrivals of goods are higher in e-markets than in non-e-markets. They do, however, add that an efficient assaying mechanism and the involvement of traders in remote markets are essential for the success of the e-NAM (Reddy and Mehjabeen, 2019). Online trade has a favorable impact on the state of Karnataka (Chand, 2017). According to the study's results, e-tendering systems offer a great deal of potential to increase competition and transparency in agricultural markets while also lowering transaction costs for both buyers and sellers

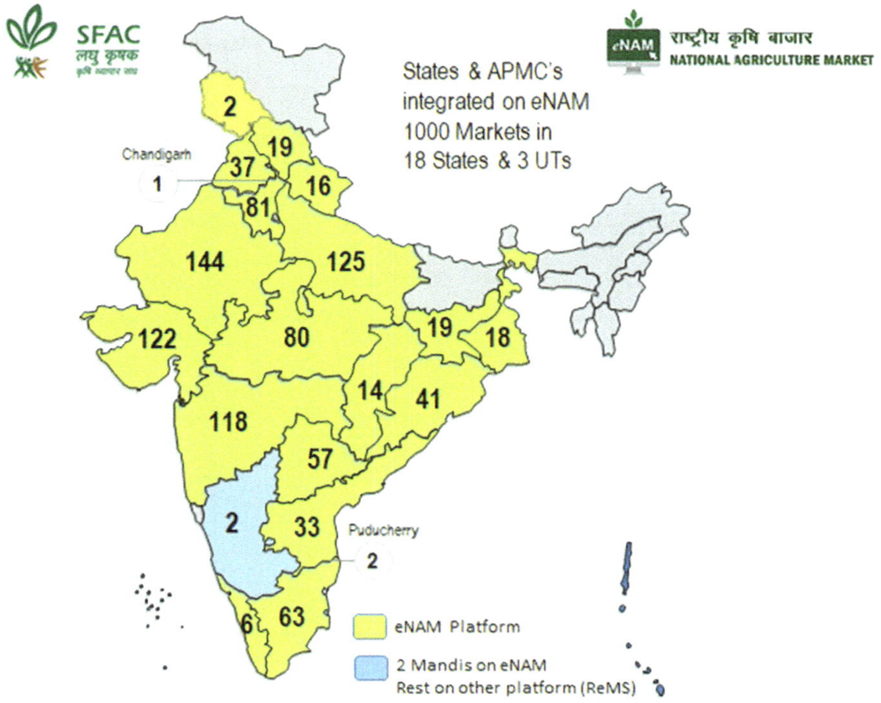

Fig. 30.1 State-wise number of APMC's integrated on e-NAM. (Source: National Agriculture Market, 2022)

without negatively affecting their business relationships or income (Pavithra et al. 2018).

The e-NAM platform now has a total of 1000 mandis spread throughout 18 States and 3 UTs. It is due to the overall success of 585 mandis in Phase 1 and further expanding its wings to incorporate 415 new mandis in Phase 2 (Fig. 30.1 and Table 30.1).

30.3 Agri-Logistics

The core of agribusiness is agri-logistics, which permits connectivity between centers of production and consumption over both geography and time with little loss of quality or quantity. When buyer and seller engage in online bidding for Inter-Mandi or Inter-State commerce on e-NAM, the demand for agri-logistics will arise. The agri-logistic services will make it possible for Farmer Producer Organizations, Traders, Processors, Exporters, and Corporates to safely store their goods and expedite the delivery to markets. Assaying, Grading, Sorting, Packaging, Agri-Transportation (Uberization), Warehousing, and Agri-logistics Professionals.

Table 30.1 Number of unified licenses (as on June 30, 2022)

Name of state/ UT	Mandis registered on e-NAM	Registered traders on e-NAM	No. of unified licenses issued by state
Andhra Pradesh	33 (3.3)	3478 (1.54)	3478 (2.93)
Chandigarh	1 (0.1)	114 (0.05)	0 (0)
Chhattisgarh	14 (1.4)	3125 (1.38)	36 (0.03)
Gujarat	122 (12.2)	9401 (4.16)	9401 (7.92)
Haryana	81 (8.1)	14,424 (6.38)	35 (0.03)
Himachal Pradesh	19 (1.9)	2010 (0.89)	0 (0)
Jammu and Kashmir	2 (0.2)	121 (0.05)	0 (0)
Jharkhand	19 (1.9)	2299 (1.02)	101 (0.09)
Karnataka	2 (0.2)	655 (0.29)	655 (0.55)
Kerala	6 (0.6)	346 (0.15)	35 (0.03)
Madhya Pradesh	80 (8)	22,337 (9.88)	1000 (0.84)
Maharashtra	118 (11.8)	21,510 (9.51)	0 (0)
Odisha	41 (4.1)	7393 (3.27)	7393 (6.23)
Puducherry	2 (0.2)	180 (0.08)	0 (0)
Punjab	37 (3.7)	2598 (1.15)	1 (0.001)
Rajasthan	144 (14.4)	82,359 (36.42)	82,359 (69.38)
Tamil Nadu	63 (6.3)	4358 (1.93)	3543 (2.98)
Telangana	57 (5.7)	5799 (2.56)	5799 (4.89)
Uttar Pradesh	125 (12.5)	35,029 (15.49)	110 (0.09)
Uttarakhand	16 (1.6)	4728 (2.09)	4728 (3.98)
West Bengal	18 (1.8)	3884 (1.72)	33 (0.03)
India	**1000 (100.00)**	**226,148 (100.00)**	**118,707 (100.00)**

Figures in parentheses are percentage to the total. Source: National Agriculture Market 2022

Stakeholders can utilize the platform for networking as well as for business development.

No state has a platform that is completely operational. There are no quality testing devices or scientific sorting/grading facilities. Another problem preventing advancement is poor internet access. According to mandi officials, a lack of technical know-how at state agricultural departments has also delayed the establishment of grading/ assaying facilities. Many states are still having difficulty implementing e-NAM because they lack the necessary infrastructure, including assaying facilities, a skilled workforce, and standardized and straightforward procedures. Still, many traders and commission brokers are hesitant to join e-NAM.

Table 30.2 Area covered under micro-irrigation (area in '000 ha)

Years	Drip irrigation	Sprinkler irrigation	Total
2015–16	347 62.86)	205 (37.14)	552 (100)
2016–17	488 (58.1)	352 (41.9)	840 (100)
2017–18	541 (51.57)	507 (48.33)	1049 (100)
2018–19	575 (49.65)	583 (50.35)	1158 (100)
2019–20	624 (53.15)	550 (46.85)	1174 (100)
2020–21	358 (38.17)	580 (61.83)	938 (100)
2021–22	358 (35.27)	657 (64.73)	1015 (100)

Figures in parentheses is percentage to the total. Source: Pradhan Mantri Krishi Sinchai Yojana, 2015–2022

30.4 Pradhan Mantri Krishi Sinchai Yojana (PMKSY)

To boost productivity, irrigation efficiency must be improved. Consequently, the funding for irrigation has risen by the government. The catchphrase is *Per Drop More Crop*. To lessen the impact of the drought and to provide *water to every farm*, the Pradhan Mantri Krishi Sinchai Yojana has been introduced. As a result, large and medium-sized projects that were in progress were also accelerated. Projects for the management and development of watersheds have been given priority. The Accelerated Irrigation Benefit Program (AIBP) of the Ministry of Water Resources, River Development, and Ganga Rejuvenation (MoWR, RD & GR), the Integrated Watershed Management Program (IWMP) of the Department of Land Resources (DoLR), and the On-Farm Water Management (OFWM) of the Department of Agriculture and Cooperation have been combined into the PMKSY (DAC). PMKSY has been authorized for nationwide implementation of a 5-year budget of Rs 50,000 crore. An expenditure of Rs 5300 crore has been made for 2015–16, of which Rs 1800 crore has been allocated to the DAC, Rs 1500 crore to the DoLR, and Rs 2000 crore to the MoWR (Rs 1000 crore for AIBP; Rs 1000 crores for PMKSY). Technology would support the promotion of micro-irrigation under the PMKSY by bringing 1 crore hectares of agricultural land under micro-irrigation, which would be coupled with "fertigation" to encourage the wise use of fertilizers. Currently, 11.4 million hectares (ha) of land have been covered by micro-irrigation (MI), of which 53.1% are covered by sprinkler systems (6.06 m ha) and 46.9% by drip systems (5.35 m ha), of which about 44% have been installed during the past 6 years (Table 30.2). Researchers from several states around the nation looked at the effects of micro-irrigation on improving yield, saving water, energy, and other resources, and boosting revenue (Table 30.3).

Table 30.3 Evidences of micro-irrigation's impact

Author(s)	Location	Crop	Impact on Yield	Income	Cost	Others
Shrivastava et al. (1994)	Soil-Water Management Project Farm, Navsari, South Gujarat	Tomato	44% increase in the drip-alone treatment as compared to flood method without mulch Drip with either of the mulches, the yield was more than 55% compared to the yield obtained by surface method			95% reduction in weed infestation, 44% saving in irrigation water when compared with the surface flood
Dalvia et al. (1999)	Akola, India	Tomato	Tomato yield increased by about 27%, with saving of 21% water and about 4% fertilizer		Approximately 50% reduction in initial capital investment	
Luhach et al. (2003)	Haryana	Wheat		The average net return per hectare was 19.53% higher than pump irrigation	The benefit–cost ratio was 1:1.97	
Kumar and Palanisami (2010)	Coimbatore, Tamil Nadu	Banana, coconut, and grape			Labor cost reduced by 69%, 69%, and 15% for banana, coconut, and grape cultivation under the drip method	Net sown area, gross cropped area, and irrigation increased by 82.0 and 98.03%, respectively, due to the drip intervention
Prasad et al. (2003)	Jodhpur	Pomegranate	Yield increased from 17.7 kg per plant to 28.2 kg per plant			

Narayanamoorthy (2004)	Maharashtra	Sugarcane	The yield difference of about 24 tons/ha for Pune district and about 27 tons/ha for Ahmednagar district		Average cost saving of nearly 14%	Water saving estimated at 44% and 1059 kWh of electricity per ha
Rajak et al. (2006)	Karnal	Cotton	The growth and yield performance of cotton irrigated through furrows, even though with good-quality canal water (ECw = 0.25 dS ml), was poor when compared drip irrigation with marginally saline water (ECw = 2.2 dS ml)	The gross income (US$ 223–690 per ha) was more with drip than furrow (US$ 67–545 per ha) irrigation		
Rao et al. (2010)	Pali-Marwar, Rajasthan	Cumin		Maximum net returns of Rs 45,059/ha and benefit–cost ratio of 3.84 was recorded under the 0.8 IW/CPE treatment in comparison to significantly low net returns of Rs 23,763/ha and benefit–cost ratio of 2.17 obtained under 0.4 IW/CPE		

(continued)

Table 30.3 (continued)

Author(s)	Location	Crop	Impact on			
			Yield	Income	Cost	Others
Madhava et al. (2016)	Kozhikode and Thrissur, Kerala	Coconut, arecanut, and nutmeg	19.11% improvement in yield for coconut, 13.3% for arecanut, and 47.1% for nutmeg than surface method of irrigation			
Surendran et al. (2016)	Tamil Nadu	Sugarcane	Single-row conventional drip irrigation recorded the highest mean yield of 120.4 tons/ha followed by single-row low-cost drip irrigation (118.6 tons/ha) found to be significantly superior to the rest of the treatments			

Table 30.4 Actuarial Premium Rate (APR) across seasonal crops

Seasons	Crops	Maximum insurance charges payable by farmer (% of sum insured)
Kharif	All foodgrain and oilseeds crops (all cereals, millets, pulses, and oilseeds crops)	2.0% of SI or actuarial rate, whichever is less
Rabi	All food grain and oilseeds crops (all cereals, millets, pulses, and oilseeds crops)	1.5% of SI or actuarial rate, whichever is less
Kharif and Rabi	Annual commercial/annual horticultural crops	5% of SI or actuarial rate, whichever is less

Pradhan Mantri Fasal Bima Yojana 2015

30.5 Risk Management Through Pradhan Mantri Fasal Bima Yojana (PMFBY)

The Pradhan Mantri Fasal Bima Yojana (PMFBY) aims to support sustainable production in the agricultural sector by: (a) compensating farmers who suffer crop loss or damage due to unforeseeable events; (b) stabilizing farmers' incomes to ensure their continued involvement in farming; (c) encouraging farmers to adopt innovative and modern agricultural practices; and (d) ensuring the flow of credit to the agriculture sector, which will contribute to food security and crop diversification. The progress is given in Table 30.4. The implementing agency (IA) under PMFBY would charge the Actuarial Premium Rate (APR). The farmer will pay insurance charges at the differential rates (Table 30.5).

30.6 Promotion of Mechanization and Technology Backstopping

With the help of the Sub-Mission on Agriculture Mechanization (SMAM), the degree of farm mechanization increased from 1.62 kW/ha in 2014 to 2.02 kW/ha in 2018, due to the distribution of agricultural machinery and the construction of custom hiring centers, high-tech hubs, and farm machinery banks. By 2022, the objective is to reach 4 kW/ha. The CHC "Uberization" app has undergone testing and is now prepared for a nationwide rollout. The SMAM offers a good platform for integrating all initiatives for the advancement of agricultural mechanization that is inclusive by offering a "single window" implementation strategy with a particular focus on small and marginal farmers. For the estimation of agricultural production, horticultural inventory, site suitability analysis for crop growth, and drought assessment, the government also uses satellite data and GIS technology. Additionally, support is given to the states under the Marketing Research & Information Network (MRIN) Scheme to establish up and maintain connectivity with the Agmarknet

Table 30.5 Pradhan Mantri Fasal Bima Yojana (PMFBY) status

	Seasons								
	Kharif					Rabi			
States	2018	2019	2020	2021	2022	2018	2019	2020	2021
States/UTs	22	20	19	19	16	21	19	18	19
Districts	475	463	390	404	370	486	445	389	410
Agriculture crops	38	37	36	26	30	40	40	36	38
Horticulture crops	57	48	46	47	45	82	83	89	92
Area coverage ('000 ha)									
Farmers (nos.)	21,663,395	20,050,883	16,868,110	15,095,926	11,582,311	14,685,273	9,660,447	10,010,514	9,837,105
Insured by PMFBY	27,789.7	29,299.6	27,173.3	23,922.5	17,182.6	19,793.9	15,420.9	15,737.4	14,850.7
Insured by RWBCIS	1554.9	2141.9	227.6	178.3	165.9	8018.0	8376.2	6927.1	6727.1
Demographic distribution (%)									
Marginal farmers	18.1	16.5	16.5	18.0	17.3	19.2	18.4	17.4	18.2
Small farmers	64.9	67.6	67.6	62.2	60.7	63.0	60.9	64.5	60.8
Others	17.1	15.9	15.8	19.8	22.0	17.8	20.7	18.1	21.0
Premium and sum insured (in lac.)									
Farmers premium	261,034	269,527	243,721	213,490	158,929	161,943	133,649	142,307	138,592
State/UTs premium	743,876	908,212	842,977	768,821	509,119	331,688	325,103	534,891	530,147
GoI premium	717,203	818,732	805,117	735,898	496,195	321,601	318,554	431,266	441,314
Gross premium	1,722,113	1,996,471	1,891,815	1,718,209	1,164,243	815,232	777,306	1,108,464	1,110,053
Sum insured	1,239,4576	13,423,613	11,024,478	9,632,577	7,208,058	9,260,486	7,186,700	8,444,725	7,883,792

Pradhan Mantri Fasal Bima Yojana 2022

portal for the collecting and distribution of market information on agri-commodity arrivals and pricing.

30.6.1 Artificial Intelligence

Finding the variables that might lead to a greater yield and harvest location is crucial for the expansion of agriculture. With the introduction of AI technologies, it is now feasible to find solutions based on data that shows weather conditions, the kind of harvest a crop would require the best sort of soil, and so on. AI technology would be welcomed by government program such as the Pradhan Mantri Fasal Bima Yojana (PMFBY) to cut down on the amount of time needed to resolve farmer claims. Additionally, the government and IBM signed an agreement to use AI to monitor the agricultural industry.

30.6.2 Mobile Applications

India saw a significant level of digitization after smartphones became a common item. The new technology increased the possibility of applications for agricultural uses as more and more activities shifted to smartphones. With the launch of *e-Nam* (National Agriculture Market), an online trading platform for agricultural commodities, the government has placed a significant emphasis on supporting technology usage in agriculture. Some states have taken independent actions that are advantageous to farmers.

30.6.3 Agro-Advisories

Farmers receive agro-advisories from a variety of service providers including the mKisan platform. The mKisan site is a platform that offers web-based mobile advice to farmers with technology help from agricultural universities and research institutes. In order to transmit SMS messages on numerous topics related to agriculture, horticulture, animal husbandry, weather forecasts, and pest and disease control, almost 5.3 crore farmers are connected to the mKisan portal.

30.6.4 Encouraging Organic Farming Through Paramparagat Krishi Vikas Yojana (PKVY)

The PGS (Participatory Guarantee System) certification program Paramparagat Krishi Vikas Yojana supports cluster-based organic farming. The program supports cluster creation, training, certification, and marketing. A farmer receives assistance of Rs 50,000 per hectare for 3 years, of which 62% (or Rs 31,000) is given as a financial incentive to use organic inputs.

30.7 Rashtriya Krishi Vikas Yojana (RKVY)

The RKVY scheme was introduced in 2007 as a general program to ensure the growth of agriculture and allied industries in a holistic manner. The states are free to select their own agricultural and allied industry development initiatives in accordance with district or state agriculture plans. The program was run as an Additional Central Assistance (ACA) to State Plan Scheme up to 2013–2014, receiving 100% of its funding from the federal government. In 2014–2015, it underwent a similar transformation into a centrally sponsored scheme with 100% central support. The scheme's financing arrangement has changed since 2015–2016, with a 60:40 split between the center and the states (90:10 for North Eastern States and Himalayan States). The Union Territories get a central grant that is distributed equally to all recipients. Pre- and postharvest infrastructure, as well as encouraging agri-entrepreneurship and innovations, are given significant attention under the Rashtriya Krishi Vikas Yojana—Remunerative Approaches for Agriculture and Allied Sector Rejuvenation (RKVY-RAFTAAR). The following streams are how the Central Government would distribute grants to the states under the RKVY-RAFTAAR.

The annual expenditure for the regular RKVY-RAFTAAR shall be divided among the states according to the following criteria:

- Infrastructure and assets account for 50% (of 70%) of the total RKVY-RAFTAAR expenditure. Preharvest infrastructure makes up 20% of the total, while postharvest infrastructure makes up 30%.
- Value-added linked production projects (agribusiness models) that guarantee or supplement farmers' income, such as Public Private Partnership for Integrated Agriculture Development (PPP-IAD) projects, which account for 30% (of 70%) of the total RKVY expenditure.
- Flexi funds make about 20% (of 70%) of the RKVY-RAFTAAR outlay. The states may utilize this fund to assist any initiatives that fit with their regional priorities, although they are encouraged to support cutting-edge initiatives in the agricultural and related industries.
- *RKVY-RAFTAAR special subschemes*, which account for 20% of total yearly outlays, are based on national objectives periodically announced by the Indian government in order to develop regions and address problem-specific areas.
- *Development of agri-entrepreneurs and innovation*—10% of yearly expenditure—for promoting agri-entrepreneurs and innovation through financial support and skill development. It will provide assistance for incubators, incubators' facilities, KVKs, prizes, and so on.

30.8 Conclusions

Fostering public–private partnerships to finance the market information infrastructure, such as electronic dashboards in market yards and rural kiosks such as e-Chowpal, can aid in supplying the farmers with the pertinent information, extensions, and other services. There is a need to aggressively include the FPOs in connecting the small farmers with the e-NAM since this networked market serves as a signal to inform India's government about national demand and supply conditions. There are several reasons why farmers visit markets. The e-NAM administrative team should assist the farmer organizations in establishing connections with organizations such as banks, warehouses, merchants in far-off marketplaces, and so on to meet their demands for finance, transportation, and better pricing. The e-NAM should have thorough institutional regulations, a grievance redressal system, and policies that may enforce the execution of the agreements established on the e-NAM platform in addition to ensuring the accuracy of product information. Specialized abilities are required for farmers, commission agents, traders, and market administrators. Farmers need to receive training on how to grade and assay their produce and conduct online transactions, whereas traders and commission agents need to receive training on how to register for an account on the e-NAM, login, link it to their Aadhaar number and bank account, participate in electronic bidding, conduct financial transactions, and other topics. In order to maintain the e-NAM, the administration must also get training.

References

Chand R (2017) Presidential address: doubling farmers' income: strategy and prospects. Indian J Agri Econ 71(1):1–23

Dalvi VB, Tiwari KN, Pawade MN, Phirke PS (1999) Response surface analysis of tomato production under micro-irrigation. Agric Water Manag 41(1):11–19

GoI (2014) Agricultural at a glance 2014, Department of Agriculture and Farmers Welfare, Ministry of Agriculture and Farmers Welfare. https://eands.dacnet.nic.in/

GoI (2018a) Agricultural at a glance 2018, Department of Agriculture and Farmers Welfare, Ministry of Agriculture and Farmers Welfare. https://eands.dacnet.nic.in/

GoI (2018b) Sub-mission on agricultural mechanization, Department of Agriculture, Cooperation & Farmers Welfare, Mechanization & Technology Division

GoI (2021) Operation Greens - Ministry of Food Processing Industries, Operation Greens: From TOP to TOTAL, Ministry of Food Processing Industries. https://pib.gov.in/FactsheetDetails.aspx?Id=148583

Kumar DS, Palanisami K (2010) Impact of drip irrigation on farming system: evidence from southern India. Agri Econ Res Rev 23(2):265–272

Luhach MS, Khatkar RK, Malik DP (2003) Economic viability of sprinkler irrigation: An empirical analysis. Indian J Agric Econ 58(3):489

Madhava CK, Surendran U (2016) Study on factors influencing the adoption of drip irrigation by farmers in humid tropical Kerala, India

Narayanamoorthy A (2004) Impact assessment of drip irrigation in India: the case of sugarcane. Deve Poli Rev 22(4):443–462

Pavithra S, Gracy CP, Saxena R, Patil GG (2018) Innovations in agricultural marketing: a case study of e-tendering system in Karnataka, India. Agri Econ Res Rev 31(347-2018-3189):53–64

Prasad RN, Bankar GJ, Vashishtha BB (2003) Effect of drip irrigation on growth, yield and quality of pomegranate in arid region. Indian J Hort 60:140–142

Rajak D, Manjunatha MV, Rajkumar GR, Hebbara M, Minhas PS (2006) Comparative effects of drip and furrow irrigation on the yield and water productivity of cotton (Gossypium hirsutum L.) in a saline and waterlogged vertisol. Agri Water Manag 83(1-2):30–36

Rao SS, Singh YV, Regar PL, Chand K (2010) Effect of micro-irrigation on productivity and water use of cumin (Cuminum cyminum) at varying fertility levels. Indian J Agric Sci 80(6):507

Reddy AA (2017) Impact study of soil health card scheme. National Institute of Agricultural Extension Management (MANAGE), Hyderabad

Reddy A, Mehjabeen. (2019) Electronic national agricultural markets, impacts, problems and way forward. IIM Kozhikode Soc Manag Rev 115:1–13

Shrivastava PK, Parikh MM, Sawani NG, Raman S (1994) Effect of drip irrigation and mulching on tomato yield. Agric Water Manag 25(2):179–184

Surendran U, Jayakumar M, Marimuthu S (2016) Low cost drip irrigation: Impact on sugarcane yield, water and energy saving in semiarid tropical agro ecosystem in India. Sci Tot Environ 573:1430–1440

The manufacturer's authorised representative in the EU is Springer Nature Customer Service Centre GmbH, Europaplatz 3, 69115 Heidelberg, Germany. If you have any concerns regarding our products, please contact ProductSafety@springernature.com

Printed and bound by CPI Group (UK) Ltd, Croydon, CR0 4YY
01/10/2026
02225868-0020